# ALGEBRAIC AND GE(

## INEQUALITIES

If $a < b$ then $a + c < b + c$

If $a < b$ then $-a > -b$

If $a < b$ and $c > 0$ then $ac < bc$

If $a < b$ and $c < 0$ then $ac > bc$

$$|x| = \begin{cases} x & \text{if } x \geq 0 \\ -x & \text{if } x < 0 \end{cases}$$

$|x| < a$ if and only if $-a < x < a$

$|x| > a$ if and only if $x > a$ or $x < -a$

$|x + y| \leq |x| + |y|$

$|xy| = |x|\,|y|$

## STRAIGHT LINES

Distance between 2 points $\quad d = \sqrt{(x_1 - x_2)^2 + (y_1 - y_2)^2}$

Slope $\quad m = \dfrac{y_2 - y_1}{x_2 - x_1}$ where $x_1 \neq x_2$

Forms of the equation of a straight line

*Standard form:* $ax + by = c$

*Slope-intercept form:* $y = mx + b$

*Point-slope form:* $y - y_1 = m(x - x_1)$

*Two-point form:* $y - y_1 = \dfrac{y_2 - y_1}{x_2 - x_1}(x - x_1)$

## CONIC SECTIONS

$(x - h)^2 + (y - k)^2 = r^2$ — **Circle with center at $(h, k)$** and radius $r$.

$\dfrac{(x - h)^2}{a^2} + \dfrac{(y - k)^2}{b^2} = 1$ — **Ellipse with center at $(h, k)$.**

The distance from the center to the foci is $c$.

If $a > b$, the ellipse is horizontal and $a^2 - c^2 = b^2$

If $a < b$, the ellipse is vertical and $a^2 + c^2 = b^2$.

$\dfrac{(x - h)^2}{a^2} - \dfrac{(y - k)^2}{b^2} = \pm 1$ — **Hyperbola with center at $(h, k)$.**

The distance from the center to the foci is $c$ and $a^2 + b^2 = c^2$.

If the right side is $+1$, the hyperbola is horizontal.

If the right side is $-1$, the hyperbola is vertical.

$(x - h)^2 = 4p(y - k)$ — **Vertical parabola.**

Vertex $(h, k)$; focus $(h, k + p)$; directrix $y = k - p$.

The parabola opens upward if $p > 0$ and downward if $p < 0$.

$(y - k)^2 = 4p(x - h)$ — **Horizontal parabola.**

Vertex $(h, k)$; focus $(h + p, k)$; directrix $x = h - p$.

The parabola opens to the right if $p > 0$ and to the left if $p < 0$.

# ALGEBRA *and* TRIGONOMETRY

# ALGEBRA *and* TRIGONOMETRY

*Second Edition*

**JOHN TOBEY, JR.**
*North Shore Community College*

**J. LOUIS NANNEY,**
**EMERITUS**
*Miami-Dade Community College*

**JOHN L. CABLE,**
**EMERITUS**
*Miami-Dade Community College*

**wcb**
Wm. C. Brown Publishers
Dubuque, Iowa

Library of Congress Catalog Card Number: 85–28615

ISBN 0–697–06898–6



Printed in the United States of America by Wm. C. Brown Publishers
2460 Kerper Boulevard, Dubuque, IA 52001

10 9 8 7 6 5 4 3 2

# Contents

# Preface

This text was written to provide a clear and readable, yet mathematically accurate, presentation of all of the content areas of a college level course in algebra and trigonometry. The text presents a carefully developed sequence of mathematical topics presented in an intuitive manner. However, the text carefully avoids a "skills only" approach by stressing concepts and properties. It also provides the student with the opportunity to examine a few mathematical proofs and to investigate a few generalized problems that will extend the student's knowledge in other topical areas. In today's high-technology society, people from all walks of life are exposed to more mathematics than ever before. A variety of interesting application problems from many academic disciplines (physics, chemistry, biology, psychology, economics, history, computer science, and electronics) and subject areas (sports, military applications, business, population growth, and space travel) serve to heighten student interest. The text emphasis on application problems helps to make the student aware of the applications of mathematics to a broad area of real world problems.

The text has been designed to provide solid content coverage for precalculus students. The level of coverage of functions and their applications and of the mathematical techniques for graphing functions efficiently is superior to competing texts. In addition, the text is specifically structured to prepare the student for further mathematics courses other than calculus. Students will find the text a solid preparation for noncalculus based courses in finite mathematics, discrete mathematics, elementary statistics, and linear programming. The text provides students with a thorough background in those topics needed to enter computer science and other high technology fields.

## FLEXIBILITY

Although this text is one of the most topically complete available, it is unusually flexible. The following features will enable the book to be used to meet a variety of student needs.

### *Flexibility of Topic Coverage*

Certain chapters may be omitted by the better-prepared student.

Chapter 1 contains a complete review of the basic properties of algebra. It may be omitted by the reasonably well-prepared class.

Chapter 2 contains a complete presentation of the skills needed to solve first and second degree equations and inequalities. A strong class may be able to omit this chapter or cover certain topics of the chapter in a brief review.

Chapters 3 and 4 provide a complete presentation of mathematical notation, the concept of a function, and the techniques for graphing functions and relations. After Chapter 4 is completed, subsequent chapters can be covered in almost any order. A more detailed discussion of topic sequences and possible coverage of chapters to meet the needs of certain types of courses is contained in the *Instructor's Manual.*

### *Flexibility in the Use of Scientific Calculators*

Mathematics instructors seem to be divided on the use of the scientific calculator in a course in algebra and trigonometry. Some prefer to avoid using a calculator and to introduce the students to the use of mathematical tables. Others integrate the scientific calculator into the entire content of the course and require each student to use a scientific calculator on homework assignments and tests. Still other instructors allow the student the choice of using or not using a calculator.

The present text encourages the use of a scientific calculator but does not require it. A complete set of mathematical tables for exponential functions, trigonometric functions, square root functions, and common logarithms is provided for those students who do not use the calculator. All homework problems can be done without the use of a calculator except those designated by the ▦ symbol.

For those students who use a scientific calculator Section 2.1 explains the use of a scientific calculator. This optional section may be covered in class or merely assigned as reading for those students needing the material. Throughout the text, those types of problems requiring special insight to solve using a scientific calculator are explained with solutions that show the calculator keystroke operations.

## SPECIAL FEATURES OF THIS TEXTBOOK

### *Challenge Problems*

Nonroutine problems and problems involving more serious effort are carefully marked with the ★ symbol. This allows the instructor freedom in assigning or avoiding such problems depending on the ability level of the class. Some instructors may find the assigning of some of these specially marked problems to be helpful in determining whether or not students should receive honors credit.

### Boxed Definitions

Key definitions, theorems, and properties are clearly marked with a shaded box so that students may more readily identify the most important concepts within each chapter.

### Chapter Review Material

The text contains an extensive chapter review section that includes

- A list of **key terms** and **concepts** covered in the chapter.
- A summary of the **key formulas** or **procedures** that are presented in the chapter.
- An optional set of **computer problems** that allows the student the opportunity to integrate his or her programming skills with the mathematical content areas of each chapter.
- A large selection of **review problems** at the end of each chapter.
- A **practice test** to help the student prepare for a classroom examination on the content of each chapter.

The text contains more sample examples than most of the competing textbooks. This wealth of sample examples includes elementary, intermediate, and advanced problems, with complete worked-out solutions. Applied sample problems are explained in detail whenever a particular problem set includes applied problems. Large detailed graphs are provided for each sample example requiring a graphical solution.

## THE STUDENT TUTORIAL

A complete learning package to accompany this book has been prepared by Marie Clarke of North Shore Community College. This *Student Tutorial* provides the student with

- **Detailed worked-out solutions** showing all the steps necessary to solve each of the odd-numbered problems in the text.
- Helpful **hints** and **suggestions** for students in those topic areas where students usually experience difficulty.
- A **complete graph** constructed as it would be normally done by a student for every odd-numbered problem in the text that requires a graph or sketch in order to complete the solution.

- An **extra Practice Test** for each chapter and the completely worked out solutions to all problems in this Practice Test.

The *Student Tutorial* is available through any college bookstore and may be ordered by the individual student if the book is not carried in stock.

## THE INSTRUCTOR'S MANUAL

The answers to all even-numbered problems in the text are contained in the *Instructor's Manual.* In addition, five versions of a chapter test ready for reproduction for each of the twelve chapters of the book are provided. Suggested chapter coverage outlines for the use of this book to meet the needs of a variety of schools are provided. Two copies of mid-term examinations covering one-half the content of the text and two copies of a final examination covering the entire content of the text are also provided in the *Instructor's Manual.*

## ACKNOWLEDGMENTS

The preparation of this textbook was significantly influenced by many individuals who gave willingly of their time and knowledge. We would like to express our deep appreciation to George Rumsey of Miami-Dade Community College who provided the major portion of the content area on vectors. We would like to personally thank several of the faculty at North Shore Community College—Tom Rourke, Marie Clarke, Hank Harmeling, Wally Hersey, Russ Sullivan, and Keith Pigott—for providing many helpful suggestions at various stages of the writing of the book.

A number of faculty reviewed various sections of the book, worked out the answers to many of the problem sets, or checked the accuracy of the content of the book. We are deeply indebted to the following people for their hard work, constructive comments, and insightful suggestions:

Helen Baril, *Quinnipiac College*
James Brown, *Northern Essex Community College*
Marie Clarke, *North Shore Community College*
Joyce Davis, *Pleasanton, Texas*
Gerald E. Gannon, *University of California-Fullerton*
Hank Harmeling, Jr., *North Shore Community College*
Arthur Hobbs, *Texas A&M University*
Jerry Karl, *Golden West College*
Richard Kennedy, *Central Connecticut State University*
Eldon Miller, *University of Mississippi*
Jacqueline Peterson, *Arizona State University*
Keith Piggott, *North Shore Community College*
Thomas Rourke, *North Shore Community College*
Shirley Sorenson, *University of Maryland-College Park*
Carol Stanton, *University of California-Riverside*
Stephanie Troyer, *University of Hartford*

Stanley Van Steenvoorte, *Northwestern Oklahoma State University*
Edward Vondrak, *Indiana Central University*
Shirley Wakin, *University of New Haven*
Carroll Wells, *Western Kentucky University*
Thomas Woods, *Central Connecticut State University*
T. Perrin Wright, Jr., *Florida State University*
Paul Young, *Kansas State University*

We also wish to acknowledge the contributions of those who reviewed the first edition:

Ignacio Bello, *Hillsboro Community College*
Rick Billstein, *University of Montana*
Janet Buck, *Seattle Pacific University*
Cal Carlson, *Brainerd Community College*
Mary Jane Causey, *University of Mississippi*
Rodney Chase, *Oakland Community College*
Charles Cook, *Tri-State University*
Vivian Dennis, *Eastfield College*
James B. Derr, *West Virginia University*
Bruce W. Earnley, *Northern Essex Community College*
Duncan O. Faus, *Monterey Peninsula College*
Charles R. Friese, *North Dakota State University*
Hank Harmeling, Jr., *North Shore Community College*
Warland Hersey, *North Shore Community College*
Calvin J. Holt, *Paul D. Camp Community College*
Jerry Karl, *Golden West College*
John G. Krupka, *Northampton County Area Community College*
Paul Lanphier, *Miami-Dade Community College*
David Lomen, *University of Arizona*
Ronald P. Morash, *University of Michigan-Dearborn*
Alan Olinsky, *Bryant College*
Chantal Shafroth, *North Carolina Central University*
Adele Shapiro, *Central Piedmont Community College*
Peter Sherman, *University of Oregon*
Charles Sinclair, *Portland State University*
John Snyder, *Sinclair Community College*
John L. Whitcomb, *University of North Dakota*
Edward Zannella, *Rhode Island Junior College*

We wish to especially thank Joan Peabody for the superb work she did in typing and re-typing the manuscript and the *Student Tutorial.*

Writing a textbook is a joint venture between authors and the members of a publishing team. We want to express our deep appreciation to Gary Folven, Ben Elderd, Carol Nolan-Fish, Jane Dahl, Louise Lindenberger; Lorraine Perrotta, Judy Fiske, and Bob Swan who provided us with much support and professional expertise from Allyn and Bacon.

A special word of love and appreciation is due to Nancy, Johnny, Marcia and Melissa Tobey who put up with the inconveniences of having an author in the house.

# ALGEBRA *and* TRIGONOMETRY

# 1 Basic Concepts of Algebra

■ *Get wisdom, get understanding: forget it not . . .*

*Proverbs* 4:5

A good foundation in the concepts of elementary algebra is essential for a student who wants to succeed in the study of more advanced mathematical topics. Such a foundation requires that a student can both perform the necessary algebraic steps (this takes a facility in algebraic skills) and understand what he or she is doing and why it is done (this takes a mastery of certain key definitions and theorems).

## 1-1 REAL NUMBERS

A **set** is a collection of objects. Sets are usually denoted by capital letters such as $A$, $B$, $C$, $Q$, $R$, or $W$. Any of the objects in a set are called the members or **elements** of a set. When listing the objects of a set, we enclose them in braces. Thus if we write $\{1, 6, 9\}$, we mean the set whose elements are the numbers 1, 6, and 9. If the set continues forever, we often use . . . to indicate that condition. If we wanted to indicate the set of all positive even numbers greater than three, we could write $\{4, 6, 8, 10, 12, \ldots\}$.

The numbers first encountered in elementary arithmetic are those used in counting, or the set of **counting numbers**, or **natural numbers**,

$$\{1, 2, 3, 4, \ldots\}$$

This set is later extended to include zero and the negatives of the counting numbers giving us the set of **integers**,

$$\{\ldots, -4, -3, -2, -1, 0, 1, 2, 3, 4, \ldots\}$$

The set of **rational numbers**—those that can be expressed as a ratio of two integers—are generally referred to as **fractions**. This set includes the integers: for example, 3 can be expressed as $\frac{3}{1}$, $\frac{6}{2}$, etc. In fact, this set includes all number expressions that involve only the operations of addition, subtraction, multiplication, and division of integers.

Next, a set of numbers called the **irrational numbers**, which cannot be expressed as the ratio of integers, is developed. This set includes such numbers as $\pi$, $\sqrt{5}$, and $\sqrt[3]{7}$. (The decimal expansion of these numbers will neither repeat nor terminate.)

The sets of rational and irrational numbers together make up a set called the **real numbers**. Elementary algebra is a study of the real numbers and their properties.

We can summarize some subsets of the set of real numbers that we have discussed as follows:

| *Type of Real Number* | *Description* | *Example* |
|---|---|---|
| Natural numbers | Those numbers used in counting | $\{1, 2, 3, \ldots\}$ |
| Integers | The counting numbers, 0, and the negative of each counting number | $\{\ldots -2, -1, 0, 1, 2, 3, \ldots\}$ |
| Rational numbers | Those numbers that *can* be written as the quotient of two integers $a$, $b$ where $b \neq 0$ | $\left\{-\frac{1}{2}, 3, 0, -\frac{1}{5}, \frac{2}{7}, 0.7, 0.\overline{12}, 4.16, \ldots\right\}$ |
| Irrational numbers | Those real numbers that *cannot* be written as the quotient of two integers | $\left\{\sqrt{2}, -3\sqrt{5}, e, \pi, -\frac{2}{5}\pi, \ldots\right\}$ |

---

The *decimal representation of a rational number* either terminates or has a pattern of repeating digits.

---

### Examples of the Decimal Representation of Rational Numbers

**1.** $\frac{1}{8} = 0.125$ — A terminating decimal

**2.** $\frac{1}{3} = 0.33333\ldots$ — A repeating decimal; the pattern or repeating 3's continues indefinitely

A repeating decimal is often denoted by a bar over the repeating digits.

**3.** $1.37373737\ldots = 1.\overline{37}$ A repeating decimal

---

The *decimal representation of an irrational number* does not end and does not have a pattern of repeating digits.

---

**Examples of the Decimal Representation of Irrational Numbers**

**4.** $\pi = 3.14159265358979\ldots$ There is no pattern of repeating digits.

**5.** $\sqrt{7} = 2.645751311\ldots$ There is no pattern of repeating digits.

The basic set of properties (axioms) of the real numbers can be used to justify all of the manipulations used in both arithmetic and algebra.

---

**Properties of Real Numbers (*Field Properties*)**

---

For real numbers $a$, $b$, $c$.

1. *Closure under Addition* If $a$ and $b$ are real numbers, then $a + b$ is a real number.
2. *Commutativity under Addition* $a + b = b + a$
3. *Associativity under Addition* $(a + b) + c = a + (b + c)$
4. *Additive Identity* $a + 0 = a$
5. *Additive Inverse* $a + (-a) = 0$
6. *Closure under Multiplication* If $a$ and $b$ are real numbers, then $ab$ is a real number.
7. *Commutativity under Multiplication* $ab = ba$
8. *Associativity under Multiplication* $(ab)c = a(bc)$
9. *Multiplicative Identity* $(a)(1) = a$
10. *Multiplicative Inverse* For all $a \neq 0$, $a\left(\frac{1}{a}\right) = 1$
11. *Distributive Property of Multiplication over Addition* $a(b + c) = ab + ac$

---

**EXAMPLE 1** Justify each of the following by stating one of the properties of real numbers.

**a.** $(5 + 7) + 3 = 5 + (7 + 3)$
**b.** $-2.3 + 2.3 = 0$
**c.** $(7)(-2.61) = (-2.61)(7)$
**d.** $(\sqrt{3})(\sqrt{5}) =$ a real number
**e.** $5(8 + 2) = 5 \cdot 8 + 5 \cdot 2$

***Solution***

**a.** Associative property of addition.
**b.** Additive inverse.
**c.** Commutative property of multiplication.
**d.** Closure property under multiplication.
**e.** Distributive property of multiplication over addition. ■

Any set of elements in which all eleven of these properties are true is called a **field**. The field of real numbers is a subset of another field, called the **complex numbers**, which will be studied in Chapter 2.

The student of algebra often encounters negative signs when working with real numbers. You should take great care that the following properties are understood and can be successfully employed in any problem situation.

### Basic Properties Involving Negatives for Real Numbers

| | *Property* | *Example* |
|---|---|---|
| **12.** | $-(-a) = a$ | $-(-18) = 18$ |
| **13.** | $(-a)b = -(ab) = a(-b)$ | $(-7)(2) = -14 = 7(-2)$ |
| **14.** | $(-1)a = -a$ | $-1(5) = -5$ |
| **15.** | $(-a)(-b) = ab$ | $(-8)(-3) = 24$ |
| **16.** | $(-a) + (-b) = -(a + b)$ | $(-7) + (-4) = -11$ |

The eleven properties of real numbers under addition and multiplication can be used to derive a number of other properties that involve addition or multiplication of real numbers. They can also be used to develop properties involving subtraction and division. To do so requires two additional definitions that are most useful.

***Definition of Subtraction***

For all real numbers $a$, $b$, $a - b = a + (-b)$.

*Definition of Division*

For all real numbers $a$, $b$, where $b \neq 0$, division of $a$ by $b$ is defined by $a \div b = (a)\left(\frac{1}{b}\right)$. $a \div b$ is often written as $\frac{a}{b}$.

EXAMPLE 2 Give an illustration of the definition of division.

***Solution*** $5 \div 3 = (5)\left(\frac{1}{3}\right)$ *5 divided by 3 is equivalent to 5 multiplied by the reciprocal of 3 (multiplicative inverse).* ■

Since all real numbers are classified as positive, negative, or zero, this set can be represented on a number line. By using a straight line from plane geometry (a straight line has infinite length) and choosing a point to be represented by 0 and another point to be represented by 1, we can, with certain agreements, establish a correspondence between the real numbers and the points of the line. We first agree to place the positive numbers to the right of 0 and the negative numbers to the left of 0. We also agree that the length of the line segment from 0 to 1 will be used as a unit measurement between all real numbers that differ by 1 and that real numbers between 0 and 1 will be represented by proportional parts of this unit. An illustration of such a number line is given in Figure 1-1.

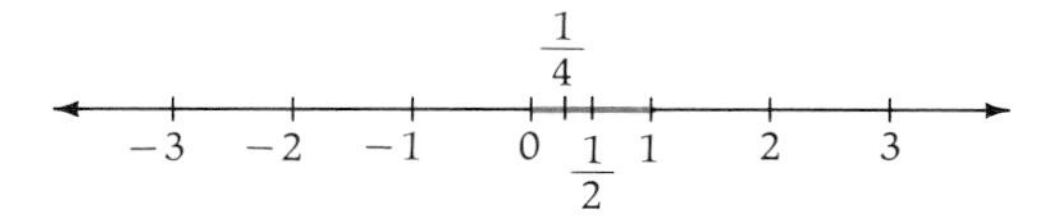


*Figure 1-1*

The number line is a very useful tool in the study of mathematics. One example of its use is in the study of inequalities. The symbol for inequality "<" can be defined in terms of the number line.

*Definition*

If $a$ and $b$ are real numbers, $a < b$ means that $a$ is to the left of $b$ on the number line.

This same symbol is sometimes turned in the other direction ">." $a < b$ is usually read "$a$ is less than $b$," and $b > a$ is usually read "$b$ is greater than $a$." Of course, both expressions mean the same thing and could be read in either direction. A good way to remember the meaning of the symbol is to recognize that the pointed end is always in the direction of the smaller number (i.e., the number which would be placed to the left of the other on the number line).

EXAMPLE 3 Replace the question mark by the proper symbol $<$ or $>$ in the following.

a. $5 \,?\, 2$ b. $-7 \,?\, -4$ c. $\frac{2}{5} \,?\, 0.43$

***Solution***

a. $5 > 2$

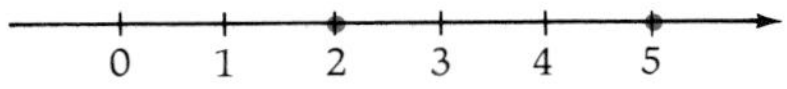

5 is greater than 2, since 2 is to the left of 5.

b. $-7 < -4$,

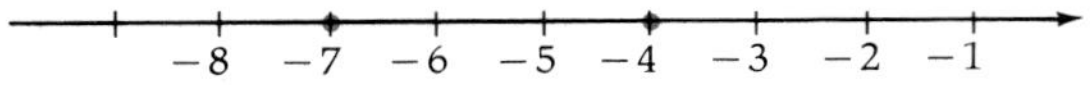

$-7$ is less than $-4$, since $-7$ is to the left of $-4$.

c. $\frac{2}{5} < 0.43$

$0.4 < 0.43$

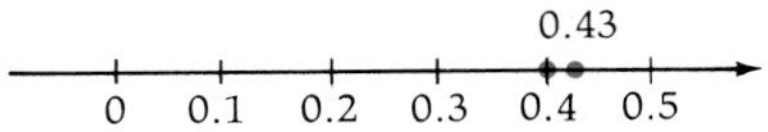

*We write $\frac{2}{5}$ in decimal form. 0.4 is to the left of 0.43.* ■

Distance, without regard to direction, is defined by the term **absolute value**. The absolute value of an expression is indicated by placing vertical bars before and after the expression. $|a - b|$ is read as "the absolute value of $a - b$." $|x|$ is read as "the absolute value of $x$." $|a - b|$ implies the distance from $a$ to $b$ or the distance from $b$ to $a$, without regard to direction. A formal definition follows.

*Definition*

$$|x| = \begin{cases} x & \text{if } x \text{ is positive} \\ 0 & \text{if } x \text{ is zero} \\ -x & \text{if } x \text{ is negative} \end{cases}$$

An immediate result of this definition is that the absolute value of an expression is always nonnegative. If $x$ is positive or zero, it is clear from the definition that $|x|$ is nonnegative. Also if $x$ is negative, then $|x| = -x$ (opposite of $x$) and is positive.

EXAMPLE 4 Evaluate the following.

a. $\left|\frac{1}{3}\right|$ b. $|-8.2|$ c. $|2 - 6|$ d. $|-3 - (-2)|$

*Solution*

a. $\left|\frac{1}{3}\right| = \frac{1}{3}$

b. $|-8.2| = 8.2$

c. $|2 - 6| = |-4| = 4$

d. $|-3 - (-2)| = |-3 + 2|$
$= |-1| = 1$ ■

## EXERCISE 1-1

In Problems 1–16, justify the equation by stating *one* of the field properties.

**1.** $4 + (2 + 3) = (4 + 2) + 3$ **2.** $5 + 0 = 5$ **3.** $(6)(1) = 6$

**4.** $(4)\left(\frac{1}{4}\right) = 1$ **5.** $5(2 + 3) = 5(2) + 5(3)$ **6.** $8 + (-8) = 0$

**7.** $(2 + 3) + 6 = (3 + 2) + 6$ **8.** $[(4)(2)](3) = (3)[(4)(2)]$ **9.** $-(8 + 6) + (8 + 6) = 0$

**10.** $8(2 + 0) = 8 \cdot 2$ **11.** $(7 \cdot 3) \cdot 5 = 7 \cdot (3 \cdot 5)$ **12.** $12 \cdot 1 = 12$

**13.** $(\pi)(\sqrt{7})$ is a real number. **14.** $(a + 6)b + (a + 6)2 = (a + 6)(b + 2)$

**15.** $x(1 + y + z) = x(y + 1 + z)$ **16.** $(3 + 4)(7 - 2) = (7 - 2)(3 + 4)$

In Problems 17–22, consider the set $A$ of numbers:

$$A = \left\{1.63,\ -2,\ 3,\ \frac{1}{5},\ -\frac{18}{3},\ \sqrt{2},\ 2\frac{1}{5},\ \sqrt{7},\ 0,\ 1.\overline{32},\ 5\right\}$$

In Problems 17–22, construct the specified set from the elements of set $A$.

**17.** List the set of natural numbers. **18.** List the set of integers.

**19.** List the set of real numbers. **20.** List the set of rational numbers.

**21.** List the set of irrational numbers. **22.** List the set of rational numbers that are not integers.

In Problems 23–26, write the fraction in decimal form. Then state whether the decimal representation terminates or has a pattern of repeating digits.

**23.** $\frac{6}{11}$ **24.** $\frac{29}{3}$ **25.** $\frac{3}{8}$ **26.** $\frac{6}{13}$

In Problems 27–34, determine which of the following sets are closed (a) under addition, (b) under multiplication.

**27.** {integers} **28.** {rational numbers} **29.** {all positive real numbers}

**30.** {all negative real numbers} **31.** {even counting numbers} **32.** {odd counting numbers}

**33.** {0} **34.** {−1, 0, 1}

**35.** The value of $\pi$ to five decimal places is 3.14159. A less accurate value, $\frac{22}{7}$, is often used for $\pi$ in elementary work. Find the decimal approximation of $\frac{22}{7}$ and then substitute $<$ or $>$ for the question mark in the following to make a true statement:

$$\frac{22}{7} ? \pi$$

In Problems 36–50, evaluate each expression.

**36.** $|12 - 6|$ **37.** $|-1.83|$ **38.** $|0|$ **39.** $|9 - 3|$

**40.** $|5 - 6|$ **41.** $|3 - 3|$ **42.** $|10 - 7|$ **43.** $|7 - 10|$

**44.** $|8 - (-5)|$ **45.** $\left|\frac{2}{5} - \frac{3}{8}\right|$ **46.** $\left|\frac{3}{7} - \frac{4}{9}\right|$ **47.** $|-3(2 + 7)|$

**48.** $|2(-3 + 4)|$ **49.** $|1.6 - 2.7 - 3.3|$ **50.** $|0.63 - 1.82 + 0.31|$

In Problems 51–60, replace the question mark by $<$ or $>$ in order to make a true statement.

**51.** $4 ? 6$ **52.** $9 ? 2$ **53.** $-3 ? 0$ **54.** $5 ? 0$ **55.** $-1 ? -5$

**56.** $3 ? -6$ **57.** $\frac{3}{2} ? 2$ **58.** $-\frac{10}{3} ? -3$ **59.** $\frac{1}{3} ? 0.33$ **60.** $-0.5 ? 0.6$

In Problems 61 and 62, write in numerical order from smallest to largest. (Table A may be used to approximate square roots.)

**61.** $-7, 3, 0, \frac{1}{5}, -2\sqrt{3}, 8, |-3|, |4 - 2|$

**62.** $\sqrt{2}, \pi, -3, 0, |5 - 7|, \frac{1}{4}, -2, \frac{7}{5}$

In Problems 63–72, show your mastery of the basic properties involving negatives for real numbers by accurately evaluating each expression.

**63.** $(-3) + (-6)$ **64.** $(-8) + (+3)$ **65.** $-2(5 + 7)$

**66.** $-8(-8 + 5)$ **67.** $(-3)(-2) + 5(-2)$ **68.** $(+3)(-8) + (-5)(-2)$

**69.** $\dfrac{-5 + 3 - 2}{-2}$ **70.** $\dfrac{(-8)(-2)}{4}$ **71.** $\dfrac{6 - (-8)}{-4}$ **72.** $\dfrac{-1(3) + 5(-2)}{2 - 4}$

**73.** What is the largest integer less than or equal to $\pi$?

★ **74.** What is the largest rational number less than or equal to $\pi$?

★ **75.** What is the largest real number less than or equal to $\pi$?

**76.** Give an example to show that division of real numbers is not commutative.

**77.** Give an example to show that subtraction of real numbers is not associative.

★ **78.** Prove for all real numbers $a, b, c$ if $c \neq 0$ that $(a + b) \div c = (a \div c) + (b \div c)$.

★ **79.** Prove for all real numbers $a, b, c$ that $a - (b - c) = a - b + c$.

★ **80.** Prove for all real numbers $a, b, c$ when $a \neq 0$ and $b \neq 0$ that $ab\left(\frac{1}{a} + \frac{1}{b}\right) = b + a$.

## 1-2 INTEGRAL EXPONENTS

Exponent notation is used throughout algebra. It provides a convenient notation to express repeated factors.

$x^2 = (x)(x)$ — *$x^2$ is read "x squared."*

$x^3 = (x)(x)(x)$ — *$x^3$ is read "x cubed."*

$x^4 = (x)(x)(x)(x)$ — *$x^4$ is read "x to the fourth power."*

We will define this for the general case.

*Definition*

$$x^n = \underbrace{(x)(x)(x)(x) \cdots (x)}_{n \text{ factors}}, \quad \text{if } n \text{ is a positive integer}$$

In words: "$x$ to the $n$th power, when $n$ is a positive integer, implies that $x$ is used as a factor $n$ times."

Thus we have

$$(-3)^5 = (-3)(-3)(-3)(-3)(-3) = -243$$

This definition gives rise to the following laws of exponents. Let $x$ be any real number and $a$, $b$, be positive integers.

*Law I*

$$x^a \cdot x^b = x^{a+b}$$

To multiply like bases, add the exponents.

***Examples*** $x^3 \cdot x^5 = x^8$ $\quad (23)^5 \cdot (23)^{10} = 23^{15}$

It should be noted that all the laws apply whether the base is a variable or a constant.

*Law II*

$$\frac{x^a}{x^b} = x^{a-b} \quad \text{where} \quad x \neq 0$$

To divide like nonzero bases, subtract the exponent of the denominator from the exponent of the numerator.

***Examples*** $x^7 \div x^5 = x^2$ $\qquad \dfrac{13^{12}}{13^7} = 13^5$ $\qquad \dfrac{y^{20}}{y^4} = y^{16}$

*Law III*

$$(x^a)^b = x^{ab}$$

To raise a power to a power, multiply exponents.

***Examples*** $(x^3)^5 = x^{15}$ $\qquad (3^8)^4 = 3^{32}$

*Law IV*

$$(xy)^a = x^a y^a$$

The power of a product is the product of the powers.

***Examples*** $(xy)^3 = x^3y^3$ $\qquad (2 \cdot 5)^{12} = 2^{12} \cdot 5^{12}$

*Law V*

$$\left(\frac{x}{y}\right)^a = \frac{x^a}{y^a} \quad \text{where} \quad y \neq 0$$

The power of a quotient is the quotient of the powers.

***Examples*** $\left(\dfrac{x}{y}\right)^4 = \dfrac{x^4}{y^4}$ $\qquad \left(\dfrac{3}{7}\right)^{16} = \dfrac{3^{16}}{7^{16}}$

Law of Exponents II gives rise to the necessity of two further definitions to expand the meaning of exponents. As written in Law II, $x^a \div x^b = x^{a-b}$, $x \neq 0$, actually would apply only when $a > b$, if we restrict ourselves to positive integers. To expand the concept of exponent, let us consider the cases when $a = b$ and $a < b$.

If $a = b$ and $x \neq 0$, then $\dfrac{x^a}{x^a} = x^{a-a} = x^0$ but $\dfrac{x^a}{x^a} = 1$ since any nonzero quantity divided by itself is 1. This observation gives rise to the following definition.

*Definition*

$$x^0 = 1, \quad \text{for all } x \neq 0.$$

*Examples* $\left(-\frac{2}{3}\right)^0 = 1 \qquad (5.2)^0 = 1$

**Note:** The symbol $0^0$ is not defined.

When $a < b$, $\frac{x^a}{x^b} = x^{a-b}$ would give rise to a negative exponent. For instance,

$$\frac{x^3}{x^5} = x^{3-5} = x^{-2}$$

From the definition of exponent, however, and a basic principle of simplifying fractions, we have

$$\frac{x^3}{x^5} = \frac{(x)(x)(x)}{(x)(x)(x)(x)(x)} = \frac{1}{x^2}$$

Hence the following definition.

*Definition*

If $a$ is a positive integer and $x \neq 0$, then

$$x^{-a} = \frac{1}{x^a}$$

EXAMPLE 1 Evaluate $3^{-3}$.

***Solution*** $3^{-3} = \frac{1}{3^3} = \frac{1}{27}$ ■

Simplify each of the following in Examples 2–6. Express each answer so that each variable appears only once.

EXAMPLE 2 $(5x^3y)(2xy^4) = (5)(2)x^{3+1} \cdot y^{1+4} = 10x^4y^5$

EXAMPLE 3 $(2ab^6)^3 = 2^3 \cdot a^3 \cdot b^{6 \cdot 3} = 8a^3b^{18}$

EXAMPLE 4 $\frac{12x^8y^6}{24x^3y^8} = \frac{12}{24} \cdot \frac{x^8}{x^3} \cdot \frac{y^6}{y^8} = \frac{1}{2} \cdot x^5 \cdot \frac{1}{y^2} = \frac{x^5}{2y^2}$

EXAMPLE 5 $\left(\frac{a^4b}{2}\right)\left(\frac{2a}{b^2}\right)^3 = \left(\frac{a^4b}{2}\right)\left(\frac{2^3a^3}{b^6}\right) = \frac{2^3a^7b}{2b^6} = \frac{2^2a^7}{b^5} = \frac{4a^7}{b^5}$

EXAMPLE 6 $(-2xy)^3(-3x^2y)^2 = [(-2)^3x^3y^3][(-3)^2x^4y^2] = (-8x^3y^3)(9x^4y^2) = -72x^7y^5$

Note carefully the use of sign rules in each step of Example 6.

The definitions we introduced for $x^0$ and $x^{-a}$ will allow us to make further use of the five laws of exponents.

Notice that the definitions are consistent with the laws of exponents, and therefore Laws I–V will still be valid when we allow exponents to be *any* integer.

To simplify wording of problems, we will assume in all examples and exercises in this text involving exponents that the variables that appear in the denominators of fractions are limited to nonzero real values.

Simplify each of the following using the laws of exponents in Examples 7–9. Express each answer so that each variable appears only once. Leave your answer with only positive exponents.

**EXAMPLE 7** $a^{-5} \cdot a^{-8} = a^{-13} = \dfrac{1}{a^{13}}$

**EXAMPLE 8** $(2x^{-4}y^3)^{-6} = 2^{-6}x^{24}y^{-18} = \dfrac{1}{2^6} \cdot \dfrac{x^{24}}{1} \cdot \dfrac{1}{y^{18}} = \dfrac{x^{24}}{2^6y^{18}} = \dfrac{x^{24}}{64y^{18}}$

**EXAMPLE 9** $\dfrac{b^{-6}d^{-8}}{b^3d^{-5}} = \dfrac{b^{-6}}{b^3} \cdot \dfrac{d^{-8}}{d^{-5}} = b^{-6-3} \cdot d^{-8-(-5)} = b^{-9}d^{-3} = \dfrac{1}{b^9d^3}$

The results of Example 9 help us to observe a useful pattern if the numerator and denominator consist only of **factors**. A factor can be moved from numerator to denominator or from denominator to numerator by changing the **sign** of its **exponent**. We will use this property in Example 10. Study the example carefully. Be sure you understand each step.

**EXAMPLE 10** Simplify $\dfrac{(2x^2y^{-3})^{-2}}{(-3x^{-4}y^{-2})^3}$.

***Solution*** $\dfrac{2^{-2}x^{-4}y^6}{(-3)^3x^{-12}y^{-6}} = \dfrac{x^{12}y^6y^6}{(-3)^3 \cdot 2^2x^4} = \dfrac{x^8y^{12}}{(-27)(4)} = -\dfrac{x^8y^{12}}{108}$ ■

***A Word of Caution!***

Some students tend to apply these rules where they should not be applied. Consider carefully the following example.

**EXAMPLE 11** Simplify $(x + y)^{-1}$.

***Solution*** $(x + y)^{-1} = \dfrac{1}{(x + y)^1} = \dfrac{1}{x + y}$

Be sure you understand that $(x + y)^{-1} \neq x^{-1} + y^{-1}$ ■

### Scientific Notation

**Definition of Scientific Notation for a Number**

A positive number is expressed in scientific notation if it is written in the form $a \times 10^n$ where $1 \leq a < 10$ and $n$ is an integer.

To write a number written in ordinary decimal form in scientific notation:

1. Move the decimal point to obtain a number between 1 and 10.
2. Multiply that number by $10^n$ or $10^{-n}$ where $n$ is the number of places the decimal point was moved to the left or to the right, respectively.

**EXAMPLE 12**

$$28 = 2.8 \times 10^1 \qquad 0.63 = 6.3 \times 10^{-1}$$
$$23{,}184 = 2.3184 \times 10^4 \qquad 0.00038 = 3.8 \times 10^{-4}$$
$$1{,}400{,}000 = 1.4 \times 10^6 \qquad 0.000000667 = 6.67 \times 10^{-7}$$

It is important to be able to apply the laws of exponents to numbers written in scientific notation. This provides an important means of checking work that is done on a scientific calculator to see if it is "in the ballpark."

Evaluate each of the following without using a calculator in Examples 13 and 14. Express your answer in scientific notation.

**EXAMPLE 13** $(1.2 \times 10^{25})^2$

***Solution*** $(1.2 \times 10^{25})^2 = (1.2)^2 \times (10^{25})^2 = 1.44 \times 10^{50}$ ■

**EXAMPLE 14** $\left(\dfrac{8 \times 10^{-18}}{2 \times 10^{-14}}\right)^3$

***Solution***

$$\left(\frac{8 \times 10^{-18}}{2 \times 10^{-14}}\right)^3 = \left(\frac{8}{2} \times \frac{10^{-18}}{10^{-14}}\right)^3 = (4 \times 10^{-4})^3$$

$$= 4^3 \times 10^{(-4)(3)} = \underline{64 \times 10^{-12}}$$

*Note that this is* not *yet in scientific notation.*

$$= 6.4 \times 10^1 \times 10^{-12}$$
$$= \underline{6.4 \times 10^{-11}}$$

*This is in scientific notation.* ■

## EXERCISE 1-2

Simplify Problems 1–44. Leave all answers with only positive exponents. Express your answer so that each variable occurs only once. Assume that all expressions represent real numbers.

1. $(2x^2)^4$
2. $(3xy^2)^3$
3. $\left(\dfrac{2}{x^2y}\right)^3$
4. $[(x^2)^3]^4$
5. $\left(\dfrac{xy^2}{x^2y}\right)^2$
6. $\left(\dfrac{2x^3}{xy^2}\right)^3$
7. $[2(xy)^2]^3$
8. $\left(\dfrac{1}{3x^2}\right)^2$
9. $\dfrac{(x^2)^3}{(x^3)^2}$
10. $\dfrac{(-3x)^3}{9x}$
11. $\dfrac{-4x^2}{(2x)^3}$
12. $\dfrac{(2x^2y^3)^3}{(2x^3y^2)^2}$
13. $2x^2y(x^2y)^3$
14. $(-x^2y)^3(-2xy^3)^2$
15. $[2(xy)^2]^5[3(x^2y)^3]^2$ (*Hint:* Do inner parentheses first.)
16. $\left(\dfrac{x}{y}\right)^3\left(\dfrac{2x}{y}\right)^4$
17. $\left(\dfrac{-2}{x}\right)^3\left(\dfrac{x}{2}\right)^2$
18. $\left(\dfrac{x^2y}{3}\right)\left(\dfrac{3}{xy^2}\right)^2$
19. $\left(\dfrac{-2x}{5y^2}\right)^3\left(\dfrac{5y}{4x^2}\right)^2$
20. $\left(\dfrac{x^3}{8}\right)^2\left(\dfrac{4}{x^2}\right)^3$
21. $x^{-3}y^{-5}$
22. $\left(\dfrac{a}{b}\right)^{-5}$
23. $\dfrac{1}{3^{-2}}$
24. $(ab)^{-1}$
25. $(a+b)^{-1}$
26. $a^{-1}+b^{-1}$
27. $x^{-2}y^4z^{-1}$
28. $(x^3)^{-2}$
29. $(x^{-2}y^4)^{-4}$
30. $\dfrac{x^{-3}}{x^5}$
31. $\dfrac{x^{-2}x^{-6}}{x^7}$
32. $\dfrac{x^{-3}y^2}{x^{-5}y^{-1}}$
33. $\dfrac{x^3y^{-7}}{x^3y^0}$
34. $(x^2)^{-5}(x^{-3})^2$
35. $(-2x^3)^{-3}(3x^{-1})^2$
36. $(3^{-1}x^{-4})^{-3}$
37. $\left(\dfrac{x^{-1}y^2}{y^{-3}}\right)^{-5}$
38. $(2x^2y^{-3})^{-3}(-2x^{-3}y^2)^2$
39. $\left(\dfrac{x}{y}\right)^{-3}\left(\dfrac{x^2}{y^{-1}}\right)^0$
40. $\left(\dfrac{x^2y^{-1}}{x^{-5}}\right)\left(\dfrac{x^{-1}y^3}{x^5}\right)^{-3}$
41. $\dfrac{(-2x^2y^{-3}z^0)^3}{(3x^{-1}y^2z^{-4})^2}$
42. $\dfrac{(4x^5yz^{-2})^2}{(-3xy^{-2}z^3)^3}$
43. $\left(\dfrac{3x^2}{y^{-3}}\right)\left(\dfrac{2x}{y^{-4}}\right)^{-3}$
44. $\left(\dfrac{x^2y^{-2}}{3x}\right)^0\left(\dfrac{-5x^{-2}y}{y^2}\right)^{-2}$

Express in scientific notation each number in Problems 45–54.

**45.** 34,560,000 **46.** 1,893,000,000 **47.** 0.00368 **48.** 0.00001217

**49.** The total wooded area of the world's largest forest is 2,700,000,000 acres.

**50.** The sun has an internal temperature of approximately 20,000,000°K.

**51.** The distance light travels in one year is known as a light year. A light year is approximately 5,878,000,000,000 miles.

**52.** Yellow light has a wavelength of 0.00000059 meter.

**53.** An electron has a charge of 0.00000000048 electrostatic unit.

**54.** An atom of oxygen weighs 0.00000000000000000000002656 gram.

Express in decimal notation each number in Problems 55–64.

**55.** $3.11 \times 10^5$ **56.** $2.38 \times 10^4$ **57.** $8 \times 10^{12}$ **58.** $5 \times 10^{15}$

**59.** $5.1836 \times 10^{-4}$ **60.** $3.11286 \times 10^{-7}$ **61.** $1.3 \times 10^{-11}$ **62.** $7.6 \times 10^{-10}$

**63.** The estimated population of the United States at one time was $2.41 \times 10^8$ people.

**64.** The average distance of the planet Neptune to the sun is $2.795 \times 10^9$ miles.

Perform the calculations in Problems 65–70 by using the laws of exponents. *Do not* use a calculator. Express your answer in scientific notation.

**65.** $\dfrac{9.3 \times 10^{12}}{3.1 \times 10^5}$ **66.** $\dfrac{6.8 \times 10^3}{1.7 \times 10^{-6}}$

**67.** $(2 \times 10^3)(1.5 \times 10^{-2})(4 \times 10^{-5})$ **68.** $(6 \times 10^{-8})(4 \times 10^{-2})(1.5 \times 10^3)$

**69.** $\dfrac{(8 \times 10^{-6})(3 \times 10^{-8})}{6 \times 10^{-3}}$ **70.** $\dfrac{(1.5 \times 10^{-12})(3 \times 10^5)}{9 \times 10^{-4}}$

Simplify Problems 71–74 for integers $m$ and $n$. In each problem, express your answer so that the variables $x$ and $y$ occur only once.

★ **71.** $(3x^m)(2x^{n+3})$ ★ **72.** $(4x^{n+m})(3x^2)$ ★ **73.** $\dfrac{12x^{m+4}}{5x^{m-2}}$ ★ **74.** $\dfrac{2x^m y^n}{10x^3 y^{2-n}}$

---

## 1-3 RADICALS

Almost all branches of mathematics require the use of expressions in radical form. Many equations in such areas as the physical sciences, economics, and psychology require the use of radicals. Any student who takes algebra in college should be able to work with radical expressions. We first consider the idea of roots.

***Definition***

Assume that $y$ and $x$ are real numbers and $n$ is a positive integer. If $y^n = x$, then $y$ is an $n$th root of $x$.

EXAMPLE 1 Since $3^2 = 9$, 3 is a square root of 9.
Since $2^4 = 16$, 2 is a fourth root of 16.
Since $(-3)^2 = 9$, $-3$ is a square root of 9.
Since $(-2)^4 = 16$, $-2$ is a fourth root of 16.

Notice from these examples that 9 has two square roots, 3 and $-3$. It can be shown by more advanced methods that every number has two square roots, three cube roots, four fourth roots, etc. Not all of these roots, however, are in the set of real numbers.

We use the radical sign to indicate the **principal root** of a number.

The symbol $\sqrt{\ }$ is called the **radical sign**. An expression containing the radical sign is called a **radical**. The radical $\sqrt[5]{32}$ is read "the principal fifth root of 32." The number under the radical sign (32 in this case) is called the **radicand**, and the integer 5 is the **index**. If the index is 2, it is usually omitted. Hence $\sqrt{8}$ and $\sqrt[2]{8}$ each indicate the principal square root of 8. If the radical sign is used to indicate any root other than the square root, the index must be written. The index number is always a positive integer greater than 1. We call $\sqrt[n]{a}$ the **principal *n*th root** of $a$.

EXAMPLE 2 $\sqrt[4]{24}$ means the principal fourth root of 24.
$\sqrt[3]{24}$ means the principal cube root of 24.
$\sqrt{24}$ means the principal square root of 24.

*Definition*

If $x$ is a real number and $n$ is a positive integer, then $\sqrt[n]{x}$, the principal $n$th root of $x$, is

- **a.** Zero if $x = 0$.
- **b.** Positive if $x > 0$.
- **c.** Negative if $x < 0$ and $n$ is an *odd* integer.
- **d.** Not a real number if $x < 0$ and $n$ is an *even* integer.

EXAMPLE 3 Evaluate the following square roots, assuming that all variables appearing inside any radical are positive.

**a.** $\sqrt{16}$ **b.** $\sqrt{\dfrac{1}{25}}$ **c.** $-\sqrt{36}$

**d.** $\sqrt{-36}$ **e.** $\sqrt{x^4}$ **f.** $\sqrt{y^{12}}$

***Solution***

**a.** $\sqrt{16} = 4$ since $4^2 = 16$

**b.** $\sqrt{\dfrac{1}{25}} = \dfrac{1}{5}$ since $\left(\dfrac{1}{5}\right)^2 = \dfrac{1}{25}$

c. $-\sqrt{36} = -6$
d. $\sqrt{-36}$ is not a real number
e. $\sqrt{x^4} = x^2$ since $(x^2)^2 = x^4$
f. $\sqrt{y^{12}} = y^6$ since $(y^6)^2 = y^{12}$ ■

**EXAMPLE 4** Evaluate the following cube roots.

**a.** $\sqrt[3]{8}$ **b.** $\sqrt[3]{-8}$ **c.** $\sqrt[3]{125}$ **d.** $\sqrt[3]{\frac{1}{64}}$ **e.** $\sqrt[3]{x^6}$ **f.** $\sqrt[3]{y^{15}}$

***Solution***

**a.** $\sqrt[3]{8} = 2$ since $2^3 = 8$
**b.** $\sqrt[3]{-8} = -2$ since $(-2)^3 = -8$
**c.** $\sqrt[3]{125} = 5$ since $5^3 = 125$
**d.** $\sqrt[3]{\frac{1}{64}} = \frac{1}{4}$ since $\left(\frac{1}{4}\right)^3 = \frac{1}{64}$
**e.** $\sqrt[3]{x^6} = x^2$ since $(x^2)^3 = x^6$
**f.** $\sqrt[3]{y^{15}} = y^5$ since $(y^5)^3 = y^{15}$ ■

**EXAMPLE 5** Evaluate the following higher-order roots. Assume all variables are positive.

**a.** $\sqrt[4]{81}$ **b.** $\sqrt[4]{-81}$ **c.** $\sqrt[4]{x^{20}}$ **d.** $\sqrt[5]{-32}$
**e.** $\sqrt[5]{x^{30}}$ **f.** $\sqrt[6]{1}$ **g.** $\sqrt[7]{x^{21}}$ **h.** $\sqrt[20]{-1}$

***Solution***

**a.** $\sqrt[4]{81} = 3$ since $3^4 = 81$
**b.** $\sqrt[4]{-81}$ is not a real number by the definition of $\sqrt[n]{x}$
**c.** $\sqrt[4]{x^{20}} = x^5$ since $(x^5)^4 = x^{20}$
**d.** $\sqrt[5]{-32} = -2$ since $(-2)^5 = -32$
**e.** $\sqrt[5]{x^{30}} = x^6$ since $(x^6)^5 = x^{30}$
**f.** $\sqrt[6]{1} = 1$ since $1^6 = 1$
**g.** $\sqrt[7]{x^{21}} = x^3$ since $(x^3)^7 = x^{21}$
**h.** $\sqrt[20]{-1}$ is not a real number by the definition of $\sqrt[n]{x}$ ■

There are certain laws of radicals that are very useful. In stating these laws we assume that the indicated roots do exist.

*Laws of Radicals*

If $a$ and $b$ are real numbers and $m$ and $n$ are positive integers, then

**I.** $(\sqrt[n]{a})^n = a$
**II.** $\sqrt[n]{ab} = \sqrt[n]{a}\,\sqrt[n]{b}$

**III.** $\sqrt[n]{\dfrac{a}{b}} = \dfrac{\sqrt[n]{a}}{\sqrt[n]{b}}$ $(b \neq 0)$

**IV.** $\sqrt[m]{\sqrt[n]{a}} = \sqrt[mn]{a}$

**Caution:** The radical symbol is always the *principal* root. $x^2 = 4$ and $x = \sqrt{4}$ are not the same. If $x^2 = 4$, then $x = 2$ or $x = -2$. *But* if $x = \sqrt{4}$, then the only solution is $x = 2$.

When students see that $\sqrt{25} = 5$ and $\sqrt{36} = 6$, they are tempted to say that $\sqrt{x^2} = x$. This is not correct unless we add a restriction. Suppose that $x = -3$; then $\sqrt{(-3)^2} = \sqrt{9} = 3$. But if we thought that $\sqrt{x^2} = x$, we would have $\sqrt{(-3)^2} = -3$. So in general, if $x$ is any real number, then $\sqrt{x^2} = |x|$. If $x \geq 0$, then $\sqrt{x^2} = x$. To avoid confusion throughout this text, we will assume that **all variables in the radicand represent positive real numbers**.

In working with radicals it is common practice to simplify radicals whenever possible. A radical in simplest form will have no factor of the radicand raised to a power equal to or greater than the index number.

**EXAMPLE 6** Simplify $\sqrt{75}$.

***Solution*** The question is: "Does 75 have a factor that is raised to the second or greater power?" To answer this question, we need only to look for factors of 75 that are perfect squares.

$$\sqrt{75} = \sqrt{(25)(3)}$$
$$= \sqrt{25}\sqrt{3} \quad \textit{Law of Radicals II.}$$
$$= 5\sqrt{3} \quad \blacksquare$$

In practice, it is usually easier to write the radical as a product of two radicals with the first radical containing all the factors for which the exact root can be obtained.

**EXAMPLE 7** Simplify $\sqrt{140x^5y^7z^8}$.

***Solution***

$\sqrt{140x^5y^7z^8}$

$= \sqrt{4x^4y^6z^8}\sqrt{35xy}$ *This step is not obvious. It will take a minute to find the factors of 140 such that one is a perfect square.* ■

$= 2x^2y^3z^4\sqrt{35xy}$

Similar measures can be used to simplify cube roots.

**EXAMPLE 8** Simplify $\sqrt[3]{16x^4y^5}$.

***Solution*** $\sqrt[3]{16x^4y^5} = \sqrt[3]{8x^3y^3}\,\sqrt[3]{2xy^2} = 2xy\sqrt[3]{2xy^2}$ ■

**EXAMPLE 9** Simplify $\sqrt[3]{-81x^8y^9z^{10}}$.

***Solution***

$$\sqrt[3]{-81x^8y^9z^{10}} = \sqrt[3]{-27x^6y^9z^9}\,\sqrt[3]{3x^2z}$$

*It is common practice when evaluating roots with an odd index to remove the negative sign from the radicand.*

$$= -3x^2y^3z^3\,\sqrt[3]{3x^2z} \quad ■$$

The distributive property is often useful in conjunction with Law of Radicals II.

By the distributive property and Law II,

$$\sqrt{6}(\sqrt{5} + \sqrt{7}) = \sqrt{30} + \sqrt{42}$$

Also by the distributive property,

$$3\sqrt{5} + 2\sqrt{5} = (3 + 2)\sqrt{5} = 5\sqrt{5}$$

In this way, like radical expressions can be combined. If the radicands are not the same, we look for opportunities to simplify the radical expressions.

**EXAMPLE 10** Simplify $3\sqrt{2} + \sqrt{50} + \sqrt{32}$.

***Solution*** We first simplify $\sqrt{50}$ and $\sqrt{32}$, obtaining

$$\begin{aligned} 3\sqrt{2} + \sqrt{50} + \sqrt{32} &= 3\sqrt{2} + 5\sqrt{2} + 4\sqrt{2} \\ &= (3 + 5 + 4)\sqrt{2} \\ &= 12\sqrt{2} \quad ■ \end{aligned}$$

**EXAMPLE 11** Simplify $\sqrt[3]{a^4} + \sqrt[3]{8a^4} - a\sqrt[3]{16a}$.

***Solution***

$$\sqrt[3]{a^4} + \sqrt[3]{8a^4} - a\sqrt[3]{16a} = a\sqrt[3]{a} + 2a\sqrt[3]{a} - 2a\sqrt[3]{2a}$$

*Again we first simplify each radical.*

$$= 3a\sqrt[3]{a} - 2a\sqrt[3]{2a} \quad ■$$

Note that since $\sqrt[3]{a}$ and $\sqrt[3]{2a}$ are not like radical terms, we cannot combine the last two quantities. The radicands must contain exactly the same expression and the index of the roots must be the same in order to add two radicals.

With the use of the distributive property we can multiply values involving radicals. In many instances, the resulting product can be further simplified.

**EXAMPLE 12** Multiply the following. Simplify your answer.

**a.** $5\sqrt{2}(\sqrt{6} + 4\sqrt{5})$ **b.** $(\sqrt{3} + \sqrt{6})(\sqrt{2} - \sqrt{6})$

***Solution***

**a.**

$$\begin{aligned} 5\sqrt{2}(\sqrt{6} + 4\sqrt{5}) &= 5\sqrt{2}\sqrt{6} + (5)(4)\sqrt{2}\sqrt{5} \\ &= 5\sqrt{12} + 20\sqrt{10} \\ &= 5(2)\sqrt{3} + 20\sqrt{10} \\ &= 10\sqrt{3} + 20\sqrt{10} \end{aligned}$$

*Observe that we must simplify* $\sqrt{12}$

**b.** $(\sqrt{3}+\sqrt{6})(\sqrt{2}-\sqrt{6}) = \sqrt{3}\sqrt{2}+\sqrt{6}\sqrt{2}-\sqrt{3}\sqrt{6}-\sqrt{6}\sqrt{6}$

$= \sqrt{6}+\sqrt{12}-\sqrt{18}-\sqrt{36}$ *Observe that we must simplify the last three radicals*

$= \sqrt{6}+2\sqrt{3}-3\sqrt{2}-6$ ■

A radical expression is not considered to be completely simplified if it contains a fraction or if the denominator of a fraction contains a radical. To summarize, we have:

---

An algebraic expression containing radicals is in simplest form when

1. Each radical in the expression is in simplest form.
2. No radical appears in the denominator of a fraction.
3. No radical contains a fraction.

---

If a fraction contains a radical in the denominator, we *rationalize the denominator.* This will remove that radical from the denominator.

EXAMPLE 13 Simplify $\frac{3}{\sqrt{8}}$.

***Solution*** Here the requirement is to find the smallest number that, when multiplied by $\sqrt{8}$, yields the square root of a perfect square. Since $(\sqrt{8})(\sqrt{2}) = \sqrt{16}$, the desired number is $\sqrt{2}$. Hence

$$\frac{3}{\sqrt{8}} = \frac{3}{\sqrt{8}}\cdot\frac{\sqrt{2}}{\sqrt{2}} = \frac{3\sqrt{2}}{\sqrt{16}} = \frac{3\sqrt{2}}{4} \quad ■$$

If the radical contains a fraction, we can simplify using Law of Radicals III.

EXAMPLE 14 Simplify $\sqrt{\frac{8a^3}{27b^7}}$.

***Solution***

$$\sqrt{\frac{8a^3}{27b^7}} = \frac{\sqrt{8a^3}}{\sqrt{27b^7}} = \frac{\sqrt{8a^3}}{\sqrt{27b^7}}\cdot\frac{\sqrt{3b}}{\sqrt{3b}} = \frac{\sqrt{24a^3b}}{\sqrt{81b^8}} = \frac{\sqrt{24a^3b}}{9b^4} = \frac{2a\sqrt{6ab}}{9b^4} \quad ■$$

Rationalizing the denominator will often be required with cube roots or higher-order radicals.

EXAMPLE 15 Simplify $\frac{2}{\sqrt[3]{5}}$.

***Solution***

$$\frac{2}{\sqrt[3]{5}} = \frac{2}{\sqrt[3]{5}} \cdot \frac{\sqrt[3]{25}}{\sqrt[3]{25}} = \frac{2\sqrt[3]{25}}{\sqrt[3]{125}} = \frac{2\sqrt[3]{25}}{5}$$ ■

The radical in the denominator may contain a binomial. However, the same basic approach is used. We wish to obtain a denominator for which we can exactly evaluate the radical.

EXAMPLE 16 Simplify $\dfrac{a-b}{\sqrt{a+b}}$.

***Solution***

$$\frac{a-b}{\sqrt{a+b}} = \frac{a-b}{\sqrt{a+b}} \cdot \frac{\sqrt{a+b}}{\sqrt{a+b}} = \frac{(a-b)\sqrt{a+b}}{\sqrt{(a+b)^2}} = \frac{(a-b)\sqrt{a+b}}{a+b}$$ ■

If the denominator is of the form $\sqrt{x} + \sqrt{y}$, we will need to multiply numerator and denominator by $\sqrt{x} - \sqrt{y}$. If the denominator is of the form $\sqrt{x} - \sqrt{y}$, then we multiply by $\sqrt{x} + \sqrt{y}$. This approach is taken to eliminate the radical from the denominator using the fact that since $(a+b)(a-b) = a^2 - b^2$, we therefore have

$$(\sqrt{x} + \sqrt{y})(\sqrt{x} - \sqrt{y}) = (\sqrt{x})^2 - (\sqrt{y})^2 = x - y$$

EXAMPLE 17 Simplify $\dfrac{3}{\sqrt{5}+1}$.

***Solution*** To rationalize the denominator of this fraction, we must multiply the numerator and denominator by $(\sqrt{5} - 1)$.

$$\frac{3}{\sqrt{5}+1} = \frac{3}{\sqrt{5}+1} \cdot \frac{\sqrt{5}-1}{\sqrt{5}-1} = \frac{3(\sqrt{5}-1)}{\sqrt{25}-1} = \frac{3\sqrt{5}-3}{4}$$ ■

EXAMPLE 18 Simplify $\dfrac{\sqrt{x} - 3\sqrt{xy}}{2\sqrt{x} + 5\sqrt{y}}$.

***Solution***

$$\frac{(\sqrt{x} - 3\sqrt{xy})}{(2\sqrt{x} + 5\sqrt{y})} \cdot \frac{(2\sqrt{x} - 5\sqrt{y})}{(2\sqrt{x} - 5\sqrt{y})}$$

*We multiply numerator and denominator by* $(2\sqrt{x} - 5\sqrt{y})$.

$$= \frac{2\sqrt{x^2} - 6\sqrt{x^2y} - 5\sqrt{xy} + 15\sqrt{xy^2}}{4\sqrt{x^2} - 25\sqrt{y^2}}$$

*Several of the radicals may be simplified.*

$$= \frac{2x - 6x\sqrt{y} - 5\sqrt{xy} + 15y\sqrt{x}}{4x - 25y}$$

*This is a final answer, since no terms in the numerator can be combined.* ■

If a radical contains another radical, it may often be simplified by Law of Radicals IV.

**EXAMPLE 19** Simplify $\sqrt[5]{\sqrt[3]{x^4}}$.

***Solution*** Since $\sqrt[m]{\sqrt[n]{a}} = \sqrt[mn]{a}$, we have $\sqrt[5]{\sqrt[3]{x^4}} = \sqrt[15]{x^4}$ ■

## EXERCISE 1-3

Simplify Problems 1–54. Assume values of $x$ and $y$ that yield real numbers in each expression.

**1.** $\sqrt{144}$ **2.** $\sqrt[4]{16}$ **3.** $\sqrt[5]{-32}$ **4.** $(\sqrt[3]{-27})^2$ **5.** $(\sqrt[5]{-1})^3$

**6.** $\sqrt{27}$ **7.** $\sqrt{125}$ **8.** $\sqrt[3]{16}$ **9.** $\sqrt[3]{x^6y^9}$ **10.** $\sqrt[8]{x^{16}y^8}$

**11.** $\sqrt[4]{x^{12}y^8}$ **12.** $\sqrt[4]{32x^5}$ **13.** $\sqrt[3]{64x^4y^6}$ **14.** $\sqrt{144x^{10}y^5}$ **15.** $\sqrt[5]{64x^6y^5}$

**16.** $\sqrt{3} + 4\sqrt{3} - 2\sqrt{3}$ **17.** $3\sqrt{5} - 2\sqrt{3} + 4\sqrt{5} - \sqrt{3}$ **18.** $2\sqrt{2} - 3\sqrt[3]{2} - \sqrt{8}$

**19.** $\sqrt{9} + \sqrt{18} - \sqrt{36}$ **20.** $\sqrt[3]{16} - \sqrt[3]{54} + \sqrt[3]{3}$ **21.** $\sqrt{50} + \sqrt[3]{2} - \sqrt{32} - \sqrt[3]{16}$

**22.** $\sqrt{2}\sqrt{5}$ **23.** $(2\sqrt{7})(3\sqrt{2})$ **24.** $3\sqrt{3}(2\sqrt{5} - 3\sqrt{7})$

**25.** $5\sqrt{3}(2\sqrt{6} + \sqrt{15})$ **26.** $\sqrt[3]{2}(4\sqrt[3]{4} - 2\sqrt[3]{32})$

**27.** $\sqrt{3}(\sqrt{2} + \sqrt{7}) + \sqrt{2}(\sqrt{5} - \sqrt{3})$ **28.** $(\sqrt{3} + \sqrt{2})(\sqrt{2} + \sqrt{5})$

**29.** $(2\sqrt{2} - 5\sqrt{7})(2\sqrt{2} + 5\sqrt{7})$ **30.** $\dfrac{1}{\sqrt{5}}$

**31.** $\dfrac{3}{\sqrt{20}}$ **32.** $\dfrac{3}{\sqrt[3]{2x}}$ **33.** $\dfrac{2y}{\sqrt[3]{4x^2y}}$ **34.** $\dfrac{1}{\sqrt{x+3}}$ **35.** $\dfrac{8}{\sqrt{x-2}}$

**36.** $\dfrac{1}{\sqrt{3}-2}$ **37.** $\dfrac{4}{3-\sqrt{5}}$ **38.** $\dfrac{1}{\sqrt{3}-\sqrt{5}}$ **39.** $\dfrac{a-1}{\sqrt{a}-1}$ **40.** $\dfrac{\sqrt{3}+\sqrt{2}}{\sqrt{3}-\sqrt{2}}$

**41.** $\sqrt[8]{\sqrt[5]{x^7}}$ **42.** $\sqrt{\sqrt{\sqrt{x}}}$ **43.** $\sqrt[3]{\dfrac{1}{5x^2}}$ **44.** $\sqrt[3]{\dfrac{1}{4xy^2}}$ **45.** $\sqrt{32x^5y^{-3}}$

**46.** $\sqrt{98x^{-3}y^4}$ **47.** $\sqrt{3xy}\sqrt{5x^3y}$ **48.** $\sqrt{2x^5y^2}\sqrt{3xy^3}$

**49.** $\sqrt[5]{3x^3y}\sqrt[5]{9x^4y^9}$ **50.** $\sqrt[4]{2xy^9}\sqrt[4]{8x^2y}$ **51.** $\dfrac{2\sqrt{x} + 3\sqrt{xy}}{3\sqrt{x} - 2\sqrt{y}}$

**52.** $\dfrac{\sqrt{x} + 2\sqrt{xy}}{4\sqrt{x} + 3\sqrt{y}}$ **53.** $(\sqrt{y+2} - \sqrt{y})(\sqrt{y+2} + \sqrt{y})$ **54.** $(\sqrt{x+b} + \sqrt{x})(\sqrt{x+b} - \sqrt{x})$

In Problems 55–58, simplify each expression *without* the use of a calculator.

**55.** $\sqrt{4 \times 10^{12}}$ **56.** $\sqrt{9 \times 10^8}$ **57.** $\sqrt{\dfrac{25 \times 10^{-7}}{36 \times 10^{-3}}}$ **58.** $\sqrt{\dfrac{16 \times 10^{-5}}{49 \times 10^7}}$

In Problems 59–62, multiply the expressions and simplify your result.

★ **59.** $\left(\sqrt{a} - \dfrac{2}{\sqrt{a}}\right)^2$ ★ **60.** $\left(\sqrt{\dfrac{a}{b}} - 2b\right)^2$

**61.** Find a specific value of $x$ and $y$ to show that

$$\sqrt{x} + \sqrt{y} \neq \sqrt{x+y}$$

**62.** Square each side of the equation to show that

$$\sqrt{8 + 2\sqrt{15}} = \sqrt{3} + \sqrt{5}$$

---

## 1-4 RATIONAL EXPONENTS

In Section 1-2 we stated five laws for exponents that hold when the exponents are positive or negative **integers**. We now extend the laws and state that the laws for exponents are valid when the **exponents are rational numbers**.

Let us examine Law III $(x^a)^b = x^{ab}$ when $a = \frac{1}{n}$ where $n$ is any positive integer and $b = n$. We see that $(x^{\frac{1}{n}})^n = x^{\frac{n}{n}} = x^1 = x$.

Since we know that $(\sqrt[n]{x})^n = x$ by Law of Radicals I, we are ready to make the following definition:

*Definition*

If $n$ is a positive integer and both $x$ and $\sqrt[n]{x}$ are real numbers, then $x^{\frac{1}{n}} = \sqrt[n]{x}$.

EXAMPLE 1 Evaluate the following.

**a.** $25^{\frac{1}{2}}$ **b.** $16^{\frac{1}{4}}$ **c.** $\left(\frac{1}{8}\right)^{\frac{1}{3}}$ **d.** $(-32)^{\frac{1}{5}}$

***Solution***

**a.** $25^{\frac{1}{2}} = \sqrt{25} = 5$ **b.** $16^{\frac{1}{4}} = \sqrt[4]{16} = 2$

**c.** $\left(\frac{1}{8}\right)^{\frac{1}{3}} = \sqrt[3]{\frac{1}{8}} = \frac{1}{2}$ **d.** $(-32)^{\frac{1}{5}} = \sqrt[5]{-32} = -2$ ■

Now we must consider what to do with expressions such as $x^{\frac{2}{3}}$ and $y^{\frac{3}{5}}$. Since the laws of exponents now hold for all rational numbers, let us examine Law III: $(x^a)^b = x^{ab}$ where $a = \frac{m}{n}$, $b = n$, and $m$ and $n$ are both integers. We now have $(x^{\frac{m}{n}})^n = x^m$. By our definition of radicals this means that $x^{\frac{m}{n}}$ is the $n$th root of $x^m$.

We are thus ready for a more general definition:

*Definition*

For all real numbers $x$ for which $\sqrt[n]{x}$ is a real number when $n$ is a positive integer and $m$ is any integer,

$$x^{\frac{m}{n}} = (\sqrt[n]{x})^m = \sqrt[n]{x^m}$$

**EXAMPLE 2** Evaluate the following.

**a.** $4^{\frac{3}{2}}$ **b.** $27^{\frac{2}{3}}$ **c.** $\left(\frac{1}{16}\right)^{\frac{3}{4}}$ **d.** $(-32)^{\frac{3}{5}}$

***Solution*** We observe that it will be easier to calculate the value if we use $x^{\frac{m}{n}} = (\sqrt[n]{x})^m$. (Note that a large cumbersome number may result in the radicand if we use $x^{\frac{m}{n}} = \sqrt[n]{x^m}$.)

**a.** $4^{\frac{3}{2}} = (\sqrt{4})^3 = (2)^3 = 8$

**b.** $27^{\frac{2}{3}} = (\sqrt[3]{27})^2 = (3)^2 = 9$

**c.** $\left(\frac{1}{16}\right)^{\frac{3}{4}} = \left(\sqrt[4]{\frac{1}{16}}\right)^3 = \left(\frac{1}{2}\right)^3 = \frac{1}{8}$

**d.** $(-32)^{\frac{3}{5}} = (\sqrt[5]{-32})^3 = (-2)^3 = -8$ ■

We will sometimes encounter negative rational exponents. We will state the following definition to deal with such cases:

*Definition*

If $x$ is a real number and $m$ and $n$ are positive integers, then

$$x^{-\frac{m}{n}} = \frac{1}{x^{\frac{m}{n}}}$$

**Warning:** You should take great care not to confuse negative exponents with negative numbers raised to a power.

Examine carefully each part of this example. Be careful of the use of the negative sign in each case. It is very important that you see how each part of the example differs from every other part.

**EXAMPLE 3** Evaluate the following.

**a.** $32^{-\frac{3}{5}}$ **b.** $(-32)^{-\frac{3}{5}}$ **c.** $(-32)^{\frac{3}{5}}$ **d.** $\left(\frac{1}{4}\right)^{-\frac{3}{2}}$

***Solution***

**a.** $32^{-\frac{3}{5}} = \frac{1}{32^{\frac{3}{5}}} = \frac{1}{(\sqrt[5]{32})^3} = \frac{1}{(2)^3} = \frac{1}{8}$

**b.** $(-32)^{-\frac{3}{5}} = \frac{1}{(-32)^{\frac{3}{5}}} = \frac{1}{(\sqrt[5]{-32})^3} = \frac{1}{(-2)^3} = -\frac{1}{8}$

**c.** $(-32)^{\frac{3}{5}} = (\sqrt[5]{-32})^3 = (-2)^3 = -8$

**d.** $\left(\frac{1}{4}\right)^{-\frac{3}{2}} = \frac{1}{\left(\frac{1}{4}\right)^{\frac{3}{2}}} = \frac{1}{\left(\sqrt{\frac{1}{4}}\right)^3} = \frac{1}{\left(\frac{1}{2}\right)^3} = \frac{1}{\frac{1}{8}} = 8$ ■

In dealing with variable quantities, sometimes the fractional exponents are retained throughout the entire problem while the laws of exponents are used to simplify the expression.

**EXAMPLE 4** Simplify the following.

**a.** $(3x^{\frac{1}{3}})(2x^{\frac{1}{2}})$ **b.** $(x^4y^2z^8)^{\frac{1}{2}}$

***Solution***

**a.** $(3x^{\frac{1}{3}})(2x^{\frac{1}{2}}) = 6x^{\frac{1}{3}+\frac{1}{2}} = 6x^{\frac{2}{6}+\frac{3}{6}} = 6x^{\frac{5}{6}}$

**b.** $(x^4y^2z^8)^{\frac{1}{2}} = x^{4\cdot\frac{1}{2}}y^{2\cdot\frac{1}{2}}z^{8\cdot\frac{1}{2}} = x^2yz^4$ ■

**EXAMPLE 5** Simplify $\frac{(x^{-\frac{1}{5}}y^{\frac{5}{2}})^{10}}{(x^{\frac{2}{3}}y^{-\frac{3}{5}})^{15}}$

***Solution*** $\frac{(x^{-\frac{1}{5}}y^{\frac{5}{2}})^{10}}{(x^{\frac{2}{3}}y^{-\frac{3}{5}})^{15}} = \frac{x^{-2}y^{25}}{x^{10}y^{-9}} = \frac{y^9y^{25}}{x^2x^{10}} = \frac{y^{34}}{x^{12}}$ ■

By changing a radical expression to exponent form, it is sometimes possible to simplify a higher-order radical by transforming it to an equivalent radical of lower order.

**EXAMPLE 6** Simplify $\sqrt[8]{x^6}$ where $x$ is a nonnegative real number.

***Solution*** $\sqrt[8]{x^6} = x^{\frac{6}{8}} = x^{\frac{3}{4}} = \sqrt[4]{x^3}$ *Note that the fraction exponent can be reduced.* ■

This general approach can be used if two or more variables are contained in the radicand.

**EXAMPLE 7** Simplify $\sqrt[6]{25x^4y^8}$.

***Solution*** $\sqrt[6]{25x^4y^8} = \sqrt[6]{5^2x^4y^8} = 5^{\frac{2}{6}}x^{\frac{4}{6}}y^{\frac{8}{6}} = 5^{\frac{1}{3}}x^{\frac{2}{3}}y^{\frac{4}{3}} = \sqrt[3]{5x^2y^4}$

We see that we still have a factor, $y$, raised to a power higher than the index number 3. We further simplify

$$\sqrt[3]{5x^2y^4} = \sqrt[3]{y^3}\sqrt[3]{5x^2y} = y\sqrt[3]{5x^2y} \quad \blacksquare$$

One further use of rational exponents should be mentioned. We cannot use the laws of radicals to multiply or divide two radicals that have a different index. However, it is possible to transform the product to new higher-order radicals. We will assume that all variables in the radicand are positive for the remainder of this chapter.

**EXAMPLE 8** Write as one radical $\sqrt{xyz}\sqrt[3]{xy^2z^0}$.

***Solution***

| | |
|---|---|
| $\sqrt{xyz}\sqrt[3]{xy^2z^0} = (xyz)^{\frac{1}{2}}(xy^2)^{\frac{1}{3}}$ | *Since* $z^0 = 1$. |
| $= (x^{\frac{1}{2}}y^{\frac{1}{2}}z^{\frac{1}{2}})(x^{\frac{1}{3}}y^{\frac{2}{3}})$ | *Using Law of Exponents IV.* |
| $= (x^{\frac{3}{6}}y^{\frac{3}{6}}z^{\frac{3}{6}})(x^{\frac{2}{6}}y^{\frac{4}{6}})$ | *Change each fractional exponent to a common denominator.* |
| $= x^{\frac{5}{6}}y^{\frac{7}{6}}z^{\frac{3}{6}}$ | *Add exponents.* |
| $= \sqrt[6]{x^5y^7z^3}$ | *Change back to a radical.* |
| $= y\sqrt[6]{x^5yz^3}$ | *Simplify the radical.* ■ |

**Note:** You should always check your final answer to ensure that no exponent in the radicand is higher than the order of the root.

## EXERCISE 1-4

In Exercise 1-4, assume that all variables are restricted to positive real numbers.

In Problems 1–8, write each expression in exponent form.

**1.** $\sqrt[5]{x^4}$ **2.** $\sqrt[7]{x^3}$ **3.** $(\sqrt[4]{ab})^3$ **4.** $(\sqrt[6]{x^2y})^2$

**5.** $\sqrt[4]{(a+2b)^3}$ **6.** $\sqrt[3]{3xy^2}$ **7.** $\dfrac{5}{\sqrt[3]{x}}$ **8.** $\dfrac{3}{\sqrt[4]{y^3}}$

In Problems 9–14, write each expression in radical form.

**9.** $x^{\frac{7}{12}}$ **10.** $y^{\frac{5}{11}}$ **11.** $(2xy^2)^{\frac{1}{5}}$ **12.** $(3a^2b)^{\frac{1}{6}}$ **13.** $(a^2+2b)^{\frac{1}{2}}$ **14.** $(2a+3b^3)^{\frac{1}{3}}$

In Problems 15–24, evaluate each expression. Do *not* use a calculator.

**15.** $4^{\frac{1}{2}}$ **16.** $9^{\frac{1}{2}}$ **17.** $27^{-\frac{1}{3}}$ **18.** $32^{-\frac{1}{5}}$ **19.** $\left(\frac{1}{16}\right)^{-\frac{1}{4}}$

**20.** $\left(\frac{8}{27}\right)^{-\frac{1}{3}}$ **21.** $(-125)^{\frac{2}{3}}$ **22.** $(-8)^{\frac{5}{3}}$ **23.** $\left(\frac{4}{49}\right)^{-\frac{3}{2}}$ **24.** $\left(\frac{1}{16}\right)^{-\frac{5}{4}}$

Simplify Problems 25–36.

**25.** $(x^4y^2z^6)^{\frac{5}{2}}$ **26.** $(a^3b^6c^{15})^{\frac{2}{3}}$ **27.** $\left(\frac{x^5}{y^{10}}\right)^{\frac{2}{5}}$ **28.** $\left(\frac{a^4}{b^{12}}\right)^{\frac{3}{4}}$

**29.** $(64x^6y^{15}z^9)^{\frac{4}{3}}$ **30.** $(36x^8y^4z^2)^{\frac{3}{2}}$ **31.** $\left(\frac{8x^3y^6}{27z^9}\right)^{\frac{2}{3}}$ **32.** $\left(\frac{16x^8y^2}{25z^4}\right)^{\frac{3}{4}}$

**33.** $(x+2y)^{\frac{1}{5}}(x+2y)^{-\frac{1}{10}}$ **34.** $\dfrac{(a+2b)^{\frac{1}{3}}}{(a+2b)^{\frac{1}{6}}}$ **35.** $\left(\frac{x^2y^7}{100x^8y}\right)^{-\frac{1}{2}}$ **36.** $(8x^{-6}y^9z^{15})^{-\frac{1}{3}}$

Simplify Problems 37–46 by reducing the order of the index.

**37.** $\sqrt[6]{4x^2}$ **38.** $\sqrt[8]{9y^6}$ **39.** $\sqrt[3]{\sqrt[4]{x^3}}$ **40.** $\sqrt[5]{\sqrt{x^5}}$ **41.** $\sqrt[8]{a^2b^8c^{14}}$

**42.** $\sqrt[6]{a^{12}b^8c^4}$ **43.** $\sqrt[12]{x^{10}y^6}$ **44.** $\sqrt[8]{64x^2y^4}$ **45.** $\sqrt[6]{25x^2y^{14}}$ **46.** $\sqrt[6]{36x^2y^{12}z^4}$

Express as a single radical in Problems 47–56. Simplify when possible.

**47.** $\sqrt{x^3}\cdot\sqrt[3]{x}$ **48.** $\sqrt[4]{x}\cdot\sqrt[3]{x^2}$ **49.** $\sqrt[4]{x^2y}\cdot\sqrt{x^2y}$ **50.** $\sqrt{x^2y^3}\cdot\sqrt[3]{xy^4}$ **51.** $\dfrac{\sqrt{xy}}{\sqrt[4]{xy^2}}$

**52.** $\dfrac{\sqrt[3]{a^3b^2}}{\sqrt[6]{ab^2}}$ **53.** $\sqrt{xy\sqrt{x}}$ **54.** $\sqrt[3]{x\sqrt{xy}}$ **55.** $\sqrt[3]{2x^2\sqrt{2y}}$ **56.** $\sqrt[4]{4b\sqrt[3]{2b}}$

Express Problem 57 without radicals, using only positive exponents.

★ **57.** $\sqrt[3]{\dfrac{(x^2y^3)^{-\frac{10}{3}}}{\sqrt[3]{x}\cdot y^2\cdot\sqrt[6]{x^3}}}$

---

## 1-5 ALGEBRAIC EXPRESSIONS AND POLYNOMIALS

Throughout this chapter, we have been using letters like $a$, $b$, $c$, $x$, $y$, $z$ to represent real numbers. We now take time to state how this specific practice relates to the overall topic of algebra.

Algebra may be considered as a mathematical system that generalizes the operations on a set of numbers. In the process of this generalization, letters are used to represent numbers. When letters are used to represent numbers, they are called **variables**.

*Definition*

> An *algebraic expression* is an indicated finite series of the operations of addition, subtraction, multiplication, division, and the extraction of roots on a set of numbers, or variables representing numbers from that set.

The following are illustrations of algebraic expressions:

$$x + 2y + \frac{1}{\sqrt{x}}, \qquad 3x^{-2}y + xy^{-2} + xy, \qquad \sqrt[3]{x} + \sqrt{x} + x + 2x^2$$

If specific numbers are substituted for the variables in an algebraic expression, the resulting number is called the **value of the expression**.

**EXAMPLE 1** Evaluate $x^2 + 2xy + y^3$ if $x = -2$ and $y = 3$.

***Solution***

$x^2 + 2xy + y^3 = (-2)^2 + 2(-2)(3) + (3)^3$ — *Raise variables to a power.*

$= 4 + 2(-2)(3) + 27$ — *Perform any products or divisions from left to right.*

$= 4 + (-12) + 27$

$= 19$ — *Perform any addition or subtraction from left to right.* ■

**EXAMPLE 2** Evaluate $\frac{3x^2\sqrt{y} + 1}{x + y}$ if $x = -4$ and $y = 9$.

***Solution***

$$\frac{3x^2\sqrt{y} + 1}{x + y} = \frac{3(-4)^2\sqrt{9} + 1}{-4 + 9} = \frac{3(16)(3) + 1}{-4 + 9} = \frac{145}{5} = 29$$ ■

The set of numbers used in this text will be the set of real numbers unless otherwise specified. There will be times when only certain real numbers may be used. The set of real numbers that is allowed for a given variable is called the domain.

*Definition*

> The *domain of a variable* is the set of real numbers that may be assigned to the variable.

**EXAMPLE 3** Give the domain of the variable $x$ in each of the following expressions.

**a.** $\frac{3}{x + 4}$ **b.** $\sqrt{x - 1}$

***Solution*** **a.** The denominator of a fraction cannot be zero. Thus $x + 4 \neq 0$. This implies that $x \neq -4$. Thus the domain of variable $x$ will be all real numbers except for $-4$.

**b.** The radicand cannot be negative. Therefore $(x - 1)$ must be zero or positive. If $x = 1$, then $(x - 1)$ has the value of 0. If $x > 1$, then $(x - 1)$ will be a positive real number. Thus the domain of the variable $x$ will be all real numbers greater than or equal to 1 $(x \geq 1)$. ■

*Definition*

In an expression that is an indicated product, the numbers (or variables) being multiplied are called *factors.*

**EXAMPLE 4** List the factors in the following expressions.

**a.** $3xy$ **b.** $5a(b + c)\sqrt{d}$

***Solution*** **a.** $3xy$ has three factors: $3$, $x$, $y$.

**b.** $5a(b + c)\sqrt{d}$ has four factors: $5$, $a$, $(b + c)$, and $\sqrt{d}$. ■

*Definition*

An expression composed entirely of factors is called a *term.*

An indicated sum may be composed of many terms.

**EXAMPLE 5** List the terms in the following expressions.

**a.** $2x + y$ **b.** $5x^3y^2 + 6xy^2 + 8xy - 2y + z$

***Solution*** **a.** $2x + y$ has two terms: $2x$ and $y$.

**b.** $5x^3y^2 + 6xy^2 + 8xy - 2y + z$ has 5 terms: $5x^3y^2$, $6xy^2$, $8xy$, $-2y$, and $z$. ■

*Definition*

An algebraic expression in which the only operation is multiplication of numbers and nonnegative integral powers of a variable is called a *monomial.* The sum of one or more monomials is called a *polynomial.*

Polynomials with two terms are called **binomials**, and polynomials with three terms are called **trinomials**. We will now introduce a definition for a polynomial with one variable.

*Definition*

A *polynomial* in $x$ is an algebraic expression of the form $a_nx^n + a_{n-1}x^{n-1} + \cdots + a_1x + a_0$ where $a_n, a_{n-1}, \ldots, a_1, a_0$ are real numbers, where the exponents of $x$ are nonnegative integers, and where $a_n \neq 0$.

The **degree** of a polynomial in $x$ is the highest exponent of $x$ in the polynomial. The polynomial $5x^3 + 3x - 1$ is of degree 3. The polynomial $2x - 7$ is of degree 1. The real numbers $a_n, a_{n-1}, \ldots, a_1, a_0$ are called the **coefficients** of the polynomial.

The word **coefficient** is used in a variety of ways in mathematics. Essentially, a coefficient means a factor. It usually refers to a constant or numerical coefficient. In the polynomial $6x + 2$, 6 is the **coefficient** of the $x$ term. It would be true (but somewhat unusual) to say that $x$ is the coefficient of 6. In an expression such as $3xy$, 3 is called the **numerical coefficient**. It is also the **coefficient** of $xy$. In discussing the term $3xy$, $x$ and $y$ would usually be called **literal factors**.

*Definition*

Two monomials, or terms of a polynomial, are *like terms* if and only if the literal factors of one are identical to the literal factors of the other.

This definition is important because **only like terms can be combined.** In general, to combine like terms, combine the numerical coefficients and use this result as the coefficient of the common literal factors of the terms.

**EXAMPLE 6** Combine like terms in each of the following.

**a.** $3x^2y - 7x^2y + 10xy$ **b.** $2(x + y) + 5(x + y)$

**c.** $2x + 3y + 3x - 7xy + 2y - 7$

***Solution***

**a.** $3x^2y - 7x^2y + 10xy = -4x^2y + 10xy$ *Note that $-4x^2y$ and $10xy$ are not like terms.*

**b.** $2(x + y) + 5(x + y) = 7(x + y)$

**c.** $2x + 3y + 3x - 7xy + 2y - 7 = 5x + 5y - 7xy - 7$ ■

A similar procedure can be used to subtract polynomials.

**EXAMPLE 7** Subtract $x^3 + 3x^2 - 2$ from $x^3 + 5x^2 - 8x - 3$.

***Solution***

$(x^3 + 5x^2 - 8x - 3) - (x^3 + 3x^2 - 2)$

$= x^3 + 5x^2 - 8x - 3 - x^3 - 3x^2 + 2$ *Remove parentheses.*

$= x^3 - x^3 + 5x^2 - 3x^2 - 8x - 3 + 2$ *Place like terms together. (This is an optional step.)*

$= 2x^2 - 8x - 1$ *Combine like terms.* ■

To find the product of a monomial and a polynomial having two or more terms, we need simply to apply the distributive property.

$$3ab(3x + 2y) = 9abx + 6aby$$

The product of two polynomials is also found by using the distributive property. We first regard one of the polynomials as a monomial.

$$(3x + 2y)(a + b + c) = (3x + 2y)a + (3x + 2y)b + (3x + 2y)c$$

We use the distributive property again to complete the process, giving

$$3ax + 2ay + 3bx + 2by + 3cx + 2cy$$

A **polynomial is simplified** if all like terms are combined.

EXAMPLE 8

Multiply each of the following.

**a.** $3x^2y(2x + 5y)$ **b.** $(x + 2y - z)(2x - 3y + z)$

***Solution***

**a.** $3x^2y(2x + 5y) = 6x^3y + 15x^2y^2$

**b.** $(x + 2y - z)(2x - 3y + z) = (x + 2y - z)(2x) + (x + 2y - z)(-3y) + (x + 2y - z)(z)$

$= 2x^2 + 4xy - 2xz - 3xy - 6y^2 + 3yz + xz + 2yz - z^2$

$= 2x^2 + xy - xz + 5yz - 6y^2 - z^2$ ■

A pattern develops when we multiply a binomial by a binomial that is very useful both in multiplying and in factoring.

$$(a + b)(c + d) = ac + ad + bc + bd$$

You should seek to perform this type of multiplication mentally without writing down intermediate steps.

EXAMPLE 9

Mentally multiply each of the following.

**a.** $(5x - 3y)(x + 2y)$ **b.** $(5x + 2y)(w + 3z)$

***Solution***

**a.** $(5x - 3y)(x + 2y) = 5x^2 + 7xy - 6y^2$

**b.** $(5x + 2y)(w + 3z) = 5wx + 15xz + 2wy + 6yz$ ■

The product of three or more binomials is accomplished by repeated multiplication.

**EXAMPLE 10** Multiply $(3x + 1)(x - 2)(x + 4)$.

***Solution***

$$\begin{aligned}(3x + 1)(x - 2)(x + 4) &= (3x^2 - 5x - 2)(x + 4) \\ &= x(3x^2 - 5x - 2) + 4(3x^2 - 5x - 2) \\ &= 3x^3 - 5x^2 - 2x + 12x^2 - 20x - 8 \\ &= 3x^3 + 7x^2 - 22x - 8 \quad \blacksquare\end{aligned}$$

## EXERCISE 1-5

In Problems 1–12, evaluate each expression by substituting the stated values:

**1.** $5x^2$, $x = 2$ **2.** $(5x)^2$, $x = 2$ **3.** $-x^2$, $x = 5$ **4.** $(-x)^2$, $x = 5$

**5.** $2x^2 + y - 1$, $x = 2$, $y = -3$ **6.** $3x^2 - 2xy + 2$, $x = -2$, $y = 3$

**7.** $\dfrac{3x\sqrt{y} + 4}{3}$, $x = 2$, $y = 4$ **8.** $\dfrac{x^2y - \sqrt{x}}{x}$, $x = 4$, $y = 3$

**9.** $\dfrac{-3x^3 + \sqrt{y}}{2y}$, $x = -2$, $y = 9$ **10.** $\dfrac{x^5y^2 - 2x^3}{\sqrt[3]{x}}$, $x = -1$, $y = 3$

**11.** $(-2x)^3\sqrt[3]{y} + \sqrt[5]{y^2}$, $x = 2$, $y = -1$ **12.** $\dfrac{x^{\frac{1}{2}}y^5 - 4xy^2}{y^{\frac{1}{2}}}$, $x = 16$, $y = 4$

In Problems 13–18, state the degree of the polynomial in $x$.

**13.** $3x^2 + 5x - 6$ **14.** $7x^3 - 3x^2 + x - 2$ **15.** $5x - 9x^4 + 3x^2 + 4$

**16.** $2x^3 + 3x^2 + 5x - 3x^5$ **17.** $-3x + (5 + 2)^2$ **18.** $2(\sqrt{7})^3 + 4x$

Simplify the following in Problems 19–32.

**19.** $3x^2 - 2xy^2 + 5xy^2 - x^2y$ **20.** $5ab - 4a^2b + 6ab - 9a^2b$ **21.** $2x(x + y) - 3y(x - 2y)$

**22.** $4x(x + 3) - (5x - 2)$ **23.** $x(a + b) - (a + b)$ **24.** $(x + 3)(x - 2)$

**25.** $(x + 3)(x^2 - 2x + 3)$ **26.** $(x + 5)(x - 5)$ **27.** $(x + 3)(x^2 - 3x + 9)$

**28.** $(2x - 5)(4x^2 + 10x + 25)$ **29.** $(a + b)(a - b)$ **30.** $(a - b)(a^2 + ab + b^2)$

**31.** $(a + b)(a^2 - ab + b^2)$ **32.** $(x + 2y)(x^2 + 4xy + 4y^2)$

In Problems 33–48, perform the indicated operations and simplify whenever possible.

**33.** $(5x^2 - 3x - 7) + (8x^2 + 7x - 4)$ **34.** $(6x^3 + 3x - 12) + (5x^2 - 4x - 3)$

**35.** $(x^3 - 3x^2 + 5x - 2) - (2x^3 - 7x - 3)$

**36.** $(2x^3 + 3x^2 - 6x - 1) - (3x^3 - 2x^2 + 5)$

**37.** $(2x^2 - 3)(x + 5)$

**38.** $(3x - 7)(x^2 + 2)$

**39.** $(x + 6)(2x - 1)(3x + 2)$

**40.** $(x - 1)(x - 3)(3x + 5)$

**41.** $(2x + 5y)^2$

**42.** $(4x - 6y)^2$

**43.** $(3x - 2)(x^2 + 5x - 4)$

**44.** $(2x + 1)(5x^2 - x - 2)$

**45.** $(x - 2)(x + 3) + (x - 5)(x + 2)$

**46.** $(2x - 1)(x + 3) + (x + 1)(x + 3)$

**47.** $(x^3 + 2x^2 + 1)(x^2 + x - 1)$

**48.** $(x^2 - 3x + 1)(2x^3 + x - 2)$

**49.** Show that $(x + y)^3 = x^3 + 3x^2y + 3xy^2 + y^3$.

**50.** Show that $(x - y)^3 = x^3 - 3x^2y + 3xy^2 - y^3$.

In Problems 51–54, let $m$ and $n$ be integers. Do the multiplication in each problem.

★ **51.** $(2x^m + 1)(3x^n + 1)$

★ **52.** $(3x^m - 1)(2x^n - 1)$

★ **53.** $(5x^m - 2y^n)(x^m - 3y^n)$

**54.** $(3x^m + y^n)(3x^m - y^n)$

## 1-6 FACTORING

When two or more polynomials are multiplied, each expression is called a **factor** of the product. The process of expressing a polynomial as a product is known as **factoring**. If a polynomial cannot be written as the product of two or more polynomials with integer coefficients, it is said to be **prime**. In this chapter we will adopt the convention that if a polynomial has integer coefficients, then the factors should be polynomials with integer coefficients.

Problems in factoring usually fall into one or more of the following categories:

### Categories of Factoring

1. Removing the greatest common factor.
2. The difference of two perfect squares.
3. The sum or difference of two perfect cubes.
4. The perfect square trinomial.
5. The general trinomial.
6. Factoring by grouping the terms so that they fall into one of the above categories.

To factor problems in categories 2, 3, and 4, it is very helpful to memorize the following formulas:

*Factoring Formulas*

For all real numbers $a$ and $b$,

$$a^2 - b^2 = (a + b)(a - b) \quad \textbf{(F-1)}$$

$$a^3 - b^3 = (a - b)(a^2 + ab + b^2) \quad \textbf{(F-2)}$$

$$a^3 + b^3 = (a + b)(a^2 - ab + b^2) \quad \textbf{(F-3)}$$

$$a^2 + 2ab + b^2 = (a + b)^2 \quad \textbf{(F-4)}$$

$$a^2 - 2ab + b^2 = (a - b)^2 \quad \textbf{(F-5)}$$

All of these formulas can be verified by using the method of multiplication of polynomials that we have previously established.

One of the most commonly encountered factoring problems is that of removing the greatest common factor. When you are doing this type of factoring problem, you are actually using the distributive property.

EXAMPLE 1 Factor the following.

**a.** $12x^2 + 16x^3$ **b.** $3x^3y + 6xyz - 18xy^2z$ **c.** $2(x + 3y) - y(x + 3y)$

***Solution*** **a.** We remove the greatest common factor, which is $4x^2$. Thus we have

$$12x^2 + 16x^3 = 4x^2(3 + 4x)$$

**b.** We remove the greatest common factor $3xy$, giving

$$3x^2y + 6xyz - 18xy^2z = 3xy(x + 2z - 6yz)$$

**c.** Here we observe that the greatest common factor is the binomial $x + 3y$. Thus we have

$$2(x + 3y) - y(x + 3y) = (x + 3y)(2 - y) \quad ■$$

Factoring problems involving the difference of two squares are fairly easy to identify, especially if the Factoring Formula (F-1), $a^2 - b^2 = (a + b)(a - b)$, is memorized.

EXAMPLE 2 Factor the following.

**a.** $x^2 - 36$ **b.** $25x^2 - 49y^2$ **c.** $16x^4 - 1$

***Solution***

a. $x^2 - 36 = (x + 6)(x - 6)$

b. $25x^2 - 49y^2 = (5x + 7y)(5x - 7y)$

c. $16x^4 - 1 = (4x^2 + 1)(4x^2 - 1)$

$= (4x^2 + 1)(2x + 1)(2x - 1)$ *Here the second factor may be factored again.* ■

In identifying factoring problems that involve the sum or difference of two perfect cubes it is necessary to identify what values are being cubed.

EXAMPLE 3 Factor the following.

a. $x^3 - 64$  b. $8x^3 + 27y^3$

***Solution***

a. We will use Factoring Formula (F-2), where $a = x$ and $b = 4$.

$$a^3 - b^3 = (a - b)(a^2 + ab + b^2)$$

$$x^3 - 64 = x^3 - (4)^3 = (x - 4)(x^2 + 4x + 16)$$

b. We will use Factoring Formula (F-3), where $a = 2x$ and $b = 3y$

$$a^3 + b^3 = (a + b)(a^2 - ab + b^2)$$

$$8x^3 + 27y^3 = (2x)^3 + (3y)^3 = (2x + 3y)(4x^2 - 6xy + 9y^2)$$ ■

A **perfect square trinomial** is a trinomial of the form $a^2 + 2ab + b^2$ or $a^2 - 2ab + b^2$. Since problems that involve a perfect square trinomial are frequently encountered, memorizing Formulas (F-4) and (F-5) is highly desirable.

EXAMPLE 4 Factor the following perfect square trinomials.

a. $16x^2 + 40x + 25$  b. $81s^2 - 90st + 25t^2$

***Solution***

a. We use Formula (F-4), where $a = 4x$ and $b = 5$.

$$16x^2 + 40x + 25 = (4x + 5)^2$$

b. We use Formula (F-5), where $a = 9s$ and $b = 5t$.

$$81s^2 - 90st + 25t^2 = (9s - 5t)^2$$ ■

For trinomials that are not perfect square trinomials we use the techniques of factoring the general trinomial. These can be described by using the formula

$$x^2 + (a + b)x + ab = (x + a)(x + b)$$

if the initial squared variable has a coefficient of 1. If the initial squared variable is not 1, we can describe the factoring pattern as

$$acx^2 + (ad + bc)x + bd = (ax + b)(cx + d)$$

However, these formulas are *not* useful to memorize, since it is far more efficient to use the methods learned in an elementary algebra course. As a check, it is very helpful to mentally multiply the two factors to see whether you obtain the original trinomial.

**EXAMPLE 5** Factor each of the following.

**a.** $x^2 + 8x + 15$ **b.** $x^2 - 7xy + 12y^2$
**c.** $3x^2 - 8x - 3$ **d.** $4x^2 + xy - 3y^2$

***Solution***

**a.** $x^2 + 8x + 15 = (x + 5)(x + 3)$
**b.** $x^2 - 7xy + 12y^2 = (x - 4y)(x - 3y)$
**c.** $3x^2 - 8x - 3 = (3x + 1)(x - 3)$
**d.** $4x^2 + xy - 3y^2 = (x + y)(4x - 3y)$ ■

Many students make an error in assuming that a trinomial that has a perfect square for both first and last terms is a perfect square trinomial. Do *not assume you have a perfect square trinomial* unless you check the middle term. A perfect square trinomial is always of the form $a^2 + 2ab + b^2$ or $a^2 - 2ab + b^2$.

**EXAMPLE 6** Factor $4x^2 + 37xy + 9y^2$.

***Solution*** We cannot use the formula $a^2 + 2ab + b^2 = (a + b)^2$, since the middle term is not $2ab = 2(2x)(3y) = 12xy$. Therefore we must use the techniques of factoring a general trinomial.

Thus $4x^2 + 37xy + 9y^2 = (4x + y)(x + 9y)$. ■

We use the sixth category, factoring by grouping, when there are four or more terms in an expression. The process of grouping is generally for the purpose of identifying the problem as one or more of the types just discussed. After the step of factoring by grouping, additional steps of factoring may be possible.

**EXAMPLE 7** Factor by grouping.

**a.** $x^3 + 2x^2 - 3x - 6$ **b.** $4x^2 + y^2 - 9 - 4xy$

***Solution***

**a.** $x^3 + 2x^2 - 3x - 6$ *We remove a common factor of $x^2$ from the first two terms.*

$= x^2(\mathbf{x + 2}) - 3(\mathbf{x + 2})$ *Notice we remove a common factor of $-3$ from the last two terms.*

$= (x + 2)(x^2 - 3)$ *Finally, we remove the common factor $(x + 2)$.*

**b.** The polynomial $4x^2 + y^2 - 9 - 4xy$ when grouped as $a^2 - b^2$ looks like

$$(4x^2 - 4xy + y^2) - 9 \text{ with } a = (2x - y) \quad \text{and} \quad b = 3$$

Thus we have

$$(2x - y)^2 - 3^2$$

which is recognized as the difference of two perfect squares and then can be factored as

$$(2x - y + 3)(2x - y - 3) \quad \blacksquare$$

Many factoring problems require first that a common factor be removed and then that the problem be factored further by other methods.

**EXAMPLE 8** Factor completely.

**a.** $16x^2 - 20x - 6$ **b.** $54x^3 - 16$

***Solution*** **a.** First we remove a common factor of 2.

$$\begin{aligned} 16x^2 - 20x - 6 &= 2(8x^2 - 10x - 3) \\ &= 2(4x + 1)(2x - 3) \end{aligned}$$

**b.** First we remove a common factor of 2.

$$54x^3 - 16 = 2(27x^3 - 8)$$

Now we have the difference of two cubes and can use Formula (F-2).

$$\begin{aligned} &= 2[(3x)^3 - (2)^3] \\ &= 2(3x - 2)(9x^2 + 6x + 4) \quad \blacksquare \end{aligned}$$

A few factoring problems are more involved. These will require some algebraic manipulations so that a pattern appears that is similar to previously introduced types of problems. Be alert. This problem is *not easy*!

**★ EXAMPLE 9** Factor completely: $4x^2 - 25y^2 - 10y - 12x + 8$.

***Solution*** If we arrange this expression as

$$4x^2 - 12x - 25y^2 - 10y + 8$$

and notice that 8 could be written as $9 - 1$, we can obtain

$$(4x^2 - 12x + 9) - (25y^2 + 10y + 1)$$

or

$$(2x - 3)^2 - (5y + 1)^2$$

which is the difference of two perfect squares giving

$$[(2x - 3) - (5y + 1)][(2x - 3) + (5y + 1)]$$

or

$$(2x - 5y - 4)(2x + 5y - 2) \quad \blacksquare$$

## EXERCISE 1-6

Factor Problems 1–58 completely.

1. $5a^2 - 10a$
2. $6a^2b + 9ab^2 - 6ab$
3. $9x^2 - 30x + 25$
4. $25x^2 - 49$
5. $5x^2 - 45$
6. $x^2 + 5x + 6$
7. $x^2 + 2x - 15$
8. $2x^2 + 5x + 2$
9. $2x^2 - x - 3$
10. $3x^2 + 13x - 10$
11. $6x^2 - 5x - 4$
12. $4x^2 + 14x + 6$
13. $6x^3 + 8x^2 - 8x$
14. $3x^2 + 9x + 15$
15. $5x^2 - 2x + 3$
16. $cx + dx + cy + dy$
17. $ax + 3ay - 3by - bx$
18. $ax - 2ay + 2xb - 4by$
19. $x^2 + 8x + 16$
20. $x^2 - 6x + 9$
21. $4x^2 - 28xy + 49y^2$
22. $9x^2 + 30xy + 25y^2$
23. $x^2 - 8x - 20$
24. $x^2 - 12x + 20$
25. $6x^2 + x - 5$
26. $6x^2 + 7x - 3$
27. $2y^2 + 7y - 30$
28. $2y^2 - 11y + 12$
29. $9y^2 - 100z^2$
30. $25x^2 - 121y^2$
31. $27s^3 - 8t^3$
32. $125s^3 - 1$
33. $ax + 20 - 5a - 4x$
34. $a^2x - 8 + 2a^2 - 4x$
35. $2a^2x - 3 - 2x + 3a^2$
36. $27a^3 + 1$
37. $x^3 - y^6$
38. $a^9 - 1$
39. $2(x + 5)^2 + 3(x + 5) - 20$
40. $3(x - 2)^2 - 4(x - 2) - 7$
41. $x^2 + 2x - y^2 - 4y - 3$
42. $x^2 + 10x - 9y^2 + 6y + 24$
43. $9a^2 - (3x + 4y)^2$
44. $x^2 - 4x - y^2 - 6yz + 4 - 9z^2$
45. $18 - 2a^2 - 27b + 3a^2b$
46. $x^4 + 2acx^2 - w^4 - 2acw^2$
47. $5x^3 - 5x^3y^3$
48. $2x^6 + 2x^3y^3$
49. $3x^2 - 36x + 108$
50. $6x^2 + 21x - 45$
51. $8x^3 - 8x^2 - 6x$
52. $8x^2 + 24x + 18$
53. $6y^3 + 21y^2 + 9y$
54. $3y^4 - 7y^5 + 2y^6$
55. ★ $x^{2n} - 1$
56. ★ $x^{n+2} - x^{n+1} + 2x^n$
57. ★ $y^{4n} - 3y^{2n} - 4$
58. ★ $y^{4n} - x^{4n}$

Some problems appear not to be factorable; however, by adding and subtracting the proper monomial they may be factored. Using the hint, factor Problems 59 and 60.

59. ★ $x^4 + x^2y^2 + 25y^4$ (*Hint:* Add and subtract $9x^2y^2$.)
60. ★ $4 + x^4$ (*Hint:* Add and subtract $4x^2$.)

## 1-7 RATIONAL EXPRESSIONS

The quotient of two polynomials is called a **rational expression**. Some examples of rational expressions are $\frac{3x + 5}{x^2 - 7}$, $\frac{x^5y^6 + x^3y^2 + 7y^8 + 3x^8}{3 + 2y}$, and $\frac{5}{3x}$. Both numerator and denominator must be polynomials in order to satisfy this definition. Thus expressions like $\dfrac{2x^{\frac{3}{4}} + 2y^{\frac{1}{4}}}{2x + 3}$ and $\dfrac{5}{2\sqrt{x}}$ are *not* rational expressions. Frequently, our work in algebra involves rational expressions. Therefore we must be able to quickly simplify or combine them.

---

The *fundamental principle of fractions* is

$$\frac{a}{b} = \frac{ax}{bx} \qquad \text{for all } x \neq 0$$

---

This is the basis of all operations on fractions.

When $a$, $b$, and $x$ are polynomials, we employ the fundamental principle of fractions to simplify rational expressions.

The simple rule, "factor completely all numerators and denominators, then cancel all like factors that are in both numerator and denominator" is sufficient for all such problems.

**Caution:** Only factors can be cancelled. Never make the common error of cancelling terms.

*It is assumed throughout this text that the denominator of any rational expression is nonzero.*

**EXAMPLE 1** Simplify: $\dfrac{ax + ay + 3x + 3y}{a^2 + 5a + 6}$.

***Solution***

$$\frac{ax + ay + 3x + 3y}{a^2 + 5a + 6} = \frac{\cancel{(a + 3)}(x + y)}{\cancel{(a + 3)}(a + 2)}$$

*We factor numerator and denominator and then cancel the like factor that appears in both.*

$$= \frac{x + y}{a + 2} \quad ■$$

**Warning:** Make sure when you are about to cancel something that it is a factor. If the expression you are about to cancel is added to or subtracted from something else, you may *not* cancel that expression.

In some cases it may be necessary to factor out a negative number or a negative sign in order to obtain common factors that may be cancelled.

**EXAMPLE 2** Simplify: $\dfrac{5x^2 + 6xy - 8y^2}{16y^2 - 25x^2}$.

***Solution***

$$\frac{5x^2 + 6xy - 8y^2}{16y^2 - 25x^2} = \frac{(5x - 4y)(x + 2y)}{(4y + 5x)(4y - 5x)}$$

*We factor numerator and denominator.*

$$= \frac{-(4y - 5x)(x + 2y)}{(4y + 5x)(4y - 5x)}$$

*Factor $5x - 4y$ as $-(-5x + 4y)$ or $-(4y - 5x)$.*

$$= \frac{-\cancel{(4y - 5x)}(x + 2y)}{(4y + 5x)\cancel{(4y - 5x)}}$$

*Cancel the common factor.*

$$= -\frac{x + 2y}{4y + 5x}$$

*Do not forget the negative sign in the answer.* ■

To multiply two or more rational expressions, we use the property of multiplication of fractions.

$$\frac{a}{b} \cdot \frac{c}{d} = \frac{ac}{bd}$$

Wherever possible, we cancel all common factors to simplify the fraction.

**EXAMPLE 3** Multiply: $\dfrac{3x^2 + 6x}{x^2 - 25} \cdot \dfrac{x^2 - x - 20}{x^2 + 7x + 12}$.

***Solution***

$$\frac{3x^2 + 6x}{x^2 - 25} \cdot \frac{x^2 - x - 20}{x^2 + 7x + 12} = \frac{3x(x + 2)\cancel{(x - 5)}\cancel{(x + 4)}}{(x + 5)\cancel{(x - 5)}\cancel{(x + 4)}(x + 3)}$$

$$= \frac{3x(x + 2)}{(x + 5)(x + 3)} \quad ■$$

To divide two or more rational expressions we use the property of division of fractions.

$$\frac{a}{b} \div \frac{d}{c} = \frac{a}{b} \cdot \frac{c}{d} = \frac{ac}{bd}$$

**EXAMPLE 4** Divide: $\dfrac{4x^2 - y^2}{x + 2y} \div (4x^2 - 2xy)$.

***Solution*** First we change to a multiplication problem by using the inverse.

$$\frac{4x^2 - y^2}{x + 2y} \div (4x^2 - 2xy) = \frac{4x^2 - y^2}{x + 2y} \cdot \frac{1}{4x^2 - 2xy}$$

Then we factor numerator and denominator and cancel common factors.

$$\frac{(2x - y)(2x + y)}{x + 2y} \cdot \frac{1}{2x(2x - y)} = \frac{2x + y}{2x(x + 2y)} \quad \blacksquare$$

If fractions to be added or subtracted have common denominators, we can quickly combine them using the following properties.

$$\frac{a}{b} + \frac{c}{b} = \frac{a + c}{b} \qquad \frac{a}{b} - \frac{c}{b} = \frac{a - c}{b}$$

If fractions do not have common denominators, we write the fractions as equivalent fractions that have the same denominator. The new denominator chosen will be the **least common denominator** (or **L.C.D.**) of the fractions.

The least common denominator of two or more rational expressions is found as follows:

**1.** Factor each denominator completely.
**2.** Form a product that contains each different factor, each taken the greatest number of times it occurs in any of the denominators.

**EXAMPLE 5** Combine the two rational expressions.

$$\frac{3}{5x} + \frac{1}{x^2}$$

***Solution*** The L.C.D. is $5x^2$. Note that the factor $x$ occurs *twice* in the expression $x^2$.

$$\frac{3}{5x} + \frac{1}{x^2} = \frac{3 \cdot x}{5x \cdot x} + \frac{1 \cdot 5}{x^2 \cdot 5} = \frac{3x}{5x^2} + \frac{5}{5x^2} = \frac{3x + 5}{5x^2} \quad \blacksquare$$

**EXAMPLE 6** Combine the two rational expressions.

$$\frac{3}{x - 2} + \frac{1}{x + 5}$$

***Solution*** The L.C.D. is $(x - 2)(x + 5)$.

$$\frac{3}{x - 2} + \frac{1}{x + 5} = \frac{3(x + 5)}{(x - 2)(x + 5)} + \frac{1(x - 2)}{(x + 5)(x - 2)}$$

*We change the fractions to equivalent fractions with the L.C.D.*

$$= \frac{3x + 15}{(x - 2)(x + 5)} + \frac{x - 2}{(x - 2)(x + 5)}$$

*Simplify numerators.*

$$= \frac{4x + 13}{(x - 2)(x + 5)}$$

*Add numerators.* ■

**EXAMPLE 7** Subtract the two rational quantities.

$$\frac{5}{x^2 + 6x + 9} - \frac{x}{x^2 + 5x + 6}$$

***Solution***

$$\frac{5}{x^2 + 6x + 9} - \frac{x}{x^2 + 5x + 6}$$

$$= \frac{5}{(x + 3)^2} - \frac{x}{(x + 3)(x + 2)}$$ *We factor the two denominators.*

The L.C.D. is $(x + 3)^2(x + 2)$. Note that the factor $(x + 3)$ occurs twice in the first denominator.

$$= \frac{5(x + 2)}{(x + 3)^2(x + 2)} - \frac{x(x + 3)}{(x + 3)^2(x + 2)}$$ *We change the two fractions to equivalent fractions with the L.C.D.*

$$= \frac{5x + 10}{(x + 3)^2(x + 2)} - \frac{x^2 + 3x}{(x + 3)^2(x + 2)}$$ *Simplify numerators.*

$$= \frac{5x + 10 - x^2 - 3x}{(x + 3)^2(x + 2)}$$ *Subtract numerators. Note the signs in $-x^2 - 3x$.*

$$= \frac{-x^2 + 2x + 10}{(x + 3)^2(x + 2)}$$ *Collect like terms.* ■

A **complex fraction** is a fractional form that contains fractions in the numerator or denominator or both.

One method that is commonly used to simplify complex fractions is to:

**1.** combine the fractions in the numerator,
**2.** combine the fractions in the denominator,
**3.** apply the rules for division of two fractions.

We will illustrate with the following example.

**EXAMPLE 8** Simplify: $\dfrac{1 + \dfrac{2}{1 + x}}{\dfrac{3}{x^2} + \dfrac{1}{x}}$.

***Solution***

**1.** $1 + \dfrac{2}{1 + x} = \dfrac{1 + x}{1 + x} + \dfrac{2}{1 + x} = \dfrac{x + 3}{1 + x}$ *Combining in the numerator.*

**2.** $\dfrac{3}{x^2} + \dfrac{1}{x} = \dfrac{3}{x^2} + \dfrac{x}{x^2} = \dfrac{x + 3}{x^2}$ *Combining in the denominator.*

**3.** $\dfrac{x+3}{1+x} \div \dfrac{x+3}{x^2} = \dfrac{\cancel{x+3}}{1+x} \cdot \dfrac{x^2}{\cancel{x+3}} = \dfrac{x^2}{1+x}$ *Divide the fractions.* ■

A second method that is often employed involves finding the L.C.D. of all fractions appearing in the numerator and the denominator. This L.C.D. is then multiplied by the numerator and denominator of the original expression. The resulting fraction will no longer be a complex fraction. We will use this method in the following example.

**EXAMPLE 9** Simplify: $\dfrac{\dfrac{-13}{x^2 - 2x - 35} - \dfrac{1}{x+5}}{\dfrac{1}{x+5} + 1}$.

***Solution*** If we factor the first denominator, we see that the *least common denominator* of all fractions in the expression is $(x + 5)(x - 7)$. Thus to eliminate all individual fractions, we multiply both numerator and denominator by this common denominator.

$$\frac{(x+5)(x-7)\left[\dfrac{-13}{(x+5)(x-7)} - \dfrac{1}{x+5}\right]}{(x+5)(x-7)\left[\dfrac{1}{x+5} + 1\right]}$$

$= \dfrac{-13 - (x - 7)}{(x - 7) + (x + 5)(x - 7)}$ *Carefully multiply each quantity by the L.C.D.*

$= \dfrac{-13 - x + 7}{x - 7 + x^2 - 2x - 35}$

$= \dfrac{-x - 6}{x^2 - x - 42}$ *Remove parentheses and simplify.*

$= \dfrac{-(x + 6)}{(x + 6)(x - 7)}$ *Note the need to factor.*

$= -\dfrac{1}{x - 7}$ or $\dfrac{1}{7 - x}$ *There are several ways to leave the answer.* ■

In general, it is a good idea to be familiar with both methods of simplifying complex fractions. Each method has its advantages in simplifying certain types of problems.

## EXERCISE 1-7

Simplify Problems 1–60.

1. $\dfrac{x^2 - 2x - 15}{x^2 + 3x - 40}$

2. $\dfrac{x^2 - 49}{x^2 + 14x + 49}$

3. $\dfrac{6x + 30}{10x^2 + 40x - 50}$

4. $\dfrac{3x + 6}{x} \cdot \dfrac{5x}{21x + 42}$

5. $\dfrac{x^2 - 1}{x + 1} \cdot \dfrac{x + 2}{x - 1}$

6. $\dfrac{x^2 + 3x + 2}{x^2 + 5x + 6} \cdot \dfrac{x^2 - 2x - 15}{x^2 - 3x - 10}$

7. $\dfrac{x + 1}{x^2 - 5x - 6} \div \dfrac{x^2 - 1}{x - 6}$

8. $\dfrac{2x - 6}{3x} \div \dfrac{x^2 - 2x - 3}{x^2 + x}$

9. $\dfrac{x^2 + 7x + 10}{x^2 + 4x - 5} \div \dfrac{x^2 + 7x + 12}{x^2 + 2x - 3}$

10. $\dfrac{x^2 - 16}{x^2 + 5x + 4} \div \dfrac{x^2 - 8x + 16}{x^2 - 1}$

11. $\dfrac{x^2 - 4x - 12}{x^2 - 7x + 6}$

12. $\dfrac{4x - 4}{3x^2 - 3}$

13. $\dfrac{12 - 2x}{x^3 - 36x}$

14. $\dfrac{(x + 2)^3}{x^3 + 8}$

15. $\dfrac{2y^2 + y - 3}{8y^2 + 2y - 15}$

16. $\dfrac{6y^2 - 7y - 3}{4y^2 - 8y + 3}$

17. $\dfrac{3s^2 - 2s}{2s^2 + 13s + 21} \cdot \dfrac{s^2 + 4s + 3}{3s^2 + s - 2}$

18. $\dfrac{s^2 + 11s + 30}{s + 6} \cdot \dfrac{3s + 18}{10 + 2s}$

19. $\dfrac{x^2 - 1}{x^3 + x} \cdot \dfrac{x^2 + 1}{x + 1} \cdot \dfrac{5x + 7xy}{3 - 3x}$

20. $\dfrac{2x + 3}{x^2 - 1} \cdot \dfrac{x^2 + 1}{4x + 6} \cdot \dfrac{5x + 10}{2x^4 + 2x^2}$

21. $\dfrac{6y^2}{y^2 - y - 6} \div \dfrac{3y}{y + 2}$

22. $\dfrac{y + 2}{3y} \div \dfrac{y^2 - 2y - 8}{15y^2}$

23. $(x^2 - x - 12) \div \dfrac{x^2 - 9}{x^2 - 3x}$

24. $\dfrac{2x^2 + 7x + 3}{4x^2 - 1} \div (x + 3)$

25. $\dfrac{4}{x + 1} + \dfrac{1}{x^2 - 4} + \dfrac{5}{x + 2}$

26. $\dfrac{4}{x^2 + 3x} + \dfrac{x}{x + 3}$

27. $\dfrac{5}{x - 1} - \dfrac{2x + 6}{x^2 + 2x - 3}$

28. $\dfrac{2x}{x^2 - 1} + \dfrac{3}{x^2 - x - 2}$

29. $\dfrac{1}{x + 2} + \dfrac{1}{x + 1} - \dfrac{2}{x - 7}$

30. $\dfrac{2x}{x^2 - 4} - \dfrac{x + 4}{x^2 + 4x - 12}$

31. $\dfrac{\dfrac{1}{x} + \dfrac{1}{y}}{\dfrac{1}{x + y}}$

32. $\dfrac{1 - \dfrac{1}{a}}{a - 1}$

**33.** $\dfrac{\dfrac{1}{x+3} - 1}{\dfrac{1}{x^2-9}}$

**34.** $\dfrac{\dfrac{6}{x^2+3x-10} - \dfrac{1}{x-2}}{\dfrac{1}{x-2} + 1}$

**35.** $\dfrac{1}{x} + \dfrac{3}{x^2} + \dfrac{2}{y}$

**36.** $\dfrac{5}{x^2y} + \dfrac{1}{y} + \dfrac{3}{x}$

**37.** $5 + \dfrac{y+1}{y-1}$

**38.** $\dfrac{y+3}{y-3} - 2$

**39.** $x - 2 + \dfrac{x-3}{x-4}$

**40.** $x - 1 + \dfrac{x+3}{x-2}$

**41.** $\dfrac{2}{2x+6} + \dfrac{x}{x+3}$

**42.** $\dfrac{x}{2x-5} + \dfrac{3}{6x-15}$

**43.** $\dfrac{5}{y} - \dfrac{2}{y+2}$

**44.** $\dfrac{8}{y+3} - \dfrac{3}{y}$

**45.** $\dfrac{2}{s-2} + \dfrac{s}{s^2-s-6}$

**46.** $\dfrac{2s}{s^2+s-2} + \dfrac{3}{s+2}$

**47.** $\dfrac{3x+2}{x-4} - \dfrac{4x+1}{5x+2}$

**48.** $\dfrac{3x}{x+3} - \dfrac{2x+1}{3x+1}$

**49.** $\dfrac{5}{x^2-1} - \dfrac{8}{x^2+2x+1}$

**50.** $\dfrac{2}{x^2-5x+6} - \dfrac{2}{x^2-4}$

★**51.** $\dfrac{x}{x^2-x-2} - \dfrac{1}{x^2+5x-14} - \dfrac{2}{x^2+8x+7}$

★**52.** $\dfrac{3}{y^2-6y+9} - \dfrac{2}{y^2-9} - \dfrac{5}{3-y}$

**53.** $\dfrac{2 - \dfrac{2}{x}}{2 + \dfrac{2}{x}}$

**54.** $\dfrac{1 - \dfrac{y^2}{x^2}}{1 - \dfrac{y}{x}}$

**55.** $\dfrac{y - \dfrac{y^2}{x-y}}{1 + \dfrac{y^2}{x^2-y^2}}$

**56.** $\dfrac{\dfrac{1}{y+2} - \dfrac{3}{y^2-4}}{\dfrac{3}{y-2}}$

★**57.** $y - \dfrac{y}{1 - \dfrac{y}{1-y}}$

★**58.** $1 - \dfrac{1}{1 - \dfrac{1}{y-2}}$

★**59.** $\left[\dfrac{y}{y^2-9} + \dfrac{2}{y-3}\right] \div \dfrac{y}{y-1}$

★**60.** $\left[1 + \dfrac{2}{y-1}\right] \div \dfrac{y^2+y}{y^2+y-2}$

## KEY TERMS AND CONCEPTS

Be sure you understand what is meant by each of these terms and can give an example of each.

- Natural numbers
- Integers
- Rational numbers
- Irrational numbers
- Real numbers
- Decimal representation of rational numbers
- Decimal representation of irrational numbers
- Absolute value
- Scientific notation
- Radical
- Radical sign
- Radicand
- Index of a radical
- Principal root of a number
- Rationalize a denominator
- Variable
- Algebraic expression
- Domain of a variable
- Factor
- Prime polynomial
- Term
- Monomial
- Polynomial
- Binomial
- Trinomial
- Degree of a polynomial
- Coefficient
- Like terms
- Literal factor
- Fundamental principle of fractions
- Cancelling factors
- Least common denominator
- Complex fraction

## SUMMARY OF PROCEDURES AND CONCEPTS

### Properties of Exponents

For all real numbers $a$, $b$ with $x \neq 0$, $y \neq 0$

$$x^a \cdot x^b = x^{a+b}$$

$$\frac{x^a}{x^b} = x^{a-b}$$

$$(x^a)^b = x^{ab}$$

$$(xy)^a = x^a y^a$$

$$\left(\frac{x}{y}\right)^a = \frac{x^a}{y^a}$$

$$x^0 = 1$$

$$x^{-a} = \frac{1}{x^a}$$

**Laws of Radicals**

If these radicals exist, then if $a$ and $b$ are real numbers and $m$ and $n$ are positive integers, then

$$(\sqrt[n]{a})^n = a$$
$$\sqrt[n]{ab} = \sqrt[n]{a}\,\sqrt[n]{b}$$
$$\sqrt[n]{\frac{a}{b}} = \frac{\sqrt[n]{a}}{\sqrt[n]{b}} \text{ if } b \neq 0$$
$$\sqrt[m]{\sqrt[n]{a}} = \sqrt[mn]{a}$$

**Properties of Rational Exponents**

For all real numbers $x$ for which $\sqrt[n]{x}$ is a real number when $n$ is a positive integer and $m$ is any integer,

$$x^{\frac{1}{n}} = \sqrt[n]{x}$$
$$x^{\frac{m}{n}} = (\sqrt[n]{x})^m = \sqrt[n]{x^m}$$
$$x^{-\frac{m}{n}} = \frac{1}{x^{\frac{m}{n}}} = \frac{1}{\sqrt[n]{x^m}}$$

**Factoring Formulas**

$$a^2 - b^2 = (a + b)(a - b)$$
$$a^3 - b^3 = (a - b)(a^2 + ab + b^2) \qquad a^3 + b^3 = (a + b)(a^2 - ab + b^2)$$
$$a^2 + 2ab + b^2 = (a + b)^2 \qquad a^2 - 2ab + b^2 = (a - b)^2$$

**Operations with Algebraic Fractions**

For polynomials $a$, $b$, $c$, $d$ where all denominators are nonzero:

$$\frac{a}{b} + \frac{c}{b} = \frac{a + c}{b} \qquad \frac{a}{b} - \frac{c}{b} = \frac{a - c}{b}$$
$$\frac{a}{b} \cdot \frac{c}{d} = \frac{ac}{bd} \qquad \frac{a}{b} \div \frac{d}{c} = \frac{a}{b} \cdot \frac{c}{d} = \frac{ac}{bd}$$

## CHAPTER 1 REVIEW EXERCISE

Evaluate Problems 1 and 2.

**1.** $|19 - 23|$ **2.** $\left|\frac{2}{3} - \frac{5}{7}\right|$

In Problems 3–5, justify the equation by stating *one* of the field properties.

**3.** $5 + (7 + 3) = (7 + 3) + 5$ **4.** $(8 \cdot 3) \cdot 2 = 8 \cdot (3 \cdot 2)$ **5.** $5 \cdot \frac{1}{5} = 1$

**6.** Is the set of nonnegative integers closed under the operation of subtraction? Explain.

Simplify Problems 7–22. Assume $x$ and $y$ represent positive real numbers. Do not leave negative exponents in your answer.

**7.** $(x^2yz^2)^3(-2xy^3z)^2$

**8.** $(x^2y^{-3})^{-2}$

**9.** $(3^0x^{-2}y^3)^{-4}$

**10.** $\left(\frac{2^{-3}}{x^4}\right)^{-2}\left(\frac{-2x^3}{x^{-1}}\right)^3$

**11.** $\sqrt[6]{-64}$

**12.** $\sqrt[12]{x^2y^8}$

**13.** $\sqrt[8]{16x^6y^{10}}$

**14.** $\sqrt{50} + \sqrt[3]{2} - \sqrt{32} - \sqrt[3]{16}$

**15.** $\sqrt{2}(3\sqrt{6} - \sqrt{10}) - 2\sqrt{3}(\sqrt{15} + 2\sqrt{12})$

**16.** $(2\sqrt{3} + \sqrt{2})(2\sqrt{3} - 5\sqrt{2})$

**17.** $\frac{1}{\sqrt{12}}$

**18.** $\frac{3}{\sqrt{2} + \sqrt{3}}$

**19.** $\frac{-3a^2b + 2ab}{b^2}$ if $a = -1, b = 3$

**20.** $21xyz + 15xy - 17xyz + 7xy^2$

**21.** $[4x - (x + 1)] - [3x + (x + 1)]$

**22.** $(x - 6)(x^2 + 6x + 36)$

Factor Problems 23–34.

**23.** $15x^4y - 20x^2y^2 + 5x^2y$ **24.** $64x^3 - 27y^3$ **25.** $x^2 + 16x + 15$

**26.** $x^2 - 20x + 96$ **27.** $x^2 + 23x - 140$ **28.** $4x^2 + 4x + 1$

**29.** $9x^2 - 12x + 4$ **30.** $15x^2 - 19x - 10$ **31.** $4x^3 + 8x^2 - 60x$

**32.** $xy - 8 - 2x + 4y$ **33.** $a^4x + 3 - x - 3a^4$ **34.** $4x^2 + 4x - y^2 - 4y - 3$

Simplify Problems 35–40. Assume the denominators do not equal zero.

**35.** $\frac{3a^2 + 17a + 10}{2a^2 + 7a - 15}$

**36.** $\frac{3x^2 + 17x + 10}{x^2 + 10x + 25} \cdot \frac{x^2 - 25}{3x^2 + 11x + 6} \cdot \frac{5x^2 + 16x + 3}{5x^2 - 24x - 5}$

**37.** $\frac{25x^2 - 1}{10x^2 + 17x + 3} \div (1 - 5x)$

**38.** $\frac{x}{x^2 - 4x - 5} - \frac{2}{x + 1} + \frac{2x - 1}{x^2 - 7x + 10}$

**39.** $\dfrac{\frac{1}{a} + \frac{1}{a + b}}{\frac{2}{a + b} + 1}$

**40.** $\dfrac{\frac{1}{x^2 + 4x + 3} + \frac{1}{x - 2}}{\frac{1}{x + 3} + \frac{1}{x^2 - x - 2}}$

**41.** Write in scientific notation: 0.0000006148

**42.** Evaluate: $\left(\frac{1}{9}\right)^{-\frac{3}{2}}$

**43.** Evaluate: $(-27)^{\frac{2}{3}}$

**44.** Simplify: $(36x^{12}y^{15})^{\frac{3}{2}}$

**45.** Simplify: $(125x^{-6}y^3)^{\frac{2}{3}}$

**46.** Write as one radical: $\sqrt[3]{\sqrt[4]{xy}}$

## PRACTICE TEST FOR CHAPTER 1

**1.** Evaluate $\left|\frac{3}{8} - \frac{4}{7}\right|$.

State which field property is illustrated by each of Problems 2–4.

**2.** $6 + (12 + 1) = 6 + (1 + 12)$ **3.** $8 + 0 = 8$ **4.** $3(2 + 9) = 3(2) + 3(9)$

Simplify Problems 5–10. Assume $x$ and $y$ represent positive real numbers.

**5.** $(3^{-2})^3(3^0)^{-1}$

**6.** $\sqrt[10]{16x^6y^8}$

**7.** $4\sqrt{3}(\sqrt{6} - \sqrt{3} + \sqrt{18})$

**8.** $\sqrt[11]{-1}$

**9.** $\sqrt{8} - 2\sqrt{18} + \sqrt{50}$

**10.** $\dfrac{2}{\sqrt{5} + 1}$

Simplify Problems 11–13.

**11.** $16a^2b - 4a^2b + 5a^2b^2 - 7a^2b$ **12.** $2(x + y) - [3x - (x + y)]$ **13.** $(x - 5)(x^2 + 5x + 25)$

Factor Problems 14–16.

**14.** $15x^2 + 5x - 70$ **15.** $3y - x - 3 + xy$ **16.** $x^2 + 8xy + 16y^2 - 25$

Simplify Problems 17–19. Assume the denominators do not equal zero.

**17.** $\dfrac{x^2 - 2x - 63}{x^2 - 81} \div \dfrac{2x^2 - 5x - 12}{2x + 3}$

**18.** $\dfrac{x + 2}{x - 5} - \dfrac{5x + 31}{x^2 - 2x - 15}$

**19.** $\dfrac{\dfrac{1}{x + y} - \dfrac{1}{y}}{\dfrac{1}{x^2 - y^2}}$

**20.** Evaluate: $\left(\dfrac{8}{27}\right)^{-\frac{2}{3}}$

# 2 *Equations and Inequalities*

> ■ *By the help of God and with His precious assistance, I say that Algebra is a scientific art . . . . The perfection of this art consists in knowledge of the scientific method by which one determines numerical and geometric unknowns.*
>
> *Omar Khayyam*

## 2-1 SCIENTIFIC CALCULATORS (OPTIONAL)

This text does *not require* the use of a calculator. However, any student taking this course is *encouraged* to consider the purchase of an inexpensive scientific calculator. It should be stressed that students are asked to avoid using a calculator for any of the exercises in which the calculations can be readily done by hand. The only problems in the text that really demand the use of a scientific calculator are marked with the ▦ symbol. Dependence on the use of the scientific calculator for regular exercises in the text will only hurt the student in the long run.

### The Two Types of Logic Used in Scientific Calculators

There are two major types of scientific calculators that are popular today. The most common type employs a type of logic known as **algebraic** logic. The calculators manufactured by Casio, Sharp, and Texas Instruments as well as many other companies employ this type of logic. An example of calculation on such a calculator would be the following. To add 14 + 26 on an algebraic logic calculator, the sequence of buttons would be:

$$14 \boxed{+} 26 \boxed{=}$$

The second type of scientific calculator requires the entry of data in **Reverse Polish Notation (RPN)**. Calculators manufactured by Hewlett-Packard and a few other specialized calculators are made to use RPN. To add 14 + 26 on a RPN calculator, the sequence of buttons would be:

14 [enter] 26 [+]

Mathematicians and scientists do not agree on which type of scientific calculator is superior. However, the clear majority of college students own calculators that employ *algebraic* logic. Therefore this section of the text is explained with reference to the sequence of steps employed by an *algebraic* logic calculator. If you already own or intend to purchase a scientific calculator that uses RPN, you are encouraged to study the instruction booklet that comes with the calculator and practice the problems shown in the booklet. After this practice you will be able to solve the calculator problems discussed in this section.

### Performing Simple Calculations

The following example will illustrate the use of the scientific calculator in doing basic arithmetic calculations.

**EXAMPLE 1** Evaluate each of the following on a scientific calculator.

**a.** $2.61 + 3.82$ **b.** $0.0006 \times 158.32$

**Note:** In the following problems your calculator may display more digits or fewer digits in the final answer.

***Solution***

**a.** 2.61 [+] 3.82 [=] 6.43

**b.** .0006 [×] 158.32 [=] 0.094992 ■

Any scientific calculator with algebraic logic uses a priority system that has a clearly defined hierarchy (or order of operations). It is the same hierarchy that is in common usage in performing arithmetic operations by hand. In either situation, calculations are performed in the following order:

1. First calculations within a parentheses are completed.
2. Then numbers are raised to a power or a root is calculated.
3. Then multiplication and division operations are performed from left to right.
4. Then addition and subtraction operations are performed from left to right.

**Note:** This hierarchy is carefully followed on *scientific calculators.* Small inexpensive calculators that do not have scientific functions often do not follow this hierarchy.

**EXAMPLE 2** Evaluate: $5.3 \times 1.62 + 1.78 \div 3.51$.

***Solution*** This problem requires that one evaluate the product and the quotient first and then add the two results. If the sequence of data is entered directly into the calculator exactly as the problem is written, the calculator will perform the calculations in the correct order.

5.3 [×] 1.62 [+] 1.78 [÷] 3.51 [=] 9.09312251 ■

In order to perform some calculations on the calculator the use of parentheses is helpful. These parentheses may or may not appear in the original problem.

EXAMPLE 3 Evaluate: $5 \times (2.123 + 5.786 - 12.063)$.

***Solution*** The problem requires that the numbers in the parentheses be combined first. By entering the parentheses on the calculator this will be accomplished.

5 [×] [(] 2.123 [+] 5.786 [−] 12.063 [)] [=] −20.77

*Note:* The result is a negative number. ■

### Notation of Numbers

#### *Negative Numbers*

EXAMPLE 4 Evaluate: $(-8.634)(5.821) + (1.634)(-16.082)$.

***Solution*** The products will be evaluated first by the calculator. Therefore parentheses are not needed as we enter the data.

8.634 [+/−] [×] 5.821 [+] 1.634 [×] 16.082 [+/−] [=] −76.536502

*Note:* The result is negative. ■

#### *Numbers in Scientific Notation*

All scientific calculators are limited to displaying numbers no larger than $9.999999999 \times 10^{99}$ and no smaller than $9.999999999 \times 10^{-99}$. If a number outside these boundaries is produced or if you try to do a problem that is not permitted (such as $6 \div 0$), the calculator will display an error message. (Common error displays are "Error," flashing display, or **E**). Be sure you are familiar with the message your calculator uses to communicate an error message.

If you wish to enter a number in scientific notation, you should use the special scientific notation button. On most calculators it is denoted as [EXP] or [EE].

**EXAMPLE 5** Multiply: $(5.31 \times 10^{16})(2.18 \times 10^{12})(1.82 \times 10^{-10})$.

***Solution*** 5.31 [EXP] 16 [×] 2.18 [EXP] 12 [×] 1.82 [EXP] 10 [+/−] [=] 2.1067956 19

Thus we know that the answer is $2.1067956 \times 10^{19}$. ■

### Special Functions

All current scientific calculators have a key for finding powers of numbers. It is usually labeled [$y^x$]. (On a few calculators the notation is [$x^y$].) To raise a number to a power, first you enter the base, then push the [$y^x$] key. Then you enter the exponent, then finally the [=] button.

**EXAMPLE 6** Evaluate the following.

**a.** $(2.16)^9$ **b.** $(1.536)^{-22}$

***Solution*** **a.** 2.16 [$y^x$] 9 [=] 1023.490369

**b.** 1.536 [$y^x$] 22 [+/−] [=] 0.000079322 ■

**EXAMPLE 7** Evaluate: $(5.163 \times 1.824 - 9.4231)^2$.

***Solution*** The quantity obtained inside the parentheses is negative. However, on most scientific calculators, when the [$x^2$] button is employed, the result becomes positive, since the square of a negative number is always positive. The use of the parentheses is needed before the [$x^2$] button is used because we are squaring the result of the operations within the parentheses.

[(] 5.163 [×] 1.824 [−] 9.4231 [)] [$x^2$] 0.00003350 ■

### Roots of Numbers

Most calculators will have a button or a series of buttons to take roots of numbers. The square root occurs so often that it is a separate button. Other roots can be evaluated by the [$\sqrt[x]{y}$] button. (On some calculators it is labelled as [$x^{\frac{1}{y}}$] or it is necessary to use a [INV] [$y^x$] or [2nd Fn] [$y^x$] sequence.) Study your calculator and its instruction manual carefully to see how roots are performed on your calculator.

**EXAMPLE 8** Evaluate: $\sqrt[5]{5618 + 2734 + 3913}$.

***Solution*** [(] 5618 [+] 2734 [+] 3913 [)] [$\sqrt[x]{y}$] 5 [=] 6.57254413 ■

Roots of numbers can also be evaluated by changing the indicated root to exponent notation and using the $\boxed{y^x}$ button.

**EXAMPLE 9** Evaluate $\sqrt[4]{0.161384}$ by using the $\boxed{y^x}$ button.

***Solution*** We know that $\sqrt[4]{0.161384} = (0.161384)^{\frac{1}{4}} = (0.161384)^{0.25}$. Thus we calculate

$$0.161384 \ \boxed{y^x} \ .25 \ \boxed{=} \ 0.63381880 \quad \blacksquare$$

This introduction to scientific calculators is brief and may not cover every situation. You are encouraged to consult the instruction booklet of your calculator for further help.

## Exercise 2-1

In Problems 1–46, perform the calculation on a scientific calculator. Do not round off your answer.

**1.** $157.81 + 98.36$

**2.** $0.00621 - 0.00563$

**3.** $125{,}361 \times 89{,}208$

**4.** $\dfrac{76.328}{5.713}$

**5.** $12{,}820 - 7{,}431 + 7.986$

**6.** $12.6 - 8.7 - 5.3 + 3.7$

**7.** $2.56 + 8.98 \times 3.14$

**8.** $1.62 + 3.81 - 5.23 \times 6.18$

**9.** $\dfrac{(2.16 \times 10^3)(1.37 \times 10^{14})}{6.39 \times 10^5}$

**10.** $\dfrac{(3.84 \times 10^{-12})(1.62 \times 10^5)}{7.78 \times 10^{-8}}$

**11.** $(3.84)(-6.2) - (4.3)(3.12)$

**12.** $5.62(5 \times 3.16 - 18.12)$

**13.** $\dfrac{12.386 \times 10^{-12}}{5.6218 - 12.3714}$

**14.** $\dfrac{5.436 \times 10^{15}}{0.0661 + 0.0872}$

**15.** $\dfrac{2.3 + 5.8 - 2.6 - 3.9}{5.3 - 8.2}$

**16.** $\dfrac{(2.6)(-3.2) + (5.8)(-0.9)}{2.614 + 5.832}$

**17.** $\sqrt{0.0713}$

**18.** $\sqrt{5.6213 - 3.7214}$

**19.** $\sqrt{56 + 83} - \sqrt{12}$

**20.** $\sqrt{(6.13)^2 + (5.28)^2}$

**21.** $\sqrt{(0.3614)^2 + (0.9217)^2}$

**22.** $\sqrt{(-5.62)^2 + 5(1.82)^2}$

**23.** $(1.78)^3 + 6.342$

**24.** $(2.26)^8 - 3.1413$

**25.** $5(4.26)^4 + (4.26)^2$

**26.** $3(2.17)^3 - 2(2.17)^2 + 5.21$

**27.** $(-5.62)^5$

**28.** $(12 - 14.63)^3$

**29.** $\sqrt[4]{156.218}$

**30.** $\sqrt[5]{37.173}$

**31.** $\sqrt[6]{(8.13 - 2.14) \times (5 + 8.13)}$

**32.** $\sqrt[3]{(2.6)^2 + (4.3)^2}$

**33.** $\sqrt[8]{(2.63)^5}$

**34.** $\sqrt[9]{(1.83)^4}$

**35.** $(2.18)^{-2} + (0.123)^3$

**36.** $(5.81)^{-4} + 0.0078 - 0.0613$

**37.** $5.816 + \pi(3.1783)$

**38.** $16.34 - \pi(2.16)^2$

**39.** $\dfrac{2.173 \times 10^{-14}}{\pi}$

**40.** $\dfrac{1.813\pi}{1.734 \times 10^{15}}$

**41.** $\dfrac{(3.68)^5(1.72)^6}{\sqrt{63.18}}$

**42.** $\dfrac{(12.31)^2(1.67)^9}{\sqrt{126.31}}$

**43.** $\dfrac{7 + \sqrt{49 - 4(2)(1.6)}}{4}$

**44.** $\dfrac{-13 + \sqrt{169 - 4(1.1)(2.8)}}{2.2}$

**45.** $2.73[1 + (2.6)^{-1} + (2.6)^{-2} + (2.6)^{-3}]$

**46.** $0.04\left[\frac{1}{2}\sqrt{3} + \sqrt{3.01} + \sqrt{3.02} + \frac{1}{2}\sqrt{3.03}\right]$

## 2-2 LINEAR EQUATIONS IN ONE UNKNOWN

An **equation** is a statement indicating that two algebraic expressions are equal. If the equation contains a variable, it is neither true nor false until a specific number is substituted for the variable. If a substitution of a particular value for the variable yields a true statement, we say that this value is the **solution** or **root** of the equation.

Two equations are said to be **equivalent** if they have exactly the same roots. We observe that $2x + 1 = 7$ is equivalent to $2x = 6$. Both equations have the root $x = 3$. (In fact, both equations have no other root but 3.)

An equation that is valid for every value of the variable is called an **identity**: $x + 3 = 5 + x - 2$ is an identity. It will yield a true statement regardless of what value is substituted for $x$. A linear equation that is valid only for one value of the variable is called a **conditional linear equation.** Clearly, $2x+1=7$ is a conditional equation. It is true when $x = 3$ but not for any other values of $x$ such as $x = 1$. The only root is 3.

The usual method of finding the solution or root of an equation is to transform it to an equivalent equation for which determination of the root is an easy matter.

---

The operations that change an equation into an equivalent equation are

1. Adding the same quantity to both sides of the equation.
2. Subtracting the same quantity from both sides of the equation.
3. Multiplying or dividing both sides of the equation by the same non-zero quantity.

4. Simplifying one or both sides of the equation by using the methods discussed in Chapter 1.

---

The **degree of a polynomial equation** with one variable is the highest power of that variable found in the equation.

In this section we will review the procedures used to solve a first-degree polynomial equation with one variable. These equations are referred to as **linear equations** in one variable. The solution of some linear equations will involve several steps. It is helpful to perform these steps in an orderly fashion.

---

The following steps will, if followed in order, yield the solution for any first-degree equation if the equation has a unique solution.

1. Remove all parentheses.
2. Multiply both sides of the equation by the least common denominator for all fractions in the equation.
3. Simplify by combining like terms on each side of the equation.
4. Add or subtract the quantities necessary to get the variable on one side and the other quantities on the other.
5. Divide both sides of the equation by the coefficient of the variable.
6. Substitute the solution into the original equation to verify that it is correct. If such a substitution makes the original statement true, then the solution is correct; otherwise it is incorrect.

---

**EXAMPLE 1** Solve for $x$: $3(x + 1) + 7 = 2x - (x + 4)$.

***Solution***

$3x + 3 + 7 = 2x - x - 4$ *Remove parentheses.*

$3x + 10 = x - 4$ *Simplify.*

$3x - x = -4 - 10$ *Subtract $x$ from each side. Subtract 10 from each side.*

$2x = -14$ *Simplify.*

$x = -7$ *Divide both sides of the equation by 2.*

*Check:*

$$3(-7 + 1) + 7 \stackrel{?}{=} 2(-7) - (-7 + 4)$$
$$3(-6) + 7 \stackrel{?}{=} -14 - (-3)$$
$$-18 + 7 \stackrel{?}{=} -14 + 3$$
$$-11 = -11 \quad \checkmark \quad \blacksquare$$

**EXAMPLE 2** Solve for $y$: $\frac{2}{3}(y-3)+\frac{1}{2}y=16-\left(\frac{1}{4}y+1\right)$.

***Solution***

$$\frac{2}{3}y-2+\frac{1}{2}y=16-\frac{1}{4}y-1 \qquad \textit{Remove parentheses.}$$

$$12\left(\frac{2}{3}y\right)-12(2)+12\left(\frac{1}{2}y\right)=12(16)-12\left(\frac{1}{4}y\right)-12(1) \qquad \textit{Multiply each side by L.C.D. = 12.}$$

$$8y-24+6y=192-3y-12 \qquad \textit{Simplify.}$$

$$14y-24=180-3y \qquad \textit{Combine like terms on each side of the equation.}$$

$$17y-24=180 \qquad \textit{Add 3y to each side.}$$

$$17y=204 \qquad \textit{Add 24 to each side.}$$

$$y=12 \qquad \textit{Divide both sides by 17.}$$ ■

Sometimes an equation that has a variable in the denominator of a fraction will reduce to a first-degree equation when multiplied by a common denominator. Such equations can then be solved by the step-by-step method discussed at the beginning of this section. However, one difficulty can arise. Since the common denominator contains the variable, we must be sure we are multiplying by a nonzero quantity.

**EXAMPLE 3** Solve for $x$: $\frac{3}{x+1}+2=8$.

***Solution*** The lowest common denominator is $x+1$, and we can multiply by this quantity only if $x \neq -1$. Doing so gives

$$(x+1)\left(\frac{3}{x+1}\right)+(x+1)(2)=(x+1)(8) \qquad \textit{Multiply each term by L.C.D.}$$

$$3+2(x+1)=8(x+1) \qquad \textit{Simplify.}$$

$$3+2x+2=8x+8 \qquad \textit{Remove parentheses.}$$

$$5+2x=8x+8 \qquad \textit{Collect like terms.}$$

$$5-6x=8 \qquad \textit{Subtract 8x from each side.}$$

$$-6x=3 \qquad \textit{Subtract 5 from each side.}$$

$$x=-\frac{1}{2} \qquad \textit{Divide both sides by } -6 \textit{ and reduce the fraction.}$$

*Check:* Can you verify the answer? ■

Probably a simpler way to arrive at the type of conclusion shown in Example 3 is to solve the equation, then try to check. *If the proposed solution makes any denominator equal to zero, it is not a solution,* since a fraction with a zero denominator is meaningless.

**EXAMPLE 4** Solve for $x$: $\dfrac{2}{x^2 - 1} - \dfrac{1}{x+1} = \dfrac{1}{x-1}$.

***Solution***

$$\frac{2}{(x+1)(x-1)} - \frac{1}{x+1} = \frac{1}{x-1} \qquad \textit{Factor first denominator.}$$

$$\cancel{(x+1)}\cancel{(x-1)}\left[\frac{2}{\cancel{(x+1)}\cancel{(x-1)}}\right] - \cancel{(x+1)}(x-1)\left(\frac{1}{\cancel{x+1}}\right)$$

$$= (x+1)\cancel{(x-1)}\left(\frac{1}{\cancel{x-1}}\right) \qquad \textit{Multiply each term by L.C.D.} = (x+1)(x-1).$$

$$2 - (x-1) = (x+1) \qquad \textit{Simplify.}$$

$$2 - x + 1 = x + 1 \qquad \textit{Remove parentheses.}$$

$$-x + 3 = x + 1 \qquad \textit{Collect like terms.}$$

$$-2x + 3 = 1 \qquad \textit{Subtract } x \textit{ from each side.}$$

$$-2x = -2 \qquad \textit{Subtract 3 from each side.}$$

$$x = 1 \qquad \textit{Divide both sides by } -2.$$

Note that $x = 1$ is only an *apparent solution.*

*Check:* When $x = 1$ is substituted into the original equation, we have

$$\frac{2}{0} - \frac{1}{2} \stackrel{?}{=} \frac{1}{0}$$

which is meaningless, since division by zero is never allowed. There is **no solution** for this problem. ■

### Equations Containing More Than One Variable

Sometimes we encounter a linear equation with more than one variable or with symbols such as $a$, $b$, and $c$, which represent constants. We will find it useful to be able to solve for one variable in terms of all other variables and constants. The six-step procedure to find the solution of a first-degree equation may be used. However, an additional step of factoring may be necessary before we divide both sides by the coefficient of the desired variable.

**EXAMPLE 5** Solve for $y$: $3y = 2x - 4(x - by)$.

***Solution***

$3y = 2x - 4x + 4by$ *Remove parentheses.*

$3y = -2x + 4by$ *Simplify.*

$3y - 4by = -2x$ *Subtract 4by from each side. We want all the y terms on one side and all terms that do not contain y on the other side.*

$y(3 - 4b) = -2x$ *Factor out the y from each term.*

$\dfrac{y(3 - 4b)}{(3 - 4b)} = \dfrac{-2x}{3 - 4b}$ *Divide each side by (3 − 4b), which is the coefficient of y.*

$y = \dfrac{-2x}{3 - 4b}$ or $\dfrac{2x}{4b - 3}$ *Note that the answer may be written in several ways.* ■

## EXERCISE 2-2

Solve Problems 1–42.

1. $5x - 8 = 0$
2. $4x + 3 = 0$
3. $3x - 3 = -3$
4. $8 - 7x = 2$
5. $2y + 3 = y - 7$
6. $5y - 7 = 3y + 2$
7. $1 + 3(x - 2) = 4$
8. $2(x - 3) = 3(1 - x)$
9. $0.5w + 0.7 = -0.8w - 0.6$
10. $0.4w - 0.3 = 0.2 - 0.5w$
11. $3.0 + 0.6x - 0.4 = 0.2x + 1.0$
12. $1.3x + 2.0 = 0.5x - 3.0$
13. $3(x - 4) = 5 + 2(x + 1)$
14. $3x + 2(x - 5) = 7 - (x + 3)$
15. $7x + 5 - 2(x - 1) = 21$
16. $\dfrac{x}{2} = \dfrac{1}{5} - x$
17. $t - \dfrac{1}{2} = \dfrac{t}{3} + 7$
18. $2t - 3(t - 2) = \dfrac{1}{2}(t + 1)$
19. $\dfrac{2}{3} + 1 = x - \dfrac{5}{2}$
20. $x - \dfrac{1}{5}x + 1 = \dfrac{1}{3}(x - 5)$
21. $\dfrac{2}{3}x - \dfrac{1}{2}x = x + \dfrac{1}{6}$
22. $\dfrac{x}{5} - \dfrac{2}{3}x + \dfrac{1}{2} = \dfrac{1}{3}(x - 4)$
23. $\dfrac{5}{x} = \dfrac{1}{2}$
24. $\dfrac{1}{x} + \dfrac{1}{2x} = 3$

25. $\frac{2}{x} + \frac{1}{3} = \frac{1}{2x} - 1$

26. $\frac{x+1}{x} - \frac{x-2}{2x} = 1$

27. $\frac{1}{x} = \frac{3}{2x}$

28. $\frac{4}{x} + \frac{1}{2x} = \frac{9}{4}$

29. $\frac{3}{x-1} = 1$

30. $\frac{x}{x-2} = \frac{4}{3}$

31. $\frac{2}{w} = \frac{1}{w+1}$

32. $\frac{w}{w+1} = \frac{w-1}{w-2}$

33. $\frac{1}{x+1} = \frac{2}{1-x^2}$

34. $\frac{x+1}{x^2+2x-3} + \frac{1}{x-1} = \frac{1}{x+3}$

35. $\frac{x+2}{x^2+7x} = \frac{1}{x+3}$

36. $\frac{1}{x^2-3x} = \frac{2}{x^2-9}$

37. $\frac{x}{x+1} + \frac{1}{2x+1} = 1$

38. $\frac{3}{x+5} = 1 - \frac{x-4}{2x+10}$

39. $(2x+3)(3x-2) = (6x+1)(x-5)$

40. $(2x+3)(x-2) = (2x-4)(x+3)$

41. $-3[x-(2x+3)-2x] = -9$

42. $-2[x-(x-1)] = -3(x+1)$

Solve for the indicated variable in Problems 43–48.

43. $V = k + gt$ for $t$

44. $A = \frac{h}{2}(b+c)$ for $c$

45. $C = \frac{5}{9}(F-32)$ for $F$

46. $L = a + (n-1)d$ for $n$

47. $5ax + by = -2(x+3y)$ for $y$

48. $\frac{1}{2}(x+ay) = 5(ax-y)$ for $x$

## Calculator Problems

Solve Problems 49–52 by using a scientific calculator. Round solutions to the nearest hundredth.

49. $7.86x - 3.21 = 0$

50. $0.17x + 5.82 = 0$

51. $3.62x - 0.59 = 1.36(2.58x - 1.18)$

52. $1.63 - 0.78x + 5.31 = \frac{1}{4}(1.23x - 6.33)$

Determine whether each of Problems 53–56 is a conditional equation or an identity.

53. $\sqrt{x^2} = x$

54. $(3x-4)^2 = 16 + 9x^2 - 24x$

55. $\frac{3}{(y+2)^2} = \frac{5+2-4}{4+4y+y^2}$

56. $y + 8 = \frac{y^2+6y-16}{y-2}$

★ **57.** **a.** Find the error in the following:

$$x^2 + 2x - 8 = x^2 - 2x$$
$$(x + 4)(x - 2) = x(x - 2)$$
$$x + 4 = x$$
$$+4 = 0$$

**b.** Now find the correct solution to the original equation.

★ **58.** **a.** Find the error in the following:

$$x^2 + 8x + 15 = x^2 + 7x + 10$$
$$(x + 5)(x + 3) = (x + 2)(x + 5)$$
$$x + 3 = x + 2$$
$$3 = 2$$

**b.** Now find the correct solution to the original equation.

★ **59.** For what value of $b$ is $x = -4$ a solution to the following equation?

$$2x + b = 3 - 4b + 5x$$

★ **60.** For what value of $b$ is $x = \frac{2}{3}$ a solution to the following equation?

$$\frac{1}{2}(x + 3b) = \frac{1}{5}$$

---

## 2-3 APPLICATIONS FOR LINEAR EQUATIONS

One of the major reasons linear equations are emphasized is that they can often be used to help solve actual problems not only in mathematics but also in business, psychology, economics, and the physical sciences. Many of these problems are stated in a few words or a short paragraph. Therefore they are often referred to as **word problems**. Actually, a more appropriate title is **applications for linear equations**. One reason that we do these types of problems in this text is to see how linear equations help us in the solution of real-world problems in various disciplines. (Later in the text we will study applications for various nonlinear equations.)

---

### Guidelines for Solving Applied Problems

---

1. Read the problem carefully and thoughtfully.
2. Determine a quantity that must be found. Label that quantity with a variable.
3. If possible, describe any other quantities that must be found in terms of this variable.
4. If possible, draw a picture or sketch that will help you to visualize the situation accurately; label the sketch.

5. Use the sketch and/or some of the sentences of the problem to determine an equation. (Sometimes a *standard formula* for perimeter, distance, interest, etc., is appropriate.)
6. Solve the equation and obtain the solution(s) to the problem.
7. Ask yourself if the answer seems reasonable. Then check to see that your solution(s) satisfy all the conditions stated in the problem.

---

### Geometry Problems

**EXAMPLE 1** The length of a rectangular-shaped garden is 3 meters longer than twice the width. The perimeter is 108 meters. What is the length and width of the garden?

***Solution*** Since the length is compared to the width, we select the width as the first quantity to be found and designate a variable for it.

Let $x =$ the width of the garden.

Since the length is 3 meters longer than twice the width, we have $2x+3=$ the length of the garden.

Now we make a helpful sketch of the object. We draw a rectangle and label its sides.

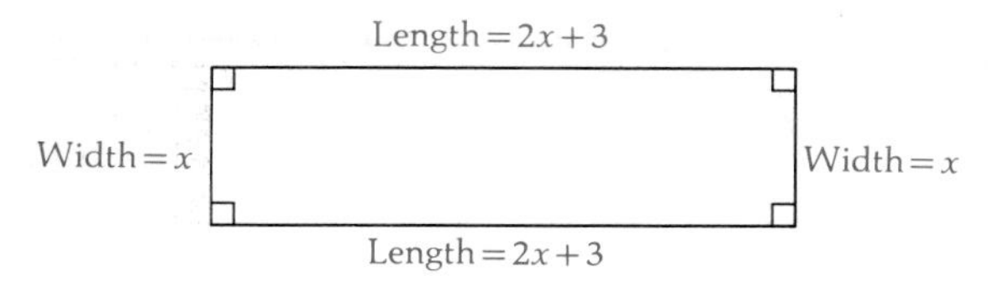


Since the perimeter is given by $P = 2l + 2w$ and we know that the perimeter is 108 meters, we can write

$$P = 2l + 2w$$
$$108 = 2(2x + 3) + 2(x)$$
$$108 = 4x + 6 + 2x$$
$$108 = 6x + 6$$
$$102 = 6x$$
$$17 = x$$

If $x = 17$, then $2x + 3 = 2(17) + 3 = 34 + 3 = 37$. Thus the width of the garden is 17 meters, and the length of the garden is 37 meters.

*Note:* Both answers are necessary. If we had stopped at $x = 17$, our problem would not have been solved. We were asked for *length and width.*

*Check:* Is the perimeter 108 meters?

$$2(17) + 2(37) \stackrel{?}{=} 108$$

$$108 = 108 \quad \checkmark$$

Is the length 3 more than double the width?

$$2(17) + 3 \stackrel{?}{=} 37$$

$$37 = 37 \quad \checkmark \quad \blacksquare$$

### Interest Problems

Simple **interest** $I$ can be calculated by the formula $I = Prt$ where $r =$ the simple interest rate, $P =$ the **principal**, and $t =$ the number of time units (usually measured in years).

If \$8000 was borrowed for 2 years at a simple interest rate of 12%, then the amount of interest after 2 years would be calculated thus:

$$I = Prt = (8000)(0.12)(2) = 1920$$

The total interest would be \$1920. A number of applied problems involve this concept of simple interest.

**EXAMPLE 2** A person invested \$8000 in two securities. One paid an annual interest rate of 4%, and the other paid 7%. The total interest on both securities for the first year was \$455. How much was invested at each rate?

***Solution*** We must first consider one of the unknown amounts invested. It does not matter which one we designate as $x$. Suppose we start with the amount of money invested at 4%. If we let

$$x = \text{the amount invested at } 4\%$$

then we can describe the second amount in terms of $x$. The two amounts total \$8000. Therefore if the amount $x$ is invested at 4%, then

$$8000 - x = \text{the amount invested at } 7\%$$

We can visualize the basic situation in this sketch:

Now if $x$ dollars is invested at 4%, then the interest for the year on this investment would be found by multiplying the investment $x$ by the rate of interest. We are using $I = Prt$ where $P = x$, $r = 0.04$, and $t = 1$. Thus $0.04x$ represents the interest on this investment. Similarly, $0.07(8000 - x)$ represents the interest on the other investment. The sum of the interests equals $455. Thus

$$0.04x + 0.07(8000 - x) = 455$$
$$0.04x + 560 - 0.07x = 455$$
$$560 - 0.03x = 455$$
$$-0.03x = -105$$
$$x = \frac{-105}{-0.03}$$

Solving, we obtain

$$x = \$3500 \qquad \text{and} \qquad 8000 - x = \$4500$$

Thus $3500 was invested at 4%, and $4500 was invested at 7%.

*Check:* Do the two investments total $8000?

$$3500 + 4500 = 8000 \checkmark$$

Does 4% of $3500 added to 7% of $4500 give a total of $455 in interest?

$$(0.04)(\$3500) + (0.07)(\$4500) \stackrel{?}{=} \$455$$
$$\$140 + \$315 = \$455 \checkmark \quad \blacksquare$$

**Rate Problems**

Problems involving distance, a uniform rate, and the traveling time will usually require the use of the formula Distance = Rate × Time or $d = rt$. The equation may be solved for any of the three variables.

**EXAMPLE 3** Two cars start at the same point and travel in opposite directions, one at 90 kilometers per hour and the other at 60 kilometers per hour. How many hours does it take them to get 450 kilometers apart?

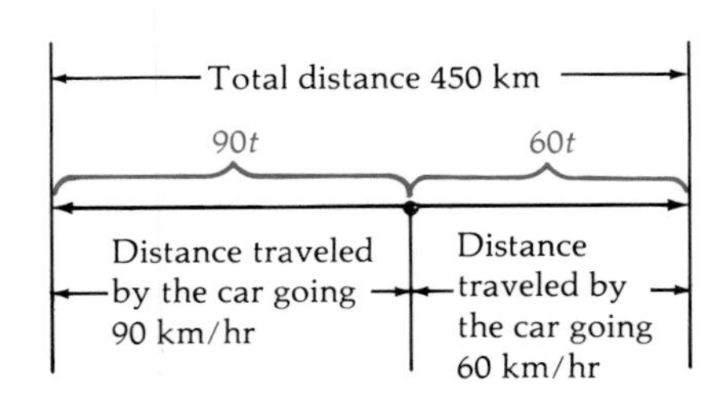


***Solution*** In this problem we make use of the formula $d = rt$ (distance = rate × time). If we let $t$ = the number of hours, then $90t$ must represent the distance traveled by the first car (distance = $90 \times t$) and $60t$ represents the distance traveled by the second car. Now, since the sum of the distances is 450 kilometers, we obtain the equation

$$90t + 60t = 450$$

Solving, we obtain $t = 3$ hours

*Check:* After 3 hours, will the cars be 450 kilometers apart? ■

### Mixture Problems

**EXAMPLE 4** A 12-liter solution of water and alcohol is 20% alcohol. How many liters of alcohol should be added to bring the solution up to 25% alcohol?

***Solution*** Let $x$ = number of liters of alcohol to be added. In a problem dealing with mixtures, concentrate on only one quantity. In this case, consider either the water or alcohol but *not* both.

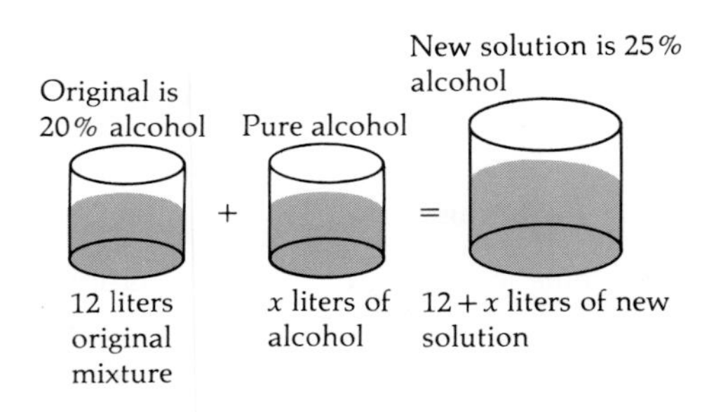

Let's consider the alcohol. How much alcohol do we start with? We have 12 liters of solution and 20% of it is alcohol. Therefore 0.20(12) represents the amount of pure alcohol.

Now we know that

(amount of alcohol in orig. mixture) + (amount of pure alcohol added)
= (amount of alcohol in new solution)

We are adding $x$ liters of pure alcohol, so $0.20(12) + x$ represents the total amount of alcohol we have in the final solution. Since we are obtaining a total of $12 + x$ liters of solution and it is 25% alcohol, then $0.25(12 + x)$ must also represent the amount of pure alcohol in the final solution. Thus

$$0.20(12) + x = 0.25(12 + x)$$

is our desired equation.

$$2.4 + x = 3.0 + 0.25x$$

$$x - 0.25x = 3.0 - 2.4$$

$$0.75x = 0.6$$

$$x = \frac{0.6}{0.75}$$

$$x = 0.8 \text{ liter} \quad \text{or} \quad \frac{4}{5} \text{ liter}$$

*Check:* Can you verify the answer? ■

**Work Problems**

EXAMPLE 5 A painter can paint a garage in 6 hours. His helper can paint the garage in 9 hours. How long will it take them if they work together to paint the garage?

***Solution*** We are asked to find how long it will take them working together to paint the garage. Let $x =$ the number of hours to do the combined job.

Let us assume that both the painter and his helper work at a constant rate and are not influenced by outside factors.

The key to work problems is to find the amount of work per unit of time (in this case, the amount of work per hour).

If a painter can do a job in 6 hours, then he can do $\frac{1}{6}$ of a job per hour. If his helper can do a job in 9 hours, the helper can do $\frac{1}{9}$ of a job per hour. If working together they can do the job in $x$ hours, then they can do $\frac{1}{x}$ of a job per hour.

Now the main idea needed is this:

$$\begin{pmatrix}\text{amount of work}\\ \text{done by painter}\\ \text{in 1 hour}\end{pmatrix} + \begin{pmatrix}\text{amount of work}\\ \text{done by helper}\\ \text{in 1 hour}\end{pmatrix} = \begin{pmatrix}\text{amount of work done}\\ \text{by both together}\\ \text{in 1 hour}\end{pmatrix}$$

$$\left(\frac{1}{6}\right) + \left(\frac{1}{9}\right) = \left(\frac{1}{x}\right)$$

$$\frac{1}{6} + \frac{1}{9} = \frac{1}{x}$$

*We observe that the L.C.D. = 18x.*

$$18x\left(\frac{1}{6}\right) + 18x\left(\frac{1}{9}\right) = 18x\left(\frac{1}{x}\right)$$

*Multiply each term by the L.C.D.*

$$3x + 2x = 18$$

$$5x = 18$$

$$x = \frac{18}{5}\text{ hours} \quad \text{or} \quad 3.6\text{ hours}$$

The amount of time together would be 3 hours and 36 minutes.

*Check:* Can you verify the solution to this problem? ■

## EXERCISE 2-3

### Comparison Problems

1. There are 9 more women chemistry majors than men chemistry majors at Jonesville College. A total of 81 chemistry majors are at the school this year. How many women are majoring in chemistry?
2. A metal rod is 98 centimeters long. It must be cut into two pieces. One piece must be 11 centimeters longer than the other. What will be the length of each piece?
3. A vitamin tablet consists only of vitamins A, B, and C. The amount of vitamin B must be triple the amount of vitamin A. The amount of vitamin C is 10 milligrams more than the amount of vitamin A. If the tablet totals 160 milligrams, how much of each vitamin should be used?
4. Fred, Alice, and Joan bought stereo sets for their rooms while enrolled in college. Alice paid double what Fred paid for his stereo set. Joan paid $80 more than Fred paid for his stereo set. The total bill for the three sets was $560.00. How much did each person pay?
5. A local college held a weight-reducing seminar. Three very large students attended the seminar for two semesters. The three of them lost a total of 215 pounds during the year. Tony lost two thirds the number of pounds that Carlos lost. Joan lost 17 pounds less than Carlos. How many pounds did each person lose?
6. The Environmental Science class at the college took measurements of three harmful pollutants in the air near a local industrial site. It was found to contain 13.5 ppm (parts per million) of the

three pollutants. There was three times as much of pollutant "A" as pollutant "B." There was five times as much of pollutant "C" as pollutant "B." How much of each type was found?

7. A father is now five times as old as his daughter. The sum of their ages in four years will be a total of 47 years. How old is each person *at the present time*?

8. A mother is now twice as old as her daughter. Eighteen years ago she was five times as old as her daughter. How old is each person *at the present time*?

## Geometry Problems

9. If the width of a rectangle is 2 meters more than half the length and the perimeter is 64 meters, find the length and width.

10. The length of a rectangle is 1 meter less than triple the width of the rectangle. The perimeter is 86 meters. Find the length and width of the rectangle.

11. The perimeter of a triangle is 51 centimeters. One side is double the second side. The third side is 3 cm longer than the second side. How long is each side?

12. The sum of the interior angles of any triangle is 180°. In a certain triangle, one angle is four times as large as a second angle. The third angle is 30° smaller than the second angle. Find the measurement of each angle.

13. A rectangular plot of ground is to be made into a garden with a path of uniform width along all four sides. That portion inside the path will be a rectangle that measures 15 meters by 30 meters. The perimeter of the entire rectangular plot of ground is 106 meters. How wide should the path be?

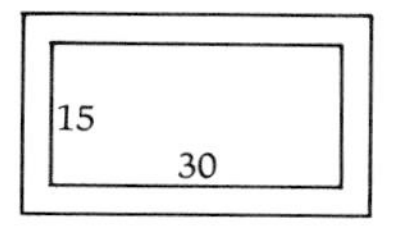


14. If the side of a square is increased by 2 feet, the perimeter of the new square is 3 feet shorter than five times the length of the side of the old square. Find the dimensions of the original square.

15. The perimeter of a rectangle is 42 meters. If the length is doubled and the width is tripled, a new rectangle is formed. The perimeter of the new rectangle is 18 meters longer than double the perimeter of the old rectangle. What are the dimensions of the old rectangle?

16. The perimeter of Sam's rectangular garden last year was 48 yards. This year Sam is thinking of expanding the area of his garden. If he has three such gardens placed side by side, the new shape of the new enlarged garden will be a square. What were the dimensions of his garden last year?

## Interest Problems

All the following problems are done using the concept of *simple interest*.

17. If part of $15,000 is invested at 6% and the remainder at 8% and the annual income from the total investment is $1072, how much is invested at each rate?

18. A woman invests part of $10,000 at 5% and the remainder at 8% per year. If her total interest for one year is $650, how much did she invest at each rate?

19. Sally earned $3000 during high school. For one year she invested part of it in the money market, which earns 14%, and the rest in an All Savers Certificate, which earns 12%. At the end of the year she had earned $394 in interest. How much did she invest at each rate?

20. Frank inherited $7000 from his uncle. For one year he invested part of it in mutual funds earning 12% interest. He left the rest of it in his special notice savings account earning 7% interest. At the end of the year he had earned $765. How much did he invest at each rate?

21. Sam received a 7% cost of living raise last year. This year his salary is $12,840. What was his salary last year before the 7% cost of living raise?

22. Susan bought a compact car last year. The dealer told her that the same model this year was $7020. This year's models have increased in price 8% over last year's models. How much did Susan pay last year for her car?

## Rate Problems

23. The Smith family took a 12-hour trip by car from one city to another. Their son Fred took the same trip by bus the next day in 8 hours. The bus traveled at an average rate of speed of 20 miles per hour faster than the car. Find the average speed of the bus and of the car on these trips.

24. A bus travels 200 miles in the same time it takes a tractor-trailer truck to travel 140 miles. The bus is traveling at an average rate of speed of 15 miles per hour faster than the truck. Find the average rate of speed of each vehicle.

25. Jim left on a trip going north of the city, traveling at 50 miles per hour. Two hours later, Robert left on the same trip, traveling at 55 miles per hour. How long will it be until Robert overtakes Jim?

26. Farmer Jones took his tractor to the tractor dealership for repairs. He left his driver's license at home. His daughter noticed this $\frac{2}{3}$ of an hour later and followed him with the car. The farmer with the tractor averaged 25 miles per hour. The daughter is traveling at 45 miles per hour. How long will it take the daughter to catch up to her father?

27. Two training jets leave the airport at the same time. One plane travels west going 64 kilometers per hour faster, while the other plane travels east. At the end of two hours the group commander reports they are 3200 kilometers apart. How fast is each jet traveling?

28. One car leaves the city traveling north at an average rate of 45 kilometers per hour. If a second car leaves the same place 20 minutes later and averages 60 kilometers per hour, how long will it take the second car to overtake the first and at what distance from the city will it occur?

## Mixture Problems

29. A solution of 10 liters of 30% alcohol must be increased in strength to a 60% solution. How many liters of pure alcohol must be added to accomplish this goal?

30. A solution of 7 liters of 10% alcohol must be increased in strength to a 30% solution. How many liters of pure alcohol must be added to accomplish this goal?

31. How much water will be required to dilute 15 liters of a 12% alcohol solution to obtain a 5% solution?

**32.** How much water will be required to dilute 1 liter of pure alcohol so that a solution is obtained that is 25% alcohol?

**33.** One container is filled with a mixture of 48% alcohol. Another container is filled with a mixture of 80% alcohol. When the two mixtures are combined, the result is 80 gallons of a mixture of 60% alcohol. How much did each original container hold?

**34.** One container is filled with a mixture of 5% alcohol. Another container holds 60 gallons of 20% alcohol. How much of the 5% alcohol mixture should be combined with 60 gallons of the 20% alcohol mixture to get a new mixture that is 10% alcohol?

### Work Problems

**35.** One worker can complete a certain job in 10 hours. A second worker can do the same job in 12 hours. How long will it take them to complete the job if they work together?

**36.** Frank can mow his lawn in 4 hours. It takes his young son Timmy 5 hours to mow the lawn. How long would it take them to do it working together?

**37.** One pipe can fill a tank in 15 minutes. A second pipe (the drain pipe) can empty the tank in 18 minutes. The drain is left open by mistake when the first pipe is turned on to fill the empty tank. How long will it take to fill the tank if the drain is left open?

**38.** One pipe can fill a tank in 1 hour and another (the drain) can empty the tank in $1\frac{1}{2}$ hours. A worker opens the valve to fill the tank and 20 minutes later notices that the drain has been left open. If the worker then closes the drain, how long will it take the first pipe to finish filling the tank?

**39.** Jean and her mother together cleaned the house in 2 days. Her mother can clean the house alone in 3 days. How long would it take Jean to clean the house alone?

**40.** A man is able to shingle his roof in 6 hours. Working together with his neighbor, Alfonse, he can do it in 3 hours and 45 minutes. How long would it take Alfonse to do the job by himself?

### Miscellaneous Problems

**41.** Brenda has maintained a 78 average in her chemistry class after taking six tests in the course. She has five of the tests with scores of 50, 71, 76, 96, and 82. She has lost her other test and cannot remember her grade. What grade did she obtain on that test?

**42.** Wong Kin has taken four tests in his math course. His grades so far are 80, 76, 70, and 90. He wants to bring his course average up to an 84 by doing well on the final examination. His final examination counts twice as much as a regular test. What is the minimum score he can obtain on the final examination and achieve his goal?

**43.** The baseball season for the local city league consists of 40 games for each team. The Giants have won 7 out of their first 10 games. *How many more games* must the Giants win in order to win 60% of their games during the season?

**44.** An executive secretary was hired for a certain starting salary at a local insurance company. The second year she was given a $600 raise. The third year she was given a $600 raise. The same thing happened during the fourth and fifth years. After five years she left the job because of low salary. Her total earnings for 5 years were $48,000. What was her salary when she started her job?

★ **45.** A park manager needs to estimate the number of bears in a wildlife refuge. He uses a technique known as the capture-mark-recapture method. He first caught and marked for identification 50 bears. He then released them. Some time later he again captured 50 bears. Of the 50, 4 had his previous mark. Let us assume that the proportion of marked bears in the second sample is the same as the proportion of all marked bears in the wildlife refuge. Can you estimate the number of bears in the refuge?

★ **46.** The combined salaries of four people in the front office at Wesson Company is $53,000. The office clerk is the lowest paid of the four. The office manager earns 20% more than the office clerk. The director earns $3000 more than the average of the office manager's salary and the office clerk's salary. The president earns twice what the director earns. What is the annual salary of each person?

---

## 2-4 QUADRATIC EQUATIONS

*Definition*

The standard form of a *quadratic equation* is

$$ax^2 + bx + c = 0$$

where $a \neq 0$ and $a, b, c$ are real numbers.

### The Factoring Method

If the expression $ax^2 + bx + c$ is factorable, the most direct method to find the roots of the equation is by the factoring method.

Solving equations by factoring is based on a theorem from algebra which says that "if a product is zero, then one of the factors is zero."

*Theorem 2-1*

If $ab = 0$, then $a = 0$ or $b = 0$.

Factorable polynomial equations can be solved by the following steps:

**1.** Arrange the equation in standard form.
**2.** Factor the polynomial completely.
**3.** Set each factor equal to zero and solve the resulting equations for the variable.
**4.*** (Optional step of checking) Substitute each obtained value of the variable into the original equation to determine if it is a solution.

* Step 4 is sometimes *required* when the original equation is not in the form of a polynomial. We will study these separately.

In some cases the terms of a quadratic expression may have a common factor.

**EXAMPLE 1** Solve for $x$: $x^2 - 5x = 0$.

***Solution***

| | | | |
|---|---|---|---|
| | $x(x - 5) = 0$ | | *Factor.* |
| $x = 0$ | or | $x - 5 = 0$ | *Set each factor equal to* 0. |
| $x = 0$ | or | $x = 5$ | *Solve for $x$.* |

*Check:* Can you verify each answer? ■

If the quadratic equation is not in standard form, then the student should perform the necessary algebraic steps to place it in standard form. Be sure to simplify and collect any like terms before attempting to use the factoring method.

**EXAMPLE 2** Solve for $x$: $6x(x + 1) = 2x - 7 - 13x$.

***Solution***

| | | | |
|---|---|---|---|
| | $6x^2 + 6x = -11x - 7$ | | *Remove parentheses on left-hand side. Collect like terms on right-hand side.* |
| | $6x^2 + 17x + 7 = 0$ | | *Add $11x + 7$ to both sides.* |
| | $(2x + 1)(3x + 7) = 0$ | | *Factor.* |
| $2x + 1 = 0$ | or | $3x + 7 = 0$ | *Set each factor equal to zero.* |
| $x = -\frac{1}{2}$ | or | $x = -\frac{7}{3}$ | *Solve for $x$.* |

*Check:* Can you check these two solutions in the original equation? ■

If the expression is factorable, the factoring method can be used when the quadratic expression contains two or more variables. In such cases the roots will usually be expressed in terms of one or more variables.

**EXAMPLE 3** Solve for $x$: $3x^2 + xy - 4y^2 = 0$.

***Solution***

| | | | |
|---|---|---|---|
| | $(3x + 4y)(x - y) = 0$ | | *Factor.* |
| $3x + 4y = 0$ | or | $x - y = 0$ | *Set each factor equal to zero.* |
| $x = -\frac{4y}{3}$ | or | $x = y$ | *Solve for $x$ in terms of $y$.* ■ |

Let us examine closely a situation in which we encounter an apparent solution that does not satisfy the original equation.

**EXAMPLE 4** Solve for $x$: $\dfrac{x^2+2}{x-2} = 4 + \dfrac{3x}{x-2}$.

***Solution*** We multiply by L.C.D. $= x - 2$.

$$\cancel{(x-2)}\left(\frac{x^2+2}{\cancel{x-2}}\right) = 4(x-2) + \cancel{(x-2)}\left(\frac{3x}{\cancel{x-2}}\right)$$

$$x^2 + 2 = 4(x-2) + 3x \qquad \textit{Simplify.}$$

$$x^2 + 2 = 4x - 8 + 3x \qquad \textit{Remove parentheses.}$$

$$x^2 + 2 = 7x - 8 \qquad \textit{Collect like terms.}$$

$$x^2 - 7x + 10 = 0 \qquad \textit{Add } -7x + 8 \textit{ to each side.}$$

$$(x-5)(x-2) = 0$$

$$x - 5 = 0 \quad \text{or} \quad x - 2 = 0$$

$$x = 5 \quad \text{or} \quad x = 2$$

*Check:* If $x = 2$, we have

$$\frac{2^2+2}{0} \stackrel{?}{=} 4 + \frac{3(2)}{0}$$

Since division by zero is impossible, this equation is not defined. We know that $x = 2$ is *not a solution*.

If $x = 5$,

$$\frac{5^2+2}{5-2} \stackrel{?}{=} 4 + \frac{3(5)}{5-2}$$

$$\frac{25+2}{3} \stackrel{?}{=} 4 + \frac{15}{3}$$

$$\frac{27}{3} \stackrel{?}{=} 4 + 5$$

$$9 = 9 \quad \checkmark$$

Thus $x = 5$ is *the only solution* to the equation $\dfrac{x^2+2}{x-2} = 4 + \dfrac{3x}{x-2}$.

■

A useful property is the following.

**Theorem 2-2**

If $x^2 = a$ where $a \geq 0$, then the solutions are $x = \pm\sqrt{a}$.

Most students remember this theorem as the way to *take the square root of each side of an equation.* However, it is important to remember when this is done to take both a positive and a negative square root, since there are *two sign possibilities.* Study the ways that this theorem can be used in each part of Example 5.

EXAMPLE 5 Find $x$ in each case.

**a.** $x^2 = 36$ **b.** $x^2 = 48$ **c.** $(x + 2)^2 = 3$

***Solution***

**a.**
$$\begin{aligned} x^2 &= 36 \\ \sqrt{x^2} &= \pm\sqrt{36} \\ x &= \pm 6 \end{aligned}$$

**b.**
$$\begin{aligned} x^2 &= 48 \\ \sqrt{x^2} &= \pm\sqrt{48} \\ x &= \pm\sqrt{48} \\ x &= \pm 4\sqrt{3} \end{aligned}$$

**c.**
$$\begin{aligned} (x + 2)^2 &= 3 \\ \sqrt{(x + 2)^2} &= \pm\sqrt{3} \\ x + 2 &= \pm\sqrt{3} \\ x &= -2 \pm \sqrt{3} \end{aligned}$$

■

### Completing the Square Method

The approach used to solve Example 5c to obtain answers for $(x + 2)^2 = 3$ can be used for any quadratic equation. It is known as the *completing the square method.*

EXAMPLE 6 Solve for $x$ by completing the square: $2x^2 - 3x - 3 = 0$.

***Solution***

$2x^2 - 3x - 3 = 0$ *We first need to make the coefficient of the $x^2$ term be one.*

$\frac{2x^2}{2} - \frac{3x}{2} - \frac{3}{2} = \frac{0}{2}$ *Divide each term by 2.*

$x^2 - \frac{3}{2}x - \frac{3}{2} = 0$ *Simplify.*

$x^2 - \frac{3}{2}x = \frac{3}{2}$ *Now we want to arrange the equation so that the $x^2$ terms and the $x$ terms are on the left and the constant term is on the right. Thus we add $\frac{3}{2}$ to each side.*

We want the left-hand side to be a perfect square trinomial. If the coefficient of $x^2$ is one, the number that is needed for the third term is always the square of one half of the coefficient of the $x$ term. Thus $\frac{1}{2}\left(-\frac{3}{2}\right) = -\frac{3}{4}$ and by squaring we obtain $\left(-\frac{3}{4}\right)^2 = \frac{9}{16}$. Hence the needed third term is $\frac{9}{16}$. [Do you see why this is necessary? We are really working backward through the formula for a perfect square trinomial, $a^2 - 2ab + b^2 = (a - b)^2$.]

$$x^2 - \frac{3}{2}x + \frac{9}{16} = +\frac{3}{2} + \frac{9}{16} \qquad \textit{Add } \frac{9}{16} \textit{ to each side.}$$

$$\left(x - \frac{3}{4}\right)^2 = +\frac{24}{16} + \frac{9}{16} \qquad \textit{Write left-hand side in factored form.}$$

$$\left(x - \frac{3}{4}\right)^2 = \frac{33}{16} \qquad \textit{Simplify.}$$

$$\sqrt{\left(x - \frac{3}{4}\right)^2} = \pm\sqrt{\frac{33}{16}}$$

$$\left(x - \frac{3}{4}\right) = \pm\frac{\sqrt{33}}{4}$$

$$x = \frac{3}{4} \pm \frac{\sqrt{33}}{4} \qquad \textit{The roots are } x = \frac{3 \pm \sqrt{33}}{4}.$$ ■

In practice, the completing the square method is rarely used to solve quadratic equations. However, it can be used to derive the useful quadratic formula, as we will see in the next section.

### The Quadratic Formula

$$ax^2 + bx + c = 0$$ *First we divide by $a$ so that the coefficient of $x^2$ will be 1. (To do this, we require $a \neq 0$.)*

$$x^2 + \frac{b}{a}x + \frac{c}{a} = 0$$

$$x^2 + \frac{b}{a}x = -\frac{c}{a}$$ *We subtract $\frac{c}{a}$ from each side.*

Now, since the coefficient of the $x$ term must be twice the square root of the constant term, we will supply a constant term to fulfill this requirement.

We take one half of the coefficient of $x$, giving $\frac{b}{2a}$, then add the square of this as a constant term to each side of the equation.

$$x^2 + \frac{b}{a}x + \frac{b^2}{4a^2} = \frac{b^2}{4a^2} - \frac{c}{a}$$

Factoring gives

$$\left(x + \frac{b}{2a}\right)^2 = \frac{b^2}{4a^2} - \frac{c}{a}$$

$$\left(x + \frac{b}{2a}\right)^2 = \frac{b^2 - 4ac}{4a^2}$$

We now use Theorem 2-2 to take the square root of each side:

$$x + \frac{b}{2a} = \pm\sqrt{\frac{b^2 - 4ac}{4a^2}}$$

or

$$x = \frac{-b \pm \sqrt{b^2 - 4ac}}{2a}$$

which is the quadratic formula. This formula is developed from the general quadratic, and therefore any equation that can be put into standard form can be solved by this formula. **You should memorize the quadratic formula.**

---

### The Quadratic Formula

---

For any rational numbers $a$, $b$, $c$, if $a \neq 0$, then the roots of the quadratic equation $ax^2 + bx + c = 0$ are given by $x = \frac{-b \pm \sqrt{b^2 - 4ac}}{2a}$

---

**Note:** Since the formula is derived from the standard form, it is *always necessary* to put a quadratic equation in the form $ax^2 + bx + c = 0$ *before* identifying the values $a$, $b$, and $c$.

**EXAMPLE 7** Solve $5x^2 - 7x = 8$ by the quadratic formula.

***Solution*** $5x^2 - 7x - 8 = 0$ *Subtract 8 from each side to obtain the standard form.*

We see that $a = 5$, $b = -7$, and $c = -8$. We substitute each of these into the quadratic formula:

$$x = \frac{-(-7) \pm \sqrt{(-7)^2 - 4(5)(-8)}}{2(5)}$$

$$= \frac{7 \pm \sqrt{49 + 160}}{10} = \frac{7 \pm \sqrt{209}}{10}$$

The two roots are $x = \dfrac{7 + \sqrt{209}}{10}$ and $x = \dfrac{7 - \sqrt{209}}{10}$. ■

Do you see that it follows from the formula that $b^2 - 4ac$ (the portion under the radical) will determine the nature of the roots of a quadratic? The term $b^2 - 4ac$ is called the **discriminant** because it indicates the nature of the roots.

Examine the following table and determine why each entry is valid. (Recall that numbers such as 1, 4, 9, 16, 25, 36, 49, 64, . . . are referred to as squares of integers or *perfect squares*.)

ROOTS OF A QUADRATIC

| *Value of the Discriminant* | *Nature of the Roots (for a, b, c, Integers)* |
|---|---|
| $b^2 - 4ac < 0$ | No real roots. |
| $b^2 - 4ac = 0$ | Roots are real and equal. (Two identical roots.) The trinomial is a perfect square. |
| $b^2 - 4ac > 0$ but not a perfect square | Roots are irrational and unequal. (Two different roots which contain a radical.) |
| $b^2 - 4ac > 0$ and a perfect square | Roots are unequal and rational. (Two different roots containing no radical.) The trinomial is factorable. |

### Complex Numbers

One entry in the "Nature of the Roots" column shows "no real roots." To the observant student this should imply that roots might exist in some other set of numbers, and such is the case.

Suppose, for example, that we want to find a solution to the equation $x^2 + 1 = 0$. This means that we must find a value for $x$ such that, when we square it, the result will be $-1$. Can you see why no real number would satisfy this condition? If we want a solution, therefore, we must look to another set of numbers.

To introduce this new set of numbers, we define the imaginary unit $i$ as a square root of $-1$.

*Definition*

$$i = \sqrt{-1} \quad \text{or} \quad i^2 = -1$$

Accepting this definition of $i$ makes it possible to find values for the square roots of negative numbers, or at least to indicate such values.

To do so, we use the following definition.

*Definition*

For all real numbers $a > 0$,

$$\sqrt{-a} = \sqrt{-1}\sqrt{a} = i\sqrt{a}$$

We will now use this definition in a few examples.

**EXAMPLE 8** Find the value of $\sqrt{-4}$ using the definition $\sqrt{-a} = \sqrt{-1}\sqrt{a} = i\sqrt{a}$.

***Solution*** $\sqrt{-4} = \sqrt{(-1)(4)} = \sqrt{-1}\sqrt{4} = i\sqrt{4} = 2i$

To check $\sqrt{-4} = 2i$ by the definition of square root, we evaluate $(2i)^2$.

$$(2i)^2 = (2i)(2i) = (2)(2)(i)(i) = 4i^2$$

Since $i^2 = -1$, then $4i^2 = -4$. ■

**EXAMPLE 9** Find the value of $\sqrt{-4}\sqrt{-4}$.

***Solution*** $\sqrt{-4}\sqrt{-4} = (2i)(2i) = 4i^2 = -4$

which is the *correct* value.

*Caution:* Many students fall into the trap of writing $\sqrt{-4}\ \sqrt{-4} = \sqrt{(-4)(-4)} = \sqrt{16} = 4$. This is *not* accurate since $\sqrt{a}\ \sqrt{b} = \sqrt{ab}$ is a true statement for *real numbers* only. In other words you can state $\sqrt{a}\ \sqrt{b} = \sqrt{ab}$ only if $a \geq 0$ and $b \geq 0$. ■

**EXAMPLE 10** Find the value of $\sqrt{-10}$ using the definition $\sqrt{-a} = \sqrt{-1}\sqrt{a} = i\sqrt{a}$.

***Solution*** $\sqrt{-10} = \sqrt{(-1)(10)} = \sqrt{-1}\sqrt{10} = i\sqrt{10}$

(The answer is left in this form, since $\sqrt{10}$ is not rational.) ■

A number such as $2i$ or $i\sqrt{10}$ is called an **imaginary number**.

*Definition*

> The indicated sum of a real number and an imaginary number $(a + bi)$ where $a$ and $b$ are real is called a *complex number.*

Some quadratic equations will yield roots that are complex numbers. Complex numbers will be studied in more detail in a later chapter.

**EXAMPLE 11** Solve for $x$ using the quadratic formula: $x^2 - 2x + 5 = 0$.

***Solution*** $a = 1 \qquad b = -2 \qquad c = 5$

$$x = \frac{-(-2) \pm \sqrt{(-2)^2 - 4(1)(5)}}{2(1)}$$

$$= \frac{2 \pm \sqrt{4 - 20}}{2} = \frac{2 \pm \sqrt{-16}}{2}$$

$$= \frac{2 \pm 4i}{2} = \frac{\not{2}(1 \pm 2i)}{\not{2}}$$

*Do not forget to factor and reduce the answer to the simplest form.*

$$x = 1 \pm 2i$$

The two roots $x = 1 + 2i$ and $x = 1 - 2i$ are both complex numbers. ■

### Applied Problems Involving Quadratic Equations

The distance $S$ an object falls during time $t$ is measured by $S = \frac{1}{2}gt^2 + v_0t$ where $g$ is the gravitation constant and $v_0$ is the initial velocity of the object.

**EXAMPLE 12** Solve for $t$ in the equation $S = \frac{1}{2}gt^2 + v_0t$.

***Solution*** $0 = \frac{1}{2}gt^2 + v_0t - S$ *Place the quadratic equation in standard form.*

We have a quadratic equation in $t$. We consider $t$ to be the variable and identify the values for $a, b, c$.

$$a = \frac{1}{2}g \qquad b = v_0 \qquad c = -S$$

$$t = \frac{-v_0 \pm \sqrt{(v_0)^2 - 4\left(\frac{1}{2}g\right)(-S)}}{2\left(\frac{1}{2}g\right)}$$

$$t = \frac{-v_0 \pm \sqrt{(v_0)^2 + 2gS}}{g}$$ ■

## EXERCISE 2-4

Solve Problems 1–12 for $x$ by factoring.

1. $2x^2 + 6x = 0$
2. $2x^2 - 3x = 0$
3. $x^2 + 2x = 3$
4. $2x^2 + 5x + 3 = 0$
5. $x^2 + 9 = 6x$
6. $x = 2 + \frac{35}{x}$
7. $5x + \frac{4}{3x} = \frac{23}{3}$
8. $6x^3 - 13x^2 = 5x$
9. $3x^2 + 5xy = 2y^2$
10. $\frac{1}{x+2} + 2 = \frac{3}{x^2 + 2x}$
11. $\frac{2x^2 + 3}{x+1} = 6 - \frac{5x}{x+1}$
12. $\frac{4}{x+2} + \frac{1}{2x-5} = 1$

Solve Problems 13–22 by the quadratic formula. Leave all solutions in simplest form.

13. $x^2 + 2x - 15 = 0$
14. $5x^2 = 7x + 6$
15. $2x^2 + 3x = 0$
16. $x^2 + 3x + 1 = 0$
17. $5x^2 + 7x + 1 = 0$
18. $3x^2 - 5 = 0$
19. $x^2 + 2x = 7$
20. $9x^2 + 12x + 4 = 0$
21. $3x^2 + 8x = 5$
22. $3x^2 + 4x - 5 = 0$

Compute $b^2 - 4ac$ for Problems 23–32 and indicate the nature of the roots.

23. $x^2 - 7x + 12 = 0$
24. $x^2 - 4x + 4 = 0$
25. $x^2 - 3x + 1 = 0$
26. $2x^2 - 7x + 3 = 0$
27. $x^2 - 3x + 5 = 0$
28. $5x^2 + 6x + 1 = 0$
29. $6x^2 + 11x - 21 = 0$
30. $4x^2 + 4x + 1 = 0$
31. $2x^2 + 5x - 3 = 0$
32. $x^2 + 5x + 7 = 0$

In Problems 33–38, find the roots to each quadratic equation. Simplify any complex numbers that are obtained.

33. $x^2 - 4x + 8 = 0$
34. $2x^2 - 2x + 1 = 0$
35. $8x^2 - 4x + 1 = 0$
36. $x^2 + x + 1 = 0$
37. $x^2 - x + 1 = 0$
38. $x^2 - 2x + 4 = 0$

In Problems 39–52, find the roots of each quadratic equation by any appropriate method.

**39.** $11x^2 - 7x + 1 = 0$ **40.** $3x^2 - 7x - 2 = 0$ **41.** $2x^2 + 4x - 5 = 0$

**42.** $3x^2 + 2x - 8 = 0$ **43.** $6x^2 - 7x - 3 = 0$ **44.** $6x^2 - 5x - 4 = 0$

**45.** $\frac{2}{3}y^2 - \frac{8}{9}y = 1$ **46.** $\frac{4}{3}y^2 - y - 1 = 0$ **47.** $\frac{8}{y^2} - \frac{4}{y} + 1 = 0$

**48.** $\frac{3}{y^2} - \frac{2}{y} + 1 = 0$ **49.** $6s(s - 1) = -1$ **50.** $t(2t - 1) = 4$

**51.** $\frac{4}{x} = 1 - \frac{4}{x + 6}$ **52.** $\frac{3}{x + 5} = 1 - \frac{3}{x - 5}$

★**53.** If $x = -5$ is one root of the equation $3x^2 + kx - 10 = 0$, then find the other root.

★**54.** If $3x^2 - 2x + k = 0$ and one root of the equation is $x = \frac{5}{3}$, then find the other root.

In Problems 55–58, solve for the indicated variable.

**55.** $F = \frac{kmv^2}{r}$ for $v$ **56.** $S = \frac{1}{2}gt^2$ for $t$

★**57.** $V = ar^2h + 2br$ for $r$ ★**58.** $V = 2\pi r(r + h)$ for $r$

**59.** A lot of land is rectangular in shape. The length is twice the width. The area is 1800 square feet. Find the length and width.

**60.** A rectangular park has a length that is 1 mile more than triple the width of the park. The area of the park is 80 square miles. Find the length and width.

**61.** One leg of a right triangle is 2 meters longer than the other. If the hypotenuse is 10 meters, find the length of the two legs.

**62.** The sum of a number and its reciprocal is $\frac{25}{12}$. Find the number.

★**63.** The length of a rectangle is twice the width. If each dimension is increased by 3, the new area would be 104 square meters. Find the original dimensions of the rectangle.

★**64.** A given square has an area of 36 square meters. The area is increased by 28 square meters to form a larger square. By how much was each side increased?

**65.** The area of a triangle is 40 square meters. The base is 2 meters shorter than the altitude. Find the base and the altitude of the triangle.

**66.** The number of diagonals $d$ of a polygon of $n$ sides is given by $\frac{n^2 - 3n}{2} = d$. A certain polygon has 44 diagonals. How many sides does it have?

## Calculator Problems

Solve for $x$ using the quadratic formula. Round your answers to the nearest hundredth.

**67.** $2.5x^2 + 1.3x - 8.5 = 0$ **68.** $1.8x^2 - 2.2x - 3.4 = 0$

**69.** The diagonal of a square is 1.52 meters longer than the side. Find the length of the side.

**70.** Under certain conditions the minimum number of feet $d$ required to stop a car traveling at speed $s$ in miles per hour is $d = 0.045\, s^2 + 1.2\, s$. Estimate the speed of the car in miles per hour if it required 150.5 feet to stop in an emergency.

---

## 2-5 OTHER TYPES OF EQUATIONS

In the preceding sections we discussed first-degree equations, equations that are factorable, and quadratic equations. There are certain other types of equations that can be solved by various methods which we will discuss in this section.

*Definition*

An equation is *quadratic in form* if a substitution can be found for the variable that will result in an equation of the form $ax^2 + bx + c = 0$.

*Rule*

*To solve an equation that is quadratic in form,* make a suitable substitution, solve the resulting quadratic equation, and then use these solutions together with the substitution to find possible roots and check them in the original equation.

**EXAMPLE 1** Solve the equation $x^{\frac{1}{2}} - 7x^{\frac{1}{4}} + 10 = 0$.

***Solution*** Let $y = x^{\frac{1}{4}}$. Then we see that $y^2 = (x^{\frac{1}{4}})^2 = x^{\frac{1}{2}}$.

$y^2 - 7y + 10 = 0$ *We make the substitution of $y$ for $x^{\frac{1}{4}}$ and $y^2$ for $x^{\frac{1}{2}}$.*

$(y - 5)(y - 2) = 0$ *Solve the resulting quadratic equation for $y$.*

$y = 5 \qquad y = 2$

$x^{\frac{1}{4}} = 5 \qquad x^{\frac{1}{4}} = 2$ *We use the $y$ values in our substitution equation to find $x$.*

$(x^{\frac{1}{4}})^4 = 5^4 \qquad (x^{\frac{1}{4}})^4 = 2^4$ *Raise each side to the fourth power.*

$x = 625 \qquad x = 16$

*Check:* Can you verify each of these solutions? ■

Certain types of equations that are quadratic in form have exponents greater than 2. Consider the next example.

**EXAMPLE 2** Solve for $x$ if $3x^4 - 2x^2 - 5 = 0$.

***Solution*** Let $y = x^2$. Then $y^2 = (x^2)^2 = x^4$. We substitute $y = x^2$ and $y^2 = x^4$ into the original equation. Then

$$3y^2 - 2y - 5 = 0$$

$$(3y - 5)(y + 1) = 0 \qquad \textit{Factor the left-hand side.}$$

Now solve by letting each factor equal zero, giving

$$y = \frac{5}{3} \qquad \text{or} \qquad y = -1$$

Thus

$$x^2 = \frac{5}{3} \qquad \text{or} \qquad x^2 = -1$$

Two roots are real numbers.

$$x^2 = \frac{5}{3}$$

$$x = \pm\sqrt{\frac{5}{3}} \qquad \textit{Take the square root of each side.}$$

$$x = \pm\sqrt{\frac{15}{9}} = \frac{\pm\sqrt{15}}{3} \qquad \textit{Rationalize the denominator.}$$

The roots for the second quadratic equation are not real.

$$x^2 = -1$$

$$x = \pm\sqrt{-1} \qquad \textit{Take the square root of each side.}$$

$$x = \pm i \qquad \textit{Use the definition } \sqrt{-1} = i.$$

Thus the four roots are: $\dfrac{\sqrt{15}}{3}, \dfrac{-\sqrt{15}}{3}, i, -i$. ■

### Radical Equations

Another type of equation that requires special treatment is the equation containing radicals. To solve such an equation, we must eliminate the radical by raising both sides of the equation to a common power. This process must be repeated until we have an equation free of radicals or in a form that we can solve by a suitable substitution. The solutions *must* be checked in the *original equation.*

The requirement of the checking step is an absolute necessity. When we raise both sides of an equation to an integral power, the new equation may have more solutions than the original equation. Any solution to the new equation that is not a solution to the original equation is called an **extraneous solution** and must be eliminated.

EXAMPLE 3 Solve $\sqrt{x+2} = x - 4$.

***Solution***

$$(\sqrt{x+2})^2 = (x-4)^2 \qquad \textit{We square both sides.}$$
$$x + 2 = x^2 - 8x + 16$$
$$x^2 - 9x + 14 = 0$$
$$(x-7)(x-2) = 0$$
$$x = 7 \quad \text{or} \quad x = 2.$$

*Check:* We must verify each potential solution into the **original equation**!

If $x = 7$,

$$\sqrt{7+2} \stackrel{?}{=} 7 - 4$$
$$\sqrt{9} \stackrel{?}{=} 3$$
$$3 = 3 \ \checkmark$$

Thus $x = 7$ is a valid solution. However, if we try $x = 2$,

$$\sqrt{2+2} \stackrel{?}{=} 2 - 4$$
$$\sqrt{4} \stackrel{?}{=} -2$$
$$2 \neq -2$$

It does not check, so $x = 2$ is an *extraneous* solution. The *only* solution to the original equation is $x = 7$. ■

In more involved problems it may be necessary to square each side of the equation more than once.

EXAMPLE 4 Solve for $x$: $\sqrt{2x-1} = 1 + \sqrt{x-1}$.

***Solution***

$$(\sqrt{2x-1})^2 = (1+\sqrt{x-1})^2 \qquad \textit{Square each side.}$$

$$2x-1 = (1)^2 + 2(1)(\sqrt{x-1}) + (\sqrt{x-1})^2$$

*Use the formula* $(a+b)^2 = a^2 + 2ab + b^2$ *where* $a = 1$ *and* $b = \sqrt{x-1}$

$$2x-1 = 1 + 2\sqrt{x-1} + x - 1$$

$$2x-1 = 2\sqrt{x-1} + x$$

$$x-1 = 2\sqrt{x-1}$$

*Now we isolate the radical on the right-hand side.*

$$(x-1)^2 = (2\sqrt{x-1})^2 \qquad \textit{Square each side.}$$

$$x^2-2x+1 = 2^2(\sqrt{x-1})^2 \qquad \textit{Note the necessity of squaring 2.}$$

$$x^2-2x+1 = 4(x-1)$$

$$x^2-2x+1 = 4x-4$$

$$x^2-6x+5 = 0 \qquad \textit{Add } -4x+4 \textit{ to each side.}$$

$$(x-5)(x-1) = 0$$

$$x = 5 \qquad x = 1$$

*Check:* If $x = 5$,

$$\sqrt{2(5)-1} \stackrel{?}{=} 1 + \sqrt{5-1}$$

$$\sqrt{9} \stackrel{?}{=} 1 + \sqrt{4}$$

$$3 = 1 + 2 \quad \checkmark$$

If $x = 1$,

$$\sqrt{2(1)-1} \stackrel{?}{=} 1 + \sqrt{1-1}$$

$$\sqrt{1} \stackrel{?}{=} 1 + \sqrt{0}$$

$$1 = 1 \quad \checkmark$$

Thus both values of $x$ satisfy the original equation. We have two solutions: $x = 5$ and $x = 1$. ■

---

**Key Procedure in Solving Radical Equations**

---

In solving equations that have one or more radicals, you should always ensure that *one radical is isolated* on one side of the equation before squaring each side. This enables you to simplify the equation.

---

Equations containing radicals other than square roots should be raised to the appropriate power.

**EXAMPLE 5** Solve for $x$: $\sqrt[3]{3x^2 - 1} = 2$.

***Solution*** $(\sqrt[3]{3x^2 - 1})^3 = 2^3$ *Cube each side.*

$3x^2 - 1 = 8$

The two roots are $x = \sqrt{3}$ and $x = -\sqrt{3}$.

*Check:* Can you verify that these two roots satisfy the original equation? ■

## EXERCISE 2-5

Solve Problems 1–38 for the variable quantity.

**1.** $x^4 - 5x^2 + 4 = 0$ **2.** $x^4 - 6x^2 + 8 = 0$ **3.** $x^4 - 7x^2 + 10 = 0$

**4.** $x^4 - 3x^2 - 4 = 0$ **5.** $x^4 + 8x^2 + 15 = 0$ **6.** $x^6 - 7x^3 - 8 = 0$

**7.** $x^{\frac{1}{2}} - 5x^{\frac{1}{4}} + 6 = 0$ **8.** $x - 4x^{\frac{1}{2}} - 21 = 0$ **9.** $x^{\frac{2}{3}} + 2x^{\frac{1}{3}} - 3 = 0$

**10.** $2x^{\frac{1}{2}} + 7x^{\frac{1}{4}} + 3 = 0$ **11.** $3x^{\frac{2}{3}} + 8x^{\frac{1}{3}} - 3 = 0$ **12.** $x^{\frac{2}{3}} - x^{\frac{1}{3}} - 6 = 0$

**13.** $x^{\frac{1}{3}} - 4x^{\frac{1}{6}} + 3 = 0$ **14.** $2x^{\frac{1}{3}} - 5x^{\frac{1}{6}} + 2 = 0$ **15.** $x^2 - 1 = 1 - \frac{1}{x^2}$

**16.** $\frac{1}{x^4} + 8 = \frac{6}{x^2}$ **17.** $(y + 2)^2 + 11(y + 2) + 18 = 0$

**18.** $(y - 4)^2 - 5(y - 4) + 6 = 0$ **19.** $\sqrt{x + 6} = 2$ **20.** $\sqrt{x - 3} = 5$

**21.** $\sqrt{x} - 7 = -2$ **22.** $\sqrt{x} - 6 = 2$ **23.** $\sqrt{3x + 4} = x$

**24.** $x + \sqrt{x} = 6$ **25.** $x - 2\sqrt{x} = 3$ **26.** $x + \sqrt{2x - 5} = 10$

**27.** $x - 2\sqrt{2x + 1} = -2$ **28.** $5 - \sqrt{5x - 1} = x$

★ **29.** $\sqrt{x + 2} + \sqrt{2x - 10} = 5$ ★ **30.** $\sqrt{x + 1} + \sqrt{2x + 3} = 5$

★ **31.** $\sqrt{3x + 4} + \sqrt{x + 2} = 8$ **32.** $\sqrt{3x - 2} - \sqrt{x + 7} = 1$

**33.** $\sqrt{x^2 + 6x} = \sqrt{2x + 21}$ **34.** $\sqrt{10x - 3} = \sqrt{3x^2 - 7x + 7}$

**35.** $\sqrt[3]{1 - 5x} = -2$ **36.** $\sqrt[3]{4x - 1} + \sqrt[3]{2x + 8} = 0$

**37.** $\sqrt{2x + 5} - \sqrt{x + 2} = \sqrt{3x - 5}$ **38.** $\sqrt{x + 1} + \sqrt{x - 7} = \sqrt{2x}$

In Problems 39–42, solve each equation for the specified variable.

**39.** Solve for $h$: $A = \pi r\sqrt{r^2 + h^2}$ **40.** Solve for $L$: $P = 2\sqrt{L \div g}$ **41.** Solve for $x$: $x^{\frac{2}{3}} + y^{\frac{2}{3}} = a^{\frac{2}{3}}$ **42.** Solve for $a$: $V = \sqrt{2g(a - b)}$

### Calculator Problems

Solve Problems 43–46 for all real values of $x$. Round off answer to the nearest thousandth.

**43.** $\sqrt{3.000x + 4.000} = 5.621$

**44.** $\sqrt{7.000x + 3.000} = 2.183$

★**45.** $3.120x + 4.180 = \sqrt{3.000x + 9.924}$

★**46.** $\sqrt{1.152x + 2.000} = 1.631x - 3.982$

## 2-6 LINEAR INEQUALITIES

To denote an interval on the number line when the endpoints are included, we will use brackets. For example $[-5, 3]$ designates the interval from $-5$ to 3 including the numbers $-5$ and 3. To say that $x$ is in $[-5, 3]$, sometimes written $x \in [-5, 3]$, would mean the same as $-5 \leq x \leq 3$.

When the endpoints of the interval are not included, we will use parentheses. For example $(-5, 3)$ designates the interval from $-5$ to 3 excluding the numbers $-5$ and 3. To say that $x$ is in $(-5, 3)$ would mean the same as $-5 < x < 3$.

When the endpoints of an interval are included in the set, the interval is said to be a **closed interval**. When the endpoints are excluded, the interval is called an **open interval**. Of course we can have **half-open** or **half-closed** intervals. In this case, one endpoint is included and the other is not.

The same notation (brackets and parentheses) can be used on the number line to graph a solution to an inequality. For instance, the statement $x \in [-5, 3]$ has the following graph:

and the statement $x \in (-5, 3)$ has the graph

The graph of $x \in (-5, 3]$ is

*Set builder notation* is sometimes used to describe inequalities. For example, if we wanted to write "the set of all numbers $x$ such that $x$ is greater than 4," we could write this in symbols as $\{x : x > 4\}$.

Similarly, "the set of all numbers $x$ such that $x$ is less than or equal to 2" could be written as $\{x : x \leq 2\}$.

The following chart may be helpful in categorizing the four types of intervals and the various forms of notation commonly used.

*Real Intervals*

| *Type of Interval* | *Interval Notation* | *Set Builder Notation* | *Graph* |
|---|---|---|---|
| Open interval | $(a, b)$ | $\{x{:}a < x < b\}$ | a b |
| Closed interval | $[a, b]$ | $\{x{:}a \leq x \leq b\}$ | a b |
| Half-open interval | $(a, b]$ | $\{x{:}a < x \leq b\}$ | a b |
| Half-open interval | $[a, b)$ | $\{x{:}a \leq x < b\}$ | a b |

If the values of $x$ can be infinitely large or infinitely small, we often use a special notation. The symbol $\infty$ is read as "infinity."

*Real Intervals Not Bounded in One Direction*

| *Interval Notation* | *Set Builder Notation* | *Graph* |
|---|---|---|
| $(a, +\infty)$ | $\{x{:}x > a\}$ | a |
| $[a, +\infty)$ | $\{x{:}x \geq a\}$ | a |
| $(-\infty, a)$ | $\{x{:}x < a\}$ | a |
| $(-\infty, a]$ | $\{x{:}x \leq a\}$ | a |

Note that we never place a closed interval notation next to the $+\infty$ or $-\infty$ symbol.

The rules for solving first-degree equations given in Section 2-2 will also solve first-degree inequalities. One difference must be noted that is extremely important. The following theorem must be taken into account as we solve first-degree inequalities.

***Theorem 2-3*** For all real numbers $a$ and $b$, if $a < b$, then $-a > -b$.

This theorem implies that if we multiply or divide each side of an inequality by a negative quantity, the sense (direction) of the inequality will be reversed. However, if we multiply or divide each side of an inequality by a positive quantity, the direction of the inequality will not be changed.

**EXAMPLE 1** Solve for $x$ and graph the solution: $8 + 5x \leq 3x - 3$.

***Solution***

$8 + 2x \leq -3$ *Subtract 3x from each side.*

$2x \leq -11$ *Subtract 8 from each side.*

$\frac{2x}{2} \leq -\frac{11}{2}$ *Divide both sides by 2.*

$x \leq -\frac{11}{2}$ or $x \leq -5\frac{1}{2}$ *The direction of the inequality is not changed.*

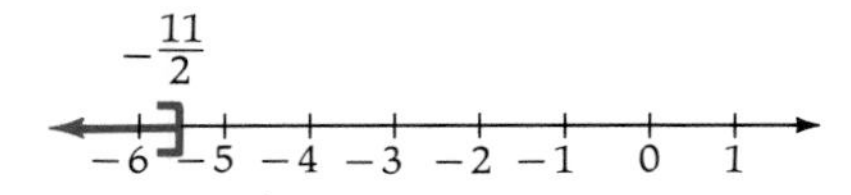


■

**EXAMPLE 2** Solve for $x$ and graph the solution: $\frac{1}{2}(x + 3) + 2x < 3x - 1$.

***Solution***

$\frac{1}{2}x + \frac{3}{2} + 2x < 3x - 1$ *Remove parentheses.*

$x + 3 + 4x < 6x - 2$ *Multiply all terms by L.C.D. = 2.*

$3 + 5x < 6x - 2$

$3 - x < -2$ *Subtract 6x from each side.*

$-x < -5$ *Subtract 3 from each side.*

$x > 5$ *The direction of the inequality is reversed by direct application of Theorem 2-3.*

The graph of the solution is

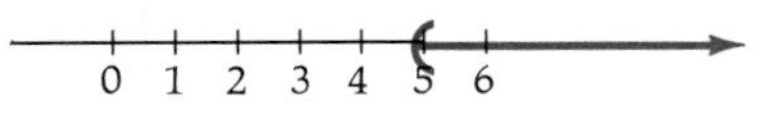


■

There are times when the graph of the solution is not appropriate or else not necessary. In such cases the answer is often expressed in interval notation.

**EXAMPLE 3** Solve for $x$ and write the solution set in interval notation.

$$3(x + 1) - 9x < x - 5(x + 4)$$

***Solution***

$$3x + 3 - 9x < x - 5x - 20$$
$$-6x + 3 < -4x - 20$$
$$-2x < -23$$
$$x > \frac{23}{2}$$ *Notice that the direction of the inequality is reversed.*

The solution in interval notation is $\left(\frac{23}{2}, +\infty\right)$. ■

Remember: The symbol $\infty$ (read as "infinity") is *not* to be considered a real number. $(-\infty, +\infty)$ designates the set of all real numbers; $(-\infty, 0)$ the negative real numbers, and $(0, +\infty)$ the positive real numbers.

In some cases it is possible to use the methods of Examples 1–3 to solve a compound inequality. The expression $3 < 5x - 2 < 18$ is a short way of writing the two inequalities $3 < 5x - 2$ and $5x - 2 < 18$. Any solution to this expression must be a solution to both of the inequalities. We can solve this expression for $x$ by working on both inequalities simultaneously.

**EXAMPLE 4** Solve for $x$ and graph the solution: $3 < 5x - 2 < 18$.

***Solution***

$3 + 2 < 5x - 2 + 2 < 18 + 2$ *We add 2 to each expression.*

$5 < 5x < 20$ *Simplify.*

$1 < x < 4$ *To solve for x, we divide each expression by 5.*

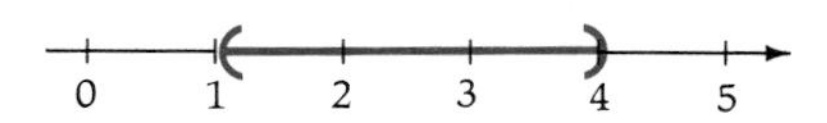


■

### Applications of Inequalities

**EXAMPLE 5** A Canadian radio station gives all its temperature readings in degrees Celsius. The meteorologist has predicted that tomorrow's temperature will range between 10°C and 20°C. Tom knows that Celsius temperature is related to Fahrenheit temperature by the equation $C = \frac{5(F - 32)}{9}$. Tom wants the temperature range if the predictions were made by using the Fahrenheit scale. Can you determine these?

***Solution*** The statement that the Celsius temperature will be between 10° and 20° can be written as: $10 \leq C \leq 20$.

$10 \leq \frac{5(F-32)}{9} \leq 20$ *We substitute the expression* $\frac{5(F-32)}{9}$ *for* $C$.

$90 \leq 5(F-32) \leq 180$ *Multiply each expression by* 9.

$90 \leq 5F - 160 \leq 180$ *Remove parentheses.*

$250 \leq 5F \leq 340$ *Add* 160 *to each expression.*

$50 \leq F \leq 68$ *Divide each expression by* 5.

Thus the temperature range will be between 50°F and 68°F. ■

The following property is useful in solving some applied problems.

*Theorem 2-4*

For all positive integers $n$ and for all positive real numbers $a$, $b$,
if $a < b$
then $a^n < b^n$ and $\sqrt[n]{a} < \sqrt[n]{b}$

## EXERCISE 2-6

Express Problems 1–5 in interval notation.

**1.** $3 \leq x \leq 8$ **2.** $-5 \leq x \leq 4$ **3.** $-1 < x < 6$ **4.** $0 \leq x < 9$ **5.** $-4 < x \leq -2$

In Problems 6–9, graph each expression on a number line

**6.** $x \in [-4, 8]$ **7.** $x \in (2, 7)$ **8.** $x \in (-2, 5]$ **9.** $x \in [-6, -3)$

**10.** How many real numbers are in the interval (1, 2)?

In Problems 11–20, solve the inequality and write the solution in interval form.

**11.** $3x - 9 \leq 0$ **12.** $5x + 15 > 0$

**13.** $6 - 2x \geq 8$ **14.** $11 - 3x < 2$

**15.** $4x + 3 < 2(x + 1)$ **16.** $6x - 5 \geq 10x + 3(x - 1)$

**17.** $\frac{1}{2}x + 3 < \frac{2}{3}x - 1$ **18.** $\frac{2}{5}(x + 1) \geq \frac{1}{4}x - \frac{1}{2}(x + 3)$

**19.** $x - 3 > 5\left(x - \frac{3}{5}\right)$ **20.** $\frac{1}{3}(3x + 5) \geq \frac{2}{3}(x + 1) - \frac{1}{2}(4x - 5)$

In Problems 21–34, solve for $x$ in each inequality and graph the solution.

**21.** $5x - 1 > 3(x + 1)$ **22.** $3(x + 3) + 2x > 3 + 7x$ **23.** $2(x + 3) \leq 7(x + 2) + 2$

**24.** $x > 4\left(\frac{1}{3}x + \frac{1}{2}\right)$ **25.** $\frac{2}{3}x + 1 \geq 2x - \left(\frac{x}{2} - 6\right)$ **26.** $5(x - 2) < 4(2x - 3) + 2$

**27.** $-4 < 3x - 1 < 5$ **28.** $6 < 2x - 6 < 12$ **29.** $9 > 5x - 1 > 4$

**30.** $5 > 2x + 3 > -7$ **31.** $4 \leq \frac{2(x - 5)}{3} \leq 6$ **32.** $-2 \leq \frac{4(2 - x)}{5} \leq 4$

**33.** $-10 < 5 - 2x \leq -5$ **34.** $-1 \leq -3 - 2x < 3$

**35.** The perimeter of a rectangular window must be less than 11 feet. The height of the window will be 3 feet. What restriction will be placed on the width of the window?

**36.** The perimeter of a rectangular sanitary landfill for a local city must be at least 2.0 kilometers. The width of the landfill is exactly 0.4 kilometers. What is the restriction on the length of the landfill?

**37.** The area of a triangle cannot exceed 46 square meters. The altitude of the triangle is 8 meters. What restriction must be placed on the base of the triangle?

**38.** In a normal day the average home may have demands of between 200 and 3100 watts of power (including both these values). The power in watts is related to the voltage and the current (measured in amps) by the equation $W = V \cdot I$. Find the daily range of current required in amps if the voltage of the house is a constant 110 volts.

**39.** The temperature in degrees Fahrenheit can be converted to degrees Celsius by the formula $C = \frac{5F - 160}{9}$. The requirements for temperature of the computer center are between 50°F and 68°F, including both these values. Find the temperature range in degrees Celsius.

**40.** A fishing boat at cruising speed can travel 4 nautical miles on a gallon of gas. The captain estimates that he has between 40 and 48 gallons of gas in his fuel tank. (He is including the values 40 and 48.) What is the range of possible distances that he can travel?

**41.** To get an A in her math course a student must obtain an average of between 90 and 100. (She is including these two averages.) She must take four tests. Her first three tests were 80, 86, and 96. What is the range of scores she can obtain on her last test and still obtain an A for the course?

**42.** A shipping box is constructed to be 6 meters long, 3 meters wide, and 2 meters high. Each measurement can be larger or smaller by 0.2 meters. What is the range of values of the volume of the box measured in cubic meters?

## Calculator Problems

Solve Problems 43–46 for $x$. Express your answers to the nearest thousandth.

**43.** $3(1.562x - 2.005) < 1.613x$ **44.** $\frac{5.368x + 8.213}{1.865} \geq 2x$

**45.** The circumference of a cylinder for an engine must be between 564.36 mm and 568.31 mm. What are the restrictions on the diameter of the circle?

**46.** A man invests $5650 in a mutual fund. The fund pays him one payment per year based on simple interest on the investment. He wants to collect between $450 and $800 per year in interest. What is the range of interest rates that will give him the desired amount of income?

## 2-7 QUADRATIC AND OTHER NONLINEAR INEQUALITIES

In Section 2-4 we solved quadratic and other equations by factoring and by using the quadratic formula. Solving quadratic and other inequalities proceeds in the same manner, until we must interpret the solution set. For instance, if we have the inequality

$$x^2 + 5x + 6 < 0$$

we factor it as $$(x + 3)(x + 2) < 0$$

We must recognize that the solution set will be those real numbers that will make the product of two factors negative. This means, of course, that one factor must be positive and the other negative. Several methods can be used to find the solution set, but all are basically the same. We must find the real numbers such that

*Case I* $x + 3 > 0$ and $x + 2 < 0$

or

*Case II* $x + 3 < 0$ and $x + 2 > 0$

The method we will present here is a graphical method that will give us a visual interpretation of the cases and perhaps make the solution set more meaningful. Be sure you understand the notation. When we write $x > 0$, it means that $x$ is positive; when we write $x < 0$, it means that $x$ is negative.

**EXAMPLE 1** Solve for $x$: $x^2 - 7x + 12 > 0$.

***Solution*** $(x - 4)(x - 3) > 0$ *Factor.*

Make a graph, showing where each factor is positive and where it is negative. We know that $x - 4$ is positive if $x > 4$ and negative if $x < 4$. Also, $x - 3$ is positive if $x > 3$ and negative if $x < 3$.

We graph these two situations thus:

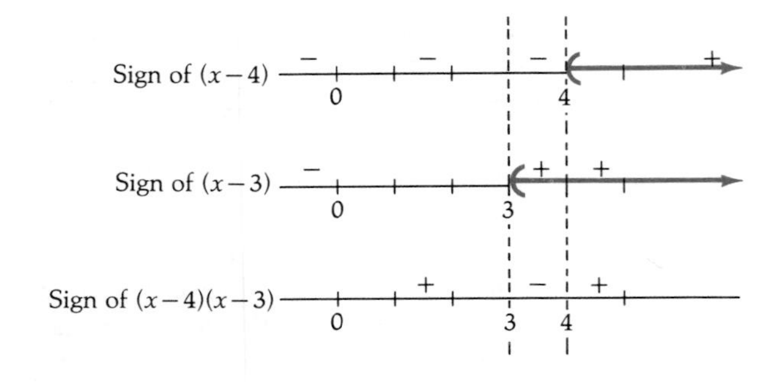

In the region $x > 4$ the value of the product is positive, since the sign of $(x - 4)(x - 3)$ is positive. Similarly, in the region $x < 3$ the value of the product is positive. Hence the graph of the solution is

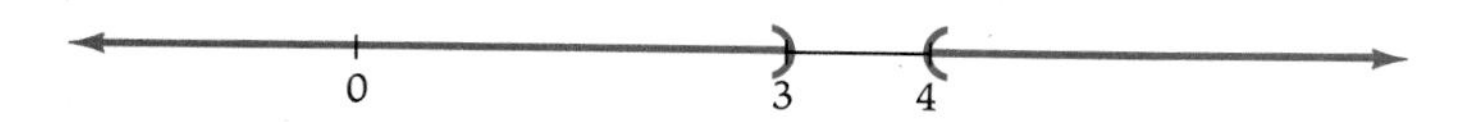


In interval notation this would be expressed as $(-\infty, 3) \cup (4, +\infty)$. ■

There may be problems that involve three or more factors.

**EXAMPLE 2** Solve for $x$: $(x + 4)(x + 2)(x - 1) \geq 0$.

***Solution*** Make a graph showing where each factor is positive and where it is negative. Notice that since the inequality is greater than or equal to zero, we include the boundary points. The intervals in our solution will be closed intervals or half-closed intervals.

Now we want to determine where $x + 4$, $x + 2$, and $x - 1$ are positive and where they are negative.

| | | | |
|---|---|---|---|
| If | $x + 4 \geq 0$ | then | $x \geq -4;$ |
| and if | $x + 2 \geq 0$ | then | $x \geq -2;$ |
| and if | $x - 1 \geq 0$ | then | $x \geq 1.$ |

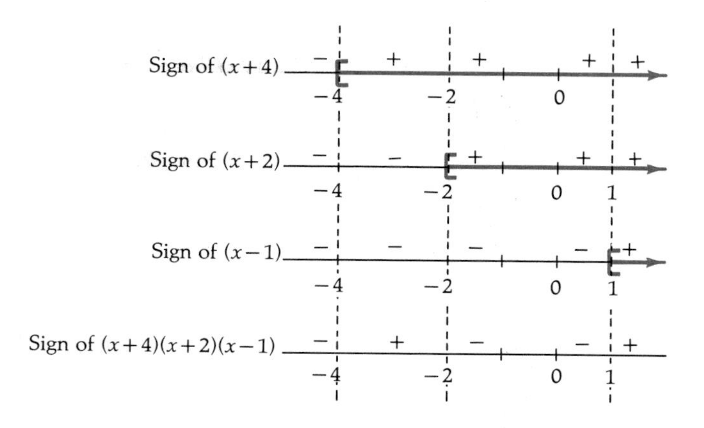


Be sure you understand clearly how to obtain the sign of the product. If all *three factors are positive,* the product is positive. This would indicate the region $x \geq 1$. If two factors are negative and only *one factor is positive,* the product is positive. This would indicate the region $-4 \leq x \leq -2$. In all other cases the product is negative. Thus the graphical solution is

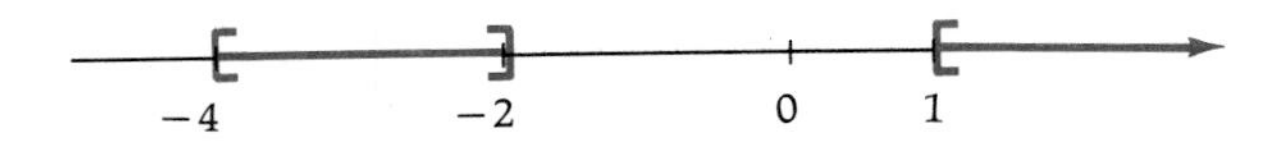


In interval notation the solution is $[-4, -2] \cup [1, +\infty]$. ■

Inequalities involving the quotient of two polynomials are solved in a similar fashion. However, it is necessary to consider separately the numerator and the denominator. Therefore *do not multiply both sides of the expression by the least common denominator.* You must examine each part of the expression in order to make a complete conclusion.

**EXAMPLE 3** Solve for $x$: $\dfrac{x+3}{x-2} < 0$

***Solution*** Make a graph showing where $(x + 3)$ and $(x - 2)$ are positive and where they are negative.

$$\begin{array}{llll} \text{If} & x + 3 > 0 & \text{then} & x > -3 \\ \text{and if} & x - 2 > 0 & \text{then} & x > \phantom{-}2 \end{array}$$

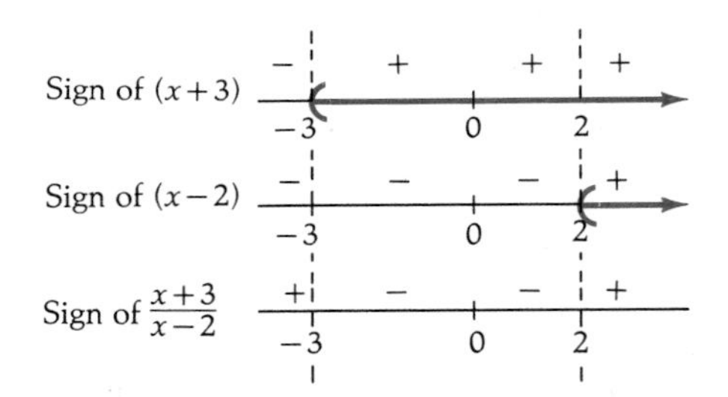


Notice here that we want to know when the quotient is negative. This will occur when one expression is positive and the other is negative. From our graph we see that this will take place when $-3 < x < 2$. The graph of the solution is

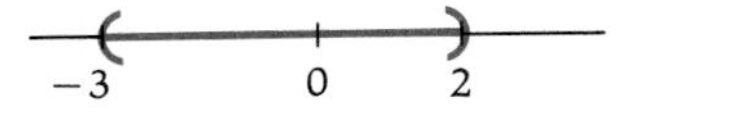


■

In order to use this method *it is necessary that the quotient of the two polynomials be greater than zero or less than zero.*

If the inequality is *not* in that form, then several algebraic operations may be necessary to place it in the proper form.

**EXAMPLE 4** Solve for $x$: $\dfrac{13x + 3}{2x + 1} > 6$.

***Solution*** Be sure you *do not* multiply both sides by $2x + 1$. We do not know whether $2x + 1$ is positive or negative, so we do not know what would happen to the inequality.

Instead we first obtain zero on the right-hand side.

$$\frac{13x + 3}{2x + 1} - 6 > 0$$

Now we would like the left-hand side to be written as one fraction. The L.C.D. is $2x + 1$; therefore we can change the left-hand side as follows:

$$\frac{13x + 3}{2x + 1} - \frac{6(2x + 1)}{2x + 1} > 0$$

$$\frac{13x + 3 - 6(2x + 1)}{2x + 1} > 0$$

$$\frac{13x + 3 - 12x - 6}{2x + 1} > 0$$

$$\frac{x - 3}{2x + 1} > 0$$

Now we continue in the usual fashion.

If $x - 3 > 0$ then $x > 3$

and if $2x + 1 > 0$ then $2x > -1$ and $x > -\dfrac{1}{2}$

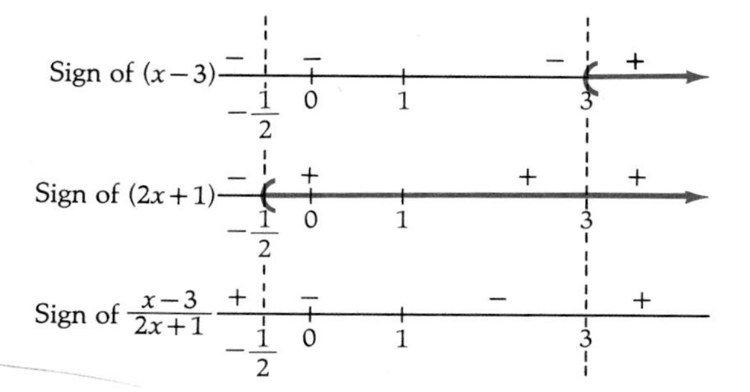

Now we want to know where $\dfrac{x-3}{2x+1} > 0$. The quotient will be positive if both numerator and denominator are positive or if both are negative. This takes place when $x > 3$ or when $x < -\dfrac{1}{2}$.

The graph of the solution set is

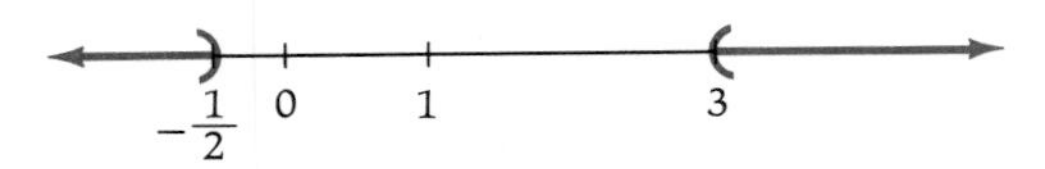


In interval notation the answer is $\left(-\infty, -\dfrac{1}{2}\right) \cup (3, +\infty)$. ■

In all the examples encountered in this section thus far, we have assumed that the left-hand side can be factored into binomial factors. In most problems of this type encountered in a mathematics course the quadratic can be factored. However, a few problems may deal with a quadratic form that cannot be factored by using rational coefficients. In such cases you should employ the quadratic formula.

## EXERCISE 2-7

In Problems 1–12, solve for $x$ and graph the solution.

**1.** $x^2 - 5x + 6 < 0$ **2.** $x^2 + 3x - 10 < 0$ **3.** $x^2 - x - 12 > 0$

**4.** $x^2 + 5x + 4 \geq 0$ **5.** $2x^2 - 9x - 18 \leq 0$ **6.** $2x^2 + 3x - 5 < 0$

**7.** $2x^2 - x - 3 > 0$ **8.** $2x^2 - 9x + 4 > 0$ **9.** $3x^2 - 5x \leq 2$

**10.** $3x^2 - 4x \leq 4$ **11.** $x(x-2) - 1 \geq 14$ **12.** $x(x+3) + 2 \leq 6$

Solve for $x$ in Problems 13–32.

**13.** $\dfrac{3x+2}{x-3} > 0$ **14.** $\dfrac{x+4}{x-1} > 0$ **15.** $\dfrac{x-5}{x+2} \leq 0$

**16.** $\dfrac{2x-1}{3x-2} \leq 0$ **17.** $\dfrac{x+2}{2x-3} > 1$ **18.** $\dfrac{3x+1}{x-2} > 2$

**19.** $(x+3)(x-2)(x+1) > 0$ **20.** $(2x+1)(x-2)(x-3) > 0$

**21.** $(x-4)(x-3)(2x-1) \leq 0$ **22.** $(3x+2)(x+5)(4x-1) \leq 0$

**23.** $\dfrac{x^2 - x - 2}{x-4} < 0$ **24.** $\dfrac{x^2 - 2x - 3}{x+5} > 0$

**25.** $\dfrac{x^2 + 2x}{x-5} > 0$ **26.** $\dfrac{x^2 + 9x + 20}{x+1} < 0$

**27.** $\dfrac{x^2 - 1}{x - 5} > -1$ ★ **28.** $x + 1 < \dfrac{x + 1}{x + 3}$ ★ **29.** $\dfrac{x^2 - 11}{x - 3} \leq 7$

★ **30.** $\dfrac{x^2 - 37}{x + 4} \geq -4$ **31.** $\dfrac{x - 5}{5 - x} \geq 0$ ★ **32.** $\dfrac{x^2 + 17}{x + 2} < \dfrac{11}{2}$

### Calculator Problems

In Problems 33–36, solve for $x$ and express your answer accurate to the nearest thousandth.

**33.** $\dfrac{x + \sqrt{2}}{2x + \pi} > 0$ **34.** $\dfrac{\pi x + 3}{x - \sqrt{7}} < 0$ **35.** $x^2 - 6x + 2 < 0$ **36.** $x^2 + 8x + 6 \leq 0$

---

## 2-8 EQUATIONS AND INEQUALITIES INVOLVING AN ABSOLUTE VALUE EXPRESSION

A special type of equation is one containing an absolute value. To solve equations involving absolute value, we must remember the definition for all real $x$

$$|x| = \begin{cases} x & \text{if } x \geq 0 \\ -x & \text{if } x < 0 \end{cases}$$

Thus if $|x| = 5$, we have $x = 5$ or $x = -5$.

EXAMPLE 1 Solve for $x$: $|2 - 3x| = 8$.

***Solution*** From the definition we have

$$2 - 3x = 8 \qquad \text{or} \qquad 2 - 3x = -8$$

We proceed to solve each equation.

$$-3x = 8 - 2 \qquad -3x = -8 - 2$$

$$x = -2 \qquad x = \frac{10}{3}$$

Thus the solution set is $\left\{-2, \dfrac{10}{3}\right\}$. ■

From the definition we can see that if $|a| = |b|$, then either $a = b$ or $a = -b$. This property is useful in solving certain absolute value equations.

EXAMPLE 2 Solve for $x$: $|5 - 6x| = |2 + 4x|$.

***Solution*** From the definition we have

$$5 - 6x = 2 + 4x \qquad \text{or} \qquad 5 - 6x = -(2 + 4x)$$

We now proceed to solve each equation.

$$5 - 6x = 2 + 4x \qquad\qquad 5 - 6x = -2 - 4x$$

$$x = \frac{3}{10} \qquad\qquad x = \frac{7}{2}$$

The solution set is $\left\{\frac{3}{10}, \frac{7}{2}\right\}$. ■

It is best to isolate the absolute value expression before attempting to use the definition to write the two equations.

**EXAMPLE 3** Solve for $x$: $\left|5x - \frac{1}{2}\right| + 2 = 6$.

***Solution*** First we isolate the absolute value expression by subtracting 2 from each side.

$$\left|5x - \frac{1}{2}\right| = 4$$

Now we apply the definition

$$5x - \frac{1}{2} = 4 \qquad \text{or} \qquad 5x - \frac{1}{2} = -4$$

Then we proceed to solve each equation. First we multiply each term by L.C.D. = 2.

$$10x - 1 = 8 \qquad\qquad 10x - 1 = -8$$

$$10x = 9 \qquad\qquad 10x = -7$$

$$x = \frac{9}{10} \qquad\qquad x = -\frac{7}{10}$$

The solution set is $\left\{-\frac{7}{10}, \frac{9}{10}\right\}$. ■

The expression within the absolute value may be a quadratic expression.

**EXAMPLE 4** Solve for $x$: $|2x^2 - x - 6| = 4$.

***Solution*** Using the definition, we have

$$2x^2 - x - 6 = 4 \qquad \text{or} \qquad 2x^2 - x - 6 = -4$$

We solve each of the resulting quadratic equations. One equation we can factor. The other requires the use of the *quadratic formula*.

$$2x^2 - x - 10 = 0$$
$$(2x - 5)(x + 2) = 0$$
$$2x - 5 = 0 \qquad x + 2 = 0$$
$$2x = 5 \qquad x = -2$$
$$x = \frac{5}{2}$$

$$2x^2 - x - 2 = 0$$
$$x = \frac{-(-1) \pm \sqrt{(-1)^2 - 4(2)(-2)}}{2(2)}$$
$$x = \frac{1 \pm \sqrt{1 + 16}}{4}$$
$$x = \frac{1 \pm \sqrt{17}}{4}$$

By solving both quadratic equations, we find the solution set of the original absolute value equation to be $\left\{-2, \frac{5}{2}, \frac{1 + \sqrt{17}}{4}, \frac{1 - \sqrt{17}}{4}\right\}$. ■

**Inequalities Involving Absolute Value**

If we think of the absolute value of a number as its distance from zero, without regard to direction, then the statement $|x| < a$ would imply that $x$ is less than $a$ units from zero and $|x| > a$ would imply that $x$ is more than $a$ units from zero.

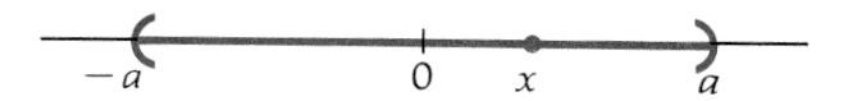


When $|x| < a$ and $x$ is some value on the number line, we can see visually that the distance between $x$ and 0 will always be less than $a$. In other words, we see that $-a < x < a$. Be sure you understand this notation. When we write $-3 < x < 3$, we are saying that $-3 < x$ *and* that $x < 3$.

We now state these ideas as a theorem.

***Theorem 2-5*** $|x| < a$ if and only if $-a < x < a$.

Thus if $|x| < 2$, we would know that $-2 < x < 2$. Consider how we use this theorem in the following example.

**EXAMPLE 5** Solve for $x$: $|2x + 1| < 5$.

***Solution*** Using Theorem 2-5, we have

$$-5 < 2x + 1 < 5$$
$$-6 < 2x < 4$$
$$-3 < x < 2$$

In interval notation the solution set is $(-3, 2)$, which graphs as

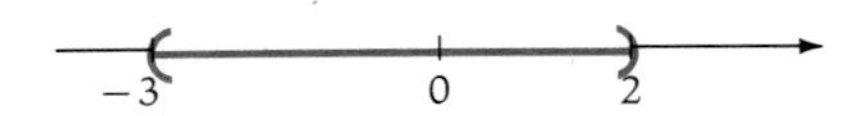


■

If an inequality of the form $|x| \leq a$ is encountered, a similar procedure would be used, but the endpoints would be included.

**EXAMPLE 6** Solve for $x$: $\left|\frac{1}{2}x - 1\right| \leq \frac{3}{4}$.

***Solution*** We employ Theorem 2-5 to obtain

$$-\frac{3}{4} \leq \frac{1}{2}x - 1 \leq \frac{3}{4}$$

$$-3 \leq 2x - 4 \leq 3 \qquad \textit{Multiply each term by L.C.D.} = 4.$$

$$1 \leq 2x \leq 7$$

$$\frac{1}{2} \leq x \leq \frac{7}{2}$$

In interval notation the solution set is $\left[\frac{1}{2}, \frac{7}{2}\right]$. ■

Sometimes we are required to express a triple inequality in the form of an absolute value inequality. To do so, we are again using Theorem 2-5.

**EXAMPLE 7** Express the interval $[-5, 3]$ as the set of all real numbers $x$ that satisfy an inequality containing an absolute value.

***Solution*** By the definition of interval notation we can immediately write $-5 \leq x \leq 3$. Now we need to have this in the form $-a \leq x \leq a$ in order to use the theorem.

$$-5 + 1 \leq x + 1 \leq 3 + 1 \qquad \textit{We add 1 to each expression.}$$

$$-4 \leq x + 1 \leq 4$$

By Theorem 2-5 this is equivalent to $|x + 1| \leq 4$. ■

We now examine a second theorem that considers the case when $|x| > a$.

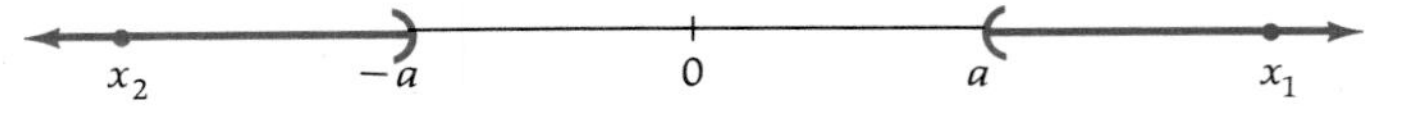


Perhaps $x$ is on the right-hand side (labeled $x_1$). Perhaps $x$ is on the left hand side (labeled $x_2$). Clearly, $x$ cannot be in both places at once, so we need separate labels $x_1$ or $x_2$. In these two cases we see that the distance from either value of $x$ to the origin is always greater than $a$. We will now state this as a theorem.

*Theorem 2-6* $|x| > a$ if and only if $x > a$ or $x < -a$.

**Caution:** Be sure you understand that there is no possible way to write this as one triple inequality. This is why there is the need of the word "or" between the two separate inequalities.

If we say $|x| > 5$, we mean *either* $x > 5$ *or* $x < -5$.

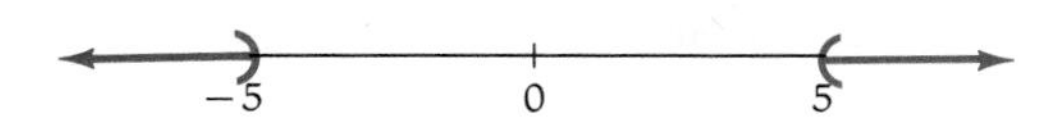


**EXAMPLE 8** Solve for $x$: $\left|\frac{1}{2}x - 1\right| > 2$.

***Solution*** By Theorem 2-6 we write the two distinct possible values for $x$.

$$\frac{1}{2}x - 1 > 2 \quad \text{or} \quad \frac{1}{2}x - 1 < -2$$

Now we solve each inequality. We first multiply each term by 2.

$$x - 2 > 4 \quad \text{or} \quad x - 2 < -4$$
$$x > 6 \quad \text{or} \quad x < -2$$

We *cannot write this as one inequality*. There are two separate inequality statments.

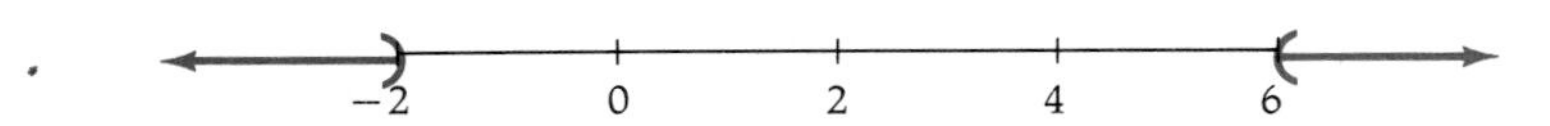


The solution is $x > 6$ or $x < -2$. In interval notation we have $(-\infty, -2) \cup (6, +\infty)$. ■

If the inequality involves $\geq$ instead of $>$, we follow the same procedure but include the endpoints.

## EXERCISE 2-8

In Problems 1–20, solve for $x$.

**1.** $|x + 3| = 8$ **2.** $|2x + 7| = 7$ **3.** $\left|\frac{1}{2}x - 2\right| = 3$

**4.** $\left|x - \frac{1}{4}\right| = 2$ **5.** $\left|\frac{x - 3}{-3}\right| = 1$ **6.** $\left|\frac{x + 5}{-4}\right| = 2$

**7.** $|2x + 1| + 3 = 8$ **8.** $|3x - 1| + 2 = 5$ **9.** $\frac{|5x - 2|}{4} = 1$

**10.** $\frac{|x + 2|}{-3} = -5$ **11.** $|3x + 4| = -2$ **12.** $|5x - 2| = 0$

**13.** $|3x - 4| = |x + 6|$ **14.** $|x - 6| = |5x + 8|$ **15.** $3|-5x + 2| = 27$

**16.** $\frac{1}{2}|3x - 2| = 4$ **17.** $|x^2 - 9| = 7$ **18.** $|x^2 + 2x + 1| = 4$

**19.** $|x^2 + 3x - 8| = 2$ **20.** $|2x^2 + x - 3| = 3$

In Problems 21–26 describe the intervals as a set of all real numbers $x$ satisfying an absolute value inequality.

**21.** $(-3, 3)$ **22.** $\left[-\frac{5}{2}, \frac{5}{2}\right]$ **23.** $[-4, 6]$

**24.** $(1, 9)$ **25.** $(-11, 5)$ **26.** $[-2.5, 4.5]$

In Problems 27–48, solve each inequality. Sketch the graph of the solutions of Problems 27–36.

**27.** $|x| < 3$ **28.** $|2x| < 5$ **29.** $|3x - 2| < 4$

**30.** $|4x + 3| < 11$ **31.** $\left|\frac{1}{2}x - 2\right| \leq 8$ **32.** $|3x + 2| \leq 0$

**33.** $|5x + 2| < 8$ **34.** $|2x + 7| \leq 1$ **35.** $\left|1 - \frac{2}{3}x\right| \leq 5$

**36.** $\left|2\left(x - \frac{1}{2}\right)\right| \leq 6$ **37.** $|2x| > 8$ **38.** $\left|\frac{1}{2}x\right| > 5$

**39.** $|4x + 3| \geq 3$ **40.** $|5x - 10| \geq 2$ **41.** $|2x - 3| > 4$

**42.** $\left|\frac{5}{2} + x\right| \geq \frac{3}{4}$ **43.** $\left|\frac{5 - x}{3}\right| > 4$ **44.** $\left|\frac{2x + 1}{-3}\right| > 2$

**45.** $\left|\frac{2x - 1}{3}\right| \geq \frac{5}{6}$ **46.** $|2(x + 3) - 1| < 9$ ★ **47.** $|3x + 8| < 5(x + 1)$

★ **48.** $|2x - 3| > 4(x - 1)$

★ **49.** Prove that $\left|x - \frac{a + b}{2}\right| < \frac{b - a}{2}$ is equivalent to the expression $a < x < b$.

★ **50.** For $a > 0$, what condition must $|x - 3|$ satisfy for it to be true that $\left|\frac{x}{2} - \frac{3}{2}\right| < a$?

---

## COMPUTER PROBLEMS FOR CHAPTER 2

If you can program a computer, you can apply your knowledge of Chapter 2 to do Problems 1 and 2.

**1.** Write a computer program to find the real solutions for any quadratic equation of the form $ax^2 + bx + c = 0$. If $b^2 - 4ac < 0$, have the program print "NO REAL SOLUTIONS." Use your program to find the real solutions for

**a.** $1.5x^2 + 6x - 7.1 = 0$

**b.** $2.8x^2 + 3.7x - 2.3 = 0$

c. $2.2x^2 - 6.8x - 9.6 = 0$
d. $3.58x^2 - 8.01x - 146.31 = 0$

2. For all positive real numbers $a$, $b$, $c$, write a computer program to find the solution for all equations of the form

$$\sqrt{ax + b} = c$$

Use your program to find the real solutions for
a. $\sqrt{5x + 3} = 8.12$
b. $\sqrt{6.1x + 2.4} = 12.36$
c. $\sqrt{\pi x + 3} = \sqrt{7}$
d. $\sqrt{2x + \sqrt{5}} = 9.3618$

## KEY TERMS AND CONCEPTS

Be sure you understand what is meant by these terms and can give an example of each.

Root of an equation
Identity
Conditional equation
Degree of an equation
Linear equation
Quadratic equation
Principal
Interest
Quadratic formula
Completing the square
Discriminant
Imaginary number
Complex number
An equation quadratic in form
Radical equation
Extraneous solution
Closed interval
Open interval
Half-open interval
Half-closed interval
Set builder notation
Interval notation

## SUMMARY OF PROCEDURES AND CONCEPTS

### Quadratic Equations

1. To solve a quadratic equation, place it in standard form, unless it is of the type $x^2 = a$.
2. If possible, factor the quadratic. Then set each factor equal to zero and solve to find the roots. (If $ab = 0$, then $a = 0$ or $b = 0$.)
3. If the quadratic cannot be factored, use the quadratic formula to find the roots: $\left(x = \dfrac{-b \pm \sqrt{b^2 - 4ac}}{2a}\right)$
4. If the quadratic equation is of type $x^2 = a$, the roots are $x = \pm\sqrt{a}$.

**Imaginary Numbers**

$$i = \sqrt{-1} \qquad \sqrt{-a} = \sqrt{-1}\sqrt{a} = i\sqrt{a} \quad (a > 0) \qquad i^2 = -1$$

**Linear Inequalities**

1. For all real numbers $a$ and $b$, if $a < b$, then $-a > -b$.
2. For all positive integers $n$ and for all positive real numbers $a$, $b$, if $a < b$, then $a^n < b^n$ and $\sqrt[n]{a} < \sqrt[n]{b}$.

**Absolute Value Equations and Inequalities**

1. Absolute value equations should be rewritten without absolute value signs by using the basic definition

$$|x| = \begin{cases} x & \text{if } x \geq 0 \\ -x & \text{if } x < 0 \end{cases}$$

to obtain two equations. Solve the resulting equations.

2. Absolute value inequalities should be rewritten in an equivalent inequality without absolute value signs by using one of the two following theorems. The resulting inequalities should then be solved.
   a. $|x| < a$ if and only if $-a < x < a$.
   b. $|x| > a$ if and only if $x > a$ or $x < -a$.

## CHAPTER 2 REVIEW EXERCISE

Solve Problems 1–10 for $x$.

1. $2x + 3 = 3(2x + 5)$
2. $\frac{x}{2} + \frac{2}{3}x = 7$
3. $\frac{3}{2x} = \frac{1}{x} + \frac{1}{2}$
4. $\frac{x}{x+7} = \frac{x-1}{x}$
5. $\frac{1}{x-2} + \frac{1}{x+2} = \frac{4}{x^2-4}$
6. $\frac{x+1}{x+2} - \frac{x-3}{x+5} = \frac{4}{x^2+7x+10}$
7. $x^2 + 10x + 21 = 0$
8. $6x^2 + 7x = 3$
9. $\frac{2}{x-3} + 1 = \frac{6}{x^2-3x}$
10. $2x^2 + 8x + 7 = 0$

Compute $b^2 - 4ac$ in Problems 11–14 and give the nature of the roots.

11. $x^2 - 3x - 18 = 0$
12. $x^2 + 6x + 9 = 0$
13. $2x^2 - 3x + 2 = 0$
14. $3x^2 + 5x + 2 = 0$

Simplify Problems 15–17, using the definition of $i$.

**15.** $\sqrt{-12}$ **16.** $\sqrt{-18}$ **17.** $\sqrt{-\frac{3}{4}}$

Solve Problems 18–40 for $x$.

**18.** $x^2 + 1 = 0$
**19.** $x^2 - 2x + 5 = 0$
**20.** $x^2 - 5x + 7 = 0$
**21.** $2x^2 - 6x + 3 = 0$
**22.** $2x^3 - 54 = 0$
**23.** $x^3 - 125 = 0$
**24.** $x^4 + 3x^2 - 10 = 0$
**25.** $x^{\frac{1}{2}} - 7x^{\frac{1}{4}} + 10 = 0$
**26.** $x - 9x^{\frac{1}{2}} + 20 = 0$
**27.** $x^{\frac{2}{3}} + 3x^{\frac{1}{3}} - 4 = 0$
**28.** $4 + \sqrt{2x} = x$
**29.** $\sqrt{x+15} - \sqrt{2x+13} = \sqrt{x+10}$
**30.** $|3x - 4| = 9$
**31.** $|x^2 + 3x - 6| = 4$
**32.** $|10x^2 - 13x + 1| = 4$
**33.** $2x + 5 \geq 6x - 7$
**34.** $\frac{1}{4}(x + 3) - 2x < \frac{2}{3}(2x - 3) + \frac{1}{2}x$
**35.** $|3x - 2| < 7$
**36.** $|6x + 5| \geq 11$
**37.** $x^2 + 4x - 21 < 0$
**38.** $3x^2 + 17x + 10 \geq 0$
**39.** $\frac{x - 5}{5 - x} \leq 0$
**40.** $\frac{x^2 - 13}{x - 3} \geq 3$

**41.** Find the angles of a triangle if the first exceeds the second by 20° and the third is 12° more than twice the second. (The sum of the angles of a triangle is 180°.)

**42.** Two positive numbers are such that one is twice the other. If the smaller number is increased by 2 and the larger is increased by 3, the product of the new numbers is 91. Find the original numbers.

**43.** Melissa earned \$2400 last summer. For one year she invested part of it in a mutual fund that earns 12% interest. The other part she invested in a special notice account that earns 10% interest. At the end of the year she had earned \$272 in interest. How much did she invest at each rate?

**44.** Two men worked together to paint a house in 20 hours. If they had worked separately, it would take Jino 9 hours longer to paint the house than it would take Wally. How long would it take each person to paint the house if each did the job working alone?

## PRACTICE TEST FOR CHAPTER 2

Compute $b^2 - 4ac$ in Problems 1–3 and give the nature of the roots.

**1.** $x^2 + 3x - 2 = 0$ **2.** $6x^2 - x - 2 = 0$ **3.** $3x^2 - 4x + 5 = 0$

Simplify Problems 4–6 by using the definition of $i$.

**4.** $\sqrt{-36}$ **5.** $\sqrt{-\frac{1}{2}}$ **6.** $\sqrt{-48}$

Solve Problems 7–20 for $x$.

**7.** $3(x + 3) = x - 5$
**8.** $\frac{x}{3} - \frac{2}{5}x + 1 = \frac{1}{3}x$
**9.** $\frac{1}{x + 1} - \frac{1}{2} = \frac{1}{3x + 3}$

**10.** $2x^2 + 3x = 2$

**11.** $\dfrac{3}{x+1} + \dfrac{3}{x^2+x} = -2$

**12.** $x^2 - 3x + 1 = 0$

**13.** $x^2 - 2x + 2 = 0$

**14.** $x^3 - 1 = 0$

**15.** $x^4 - 10x^2 + 9 = 0$

**16.** $\sqrt{3x+1} + \sqrt{2x-1} = 7$

**17.** $|x^2 + 5x + 7| = 2$

**18.** $\dfrac{2}{5}(x-1) > \dfrac{1}{3}(2x+5)$

**19.** $|5x - 1| \leq 4$

**20.** $\dfrac{x^2 + x - 6}{x - 1} > 0$

**21.** A customer buys $x$ articles at $0.20 each and twice as many articles at $0.25 each. If the cost of all the articles is $4.90, find the number bought at each price.

**22.** Bob and Michelle drove to the university, traveling at a constant speed. Bob complained to Michelle that if they had gone 5 miles per hour faster, the trip would have taken one hour less! How fast did they drive if the total distance driven was 280 miles?

$(x+5)\text{mph} = \dfrac{280}{x} - 1$

# 3 *Functions and Their Graphs*

> *The word function (or its Latin equivalent) seems to have been introduced into mathematics by Leibniz in 1694; the concept now dominates much of mathematics and is indispensable.*
>
> *E. T. Bell*

Frequently, relationships between quantities can best be seen visually. One of the most important ways of expressing a functional relationship is to draw a picture that shows how the change in one quantity produces a change in the other. In mathematics the way we picture a function is by means of a graph. Thus we will begin our study of functions with the Cartesian coordinate system. This system forms the basis for displaying elementary mathematical functions.

## 3-1 THE CARTESIAN COORDINATE SYSTEM

One method of uniquely naming each point in the plane is the Cartesian coordinate system (so named for the French mathematician René Descartes, 1596–1650).

This rectangular coordinate system is constructed with two perpendicular number lines in the plane that intersect at zero on each line. One line is called vertical and the other is called horizontal, with positive direction upward and to the right, respectively. The two lines are called **coordinate axes**, and their point of intersection is called the **origin**. The horizontal axis is usually referred to as the ***x*-axis**, and the vertical axis is usually referred to as the ***y*-axis**.

The four regions of the plane formed by the $x$- and $y$-axes are called **quadrants** and are numbered in a counterclockwise direction starting with the upper right. (See Figure 3-1.)

Positive $x$ values lie to the right of the origin, and negative $x$ values lie to the left. Positive $y$ values lie above the origin, and negative $y$ values lie below the origin.

In the Cartesian coordinate system, each point on the plane is always represented by a unique ordered pair $(x, y)$. The real numbers $x$ and $y$ of this ordered

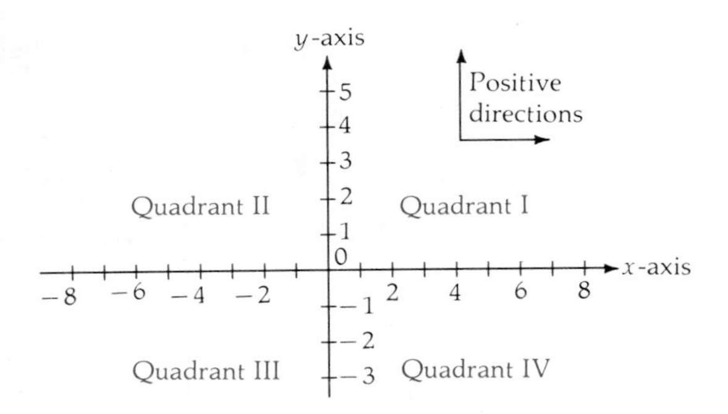


*Figure 3-1*

pair are called the **coordinates** of the point. The $x$-coordinate is called the **abscissa**. The $y$-coordinate is called the **ordinate**.

**EXAMPLE 1** Plot the points $A\ (5, 7)$, $B\ (-7, -3)$, $C\ (-2, 6)$, $D\ (0, -4)$, and $E\ (3, -8)$ using the Cartesian coordinate system.

***Solution*** Refer to Figure 3-2.

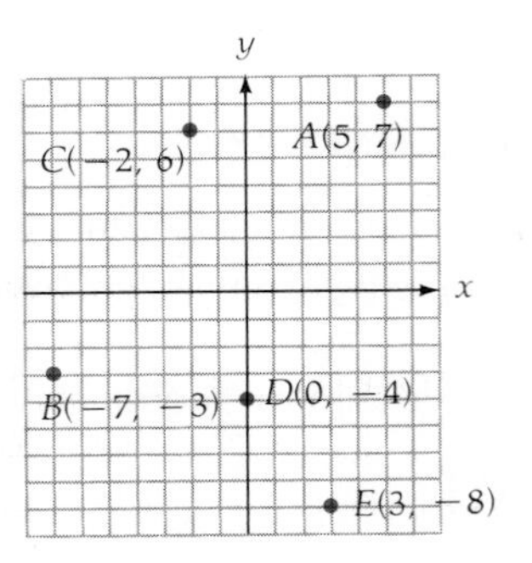


*Figure 3-2*

■

The length of a line segment on the coordinate plane, or the distance from point $P_1$ to point $P_2$, is a useful tool in many instances.

---

**Distance Formula**

---

The distance between points $P_1\ (x_1, y_1)$ and $P_2\ (x_2, y_2)$ in the coordinate plane is given by $\overline{P_1P_2} = \sqrt{(x_1 - x_2)^2 + (y_1 - y_2)^2}$.

---

**EXAMPLE 2** Find the distance from $P_1$ (3, 7) to $P_2$ (−4, 6).

***Solution***

$$\begin{aligned}\overline{P_1P_2} &= \sqrt{[3-(-4)]^2 + (7-6)^2} \\ &= \sqrt{49+1} \\ &= \sqrt{50} \\ &= 5\sqrt{2} \quad ■\end{aligned}$$

Another useful formula is the *midpoint formula.* The **midpoint** of a line segment $P_1P_2$ would be a point on the segment midway between $P_1$ and $P_2$. The coordinates of this point would therefore be the averages of the coordinates of the endpoints.

---

**Midpoint Formula**

---

The midpoint of a line segment from $P_1(x_1, y_1)$ to $P_2$ $(x_2, y_2)$ is

$$\left(\frac{x_1 + x_2}{2}, \frac{y_1 + y_2}{2}\right)$$

---

**EXAMPLE 3**

**a.** Find the midpoint of the line segment joining the points $P_1$ (5, −7) and $P_2$ (3, 1).

**b.** Assuming $M$ is on the line segment, verify that it is in fact the midpoint.

***Solution***

**a.** The midpoint is obtained by the formula

$$\left(\frac{5+3}{2}, \frac{-7+1}{2}\right) = (4, -3) = M$$

**b.** Assuming that (4, −3) is on the line segment, you can verify that (4, −3) is the midpoint by finding its distance from each endpoint of the line segment.

Using the distance formula:

$$\overline{P_1M} = \sqrt{(5-4)^2 + [-7-(-3)]^2} = \sqrt{(1)^2 + (-7+3)^2}$$
$$= \sqrt{1+16} = \sqrt{17}$$
$$\overline{MP_2} = \sqrt{(4-3)^2 + (-3-1)^2} = \sqrt{(1)^2 + (-4)^2} = \sqrt{1+16} = \sqrt{17}$$

Since there exists only one point $M$ on $P_1P_2$ where $\overline{P_1M} = \overline{MP_2}$ and since $\overline{P_1M} = \overline{MP_2}$, we have verified that $M$ is in fact the midpoint of line segment $P_1P_2$. These points are plotted in Figure 3-3. ■

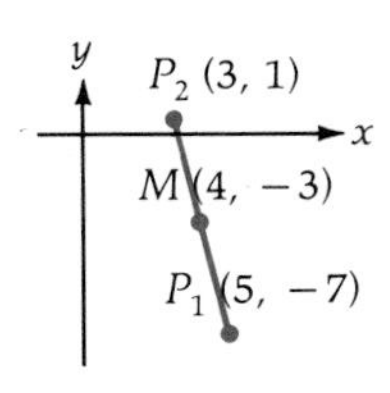


*Figure 3-3*

**EXAMPLE 4** Plot the points $P_1(-5, 2)$, $P_2(2, 9)$, and $P_3(-1, -2)$ and prove that the triangle created by connecting those three points by straight lines is a right triangle.

***Solution*** We will plot $P_1$, $P_2$, and $P_3$ in Figure 3-4 and connect each pair of points by line segments.

By examining Figure 3-4 we are inclined to believe that $P_1P_2P_3$ is a right triangle. However, we cannot be sure on the basis of appearances!

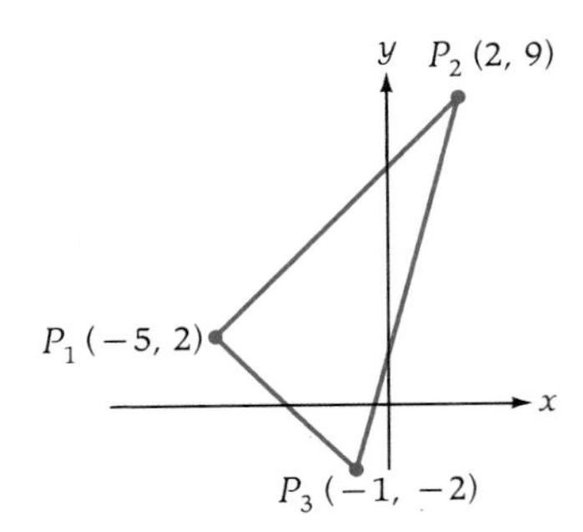


*Figure 3-4*

To prove that $P_1P_2P_3$ is a right triangle, we must first find the length of each line segment.

We first use the distance formula for each line segment.

$$\overline{P_1P_2} = \sqrt{(-5-2)^2 + (2-9)^2} = \sqrt{(-7)^2 + (-7)^2}$$
$$= \sqrt{49 + 49} = \sqrt{98} = 7\sqrt{2}$$
$$\overline{P_2P_3} = \sqrt{[2-(-1)]^2 + [9-(-2)]^2} = \sqrt{(3)^2 + (11)^2}$$
$$= \sqrt{9 + 121} = \sqrt{130}$$
$$\overline{P_3P_1} = \sqrt{[-1-(-5)]^2 + (-2-2)^2} = \sqrt{(4)^2 + (-4)^2}$$
$$= \sqrt{16 + 16} = \sqrt{32} = 4\sqrt{2}$$

We will use the converse of the Pythagorean Theorem to prove that the triangle is a right triangle.

$$(\overline{P_1P_2})^2 = (7\sqrt{2})^2 = 98$$
$$(\overline{P_2P_3})^2 = (\sqrt{130})^2 = 130$$
$$(\overline{P_3P_1})^2 = (4\sqrt{2})^2 = 32$$

Finally, we see that

$$(\overline{P_2P_3})^2 = (\overline{P_1P_2})^2 + (\overline{P_3P_1})^2$$
$$130 = 98 + 32$$

Thus we have shown that the triangle is a right triangle. ■

## EXERCISE 3-1

1. Plot and label the points on a rectangular coordinate system: $A(3, 1)$, $B(-6, -2)$, $C(-5, 1)$, $D(0, -5)$.
2. Plot and label the points on a rectangular coordinate system: $A(5, 1)$, $B(3, -7)$, $C(-5, 0)$, $D(-8, -4)$.
3. Choose *a convenient scale* and plot the following points on a rectangular coordinate system: $A(30, 65)$, $B(-120, 20)$, and $C(-80, -60)$.
4. Choose *a convenient scale* and plot the following points on a rectangular coordinate system: $A(0.25, -0.40)$, $B(-0.35, -0.20)$, $C(-0.40, 0.15)$.

Illustrate with a diagram the set of all points that satisfy the conditions in Problems 5–12 for ordered pairs $(x, y)$ in a coordinate plane.

5. $x = y$ 6. $y > 0$ 7. $x \le 0$ 8. $xy > 0$
9. $x = -4$ 10. $y = -1.5$ 11. $xy < 0$ 12. $x = -y$

In Problems 13–22, find the distance between points $C$ and $D$ and find the midpoint of line segment $CD$.

13. $C(0, -3)$, $D(0, 5)$ 14. $C(-2, 8)$, $D(6, 0)$ 15. $C(3, -1)$, $D(5, 2)$
16. $C(5, -2)$, $D(-1, -3)$ 17. $C(-4, -6)$, $D(6, 7)$ 18. $C(-5, 1)$, $D(9, -4)$
19. $C(6, -5)$, $D(-2, -3)$ 20. $C(-2, -1)$, $D(0, -7)$ 21. $C\left(\frac{1}{2}, -3\right)$, $D\left(\frac{1}{3}, -2\right)$
22. $C\left(5, -\frac{5}{2}\right)$, $D\left(-1, -\frac{1}{3}\right)$

### Calculator Problems

In Problems 23–26, express your answer to the nearest thousandth. Use a scientific calculator.

Find the distance between points $E$ and $F$ and find the midpoint of line segment $EF$.

23. $E(\sqrt{3}, -\sqrt{5})$, $F(\sqrt{2}, 3\sqrt{2})$ 24. $E(\sqrt{7}, 2\sqrt{3})$, $F(-\sqrt{6}, -\sqrt{7})$
25. $E(1.623, 8.118)$, $F(-1.001, 7.123)$ 26. $E(-2.198, 3.6111)$, $F(15.012, 4.685)$
27. Line segment $AB$ has a midpoint at $(-1, 6)$, and the coordinates of $A$ are $(3, -4)$. Find the coordinates of $B$.
28. Line segment $AB$ has a midpoint at $(-2, -3)$, and the coordinates of $B$ are $(7, 1)$. Find the coordinates of $A$.

In Problems 29 and 30, show that the given set of three points are the vertices of a right triangle.

**29.** $(-2, 1)$, $(3, -4)$, and $(5, -2)$ **30.** $(-6, 1)$, $(-2, 9)$, and $(4, -4)$

In Problems 31 and 32, show that the given set of four points are the vertices of a square.

**31.** $(-1, 7)$, $(-4, 3)$, $(3, 4)$, and $(0, 0)$ **32.** $(-2, -2)$, $(2, -1)$, $(1, 3)$, and $(-3, 2)$

Three or more points are *collinear* if they lie on the same straight line. Show that the points in Problems 33 and 34 are collinear. (*Hint:* Use the property that the shortest distance between two points is a straight line.)

**33.** $A\ (-4, 5)$, $B\ (0, 1)$, and $C\ (3, -2)$ **34.** $A\ (-3, -3)$, $B\ (1, -1)$, and $C\ (5, 1)$

**35.** Find the distance between points whose coordinates are $(a, 3a)$ and $(5a, 2a)$. Find the midpoint of the line segment connecting these points.

**36.** Find the distance between points whose coordinates are $(7, 4b)$ and $(-3, 2b)$. Find the midpoint of the line segment connecting these points.

**37.** The distance between $A\ (1, 3)$ and $B\ (x, 7)$ is 8. Find the value of $x$.

**38.** The distance between $A\ (5, y)$ and $B\ (1, 2)$ is 10. Find the value of $y$.

Using the procedures developed in this section as well as the distance formula and the midpoint formula, see whether you can prove the statements in Problems 39 and 40.

**★39.** The diagonals of any given rectangle are equal in length.

**★40.** Construct the midpoint of the hypotenuse of a right triangle. Prove that the midpoint is equally distant from the three vertices of the right triangle.

---

## 3-2 GRAPHS

Suppose we consider a set of ordered pairs. The graph of a set of ordered pairs is a set of points $(x, y)$ located on the coordinate plane. When we are asked to "graph an equation," we are being asked to show a sketch of how the collection of ordered pairs appears on the coordinate plane.

**EXAMPLE 1** Sketch the graph of $y = 3x - 1$.

***Solution*** Our first task is to find ordered pairs $(x, y)$ that are solutions to the equation. We accomplish this by assigning arbitrary values to $x$. Since we are free to assign values to $x$, we will call $x$ the **independent variable**. We now find the corresponding values of $y$. Since each value of $y$ is determined by the value of $x$, we call $y$ the **dependent variable**. First, we note that the set of possible values for $x$ is the set of all real numbers. We can therefore assign any real number for $x$. For instance, if $x = 1$, then $y = 3(1) - 1 = 2$, so the pair $(1, 2)$ is a solution to the equation and hence a point on the graph.

It is convenient to place these pairs in a "table of values." We will let $x$ take the values $-1$, 0, 1. We find the corresponding values of $y$ and set up the table as follows.

| $x$ | $-1$ | 0 | 1 |
|---|---|---|---|
| $y$ | $-4$ | $-1$ | 2 |

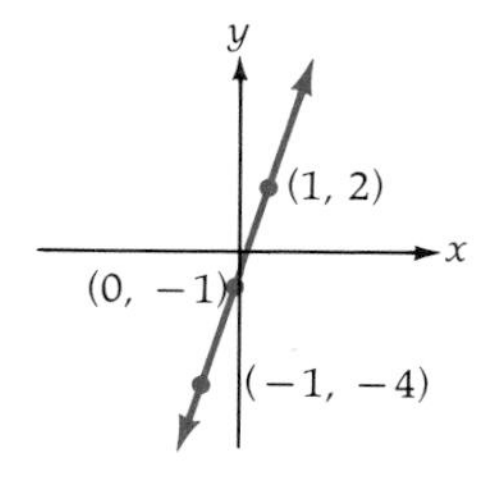


*Figure 3-5*

We now locate the points $(-1, -4)$, $(0, -1)$, and $(1, 2)$ on the coordinate plane. Since the set of possible values for both $x$ and $y$ consists of all real numbers, there will be infinitely many points on the graph. We can see from plotting these three points that the graph in Figure 3-5 appears to be a straight line.

These points establish a pattern, and it is true that the graph of $y = 3x - 1$ is a straight line. It can be shown that the graph of any first-degree polynomial equation in two variables will form a straight line. It is common practice to plot three points when graphing a straight line. (Although two points would be sufficient, the third point provides an accuracy check.) ■

**EXAMPLE 2** Sketch the graph of $y = x^2 - 3x - 4$.

***Solution*** We again set up a table of values and plot enough points to establish a pattern.

| $x$ | $-2$ | $-1$ | 0 | 1 | 2 | 3 | 4 | 5 |
|---|---|---|---|---|---|---|---|---|
| $y$ | 6 | 0 | $-4$ | $-6$ | $-6$ | $-4$ | 0 | 6 |

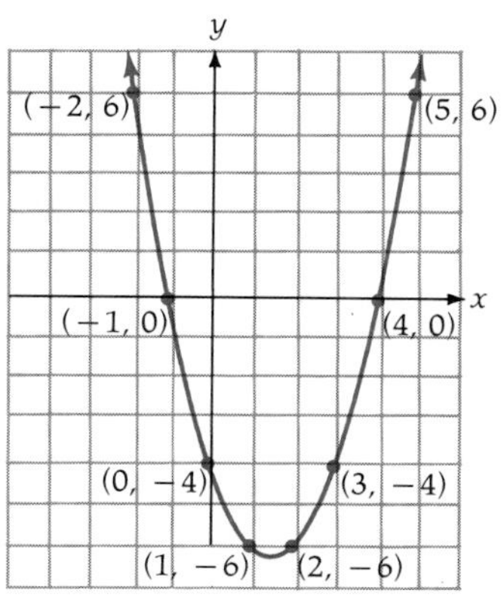


Graph of $y = x^2 - 3x - 4$

*Figure 3-6*

The resulting curve shown in Figure 3-6 is called a *parabola* and will be studied in more detail in Chapter 4.

Here, we are plotting enough points that we see a general pattern and then connecting the points with a smooth line. Later in the text, we will develop more powerful mathematical tools to obtain graphs more accurately and quickly. However, these tools will be meaningful only if you are proficient in accurately finding a few ordered pairs that satisfy a given equation and carefully plotting those points. ■

**EXAMPLE 3** Sketch the graph of $y = x^3 + x^2 - 6x$.

***Solution*** Again we choose several values of $x$ and obtain the corresponding values for $y$. Some of the values we obtain are given in the following table.

| $x$ | $-4$ | $-3$ | $-2$ | $-1$ | 0 | 1 | 2 | 3 |
|---|---|---|---|---|---|---|---|---|
| $y$ | $-24$ | 0 | 8 | 6 | 0 | $-4$ | 0 | 18 |

We plot several of these points and connect them by a smooth curve, as shown in Figure 3-7.

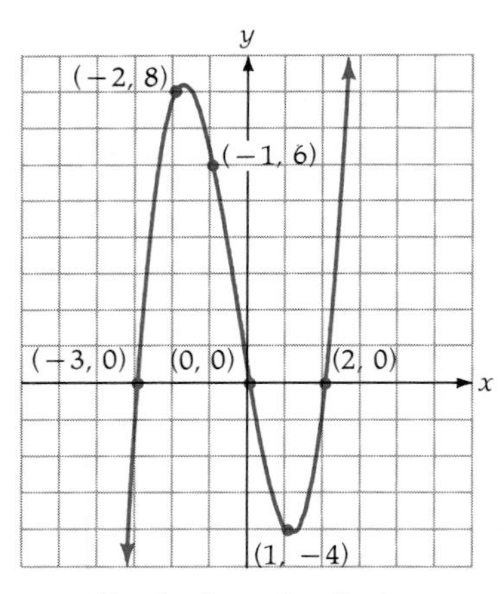


Graph of $y=x^3+x^2-6x$ ■

*Figure 3-7*

Many equations include the absolute value of a given expression. These are somewhat different from the ones we have studied so far.

You should not assume that the lines extend endlessly in any given graph. The variable $x$ sometimes cannot take on certain values. We observe such a limitation in Example 4.

**EXAMPLE 4** Sketch the graph of $y = \sqrt{x - 2}$.

***Solution*** We first should notice that $x$ cannot be less than 2. We will not obtain a real number if we have the square root of a negative quantity. Thus we will try only values of $x$ such that $x \geq 2$. Some of these values appear in the following table.

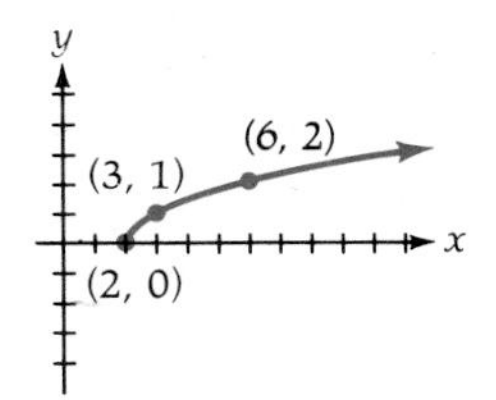


*Figure 3-8*

| $x$ | 2 | 3 | 4 | 5 | 6 | 11 |
|---|---|---|---|---|---|---|
| $y$ | 0 | 1 | $\sqrt{2}$ | $\sqrt{3}$ | 2 | 3 |

Notice in Figure 3-8 that the graph of $y = \sqrt{x - 2}$ does not go to the left of 2 on the $x$-axis and it does not go below the $x$-axis. ■

### Using Symmetry in Sketching Graphs of Equations

One of the most useful mathematical tools in sketching graphs is the property of symmetry. This allows us to sketch one half of the graph by plotting points and to reproduce the other half by comparing it to the part of the graph previously sketched.

Figure 3-9(a) shows a graph that is **symmetric to the $y$-axis**. If the graph were folded in half along the $y$-axis, the right-hand branch of the curve would coincide exactly with the left-hand branch. The point $(x, y)$ would coincide with $(-x, y)$.

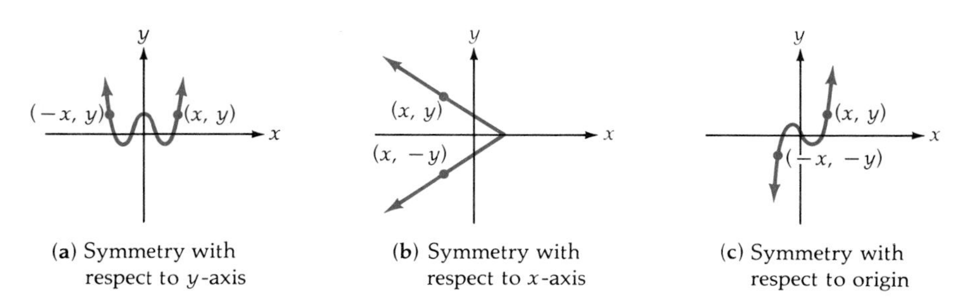


*Figure 3-9*

Figure 3-9(b) shows a graph that is **symmetric to the $x$-axis**. If the graph were folded in half along the $x$-axis, the upper branch of the curve would coincide exactly with the lower branch. The point $(x, y)$ would coincide with $(x, -y)$.

Figure 3-9(c) shows a graph that is **symmetric to the origin**. If the graph were rotated 180°, it would coincide with the original graph. The point $(x, y)$ would coincide with $(-x, -y)$.

We will find an understanding of symmetry helpful in the next two examples.

But the question occurs: How can we know in advance if a graph will be symmetric? We could say that the graph of an equation is symmetric to the $x$-axis if replacing $y$ by $-y$ produces an equivalent equation. In the equation $x = y^2$, if we replace $y$ by $-y$, we have $x = (-y)^2 = y^2$, so we see that we have an equivalent equation. Thus the graph of $x = y^2$ is symmetric with respect to the $x$-axis.

It is also useful to study symmetry with respect to the $y$-axis and with respect to the origin. We list these three useful tests.

### Tests for Symmetry

1. The graph of an equation is symmetric with respect to the $x$-axis if replacing $y$ by $-y$ produces an equivalent equation.
2. The graph of an equation is symmetric with respect to the $y$-axis if replacing $x$ by $-x$ produces an equivalent equation.
3. The graph of an equation is symmetric with respect to the origin if the replacing of both $x$ by $-x$ and $y$ by $-y$ produces an equivalent equation.

Let us examine a curve that is symmetric with respect to the $y$-axis and see how this property of symmetry aids in graphing.

**EXAMPLE 5** Sketch the graph of $y = x^2 - 2$, making use of symmetry.

***Solution*** We can replace $x$ by $-x$ and obtain an equivalent equation:

$$y = (-x)^2 - 2 = x^2 - 2$$

Thus we have satisfied the test for symmetry with respect to the $y$-axis. (Do you see that $y = x^2 - 2$ does not satisfy the test for symmetry with respect to the $x$-axis and also does not satisfy the test for symmetry with respect to the origin?)

We obtain a table of values for positive values of $x$ and $x = 0$.

| $x$ | 0 | 1 | 2 | 3 | 4 |
|---|---|---|---|---|---|
| $y$ | $-2$ | $-1$ | 2 | 7 | 14 |

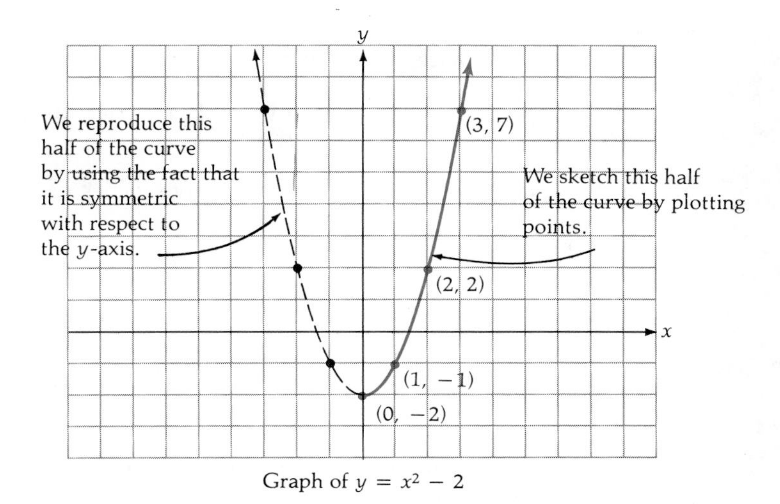


*Figure 3-10*

By symmetry we can sketch the other side of the curve. (See Figure 3-10.) This curve is symmetric with respect to the $y$-axis. ■

Finally, we examine a curve that is symmetric with respect to the origin. We do not always have one branch of the curve in the first quadrant. Consider the following example.

**EXAMPLE 6** Sketch the graph of $y = -\frac{4}{x}$, making use of symmetry.

***Solution***

1. If we replace $x$ by $-x$, we do not obtain an equivalent equation.
2. If we replace $y$ by $-y$, we do not obtain an equivalent equation.

(Take a minute to make sure you agree with these two statements.)

3. If we replace both $y$ by $-y$ and $x$ by $-x$ with equation $y = -\frac{4}{x}$, we obtain the following:

$$-y = -\frac{4}{-x} = \frac{-4}{-x} = \frac{4}{x}$$

If $-y = \frac{4}{x}$ and we multiply both sides of the equation by $-1$, we obtain the original equation $y = -\frac{4}{x}$. *This means that the curve is symmetric with respect to the origin.* If any point $(x, y)$ lies on the curve, then so does the point $(-x, -y)$.

We obtain a table of values for positive values of $x$, but we cannot use $x = 0$, since division by 0 is not allowed.

| $x$ | $\frac{1}{2}$ | 1 | 2 | 4 | 6 | 8 |
|---|---|---|---|---|---|---|
| $y$ | $-8$ | $-4$ | $-2$ | $-1$ | $-\frac{2}{3}$ | $-\frac{1}{2}$ |

The curve is symmetric with respect to the origin. We will use this fact to sketch the curve for negative values of $x$. (See Figure 3-11.)

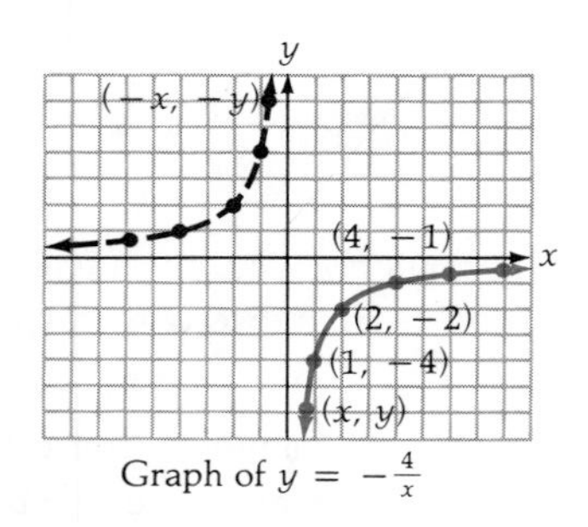


Graph of $y = -\frac{4}{x}$ ■

*Figure 3-11*

The equation could be written as $xy = -4$. Such curves of the form $xy =$ constant are included in the class of *hyperbolas*. We will study hyperbolas more extensively later in Chapter 4.

## EXERCISE 3-2

Sketch the graph of each equation in Problems 1–38. Use the tests for symmetry to aid in your work.

**1.** $y = 2x - 1$
**2.** $y = 6 - 3x$
**3.** $3x - 4y = 12$
**4.** $-3x + 5y = 15$
**5.** $y = x^2 + 4x - 1$
**6.** $y = 2x^2 - 3$
**7.** $y = x^3 - 3x^2 + 1$
**8.** $y = 1 - x^3$
**9.** $y = |x + 5|$
**10.** $y = |x - 4|$
**11.** $y = |x| - 2$
**12.** $y = |x| + 3$
**13.** $y = |2x + 1|$
**14.** $y = |3x - 2|$
**15.** $y = \frac{1}{4}x^4$
**16.** $y = \frac{1}{6}x^4$
**17.** $9x = y^2$
**18.** $4x = y^2$
**19.** $y = \sqrt{x + 7}$
**20.** $y = \sqrt{4 + x}$
**21.** $x^2 = y - 6$
**22.** $x^2 = y + 8$
**23.** $y = |5 - 2x|$
**24.** $y = |7 - 3x|$

25. $y = \frac{3}{x}$ 26. $y = -\frac{5}{x}$ 27. $xy = -8$ 28. $xy = 12$

29. $y^3 = 8 - x$ 30. $y^3 = x - 2$ 31. $y = x^3 + x$ 32. $y - 1 = x^3$

33. $\sqrt{x} = y + 2$ 34. $\sqrt{x} = y - 4$ 35. $x = y^2 + 2$ 36. $x = y^2 - 2y + 2$

37. $x = y^2 - 2y + 4$ 38. $x = y^2 - 4y$

### Calculator Problems

In Problems 39–44, enter several values for $x$ on your calculator in order to obtain sufficient points to sketch the graph. (You can still use your tests for symmetry to help you.) Choose appropriate units for your axes.

★39. $y = \sqrt[5]{x} + 3$ ★40. $y = \sqrt[5]{x} - 2$ ★41. $y = \sqrt[4]{x + 3}$

★42. $y = \sqrt[4]{\frac{1}{3}x}$ ★43. $y = 0.12x^2 + 0.36x^4$ ★44. $y = 0.52x^2 - 0.12x^3$

## 3-3 THE DEFINITION OF A FUNCTION

We frequently encounter the notion of an element in one set corresponding to an element in another set. For example, let us take the set of students in a campus club. As a second set, take the weight of each student in this club in pounds. (See Figure 3-12.)

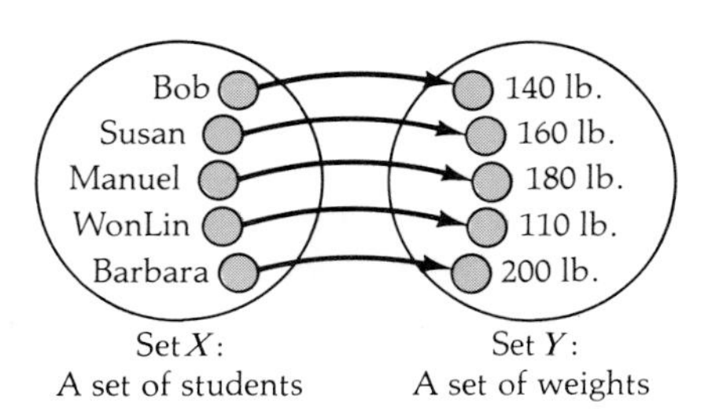


*Figure 3-12*

For each element of the first set there corresponds an element in the second set. The rule that assigns an element in the first set to an element in the second set is ". . . has a weight in pounds of . . . ." The idea of a function is a rule that defines the correspondence from one set to another.

***Definition***

A *function* is a rule that assigns to each element in one set $X$ exactly one element of another set $Y$. The set $X$ is called the *domain* of the function. The set $Y$ is called the *range* of the function.

**EXAMPLE 1** Determine whether the diagram in Figure 3-13 represents a function.

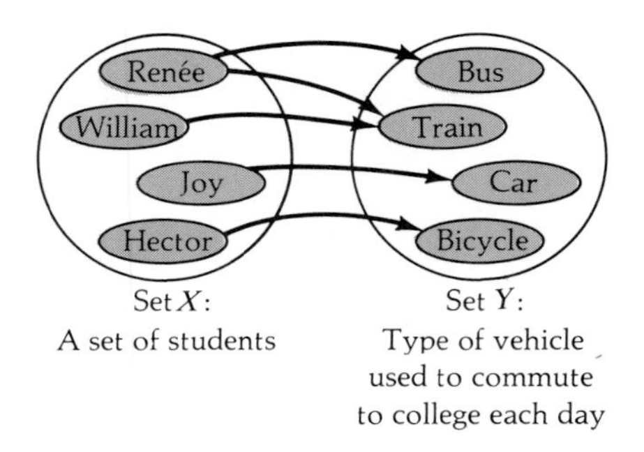


*Figure 3-13*

***Solution*** No, it is not a function. Each element in set $X$ must be assigned exactly *one* value in set $Y$. Since Renée uses both train and bus to travel to college each day, Renée is assigned *two* values in $Y$. Thus the function definition is *not* satisfied. ■

### Functional Notation

Mathematicians usually use a single letter such as $f$ or $g$ to denote a function. If $f$ is a function and $x$ is an element of the domain $X$, then $f(x)$ is the element in $Y$ that corresponds to $x$. The symbol $f(x)$ is read *f of x* or *f at x* (not "$f$ times $x$"!).

Be sure you understand the essential idea behind the notation $f(x)$. The letter $f$ is used to name the function. Then $f(x)$ denotes the value that $f$ assigns to $x$. (A similar understanding is maintained for $g(x)$, $h(x)$, etc.) For a given element $x$, $f(x)$ is known as *the value of f at x* or *the image of x.*

**EXAMPLE 2** Write the correspondence from $x$ values to $y$ values implied by the equation $y = 3x - 1$ in functional notation. Find the image corresponding to $x = 3$ and $x = -2$.

***Solution***

1. We write $f(x) = 3x - 1$ as the functional expression.
2. Then to find the image corresponding to $x = 3$, we write $f(3) = 3(3) - 1 = 9 - 1 = 8$. The *image* of $x = 3$ is 8.
3. Finally, to find the image corresponding to $x = -2$, we evaluate $f(-2) = 3(-2) - 1 = -6 - 1 = -7$. The *image* of $x = -2$ is $-7$. ■

It is important to be able to use this concept of a function when the variable $x$ is replaced by a number or when it is replaced by a variable quantity.

**EXAMPLE 3** Given the function $g(x) = 3x^2 - 2x$, find $g(-2)$, $g(5)$, $g(a)$, $g(a^2)$, and $g(a + 2)$.

***Solution***

$$\begin{aligned} g(-2) &= 3(-2)^2 - 2(-2) \\ &= 3(4) + 4 = 12 + 4 = 16 \end{aligned}$$

*To find $g(-2)$, we replace every $x$ in $g(x)$ by the number $-2$.*

$$\begin{aligned} g(5) &= 3(5)^2 - 2(5) \\ &= 3(25) - 10 = 75 - 10 = 65 \end{aligned}$$

*To find $g(5)$, we replace every $x$ in $g(x)$ by the number 5.*

$$\begin{aligned} g(a) &= 3(a)^2 - 2(a) \\ &= 3a^2 - 2a \end{aligned}$$

*To find $g(a)$, we replace every $x$ in $g(x)$ by the variable $a$.*

$$\begin{aligned} g(a^2) &= 3(a^2)^2 - 2(a^2) \\ &= 3a^4 - 2a^2 \end{aligned}$$

*To find $g(a^2)$, we replace every $x$ in $g(x)$ by the quantity $a^2$.*

$$\begin{aligned} g(a + 2) &= 3(a + 2)^2 - 2(a + 2) \\ &= 3(a^2 + 4a + 4) - 2(a + 2) \\ &= 3a^2 + 12a + 12 - 2a - 4 \\ &= 3a^2 + 10a + 8 \end{aligned}$$

*To find $g(a + 2)$, we replace every $x$ in $g(x)$ by the quantity $(a + 2)$.* ■

### Determining the Domain and Range of a Function

The **domain** of a function is the set of elements to which the function assigns values. The set of values of the function is called the **range**.

**EXAMPLE 4** Find the domain and range of the function $g(x) = \sqrt{x + 4}$.

***Solution*** The square root of a negative number is not real, so we want $x + 4 \geq 0$. Thus the domain of $g(x)$ is $x \geq -4$.

The notation $\sqrt{x + 4}$ indicates the principal root of the quantity. It will always be positive or zero. Thus the range is all nonnegative real numbers. We express this by saying $g(x) \geq 0$. ■

**EXAMPLE 5** Find the domain and range of the function $h(x) = \dfrac{1}{(4 - x)^2}$.

***Solution*** We cannot allow $x$ to be 4 or we will have division 0 (which is not defined). The domain is all values of $x$ except 4. This is often written as follows $\{x\colon \ x \neq 4\}$.

Clearly, since the denominator of $h(x)$ is squared, it cannot be negative. There is no value of $x$ that will make $h(x)$ be zero. (Do you see why?) Thus the range of $h(x)$ is all positive real numbers. ■

Two particular types of functions are quite useful in mathematics. These are known as even functions and odd functions. We will study further properties of even functions and odd functions in Section 3-4.

*Definition*

For all $x$ in the domain of any function $f(x)$:

1. The function is *even* if $f(-x) = f(x)$.
2. The function is *odd* if $f(-x) = -f(x)$.

**EXAMPLE 6** Test to see whether the following functions are *even, odd,* or *neither.*

**a.** $f(x) = x^4 + 5x^2$ **b.** $f(x) = 3x^3 + 2x$ **c.** $f(x) = 2x^2 - 2x + 1$

***Solution***

**a.** We replace $x$ by $-x$ and obtain $f(-x) = (-x)^4 + 5(-x)^2 = x^4 + 5x^2 = f(x)$. Thus $f(x) = x^4 + 5x^2$ is an even function.

**b.** We replace $x$ by $-x$ to obtain $f(-x) = 3(-x)^3 + 2(-x) = -3x^3 - 2x = -(3x^3 + 2x) = -f(x)$. Thus $f(x) = 3x^3 + 2x$ is an odd function.

**c.** We replace $x$ by $-x$ and obtain $f(-x) = 2(-x)^2 - 2(-x) + 1 = 2x^2 + 2x + 1$. This is not equal to $f(x)$, or $-f(x)$, so $f(x) = 2x^2 - 2x + 1$ is neither an even function nor an odd function. ■

### The Conversion of Equations to Functions

Since we use equations so extensively in mathematics and since most functions that we encounter are defined by equations, the question naturally arises: "Can all equations be expressed as functions?" The answer is no. We shall see this in the following discussion.

If a variable can take on any value in the domain, this variable is called the **independent variable**. If the value of a second variable is determined by this choice, then that variable is called the **dependent variable**. It is possible that one equation containing two variables may be used to determine separate functions with different letters designated to represent the independent variable in each case. Consider the following two examples.

**EXAMPLE 7** Can $y = 5x - 2$ be written as:

**a.** A function where $x$ is the independent variable?
**b.** A function where $y$ is the independent variable?

***Solution***

**a.** If we use the equation in its original form, it is already solved for the dependent variable ($y$) in terms of the independent variable ($x$).

We can write $f(x) = 5x - 2$. For each element $x$ in the domain there corresponds exactly one element $f(x)$. Thus the equation can be written as a function where $x$ is the independent variable.

**b.** $y = 5x - 2$ *We need to solve the equation for $x$ in terms of the dependent variable $y$.*

$$y + 2 = 5x$$

$$\frac{y + 2}{5} = x$$

We can write $g(y) = \dfrac{y + 2}{5}$.

For each element $y$ in the domain there corresponds exactly one element $g(y)$. Thus the equation can be written as a function where $y$ is the independent variable. ■

However, some frequently encountered equations cannot be expressed as functions because the definition of function is not satisfied.

**EXAMPLE 8** Can $y - 1 = |2 - x|$ be written as:

**a.** A function where $x$ is the independent variable?
**b.** A function where $y$ is the independent variable?

***Solution*** **a.** $y = |2 - x| + 1$ *We need to solve the equation for $y$ in terms of the dependent variable $x$.*

We can write $f(x) = |2 - x| + 1$, since for each element $x$ in the domain there is exactly one value of $y$. This equation is a function where $x$ is the dependent variable.

**b.** $y - 1 = |2 - x|$ is a more difficult equation to solve for $x$. The absolute value expression indicates two possibilities. To remove the absolute value symbol, we will obtain two equations.

If $\qquad y - 1 = |2 - x|$

then

$$y - 1 = 2 - x \qquad \text{or} \qquad -(y - 1) = 2 - x$$

$$y - 3 = -x \qquad\qquad -y + 1 = 2 - x$$

$$x = -y + 3 \qquad\qquad -y - 1 = -x$$

$$x = y + 1$$

Thus for a given value of $y$ we will obtain *two different values of $x$* (one from each equation). Therefore the definition of function is *not* satisfied. A function must have *exactly one* value in the range corresponding to any *one value* in the domain. This equation is *not* a function where $y$ is the independent variable. ■

**EXAMPLE 9** Can we write $y = \pi x^2$ as a function of $y$?

***Solution*** First we obtain

$$\frac{y}{\pi} = x^2$$

Taking the square root of both sides, we have

$$\pm\sqrt{\frac{y}{\pi}} = x$$

Since each positive value of $y$ would yield two values for $x$, the expression cannot be written as a function of $y$.

**EXAMPLE 10** Express the radius $r$ of a circle as a function of the area $A$.

***Solution*** We will use the formula $A = \pi r^2$. We know, however, that the radius of a circle cannot be negative. Thus if we solve the equation for $r$, we have

$$\frac{A}{\pi} = r^2 \qquad \sqrt{\frac{A}{\pi}} = r$$

*We select only the principal square root of* $\frac{A}{\pi}$.

Here we want only the positive square root of $\frac{A}{\pi}$, since only nonnegative values of $r$ are possible. Thus $h(A) = \sqrt{\frac{A}{\pi}}$ is a function, since it will determine exactly one radius for each element in the domain (each value of area). ■

## EXERCISE 3-3

**1.** If $f(x) = 4 - 3x$, find $f(2)$, $f(-3)$ and $f\left(\frac{1}{2}\right)$.

**2.** If $f(x) = \pi x + 6$, find $f(-1)$, $f(3)$, and $f\left(\frac{1}{2}\right)$.

**3.** If $f(x) = \frac{1}{x - 5}$, find $f(2)$, $f(-3)$, $f(0)$, and $f(6)$.

**4.** If $f(x) = (2x)(x + 1)$, find $f(0)$, $f(-3)$, $f(4)$, and $f\left(\frac{1}{3}\right)$.

**5.** If $f(x) = 2x^2 - 4x + 1$, find $f(-1)$, $f(-2)$, $f(0)$, and $f(3)$.

**6.** If $f(x) = x^3 + 2x^2 - x + 1$, find $f(1)$, $f(-1)$, $f(2)$, and $f(-2)$.

**7.** If $f(x) = x^2 - 2x$, find $f(a)$, $f(2a)$, and $f(a + 2)$.

**8.** If $f(x) = \frac{2x}{x + 2}$, find $f(b)$, $f(3b)$, and $f(b + 2)$.

**9.** If $g(x) = x^2 + 2$, find $g(b)$, $g(b + c)$, $g(b) + g(c)$ and $g(b + c) - g(b)$.

**10.** If $h(x) = \sqrt{x + 2}$, find $h(a)$, $h(b + a)$, $h(a) + h(b)$, and $h(b + a) - h(b)$.

Find the *domain* and the *range* of the functions in Problems 11–18.

**11.** $f(x) = 2x^2$ **12.** $f(x) = 2x + 1$ **13.** $f(x) = |x + 4|$ **14.** $f(x) = |3 - x|$

**15.** $f(x) = \sqrt{4 - x}$ **16.** $f(x) = \sqrt{2x + 3}$ **17.** $f(x) = 2x^3$ **18.** $f(x) = x^4$

Find the *domain* of the functions in Problems 19–28.

**19.** $f(x) = \dfrac{2}{\sqrt{x + 4}}$ **20.** $f(x) = \dfrac{x}{|x + 2|}$ **21.** $g(x) = \dfrac{5}{\sqrt{x - 2}}$ **22.** $g(x) = \sqrt{x^2 - 4}$

**23.** $g(x) = \dfrac{2x}{(x + 1)(x - 2)}$ **24.** $g(x) = \dfrac{1}{9 - x^2}$ **25.** $h(x) = \dfrac{3}{4 - \sqrt{x}}$ **26.** $h(x) = \dfrac{2}{\sqrt{x + 1} - 3}$

**27.** $b(x) = \dfrac{\sqrt{x}}{x^2 + 5x + 6}$ **28.** $c(x) = \dfrac{x^2}{x - 8}$

Determine whether the functions are *even*, *odd*, or *neither* in Problems 29–36.

**29.** $f(x) = 3x - 9$ **30.** $f(x) = x^4 + 2x^2$ **31.** $f(x) = x^5 + 3x^3$

**32.** $f(x) = \dfrac{5x}{x + 2}$ **33.** $f(x) = \sqrt{x^2 + 3}$ **34.** $f(x) = 2x^5 - x^3 + x$

**35.** $f(x) = |x| - 3$ **36.** $f(x) = x^3 + 3x^2 + x$

For each of the expressions in Problems 37–46, determine whether the specified equation can be written

**a.** As a function with $x$ as the independent variable.

**b.** As a function with $y$ as the independent variable.

In each case where it is possible, write the equation using functional notation.

**37.** $y = \dfrac{3}{x}$ **38.** $y = x^3$ **39.** $y = 2x^2 + x$ **40.** $y = \sqrt{x + 2}$ **41.** $|y + 1| = x$

**42.** $2y = |x + 3|$ **43.** $x = y^3$ **44.** $x^2 + y^2 = 9$ **45.** $x = \sqrt{9 - y^2}$ **46.** $y = \dfrac{x + 2}{x - 3}$

## Calculator Problems

Evaluate Problems 47–50, using a scientific calculator. Do *not* round off your answer.

**47.** $f(x) = 2x^2 + x - 3$
$f(3.123)$ and $f(-0.00026)$

**48.** $f(x) = \sqrt{5x - 6}$
$f(2.897)$ and $f(12{,}674)$

**49.** $g(r) = \dfrac{4}{3}\pi r^3$
$g(1.0008)$ and $g(1576)$

**50.** $f(x) = x^3 + 3x^2 - 2x$
$f(56.12)$ and $f(0.00361)$

★**51.** Express the side $s$ of an equilateral triangle as a function of its area $A$.

**52.** Express the radius $r$ of a sphere as a function of its volume $V$.

**★53.** Let $f(x) = \dfrac{1}{x+1}$; show that $f\left(\dfrac{1}{x}\right) = x \cdot f(x)$.

**★54.** Let $f(x) = \dfrac{x+1}{-x+1}$; show that $f\left(\dfrac{1}{x}\right) = -f(x)$.

---

## 3-4 THE GRAPH OF A FUNCTION

Basically, when we graph a function, we are drawing a graph of the equation $y = f(x)$. We will make this more exact by the following definition of the graph of a function.

*Definition*

> Let $f(x)$ be any function. The graph of the function is the set of all points $(x, f(x))$ such that $x$ is in the domain of the function.

For any given point $(a, b)$ on the graph of the function, $a$ is the value of the abscissa and $f(a) = b$ is the value of the ordinate $b$. In general, for any value $a$ in the domain there corresponds an image $b$, which is $f(a)$.

**EXAMPLE 1** Sketch the graph of $f(x) = 4 + x - x^2$.

***Solution*** We will list a table of values for $(x, f(x))$ for some of the points of the graph of $f(x)$.

| $x$ | $-3$ | $-2$ | $-1$ | $0$ | $0.5$ | $1$ | $2$ | $3$ | $4$ |
|---|---|---|---|---|---|---|---|---|---|
| $f(x)$ | $-8$ | $-2$ | $2$ | $4$ | $4.25$ | $4$ | $2$ | $-2$ | $-8$ |

■

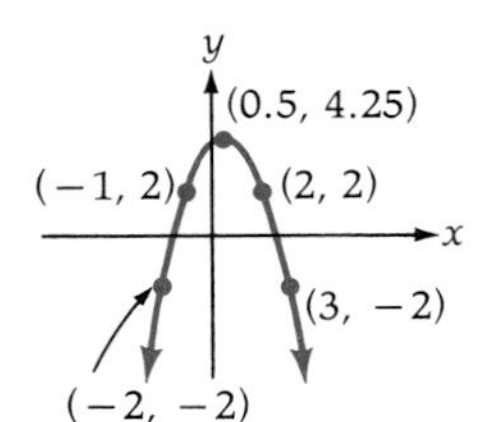


*Figure 3-14*

An important property of the graph of a function is the location of the points where the graph $f(x)$ crosses the $x$-axis. These are clearly the places where $f(x) = 0$. Any number $a$ for which $f(a) = 0$ is known as a **zero of the function.**

From Figure 3-14 we could make an estimate of the zeros of the function located at approximately $x = 2\frac{2}{3}$ and $x = -1\frac{2}{3}$. By using the quadratic formula we find the zeros exactly in the following way:

$0 = 4 + x - x^2$ *Set $f(x) = 0$*

$0 = -x^2 + x + 4$ *We rearrange in descending powers of $x$. The quadratic expression does not factor, so we will use the quadratic formula with $a = -1$, $b = +1$, and $c = 4$.*

$$x = \frac{-1 \pm \sqrt{(1)^2 - 4(-1)(4)}}{2(-1)} = \frac{-1 \pm \sqrt{1 + 16}}{-2} = \frac{-1 \pm \sqrt{17}}{-2}$$

Thus the two zeros of $f(x)$ are at $x = \dfrac{-1 + \sqrt{17}}{-2}$ and at $x = \dfrac{-1 - \sqrt{17}}{-2}$.

Entering this on a calculator, we obtain an approximation (to the nearest thousandth) of $x = -1.562$ and $x = 2.562$, which is close to our original estimate of $x = -1\frac{2}{3}$ (approximately $-1.667$) and $x = 2\frac{2}{3}$ (approximately 2.667).

The zeros of a function cannot always be obtained by algebraic methods, so the ability to estimate zeros of a function from a graph is often useful in mathematics.

If for a given set of increasing values for $x$ the values of $f(x)$ are becoming larger, we say that the function is increasing. If over the same set of $x$ values the $f(x)$ values are becoming smaller, we say the function is decreasing.

To be more precise we state the following definition of increasing and decreasing functions.

***Definition***

If $f(x)$ is a function in the interval $[a, b]$, then whenever $x_1 < x_2$ and both $x_1$ and $x_2$ are points in $[a, b]$

**1.** $f(x)$ is *increasing* in the interval $[a, b]$ if $f(x_1) < f(x_2)$.
**2.** $f(x)$ is *decreasing* in the interval $[a, b]$ if $f(x_1) > f(x_2)$.

When we apply the definition to some functions, we find that some functions are always increasing or always decreasing. (See Figure 3-15.)

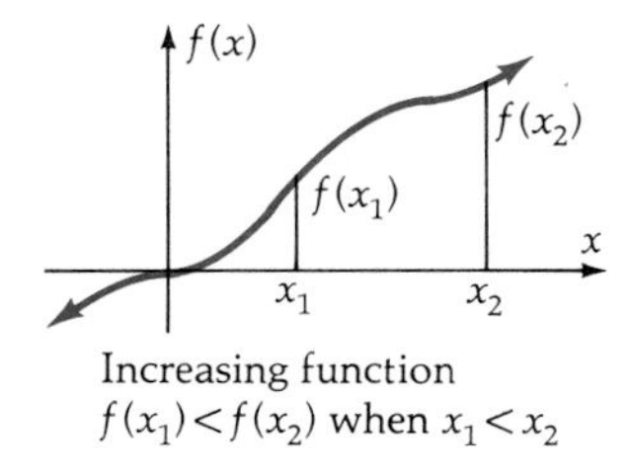

Increasing function
$f(x_1) < f(x_2)$ when $x_1 < x_2$

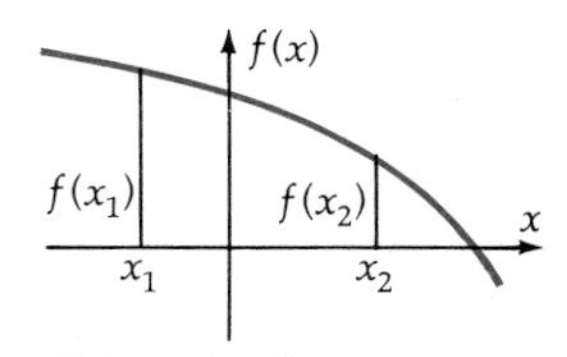

Decreasing function
$f(x_1) > f(x_2)$ when $x_1 < x_2$

***Figure 3-15***

Some functions are increasing in a certain interval and decreasing in another interval. In our graph of the function $f(x) = 4 + x - x^2$ displayed in

Figure 3-14, the function is increasing in the interval $(-\infty, 0.5]$ and decreasing in the interval $[0.5, +\infty)$.

The sketching of functions like the sketching of equations is aided by the careful use of properties of symmetry.

---

**Properties of Symmetry for Functions**

---

1. The graph of an *even* function is symmetric with respect to the $y$-axis.
2. The graph of an *odd* function is symmetric with respect to the origin.
3. A function of $x$ cannot be symmetric with respect to the $x$-axis. Any curve that is symmetric to the $x$-axis is not a function of the form $y = f(x)$.

---

We will examine some of these properties in the following example.

**EXAMPLE 2** Graph the function $f(x) = \sqrt[3]{x}$ and state what properties it has as a function.

***Solution*** We see that since $\sqrt[3]{-x} = -\sqrt[3]{x}$ for all $x$, $f(-x) = -f(x)$. *The function is an odd function* and is symmetric with respect to the origin. For every point $(x, y)$ on the curve the point $(-x, -y)$ is on the curve. We will construct a table of values for only nonnegative values of $x$ such as $x = 0, 1, 8$ and obtain integer values for $f(x)$ of 0, 1, 2, respectively.

***Calculator Values***

We find that integer values are not sufficient for a complete graph, so we obtain a few calculator values. (The cube root can be obtained by using the $\boxed{\sqrt[x]{y}}$ button.)

If $x = 4$, 4 $\boxed{\sqrt[x]{y}}$ 3 $\boxed{=}$ 1.5874011

If $x = 12$, 12 $\boxed{\sqrt[x]{y}}$ 3 $\boxed{=}$ 2.2894285

We will round these to the nearest tenth and include them in our table of values.

| $x$ | 0 | 1 | 4 | 8 | 12 |
|---|---|---|---|---|---|
| $f(x)$ | 0 | 1 | 1.6 | 2 | 2.3 |

Now we will graph the function. (See Figure 3-16.)

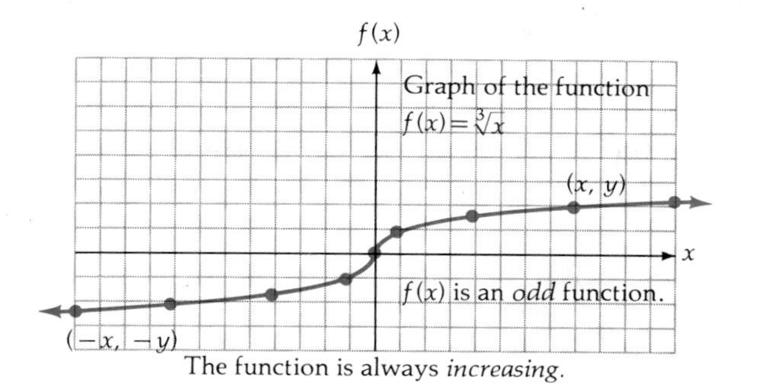


*Figure 3-16*

**EXAMPLE 3**

a. Graph the equation $x - 2 + y^2 = 0$.

b. Can the graph of $x - 2 + y^2 = 0$ be considered the graph of a function of the form $y = f(x)$?

***Solution***

a. To graph the equation, we first solve the equation for $x$, letting $y$ be the independent variable. This will be the easiest form in which to obtain ordered pairs $(x, y)$.

$$x = 2 - y^2$$

We let $y = 0, 1, 2, 3, 4$ and obtain the following table of values.

| $y$ | 0 | 1 | 2 | 3 | 4 |
|---|---|---|---|---|---|
| $x$ | 2 | 1 | −2 | −7 | −14 |

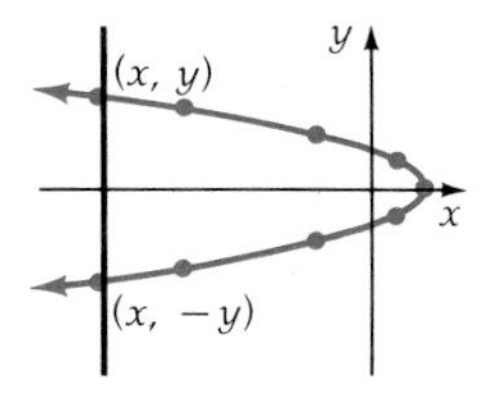


*Figure 3-17*

Since when we replace $y$ by $-y$ in the equation we have an equivalent equation, the curve is symmetric to the $x$-axis. We therefore do not need negative $y$ values in our table, since for every point $(x, y)$ on the curve, $(x, -y)$ is also on the curve. (See Figure 3-17.)

b. In a variety of ways we can prove that this is not a function. For example, by the third property of symmetry we see that this equation cannot be considered a function of $x$, since the equation is symmetric to the $x$-axis. Without referring to the graph we can verify that the equation is not a function of $x$ in the following way:

When we solve the equation for $y$, we obtain $y^2 = 2 - x$.

$$y = \pm\sqrt{2 - x} \qquad \textit{Taking the square root of both sides.}$$

Since for every value of $x$ less than 2 we obtain *two* values for $y$, we cannot consider the graph a function of the form $y = f(x)$. ■

It is obvious from Figure 3-17 that for any point $(x, y)$ on the curve the point $(x, -y)$ is on the curve, and therefore the equation cannot be considered a function of $x$. Whenever you can draw even one vertical line through the curve and have it intersect the curve at two points, you know you do not have a function. This property leads to a useful test.

---

**Vertical Line Test**

---

An equation determines a function if and only if no vertical line meets the graph of the equation at more than one point.

---

### Vertical and Horizontal Shifts of Graphs

If you know how to graph a function $f(x)$, you can easily find the graph of a function of the form $f(x) + a$ by moving the graph upward or downward $a$ units without changing the shape or orientation of the graph. We can see this clearly in the following example.

**EXAMPLE 4** Graph the functions $f(x) = x^2$, $f(x) = x^2 + 5$, and $f(x) = x^2 - 6$ on one axis in the region $-3 \leq x \leq 3$.

***Solution***

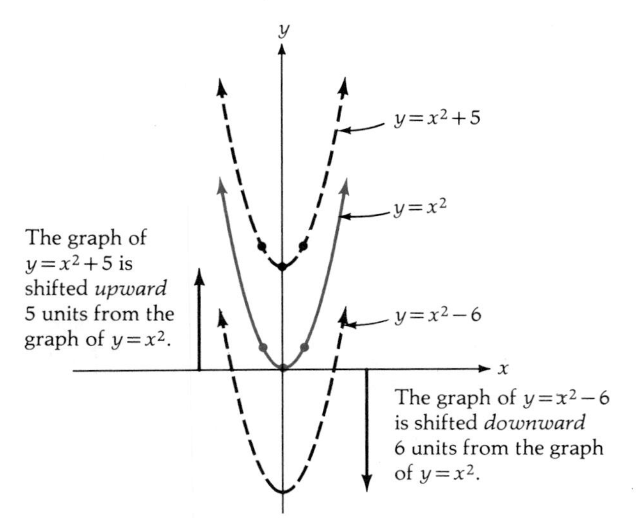
y
y=x²+5
y=x²
The graph of y=x²+5 is shifted upward 5 units from the graph of y=x².
y=x²−6
x
The graph of y=x²−6 is shifted downward 6 units from the graph of y=x².

*Figure 3-18*

Figure 3-18 shows the graph of each function of the form $y = f(x)$. ■

We can generalize the pattern observed in Example 4 to formulate the following rule that is helpful in graphing functions.

---

**Vertical Shifts of Graphs**

---

If you have the graph of $f(x)$, to obtain a graph of the function (for any positive number $a$):

1. $f(x) + a$ You should shift the graph of $f(x)$ a total of $a$ units **upward**.
2. $f(x) - a$ You should shift the graph of $f(x)$ a total of $a$ units **downward**.

---

A similar pattern is apparent for horizontal shifts.

EXAMPLE 5 Graph the functions $f(x) = x^3$, $f(x) = (x + 6)^3$ and $f(x) = (x - 5)^3$ on one set of coordinate axes.

***Solution*** Figure 3-19 shows the graph of each function of the form $y = f(x)$.

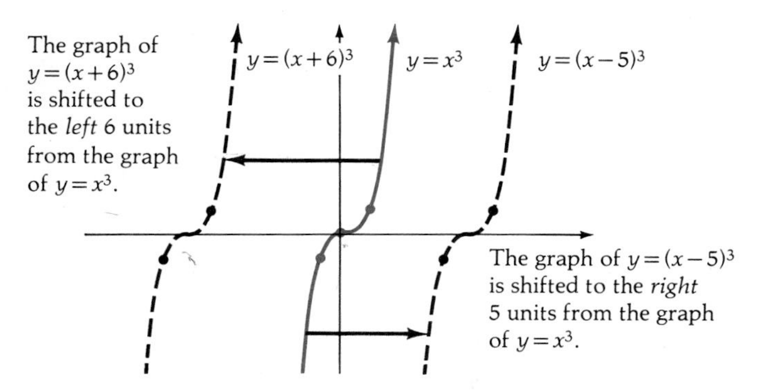


*Figure 3-19*

Note carefully the direction of the shift. The graph of $f(x) = (x - 5)^3$ is 5 units to the *right* of the graph of $f(x)$. Students sometimes incorrectly go to the left when they encounter the $-5$ quantity. ■

We can generalize the pattern observed in Example 5 to formulate the following rule.

### Horizontal Shifts of Graphs

If you have the graph of $f(x)$, to obtain a graph of the function (for any positive number $b$):

1. $f(x+b)$ You should shift the graph of $f(x)$ a total of $b$ units to the **left**.
2. $f(x-b)$ You should shift the graph of $f(x)$ a total of $b$ units to the **right**.

## Stretching and Shrinking of Graphs

There is one further situation that is helpful to analyze. If you know how to graph a function $y = f(x)$, you can easily find the graph of $y = c \cdot f(x)$ where $c$ is a real number. The general shape of the curve will be similar but not exactly the same.

### Stretching and Shrinking of Graphs

If you have the graph of $f(x)$, to obtain a graph of the function (for any real number $c$):

$c \cdot f(x)$ You should multiply the ordinates of points on the graph $y = f(x)$ by $c$.

EXAMPLE 6

**a.** Sketch the graph of $f(x) = \sqrt{x-1}$ and $g(x) = 3\sqrt{x-1}$.

**b.** On a separate axis, sketch the graph of $f(x) = \sqrt{x-1}$ and $h(x) = \frac{1}{2}\sqrt{x-1}$.

***Solution*** The radicand must be nonnegative. This means $x - 1 \geq 0$ or $x \geq 1$. Therefore the domain of each function is $\{x\colon \; x \geq 1\}$.

We plot a few points for each function and sketch the graphs in Figure 3-20.

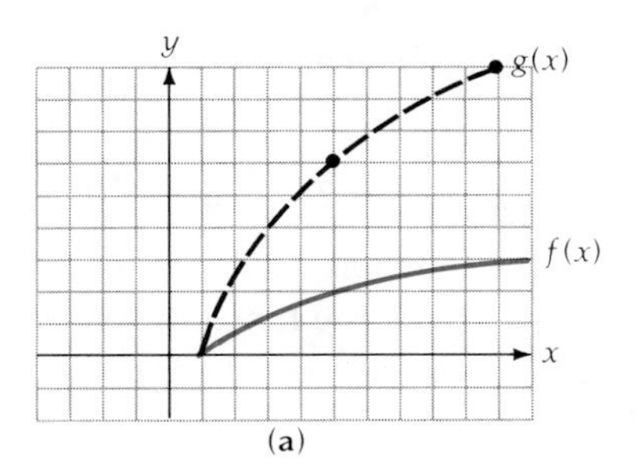


(a)

(b)

*Figure 3-20*

**a.** The graph of $f(x) = \sqrt{x - 1}$ and $g(x) = 3\sqrt{x - 1}$

The curve of $g(x)$ is "stretched" by a factor of 3. For any value of $x$ the ordinate of $g(x)$ is 3 times as tall as the ordinate of $f(x)$.

**b.** The graph of $f(x) = \sqrt{x - 1}$ and $h(x) = \frac{1}{2}\sqrt{x - 1}$.

The curve of $h(x)$ is "shrunk" by a factor of 2. For any value of $x$ the ordinate of $h(x)$ is only one half as tall as the ordinate of $f(x)$. ■

If the number $c$ is negative, then the curve of the function $c \cdot f(x)$ will be inverted and will be on the opposite side of the $x$-axis. We call the graph of $y = -f(x)$ a *reflection* of the graph of $y = f(x)$ through the $x$-axis.

EXAMPLE 7 Sketch the graph of $f(x) = x^2$, $f(x) = -x^2$, and $f(x) = -\frac{1}{3}x^2$ on the same axis.

***Solution*** We will graph each function of the form $y = f(x)$ as in Figure 3-21.

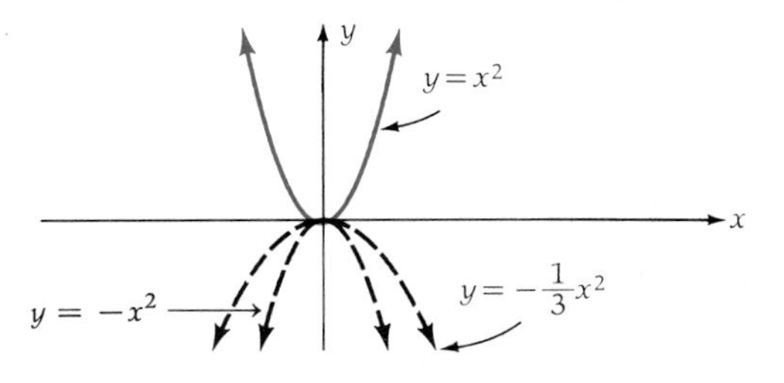


■

*Figure 3-21*

Sometimes the graph of a function is defined differently for different regions of $x$.

**EXAMPLE 8** Sketch the graph of the function $f(x)$ that is defined in the following way.

$$f(x) = \begin{cases} 2 & \text{if } x < 0 \\ x^2 + 2 & \text{if } x \geq 0 \end{cases}$$

***Solution*** For all negative values of $x$ the function value $f(x)$ is 2. This is a constant function for all values of $x < 0$. If $x \geq 0$, the function values are defined by $x^2 + 2$.

In actuality, we proceed as if we were doing two problems. For negative $x$ we graph $f(x) = 2$. This is a straight line, the line $y = 2$. For $x \geq 0$ we graph $f(x) = x^2 + 2$. This is one half of a parabola opening upward. We obtain a few points on the parabola (0, 2), (1, 3), (2, 6) and plot them in Figure 3-22.

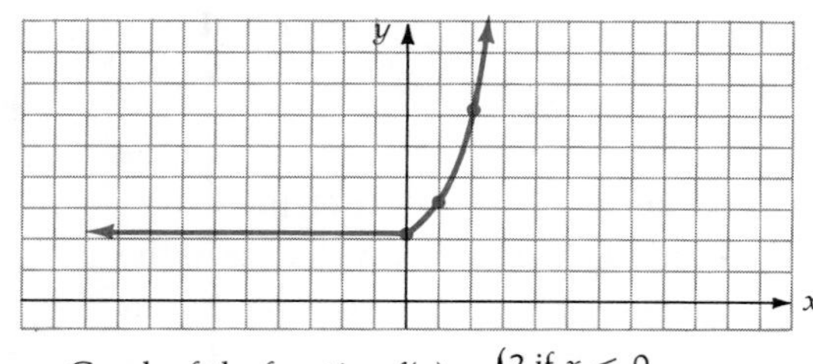


Graph of the function $f(x) = \begin{cases} 2 \text{ if } x < 0 \\ x^2 + 2 \text{ if } x \geq 0 \end{cases}$ ■

*Figure 3-22*

## EXERCISE 3-4

For each of the functions in Problems 1–14:

**a.** State what types of symmetry each function has, if any.
**b.** Graph the function.
**c.** State the intervals for which the function is *increasing* and the intervals for which the function is *decreasing*.

**1.** $f(x) = x + 5$ **2.** $f(x) = 3 - 2x$ **3.** $f(x) = -\frac{1}{3}x + 4$ **4.** $f(x) = 4x - 7$

**5.** $f(x) = -3x^2$ **6.** $f(x) = 2 - x^2$ **7.** $f(x) = \frac{1}{2}x^2 - 3$ **8.** $f(x) = \frac{1}{4}x^2 - 2$

**9.** $f(x) = \frac{2}{x} + 1$ **10.** $f(x) = \frac{3}{x} - 2$ **11.** $f(x) = 5 - |x|$ **12.** $f(x) = |5 - x|$

**13.** $f(x) = \sqrt{x + 9}$ **14.** $f(x) = 2\sqrt{x} + 1$

Sketch the graphs of the functions for each problem in Problems 15–22 on the same coordinate axes.

**15.** $f(x) = x^2$
$g(x) = 3x^2$
$h(x) = \frac{1}{3}x^2$

**16.** $f(x) = x^2$
$g(x) = 4x^2$
$h(x) = \frac{1}{4}x^2$

**17.** $f(x) = \frac{1}{3}x^3$
$g(x) = -\frac{1}{3}x^3$

**18.** $f(x) = (x - 2)^2$
$g(x) = -1(x - 2)^2$

**19.** $f(x) = 2\sqrt{x}$
$g(x) = 2\sqrt{x} + 3$
$h(x) = 2\sqrt{x} - 4$

**20.** $f(x) = x^3 + 2$
$g(x) = x^3$
$h(x) = x^3 - 3$

**21.** $f(x) = |x|$
$g(x) = |x + 2|$
$h(x) = |x - 2|$

**22.** $f(x) = 3 + |x|$
$g(x) = 3 + |x + 1|$
$h(x) = |x + 1|$

Graph each of the equations in Problems 23–26. In each case, explain why the graph of the equation is *not* the graph of a function of the form $y = f(x)$.

**23.** $y^4 = x$ **24.** $x^2 + y^2 = 4$ **25.** $y^2 + 2y - x = 0$ **26.** $y^2 - 2x + 3 = 0$

Graph the functions in Problems 27–30.

**27.** $g(x) = \begin{cases} x^2 & \text{for } x \geq 0 \\ x & \text{for } x < 0 \end{cases}$

**28.** $h(x) = \begin{cases} 4 & \text{for } x \geq 0 \\ x - 2 & \text{for } x < 0 \end{cases}$

**29.** $f(x) = \begin{cases} 3 & \text{for } x < -4 \\ x + 2 & \text{for } -4 \leq x < 0 \\ 2x & \text{for } x \geq 0 \end{cases}$

**30.** $b(x) = \begin{cases} 2 & \text{for } x < 0 \\ x^2 + 5 & \text{for } 0 \leq x < 2 \\ 2 - x & \text{for } x \geq 2 \end{cases}$

In Problems 31–35 we will refer to the greatest integer function. The *greatest integer function* denoted by $[x]$ is defined to be the greatest integer less than or equal to $x$. Examples of values of this function for certain $x$ values are:

$$[1] = 1 \qquad [\pi] = 3 \qquad [2] = 2$$
$$[2.2] = 2 \qquad [-1.6] = -2 \qquad [\sqrt{17}] = 4$$

**★ 31.** Evaluate the following: $[0]$, $[5.8]$, $[\sqrt[3]{7}]$, $[-17.6]$, $\left[-\frac{\pi}{2}\right]$, $[\pi^2]$.

**★ 32.** Graph $f(x) = [x]$. **★ 33.** Graph $f(x) = 2[x]$.

**★ 34.** Graph $f(x) = [x - 3]$. **★ 35.** Graph $f(x) = [x] - 3$.

**★ 36.** The voltage in a certain circuit can be measured by the following function where the independent variable $t$ is measured in seconds:

$$f(t) = \begin{cases} 0 \text{ volts, } 0 \leq t < 2 \\ t - 2 \text{ volts, } 2 \leq t < 5 \\ 3 \text{ volts, } t \geq 5 \end{cases}$$

Sketch the graph of the voltage function for $0 \leq t \leq 10$.

**★ 37.** The cost to manufacture a quantity of a certain type of chip for digital computers is given by the following function where the independent variable $n$ is the number of chips manufactured.

$$C(n) = \begin{cases} 5000 \text{ dollars, for } 0 \leq n < 100 \\ 5000 + 25(n - 100) \text{ dollars for } 100 \leq n < 200 \\ 7500 + 10(n - 200) \text{ dollars for } n \geq 200 \end{cases}$$

Evaluate $C(0)$, $C(50)$, $C(100)$, $C(150)$, $C(200)$, $C(250)$, and $C(300)$. Then sketch the graph of function for $0 \leq n \leq 300$.

## 3-5 LINEAR FUNCTIONS

One of the most frequently encountered functions is a linear function. A **linear function** is a function of the form $f(x) = ax + b$ for all real numbers $a$, $b$ with $a \neq 0$. We will see that any straight line that is neither vertical nor horizontal can be associated with a linear function. A horizontal line can be associated with a constant function of the form $f(x) = b$. Can we represent a vertical line by a function?

Do you see that the graph of a vertical line does not represent a function, since one $x$ value has many $y$ values? All other straight lines represent functions. Remember that the graph of any first-degree equation in two variables will form a straight line.

*Definition*

For real numbers $a$, $b$, $c$ (where $a$ and $b$ are not both 0) the *standard form of the equation of a straight line* is given as

$$ax + by = c$$

An important concept, related to the equation of a straight line, is that of slope. Intuitively, we think of the slope of a line as the "steepness" of the line. The following definition gives us a more precise meaning.

*Definition*

The *slope* ($m$) of a line through the two distinct points $(x_1, y_1)$ and $(x_2, y_2)$ is given by the ratio

$$m = \frac{y_2 - y_1}{x_2 - x_1}, \qquad x_1 \neq x_2$$

**EXAMPLE 1** Find the slope of a line through the points $(1, -5)$ and $(4, 0)$.

***Solution*** Refer to Figure 3-23. If we let $(1, -5)$ be $(x_1, y_1)$ and $(4, 0)$ be $(x_2, y_2)$, then applying the formula we obtain

$$m = \frac{0 - (-5)}{4 - 1} = \frac{0 + 5}{4 - 1} = \frac{5}{3}$$

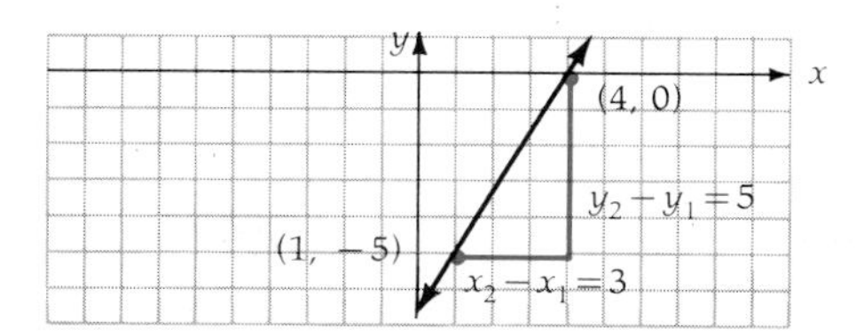


*Figure 3-23*

The ratio $\dfrac{y_2 - y_1}{x_2 - x_1}$ is not dependent on the points chosen as long as they are two different points on the line. A proof of this statement can be developed by using an argument involving similar triangles. A given line has only one slope.

The definition of slope, and the fact that the slope of a given line is constant, is the basis for the solution of many problems concerning linear functions.

For instance, if we are given that a line contains the points $(x_1, y_1)$ and $(x_2, y_2)$ and are asked for the equation of the line, we proceed in the following manner.

First we choose any other point on the line and call it $(x, y)$. The slope is now given by the ratio $\dfrac{y_2 - y_1}{x_2 - x_1}$ or by the ratio $\dfrac{y - y_1}{x - x_1}$; and since the slope is constant, we can equate these two ratios, giving

$$\frac{y - y_1}{x - x_1} = \frac{y_2 - y_1}{x_2 - x_1}$$

which becomes the **two-point form** of the equation of a straight line. We will solve this equation for the expression $y - y_1$ and state the definition.

*Definition*

> For a line through points $(x_1, y_1)$ and $(x_2, y_2)$ the *two-point form* of the equation of a straight line is given by
>
> $$y - y_1 = \left(\frac{y_2 - y_1}{x_2 - x_1}\right)(x - x_1), \qquad x_1 \neq x_2$$

EXAMPLE 2 Write the equation in standard form of the line through the points (3, 5) and (−2, 7).

***Solution*** Using the two-point form, we have

$$y - 5 = \left(\frac{7 - 5}{-2 - 3}\right)(x - 3)$$

Simplifying, we obtain

$$y - 5 = -\frac{2}{5}(x - 3)$$

or

$$2x + 5y = 31$$

The student should check to see that both points are on this line by substituting into the equation. ■

If we are given the slope $m$ of a line and a point $(x_1, y_1)$ and are asked to write the equation, we proceed as follows.

We first choose some other point on the line and call it $(x, y)$. The definition of the slope gives us the ratio $\frac{y - y_1}{x - x_1}$. But since we are given the slope to be $m$, we can write the equation

$$m = \frac{y - y_1}{x - x_1}$$

We can write this as follows: $y - y_1 = m(x - x_1)$. This form is very useful and is known as the **point-slope** form.

***Definition***

For a line through the point $(x_1, y_1)$ with slope $m$ the *point-slope form* of the equation of a straight line is given by

$$y - y_1 = m(x - x_1)$$

**EXAMPLE 3** Write the equation in standard form of the line through the point (2, 5) that has a slope of $\frac{2}{5}$.

***Solution*** Using the point-slope form gives us $y - 5 = \frac{2}{5}(x - 2)$, which in standard form is $2x - 5y = -21$. ■

***Definition***

The *y-intercept* of a straight line is the ordinate of the point where the line intersects the $y$-axis, and the *x-intercept* of a straight line is the abscissa of the point where the line intersects the $x$-axis.

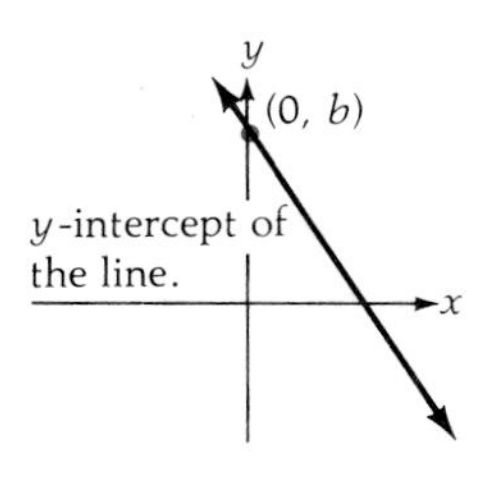


*Figure 3-24*

In other words, the $y$-intercept is the value of $y$ when $x = 0$. We will designate the $y$-intercept as $b$. Refer to Figure 3-24.

If we are given the slope $m$ of a line and the point $(0, b)$ on the line, then the *point-slope form* gives us

$$y - b = m(x - 0)$$

or

$$y = mx + b$$

**Definition**

> For a line of slope $m$ that has a $y$-intercept of $b$ the *slope-intercept form* of the equation of a straight line is
>
> $$y = mx + b$$

The *slope-intercept form* is quite useful, as we can see in the following example.

**EXAMPLE 4** If the slope of a line is $-\frac{5}{8}$ and the $y$-intercept is 6, write the equation in standard form.

***Solution*** We are using the equation $y = mx + b$ with $m = -\frac{5}{8}$ and $b = 6$. The slope-intercept form gives us $y = -\frac{5}{8}x + 6$. By using a few algebraic steps to obtain the standard form this becomes $5x + 8y = 48$. ■

**EXAMPLE 5** Find the slope and the $y$-intercept of the line given by the equation $3x - 2y = 9$.

***Solution*** Solving for $y$ in terms of $x$ gives us $y = \frac{3}{2}x - \frac{9}{2}$, so the slope is $m = \frac{3}{2}$ and the $y$-intercept is $b = -\frac{9}{2}$. ■

For two nonvertical lines to be parallel they must have the same slope. We can state this formally.

**Definition**

> *Slopes of Parallel Lines.* Two nonvertical lines are *parallel* if and only if they have exactly the same slope $m$.

This definition is useful if we need to determine the equation of a line that is parallel to a given line. Consider this example.

**EXAMPLE 6** Find the equation of a line that passes through (1, 2) that is parallel to the line $3x - 5y - 15 = 0$.

***Solution*** The line $3x - 5y - 15 = 0$ should first be written in the slope-intercept form.

$-5y = -3x + 15$ *Add 15 to both sides and subtract 3x from each side of equation.*

$y = \frac{3}{5}x - 3$ *Divide both sides by −5 and simplify.*

Since this is in the form $y = mx + b$, we see that the slope $m = \frac{3}{5}$. Thus we want to find the equation of a line through (1, 2) with slope $= \frac{3}{5}$. We will use the point-slope form now:

$$y - y_1 = m(x - x_1)$$

$$y - 2 = \frac{3}{5}(x - 1) \quad \text{Substitute } m = \frac{3}{5},\ x_1 = 1, \text{ and } y_1 = 2.$$

$$y - 2 = \frac{3}{5}x - \frac{3}{5} \quad \text{Remove parentheses.}$$

$$5y - 10 = 3x - 3 \quad \text{Multiply each term by 5.}$$

$$-3x + 5y - 10 + 3 = 0 \quad \text{Collect terms on left.}$$

$$-3x + 5y - 7 = 0 \quad \text{Simplify.}$$

$$3x - 5y + 7 = 0$$

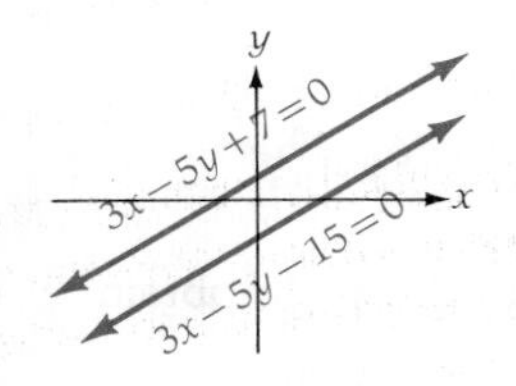

*Figure 3-25*

Thus we can see from the graphs of the two lines in Figure 3-25 that the two lines $3x - 5y + 7 = 0$ and $3x - 5y - 15 = 0$ are parallel. ■

Nonvertical *perpendicular lines* always have *slopes* that are *negative reciprocals of each other.* Let the slope of one line be $m_1$ and the slope of the other be $m_2$. If $m_1 = \frac{-1}{m_2}$, then obviously $m_1 m_2 = -1$. We can state this in a useful way as follows:

***Definition***

> *Slopes of Nonvertical Perpendicular Lines.* Two nonvertical lines are perpendicular if their slopes $m_1$ and $m_2$ satisfy the equation $m_1 m_2 = -1$.

There is one special case we should note. *If one line is vertical and one line is horizontal, the lines are perpendicular.* This situation is not contained in our definition, since a vertical line has *no slope* and a horizontal line has *zero slope.*

We can use the properties of slopes of perpendicular lines to find the equation of a line perpendicular to a given line.

**EXAMPLE 7** Find the equation of a line through the point $(-3, 4)$ that is perpendicular to the line $2x + 5y + 6 = 0$.

***Solution***

$$5y = -2x - 6$$

$$y = -\frac{2}{5}x - \frac{6}{5} \qquad \textit{We transform to the form } y = mx + b.$$

The slope $m_1$ of the given line is $-\frac{2}{5}$. For a line perpendicular to this we want

$$m_2 = -\frac{1}{m_1}, \text{ so } m_2 = \frac{-1}{-\frac{2}{5}} = +\frac{5}{2}.$$

We now use the point slope form for $m_2 = \frac{5}{2}$, $x_1 = -3$, and $y_1 = 4$.

$$y - y_1 = m_2(x - x_1)$$

$$y - 4 = \frac{5}{2}(x + 3)$$

$$2y - 8 = 5x + 15 \qquad \textit{Remove the parentheses and multiply each term by 2.}$$

$$5x - 2y = -23 \qquad \textit{Simplify to obtain the standard form.}$$

■

## EXERCISE 3-5

Find an equation, in standard form, of the line through each of the pairs of points in Problems 1–6.

**1.** $(2, 1)$ and $(-3, 9)$
**2.** $(3, 4)$ and $(1, 2)$
**3.** $(-6, 2)$ and $(4, -3)$
**4.** $(-5, 1)$ and $(-1, -2)$
**5.** $(16, -5)$ and $(4, 0)$
**6.** $\left(\frac{1}{2}, \frac{1}{3}\right)$ and $(-1, 0)$

In Problems 7–16, find an equation, in standard form, of the line through each of the given points and having the given slope.

**7.** $(1, 7)$, $m = 3$
**8.** $(1, -3)$, $m = 2$
**9.** $(-2, 0)$, $m = \frac{1}{2}$
**10.** $(3, -4)$, $m = -\frac{1}{2}$

**11.** $(0, 4), m = 5$ **12.** $(0, 0), m = -1$ **13.** $(3, -1), m = -\frac{2}{3}$ **14.** $\left(1, -\frac{1}{2}\right), m = -3$

**15.** $(-2, 3)$, parallel to the $y$-axis **16.** $(-3, -5)$, parallel to the $x$-axis.

Write each of the equations in Problems 17–24 in slope-intercept form. Specify the slope and $y$-intercept. Draw a sketch for each.

**17.** $2x + y = 5$ **18.** $\frac{1}{3}x + y = 7$ **19.** $3x - 6y = 4$ **20.** $x - \frac{1}{3}y + 4 = 0$

**21.** $5x - 2y = 0$ **22.** $5x = -8y$ **23.** $\frac{1}{5}x - \frac{1}{2}y = \frac{3}{4}$ **24.** $\frac{1}{2}x + \frac{1}{3}y = -6$

**25.** A line has an $x$-intercept of $x = -2$ and a $y$-intercept of $y = 7$. Find an equation of the line in standard form.

**26.** A line has an $x$-intercept of $x = 5$ and a $y$-intercept of $y = -2$. Find an equation of the line in standard form.

**27.** A line has a $y$-intercept of $-\frac{5}{6}$ and a slope of 3. Find an equation of the line in standard form.

**28.** A line has a $y$-intercept of $\frac{3}{4}$ and a slope of $-2$. Find an equation of the line in standard form.

In Problems 29–34, write an equation in standard form of the line passing through the given point and *parallel* to the given line.

**29.** $(3, 1), y = 2x + 1$ **30.** $(2, 0), 3x + y = 5$ **31.** $(5, -1), 2y - 5x = 1$

**32.** $(3, -10), 2x - y = 4$ **33.** $(-8, 1), y = -3$ **34.** $(4, -9), x = 2$

In Problems 35–40, write an equation in standard form of the line passing through the given point and *perpendicular* to the given line.

**35.** $(2, 1), y = 3x + 1$ **36.** $(1, -2), y = 2x - 1$ **37.** $(4, -3), 2x + y + 6 = 0$

**38.** $(-3, -1), 3x - y + 1 = 0$ **39.** $(-5, 8), 4x + y = 3$ **40.** $(2, -5), x = -2y + 3$

**41.** Find an equation of a line with $x$-intercept 4 and $y$-intercept $-3$.

**42.** Find an equation of a line that passes through $(-3, -3)$ and is perpendicular to the line passing through $(5, 1)$ and $(-2, 3)$.

**43.** Find the number $d$ such that the point $(-3, 1)$ is on the line $dx - 8y + 5 = 0$.

**44.** Find the number $c$ such that the line $2x + cy + 1 = 0$ is parallel to the line $4x = 5y + 3$.

**45.** Find the number $b$ such that the line $5x + by = 2$ is perpendicular to the line $7y = 2x - 3$.

★ **46.** Show that the figure determined by the four points $(-7, 2)$, $(-5, 5)$, $(1, 1)$ and $(-1, -2)$ is a rectangle by showing that the adjacent sides are perpendicular.

★ **47.** Show that the figure determined by the four points $(2, -1)$, $(3, 3)$, $(2, 7)$ and $(1, 3)$ is a parallelogram by showing that the opposite sides are parallel.

**48.** Graph the following function: $f(x) = \begin{cases} 3x + 2 & \text{if } x \leq 1 \\ -2x + 7 & \text{if } x > 1 \end{cases}$

**49.** Graph the following function: $f(x) = \begin{cases} \frac{1}{3}x - 2 & \text{if } x < -1 \\ 2x - \frac{1}{3} & \text{if } x \geq -1 \end{cases}$

★**50.** Consider the two lines $dx + ey = f$ and $ex - dy = g$. If $de \neq 0$, show that the two lines are perpendicular.

★**51.** Show that the equation of any line with $x$-intercept $a$ and $y$-intercept $b$ can be written in the form $\frac{x}{a} + \frac{y}{b} = 1$.

★**52.** Suppose $f$ is a linear function. Write an equation for $f(x)$ if $f(4x) = 4f(x)$.

★**53.** Suppose $f$ is a linear function. Write an equation for $f(x)$ if $f(x - 1) = f(x) - 1$.

### Calculator Problems

Find the equations of the lines in Problems 54–57. Write the equation in the slope-intercept form.

**54.** A line through (1.623, 1.058) with slope = 0.002.

**55.** A line through (152.61, −137.23) with slope = 5.723.

★**56.** A line that passes through (−1, 2) that is parallel to the line $1.563x - 2.086y = 1.8734$.

★**57.** A line that passes through (3, 1.5) that is perpendicular to the line $0.0026x + 0.0084y = 0.1673$.

## 3-6 COMPOSITE AND INVERSE FUNCTIONS

In using functions in areas of mathematics, science, and business it is frequently necessary to combine two functions to obtain a third function.

The set of all functions, together with operations on this set, forms an algebra of functions. In this section we will discuss a few of the basic ideas of this algebra.

*Definition*

*Combining Functions.* If $f$ and $g$ are functions, then

1. $(f + g)(x) = f(x) + g(x)$ is the sum function.
2. $(f - g)(x) = f(x) - g(x)$ is the difference function.
3. $\left(\frac{f}{g}\right)(x) = \frac{f(x)}{g(x)}$ is the quotient function, if $g(x) \neq 0$.
4. $(f \cdot g)(x) = [f(x)][g(x)]$ is the product function.

Take note of the fact that in this definition the domain of the sum, difference, quotient, or product function would contain only those numbers that are in both the domain of $f$ and in the domain of $g$. The symbol $\cap$, which is read as "intersection," is used to describe elements that are common to two sets. If $D_f$ represents the domain of $f$ and $D_g$ represents the domain of $g$, then $D_f \cap D_g$ would represent the intersection of the two domains. For example, if

$$D_f = \{\text{all real numbers}\}$$

and

$$D_g = \{\text{all positive real numbers}\}$$

then

$$D_f \cap D_g = \{\text{all positive real numbers}\}$$

This intersection is the domain of $f + g$, $f - g$, $\frac{f}{g}$, and $f \cdot g$ with the further stipulation that in the quotient function the denominator $g$ cannot be zero.

**EXAMPLE 1** If $f(x) = 2x - 1$ and $g(x) = 3x + 4$, then find $f + g$, $f - g$, $\frac{f}{g}$, and $f \cdot g$. Give the domain of each.

***Solution***

$$(f + g)(x) = (2x - 1) + (3x + 4)$$
$$= 5x + 3 \qquad \text{domain: } (-\infty, +\infty)$$

$$(f - g)(x) = (2x - 1) - (3x + 4)$$
$$= -x - 5 \qquad \text{domain: } (-\infty, +\infty)$$

$$\left(\frac{f}{g}\right)(x) = \frac{2x - 1}{3x + 4} \qquad \text{domain: } \left(-\infty, -\frac{4}{3}\right) \cup \left(-\frac{4}{3}, +\infty\right)$$

$$(f \cdot g)(x) = (2x - 1)(3x + 4)$$
$$= 6x^2 + 5x - 4 \qquad \text{domain: } (-\infty, +\infty)$$ ■

There is another important way that two functions $f$ and $g$ can be combined. Suppose we want to determine the values obtained by function $f$ and use them as the domain for function $g$. That is, we want to find $g[f(x)]$.

**EXAMPLE 2** If $f(x) = x^2$ and $g(x) = x + 3$, find a formula $h(x)$ that expresses the result of finding $g[f(x)]$.

***Solution***

$$g[f(x)] = g[x^2] \qquad \textit{Substituting } x^2 \textit{ for } f(x).$$
$$= (x^2) + 3 \qquad \textit{Replacing } x^2 \textit{ for } x \textit{ in the formula for } g(x).$$

$$h(x) = x^2 + 3$$

The domain of $h(x)$ is all real numbers. ■

Usually, $f[g(x)] \neq g[f(x)]$ except in certain special cases (which we will study later). We can see an illustration of this by looking again at the functions $f$ and $g$ used in Example 2 and obtaining $f[g(x)]$.

**EXAMPLE 3** If $f(x) = x^2$ and $g(x) = x + 3$, find a formula $j(x)$ that expresses the result of finding $f[g(x)]$.

***Solution***

$$j(x) = f[g(x)]$$

$f[g(x)] = f[x + 3]$ — *Substituting* $x + 3$ for $g(x)$.

$= (x + 3)^2$ — *Replacing* $x + 3$ *for* $x$ *in the formula for* $f(x)$.

$= x^2 + 6x + 9$ — *Multiplying the binomial.*

Therefore $j(x) = x^2 + 6x + 9$

The domain of $j(x)$ is all real numbers. Note that $j(x) \neq h(x)$, so we have an instance in which

$$f[g(x)] \neq g[f(x)] \quad \blacksquare$$

In Examples 2 and 3 we have formed what is known as a *composite function.* In mathematics, rather than writing $f[g(x)]$ when expressing a composite function, we often use the notation $(f \circ g)(x)$. We will state this more formally as follows:

*Definition*

> If $f$ and $g$ are functions, the composition of $f$ and $g$ or the *composite function* denoted by $(f \circ g)(x)$ is defined as $(f \circ g)(x) = f[g(x)]$ for every $x$ in the domain of $g$ for which $g(x)$ is also in the domain of $f$.

The domain of $(f \circ g)$ requires special attention. The definition implies that for $(f \circ g)(x)$ we would substitute $x$ in $g$ and then $g(x)$ in $f$. This means that $g(x)$ must be in the domain of $f$ and $x$ must be in the domain of $g$.

**EXAMPLE 4** If $f(x) = \sqrt{x + 1}$ and $g(x) = 2x - 5$, find $(f \circ g)(x)$ and give its domain.

***Solution***

$$\begin{aligned}(f \circ g)(x) &= f[g(x)] \\ &= f(2x - 5) \\ &= \sqrt{(2x - 5) + 1} \\ &= \sqrt{2x - 4}\end{aligned}$$

The domain of $f$ is $[-1, +\infty)$, and the range of $g$ is all reals. Therefore the domain of $(f \circ g)$ is the set of $x$ such that $g(x) \geq -1$ or $2x - 5 \geq -1$, giving $x \geq 2$. Thus the domain of $(f \circ g)(x)$ is $[2, +\infty)$. $\blacksquare$

**Caution:** The domain of $(f \circ g)$ cannot be determined only from looking at the final equation.

Study the following example carefully.

**EXAMPLE 5** If $f(x) = x^2$ and $g(x) = \sqrt{x + 2}$, find $(f \circ g)(x)$ and give its domain.

***Solution***

$$
\begin{aligned}
(f \circ g)(x) &= f[g(x)] \\
&= f[\sqrt{x + 2}] \\
&= (\sqrt{x + 2})^2 \\
&= x + 2
\end{aligned}
$$

The domain of $f$ is all reals. However, the domain of $g$ is $[-2, \infty)$. Therefore the domain of $(f \circ g)$ is $[-2, \infty)$. ■

### Inverse Functions

We now consider a special type of function. A **one-to-one function** has the property that each element in the domain corresponds to exactly one element in the range and also that the function never assigns the same value to the different values of $x$. This can be stated formally as follows:

*Definition*

A function $f(x)$ is said to be a *one-to-one function* for all real numbers $a$, $b$ if

$$f(a) = f(b) \qquad \text{only when } a = b$$

We can quickly identify a one-to-one function by examining its graph.

If no horizontal line meets the graph of a function in more than one point, then the function is one-to-one.

**EXAMPLE 6** Which of the graphs in Figure 3-26 are one-to-one functions?

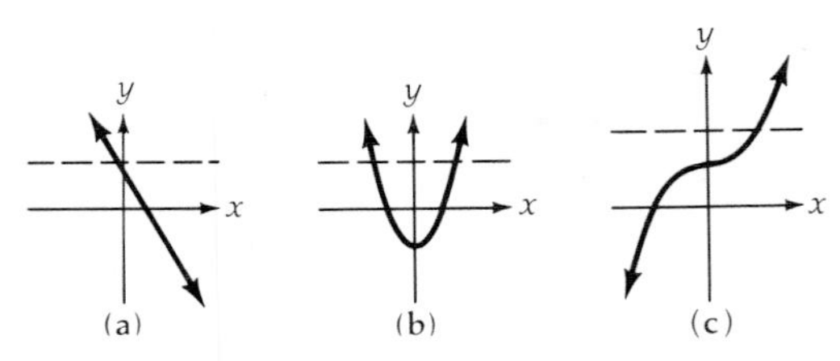


*Figure 3-26*

***Solution*** The graphs pictured in (a) and (c) are one-to-one functions. No horizontal line could meet the graph of the function in more than one point.

The graph pictured in (b) is not a one-to-one function. The horizontal line drawn meets the graph in two points. ■

In general an *increasing* function or a *decreasing* function is always *one-to-one.*

Consider a one-to-one function $f$ that assigns each value in set $X$ to a value in set $Y$. Then consider a function $g$ that assigns each value in set $Y$ back to the original value in set $X$. (See Figure 3-27.)

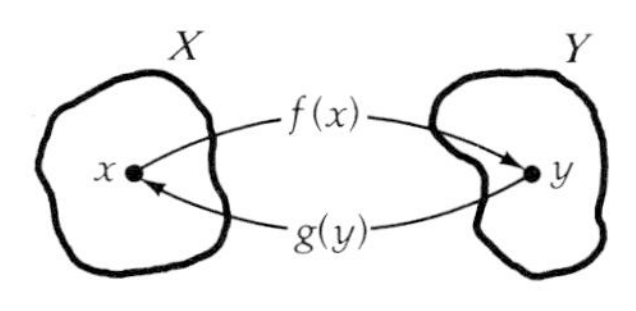


*Figure 3-27*

In this case, $y = f(x)$ for every $x$ in $X$ and $g(y) = x$ for every $y$ in $Y$. These functions $f(x)$ and $g(y)$ are called **inverse functions**.

***Definition***

If $f$ is a one-to-one function with domain $X$ and range $Y$, then the function $g$ with domain $Y$ and range $X$ satisfying

$$g[f(x)] = x \qquad \text{for every } x \text{ in } X$$

and

$$f[g(y)] = y \qquad \text{for every } y \text{ in } Y$$

is called the *inverse function* of $f$.

The notation of the inverse function of $f$ is written $f^{-1}$. Thus we see that

$$f^{-1}[f(x)] = x \qquad \text{for every } x \text{ in } X$$

and

$$f[f^{-1}(y)] = y \qquad \text{for every } y \text{ in } Y$$

**Note:** Do not confuse $f^{-1}(x)$ with exponent notation; $f^{-1}(x)$ does not mean $\frac{1}{f(x)}$. The $-1$ notation is written to indicate the inverse of a function.

To find the inverse $f^{-1}(x)$ of a one-to-one function $f(x)$:

1. Replace $f(x)$ with $y$.
2. Interchange the variables $x$ and $y$.
3. Solve for $y$ in terms of $x$.
4. Replace $y$ with $f^{-1}(x)$.

**EXAMPLE 7**

a. Find the inverse of the one-to-one function $f(x) = 2x - 1$.
b. Verify that your answer is correct by showing that $f^{-1}[f(x)] = f[f^{-1}(x)] = x$.

***Solution***

a.

$$y = 2x - 1 \qquad \textit{Replace } f(x) \textit{ with } y.$$

$$x = 2y - 1 \qquad \textit{Interchange the variables } x \textit{ and } y.$$

$$\frac{x+1}{2} = y \qquad \textit{Solve for } y \textit{ in terms of } x.$$

$$f^{-1}(x) = \frac{x+1}{2} \qquad \textit{Replace } y \textit{ with } f^{-1}(x).$$

b. $$f^{-1}[f(x)] = f^{-1}[(2x-1)] = \frac{(2x-1)+1}{2} = \frac{2x}{2} = x$$

Thus $f^{-1}[f(x)] = x$.

$$f[f^{-1}(x)] = f\left[\frac{x+1}{2}\right] = 2\left(\frac{x+1}{2}\right) - 1 = (x+1) - 1 = x$$

Thus $f[f^{-1}(x)] = x$. ■

Study Example 7 carefully. It will help you to understand the idea of an inverse function. Do you see that if $g[f(x)] = f[g(x)]$ for all $x$, it follows that $g(x) = f^{-1}(x)$?

**EXAMPLE 8**

Let $f(x) = \sqrt[3]{x}$.

a. Does $f^{-1}(x)$ exist?
b. If so, find it.

***Solution***

a. First we obtain a table of values for $f(x) = \sqrt[3]{x}$

| $x$ | $-8$ | $-1$ | 0 | 1 | 8 | 27 |
|---|---|---|---|---|---|---|
| $f(x)$ | $-2$ | $-1$ | 0 | 1 | 2 | 3 |

and plot those ordered pairs.

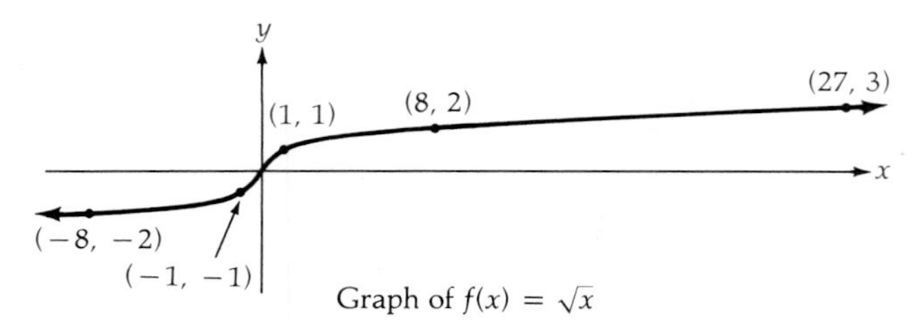


*Figure 3-28*

We see that the function is always increasing and therefore is a one-to-one function. Therefore $f^{-1}(x)$ does exist.

**b.** $y = \sqrt[3]{x}$ *Replace $f(x)$ with $y$.*

$x = \sqrt[3]{y}$ *Interchange the variables $x$ and $y$.*

$x^3 = y$ *Solve for $y$ in terms of $x$.*

$f^{-1}(x) = x^3$ *Replace $y$ with $f^{-1}(x)$.* ■

Sometimes it is necessary to restrict the domain of a function so that it is a one-to-one function in a particular region. The function will then have an inverse function as long as the original function has the appropriate restriction.

**EXAMPLE 9** **a.** Let $f(x) = x^2$ in the interval $[0, \infty)$. Find $f^{-1}(x)$.

**b.** Graph $f(x)$ and $f^{-1}(x)$ on the same axes.

***Solution*** **a.** $y = x^2$ *Replace $f(x)$ with $y$.*

$x = y^2$ *Interchange the variable $x$ and $y$.*

$\pm\sqrt{x} = y$ *Solve for $y$ in terms of $x$.*

Now $y = \pm\sqrt{x}$ is *not* a function in $x$. However, since the original interval for the function was $[0, \infty)$ we know that $x \geq 0$ and therefore $y \geq 0$. We can say that $y = \pm\sqrt{x}$ when $y \geq 0$ is the same as $y = \sqrt{x}$. Thus we have $f^{-1}(x) = \sqrt{x}$.

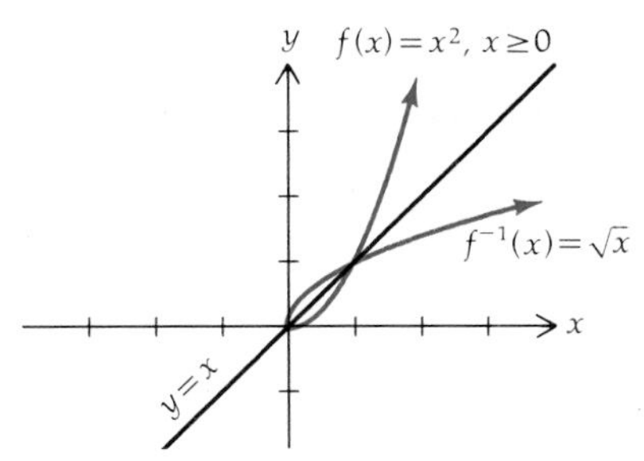


***Figure 3-29***

We see from the graph in Figure 3-29 that $f(x)$ and $f^{-1}(x)$ are reflections about the line $y = x$. If you folded the graph paper on the line $y = x$, the graphs of $f(x)$ and $f^{-1}(x)$ would coincide. We will examine this relationship between a function and its inverse again as we study logarithmic functions. ■

## EXERCISE 3-6

For each of the pairs of functions in Problems 1–4, find

a. $(f+g)(x)$ b. $(f-g)(x)$ c. $(f \cdot g)(x)$ d. $\left(\frac{f}{g}\right)(x)$ e. $\left(\frac{g}{f}\right)(x)$

and, in each case, state the domain.

**1.** $f(x) = x^2$, $g(x) = 3x + 2$
**2.** $f(x) = x^2 + 2x - 15$, $g(x) = x - 3$
**3.** $f(x) = \sqrt{x}$, $g(x) = x^2$
**4.** $f(x) = \sqrt{x+5}$, $g(x) = \sqrt{x-5}$

Find $(f \circ g)(x)$ and $(g \circ f)(x)$ in each of Problems 5–11. State the domain in each case.

**5.** $f(x) = 5x + 1$, $g(x) = 2x - 3$
**6.** $f(x) = \frac{1}{x}$, $g(x) = 3x + 1$
**7.** $f(x) = 2x^2 - x + 1$, $g(x) = 2x + 1$
**8.** $f(x) = \sqrt{x+2}$, $g(x) = 5x - 2$
**9.** $f(x) = \sqrt{3x-1}$, $g(x) = x^2 + 3$
**10.** $f(x) = 8$, $g(x) = 2$
**11.** $f(x) = x^2$, $g(x) = \sqrt{x}$

For each of Problems 12–18:
**a.** Does the equation define a function of $x$?
**b.** Is it a one-to-one function for all $x$ in the domain?
**c.** If the answers to a and b are yes, find the inverse function.

**12.** $y = x + 1$
**13.** $y = 5x - 2$
**14.** $y = x^2 + 1$
**15.** $y = 3x^2 - 4$
**16.** $y = x^3 - 4$
**17.** $y = (x-2)^2 + 1$
**18.** $3x + 2y = 7$

For Problems 19–24:
**a.** Impose any necessary restrictions so that the given function has an inverse function.
**b.** Find the inverse of the given function.

**19.** $f(x) = 2x - 7$
**20.** $f(x) = \frac{1}{2}x + \frac{2}{3}$
**21.** $f(x) = \frac{1}{x}$, $x \neq 0$
**22.** $f(x) = \frac{x+3}{x}$, $x \neq 0$
**23.** $f(x) = x^2 - 3$
**24.** $f(x) = \sqrt{x^2 + 4}$

★ **25.** In Example 9 discussed in this section, could we have used $f(x) = x^2$ and $f^{-1}(x) = -\sqrt{x}$ to obtain inverse functions? Explain.

In Problems 26 and 27, **a.** show that $f[g(x)] = g[f(x)]$ and **b.** sketch the graphs of $f$ and $g$ on the same axis and graph the line of symmetry $y = x$. Do you see that $g(x) = f^{-1}(x)$?

**26.** $f(x) = 2x - 3$
$g(x) = \frac{x+3}{2}$

**27.** $f(x) = x^2 - 5$, $x \geq 0$
$g(x) = \sqrt{x+5}$, $x \geq -5$

## COMPUTER PROBLEMS FOR CHAPTER 3

If you can program a computer, see whether you can write a program to solve Problems 1 and 2.

1. Write a computer program to print out all function values of $f(x)$ where $f(x) = |2x^2 - 4x - 3|$ for all values of $x$ in the set $\{-8.0, -7.8, -7.6, -7.4, \ldots, 7.0, 7.2, 7.4, 7.6, 7.8, 8.0\}$.
2. Write a computer program to print out the function values of $g(x)$ where

$$g(x) = 2.12x^3 + 3.41x^2 - 8.61x + 2.75$$

for all values of $x$ in the set $\{-4.0, -3.9, -3.8, -3.7, \ldots, 3.4, 3.5, 3.6, 3.7, 3.8, 3.9\}$. From the results of your printout, estimate to the nearest tenth the intervals where $f(x)$ is increasing and decreasing.

## KEY TERMS AND CONCEPTS

Be sure you understand and can explain these terms and can use them in solving a problem.

Function
Cartesian coordinate system
The graph of a function
Symmetry
Domain
Range
Even and odd functions
Dependent variable
Independent variable
Decreasing and increasing functions
The vertical line test
Horizontal shifts of graphs
Vertical shifts of graphs
The distance between two points
The midpoint of a line segment
Stretching and shrinking of graphs
Reflection of a graph
Slope
Parallel and perpendicular lines
Standard form of equation of a line
The slope-intercept equation of a line
The point-slope equation of a line
Intercepts
Composite functions
Inverse of a function

## SUMMARY OF PROCEDURES AND CONCEPTS

### Properties Relating to Straight Lines

1. The *distance* between two points $(x_1, y_1)$ and $(x_2, y_2)$ is given by $\sqrt{(x_1 - x_2)^2 + (y_1 - y_2)^2}$.
2. The *midpoint* of a line segment connecting $(x_1, y_1)$ and $(x_2, y_2)$ is given by $\left(\frac{x_1 + x_2}{2}, \frac{y_1 + y_2}{2}\right)$.

3. The *slope* of a line containing the points $(x_1, y_1)$ and $(x_2, y_2)$ is given by $m = \dfrac{y_2 - y_1}{x_2 - x_1}$ where $x_1 \neq x_2$.
4. Two nonvertical *parallel lines* have the same slope $m$.
5. Two nonvertical *perpendicular lines* have slopes that are negative reciprocals.
6. The *equation of a straight line* can be expressed in the following *forms*. (Assume that $m$ is the slope of the line and $(x_1, y_1)$ and $(x_2, y_2)$ are points on the line such that $x_1 \neq x_2$.)
   a. *Standard form:* $ax + by = c$.
   b. *Slope-intercept form:* $y = mx + b$.
   c. *Point-slope form:* $y - y_1 = m(x - x_1)$.
   d. *Two-point form:* $y - y_1 = \dfrac{y_2 - y_1}{x_2 - x_1}(x - x_1)$.

### Basic Steps to Follow in Graphing Equations

1. Determine the domain and (if possible) the range of the equation.
2. Test for symmetry. If the curve is symmetric, half of it can be sketched from the other half.
   a. The graph of an equation is *symmetric with respect to the x-axis* if replacing $y$ by $-y$ produces an equivalent equation.
   b. The graph of an equation is *symmetric with respect to the y-axis* if replacing $x$ by $-x$ produces an equivalent equation.
   c. The graph of an equation is *symmetric with respect to the origin* if the replacing of both $x$ by $-x$ and $y$ by $-y$ produces an equivalent equation.
3. Find where the curve crosses the coordinate axes (if possible).
4. Plot a few points and connect them by a smooth line. Be careful not to plot points where the equation is not defined.

## CHAPTER 3 REVIEW EXERCISE

1. Find the distance between the points $(-9, 3)$ and $(5, -7)$.
2. If the line segment $AB$ has midpoint $(5, -2)$ and the coordinates of $B$ are $(-8, 4)$, find the coordinates of $A$.
3. Find the midpoint of the line joining $(-4, 5)$ and $\left(\dfrac{1}{2}, -6\right)$.
4. Determine algebraically if the following points are collinear: $(-4, 5)$, $(3, 2)$, and $(-9, 7)$.
5. Find the distance between the points whose coordinates are $(3, 2b)$ and $(-5, 4b)$.

Graph each of the equations given in Problems 6–11. Give the domain and range. State whether or not the equation can be expressed as a function of $x$.

**6.** $y = 5x - 2$ **7.** $y = x^2 - 2x + 3$ **8.** $y^2 - 4x = 0$

**9.** $x = 4$ **10.** $y = |x - 2| + 3$ **11.** $y = \sqrt{x^2 - 1}$

**12.** If $f(x) = x^3 - 2x^2 - 3x + 1$, find $f(-1)$, $f(2)$, $f(a)$, and $f(a + b)$.

**13.** If $f(x) = \sqrt{x^2 - 2x}$, find $f(0)$, $f(-1)$, $f(2)$, and $f(a + 1)$.

Find the domain of each of the equations in Problems 14–23. In each case, state whether the equation can be considered a function of $x$. [That is, can we write $y = f(x)$?]

**14.** $y = 2x^2$ **15.** $y = \frac{1}{3}x$ **16.** $x = y^2 + 1$ **17.** $x = \frac{3}{y}$

**18.** $y = \sqrt{4 - x^2}$ **19.** $y = |x^2 - 1|$ **20.** $y = x^5 - 3x^2 + \sqrt{x}$ **21.** $y = \frac{3}{\sqrt{x + 1}}$

**22.** $y = \frac{x + 1}{x - 5}$ **23.** $y = \frac{2}{|x| - 1}$

State whether the functions in Problems 24–27 are *even* or *odd* or neither.

**24.** $f(x) = x^3 - 3x$ **25.** $f(x) = \frac{4x}{x + 1}$ **26.** $f(x) = x^2 + 2x$ **27.** $f(x) = 3x^4 - 2x^2$

For each of the functions in Problems 28–31:

**a.** State whether the function has symmetry. If it does, identify the type of symmetry.
**b.** Graph the function.
**c.** State the interval when the function is increasing or decreasing.

**28.** $f(x) = 3x^2 - 2$ **29.** $f(x) = 2 - 4x^2$ **30.** $f(x) = \sqrt{2x}$ **31.** $f(x) = \sqrt{x - 4}$

**32.** Sketch the graph of the following on the same coordinate axes:

$$f(x) = 3\sqrt{x} \qquad g(x) = 3\sqrt{x} - 2$$

**33.** Sketch the graph of the following on the same coordinate axes:

$$f(x) = x^2 - 3 \qquad g(x) = (x - 1)^2 - 3$$

**34.** Graph $f(x) = \begin{cases} 3 & \text{for } x < 0 \\ 3 - x^2 & \text{for } x \geq 0 \end{cases}$ **35.** Graph $g(x) = \begin{cases} 2x - 3 & x \leq 1 \\ 3 - 4x & x > 1 \end{cases}$

**36.** Find the equation, in standard form, of the line through the point $(2, -9)$ and having a slope of $-\frac{3}{5}$.

**37.** Write the equation $2x - 5y = 10$ in slope-intercept form.

**38.** Find the equation of a line, in standard form, whose slope is 5 and whose $y$-intercept is $-3$.

**39.** Write the equation, in slope-intercept form, of the line passing through the point $(-2, 7)$ that is parallel to the line $3x + 8y = 1$.

**40.** Write the standard form of the equation of a line that passes through $(-1, -2)$ and is perpendicular to the line $5x + y = 3$.

**41.** Find the intercepts of the line $-7x + 3y - 8 = 0$

Given $f(x) = 3x^2 + 2x$ and $g(x) = x - 1$, in Problems 42–46, find the indicated function and give its domain.

**42.** $(f + g)(x)$ **43.** $(f - g)(x)$ **44.** $(f \cdot g)(x)$ **45.** $\left(\frac{f}{g}\right)(x)$ **46.** $(f \circ g)(x)$

Given $f(x) = \dfrac{1}{x + 1}$ and $g(x) = \dfrac{2}{x}$, in Problems 47–51, find the indicated function and give its domain.

**47.** $(f + g)(x)$ **48.** $(f - g)(x)$ **49.** $(f \cdot g)(x)$ **50.** $\left(\frac{f}{g}\right)(x)$ **51.** $(f \circ g)(x)$

**52.** If $f(x) = 2x^2 + 5$ and $g(x) = 3x - 1$, find $f[g(x)]$ and $g[f(x)]$.

**53.** If $f(x) = \sqrt{3x + 1}$ and $g(x) = 3 - x$, find $f[g(x)]$ and $g[f(x)]$.

In Problems 54 and 55, prove that the pairs of functions are inverses of each other by showing that $f[g(x)] = g[f(x)]$.

**54.** $f(x) = 7x + 3, \quad g(x) = \dfrac{x - 3}{7}$ **55.** $f(x) = \sqrt{5x + 1}, \quad x \geq -\dfrac{1}{5}; \quad g(x) = \dfrac{x^2 - 1}{5}, \quad x \geq 0$

## PRACTICE TEST FOR CHAPTER 3

**1.** Give the domain and range of each of the following equations. In each case, state whether the equation is a function. Assume that $x$ is the independent variable.

**a.** $y = \sqrt{2x - 5}$ **b.** $x^2 + y^2 = 9$ **c.** $y = |x - 5|$

**2.** Given the points $A$: $(2, 8)$ and $B$: $(-7, 3)$,

**a.** Find the distance between $A$ and $B$.

**b.** Find the midpoint of $AB$.

**c.** Find the equation, in standard form, of the line that passes through points $A$ and $B$.

**3.** If $f(x) = \begin{cases} x^2 - 2 & x < -1 \\ -1 & -1 \leq x < 2 \\ 3x - 7 & x \geq 2 \end{cases}$

find $f(-3)$, $f(0)$, $f(1)$, $f(3)$ and then graph the function.

Graph the following equations in Problems 4 and 5. Give the domain and range of each and state whether or not the equation can be considered as a function of $x$.

**4.** $y = x^2 - 4$ **5.** $y = \sqrt{x + 2}$

**6.** Graph the function $f(x) = x^3 - 3x + 1$.

**7.** Write the equation $8x - 3y = 12$ in slope-intercept form.

**8.** Given $f(x) = \dfrac{1}{2x - 3}$ and $g(x) = 3x - 1$, find the indicated function and give its domain:

**a.** $(f + g)(x)$ **b.** $\left(\dfrac{f}{g}\right)(x)$ **c.** $(f \circ g)(x)$

**9.** Find the inverse function of each of the following functions:

**a.** $f(x) = 6x - 5$ **b.** $f(x) = \sqrt{x + 4},\ x \geq -4$

**10.** Graph the equation and discuss the symmetry of $x = 2y^2 - 3$. Can this equation determine a function of the form $y = f(x)$?

**11.** Sketch on one coordinate axis $f(x) = (x + 3)^2$, $g(x) = x^2$, and $h(x) = x^2 + 3$.

**12.** Write the equation of a line passing through $\left(\dfrac{1}{2}, 2\right)$ that is perpendicular to $3x + 5y = 15$. Express your answer in the slope-intercept form.

# 4 Polynomial Functions, Rational Functions, and Conic Sections

■ *We are not to imagine or suppose, but to discover, what nature does or may be made to do.*

*Sir Francis Bacon*

Having learned the basic definitions of functions and the general approaches to graphing functions in Chapter 3, we now wish to examine in detail certain specialized functions and certain important equations that are not functions. It is interesting to note that our knowledge of certain types of equations developed as a result of such things as the desire to understand the orbiting of a planet about the sun or the path of an object thrown upward against the pull of gravity. The spirit of discovery alluded to in the quotation from Sir Francis Bacon became a catalyst that helped speed the knowledge of certain functions and other equations and the manner in which they can be used to model the patterns observed in nature.

## 4-1 QUADRATIC FUNCTIONS

If $n$ is a nonnegative integer and $a_n, a_{n-1}, a_{n-2}, \ldots, a_1, a_0$ are real numbers, then a function $f$ is called a **polynomial function** in $x$ if

$$f(x) = a_nx^n + a_{n-1}x^{n-1} + a_{n-2}x^{n-2} + \cdots + a_1x + a_0$$

If the degree of the polynomial is one, we have a linear function, which was covered extensively in Section 3-5. If the polynomial is of degree two, then we obtain a **quadratic function**. The definition is usually stated thus:

*Definition*

A function $f$ is called a *quadratic function* if $f(x) = ax^2 + bx + c$ where $a$, $b$, $c$ are real numbers and $a \neq 0$.

The graph of a quadratic function is a **parabola**. We previously graphed parabolas in Sections 3-2 and 3-4. The parabola opens upward if $a > 0$ and downward if $a < 0$. The lowest or highest point on the parabola is called the **vertex**. The graph of a quadratic function is always a parabola that is symmetric to the $y$-axis or another line parallel to the $y$-axis. The line to which the parabola is symmetric is known as the axis of symmetry.

In Section 3-4 we studied vertical and horizontal shifts of graphs. For example assume that $h, k$ are positive real numbers. If we know the shape of the graph of $f(x)$, then we can obtain the graph of $f(x - h)$ by shifting the graph $h$ units to the right. If we have the graph of $f(x)$, to obtain the graph $f(x) + k$ we shift the graph $k$ units upward.

Thus if we have a graph of a parabola defined by $f(x) = x^2$, then the graph of $f(x) = (x - h)^2 + k$ has the same general shape, but the vertex is $h$ units to the right and $k$ units above the vertex of $f(x) = x^2$. This is illustrated in Figure 4-1.

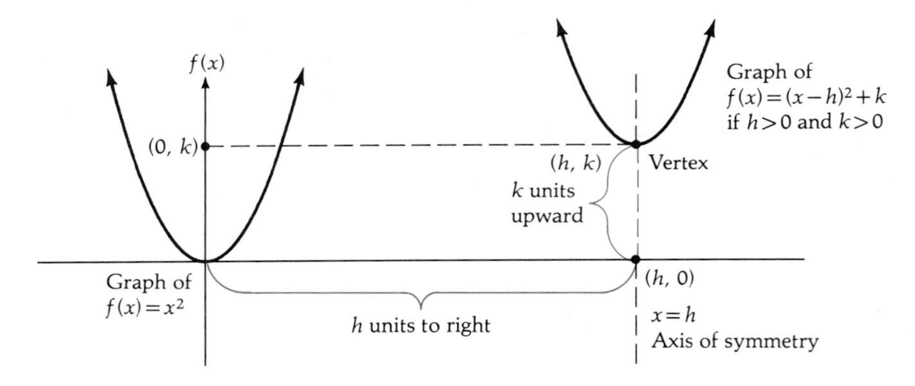


*Figure 4-1*

We do not need to restrict $h$ and $k$ to positive values. In a more general case we could say for *any real numbers h, k* that the graph of $f(x) = a(x - h)^2 + k$ can be obtained by shifting the graph of the parabola $f(x) = ax^2$ by $|h|$ units in a horizontal direction (right if $h$ is positive, left if $h$ is negative) and by $|k|$ units in a vertical direction (up if $k$ is positive, down if $k$ is negative).

The intercepts of the quadratic function $f(x) = ax^2 + bx + c$ are useful for graphing. If $x = 0$, then $f(0) = a(0)^2 + b(0) + c$ so $f(0) = c$. The $y$-intercept of the quadratic function is $c$. By solving the equation $f(x) = 0$, we obtain the $x$-intercepts (if any exist). These can be obtained by factoring the quadratic or using the quadratic formula to obtain the roots. A quadratic function may have two real roots, one real root, or no real roots. Therefore there are three possibilities for $x$-intercepts. This is illustrated in Figure 4-2. However, the *quadratic*

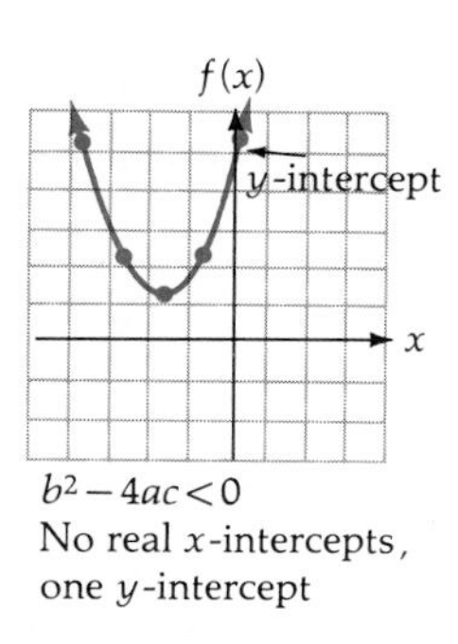


$b^2 - 4ac < 0$
No real x-intercepts,
one y-intercept

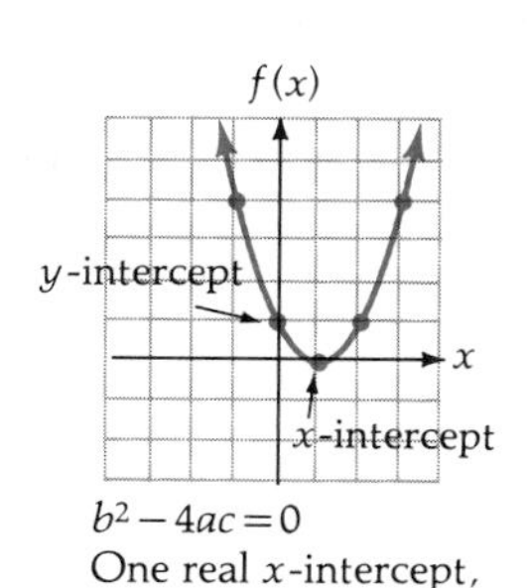


$b^2 - 4ac = 0$
One real x-intercept,
one y-intercept

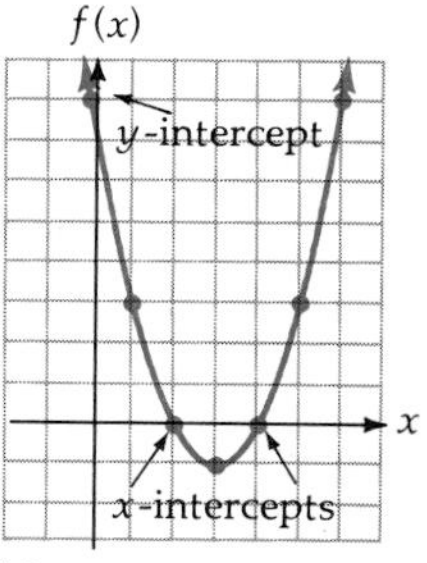


$b^2 - 4ac > 0$
Two real x-intercepts,
one y-intercept

*Figure 4-2*

*function will always have one y-intercept,* since $f(0) = c$ for all quadratic functions. Graphically, you can always extend the parabola until it touches the $y$-axis.

Be sure you understand the terms *roots, x-intercepts,* and *zeros. Roots* are solutions to an equation. The *x-intercepts* are those values of $x$ where an equation crosses the $x$-axis. *Zeros* of a function are those values of the independent variable that make the function value zero.

When we graphed parabolas in Sections 3-2 and 3-4, we found it necessary to plot six to eight points in order to have a reasonably accurate graph. However, we are now ready to summarize our findings so that quadratic functions may be graphed more rapidly.

---

**Procedure for Graphing Quadratic Functions:**
$$f(x) = ax^2 + bx + c$$

---

**1.** Determine the direction of the parabola.

If $a > 0$, it opens upward; if $a < 0$, it opens downward.

**2.** Plot the $y$-intercept. Since $f(0) = c$, the $y$-intercept is at $(0, c)$.

**3.** Plot any $x$-intercepts. The roots may be found by the quadratic formula or by factoring the equation $f(x) = 0$.

If $b^2 - 4ac > 0$, the function has two real roots.
If $b^2 - 4ac = 0$, the function has one real root.
If $b^2 - 4ac < 0$, the function has no real roots.

**4.** Plot the vertex. It is located at $\left[\dfrac{-b}{2a}, f\left(\dfrac{-b}{2a}\right)\right]$

5. Draw the axis of symmetry. It is the line $x = \frac{-b}{2a}$.
6. If necessary, plot a couple of additional points on one side of the axis of symmetry. Use the property of symmetry to find corresponding points on opposite side of $x = \frac{-b}{2a}$.

---

EXAMPLE 1 Graph the quadratic function $f(x) = x^2 + 2x - 8$.

***Solution***

1. $a = 1$. Since $a > 0$, the parabola opens upward.
2. The $y$-intercept is $-8$.
3. We set $f(x) = 0$ to find $x$-intercepts.

$$x^2 + 2x - 8 = 0$$

$$(x + 4)(x - 2) = 0$$

The $x$-intercepts are $x = -4$ and $x = 2$.

4. The vertex is at $\left[\frac{-b}{2a}, f\left(\frac{-b}{2a}\right)\right]$

$$\frac{-b}{2a} = \frac{-2}{2(1)} = -1$$

$$f(-1) = (-1)^2 + 2(-1) - 8 = 1 - 2 - 8 = -9$$

Thus the vertex is at $(-1, -9)$.

5. The axis of symmetry is $x = -1$.

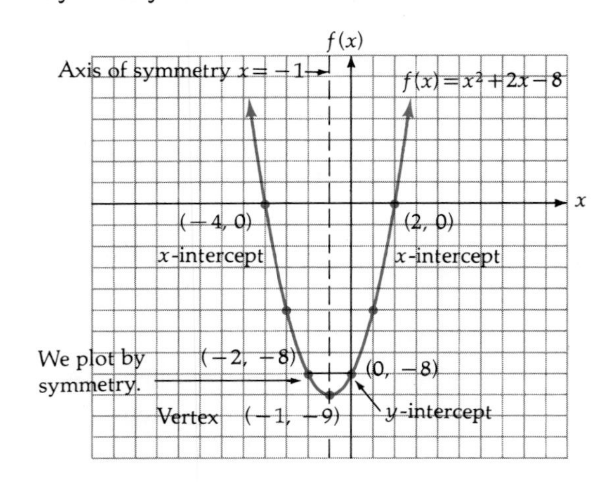


*Figure 4-3*

6. We can plot the point $(-2, -8)$ by symmetry, since it is on the opposite side of $x = -1$ from the $y$-intercept at $(0, -8)$. No further points are needed. (See Figure 4-3.) ■

**EXAMPLE 2** Graph the quadratic function $f(x) = -2x^2 + 4x + 1$.

***Solution***

1. $a = -2$. Since $a < 0$, the parabola opens downward.
2. The $y$-intercept is 1.
3. We cannot factor $f(x)$, so we use the quadratic formula:

$$x = \frac{-4 \pm \sqrt{16 - 4(-2)(1)}}{2(-2)} = \frac{-4 \pm \sqrt{16 + 8}}{-4}$$

$$= \frac{-4 \pm \sqrt{24}}{-4} = \frac{-4 \pm 2\sqrt{6}}{-4} = \frac{-2 \pm \sqrt{6}}{-2}$$

It will aid us in graphing to approximate these two roots to the nearest tenth. Thus $x \doteq -0.2$ and $x \doteq 2.2$.

4. $\dfrac{-b}{2a} = \dfrac{-4}{2(-2)} = +1; \qquad f(1) = -2 + 4 + 1 = 3.$

Thus the vertex is at $\left[\dfrac{-b}{2a}, f\left(\dfrac{-b}{2a}\right)\right] = (1, 3).$

5. The axis of symmetry is $x = 1$.
6. If $x = 3$, then $f(3) = -2(3)^2 + 4(3) + 1 = -18 + 12 + 1 = -5$.

Thus one point on the curve is $(3, -5)$. By symmetry we know that another point is $(-1, -5)$. The graph is shown in Figure 4-4. ■

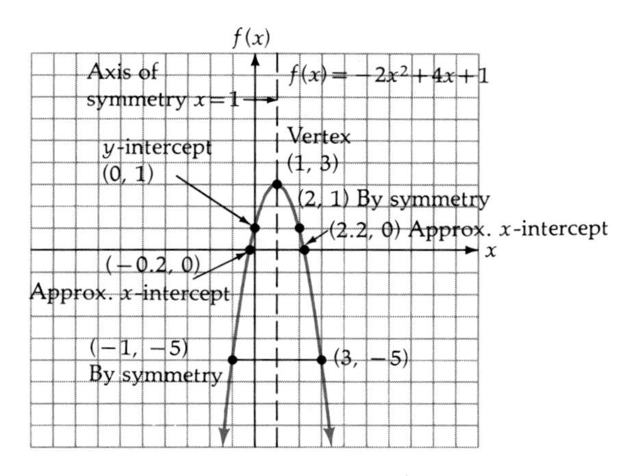


*Figure 4-4*

### Maximum and Minimum Problems

For a parabola opening upward, the vertex is the lowest point; therefore for any quadratic function $f(x) = ax^2 + bx + c$ where $a > 0$, $f\left(\frac{-b}{2a}\right) \leq f(x)$ for all $x$.

Similarly, for a parabola opening downward the vertex is the highest point. Thus for any quadratic function where $a < 0$, $f\left(\frac{-b}{2a}\right) \geq f(x)$ for all $x$. These facts yield the following definition:

*Definition*

For the quadratic function $f(x) = ax^2 + bx + c$:

If $a > 0$, then $f\left(\frac{-b}{2a}\right)$ is the *minimum* value of the function and there is no maximum value.

If $a < 0$, then $f\left(\frac{-b}{2a}\right)$ is the *maximum* value of the function and there is no minimum value.

In calculus, more general methods of finding minimum and maximum values of functions are developed. However, even the methods of this section are very useful. Since quadratic functions often describe area, volume, profit, cost, velocity, and distance, the techniques of this section provide us with a tool to find maximum or minimum values for many of these applied situations.

### Applications

EXAMPLE 3 A farmer wishes to fence in a rectangular grazing area. He has 400 meters of fencing available. One side of the rectangle will border a river, so he will not need to fence in that side. What dimensions should he use in order to maximize the grazing area?

***Solution*** Let us call the width of the rectangle $x$. Then $x + \text{length} + x = 400$, since there are 400 meters of fencing. Thus length $= 400 - 2x$. Refer to Figure 4-5.

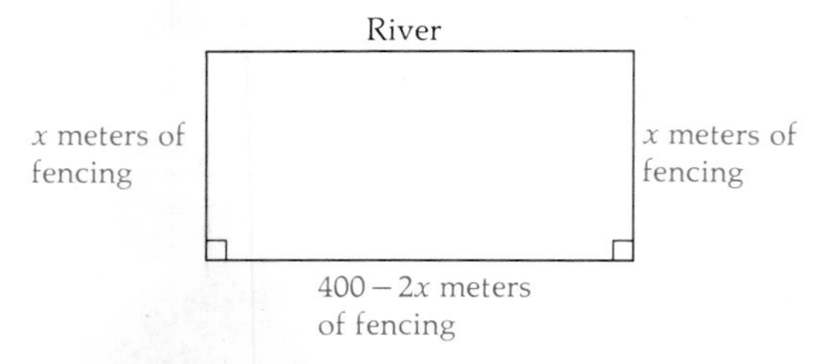


*Figure 4-5*

We wish to maximize the area.

$$A = (\text{length})(\text{width})$$
$$A = (400 - 2x)(x) = 400x - 2x^2$$

Here the area function is a function with $x$ as the independent variable.

$$A(x) = -2x^2 + 400x$$
$$a = -2, \qquad b = 400, \qquad \text{and} \qquad c = 0$$

By definition, since $a < 0$, $f\left(\frac{-b}{2a}\right)$ is the maximum area. In other words, the maximum area will take place when $x = \frac{-b}{2a}$:

$$x = \frac{-b}{2a} = \frac{-400}{2(-2)} = \frac{-400}{-4} = 100$$

Thus when the width is 100 meters, the maximum area will result. If the width is 100, the length $= 400 - 2x = 400 - 2(100) = 200$. The dimensions required to achieve maximum grazing area are a width of 100 meters (sides perpendicular to the river) and a length of 200 meters (side parallel to the river). ■

When a ball is thrown upward with initial velocity $v_0$ in meters per second, its height $S$ above the ground is given approximately by the quadratic equation $S = -4.9t^2 + v_0 t + h$ where $t$ is the time measured in seconds from the moment the ball is thrown and $h$ is the starting height measured in meters.

**EXAMPLE 4** Find the maximum height achieved by a baseball thrown upward from a building at a point 126 meters high with an initial velocity of 25 meters per second. Find the number of seconds it takes to achieve this height. Round your answers to nearest tenth.

***Solution***

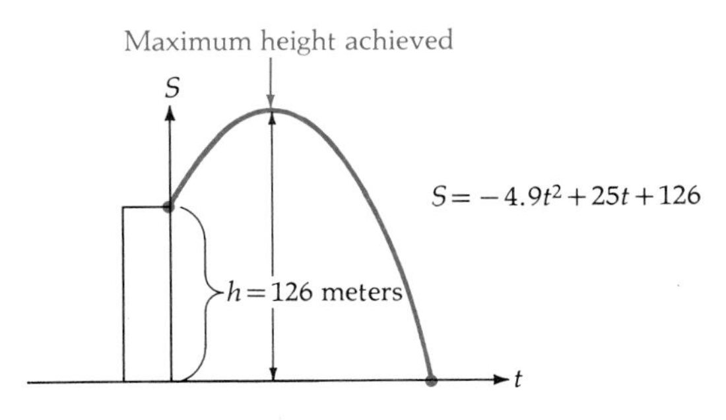


*Figure 4-6*

Referring to Figure 4-6, since $v_0 = 25$ and $h = 126$, we have the quadratic function $S = -4.9t^2 + 25t + 126$. Maximum height will be achieved when $t = \frac{-b}{2a}$. Thus

$$t = \frac{-25}{2(-4.9)} = \frac{-25}{-9.8} = 2.551020408$$

*To the nearest tenth of a second we have $t = 2.6$ seconds.*

The maximum height will be approximated by $f(2.6)$:

$$f(2.6) = -4.9(2.6)^2 + 25(2.6) + 126$$

This may be accomplished directly on a scientific calculator:

4.9 [+/−] [×] 2.6 [$x^2$] [+] 25 [×] 2.6 [+] 126 [=] 157.876

*To the nearest tenth the maximum height is 157.9 meters.* ■

## EXERCISE 4-1

Graph the quadratic functions in Problems 1–20. Label intercepts and the vertex.

**1.** $f(x) = x^2 + 3$
**2.** $f(x) = x^2 - 2$
**3.** $f(x) = x^2 - 6x$
**4.** $f(x) = 2x^2 + x$
**5.** $f(x) = x^2 + 7x + 12$
**6.** $f(x) = x^2 - 5x + 6$
**7.** $f(x) = 2x^2 - 12x + 9$
**8.** $f(x) = 3x^2 - 6x + 2$
**9.** $f(x) = -x^2 + 6x - 5$
**10.** $f(x) = -x^2 + 10x - 15$
**11.** $f(x) = -3x^2 - 6x + 2$
**12.** $f(x) = -2x^2 + 8x - 3$
**13.** $f(x) = \frac{1}{2}x^2 - 7x + 20$
**14.** $f(x) = \frac{1}{2}x^2 + 2x - 2$
**15.** $f(x) = 2x^2 - 6x - 3$
**16.** $f(x) = 3x^2 - 10x - 2$
**17.** $f(x) = \frac{x^2 + 2x - 3}{2}$
**18.** $f(x) = x - \frac{2}{3}x^2 - \frac{1}{3}$
**19.** $f(x) = 1 - 4x - 3x^2$
**20.** $f(x) = 5 - 8x + 2x^2$

**21.** A farmer wishes to fence in a rectangular grazing area. He has 160 meters of fencing available. One side of the rectangle will be bordered by a barn, so no fencing is needed on that side. What dimensions should he use in order to maximize the grazing area?

**22.** An elementary school principal wants to enclose a small rectangular play area with 120 meters of fencing. One side of the rectangle will be the back wall on the school. What dimensions should she use to maximize the play area?

**23.** Two numbers have a sum of 36. The sum of the squares of those numbers is as small as possible. What are the numbers?

**24.** Two numbers have a sum of 100. Their product is as large as possible. What are the numbers?

**25.** A rectangular plot of land has a perimeter of 600 yards. A man stated that the area in the plot of land is the maximum possible with that perimeter. If he is correct, what are the dimensions of the rectangle?

**26.** The cost to manufacture electronic test kits is given by the equation $c(x) = \frac{1}{2}x^2 - 10x + 200$ where $x$ is the number of items produced in one day. What value of $x$ will result in the minimum cost?

**27.** The profit achieved by a local company is given by $P(x) = 240 + 96x - 2x^2$ where $x$ is the number of items produced in a given day. How many items should be produced each day to maximize profits?

**★28.** Frank and William have a car-waxing service just outside the college campus. They charge $10.00 to wax any car. They average 120 customers per week. They have tried charging different prices. For every 50¢ increase in price they average four fewer customers per week. What price should they charge to maximize their profits?

**★29.** When the university gymnastics team played home games, the admission charge was $4.00 per person and the average attendance was 300 students per game. The Athletic Director tried different admission rates. She found that for each decrease in price of 20¢ the number of students coming to see the game increased by 30. What admission price should she charge to maximize revenue?

## Calculator Problems

Graph the functions in Problems 30 and 31. Use a scientific calculator.

**★30.** $f(x) = 2.3x^2 - 4.6x + 5.8$ **★31.** $f(x) = -1.1x^2 + 5.7x - 6.1$

In Problems 32 and 33, use $S = -4.9t^2 + v_0t + h$ and refer to Example 4. Round each answer to the nearest tenth. Use a scientific calculator.

**32.** Find the maximum height achieved by a projectile fired upward at a rate of 143 meters per second from a silo that is 100 meters below the ground surface. Find the number of seconds it takes to achieve this height.

**33.** Find the maximum height achieved by a projectile fired upward at a rate of 120 meters per second from a mountain peak 1260 meters above level ground. Find the number of seconds it takes to achieve this height.

**34.** Find the dimensions of the largest rectangle that can be inscribed in the right triangle shown in Figure 4-7. (The figure is not drawn to scale.)

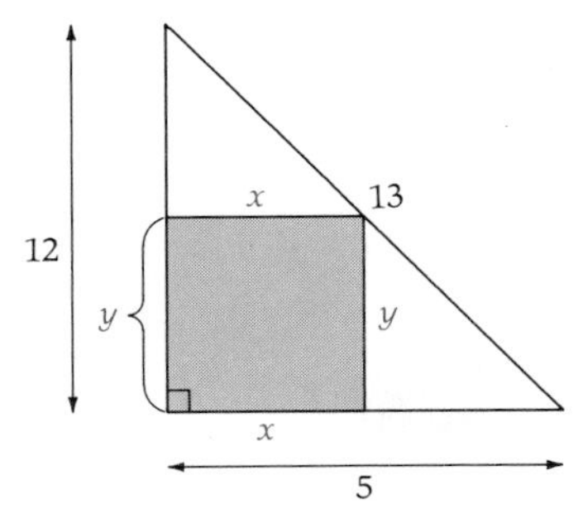


*Figure 4-7*

## 4-2 GRAPHS OF HIGHER-DEGREE POLYNOMIAL FUNCTIONS

The graphs of polynomial functions of degree 3 or higher are more complicated than quadratic functions. We will study some involved procedures to determine useful facts about the polynomial in Chapter 11. However, a few basic facts are helpful at this point in sketching the polynomial. Some of the properties can be observed by looking at the rough sketch of polynomials of specific degrees.

### Cubic Function

Figures 4-8, 4-9, 4-10, and 4-11 illustrate some representative graphs of the **cubic function** $f(x) = ax^3 + bx^2 + cx + d$ where $a \neq 0$.

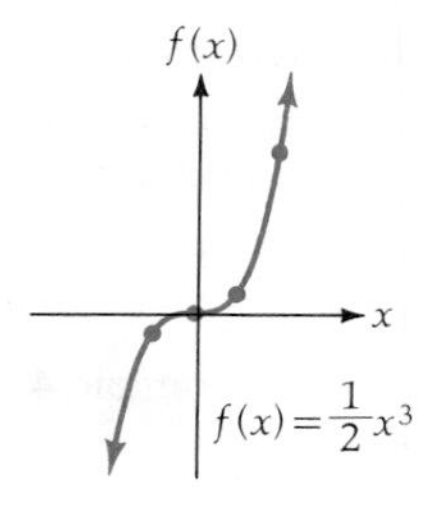


*Figure 4-8*

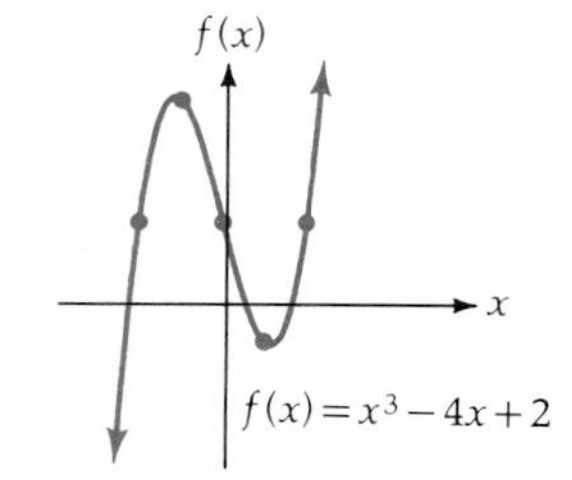


*Figure 4-9*

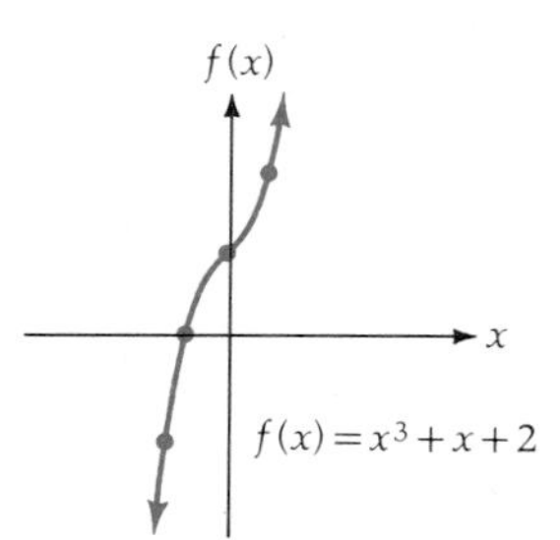


*Figure 4-10*

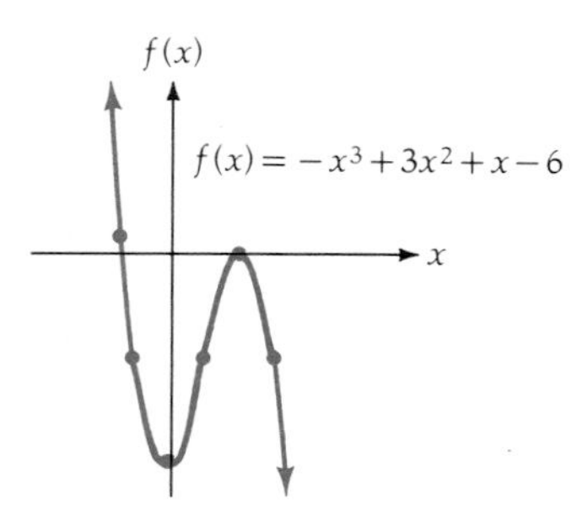


*Figure 4-11*

### Fourth-Degree Polynomial Function

Figures 4-12, 4-13, 4-14, and 4-15 illustrate some representative graphs of the fourth-degree polynomial function $f(x) = ax^4 + bx^3 + cx^2 + dx + e$ where $a \neq 0$.

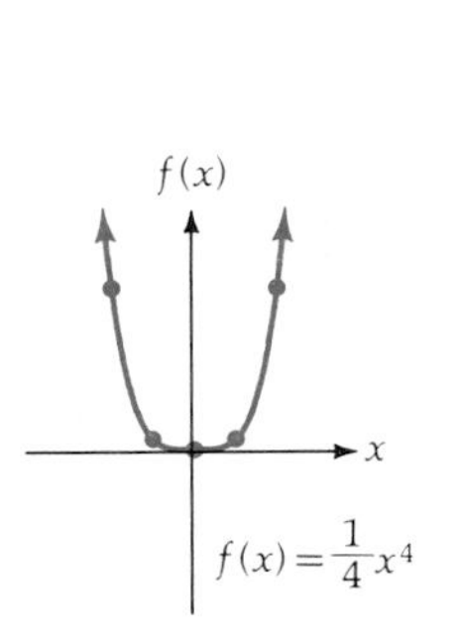


*Figure 4-12*

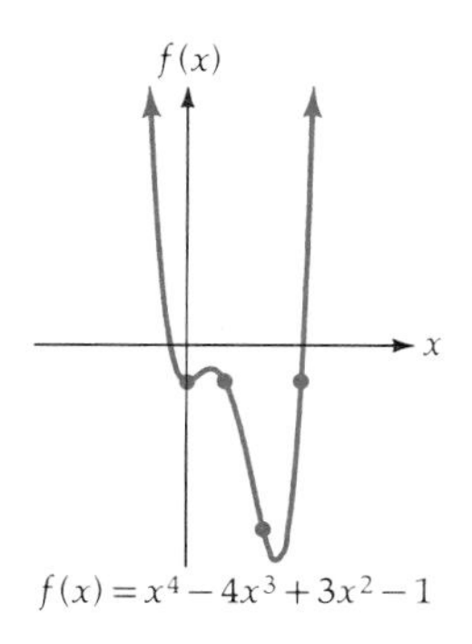


*Figure 4-13*

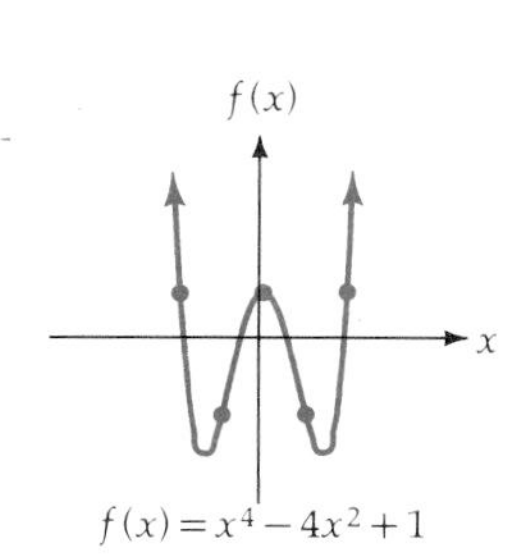


*Figure 4-14*

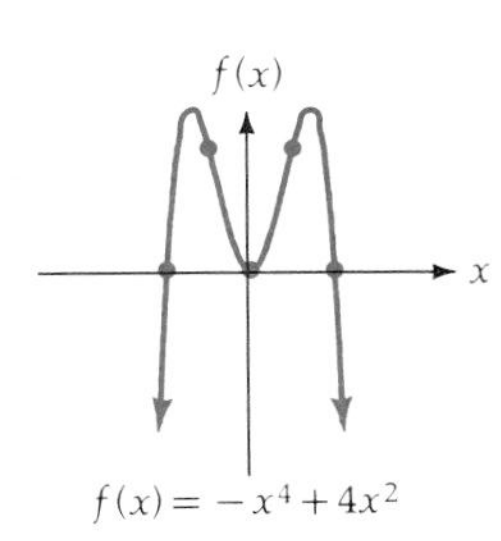


*Figure 4-15*

The graph of a polynomial is a smooth continuous curve.

From examining Figures 4-8, 4-9, 4-10, and 4-11 we observe that the cubic function can have $f(x) = 0$ at one, two, or three places. Thus it would appear that the cubic polynomial of the form $f(x) = 0$ can have one, two, or three real roots.

From examining Figures 4-12, 4-13, 4-14, and 4-15 we observe that the fourth-degree polynomial function can have $f(x) = 0$ at one, two, three, or four places. Clearly, if we graphed $f(x) = \frac{1}{4}x^4 + 1$, we would see a curve that was raised vertically 1 unit from Figure 4-12 and did not cross the $x$-axis at all. Thus we would assume that the fourth-degree polynomial of the form $f(x) = 0$ can have zero, one, two, three, or four real roots.

In a higher-level mathematics course, procedures are developed that allow the proof of the following theorem:

***Theorem 4-1***

If $P(x) = a_nx^n + a_{n-1}x^{n-1} + a_{n-2}x^{n-2} + \cdots + a_1x + a_0$ is a polynomial of degree $n$:

1. Then the graph of any $P(x)$ is a smooth curve that has at most $n - 1$ *turning points.*
2. The graph of any $P(x)$ crosses the $x$-axis no more than $n$ times. Thus the equation $P(x) = 0$ has at most $n$ real roots.
3. If $P(x)$ is of *odd* degree, it will cross the $x$-axis at least once. Thus if $P(x)$ is of *odd* degree, the equation $P(x) = 0$ has at least *one* real root.

It is quite difficult to graph most polynomial functions without the use of the skills developed in calculus. In fact, the formal concept of *turning point* depends on the idea of a relative maximum or minimum. We will *informally* use the idea of turning point as a turn of the curve to a different direction from upward to downward or downward to upward. If the polynomial function can be factored, then we can make a rough sketch without plotting a large number of points. It is wise to factor the polynomial (whenever possible) before you graph it.

**EXAMPLE 1** Make a rough sketch of the polynomial function $f(x) = x^4 - 9x^2$.

***Solution*** We factor to obtain $x^2(x^2 - 9)$, which we further factor as $x^2(x + 3)(x - 3)$. Thus $f(x) = 0$ when $x$ has the following values:

$$x^2 = 0 \qquad x + 3 = 0 \qquad x - 3 = 0$$

so

$$x = 0 \qquad x = -3 \qquad x = 3$$

Suppose we would like to find two points between (0, 0) and (3, 0). Let us try $x = 1$ and $x = 2$.

$$f(x) = x^4 - 9x^2$$

$$f(1) = 1^4 - 9(1)^2 = 1 - 9 = -8$$

$$f(2) = 2^4 - 9(2)^2 = 16 - 9(4) = 16 - 36 = -20$$

Now $f(x)$ is an *even function,* since the only powers of $x$ are even powers. Therefore the function is symmetric to the $y$-axis. By symmetry we have the values $f(-1) = -8$ and $f(-2) = -20$.

It will be helpful to use a different scale on the $y$-axis, since the magnitude of the values for $f(x)$ are significantly larger. The polynomial is of degree 4, and can have at most three turning points. From our graph in Figure 4-16 the approximate values of the turning points are at (0, 0), $(-2, -20)$, and $(2, -20)$. ■

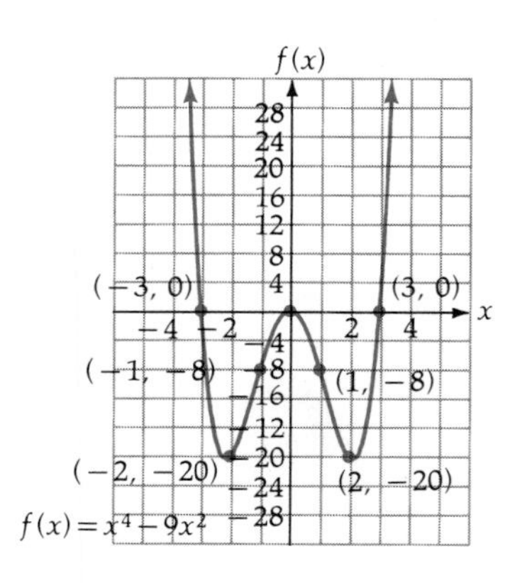


*Figure 4-16*

To determine the possible factors of some polynomials, we will need the skills developed in Chapter 11. However, by using the method of grouping, it is possible to factor several polynomials that do not immediately appear to be factorable.

**EXAMPLE 2** Make a rough sketch of the polynomial function $f(x) = x^3 + 5x^2 - x - 5$.

***Solution*** We may factor $f(x)$ by grouping.

$f(x) = (x^3 + 5x^2) - (x + 5)$ *Separate into two groups.*

$= x^2(x + 5) - 1(x + 5)$ *Remove a common factor in each of the two groups.*

$= (x + 5)(x^2 - 1)$ *Remove the common factor of* $(x + 5)$.

$= (x + 5)(x + 1)(x - 1)$ *Factor by the difference of two squares.*

Thus $f(x) = 0$ when $x$ has the following values:

$$x + 5 = 0 \qquad x + 1 = 0 \qquad x - 1 = 0$$

$$x = -5 \qquad x = -1 \qquad x = 1$$

We wish to find $f(x)$ for a value of $x$ in the interval $-5 < x < -1$ and another in the interval $-1 < x < 1$. Let us use $x = -3$ and $x = 0$.

$$f(x) = x^3 + 5x^2 - x - 5$$

$$f(-3) = (-3)^3 + 5(-3)^2 - (-3) - 5 = -27 + 5(9) + 3 - 5$$

$$= -27 + 45 + 3 - 5 = 16$$

$$f(0) = (0)^3 + 5(0)^2 - (0) - 5 = -5$$

A rough sketch of this polynomial is given in Figure 4-17. We can approximate from the graph that the turning points are at $(-3, 16)$ and $(0, -5)$. ■

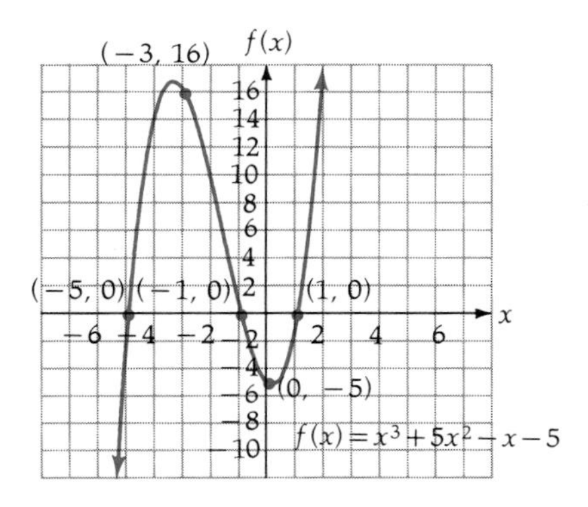


*Figure 4-17*

**Caution:** In general we will not always be able to find the *turning points* without methods developed in calculus. The major benefit of Theorem 4-1, paragraph 1, is that we know *the maximum number* of turning points of any polynomial. There may be fewer turning points, and we cannot always find them without more advanced methods. In most cases we only *approximate* the locations of the turning points.

## EXERCISE 4-2

In Problems 1–18, plot four to six points and make a rough sketch of each of the following polynomials. Choose a convenient scale on each axis.

**1.** $f(x) = \frac{1}{3}x^3$ **2.** $f(x) = \frac{x^3}{5}$ **3.** $f(x) = -2x^3$

**4.** $f(x) = -3x^3$ **5.** $f(x) = \frac{1}{2}x^4$ **6.** $f(x) = \frac{1}{3}x^4$

**7.** $f(x) = 2 - x^4$ **8.** $f(x) = 3 - \frac{1}{2}x^4$ **9.** $f(x) = x^3 + 5$

**10.** $f(x) = x^3 + 2x$ **11.** $f(x) = \frac{1}{6}x^5$ **12.** $f(x) = \frac{1}{8}x^5$

**13.** $f(x) = -\frac{1}{2}x^5 + 1$ **14.** $f(x) = -\frac{1}{4}x^5 + 4$ **15.** $f(x) = \frac{1}{2}(x - 3)^3$

**16.** $f(x) = \frac{1}{3}(x + 2)^2$ **17.** $f(x) = (x + 2)^4 - 2$ **18.** $f(x) = (x - 1)^4 + 1$

In Problems 19–36, find the roots of the polynomial. (Factor the polynomial if necessary to find the roots.) Then use the $x$-intercepts as well as a few other points to make a rough sketch of the polynomial.

**19.** $f(x) = x(x + 2)(x - 2)$ **20.** $f(x) = x(x + 3)(x - 1)$

**21.** $f(x) = x(x - 4)(x - 2)$ **22.** $f(x) = x(x - 2)(x - 1)$

**23.** $f(x) = (x + 3)(x + 1)(x - 2)$ **24.** $f(x) = (x - 4)(x - 2)(x + 1)$

**25.** $f(x) = -2x(x - 2)^2(x + 1)$ **26.** $f(x) = -2x(x + 3)^2(x - 1)$

**27.** $f(x) = 9x - x^3$ **28.** $f(x) = -x^3 - x^2 + 2x$

**29.** $f(x) = -4x^3 - 2x^2 + 12x$ **30.** $f(x) = x^3 - 4x^2 + 4x$

**31.** $f(x) = x^3 + x^2 - 4x - 4$ **32.** $f(x) = x^3 + 3x^2 - x - 3$

**★33.** $f(x) = (x^2 - 9)(x^2 + 3x + 2)$ **★34.** $f(x) = (x^2 - 4)(x^2 + 4x + 4)$

**★35.** $f(x) = \frac{(2x + 1)(x - 2)^2}{-2}$ **★36.** $f(x) = \frac{(3x - 2)(x - 3)^2}{-2}$

### Calculator Problems

With the aid of a scientific calculator graph each of the polynomial equations in Problems 37 and 38 for $\{x: -2 \leqslant x \leqslant 3\}$. Use an appropriate scale for your axes.

**★37.** $f(x) = 1.6x^3 - 3.4x^2 + 3.6$ **★38.** $f(x) = 1.2x^3 - 2.9x - 3.8$

In Problems 39 and 40, use your calculator to approximate where the zero is for the specified function $g(x)$.

**★39.** $g(x) = -2x^3 + x^2 + 2$. Show that $g(x)$ has a zero between $x = 1$ and $x = 2$ by evaluating $f(x)$ for the values $x = 1.0, 1.1, 1.2, 1.3, \ldots, 1.9, 2.0$.

**★40.** $g(x) = x^3 - x^2 - 2x + 3$. Show that $g(x)$ has a zero between $x = -2$ and $x = -1$ by evaluating $f(x)$ for the values $x = -2.0, -1.9, -1.8, \ldots, -1.1, -1.0$.

---

## 4-3 RATIONAL FUNCTIONS

Rational functions are formed by dividing one polynomial by another polynomial. We define it more carefully as follows:

*Definition*

If $g(x)$ and $h(x)$ are polynomials without common factors and $h(x) \neq 0$, then $f(x) = \dfrac{g(x)}{h(x)}$ is called a *rational function* of $x$.

Any values of $x$ for which $h(x) = 0$ are excluded from the domain of the rational function. Thus a rational function is not in general continuous but may have one or more breaks in it.

EXAMPLE 1 Graph the rational function $f(x) = \dfrac{4}{x}$.

***Solution*** We plot several points in order to see the general behavior of the curve.

| $x$ | $\frac{1}{2}$ | 1 | 2 | 4 | 8 |
|---|---|---|---|---|---|
| $f(x)$ | 8 | 4 | 2 | 1 | $\frac{1}{2}$ |

On the basis of our discussion in Section 3-2 we see that this curve is symmetric to the origin, since if we replace $x$ by $-x$ and $f(x)$ by $-f(x)$ we obtain an equivalent equation. Thus we immediately have the following points:

| $x$ | $-\frac{1}{2}$ | $-1$ | $-2$ | $-4$ | $-8$ |
|---|---|---|---|---|---|
| $f(x)$ | $-8$ | $-4$ | $-2$ | $-1$ | $-\frac{1}{2}$ |

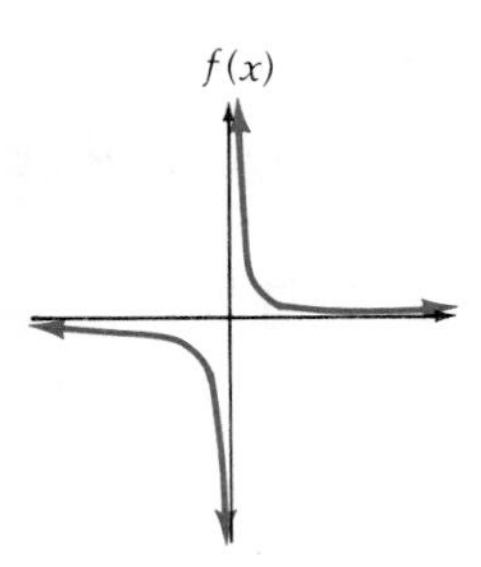


*Figure 4-18*

Now we sketch the graph of $f(x)$ in Figure 4-18.

We observe that $f(x)$ is not defined where $x = 0$. Therefore we do not have a value on our graph for $(0, f(0))$, since the function does not exist at that point.

If we plotted additional points, we would also observe that the curve gets very close to the lines $x = 0$ and $y = 0$ but never touches those lines. When a curve gets closer and closer to a line but does not touch it, we say that the line is an **asymptote** of the curve. We shall study two types of asymptotes.

1. **Vertical asymptotes:** asymptotic lines that are either the $y$-axis or parallel to the $y$-axis.
2. **Horizontal asymptotes:** asymptotic lines that are either the $x$-axis or parallel to the $x$-axis.

The graph of $f(x) = \frac{4}{x}$ has a horizontal asymptote of the $x$-axis and a vertical asymptote of the $y$-axis. This can be easily seen in Figure 4-18. ■

The general properties of graphing functions in Section 3-4 using vertical shifts of graphs and horizontal shifts of graphs are useful in graphing rational functions.

**EXAMPLE 2** Graph the rational function $k(x) = \dfrac{4}{x - 2} + 3$.

***Solution*** We observe that the graph of $k(x)$ has been shifted 2 units to the right of the graph of $f(x)$ in Example 1. We also observe the graph of $k(x)$ has been shifted 3 units upward from the graph of $f(x)$. The sketch of $k(x)$ is given in Figure 4-19.

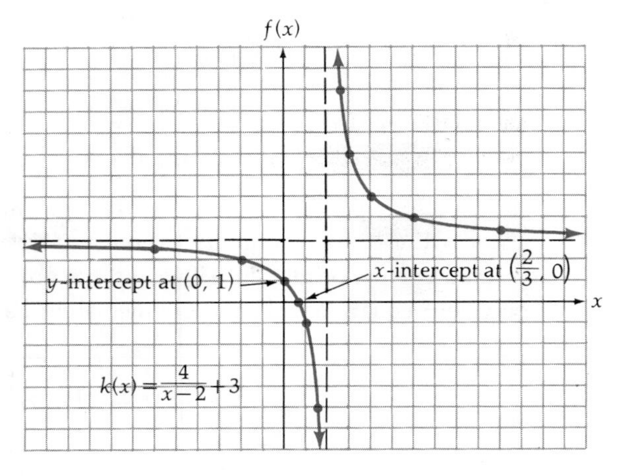


*Figure 4-19*

*Note*: Although $k(x)$ is a rational function, it is not written in the form $= \frac{g(x)}{h(x)}$. However, it could be readily written in that form by simple algebraic operations:

$$k(x) = \frac{4}{x-2} + 3 = \frac{4}{x-2} + \frac{3(x-2)}{x-2} = \frac{4+3x-6}{x-2} = \frac{3x-2}{x-2}.$$

So an alternative form for $k(x)$ that indicates one polynomial divided by another is $k(x) = \frac{3x-2}{x-2}$.

Here we observe that $k(x)$ has a vertical asymptote at $x = 2$ and a horizontal asymptote at $y = 3$. We note that $k(x)$ is not defined at $x = 2$. The left branch of $k(x)$ has a $y$-intercept at $y = 1$. It also has an $x$-intercept. If $k(x) = 0$, then $0 = \frac{4}{x-2} + 3$. Solving for $x$ we have:

$$-3 = \frac{4}{x-2}$$

$$-3(x-2) = 4 \qquad \textit{As long as } x \neq 2.$$

$$-3x + 6 = 4$$

$$-3x = -2$$

$$x = \frac{2}{3}$$

Since $k\left(\frac{2}{3}\right) = 0$, we have an $x$-intercept at $x = \frac{2}{3}$. ■

Symmetry about the $y$-axis or a line parallel to the $y$-axis is a useful property in graphing certain rational functions.

**EXAMPLE 3** Graph the rational function $f(x) = \frac{-0.5}{x^2}$.

***Solution*** We obtain a few function values for positive values of $x$.

| $x$ | $\frac{1}{4}$ | $\frac{1}{2}$ | 1 | 2 | 3 |
|---|---|---|---|---|---|
| $f(x)$ | $-8$ | $-2$ | $-0.5$ | $-0.125$ | $-0.055\ldots$ |

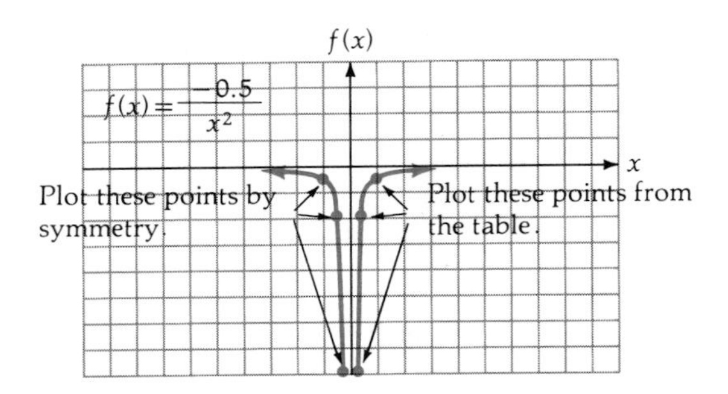


*Figure 4-20*

Since $f(-x) = f(x)$, we have an *even function* symmetrical to the $y$-axis. The curve is sketched in Figure 4-20. The curve has a vertical asymptote of the $y$-axis and a horizontal asymptote of the $x$-axis. ■

All of the rational functions sketched so far have been elementary. Other rational functions become more difficult to sketch. It will be helpful to establish an organized procedure so that we can identify the location of vertical and horizontal asymptotes.

Vertical asymptotes have occurred at points $x = a$ where that value of $x$ makes the polynomial in the denominator become zero. The proof involves a higher level of mathematics, but we shall state it as a theorem.

*Theorem 4-2*

If $f(x) = \dfrac{g(x)}{h(x)}$ is a rational function, then $f(x)$ will have a vertical asymptote at $x = a$ if and only if $h(x)$, the polynomial in the denominator, has a zero at $x = a$.

A rational function can have at most one horizontal asymptote. If the degree of the numerator of the rational function is less than or equal to the degree of the denominator, the horizontal asymptote can be found quickly by the following theorem:

*Theorem 4-3*

Let $f(x)$ be a rational function of the form

$$f(x) = \frac{a_nx^n + a_{n-1}x^{n-1} + a_{n-2}x^{n-2} + \cdots + a_1x + a_0}{b_mx^m + b_{m-1}x^{m-1} + b_{m-2}x^{m-2} + \cdots + b_1x + b_0}$$

Then if $n < m$, the line $y = 0$ is a *horizontal asymptote*. If $n = m$, the line $y = \dfrac{a_n}{b_m}$ is a *horizontal asymptote.*

The addition of these two theorems gives us several tools to quickly sketch the graph of a rational function. It is suggested that a procedure such as the following be followed when sketching rational functions.

---

**Procedure to Sketch Rational Functions**

---

1. Examine the function for symmetry.
2. Solve to find any intercepts.
3. Locate any horizontal asymptotes by Theorem 4-3.
4. Locate any vertical asymptotes by Theorem 4-2.
5. Plot one or two points in each of the intervals into which the $x$-intercepts and the vertical asymptotes divide the axis.
6. Sketch the curve.

---

We will employ this procedure in the following example.

EXAMPLE 4 Graph the rational function $f(x) = \dfrac{2x^2 + 2}{x^2 - 4}$.

***Solution***

1. Since $f(x) = f(-x)$, we know that $f(x)$ is an even function and the curve is *symmetric to the y-axis.*

2. Intercepts: if $x = 0$, then $f(0) = \dfrac{2(0)^2 + 2}{0 - 4} = \dfrac{2}{-4} = -\dfrac{1}{2}$ and thus the $y$-intercept is $-\dfrac{1}{2}$. To find the $x$-intercept, if one exists, we know $f(x)$ must be zero. Thus the only possibility is if

$$2x^2 + 2 = 0$$
$$2x^2 = -2$$
$$x^2 = -1$$

but this has no real roots, so there is no $x$-intercept.

3. By Theorem 4-3 we see that since the degree of $2x^2 + 2$ is equal to the degree of $x^2 - 4$, the horizontal asymptote is at

$$y = \frac{a_n}{b_m} = \frac{2}{1} = 2$$

4. By Theorem 4-2 any vertical asymptotes will occur if $x^2 - 4 = 0$, which yields

$$x^2 = 4$$

$$x = \pm 2$$

So there are two vertical asymptotes. One is $x = 2$ and the other $x = -2$.

5. We need to plot points only for $x \geq 0$. We can complete the graph by symmetry. We obtain a few function values (rounded to the nearest hundredth).

| $x$ | 0.5 | 1 | 1.5 | 2.5 | 3 | 6 |
|---|---|---|---|---|---|---|
| $f(x)$ | $-0.67$ | $-1.33$ | $-3.71$ | 6.44 | 4 | 2.31 |

6. The graph of the curve is shown in Figure 4-21. ■

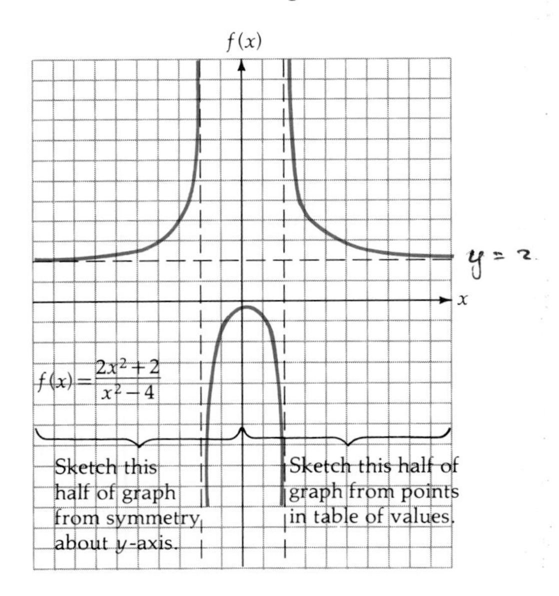


*Figure 4-21*

## EXERCISE 4-3

Graph the rational functions in Problems 1–22. Include any properties of symmetry, the intercepts, and the existence of horizontal or vertical asymptotes.

1. $f(x) = \dfrac{2}{x - 2}$ 2. $f(x) = \dfrac{3}{x + 1}$ 3. $g(x) = \dfrac{-4}{x + 3}$

**4.** $g(x) = \dfrac{-2}{x-4}$ **5.** $h(x) = \dfrac{2}{x} + 4$ **6.** $h(x) = \dfrac{3}{x-1} + 2$

**7.** $f(x) = \dfrac{1}{x^2 - 5x + 6}$ **8.** $f(x) = \dfrac{1}{x^2 - 2x - 8}$ **9.** $f(x) = \dfrac{2}{(x-3)^2}$

**10.** $f(x) = \dfrac{3}{(x+2)^2}$ **11.** $f(x) = \dfrac{x+2}{x-3}$ **12.** $f(x) = \dfrac{x-2}{x+4}$

**13.** $g(x) = \dfrac{2x}{x^2 - 16}$ **14.** $g(x) = \dfrac{4x}{x^2 - 9}$ **15.** $h(x) = \dfrac{x+1}{x^2 - 1}$

**16.** $h(x) = \dfrac{2x^2}{4 - x^2}$ **17.** $f(x) = \dfrac{2x^2}{x^2 - 3x}$ **18.** $f(x) = \dfrac{2x+1}{x^2 + x}$

**★19.** $f(x) = \dfrac{x^2 + 1}{2x^2 - x - 3}$ **★20.** $f(x) = \dfrac{x^2}{x^2 - x - 2}$ **★21.** $g(x) = \dfrac{1}{x^3 + x^2 - 6x}$

**★22.** $g(x) = \dfrac{1}{x^3 + 2x^2 - 3x}$

### Calculator Problems

**23.** Wein's Law of measuring temperatures of astronomical bodies is given by the equation $w = \dfrac{0.00290}{T}$ where the wavelength $w$ is the maximum amount for each temperature $T$. The temperature $T$ is measured in degrees Kelvin. Plot a graph for a temperature range of 4000–7000°K (which would correspond to solar bodies with a temperature in the range of our own sun). Label your axes with a convenient scale.

**24.** Ceramic tiles are used to make a new type of tile stove that retains heat longer than a conventional cast iron stove. The heat loss from a test piece of tile is measured in BTU's (British thermal units). The time period is measured in hours. The theoretical equation is $H = \dfrac{5000t}{t^2 + 5}$. Graph the heat loss function for $0 \leq t < 10$ hours.

## 4-4 THE CIRCLE AND THE ELLIPSE

The curves formed by second-degree equations in two variables are the **circle**, **ellipse**, **hyperbola**, and **parabola**. These curves are studied in detail in the subject of analytic geometry; but because their equations are also important in algebra, we will briefly discuss them in Sections 4-4 to 4-6.

The geometric interpretation of these curves was studied extensively by the ancient Greeks. In the third century B.C. a famous Greek by the name of Apollonius wrote eight treatises about the intersection of a plane and a right

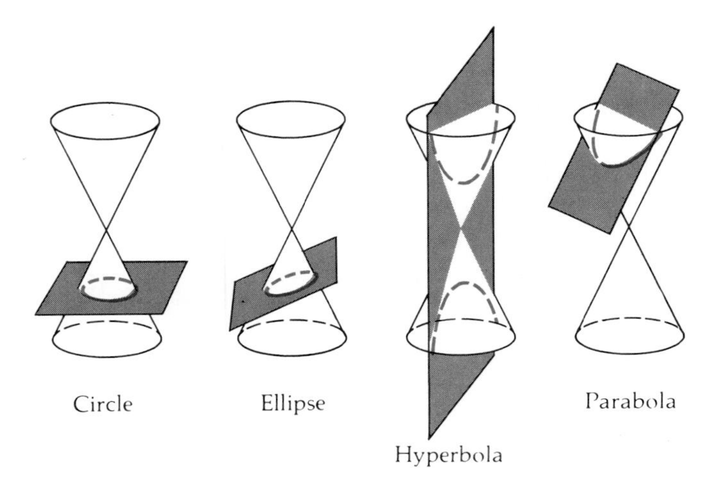


*Figure 4-22*

circular cone. The curves that result from these intersections have borne the name **conics** up to this present time and are shown in Figure 4-22.

A powerful discovery in mathematics was that the equation for all of the conics could be obtained by considering special cases of the general second-degree equation in two unknowns $Ax^2 + By^2 + Cx + Dy + E = 0$. The first case we shall consider is that of a circle. If $A = B \neq 0$, we obtain the equation of a circle. We will shortly see why this is so. But first we must study some important properties of the circle.

The essential definition commonly used for a circle is the following:

***Definition***

A *circle* is the set of all points in a plane that are a given distance from a fixed point. The given distance is the length of any line segment from the fixed point (called the center) to a point on the circle. All of these line segments are called radii (singular radius). Often $r$ is used to represent the radius, and the center is called $(h, k)$.

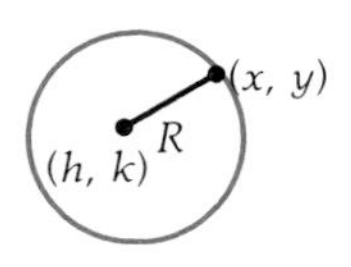


*Figure 4-23*

Let us consider a circle of radius $r$ with its center at the point $(h, k)$, as shown in Figure 4-23.

Using this definition and letting $(x, y)$ represent any point in the set so described, the distance formula gives

$$r = \sqrt{(x - h)^2 + (y - k)^2} \qquad \text{or} \qquad (x - h)^2 + (y - k)^2 = r^2$$

*Definition*

The *standard form* of the *equation of a circle* having center $(h, k)$ and radius $r$ is given by

$$(x - h)^2 + (y - k)^2 = r^2$$

The graph of a circle is easy to sketch when we know its center and radius.

EXAMPLE 1

Sketch the graph of a circle with center $(2, -1)$ and a radius of 3. Write its equation in standard form.

***Solution***

We can easily sketch the circle. Refer to Figure 4-24.
The equation in standard form is

$$(x - 2)^2 + [y - (-1)]^2 = 3^2$$

or

$$(x - 2)^2 + (y + 1)^2 = 9 \quad ■$$

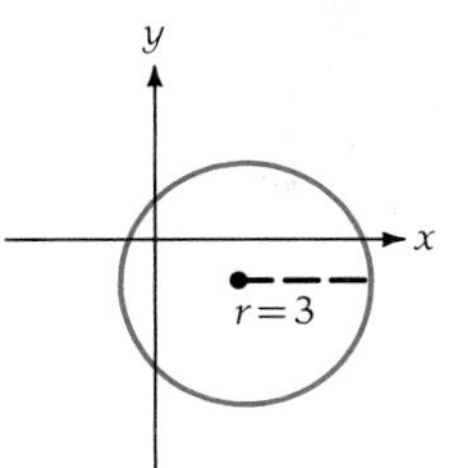


*Figure 4-24*

If we have the equation of a circle in a different form, we perform algebraic steps to place it in standard form before we attempt to complete the sketch.

EXAMPLE 2

$x^2 + y^2 + 4x + 8y = -4$ is the equation of a circle. Find its center and radius, and sketch its graph.

***Solution***

First we wish to put the equation in standard form. To do this, we complete the square on each part by adding the correct number to both sides of the equation.

$$x^2 + 4x + \_\_ + y^2 + 8y + \_\_ = -4$$

$$x^2 + 4x + 4 + y^2 + 8y + 16 = -4 + 4 + 16$$

or

$$(x + 2)^2 + (y + 4)^2 = 16$$

In this form we see that the center is $(-2, -4)$ and the radius is 4. We now sketch the graph, as shown in Figure 4-25.

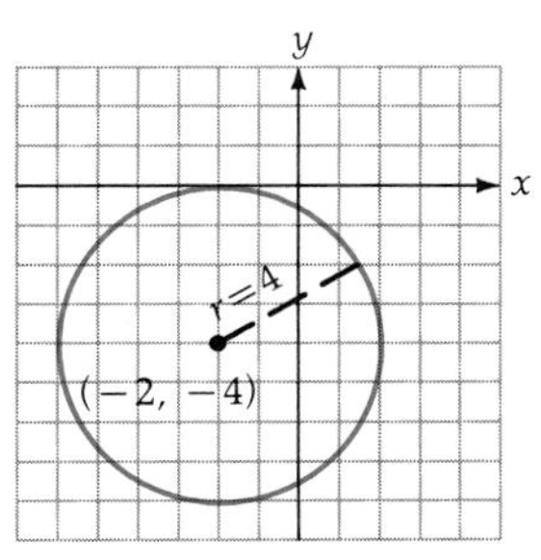


*Figure 4-25*

The domain is $[-6, 2]$, and the range is $[-8, 0]$. Is this a function? ■

On the basis of a careful examination of Example 2 we can conclude that if $A = B$ in the equation $Ax^2 + By^2 + Cx + Dy + E = 0$, the equation represents a circle. Do you see why this is so?

### The Ellipse

In the equation $Ax^2 + By^2 + Cx + Dy + E = 0$ if $A \neq B$ and both $A > 0$ and $B > 0$, we obtain an ellipse. An ellipse looks like a "flattened" circle. The study of the ellipse is quite important. For example, all of the planets and even some comets orbit the sun in elliptical orbits. (See Figure 4-26.)

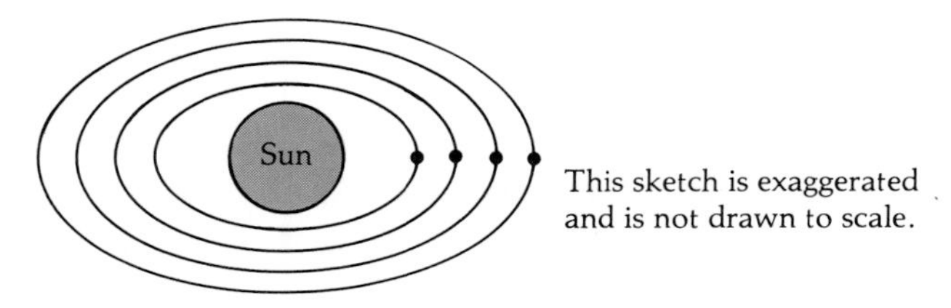
Sun
This sketch is exaggerated
and is not drawn to scale.

*Figure 4-26*

We will now state the classical Greek definition of an ellipse.

*Definition*

> An *ellipse* is the set of all points in a plane, each such that the sum of its distances from two fixed points, called *foci* (singular: focus), is a constant.

To obtain the equation of an ellipse, we place the ellipse on the coordinate plane so that the center of the ellipse is at $(h, k)$. We will place the foci on a line parallel to the $x$-axis.

We shall call the distance from the center to the foci "$c$"; thus the coordinates of the foci are $(h - c, k)$ and $(h + c, k)$. If we now pick any point on the ellipse and apply the definition, we see from Figure 4-27 that the sum of its distances to the two foci must be a constant. We shall call this constant $2a$. The definition thus states

$$\sqrt{[x - (h - c)]^2 + (y - k)^2} + \sqrt{[x - (h + c)]^2 + (y - k)^2} = 2a$$

It is left as an exercise to show that this equation becomes

$$\frac{(x - h)^2}{a^2} + \frac{(y - k)^2}{a^2 - c^2} = 1$$

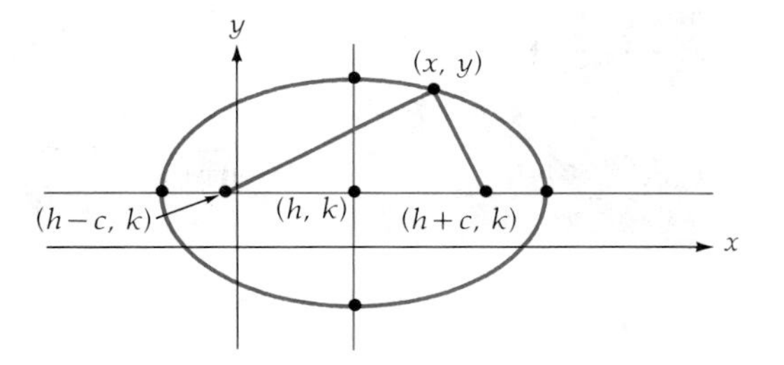


*Figure 4-27*

If we let $a^2 - c^2 = b^2$, then the equation can be stated as

$$\frac{(x-h)^2}{a^2} + \frac{(y-k)^2}{b^2} = 1$$

***Definition***

The standard form of the equation of an ellipse whose center is $(h, k)$ is given by

$$\frac{(x-h)^2}{a^2} + \frac{(y-k)^2}{b^2} = 1$$

The graph of an ellipse in standard form may appear in two different shapes, as seen in Figure 4-28.

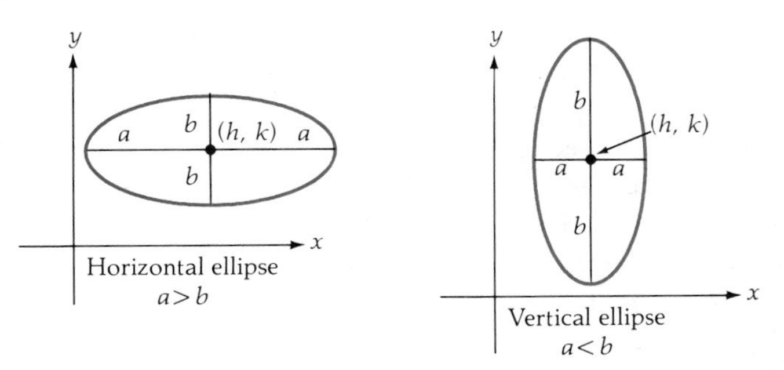


*Figure 4-28*

In each case, $2a$ is the length of the horizontal axis, and $2b$ is the length of the vertical axis. Do you see why this is so?

Now, one important characteristic differs in the two cases. If $a > b$, the ellipse is called a **horizontal ellipse** because its foci are on the horizontal axis. In this case, $a^2 - c^2 = b^2$. If $a < b$, we have a **vertical ellipse**, and the foci are on the vertical axis. In this case, $a^2 + c^2 = b^2$. The reason for this is left as an exercise.

If we know $a$ and $b$ for any given ellipse, we can readily find the distance $c$ by solving the appropriate equation. The value of $c$ is always the distance of each focus from the center of the ellipse at $(h, k)$. Refer to Figure 4-29.

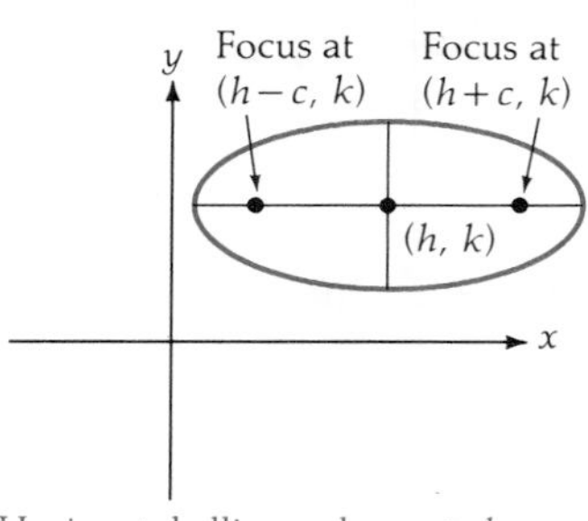


Horizontal ellipse where $a > b$

Since $a^2 - c^2 = b^2$, $a^2 - b^2 = c^2$

To obtain $c$, we have $c = \sqrt{a^2 - b^2}$

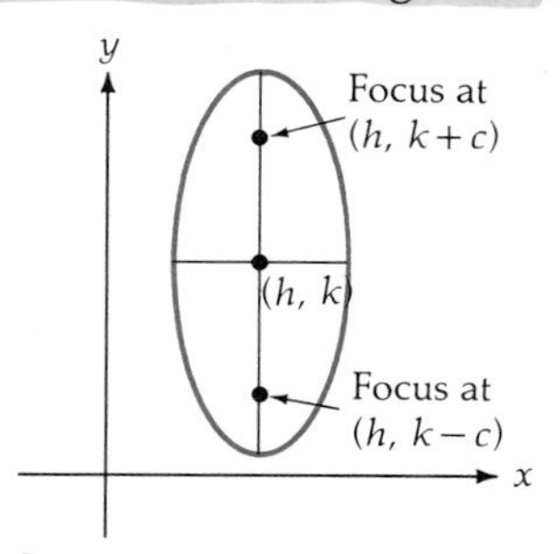


Vertical ellipse where $a < b$

Since $a^2 + c^2 = b^2$, $c^2 = b^2 - a^2$

To obtain $c$, we have $c = \sqrt{b^2 - a^2}$

*Figure 4-29*

The longer axis of an ellipse is referred to as the **major axis**. The shorter axis is referred to as the **minor axis**.

If the center of the ellipse is at (0, 0), then the equation simplifies to

$$\frac{x^2}{a^2} + \frac{y^2}{b^2} = 1.$$

**EXAMPLE 3** Sketch the graph of $\dfrac{x^2}{9} + \dfrac{y^2}{4} = 1$.

***Solution*** We see that the center is at (0, 0), since $h = 0$ and $k = 0$. If $a^2 = 9$, then $a = 3$; and if $b^2 = 4$, then $b = 2$. (Remember $a$ and $b$ are always positive.)

Since $a > b$, we have a horizontal ellipse. We know that the major axis is the $x$-axis. The $x$-intercepts will be 3 units to the right and left of (0, 0). The $y$-intercepts will be 2 units above and below (0, 0).

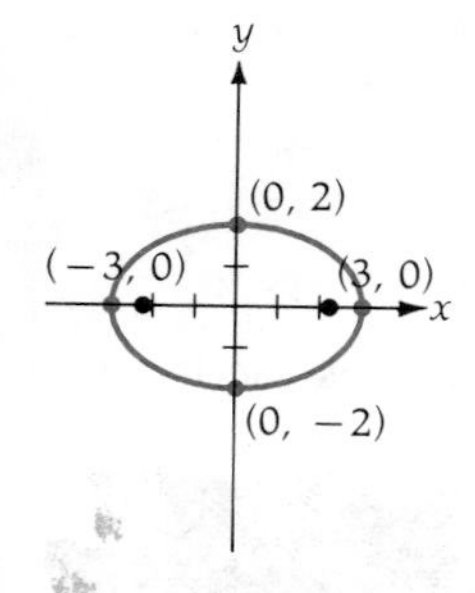


*Figure 4-30*

For a horizontal ellipse, $a^2 - c^2 = b^2$

Solving for $c^2$, we have $a^2 - b^2 = c^2$

Since $a = 3$ and $b = 2$, $9 - 4 = c^2$

$$\sqrt{5} = c$$

The foci are $c$ units from (0, 0). Thus we have foci at $(-\sqrt{5}, 0)$ and $(\sqrt{5}, 0)$, as shown in Figure 4-30. ■

The values of $a$, $b$ are not always rational numbers.

EXAMPLE 4 Sketch the graph of $6x^2 + 4y^2 = 24$ and label the foci.

***Solution*** We divide each side of the equation by 24 to place the equation in *standard form*:

$$\frac{6x^2}{24} + \frac{4y^2}{24} = \frac{24}{24}$$

$$\frac{x^2}{4} + \frac{y^2}{6} = 1 \qquad \textit{Simplify.}$$

Since $a^2 = 4$, $a = 2$; and since $b^2 = 6$, $b = \sqrt{6}$. (To assist in our sketch we approximate $b \doteq 2.45$.)

Since $b > a$, we have a vertical ellipse. Since $(h, k) = (0, 0)$, the center is at the origin. The foci will be $(0, c)$ and $(0, -c)$.

Since for a vertical ellipse $a^2 + c^2 = b^2$, then $c = \sqrt{2}$.

Thus the foci are at $(0, \sqrt{2})$, $(0, -\sqrt{2})$. To aid our graph in Figure 4-31, we approximate these values to obtain $(0, 1.41)$, $(0, -1.41)$. ■

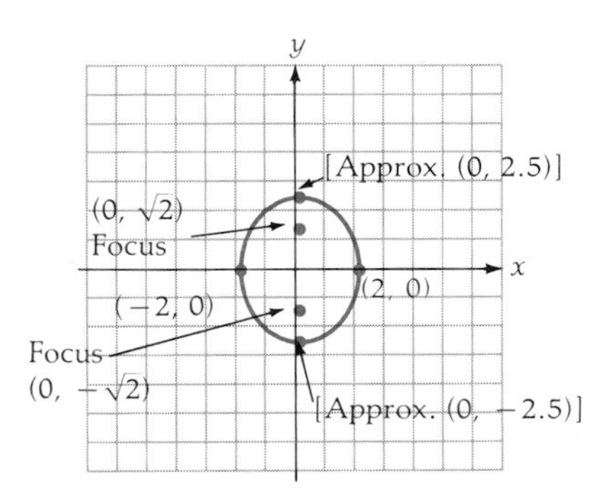


Graph of $6x^2 + 4y^2 = 24$

*Figure 4-31*

Some equations will involve several steps of algebra to obtain the desired standard form. Observe carefully the following example.

EXAMPLE 5 $9x^2 + 18x + 16y^2 - 64y = 71$ is the equation of an ellipse. Find its center, foci, and vertices (endpoints of the axes). Sketch the graph.

***Solution*** We first wish to get the equation in standard form. To do this we factor 9 from the $x$ terms and 16 from the $y$ terms.

$$9(x^2 + 2x + \_\_) + 16(y^2 - 4y + \_\_) = 71$$

We next complete the squares.

$$9(x^2 + 2x + 1) + 16(y^2 - 4y + 4) = 71 + 9 + 64$$

$$9(x + 1)^2 + 16(y - 2)^2 = 144$$

or

$$\frac{(x + 1)^2}{16} + \frac{(y - 2)^2}{9} = 1$$

*Be careful here. Do you see why we added 9 and 64 to the right-hand side?*

The equation is now in standard form, and we see that the center is $(-1, 2)$. We also note that it is a horizontal ellipse, since $a = 4$ is greater than $b = 3$.

Since it is a horizontal ellipse,

$$a^2 - c^2 = b^2$$

or

$$\begin{aligned} c^2 &= a^2 - b^2 \\ &= 16 - 9 \\ &= 7 \end{aligned}$$

Thus

$$c = \sqrt{7}$$

Since the foci are on the horizontal axis, we add and subtract the $c$ value to the $x$-coordinate of the center point. This yields $(-1 \pm \sqrt{7}, 2)$. Thus the coordinates of the foci are $(-1 - \sqrt{7}, 2)$ and $(-1 + \sqrt{7}, 2)$. The vertices of the horizontal axis are $(-5, 2)$ and $(3, 2)$, and the vertices of the vertical axis are $(-1, -1)$ and $(-1, 5)$. Refer to Figure 4-32.

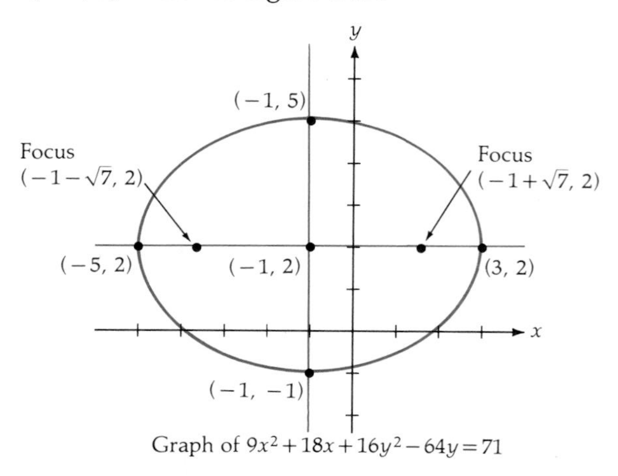


*Figure 4-32*

The domain of this equation is $[-5, 3]$, and the range is $[-1, 5]$. ■

The ancient Greeks sketched an ellipse by means of a specific construction technique. You can do the same by means of some string, a pencil and 2 thumbtacks mounted on a piece of cardboard. The points on an ellipse are those points such that the sum of the distances from two fixed points $F_1$ and $F_2$ (where we place thumbtacks) is a constant. The thumbtacks are located at the two foci. Keeping the string tight, move the pencil slowly and allow it to trace out the curve. The sketch is shown in Figure 4-33.

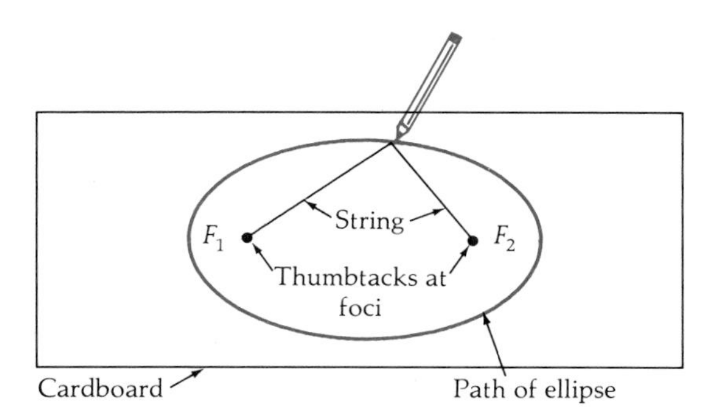


*Figure 4-33*

## EXERCISE 4-4

1. Find the equation of a circle in standard form with center $(-2, 3)$ and radius 6.
2. Find the equation of a circle in standard form with center $(0, 0)$ and radius 7.

In Problems 3–6, each equation is the equation of a circle. Put the equation in standard form and give the center and radius. Sketch the graph.

3. $x^2 + y^2 + 2x - 12y = -28$
4. $x^2 + y^2 - 6x - 4y = 36$
5. $x^2 + y^2 + 2x + 10y = -25$
6. $4x^2 + 4y^2 - 4x + 32y = 35$
7. Find the equation of a circle in standard form if the center is $(3, -4)$ and the circle passes through the origin.
8. Find the equation of a circle in standard form if the endpoints of the diameter are $(3, -9)$ and $(-5, 13)$. (*Hint:* Use the midpoint formula to determine the center.)

In Problems 9–21, each equation represents the equation of an ellipse. If necessary, place each equation in standard form. Then find the center, vertices, and foci of each one. Graph each ellipse.

9. $\dfrac{x^2}{49} + \dfrac{y^2}{25} = 1$
10. $\dfrac{x^2}{16} + \dfrac{y^2}{4} = 1$
11. $\dfrac{x^2}{9} + \dfrac{y^2}{25} = 1$
12. $\dfrac{x^2}{6} + \dfrac{y^2}{36} = 1$
13. $5x^2 + y^2 = 25$
14. $8x^2 + y^2 = 8$
15. $\dfrac{(x-3)^2}{25} + \dfrac{(y-1)^2}{9} = 1$
16. $\dfrac{(x+1)^2}{16} + \dfrac{(y-4)^2}{36} = 1$
17. $\dfrac{(x+5)^2}{1} + \dfrac{(y+2)^2}{4} = 1$

**18.** $\dfrac{(x-2)^2}{25}+\dfrac{(y-4)^2}{9}=1$

**19.** $4x^2+25y^2-24x+50y-39=0$

**20.** $4x^2+16y^2-16x+160y=-352$

**21.** $25x^2+150x+16y^2-96y=31$

**22.** Write the equation of the ellipse whose vertices are $(-10, 2)$, $(4, 2)$, $(-3, -1)$, and $(-3, 5)$.

**23.** Write an equation that describes the set of all points in a plane satisfying the property that the sum of the distances from each to the points $(-4, 3)$ and $(2, 3)$ is 10.

**★24.** Show that the equation

$$\sqrt{[x-(h-c)]^2+(y-k)^2}+\sqrt{[x-(h+c)]^2+(y-k)^2}=2a$$

becomes $\dfrac{(x-h)^2}{a^2}+\dfrac{(y-k)^2}{b^2}=1$ if we substitute $b^2$ for $a^2-c^2$.

**★25.** Suppose we have a vertical ellipse as shown in Figure 4-34. The definition of the ellipse tells us that $\overline{PF_1}+\overline{PF_2}$ is a constant, which we may arbitrarily call $2b$. Thus $\overline{PF_1}+\overline{PF_2}=2b$. Using the distance formula and letting $b^2=a^2+c^2$, derive the standard form of the equation of the ellipse.

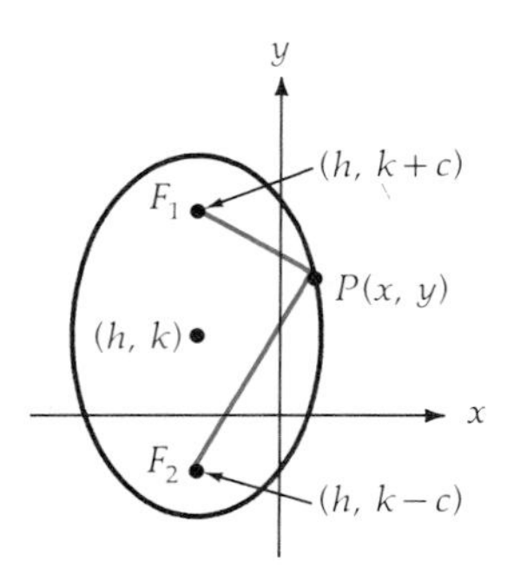


*Figure 4-34*

Each of the equations in Problems 26–32 is either a circle or an ellipse. Place the equation in standard form, identify it, and sketch the graph.

**26.** $3x^2+2+3y^2-12x=2$

**27.** $x^2-2x+15+8y+y^2=0$

**28.** $x^2+4y^2-8x+16y=-28$

**29.** $4x^2+y^2+24x-10y=-45$

**30.** $9y^2+3+4x^2-16x-20=3$

**31.** $x^2+2y^2+6x+7=0$

**32.** $4x^2+y^2+4x+4=4y$

## 4-5 THE HYPERBOLA

In the equation $Ax^2+By^2+Cx+Dy+E=0$ if $A\neq 0$ and $B\neq 0$ and $A$ is opposite in sign to $B$, we obtain the equation of a hyperbola. A hyperbola has two branches that are not connected. The study of the hyperbola is particularly important in certain areas of physics and astronomy. For example, some comets that do not orbit the sun travel in a hyperbolic orbit. If a space vehicle is launched from orbit or from a multi-stage rocket high above the earth's surface, it will travel in a hyperbolic path (as long as the velocity of travel is greater than the minimum escape velocity required to leave the earth's gravitational field).

We will now state the classical Greek definition for a hyperbola.

***Definition***

A *hyperbola* is the set of all points in a plane, each such that the absolute value of the difference of its distances to two fixed points (foci) is a constant. Refer to Figure 4-35.

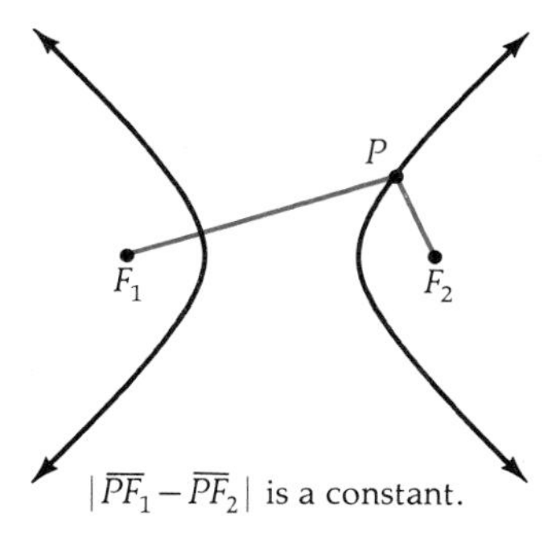


*Figure 4-35*

Using methods similar to those for the ellipse, we can establish the *standard form* for the equation of the hyperbola. This form is

$$\frac{(x-h)^2}{a^2} - \frac{(y-k)^2}{b^2} = \pm 1$$

where $(h, k)$ is the *center*. The foci are located at a distance $c$ from the center, and $a^2 + b^2 = c^2$. If the right-hand side of the equation is $+1$, we have a *horizontal* hyperbola; and if the right-hand side of the equation is $-1$, the hyperbola is *vertical*.

Using direct substitution of variable values to obtain points is not the best approach for sketching the hyperbola. However, if we sketch a small rectangle using the values of $a$ and $b$, then the hyperbola is one of the easiest curves to sketch. Using $(h, k)$ as the center, we construct a rectangle that is $2a$ units wide and $2b$ units high. We then draw and extend the diagonals of the rectangle. These extended diagonals are the *asymptotes* for the branches of the hyperbola. The *asymptote* has the characteristic that as we move out on any branch of the hyperbola, the curve continually approaches the asymptote but never reaches it. The hyperbola can then be easily sketched.

Suppose we have a hyperbola with a center at (0, 0) that has the equation $\frac{x^2}{a^2} - \frac{y^2}{b^2} = 1$ where $a > b$. The diagonal lines of the rectangle would have the equations $y = \frac{b}{a}x$ and $y = -\frac{b}{a}x$. The graph of this hyperbola appears in Figure 4-36.

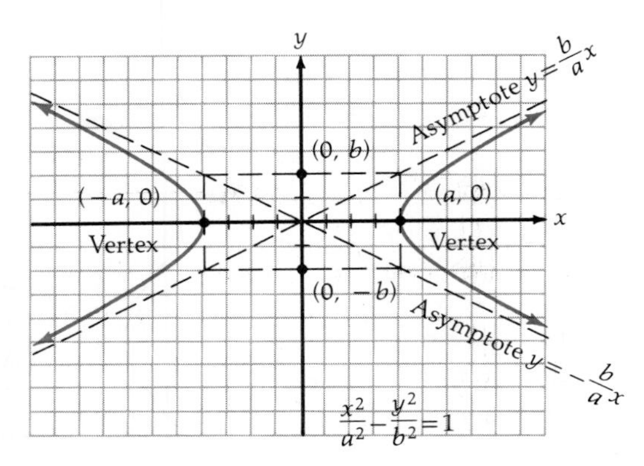


*Figure 4-36*

As any point $(x, y)$ on the hyperbola moves farther and farther from the center, it comes closer and closer to one of these asymptotes. The asymptotes are the diagonals of the rectangle and are obviously not part of the hyperbola—they are merely aids to graphing the hyperbola.

When sketching and identifying the hyperbola, we usually refer to one of two standard forms.

***Definition***

The standard equation of a *horizontal hyperbola* with a center at $(h, k)$ is given by

$$\frac{(x-h)^2}{a^2} - \frac{(y-k)^2}{b^2} = 1$$

The two vertices of a horizontal hyperbola will be $a$ units to the right and the left of the center at $(h, k)$. Thus the two vertices of a horizontal hyperbola are located at $(h - a, k)$ and $(h + a, k)$. The line segment connecting the two **vertices** is called the **transverse axis** of the hyperbola. The two foci of a horizontal hyperbola will be $c$ units to the right and the left of the center at

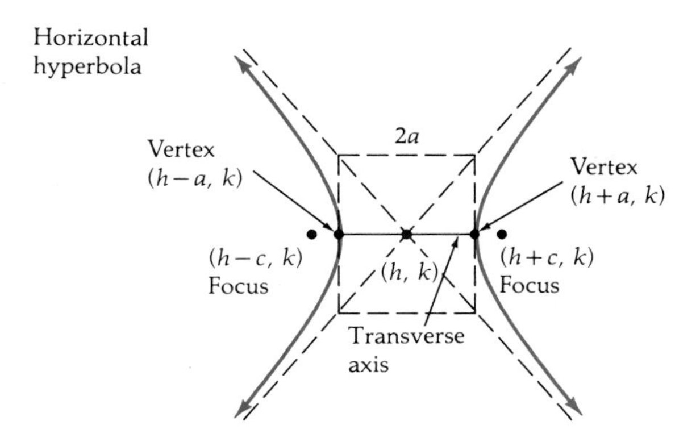


*Figure 4-37*

$(h, k)$. Thus the two foci of a horizontal hyperbola are at $(h - c, k)$ and $(h + c, k)$. Refer to Figure 4-37.

If the transverse axis is parallel to the $y$-axis, we have a vertical hyperbola.

***Definition***

The standard equation of a *vertical hyperbola* with a center at $(h, k)$ is given by

$$\frac{(x-h)^2}{a^2} - \frac{(y-k)^2}{b^2} = -1$$

In similar fashion we would find that the vertices of a vertical hyperbola are located at $(h, k - b)$ and $(h, k + b)$. The foci of a vertical hyperbola are located at $(h, k - c)$ and $(h, k + c)$. Refer to Figure 4-38.

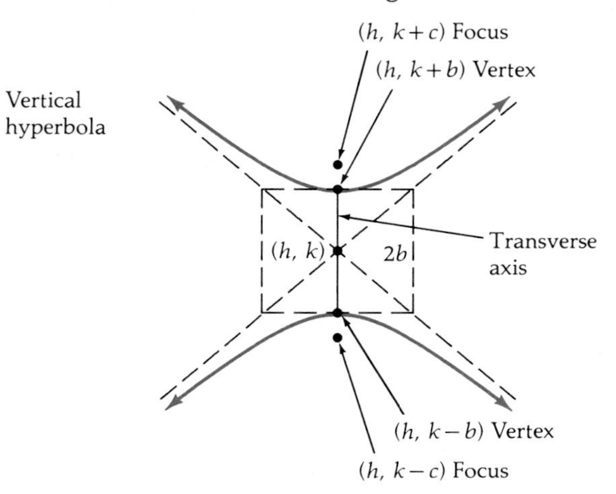


*Figure 4-38*

It is important for a student to remember that there are three possibilities. We can have $a > b$, $a < b$, or $a = b$ for either the horizontal or vertical hyperbolas. The equation $a^2 + b^2 = c^2$ is true for both *horizontal and vertical hyperbolas.*

**EXAMPLE 1** Sketch the graph of $\frac{x^2}{4} - \frac{y^2}{1} = 1$. Locate and plot the vertices and the foci.

***Solution*** The equation is in the standard form of the equation of a horizontal hyperbola. Since $(h, k)$ is $(0, 0)$, the center is at the origin. Since $a^2 = 4$, we have $a = 2$. Since $b^2 = 1$, we have $b = 1$.

We now construct a rectangle 4 units ($2a$) long and 2 units ($2b$) high. We extend the diagonals of the rectangle to form the asymptotes for the hyperbola, as shown in Figure 4-39. One vertex is $a$ units to the right of the origin. It is located at $(2, 0)$. The other is $a$ units to the left. It is located at $(-2, 0)$.

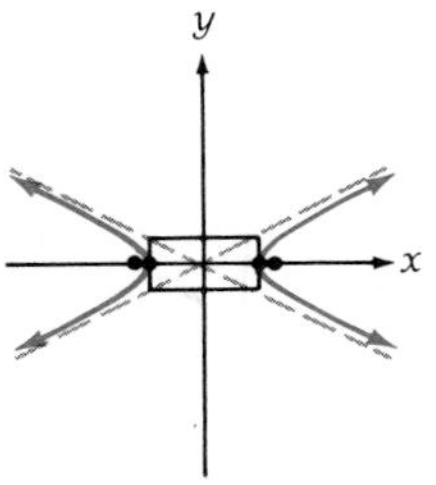

We find $c$ by the equation:

$$c^2 = a^2 + b^2$$

$$c^2 = 4 + 1$$

$$c^2 = 5 \qquad c = \sqrt{5} \quad (\text{or } c \doteq 2.24 \text{ to nearest hundredth})$$

*Figure 4-39*

The foci are $c$ units to the right and left of the origin. The foci are $(-\sqrt{5}, 0)$ and $(\sqrt{5}, 0)$. The domain is $(-\infty, -2] \cup [2, +\infty)$. The range is $(-\infty, +\infty)$. ■

We will now consider two examples in which the center of a hyperbola is not at the origin.

**EXAMPLE 2** Sketch the graph of $\frac{(x-2)^2}{9} - \frac{(y+1)^2}{16} = 1$. Label the vertices and the foci.

***Solution*** The equation is in standard form. We obtain the values $a = 3$ and $b = 4$.

We observe that it is a horizontal hyperbola. (Why?) We next locate the center at $(2, -1)$. Using this as the center, we construct a rectangle 6 units ($2a$) long and 8 units ($2b$) high. We extend the diagonals of the rectangle to form the asymptotes for the hyperbola. Using $(-1, -1)$ and $(5, -1)$ as the vertices of the hyperbola, we may now sketch the graph, which is shown in Figure 4-40. To plot the foci, we need the value of $c$.

$$c^2 = a^2 + b^2$$

$$c^2 = 9 + 16 = 25$$

$$c = 5$$

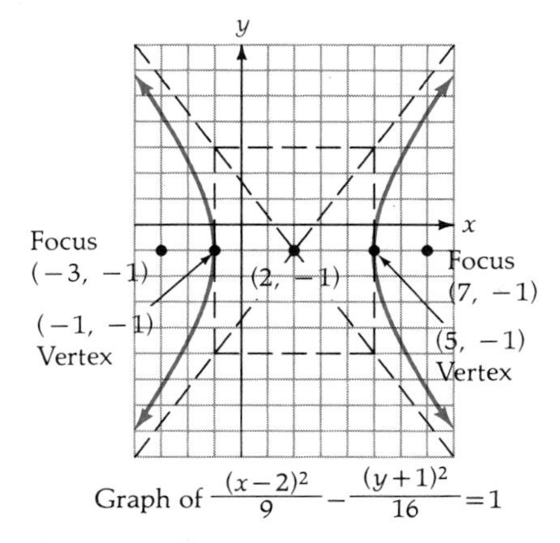


*Figure 4-40*

The foci are 5 units to the right and to the left of the center at $(2, -1)$, which we can write as $(2 + 5, -1)$ and $(2 - 5, -1)$. Thus the foci are at $(7, -1)$ and $(-3, -1)$. The domain of this equation is $(-\infty, -1] \cup [5, +\infty)$. The range is $(-\infty, +\infty)$. ■

Some work may be necessary to transform the equation into the standard form.

**EXAMPLE 3** Sketch the graph of $x^2 + 12x - y^2 + 10y = -15$. Label the vertices and foci.

***Solution*** We rewrite the equation in a more convenient form:

$$x^2 + 12x + \underline{\quad} - (y^2 - 10y + \underline{\quad}) = -15$$

$$x^2 + 12x + \underline{36} - (y^2 - 10y + \underline{25}) = -15 + 36 - 25$$

Notice that we add 36 and subtract 25 from each side to complete the square.

$$(x + 6)^2 - (y - 5)^2 = -4 \qquad \textit{Write in factored form.}$$

$$\frac{(x + 6)^2}{4} - \frac{(y - 5)^2}{4} = -1 \qquad \textit{Divide each side by 4.}$$

This is in standard form. The hyperbola is a vertical hyperbola with a center at $(-6, 5)$. Note that $a = b = 2$. (When working on the previous section, we found that if $a = b$, we did not really have an ellipse but a circle.) The hyperbola may have $a = b$.

As shown in Figure 4-41, we draw a rectangle 4 units long and 4 units high. The vertices are $b$ units below and above $(-6, 5)$. Thus the vertices are at $(-6, 3)$ and $(-6, 7)$.

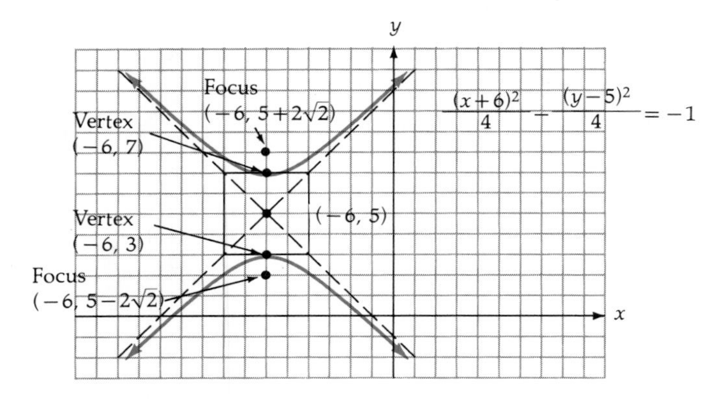


*Figure 4-41*

$$c^2 = a^2 + b^2 = 4 + 4$$
$$c^2 = 8$$
$$c = \sqrt{8} = 2\sqrt{2} \quad (c \doteq 2.83 \text{ to nearest hundredth})$$

The foci are at $(-6, 5 + 2\sqrt{2})$ and $(-6, 5 - 2\sqrt{2})$. Using our calculator we approximate the foci at $(-6, 7.83)$ and $(-6, 2.17)$. ■

### Other Forms of the Equation of the Hyperbola

Not all hyperbolas can be placed in the standard form of the equation of a hyperbola developed in this chapter.

We have previously graphed equations of the form $xy = c$ where $c$ is a real number and $c \neq 0$. We have indicated that all equations of this type are hyperbolas. In general it can be shown by using more advanced methods that for all real numbers $a$, $b$ and all real numbers $c$ where $c \neq 0$, the equation of the form $(x - a)(y - b) = c$ is the equation of a hyperbola.

If this type of equation is solved for the variable $y$, we have

$$y - b = \frac{c}{x - a}$$

$$y = \frac{c}{x - a} + b$$

Such equations can be readily graphed by using the techniques of graphing rational functions.

## EXERCISE 4-5

In Problems 1–22, sketch the hyperbola for each equation. Find and label the center, vertices, and foci.

**1.** $\dfrac{x^2}{4} - \dfrac{y^2}{25} = 1$ **2.** $\dfrac{x^2}{4} - \dfrac{y^2}{9} = 1$ **3.** $\dfrac{x^2}{36} - \dfrac{y^2}{25} = -1$ **4.** $\dfrac{x^2}{9} - \dfrac{y^2}{49} = -1$

**5.** $9y^2 - x^2 = 36$ **6.** $9y^2 - 25x^2 = 225$ **7.** $9x^2 - y^2 = 9$ **8.** $x^2 - y^2 = 9$

**9.** $y^2 - 4x^2 - 16 = 0$ **10.** $4y^2 - 25x^2 - 100 = 0$

**11.** $\dfrac{(x-2)^2}{9} - \dfrac{(y-3)^2}{16} = 1$ **12.** $\dfrac{(x+4)^2}{25} - \dfrac{(y-2)^2}{9} = -1$

**13.** $\dfrac{(x+1)^2}{4} - \dfrac{(y-2)^2}{16} = -1$ **14.** $\dfrac{(x-2)^2}{4} - \dfrac{(y-3)^2}{25} = 1$

**15.** $16(x-3)^2 - 25(y+2)^2 = 400$ **16.** $25(x+2)^2 - 16(y-3)^2 = 400$

**17.** $9(y-2)^2 - (x+5)^2 - 9 = 0$ **18.** $4(y+3)^2 - (x+1)^2 - 16 = 0$

**19.** $9x^2 - 4y^2 - 54x - 32y = 19$ **20.** $x^2 - 4y^2 - 4x - 8y = 4$

**21.** $25y^2 - 9x^2 - 100y - 54x = -10$ **22.** $4y^2 - x^2 + 32y - 8x = -49$

**23.** Write the equation of the hyperbola whose vertices are $(1, -1)$ and $(1, -9)$ and whose foci are $(1, 0)$ and $(1, -10)$.

**24.** Determine the equation of the hyperbola with vertices at $(-2, 0)$ and $(2, 0)$ and foci at $(-3, 0)$ and $(3, 0)$.

**25.** Write an equation that describes the set of all points in a plane for which the difference of the distances from each point to the points $(-3, 2)$ and $(5, 2)$ is 6.

**★26.** Using Figure 4-42 along with the distance formula and the definition, establish the standard form for the equation of a horizontal hyperbola. Let $b^2 = c^2 - a^2$.

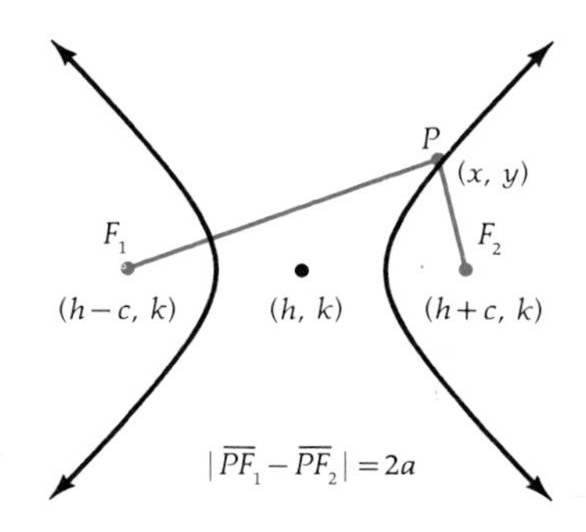


*Figure 4-42*

**★27.** Using a similar method to that used in Problem 26, establish the standard form for the equation of a vertical hyperbola. (Use $|\overline{PF}_1 - \overline{PF}_2| = 2b$ and $a^2 + b^2 = c^2$.)

**28.** Graph the hyperbola $xy = -8$. **29.** Graph the hyperbola $(x-5)y = 3$.

**30.** Graph the hyperbola $(x+2)(y-3) = 4$.

## 4-6 THE PARABOLA

In Section 4-1 we discussed the graphs of the quadratic function. The graph of every quadratic function is a parabola. However, there are a number of equations of parabolas that are *not functions*.

Thus we find a need to discuss parabolas in a more general sense. The general equation for conic sections is $Ax^2 + By^2 + Cx + Dy + E = 0$. If either $A = 0$ or $B = 0$, but not both, then we will obtain the equation of a parabola.

We now proceed to the classic Greek definition of a parabola.

*Definition*

A *parabola* is the set of all points in a plane equidistant from a fixed point (focus) and a fixed line (*directrix*) not containing the fixed point.

The midpoint between the focus and the directrix on a line segment drawn perpendicular to the directrix is called the **vertex**. The line drawn through the vertex and the focus is the **axis of symmetry**. (See Figure 4-43.)

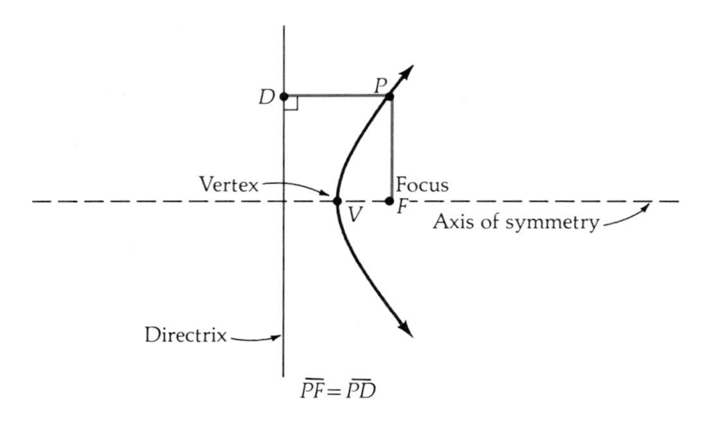


*Figure 4-43*

Using methods similar to those discussed earlier, we can establish the *standard form* for the equation two types of parabolas. A *vertical* parabola opens upward or downward. A *horizontal* parabola opens to the right or to the left.

*Definition*

The equation for a *vertical* parabola in standard form is

$$(x - h)^2 = 4p(y - k)$$

where $(h, k)$ is the vertex of the parabola. The focus is $(h, k + p)$, and the directrix is $y = k - p$. The parabola opens upward if $p > 0$ and downward if $p < 0$. (See Figure 4-44.)

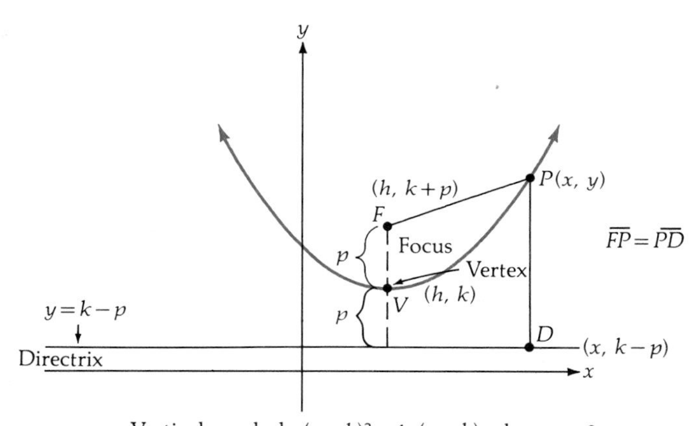


Vertical parabola $(x-h)^2 = 4p(y-k)$ where $p>0$

*Figure 4-44*

The derivation of the equation of the vertical parabola will be left as an exercise.

In a similar fashion we can examine a *horizontal* parabola.

*Definition*

The equation for a *horizontal* parabola in standard form is

$$(y - k)^2 = 4p(x - h)$$

where $(h, k)$ is the vertex. The focus is $(h + p, k)$, and the directrix is $x = h - p$. The parabola opens to the right if $p > 0$ and to the left if $p < 0$.

The derivation of the equation of the horizontal parabola will be left as an exercise.

It should be mentioned immediately that a parabola is not one branch of a hyperbola as might be thought if they are carelessly drawn. They have different definitions, and a parabola does not have any asymptotes.

This difference is evident if we quickly sketch a parabola and the right branch of a hyperbola that have the same focus $F$ and the same vertex $V$. Refer to Figure 4-45.

The more we extend the graphs to the right, the more evident this difference becomes.

In the next few examples we will illustrate the use of the standard form for horizontal and vertical parabolas to enable us to quickly sketch the graph of any given parabola.

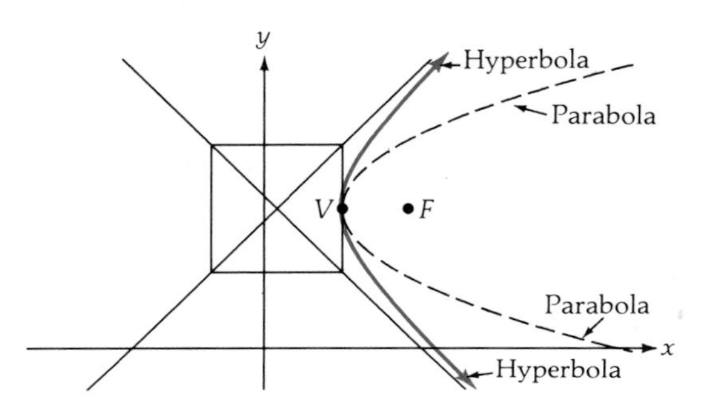


*Figure 4-45*

**EXAMPLE 1** Sketch the graph of $y^2 = -16x$. Find and label the vertex, the focus, and the directrix.

***Solution*** We see that the $y$-variable is squared and the $x$-variable is not. Thus we immediately know that we have a horizontal parabola. Since $(h, k) = (0, 0)$, we know that the vertex is at the origin.

$$y^2 = \underbrace{4px} \qquad\qquad y^2 = \underbrace{-16x}$$

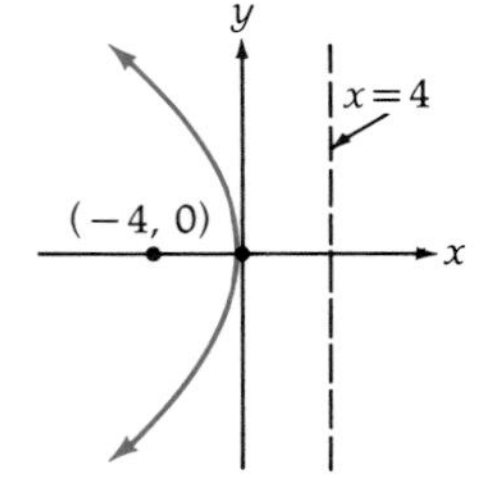


*Figure 4-46*

If we set $4p = -16$, we know that $p = -4$. Since $p < 0$, we know that the parabola opens to the left. The focus is at $(h + p, k)$, so we have $(0 - 4, 0)$ or $(-4, 0)$ as the focus. The equation of the directrix is $x = h - p$. So we have $x = 0 - (-4)$ or $x = 4$ as the directrix. The sketch is shown in Figure 4-46.

The accuracy of our sketch can be improved by plotting two points. Let $x = -4$; then $y^2 = (-16)(-4) = 64$, and $y = \pm 8$. So we plot $(-4, 8)$ and $(-4, -8)$.

The axis of symmetry is the $x$-axis. ■

**EXAMPLE 2** Sketch the graph of $(x - 4)^2 = 8(y - 3)$. Find and label the vertex, the focus, and the directrix.

***Solution*** The equation is in standard form. This is a vertical parabola with vertex (4, 3), as shown in Figure 4-47. Since $4p = 8$, we know that $p = 2$, and therefore the parabola opens upward. The focus is at $(h, k + p)$, so we have $(4, 3 + 2)$. Thus the focus is at (4, 5). The directrix is $y = k - p$. This would yield $y = 3 - 2$ or $y = 1$ for the equation of the directrix.

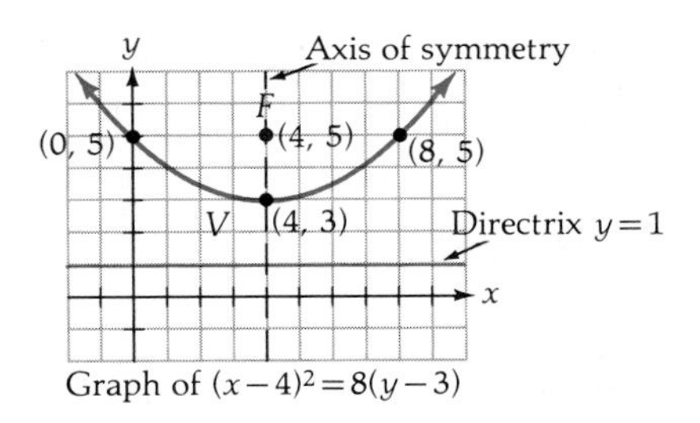


*Figure 4-47*

The graph is more accurate if we plot two additional points. Let $x = 0$; then $(-4)^2 = 8(y - 3)$ and thus $y = 5$.

We plot (0, 5) on the graph. Since the parabola is symmetric to the line $x = 4$ that passes through points $F$ and $V$, we can also plot (8, 5). ■

EXAMPLE 3 Sketch the graph of $(y + 1)^2 = -12(x - 3)$. Find and label the vertex, the focus, and the directrix.

***Solution*** This is a horizontal parabola with vertex $(3, -1)$, $p = -3$. The parabola opens to the left. (Why?) The focus is $(0, -1)$ and the directrix is $x = 6$. Refer to Figure 4-48.

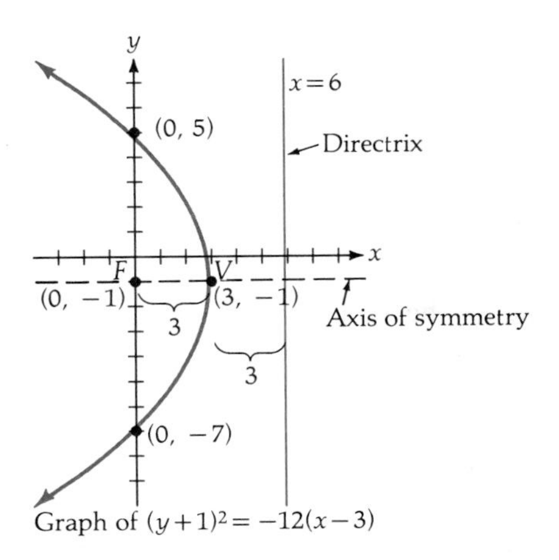


*Figure 4-48*

We note that the distance from $F$ to $V$ is always $|p|$ units or 3. Similarly the distance from $V$ to the directrix along the axis of symmetry is 3 units.

Our sketch will be more accurate if we locate the $y$-intercepts. If $x = 0$, we have $(y + 1)^2 = -12(-3)$ and thus $y = -7$ and $y = 5$. So we plot the points $(0, -7)$ and $(0, 5)$. ■

Since it is usually necessary to plot two other points on the parabola after we find the vertex, focus, and directrix, the student is probably wondering if there is a quick way to find these two additional points. There is if we use a property of the **latus rectum** of the parabola.

*Definition*

The *latus rectum* of a parabola is a line segment connecting two points on the parabola and passing through the focus perpendicular to the axis of symmetry. The length of the latus rectum is $4|p|$. Refer to Figure 4-49.

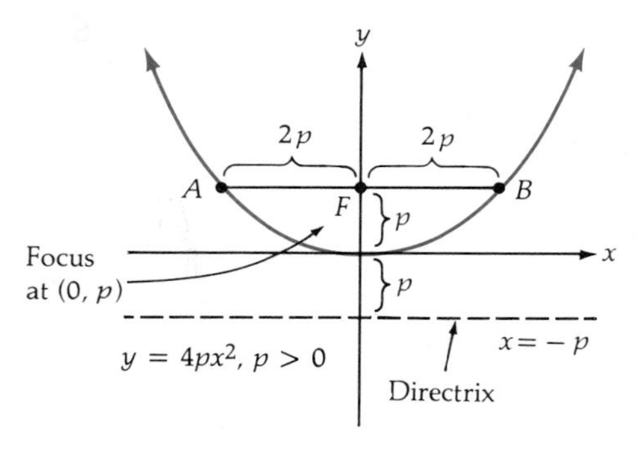


*Figure 4-49*

In the next example it is necessary to complete the square to obtain the standard form of the parabola.

**EXAMPLE 4** Sketch the graph of $x^2 + 8x + 6y + 22 = 0$. Find and label the vertex, the focus, and the directrix.

***Solution***

$$x^2 + 8x = -6y - 22 \qquad \textit{Add } -6y - 22 \textit{ to each side.}$$

$$x^2 + 8x + 16 = -6y - 22 + 16 \qquad \textit{Complete the square by adding 16 to each side.}$$

$$(x + 4)^2 = -6y - 6 \qquad \textit{Factor the left-hand side.}$$

$$(x + 4)^2 = -6(y + 1) \qquad \textit{Factor the right-hand side.}$$

The equation is now in the standard form for a vertical parabola. The vertex is at $(-4, -1)$. Since $4p = -6$, we know that $p = -\frac{3}{2}$. Since $p < 0$, the parabola opens downward. The focus is located at $(h, k + p)$, which becomes $\left(-4, -1 - \frac{3}{2}\right)$ or $(-4, -2\frac{1}{2})$. The directrix is $y = k - p$. Substituting, we have

$$y = -1 - \left(-\frac{3}{2}\right)$$

$$y = -1 + \frac{3}{2}$$

$$y = \frac{1}{2}$$

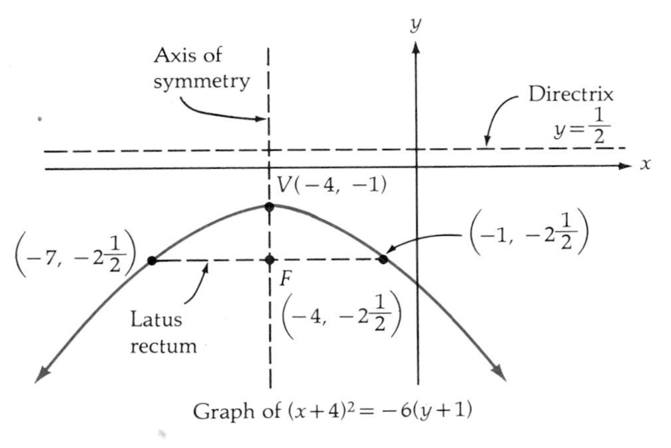


*Figure 4-50*

The graph in Figure 4-50 will be more accurate if we plot two additional points. The *latus rectum* passes through $F$ perpendicular to the axis of symmetry. It intersects the parabola $2|p|$ units to the right and $2|p|$ units to the left of the focus. Since $|p| = \left|-\frac{3}{2}\right| = \frac{3}{2}$, we have $2|p| = 3$. The two points on the parabola are $(-4 - 3, -2\frac{1}{2})$ or $(-7, -2\frac{1}{2})$ and $(-4 + 3, -2\frac{1}{2})$ or $(-1, -2\frac{1}{2})$. ■

### Degenerate Conics

We illustrated the **curves** that are called **conics** when they result in the intersection of a plane and a right circular cone. Refer to Figure 4-22 on page 180.

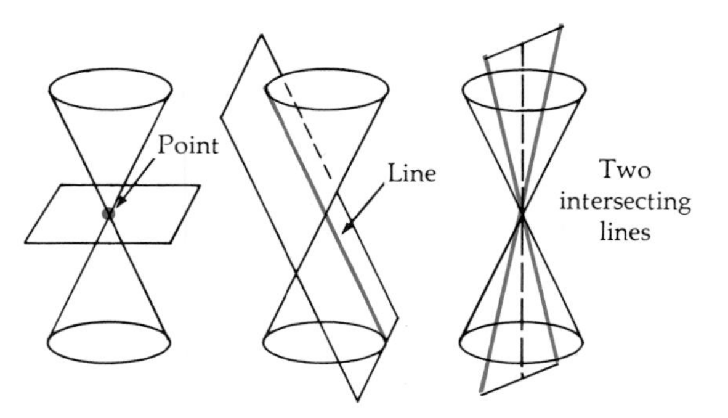


*Figure 4-51*

In a few special cases we do not obtain a curve at all. When we obtain a point, a line, or two intersecting lines, we say that these are **degenerate conic sections**. (See Figure 4-51.)

Therefore it is not correct to assume that a general second-degree equation will always yield a curve.

To illustrate, when we *attempt* to place $4x^2 + 32x - y^2 + 2y = -63$ in the standard form for a hyperbola, we obtain

$$\frac{(x+4)^2}{1} - \frac{(y-1)^2}{4} = 0$$

Since the right-hand side equals zero, we see that it is impossible to obtain the standard form of the equation of a hyperbola. The graph of this equation yields two intersecting straight lines and is called a **degenerate hyperbola**.

When we *attempt* to place $4y^2 + 16y + 100x^2 - 300x = -241$ in standard form for an ellipse, we obtain $\left(x - \frac{3}{2}\right)^2 + (y+2)^2 = 0$. The only solution to this equation is $\left(\frac{3}{2}, -2\right)$ so this degenerate case is called a **point ellipse**.

## EXERCISE 4-6

For each of the parabolas in Problems 1–20, sketch the graph and label the vertex, the focus, and the directrix.

1. $x^2 = -8y$
2. $y^2 = 20x$
3. $(y-1)^2 = 4x$
4. $(x+2)^2 = -12y$
5. $(x-3)^2 = 12(y+1)$
6. $(y-2)^2 = 4(x+3)$
7. $(y-1)^2 = -2(x+3)$
8. $(x+1)^2 = -8(y-3)$
9. $\left(y - \frac{1}{2}\right)^2 = 6\left(x - \frac{3}{2}\right)$
10. $\left(y + \frac{3}{2}\right)^2 = 2\left(x - \frac{3}{2}\right)$
11. $x^2 = y - x$
12. $2x^2 = \frac{1}{2}y + 1$

**13.** $y^2 - 2y + 8x - 15 = 0$ **14.** $y^2 - 6y - 2x + 7 = 0$ **15.** $y^2 + 6y - x + 5 = 0$

**16.** $y^2 - 4y - x - 6 = 0$ **17.** $x^2 + 4y - 6x + 15 = 0$ **18.** $x^2 - y + 6x + 10 = 0$

**19.** $x^2 - 10x + 4y = -33$ **20.** $x^2 + 6y = 9 - 6x$

**21.** Write an equation that describes the set of all points in a plane that are equidistant from the point (3, 4) and the line $y = -2$.

**22.** Write the equation of a parabola whose focus is (5, −1) and whose directrix is $y = 3$.

**23.** Write the equation of a parabola whose focus is (−2, 3) and whose directrix is $y = -3$.

**24.** Write an equation that describes the set of points in a plane that are equidistant from the point (−1, 1) and the line $x = -5$.

**25.** Write the equation of a parabola that passes through (−2, 6) and (0, 0) and that has the $y$-axis as the axis of symmetry.

**★26.** Using Figure 4-52, along with the distance formula and the definition, establish the equation of the standard form of a vertical parabola.

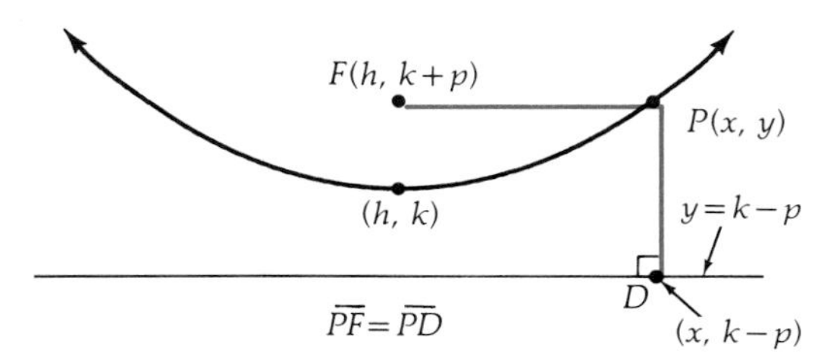


*Figure 4-52*

**★27.** Using a similar method to that used in Problem 26, establish the standard form of the equation of a horizontal parabola.

## Calculator Problems

Use a calculator to obtain the required equations in Problems 28–29.

**★28.** A Boy Scout troop constructed a suspension bridge from some surplus cable. The cable is hung between the two supports and is in the shape of a parabola. The distance labelled $y$ in Figure 4-53 is actually 30 feet. (The distance labelled $y$ measures the distance from the top of the supports to the vertex.) The distance labelled $x$ on the diagram is actually 350 feet. Can you obtain the equation of the parabola that describes the path of the cable that is suspended between the two supports?

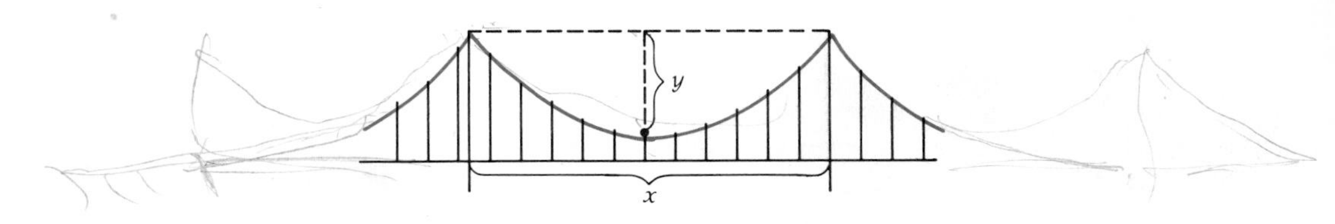


*Figure 4-53*

★**29.** A parabolic solar heating device is being built that will achieve temperatures of 3000°C at the focal point. The focal distance (the distance from the vertex of the parabolic bowl to the focal point) is 8.5 feet. The parabolic bowl is 3.0 feet deep, as shown in Figure 4-54. How large a diameter will be needed for the parabolic bowl?

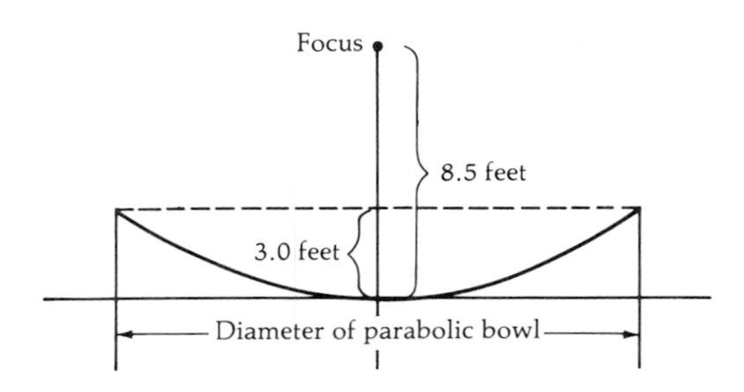


*Figure 4-54*

### Review Problems for the Conic Sections

Put each of the equations in Problems 30–41 in standard form. Identify the conic and give important data such as center, vertices, foci, directrix, etc. Sketch the graph.

**30.** $x^2 + 9y^2 - 10x + 36y + 52 = 0$

**31.** $x^2 - 8x - 8y + 8 = 0$

**32.** $x^2 + y^2 + 8x - 6y - 5 = 0$

**33.** $9x^2 - 16y^2 + 72x + 96y - 144 = 0$

**34.** $9x^2 + 4y^2 + 36x + 32y + 64 = 0$

**35.** $y^2 - 12x + 8y + 40 = 0$

**36.** $x^2 + 25y^2 - 2x - 150y + 201 = 0$

**37.** $x^2 - 12x - 4y + 48 = 0$

**38.** $x^2 + y^2 + 6x - 8y - 9 = 0$

**39.** $9x^2 - 4y^2 + 36x + 40y - 100 = 0$

**40.** $4x^2 - y^2 - 24x + 4y + 36 = 0$

**41.** $4x^2 + y^2 - 8x - 8y + 4 = 0$

## COMPUTER PROBLEMS FOR CHAPTER 4

If you can program on a computer, solve the following problems.

**1.** The profit in dollars for manufacturing $x$ items daily in a factory is given by the equation $P(x) = 3286.111 + 1.2777778x - 0.0013889x^2$. Write a computer program that will print out a table of profit values for each given value of $x$. (Round profit values to the nearest cent.) Let $x$ assume each value in the set $\{20, 40, 60, 80, 100, \ldots, 960, 980, 1000\}$.

**2.** A polynomial function is defined by $P(x) = 3x^7 + 5x^6 - 2x^5 + 3x^4 - 2x^3 + 8x - 4$.

**a.** Write a computer program that will print out a table of function values of $P(x)$ for each given value of $x$. Let $x$ assume each value in the set $\{0, 0.2, 0.4, 0.6, \ldots, 9.2, 9.4, 9.6, 9.8, 10.0\}$.

**b.** Based on your computer printout, for what values of $x$ is $P(x)$ close to zero? (In other words, what $x$ values are closest to the zeros of $P(x)$?)

## KEY TERMS AND CONCEPTS

Be sure you understand the following terms and can use your knowledge in solving various mathematical problems.

- Polynomial function
- Quadratic function
- Parabola
- Vertex
- Minimum value of a quadratic function
- Maximum value of a quadratic function
- Cubic function
- Rational function
- Asymptote
- Vertical asymptote
- Horizontal asymptote
- Conic section
- Circle
- Ellipse
- Focus
- Horizontal ellipse
- Vertical ellipse
- Major axis
- Minor axis
- Vertices
- Hyperbola
- Transverse axis
- Directrix
- Axis of symmetry
- Horizontal parabola
- Vertical parabola
- Latus rectum

## SUMMARY OF PROCEDURES AND CONCEPTS

### The Quadratic Function

The graph of the quadratic function $f(x) = ax^2 + bx + c$ where $a \neq 0$ is a parabola.

1. It opens upward if $a > 0$ and downward if $a < 0$.
2. It has a vertex at $\left[-\dfrac{b}{2a}, f\left(-\dfrac{b}{2a}\right)\right]$.
3. The axis of symmetry is $x = \dfrac{-b}{2a}$.
4. If $a > 0$, then the function has a minimum value at the vertex. If $a < 0$, then the function has a maximum value at the vertex.

### Rational Functions

If rational functions are placed in the form $f(x) = \dfrac{g(x)}{h(x)}$ where $g(x)$ and $h(x)$ are polynomials without common factors, then:

1. The function will have a vertical asymptote at $x = a$ only if $h(x)$ has a zero at $x = a$ where $a$ is a real number.

2. If the degree of $g(x)$ is less than the degree of $h(x)$, then the function will have a horizontal asymptote at $y = 0$.
3. If the degree of $g(x)$ is equal to the degree of $h(x)$, then the function will have a horizontal asymptote at $y = \frac{a_n}{b_n}$ where $a_n$ is the leading coefficient $g(x)$ and $b_n$ is the leading coefficient of $h(x)$.

## The Conic Sections

The conic sections are curves that result from the intersection of a plane and a right circular cone. They are the circle, the ellipse, the hyperbola, and the parabola.

### *The Circle*

The standard form of the equation of a circle having a center at $(h, k)$ and a radius $r$ is given by $(x - h)^2 + (y - k)^2 = r^2$.

### *The Ellipse*

The standard form of the equation of an ellipse whose center is at $(h, k)$ is given by $\frac{(x-h)^2}{a^2} + \frac{(y-k)^2}{b^2} = 1$.

If $a > b$, the major axis is parallel to the $x$-axis and $a^2 - c^2 = b^2$. This is a horizontal ellipse.

If $a < b$, the major axis is parallel to the $y$-axis and $a^2 + c^2 = b^2$. This is a vertical ellipse.

The foci are $c$ units along the major axis from the center at $(h,k)$.

### *The Hyperbola*

1. The standard equation for a *horizontal hyperbola* with center at $(h, k)$ is given by $\frac{(x-h)^2}{a^2} - \frac{(y-k)^2}{b^2} = 1$. The transverse axis is parallel to the $x$-axis.
2. The standard equation for a *vertical hyperbola* with center at $(h, k)$ is given by $\frac{(x-h)^2}{a^2} - \frac{(y-k)^2}{b^2} = -1$. The transverse axis is parallel to the $y$-axis.
3. The foci for either a horizontal hyperbola or a vertical hyperbola are $c$ units along the transverse axis from the center at $(h, k)$.
4. The equation $a^2 + b^2 = c^2$ is true for both horizontal and vertical hyperbolas.

### The Parabola

1. The equation for a *vertical parabola* in standard form is $(x - h)^2 = 4p(y - k)$ where $(h, k)$ is the vertex. The focus is at $(h, k + p)$, and the directrix is $y = k - p$. The parabola opens upward if $p > 0$ and downward if $p < 0$. The axis of symmetry passes through the vertex parallel to the $y$-axis.
2. The equation for a *horizontal parabola* in standard form is $(y - k)^2 = 4p(x - h)$ where $(h, k)$ is the vertex. The focus is at $(h + p, k)$, and the directrix is $x = h - p$. The parabola opens to the right if $p > 0$ and to the left if $p < 0$. The axis of symmetry passes through the vertex parallel to the $x$-axis.

## CHAPTER 4 REVIEW EXERCISE

In Problems 1–4, graph the quadratic functions. Label any intercepts and label the vertex. Write the equation of the line of the axis of symmetry.

**1.** $f(x) = 3x^2 - 7x + 2$ **2.** $f(x) = 5x^2 + 8x - 4$

**3.** $f(x) = 2 - 4x - \frac{1}{2}x^2$ **4.** $f(x) = 3 + 4x - \frac{1}{2}x^2$

**5.** A man wishes to enclose a rectangular area for a garden. He will be using a barn as a barrier for one side. He will fence in the other three sides. He has 86 meters of fencing available. What dimensions should he use for the garden to maximize the area of the garden?

**6.** Two numbers have a sum of 30. The sum of the squares of those numbers is as small as possible. What are the numbers?

In Problems 7 and 8, plot four to six points and make a rough sketch of the polynomial. Choose a convenient scale on each axis.

**7.** $f(x) = \frac{1}{2}x^5 + 3x^2 - 2$ **8.** $f(x) = 4x^4 - 2x^2 + 3x + 1$

In Problems 9 and 10, find the roots of the polynomial. Then use the $x$-intercepts as well as a few other points to make a rough sketch of the polynomial.

**9.** $f(x) = (x - 3)(x + 2)(x + 4)$ **10.** $f(x) = -3x(x - 2)(x + 1)$

In Problems 11–14, graph each rational function. Examine each function for symmetry. Locate and label any intercepts or asymptotes.

**11.** $f(x) = \dfrac{4}{x - 3} + 2$ **12.** $f(x) = \dfrac{-2}{x + 3} - 1$ **13.** $f(x) = \dfrac{3x^2}{9 - x^2}$ **14.** $f(x) = \dfrac{3x + 1}{x^2 + x}$

**15.** Put the equation of the circle $x^2 + y^2 - 6x + 2y = 15$ in standard form. Give the center and radius.

**16.** Classify the curve represented by the equation $x^2 + y^2 + 6x - 8y + 25 = 0$.

In Problems 17 and 18, graph each ellipse. Find and label the center, vertices, and foci.

**17.** $\dfrac{(x-4)^2}{16} + \dfrac{(y+2)^2}{9} = 1$ **18.** $\dfrac{(x+1)^2}{\frac{1}{4}} + \dfrac{(y-2)^2}{4} = 1$

**19.** Write the equation of the ellipse whose vertices are $(-5, 1)$, $(-2, 5)$, $(1, 1)$, and $(-2, -3)$.

In Problems 20 and 21, graph each hyperbola. Find and label the center, vertices, and foci.

**20.** $\dfrac{(x+2)^2}{16} - \dfrac{(y+3)^2}{25} = 1$ **21.** $\dfrac{(x-4)^2}{4} - \dfrac{(y-2)^2}{16} = -1$

**22.** Find the equation of the hyperbola whose vertices are at $(-3, 1)$ and $(3, 1)$ and whose foci are $(-4, 1)$ and $(4, 1)$.

In Problems 23 and 24, graph each parabola. Find and label the vertex, the focus, and the directrix.

**23.** $(y-2)^2 = -3(x+1)$ **24.** $(x+4)^2 = -10(y-3)$

In Problems 25–34, put the equation in standard form and classify the curve. Sketch the graph.

**25.** $25x^2 + 9y^2 + 150x - 90y + 225 = 0$ **26.** $4x^2 + 4y^2 - 4x - 48y + 45 = 0$

**27.** $25y^2 - 16x^2 + 50y + 64x - 439 = 0$ **28.** $y^2 - 8y - 2x + 6 = 0$

**29.** $x^2 + 4y^2 + 6x = 16y - 21$ **30.** $4x^2 - 9y^2 - 16x = 18y - 43$

**31.** $2y^2 - 16y = -35 - x$ **32.** $25x^2 + 100x - 284 = 32y - 16y^2$

**33.** $9x^2 - y^2 - 9 = 36x - 12y$ **★34.** $x - 4x^2 = 3y^2 + 8y - 4x^2 - 3$

## PRACTICE TEST FOR CHAPTER 4

**1.** Put the equation of the circle $x^2 + y^2 - 10x + 4y = 7$ in standard form. Give the center and radius.

**2.** Graph the function $f(x) = \dfrac{x^2 - 4x - 3}{2}$. Label the vertex and axis of symmetry.

**3.** Plot four to six points and make a rough sketch of the polynomial $P(x) = x^4 - 2x^3 - 3x + 4$.

**4.** Graph the function $f(x) = \dfrac{2x^2 + x - 1}{x^2 - 16}$. Label any asymptotes.

In Problems 5–10, put the equation in standard form. Classify the curve. Sketch the graph.

**5.** $16x^2 - 4y^2 - 64x - 32y - 64 = 0$ **6.** $4x^2 + y^2 + 4y + 36 = 24x$

**7.** $2x^2 + 2y^2 + 2 = -4x + 8y$ **8.** $y^2 - 2x - 6y + 3y^2 = 3y^2 - 13$

**9.** $25x^2 - 9y^2 - 100x - 54y = -10$ **10.** $2 + 3x^2 + 12x + 3y^2 - 4y = 65 - 4y$

# 5 Exponential and Logarithmic Functions

■ *"Seeing there is nothing (right well-beloved Students of Mathematics) that is so troublesome to mathematical practice, nor doth more molest and hinder calculations, than the multiplications, divisions, square and cubical extractions of great numbers, which besides the tedious expense of time are for the most part subject to many slippery errors, I began therefore to consider in my mind by what certain and ready art I might remove those hindrances."*

*John Napier*

The history of mathematics has had many chapters written by men who desired to find a means to perform difficult or seemingly impossible calculations. Logarithms are the invention of John Napier (1550–1617). Napier worked on the concept for many years in an attempt to simplify certain difficult calculations. The development of the theory of logarithms by Napier created an intense interest on the part of many mathematicians and scientists. They saw logarithms as a convenient approach to performing the difficult calculations required in that day to study astronomy in general, and more specifically the orbits of heavenly bodies.

Similarly, the desire of men and women to understand patterns of growth and decay in the natural world have fostered the advance of our knowledge of exponential functions. Since exponential and logarithmic functions are so closely related, we shall study them both in this chapter.

## 5-1 EXPONENTIAL FUNCTIONS AND THEIR APPLICATIONS

In the previous two chapters we devoted much attention to a variety of algebraic and rational functions that have a variable base and a constant integral exponent. We now turn to the case in which the base is a constant and the exponent is a variable.

Exponential functions occur in many areas of mathematics but in a number of other disciplines as well. The exponential function is particularly useful in the natural sciences, to examine rates of growth and decay of both organic and inorganic matter. The exponential function is also useful in economics, history, and business to examine the properties of compound interest, inflation, and the growth of human population.

The laws of exponents are not limited to integers. We shall therefore state that the laws of exponents discussed in Chapter 1 are true for all real numbers. These laws will prove useful in dealing with exponential functions.

*Notation*

For any given real number $a$, where $a > 0$, the *exponential function* $f$ with base $a$ is defined by $f(x) = a^x$ where $x$ is a variable that can assume the value of any real number.

By examining the graphs of a few exponential functions we will observe a few properties of such functions.

**Note:** You will observe that we really have not defined the function $a^x$, but rather introduced the notation. Suppose that $a = 2$. What would $2^{\sqrt{2}}$ indicate? It is best to think of it by approximating $\sqrt{2}$ as 1.4, 1.41, 1.414, 1.4142, . . . . Thus $2^{\sqrt{2}}$ is the value we approach as we take successively more accurate approximations for the irrational exponent. A precise definition for $a^x$ where $x$ is any rational or irrational number will be encountered in a calculus course.

**EXAMPLE 1** Graph the exponential function $f(x) = 2^x$.

***Solution*** If we let $x$ assume a few values and find the corresponding function values, we can quickly set up a table. The graph appears in Figure 5-1.

| $x$ | $-2$ | $-1$ | 0 | 1 | 2 | 3 |
|---|---|---|---|---|---|---|
| $f(x)$ | $\frac{1}{4}$ | $\frac{1}{2}$ | 1 | 2 | 4 | 8 |

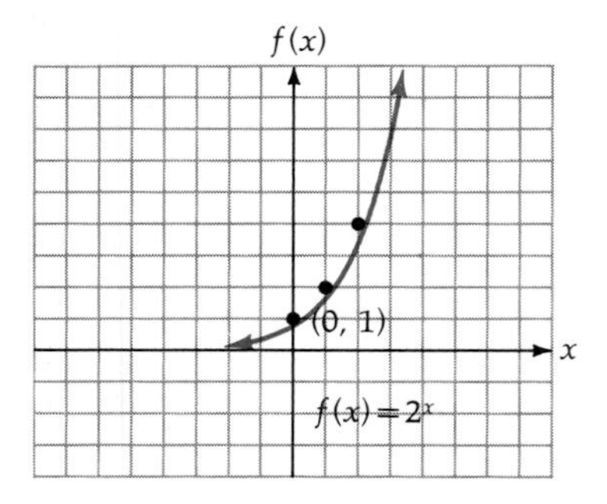


*Figure 5-1*

**EXAMPLE 2** Graph the exponential function $f(x) = \left(\frac{1}{3}\right)^x$.

***Solution*** Using our laws of exponents, we can write $f(x) = \left(\frac{1}{3}\right)^x = \frac{1^x}{3^x} = \frac{1}{3^x}$ as an equivalent form. (Can you state a reason for each step?) Now we obtain a few function values and list them in a table.

| $x$ | $-2$ | $-1$ | $0$ | $1$ | $2$ |
|---|---|---|---|---|---|
| $f(x)$ | $9$ | $3$ | $1$ | $\frac{1}{3}$ | $\frac{1}{9}$ |

In many cases of graphing exponential functions a convenient scale should be chosen for the axis before plotting the points. (See Figure 5-2.)

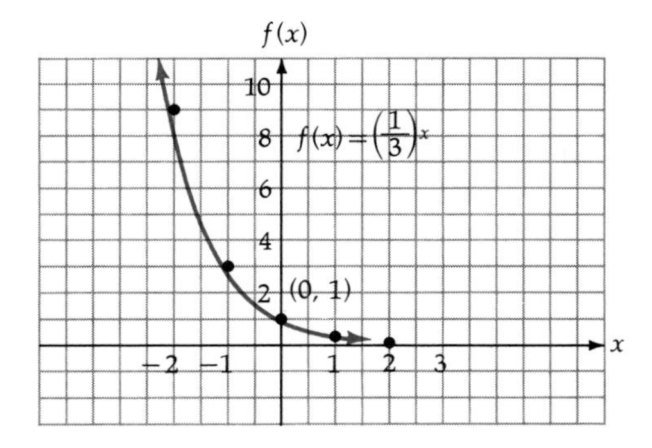


*Figure 5-2*

We are not limited to integer values of $x$. In some cases we will need to obtain approximations on a calculator. Suppose we let $x = -1.5$. Then we want $f(-1.5) = \frac{1}{3^{-1.5}}$. On the scientific calculator we obtain the approximation thus:

$$3 \ \boxed{y^x} \ 1.5 \ \boxed{+/-} \ \boxed{=} \ \boxed{\frac{1}{x}} \ 5.1961524$$

We round this to the nearest tenth and plot the point $(-1.5, 5.2)$. ■

Even a rapid glance at these two examples allows us to observe certain characteristics of exponential functions. We will state these properties without proof.

**Properties of the Graph of Exponential Functions $f(x) = a^x$**

1. The graph passes through (0, 1).
2. The graph has a horizontal asymptote of the $x$-axis.
3. If $a > 1$, the function is always increasing.
4. If $0 < a < 1$, the function is always decreasing.

Some functions of a similar nature consist of the form $f(x) = k(a^x)$ where $k =$ some real number. They can be graphed readily, since they are similar to the exponential functions discussed previously.

### Exponential Functions with Base e

The base of an exponential function may be an irrational number. The three exponential functions $f(x) = (\sqrt{3})^x$, $f(x) = \pi^x$, and $f(x) = (\sqrt[3]{7})^x$ all have a base that is irrational. We can approximate the irrational numbers fairly easily on a scientific calculator.

One of the most useful exponential functions is $f(x) = e^x$ where $e$ is an irrational number that is approximated by

$$e \doteq 2.718281828459\ldots \qquad \textit{Accurate to 12 decimal places.}$$

The famous Swiss mathematician Leonhard Euler (1707–1783) was the first person to denote this irrational number by the letter "$e$". Euler was able to develop a number of powerful mathematical procedures involving the use of this irrational number.

The function $f(x) = e^x$ is called the **natural exponential function**. It is so named because the base $e$ occurs naturally in describing many of the physical phenomena of our world. Tables of values of $e^x$ have been compiled to find values such as $e^{1.5}$ and $e^2$. Table B at the end of this text provides such values. To obtain values for $e^x$ on a scientific calculator, you enter the exponent then press the $\boxed{e^x}$ button. To obtain values of $e^x$ without a calculator, use Table B.

**Note:** If your scientific calculator does not have an $\boxed{e^x}$ button, use the sequence $\boxed{\text{INV}}\boxed{\text{LN X}}$ or $\boxed{\text{2nd FN}}\boxed{\text{LN X}}$ in place of $e^x$. For further directions, consult the instruction manual of your calculator.

EXAMPLE 3 Graph the function $f(x) = e^x$.

***Solution*** Let us obtain the function values $f(-2)$, $f(-1)$, $f(0)$, $f(1)$, $f(1.5)$, $f(2)$ on the calculator. (If no calculator is available, use Table B.)

$f(-2)$: 2 $\boxed{+/-}$ $\boxed{e^x}$ 0.1353353

$f(-1)$: 1 $\boxed{+/-}$ $\boxed{e^x}$ 0.3678794

$f(0)$: $= 1$ *Obvious problems like this should be done in your head, not on a calculator!*

$f(1)$: 1 $\boxed{e^x}$ 2.7182818

$f(1.5)$: 1.5 $\boxed{e^x}$ 4.4816891

$f(2)$: 2 $\boxed{e^x}$ 7.3890561

The results above are recorded in the following table (values rounded to nearest hundredth) and the graph is shown in Figure 5-3.

| $x$ | $-2$ | $-1$ | 0 | 1 | 1.5 | 2 |
|---|---|---|---|---|---|---|
| $f(x)$ | 0.14 | 0.37 | 1 | 2.72 | 4.48 | 7.39 |

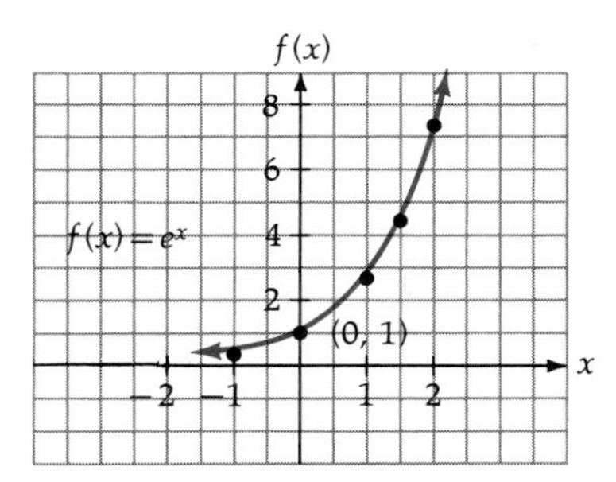


*Figure 5-3*

■

**EXAMPLE 4** Graph the function $f(x) = e^{-x^2}$.

***Solution*** It is important to enter the sequence correctly on the calculator.

$f(2)$: 2 $\boxed{x^2}$ $\boxed{+/-}$ $\boxed{e^x}$ 0.0183156

We follow a similar procedure to find other points and list in this table. (We round values to the nearest hundredth.)

| $x$ | 0 | 0.3 | 0.6 | 1 | 1.5 | 2 |
|---|---|---|---|---|---|---|
| $f(x)$ | 1 | 0.91 | 0.70 | 0.37 | 0.11 | 0.02 |

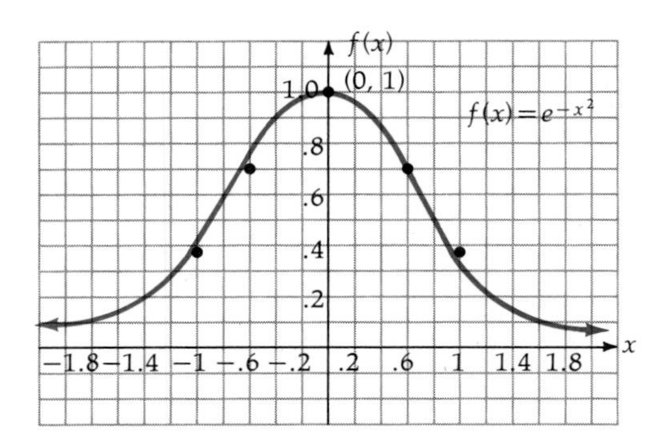


*Figure 5-4*

Since $f(x) = f(-x)$, the function is an even function. Therefore the curve is symmetric to the $y$-axis, as shown in Figure 5-4. We will not need to calculate function values for negative values of $x$, since we can sketch the left-hand side of the curve by symmetry.

We choose a convenient scale for each axis. ■

You may recall seeing a similar curve called a normal curve or a bell curve. This type of curve is used extensively in statistics.

### Compound Interest

Let us assume that a certain amount $P$ of money (called the principal) is invested at an interest rate $r$. The interest is compounded annually and added to the principal. The amount $A$ of money after $t$ years will be determined by the equation

$$\boxed{A = P(1 + r)^t}$$

**EXAMPLE 5** Determine how much money a man will have after three years if he invests \$3000 in a mutual fund that pays 16% interest compounded annually.

***Solution*** We use the equation $A = P(1 + r)^t$, where $P = \$3000$, $r = 0.16$, and $t = 3$.

$$A = 3000(1 + 0.16)^3 = 3000(1.16)^3$$

This can be quickly evaluated on a calculator as follows:

$$3000 \boxed{\times} 1.16 \boxed{y^x} 3 \boxed{=} 4682.688$$

Rounded to the nearest cent, the man will have \$4682.69. ■

Use of the formula is not limited to years. Any time period can be used as long as the rate $r$ corresponds to the same unit of time as the time measurement $t$.

**EXAMPLE 6** The man in Example 5 is considering moving his investment to a different mutual fund that only pays 15.6% annual interest, but the interest is compounded monthly. Would this be a better investment?

***Solution*** We want to base all our time units on months. The annual rate of 15.6% corresponds to $\frac{15.6}{12}$ or 1.3% per month. Thus for a month, $r = 0.013$.

The period of 3 years on a monthly basis corresponds to 36 months. Thus $t = 36$.

$$A = P(1 + r)^t = 3000(1 + 0.013)^{36} = 3000(1.013)^{36}$$

On the calculator we have

3000 [×] 1.013 [$y^x$] 36 [=] 4775.9667

Rounding to the nearest cent, we see that the man will have a total amount of \$4775.97 by investing in the second location.

Thus we see that in terms of total amount earned over the three-year period, the mutual fund that compounds monthly at a slightly lower interest rate is a better investment by \$93.28. ■

### Population Growth

In a similar fashion the equation can be used to provide a model for population growth. In this case the amount $A$ of people is given by

$$A = P(1 + r)^t$$

where $P =$ the original population at some initial time indicated by $t = 0$, $r$ is the rate of population increase per unit of time, and $t$ is the number of time units. It is assumed in this case that the *population rate of increase is constant during the time period.*

Some growth rates are very rapid. Certain bacteria growing in optimum conditions double every hour. Consider the following example.

**EXAMPLE 7** A growth experiment is being conducted in a sealed container that holds 100 bacteria. The number of bacteria is observed to double every three hours. How many bacteria will be present two days (48 hours) after the experiment begins? (Assume that the bacteria continue to double every three hours.)

***Solution*** The basic time unit is a three-hour period. The rate of increase is 100%, so $r = 1.00$. In 48 hours we have $t = \frac{48}{3} = 16$ of these three-hour time periods. The initial population of the bacteria is 100, so $P = 100$.

$$A = P(1 + r)^t = 100(1 + 1)^{16} = 100(2)^{16}$$

On a calculator we obtain

$$100 \boxed{\times} 2 \boxed{y^x} 16 \boxed{=} 6{,}553{,}600$$

Thus if the bacteria continue to double every three hours, in two days the bacteria count will be 6,553,600. ■

### Continuous Growth

We have discussed problems where interest is compounded monthly. Some banks claim to compound interest every day. A few advertise **continuous compounding.** What happens as the interest is compounded more frequently? Is there a limit to how much you will gain by using smaller and smaller time periods?

Let us examine a simple case for $A = P(1 + r)^t$ where the rate of interest is 100% (so $r = 1$). Assume that $P = 1$ and the time period is one year. We thus have $A = 1(1 + 1)^1$ for *compounding yearly.*

If we *compound interest quarterly,* the rate of interest for each quarter is $\frac{1}{4}$ of the yearly rate. Since $r = 1$, we see that in this case $\frac{r}{4} = \frac{1}{4}$. Since the one-year time period is measured in quarters of a year, we would have $t = 4$. Thus $A = 1\left(1 + \frac{1}{4}\right)^4$.

We will now construct a table and evaluate these expressions. The table will be expanded to consider even a larger number of periods per year.

| *How Often Interest Is Compounded* | *Number of Periods per Year* | *Amount after One Year* | *Approximate Value* |
|---|---|---|---|
| Yearly | 1 | $(1 + 1)^1$ | 2.000000 |
| Quarterly | 4 | $\left(1 + \frac{1}{4}\right)^4$ | 2.4414063 |
| Monthly | 12 | $\left(1 + \frac{1}{12}\right)^{12}$ | 2.6130353 |
| Every minute | 525,600 | $\left(1 + \frac{1}{525600}\right)^{525600}$ | 2.7182997 |
| A million times per year | 1,000,000 | $\left(1 + \frac{1}{1{,}000{,}000}\right)^{1{,}000{,}000}$ | 2.7182818 |

Observe that the number in the approximate value column keeps getting closer and closer to $e$ where $e \doteq 2.718281828459 \ldots$.

The expression evaluated is $\left(1 + \frac{1}{x}\right)^x$ where $x$ is becoming larger and larger. The limit of $\left(1 + \frac{1}{x}\right)^x$ as $x$ becomes infinitely large appears to be $e$. In calculus this is proved. The conclusion is stated in this way:

$$\lim \left(1 + \frac{1}{x}\right)^x = e \qquad x \to \infty$$

In general, continuous compounding can be expressed by the formula $A = Pe^{rt}$ where $P =$ the principal, $r =$ the rate per time unit, and $t =$ the number of time units. The time unit is years in most problems in this text.

**EXAMPLE 8** A woman invests \$3000 in a high-interest fund that pays 15.6% annual interest. This fund advertises continuous compounding of interest. How much will she have after three years? Compare the result to Example 6.

***Solution*** The compounding is done continuously, and we will use a time unit of years. Thus

$$A = Pe^{rt} \quad \text{where } P = 3000,\ r = 0.156, \text{ and } t = 3.$$
$$A = 3000e^{(0.156)(3)} = 3000e^{0.468}$$

On a calculator we have

3000 [×] .468 [$e^x$] [=] 4790.3922

We round to the nearest cent. The woman will have a total amount of \$4790.39 in three years with continuous compounding.

From Example 6 with the same principal and rate the man had a total of \$4775.97 in three years with compounding done monthly. The continuous compounding of interest increased the amount by only \$14.42 over the monthly compounding of interest. ■

Examples 5–8 illustrate **exponential increase**. We now turn to consider some examples of exponential decrease. Some physical quantities *decrease* at a rate proportional to the amount of the quantity present at that time. The decrease of electrical current in certain electrical circuits and the decay of a radioactive element are examples of **exponential decrease**.

### Radioactive Decay Using the Half-Life Equation $A = A_0 \left(\frac{1}{2}\right)^{\frac{t}{h}}$

In any radioactive substance, some of the atoms are splitting, leaving non-radioactive by-products. In this way the amount of the substance that is actually radioactive is always decreasing. This decrease is called radioactive decay. The decay is often described in terms of half-life. The **half-life** of a radioactive substance is the amount of time it takes for half of the substance to decay into another element or isotope. Many elements have various forms or isotopes. The number following the name of the element indicates which isotope is being discussed.

It is known that krypton 92 has a half-life of approximately 3 seconds. The amount of krypton 92 remaining if we start with 10 mg of krypton 92 at time $t = 0$ is given in Figure 5-5.

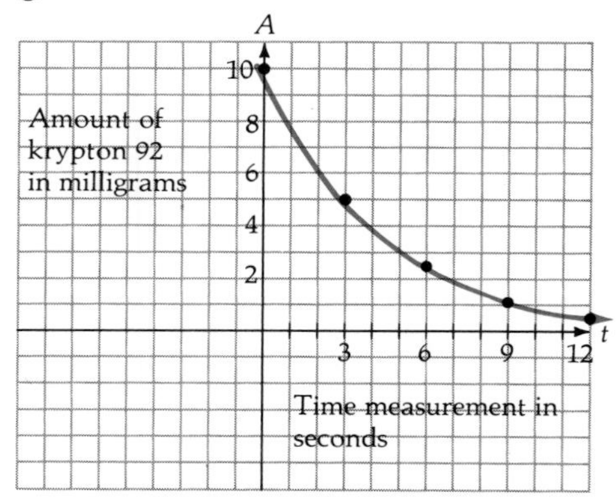


*Figure 5-5*

We could express this equation exactly by $A = 10.0\left(\frac{1}{2}\right)^{\frac{t}{3}}$.

- After 3 seconds we have $10\left(\frac{1}{2}\right)^1$ or 5 mg left.
- After 6 seconds we have $10\left(\frac{1}{2}\right)^2$ or 2.5 mg left.

After $t$ seconds we have $A = 10.0\left(\frac{1}{2}\right)^{\frac{t}{3}}$ mg of krypton 92 left.

For any radioactive substance with a half life of $h$ times units, the amount $A$ of the substance remaining of the original amount $A_0$ after $t$ time units is given by $A = A_0\left(\frac{1}{2}\right)^{\frac{t}{h}}$.

**EXAMPLE 9** How much krypton 92 is left after one minute if we start with 160 mg of the substance?

***Solution*** We use $A = 160\left(\frac{1}{2}\right)^{\frac{t}{3}}$ to represent the amount of Krypton 92 remaining after $t$ seconds. At $t = 60$ seconds (one minute) we have

$$A = 160\left(\frac{1}{2}\right)^{\frac{60}{3}} = 160\left(\frac{1}{2}\right)^{20}$$

On a calculator we have

$$160 \boxed{\times} .5 \boxed{y^x} 20 \boxed{=} 0.0001526$$

Thus if we start with 160 mg of krypton 92, in one minute only 0.0001526 mg of krypton 92 remains. ■

### Radioactive Decay Using the Natural Exponential Equation $A = Ce^{kt}$

The radioactive decay equation is often described in the form

$$A = Ce^{kt}$$

where $C$ is the original amount of the radioactive substance and $k$ is the radioactive decay constant for the radioactive substance. Finding $k$ for a given element is an involved process, and we will not discuss it at this time.

**EXAMPLE 10** Radon 222 has a half-life of 3.823 days. The amount of radon 222 left in $t$ days is given by $A = Ce^{-0.1813t}$. If we had 4.2 mg of radon 222 present in a laboratory 50 days ago, how much would be left now? Use $A = Ce^{-0.1813t}$.

***Solution*** If $t = 50$ and $C = 4.2$, then $A = 4.2e^{-0.1813(50)}$.

*Note:* We can avoid extra steps on the calculator by using the parentheses.

$$4.2 \boxed{\times} \boxed{(} 0.1813 \boxed{+/-} \boxed{\times} 50 \boxed{)} \boxed{e^x} \boxed{=} 0.000485702$$

Rounding to the nearest hundred thousandth of a milligram after 50 days, we would have only 0.00049 mg left of the original 4.2 mg. ■

## EXERCISE 5-1

Sketch the graph of the functions in Problems 1–12. Choose an appropriate scale for the $y$-axis.

**1.** $f(x) = 3^x$ **2.** $f(x) = 4^x$ **3.** $f(x) = 2^{x+2}$ **4.** $f(x) = 3^{x-1}$

**5.** $f(x) = \frac{1}{4}(2^x)$ **6.** $f(x) = \frac{1}{9}(3^x)$ **7.** $f(x) = \left(\frac{1}{4}\right)^x$ **8.** $f(x) = \left(\frac{1}{2}\right)^x$

**9.** $f(x) = \left(\frac{2}{3}\right)^x$ **10.** $f(x) = \left(\frac{3}{4}\right)^x$ **11.** $f(x) = 3^{-x}$ **12.** $f(x) = 4^{-x}$

### Calculator Problems

Graph each of the functions in Problems 13–22. Use Table B if you do not have a calculator.

**13.** $f(x) = 3e^x$ **14.** $f(x) = 2e^x$ **15.** $f(x) = e^{-x}$ **16.** $f(x) = e^{1-x}$

**17.** $f(x) = 5^{x^2}$ **18.** $f(x) = 3^{x^2} + 1$ **19.** $f(x) = 10e^{-0.25x}$ **20.** $f(x) = -15e^{0.40x}$

★**21.** $g(x) = \dfrac{e^x + e^{-x}}{2}$ (This is a specialized function called the hyperbolic cosine that is studied in more advanced mathematics courses.)

★**22.** $h(x) = \dfrac{e^x - e^{-x}}{2}$ (This is a specialized function called the hyperbolic sine that is studied in more advanced math courses.)

### Exponential Increase Problems

A calculator is very helpful in completing these problems.

**23.** How much money will a woman have in five years if she invests $4000 at 14% annual rate of interest compounded annually?

**24.** How much money will a woman have in six years if she invests $5000 at 12% annual rate of interest compounded annually?

**25.** How much money will a man have in three years if he invests $3000 at 15% annual rate of interest compounded quarterly? How much will he have if it is compounded monthly?

**26.** How much money will a man have in four years if he invests $2000 at 17% annual rate of interest compounded quarterly? How much will he have if it is compounded monthly?

**27.** Under optimum conditions the number of bacteria in a certain culture double every 40 minutes. If the initial bacteria count is 500, what will the count be in three hours?

**28.** In a certain developing country the number of households using electricity has been observed to double every 10 years. If 50,000 households have electricity now, what is the predicted number of households that will have electricity 25 years from now?

A certain bank has a special notice account that yields 9.5% annual rate of interest. The bank is considering changing their special notice accounts from having interest compounded *daily* to having interest compounded *continuously*. In Problems 29 and 30, calculate the impact on the specified investment.

**29.** A man plans to invest $4000 for five years. How much more will he earn if the change is implemented?

**30.** A woman plans to invest $6000 for four years. How much more will she earn if the change is implemented?

### Exponential Decrease Problems

**31.** The half-life of curium 242 is 163 days. If a laboratory has 4 mg of the substance now and it is not disturbed, how much will remain one year from now?

**32.** The half-life of einsteinium 253 is 20.7 days. If a laboratory has 2 mg of the substance now and it is not disturbed, how much will remain 60 days from now?

**33.** The radioactive decay of americium 241 can be described by the following equation: The amount $A$ remaining is $A = Ce^{-0.0016008t}$ where $C$ is the original amount and $t$ is the time elapsed in years. If 5 mg are present in a sealed container now, how much would theoretically be present in 1000 years?

**34.** The radioactive decay of radium 226 can be described by the following equation: The amount $A$ remaining is $A = Ce^{-0.0004279t}$ where $C$ is the original amount and $t$ is the time elapsed in years. If 3 mg are present in a sealed container now, how much would theoretically be present in 2000 years?

★ **35.** The charge $Q$ in microcoulombs on a certain discharging capacitor is given by $Q = 7.6e^{-2.8t}$ where $t$ is time measured in seconds. How much charge is lost in the time period from $t = 0.4$ second to $t = 0.8$ second?

★ **36.** The atmospheric pressure measured in pounds per square inch is given by the equation $P = 14.7e^{-0.21d}$ where $d$ is the distance in miles above sea level. What is the decrease in atmospheric pressure if you go from a level valley 100 feet above sea level to a nearby mountain peak that is 8564 feet above sea level?

---

## 5-2 LOGARITHMIC FUNCTIONS

The general topic of logarithms has been a part of the study of mathematics since their invention by John Napier (1550–1617). At that time their main use was to simplify certain numerical computations. They were used as the basis for early computers—the slide rule is a classic example of the use of the **logarithmic scale**.

In our age, people very seldom use logarithms to perform calculations. The advent of the modern computer and the availability of the scientific calculator have diminished the need for a person to know how to calculate with logarithms. Yet the use of logarithms provides the basis for raising numbers to a power on a scientific calculator. For example, when you evaluate $(2.6)^{35}$ on a scientific calculator, the calculator does *not* multiply 2.6 by itself 35 times! Rather it finds the logarithm of 2.6 and then performs two quick operations. Thus the power and speed of a scientific calculator are partially due to the amazing properties of logarithms.

Just what are these logarithms and how can we use them?

***Definition***

The *logarithm* of a positive number $x$ to the base $b$ is the exponent $y$ to which the base $b$ is raised to obtain $x$. ($b$ can represent any positive number except 1.) In symbols we may write

$$\text{if } x = b^y, \quad \text{then} \quad y = \log_b x$$

We read $\log_b x$ as "the logarithm, base $b$, of $x$" or usually as "log, base $b$, of $x$." From the definition we see that

$$8 = 2^3 \qquad \text{and} \qquad \log_2 8 = 3$$

are equivalent statements. $8 = 2^3$ is the **exponential form** and $\log_2 8 = 3$ is the **logarithmic form**.

This may seem a bit confusing at first. It will help to practice changing exponential form to logarithmic form using the definition of the logarithm of a number.

You should memorize the following statement.

$$\boxed{\text{If } x = b^y, \quad \text{then} \quad y = \log_b x}$$

Then study the following examples.

**EXAMPLE 1** Change the following equation to logarithmic form.

$16 = 4^2$

***Solution*** If $x = b^y$, then $y = \log_b x$.

Since $16 = 4^2$ has $x = 16$, $b = 4$, and $y = 2$, then the logarithmic equation is $2 = \log_4 16$. ■

**EXAMPLE 2** Change the following equation to exponential form.

$-2 = \log_5 \dfrac{1}{25}$

***Solution*** The definition can be used in reverse. If $y = \log_b x$, then $x = b^y$.

Since $y = -2$, $b = 5$, and $x = \dfrac{1}{25}$, then the exponential equation is $\dfrac{1}{25} = 5^{-2}$. ■

We can now use the definition to find the value of an unknown in certain logarithmic equations.

**EXAMPLE 3** Find the value of the variable in the given equation.

$\log_b 36 = 2$

***Solution*** If $\log_b 36 = 2$, we have $b^2 = 36$. Since by our definition $b$ is a positive number, the only solution is $b = 6$, since $6^2 = 36$. ■

Now we want to use the definition to answer one additional question: How do I find the logarithm of a number to a given base? In this section we will deal with certain special cases. Consider the following:

**EXAMPLE 4** **a.** Find $\log_3 \frac{1}{9}$ **b.** Find $\log_2 0.25$.

***Solution*** **a.** Let $y = \log_3 \frac{1}{9}$. Then we have the corresponding exponential equation $3^y = \frac{1}{9}$. Now we know that $3^{-2} = \frac{1}{3^2} = \frac{1}{9}$, so therefore we have the solution $y = -2$. Thus $\log_3 \frac{1}{9} = -2$.

**b.** Let $y = \log_2 0.25$. Then $2^y = 0.25$. Now we know that $2^{-2} = \frac{1}{4} = 0.25$; therefore we have the solution $y = -2$. Thus $\log_2 0.25 = -2$. ■

Thus we see that in certain special cases we can "find a logarithm" by setting up a logarithmic equation but solving the corresponding exponential equation.

In order to get an idea of what pattern occurs by finding values of logarithms we need a picture. A sketch of the graph of the logarithmic curve is helpful.

**EXAMPLE 5** Sketch the graph of $y = \log_2 x$.

***Solution*** If we first change from logarithmic form to exponential form, the equation becomes

$$x = 2^y$$

We now set up a table of values choosing values for $y$ and then finding the corresponding values for $x$.

| $x$ | $\frac{1}{8}$ | $\frac{1}{4}$ | $\frac{1}{2}$ | 1 | 2 | 4 | 8 |
|---|---|---|---|---|---|---|---|
| $y$ | $-3$ | $-2$ | $-1$ | 0 | 1 | 2 | 3 |

We then sketch the graph of $y = \log_2 x$.

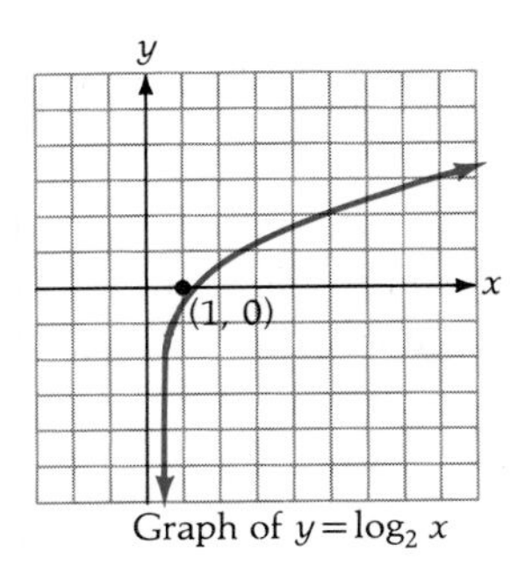

Graph of $y = \log_2 x$

Some observations from the graph regarding the properties of $y = \log_2 x$:

1. $y = \log_2 x$ is a *function*, since each value of $x$ is associated with only one $y$ value.
2. Each $y$ value is associated with only one $x$ value.
3. The point (1, 0) is on the graph for the logarithm to any base.
4. Numbers less than or equal to zero have no logarithm. The domain of the function is $(0, +\infty]$.
5. Numbers greater than zero but less than 1 have negative logarithms.
6. Numbers greater than 1 have positive logarithms.
7. As the value of $x$ increases, the value of $y$ increases.
8. The graph is asymptotic to the $y$-axis.

These facts would hold true for any base greater than 1. A different base would change the rate at which the curve changes direction but not its general shape.

We will find it useful to derive some equations that involve logarithms. First we let

$$y = a^x \quad \text{where} \quad a > 0 \quad \text{and} \quad a \neq 1 \tag{1}$$

By the definition of a logarithm an equivalent equation is

$$x = \log_a y \tag{2}$$

Now take equation (1) and replace $x$ by the expression for $x$ in equation (2). We have

$$\boxed{y = a^{\log_a y}} \tag{3}$$

Now we take equation (3), and in each instance we encounter $y$ we replace it by $a^x$ from equation (1). We have

$$a^x = a^{\log_a a^x} \tag{4}$$

Now a property of exponential functions is that if $a^m = a^n$ then $m = n$.

Applying this property to equation (4) yields

$$x = \log_a a^x \tag{5}$$

Finally, there are two special cases that are most helpful. What is $\log_b 1$? Let $\log_b 1 = y$; then $b^y = 1$. We know for any $b \neq 0$ that $b^0 = 1$, so we know that in this case $y = 0$. We state this as a formal equation.

$$\log_b 1 = 0 \tag{6}$$

Finally, what is $\log_b b$? Let $\log_b b = y$. Then we have $b^y = b$. Clearly, $b^y = b^1$, so $y = 1$. Thus we have the equation

$$\log_b b = 1 \tag{7}$$

The student will find several instances in which a quick recall of equations (3), (5), (6), and (7) will be needed in more advanced work.

The logarithmic function is the inverse of the exponential function studied in Section 5-1. We will state this formally.

***Theorem 5-1*** The exponential and logarithmic functions with base $a$ are *inverse functions.*

***Proof*** Consider any logarithmic function $f(x) = \log_a x$ and an exponential function $g(x) = a^x$ for all $a > 0$, $a \neq 1$ and $x > 0$.

1. We now find $g[f(x)]$ by using the techniques of Section 3-6: $g[f(x)] = a^{f(x)} = a^{\log_a x}$, but $a^{\log_a x} = x$ by equation (3). Thus $g[f(x)] = x$.
2. Next we find $f[g(x)]$:
   $f[g(x)] = \log_a g(x) = \log_a a^x$ but $\log_a a^x = x$ by equation (5). Thus we have $f[g(x)] = x$.
3. Now our definition of the inverse of a function in Section 3-6 implies that $f^{-1}[f(x)] = f[f^{-1}(x)] = x$. Since $g[f(x)] = f[g(x)] = x$ above, we have shown that $g(x)$ is the inverse of $f(x)$ and vice versa. This completes the proof. ■

There is a geometric confirmation that the exponential function and the logarithmic function are inverse functions. The graph of a function and the inverse of that function are symmetric about the line $y = x$.

If we graph for $a > 1$, $y = a^x$, and $y = \log_a x$, we see this property. For points $P$ and $Q$ on the graph of $y = \log_a x$ there correspond points $P'$ and $Q'$ that are on the curve $y = a^x$. If we folded the graph paper along the line $y = x$, the two curves would coincide. (See Figure 5-6.)

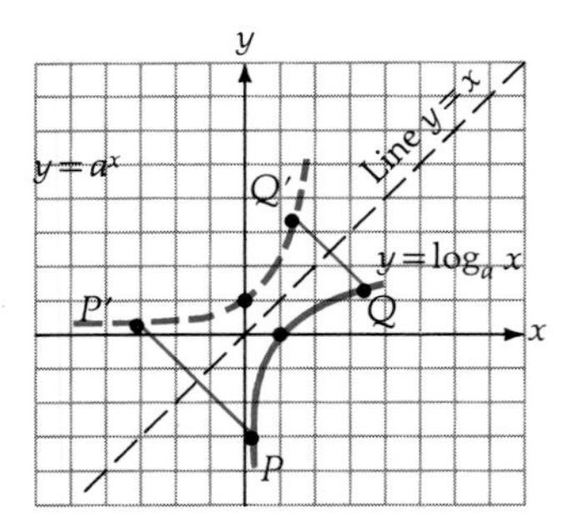


*Figure 5-6* ***Geometrical interpretation showing that $y = a^x$ and $y = \log_a x$ are inverses.***

We now establish what are commonly called the three laws of logarithms.

From the definition of a logarithm we observe that a *logarithm is an exponent*. We then should expect that logarithms would have the properties of exponents and would follow the operational laws of exponents.

There are three elementary laws of logarithms which are important to us. They will allow us to multiply, divide, and find powers by logarithms.

*First Law of Logarithms*

$$\log_b (xy) = \log_b x + \log_b y$$

*Second Law of Logarithms*

$$\log_b \left(\frac{x}{y}\right) = \log_b x - \log_b y$$

*Third Law of Logarithms*

$$\log_b (x)^n = n \log_b x$$

These three laws are consequences of the laws of exponents. We will demonstrate how the first law is derived. Let

$$\log_b x = k \qquad \text{and} \qquad \log_b y = n$$

Then $\log_b x = k$ may be written as $b^k = x$, and $\log_b y = n$ may be written as $b^n = y$. Then

$$xy = b^k b^n = b^{k+n}$$

But $xy = b^{k+n}$ may be written as $\log_b xy = k + n$. And since $\log_b x = k$ and $\log_b y = n$,

$$\log_b xy = \log_b x + \log_b y$$

As an exercise, you may construct a similar proof for the second and third laws. We now look at some applications of these laws.

**EXAMPLE 6** If $\log_b 5 = x$ and $\log_b 3 = y$, find $\log_b 15$.

***Solution***

$$\begin{aligned} \log_b 15 &= \log_b (5)(3) \\ &= \log_b 5 + \log_b 3 \qquad \textit{By the First Law of Logarithms.} \\ &= x + y \quad \blacksquare \end{aligned}$$

**EXAMPLE 7** If $\log_b 5 = x$ and $\log_b 3 = y$, find $\log_b \frac{3}{5}$.

***Solution***

$$\begin{aligned} \log_b \frac{3}{5} &= \log_b 3 - \log_b 5 \qquad \textit{By the Second Law of Logarithms.} \\ &= y - x \quad \blacksquare \end{aligned}$$

**EXAMPLE 8** If $\log_b 5 = x$, find $\log_b 125$.

***Solution***

$$\begin{aligned} \log_b 125 &= \log_b (5)^3 \qquad \textit{Since } 125 = 5^3. \\ &= 3 \log_b 5 \qquad \textit{By the Third Law of Logarithms.} \\ &= 3x \quad \blacksquare \end{aligned}$$

From equation (6) we know that $\log_b 1 = 0$ for any $b$. Thus

$$\boxed{\log_{10} 1 = 0}$$

What is $\log_{10} 10$? Let $\log_{10} 10 = y$. The exponential equation is $10^y = 10$, so $y = 1$. Thus

$$\boxed{\log_{10} 10 = 1}$$

We employ these facts to evaluate a logarithm with a base of ten.

**EXAMPLE 9** Using $\log_{10} 1 = 0$ and $\log_{10} 10 = 1$, evaluate $\log_{10} 100{,}000$.

***Solution***

$\log_{10} 100{,}000 = \log_{10}(1.0 \times 10^5)$ — *Scientific notation.*

$= \log_{10} 1 + \log_{10} 10^5$ — *By the First Law of Logarithms.*

$= \log_{10} 1 + 5 \log_{10} 10$ — *By the Third Law of Logarithms.*

$= 0 + 5(1) = 5$ — *Since* $\log_{10} 1 = 0$ *and* $\log_{10} 10 = 1$. ■

Review the properties of logarithms before you do the exercise set. In this section, do *not* use a calculator. It is important for any student of mathematics to *understand* what logarithms are and how to use the proper laws and properties of logarithms *before* he or she begins to push [log $x$] and [ln $x$] buttons.

## EXERCISE 5-2

Change to logarithmic form in Problems 1–10.

**1.** $3^2 = 9$ **2.** $5^2 = 25$ **3.** $5^3 = 125$ **4.** $49 = 7^2$ **5.** $27 = 3^3$

**6.** $64 = 2^6$ **7.** $\frac{1}{8} = 2^{-3}$ **8.** $3^0 = 1$ **9.** $\frac{1}{16} = 2^{-4}$ **10.** $3^1 = 3$

Change to exponential form in Problems 11–18.

**11.** $\log_2 4 = 2$ **12.** $\log_3 9 = 2$ **13.** $\log_4 16 = 2$ **14.** $\log_3 27 = 3$

**15.** $\log_3 81 = 4$ **16.** $\log_2 \frac{1}{2} = -1$ **17.** $\log_5 125 = 3$ **18.** $\log_2 32 = 5$

Find the value of the variable in Problems 19–34.

**19.** $\log_3 x = 4$ **20.** $\log_3 \frac{1}{3} = y$ **21.** $\log_b 16 = 4$ **22.** $\log_2 \frac{1}{8} = y$

**23.** $\log_b 125 = 3$ **24.** $\log_2 \frac{1}{32} = y$ **25.** $\log_3 x = -2$ **26.** $\log_2 x = -4$

**27.** $\log_7 x = 2$ **28.** $\log_{10} 1 = y$ **29.** $\log_{10} 10 = y$ **30.** $\log_{10} 100 = y$

**31.** $\log_{10} 1{,}000 = y$ **32.** $\log_{10} \frac{1}{10} = y$ **33.** $\log_{10} \frac{1}{100} = y$ **34.** $\log_{10} \frac{1}{1{,}000} = y$

Sketch the graphs of the equations in Problems 35–38.

**35.** $y = \log_3 x$ **36.** $y = \log_{10} x$ **37.** $y = \log_{\frac{1}{4}} x$ **38.** $y = \log_{\frac{1}{2}} x$

**39.** Answer in your own words (do not quote a definition from this textbook): What is a *logarithm*?

**40.** Why can't you take the *logarithm* of a *negative number*?

Evaluate the expressions in Problems 41–46.

**41.** $3^{\log_3 8}$ **42.** $5^{4\log_5 4}$ **43.** $2^{6\log_2 6}$ **44.** $8^{\log_8 2}$ **45.** $\log_5 \sqrt[3]{5}$ **46.** $\log_2 \sqrt[4]{2}$

Use the laws of logarithms to find an expression for each of the Problems 47–54 if $\log_b 5 = x$ and $\log_b 3 = y$.

**47.** $\log_b 9$ **48.** $\log_b \frac{3}{25}$ **49.** $\log_b 75$ **50.** $\log_b 45$

**51.** $\log_b \frac{5}{9}$ **52.** $\log_b \frac{5}{3}$ **53.** $\log_b \frac{9}{25}$ **54.** $\log_b 375$

Using the facts that $\log_{10} 1 = 0$ and $\log_{10} 10 = 1$, evaluate in Problems 55–62.

**55.** $\log_{10} 100$ **56.** $\log_{10} 1{,}000$ **57.** $\log_{10} 10{,}000$ **58.** $\log_{10} \frac{1}{10}$

**59.** $\log_{10} 0.01$ **60.** $\log_{10} 10^4$ **61.** $\log_{10} 10^5$ **62.** $\log_{10} 10^n$

Using the approximation $\log_{10} 3.56 = 0.5514$, evaluate in Problems 63–68.

**63.** $\log_{10} (3.56)(10^2)$ **64.** $\log_{10} (3.56)(10^{-1})$ **65.** $\log_{10} (3.56)(10^{-2})$

**66.** $\log_{10} (3.56)^2$ **67.** $\log_{10} (3.56)(10^{-3})$ **68.** $\log_{10} (3.56)^{\frac{1}{3}}$

★ **69.** Prove the second law of logarithms. ★ **70.** Prove the third law of logarithms.

## 5-3 LOGARITHMS TO VARIOUS BASES FROM A CALCULATOR

### Common Logarithms

Although we can find a logarithm to any base of a given number, the most common bases are a base of 10 and a base of $e$.

Base 10 has commonly been used for hundreds of years because of ease of computation with logarithms to base 10. Logarithms with base 10 are known as **common logarithms**.

Common logarithms are often specified by the expression $\log x$ without any subscript. Thus we state the following:

***Definition***

The *common logarithm* of a number $x$ is given by

$$\log x = \log_{10} x \text{ for all } x > 0$$

A few common logarithms can be found instantly without the aid of a calculator.

**EXAMPLE 1** Without the aid of a calculator, find

**a.** log 100 **b.** log 10,000

***Solution***

**a.** Let $\log 100 = y$; then by the definition of common logarithm, $10^y = 100$. Since $10^2 = 100$, we have $y = 2$. Thus $\log 100 = 2$.

**b.** Let $\log 10{,}000 = y$. We can write $10^y = 10{,}000$. Since $10^4 = 10{,}000$, we have $y = 4$. Therefore $\log 10{,}000 = 4$. ■

If you review the solution to Example 1 briefly, you will begin to see why 10 is such a convenient choice for the base of a logarithm.

Determining most common logarithms requires finding an approximation on a calculator or the use of a common logarithm table, such as Table C.

To find the common logarithm of a number on a scientific calculator, enter the number and then depress the [log x] or [log] button. (On some calculators this requires that two buttons be depressed. In such cases the first one is labelled [2nd Fn] or [INV]. Consult your owner's manual for further information.)

**EXAMPLE 2** On a scientific calculator, find a decimal approximation for:

**a.** log 3.58 **b.** log 8.86 **c.** log 88.6 **d.** log 886

***Solution***

**a.** 3.58 [log x] 0.553883026

**b.** 8.86 [log x] 0.947433722

**c.** 88.6 [log x] 1.947433722

**d.** 886 [log x] 2.947433722 ■

You will observe a pattern in comparing the answers to parts b, c, and d of Example 2. This pattern can be explained easily. Since $\log 8.86 = 0.947433722$, we could write

$$\log 88.6 = \log(8.86 \times 10^1)$$

$$= \log 8.86 + \log 10 \qquad \textit{By the First Law of Logarithms.}$$

$$= 0.947433722 + 1 \qquad \textit{We have previously shown that } \log_{10} 10 = 1.$$

$$= 1.947433722$$

In general, for any number $N$ there exists a real number $r$ such that $1 \leq r \leq 10$ and an integer $k$ such that

$$\log N = \log (r \times 10^k) = \log r + \log 10^k = \log r + k$$

In the examples we have considered so far in this section we have evaluated $\log x$ only when $x > 1$. We have always obtained a positive value.

The common logarithm of a number $x$ such that $0 < x < 1$ will yield a negative number. We would expect such a result from our previous work of graphing log functions, in particular the graphs with base 10. (See Figure 5-7.)

**EXAMPLE 3** On a scientific calculator, find a decimal approximation for:

**a.** log 0.623 **b.** log 0.128 **c.** log 0.002614

***Solution***

**a.** 0.623 [log x] −0.205511953

**b.** 0.128 [log x] −0.892790030

**c.** 0.002614 [log x] −2.582694417

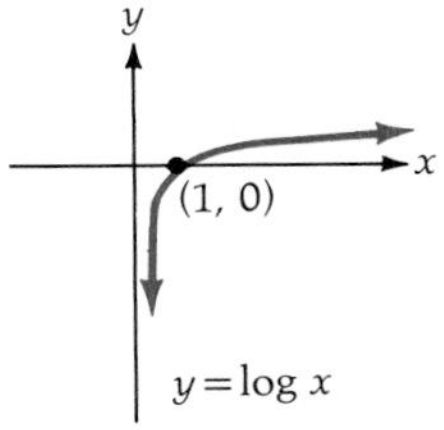

The obtained values on the calculator are consistent with our graph of $y = \log_{10} x$. ■

y
x
(1, 0)
$y = \log x$

*Figure 5-7*

Sometimes the common logarithm of a number is described as having two main parts: a nonnegative decimal fraction between 0 and 1 called the **mantissa** and an integral part called the **characteristic**.

In part c of Example 2 we found that log 88.6 = 1.947433722.

| | |
|---|---|
| In this case, | 1 is the *characteristic* |
| and | 0.947433722 is the *mantissa* |

If a common logarithm is not in that form, we can perform a few mathematical steps to place it in that form. Consider the result from part a of Example 3.

$$\begin{aligned}\log 0.623 &= -0.205511953\\ &= 1 - 0.205511953 - 1 \quad &&\textit{Since } +1 - 1 = 0, \textit{ we have not changed the value.}\\ &= 0.794488047 - 1 \quad &&\textit{Now we combine the first two terms.}\end{aligned}$$

| | |
|---|---|
| In this case, | −1 is the *characteristic* |
| and | 0.794488047 is the *mantissa* |

### Natural Logarithms

For most theoretical work in mathematics and other sciences the most useful base for logarithms is the base $e$. Logarithms with base $e$ are known as **natural logarithms**.

Natural logarithms are often specified by the expression $\ln x$ without any subscript. Thus we will state the following:

*Definition*

The *natural logarithm* of a number $x$ is given by

$$\ln x = \log_e x \quad \text{for all } x > 0$$

Since $e$ is an irrational number that we approximate by $e \doteq 2.718281828$, it is necessary to use a scientific calculator or a natural logarithm table to find the natural logarithm of a number. On a scientific calculator this is usually accomplished by entering the number and depressing the $\boxed{\ln x}$ or $\boxed{\ln}$ button.

**EXAMPLE 4** On a scientific calculator, find

**a.** $\ln 5.62$ **b.** $\ln 56.2$ **c.** $\ln 186{,}324$ **d.** $\ln 0.00264$

***Solution***

**a.** 5.62 $\boxed{\ln x}$ 1.726331664

**b.** 56.2 $\boxed{\ln x}$ 4.028916757

**c.** 186324 $\boxed{\ln x}$ 12.13524237

**d.** 0.00264 $\boxed{\ln x}$ −5.936976362 ■

Whether you are dealing with common logarithms, natural logarithms, or logarithms to any base, you should remember that $\log x$, $\ln x$, and $\log_b x$ are *defined* only *for* $x > 0$. You *cannot* take the logarithm of a negative number or take the logarithm of zero.

**EXAMPLE 5** Attempt to evaluate these impossible expressions on a scientific calculator.

**a.** $\ln 0$ **b.** $\ln(-2.52)$

***Solution*** When you enter the appropriate number and then $\boxed{\ln x}$, you will obtain an error message. ■

It can be shown that $\ln x$ and $e^x$ are inverse functions. Similarly, we could show that $\log x$ and $10^x$ are inverse functions. In general, $\log_b x$ and $b^x$ are inverse functions for all $x > 0$, as we stated in Theorem 5-1.

### Inverse Logarithms

We know that for any base $a$,

$$a^{\log_a y} = y \qquad \textit{From equation 3 in Section 5-2.}$$

We can use this property to find *inverse logarithms* or *antilogarithms.*

**EXAMPLE 6** Find $x$ if $\log x = 2.31$.

***Solution*** We know that the common log equation $\log x = 2.31$ is equivalent to $10^{2.31} = x$. On a calculator we have

2.31 [$10^x$] 204.1737945

(If your scientific calculator does not have a $10^x$ button, you can usually use [2nd] [$\log x$] or [INV] [$\log x$] as an equivalent.) ■

**EXAMPLE 7** Find $x$ if $\ln x = 0.8614$.

***Solution*** We know that the natural log equation $\ln x = 0.8614$ is equivalent to $e^{0.8614} = x$. Thus we evaluate on the calculator

0.8614 [$e^x$] 2.366471436

(If your scientific calculator does not have an $e^x$ button, you can usually use [2nd] [$\ln x$] or [INV] [$\ln x$] as an equivalent.) ■

### Logarithms to Bases Other Than 10 or e

A student will not frequently encounter logarithms to bases other than 10 or $e$. If such an expression is encountered, it can be evaluated by the *change of base formula.*

***Change of Base Formula***

$$\log_b x = \frac{\log_a x}{\log_a b}$$

The change of base formula $\log_b x = \dfrac{\log_a x}{\log_a b}$ is usually used in terms of common logarithms or in terms of natural logarithms.

***Change of Base Formula for Common Logarithms***

$$\log_b x = \frac{\log x}{\log b}$$

***Change of Base Formula for Natural Logarithms***

$$\log_b x = \frac{\ln x}{\ln b}$$

**EXAMPLE 8** Evaluate on a calculator using common logarithms.

**a.** $\log_5 68$ **b.** $\log_2 3.1642$

***Solution*** **a.** $\log_5 68 = \dfrac{\log 68}{\log 5}$ *The change of base formula for common logarithms.*

On a calculator we have

68 [log x] [÷] 5 [log x] [=] 2.621727544

**b.** $\log_2 3.1642 = \dfrac{\log 3.1642}{\log 2}$ *The change of base formula for common logarithms.*

3.1642 [log x] [÷] 2 [log x] [=] 1.661840791 ■

## EXERCISE 5-3

Use a scientific calculator to approximate the following logarithms. (Your answers may differ slightly from those in the answer key at the back of the book, depending on the number of digits displayed on your calculator and the internal procedure used in your calculator to obtain logarithms.)

Evaluate the common logarithms in Problems 1–10.

**1.** $\log 5.63$ **2.** $\log 8.15$ **3.** $\log 127.13$ **4.** $\log 318.24$ **5.** $\log 0.1683$
**6.** $\log 0.4251$ **7.** $\log 0.0001683$ **8.** $\log 0.0004251$ **9.** $\log 13643$ **10.** $\log 76182$

Evaluate the natural logarithms in Problems 11–20.

**11.** $\ln 7.062$ **12.** $\ln 5.135$ **13.** $\ln 0.7413$ **14.** $\ln 0.8613$ **15.** $\ln 3418$
**16.** $\ln 2983$ **17.** $\ln 341800$ **18.** $\ln 298300$ **19.** $\ln 0.00536$ **20.** $\ln 0.00247$

Evaluate the inverse logarithms in Problems 21–28 by changing to exponent form and solving for $x$.

**21.** $\log x = 0.3712$ **22.** $\log x = 0.5148$ **23.** $\log x = 3.6413$
**24.** $\log x = 4.0193$ **25.** $\log x = -1.7418$ **26.** $\log x = -2.4733$
**27.** $\log x = 0.6124 - 3$ **28.** $\log x = 0.1527 - 2$

Evaluate the inverse logarithms in Problems 29–36 by changing to exponent form and solving for $x$.

**29.** $\ln x = 0.8913$ **30.** $\ln x = 0.5617$ **31.** $\ln x = 3.621843$ **32.** $\ln x = 4.712463$
**33.** $\ln x = 7.3182006$ **34.** $\ln x = 9.0615428$ **35.** $\ln x = -4.4712$ **36.** $\ln x = -5.0892$

Use the change of base formula and common logarithms to evaluate in Problems 37–40.

**37.** $\log_3 58.2$ **38.** $\log_4 77.6$ **39.** $\log_7 1.3684$ **40.** $\log_9 2.1641$

Use the change of base formula and natural logarithms to evaluate in Problems 41–46.

**41.** $\log_5 12683$ **42.** $\log_6 57122$ **43.** $\log_2 0.0618$

**44.** $\log_3 0.1386$ **45.** $\log_8 (41.3217)$ **46.** $\log_{12} (82.0186)$

Pick a convenient numerical value for $x$, $y$, and $b$ and verify on the calculator that the properties of logarithms in Problems 47–53 are true.

**47.** $\log_b (xy) = \log_b x + \log_b y$ **48.** $\log_b \left(\frac{x}{y}\right) = \log_b x - \log_b y$ **49.** $\log_b x^n = n \log_b x$

**50.** $\log_b 1 = 0$ **51.** $\log_b b^x = x$ **52.** $\log_b b = 1$ **53.** $a^x = b^{x \log_b a}$

★ **54.** Let $x = \pi$ and let $a = 3$. Verify on your calculator that $\log_a \left(\frac{1}{x}\right) = \log_{(\frac{1}{a})} x$.

★ **55.** Let $x = \sqrt{7}$ and let $a = 10$. Verify on your calculator that $a^{\log_a x} = x$.

---

## 5-4 COMMON LOGARITHMS, INTERPOLATION, AND COMPUTATION FROM A TABLE (OPTIONAL)

### Finding Common Logarithms from a Table

When finding common logarithms from a table of common logarithms, it is absolutely essential to be able to rapidly write numbers in scientific notation and likewise rapidly convert numbers from scientific notation to ordinary notation.

Be sure you are absolutely clear what is meant by scientific notation. A number is expressed in **scientific notation** if it is written as a product of two factors, one of which is a number equal to or greater than 1 but less than 10 and the other is an integer power of 10.

EXAMPLE 1 Write each of the following in scientific notation.

**a.** 26,800 **b.** 268 **c.** 0.0268 **d.** 2.68

***Solution***

**a.** $26{,}800 = 2.68 \times 10^4$

**b.** $268 = 2.68 \times 10^2$

**c.** $0.0268 = 2.68 \times 10^{-2}$

**d.** $2.68 = 2.68 \times 10^0$ ■

We now have three facts that will aid us in finding logarithms of numbers from a table of common logarithms. (Remember that common logarithms have a base of ten.)

1. Every decimal number can be written in scientific notation.
2. $\log 10^n = n$.
3. $\log xy = \log x + \log y$.

We will now discuss finding the value of a common logarithm using Table C. Remember these facts about the table.

1. Every entry in column $N$ represents a number $x$ such that $1 \leq x < 10$.
2. Every logarithm given is a number $y$ such that $0 \leq y < 1$.

The entry in the table (logarithm of a number from 1 to 10) is referred to as the **mantissa**. The integer part of the logarithm (the exponent of 10 when the number is in scientific notation) is called the **characteristic**.

A step-by-step method of finding the logarithm of any number follows:

*Step 1.* Write the number in scientific notation.

*Step 2.* Use the table to find the mantissa.

*Step 3.* Add the characteristic (exponent of 10).

**EXAMPLE 2** Find log 5280 in Table C.

***Solution*** $\log 5280 = \log (5.28 \times 10^3) = \log 5.28 + \log 10^3 = \log 5.28 + 3$

Now we look up log 5.28 in Table C and obtain

$$= 0.7226 + 3$$

This answer may be left as $0.7226 + 3$ or written as 3.7226. (Notice that this answer is consistent with a calculator value $\log 5280 = 3.722633923$ if the calculator value is rounded to four decimal places.) ■

**EXAMPLE 3** Find log 0.0184 by using Table C.

***Solution*** $\log 0.0184 = \log (1.84 \times 10^{-2})$

$= \log 1.84 - 2 = 0.2648 - 2$ *(by evaluating* log 1.84 *in Table C)*

This answer may be left as $0.2648 - 2$ or written as $-1.7352$. Usually, it is more convenient to leave it as $0.2648 - 2$ if you are going to do computations with logarithms. (Notice that if you find log 0.0184 on a calculator, you will obtain

$$\log 0.0184 = -1.735182177$$

If we round that to the nearest ten thousandth, we obtain $-1.7352$, which agrees with our work above.) ■

### Finding the Inverse of a Common Logarithm from a Table (Finding Antilogarithms)

*Definition*

> The number having a certain logarithm is called the *antilogarithm* of the given logarithm.

The problem "Find the antilogarithm of 0.8235" is the same as the problem "Find $N$ if $\log N = 0.8235$."

To find the antilogarithms, we must reverse the procedure for finding logarithms. A step-by-step procedure follows.

*Step 1.* Separate the logarithm into mantissa and characteristic. (Remember that all mantissas are positive and between 0 and 1.)

*Step 2.* Use the table to find the antilogarithm of the mantissa.

*Step 3.* Use the characteristic as the exponent of 10 and multiply by the antilogarithm of the mantissa found in Step 2.

**EXAMPLE 4** Find the antilogarithm of 0.0899 using Table C.

***Solution*** We look on the inside of the common logarithm table (Table C) and find .0899 We see that .0899 lies in the row where $N = 1.2$ and under the 3 column.

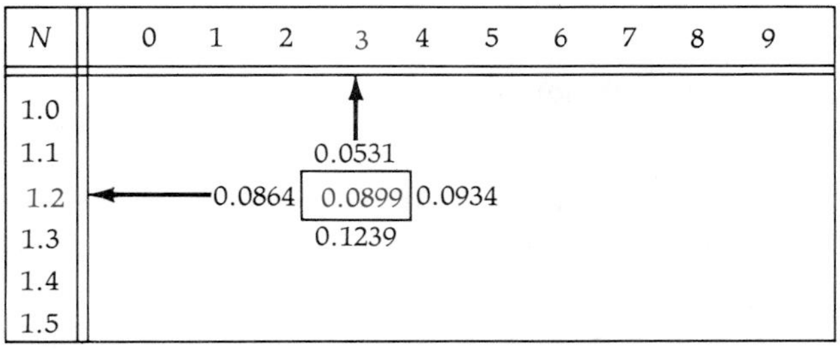

| $N$ | 0 | 1 | 2 | 3 | 4 | 5 | 6 | 7 | 8 | 9 |
|---|---|---|---|---|---|---|---|---|---|---|
| 1.0 | | | | | | | | | | |
| 1.1 | | | | 0.0531 | | | | | | |
| 1.2 | | | 0.0864 | 0.0899 | 0.0934 | | | | | |
| 1.3 | | | | 0.1239 | | | | | | |
| 1.4 | | | | | | | | | | |
| 1.5 | | | | | | | | | | |

Thus the antilogarithm of 0.0899 is 1.23. ■

The logarithm should be separated into the two parts: the decimal part (*mantissa*) and then the integer part (*characteristic*) if the characteristic is not zero. This additional step is shown in the next three examples.

**EXAMPLE 5** Find the antilogarithm of 3.4393 using Table C.

***Solution***

$$3.4393 = 0.4393 + 3$$

The antilogarithm of 0.4393 is 2.75 (from Table C); therefore

$$\text{antilog } 3.4393 = 2.75 \times 10^3 \qquad \textit{Characteristic is 3.}$$
$$= 2750 \quad \blacksquare$$

**EXAMPLE 6** Find $N$ if $\log N = 0.3284 - 3$ using Table C.

***Solution***

$$\text{antilog } 0.3284 = 2.13$$

Therefore $N = 2.13 \times 10^{-3} = 0.00213$ ■

**EXAMPLE 7** Find the antilogarithm of $-2.2984$ using Table C.

***Solution*** Step 1 is a little more difficult in this case, since the given logarithm is negative. It is necessary to rename $-2.2984$ so that the mantissa is positive. This can be accomplished by adding and then subtracting 3. Thus

$$-2.2984 = (-2.2984 + 3) - 3 = 0.7016 - 3$$

Then $\text{antilog } 0.7016 = 5.03$

Therefore $\text{antilog } (0.7016 - 3) = 5.03 \times 10^{-3} = 0.00503$ ■

The inverse of taking a common log of $x$ is raising 10 to the $x$ power. Thus to find an antilogarithm on a calculator, we would use the $\boxed{10^x}$ button. (If the $\boxed{10^x}$ button is not on your calculator, use the $\boxed{\text{INV}}$ $\boxed{\log x}$ or the $\boxed{\text{2nd Fn}}$ $\boxed{\log x}$ sequence.) Thus to verify Example 7 on a calculator, we would find $10^{-2.2984}$ and obtain 0.005030371. If we round the calculator value to five decimal places, it agrees with our table result from Table C.

### Interpolation of Table Values

In general, Table C that we are using gives the logarithms of numbers between 1 and 10 if the numbers have no more than two decimal places (i.e., it is written correct to hundredths). If a number has more than two decimal places, we must find the mantissa by a procedure called **interpolation**.

If a number has more than three decimal places, it should be rounded off to three places, since interpolation for more than one place beyond that given in a table is not accurate enough to be meaningful. Finding a logarithm by interpolation from Table C will be correct to four decimal places or within $\frac{1}{10{,}000}$.

**EXAMPLE 8** Find log 1.236 by using interpolation with Table C.

***Solution*** 1.236 is between 1.23 and 1.24, each of which can be found in the table.

$$\log 1.23 = 0.0899$$
$$\log 1.24 = 0.0934$$

A pattern such as given below might be helpful.

$$10\left[\begin{array}{l} 6\left[\begin{array}{l}\log 1.230 = 0.0899\\ \log 1.236 = ?\end{array}\right]x \\ \quad\log 1.240 = 0.0934\end{array}\right]35$$

Since we are dealing with proportional parts, it is sufficient to ignore the decimals in interpolation. We actually have the proportion

$$\frac{6}{10} = \frac{x}{35}$$

Solving this, we obtain

$$x = \frac{6}{10}(35) = 21$$

Adding the 21 to 899 gives us the 920, which is, of course, the 0.0920 we seek. Thus $\log 1.236 = 0.0920$. ■

**Note:** If we evaluated log 1.236 on a calculator, we would have $\log 1.236 =$ **0.092018471**. We see our approximation above is correct to four decimal places.

In general, linear interpolation is used frequently in higher mathematics and applied mathematical applications. It is a very good way to approximate as long as that curve we are approximating is very close to a straight line for an interval involving small changes in the values of $x$.

**EXAMPLE 9** Find log 5384 by using interpolation.

***Solution***

$$\log 5384 = \log (5.384 \times 10^3) = \log 5.384 + 3$$

$$10\left[\begin{array}{l} 4\left[\begin{array}{l}\log 5.380 = 0.7308\\ \log 5.384 = ?\end{array}\right]x \\ \quad\log 5.390 = 0.7316\end{array}\right]8$$

$$\frac{4}{10} = \frac{x}{8}$$

$$x = \frac{4}{10}(8)$$

$$= 3.2 \quad \text{which rounds to 3}$$

(Since we are ignoring the decimals in this process, we will always round to the nearest whole number in this step.) Adding 3 to 7308, we obtain 7311. Therefore

$$\log 5.384 = 0.7311$$

and

$$\log 5384 = 0.7311 + 3 = 3.7311 \quad ■$$

The need for interpolation of course is avoided when we use a calculator, which will give a higher degree of accuracy than our interpolation process.

**EXAMPLE 10** Find the antilog of 0.8714 using interpolation.

***Solution*** We look in the table for a mantissa of 0.8714 and find that it is between the two entries 0.8710 and 0.8716.

$$10\left[\begin{array}{l} x\left[\begin{array}{l}\log 7.430 = 0.8710 \\ \log ? \quad\;\; = 0.8714\end{array}\right] 4 \\ \quad \log 7.440 = 0.8716 \end{array}\right] 6$$

We solve the proportion

$$\frac{x}{10} = \frac{4}{6}$$

$$x = \frac{4}{6}(10)$$

$$\doteq 6.7 \quad \text{which rounds to 7}$$

So $$0.8714 = \log 7.437$$

or $$\text{antilog of } 0.8714 = 7.437$$

*Note:* To verify that this extra work of interpolation is really providing more accurate information, note that the calculator value of $10^{0.8714} =$ 7.437037990. Our answer above is *correct to the nearest thousandth!* ■

**EXAMPLE 11** Find $N$ if $\log N = -3.9518$. Use interpolation with Table C.

***Solution*** First we must change the form of $-3.9518$ to $0.0482 - 4$ by adding and subtracting 4. This gives us a positive mantissa of 0.0482. We then proceed as before.

$$10\left[\begin{array}{l} x\left[\begin{array}{l}\log 1.110 = 0.0453 \\ \log ? \quad\;\; = 0.0482\end{array}\right] 29 \\ \quad \log 1.120 = 0.0492 \end{array}\right] 39$$

$$\frac{x}{10} = \frac{29}{39}$$

$$x = \frac{29}{39}(10)$$

$$\doteq 7.4 \quad \text{which rounds to 7}$$

Hence $\quad 0.0482 = \log 1.117$

and $\quad 0.0482 - 4 = \log (1.117 \times 10^{-4})$

So $\quad N = 0.0001117$ ■

### Computations

Logarithms were originally invented to do tedious calculations long before calculators and computers were invented. In this section we shall see how various products, divisions, powers, and roots can be done by using logarithms.

Two important properties will be used in much of our work with logarithms.

1. If $M = N$ (both $M$ and $N$ being positive), then $\log_b M = \log_b N$.
2. If $\log_b M = \log_b N$, then $M = N$.

EXAMPLE 12 Evaluate (5.280)(361.0) using logarithms. (Use Table C.)

***Solution*** Since the log of a product is the sum of the logs (the First Law of Logarithms), we may write

$$\begin{aligned} \log (5.280)(361.0) &= \log 5.28 + \log 361 = \log 5.28 + \log (3.61 \times 10^2) \\ &= 0.7226 + 2.5575 = 3.2801 = 0.2801 + 3 \\ &= \log (1.906 \times 10^3) \qquad \textit{By interpolation.} \\ &= \log 1906 \end{aligned}$$

Hence (5.280)(361.0) = 1906.

*Note:* The exact product is (5.280)(361.0) = 1906.08. Clearly, our answer is correct to four significant digits. ■

The Third Law of Logarithms states

$$\log (x)^n = n \log x$$

This law is used where $n$ is any rational number.

EXAMPLE 13 Evaluate $(0.0432)^8$ using logarithms.

***Solution***

$$\begin{aligned} \log (0.0432)^8 &= 8 \log 0.0432 = 8(0.6355 - 2) \\ &= 5.0840 - 16 = 0.0840 - 11 \\ &= \log (1.213 \times 10^{-11}) \qquad \textit{By interpolation.} \\ &= \log 0.00000000001213 \end{aligned}$$

Therefore $\quad (0.0432)^8 = 0.00000000001213$ ■

**Note:** A noteworthy thing about Example 13 is that we have done the problem in exactly the fashion of a scientific calculator. When you use the $\boxed{y^x}$ key on a scientific calculator, the calculator uses logarithms rather than repeated multiplication. What you have calculated by hand above is done in microseconds in the circuits of a scientific calculator when the problem $(0.0432)^8$ is entered on the calculator.

**EXAMPLE 14** Evaluate $\sqrt[3]{0.0357}$ using logarithms.

***Solution*** $\log \sqrt[3]{0.0357} = \log (0.0357)^{\frac{1}{3}} = \frac{1}{3} \log 0.0357 = \frac{1}{3}(0.5527 - 2)$

Since $0.5527 - 2$ divided by 3 would give a fractional characteristic, it is convenient to rewrite the expression as $1.5527 - 3$. Then

$$\log \sqrt[3]{0.0357} = \frac{1}{3}(1.5527 - 3) = 0.5176 - 1$$

$$= \log (3.293 \times 10^{-1}) \qquad \textit{By interpolation.}$$

$$= \log 0.3293$$

Thus $\sqrt[3]{0.0357} = 0.3293$ ■

### Change of Bases Using a Common Logarithm Table

We can use Table C to find the logarithm to any other base. In such cases we will be using the change of base formula

$$\log_b x = \frac{\log_a x}{\log_a b}$$

where $a = 10$, since Table C is a common log table.

**EXAMPLE 15** Find $\log_5 45$ using only Table C.

***Solution*** We may use our base 10 table by applying the formula.

$$\log_5 45 = \frac{\log 45}{\log 5}$$

Find the values of log 45 and log 5 in Table C. We have

$$\log_5 45 = \frac{1.6532}{0.6990}$$

We are now faced with the problem of dividing 1.6532 by 0.6990.

We must either perform this computation by long division or by using logarithms. We will use logarithms. We will evaluate each logarithm using Table C and interpolation.

$$\log \frac{1.6532}{0.6990} = \log 1.6532 - \log 0.6990$$

$$= 0.2183 - (0.8445 - 1) = 0.3738 = \log 2.365$$

Therefore $\dfrac{1.6532}{0.6990} = 2.365$

Hence $\log_5 45 = 2.365$

Note that the original problem in the example, find $\log_5 45$, could have been stated "Find $x$ if $5^x = 45$," since

$$\log_5 45 = 2.365$$

in exponential form is $5^{2.365} = 45$ ■

In order to find natural logarithms using a common logarithm table (such as Table C) we will need the fact that a four-digit approximation of $e$ is 2.718. Using interpolation, we can calculate that $\log 2.718 = 0.4343$. Thus we can change from common logarithms to natural logarithms. (We will always use the notation $\ln x$ for natural logarithms rather than $\log_e x$.)

**EXAMPLE 16** Find ln 25 by using only Table C.

***Solution*** $\ln 25 = \dfrac{\log 25}{\log e} = \dfrac{1.3979}{0.4343}$

Again using logarithms,

$$\log \frac{1.3979}{0.4343} = \log 1.3979 - \log 0.4343$$

$$= 0.1455 - (0.6378 - 1) \qquad \textit{By interpolation.}$$

$$= 0.5077 = \log 3.219$$

Therefore $\dfrac{1.3979}{0.4343} = 3.219$

Hence $\ln 25 = 3.219$

*Note:* On a scientific calculator we obtain $\ln 25 = 3.218875825$. We see that our answer to Example 16 is correct to the nearest thousandth. ■

## EXERCISE 5-4

Use Table C for all problems in Exercise 5-4.

Find the common logarithms of the numbers in Problems 1–10. Check your answers using a calculator having the logarithmic function.

**1.** 2.35 **2.** 6.82 **3.** 1.93 **4.** 2.3 **5.** 7.9

**6.** 28.3 **7.** 37.1 **8.** 468 **9.** 329 **10.** 8040

★ **11.** In chemistry the pH (hydrogen potential) of a solution is defined to be pH $= -\log [H^+]$ where $[H^+]$ is the concentration of hydrogen ions in moles per liter. Find the pH of a solution in which the concentration of hydrogen ions is $3.0 \times 10^{-4}$ moles per liter.

★ **12.** From the information in Problem 11, what is the pH of a solution if the concentration of hydrogen ions is $2.78 \times 10^{-4}$ moles per liter?

Find the antilogarithms in Problems 13–16. Check answers using a calculator.

**13.** 3.8797 **14.** 5.4843 **15.** −1.3468 **16.** −3.0958

Find $N$ in each of Problems 17–20.

**17.** $\log N = 0.7789 - 3$ **18.** $\log N = 0.8476 - 5$

**19.** $\log N = 7.5378$ **20.** $\log N = 4.6702$

Use interpolation to find the common logarithms of the numbers in Problems 21–24. Check answers using a calculator.

**21.** 9282 **22.** 0.01434 **23.** 0.2986 **24.** 367.4

**25.** It was mentioned that using interpolation to find the square of a number or the logarithm of a number is not totally accurate. Why is this so?

**26.** What is the most accuracy we can expect from interpolation in the table of common logarithms (Table C)?

Use interpolation to find $N$ for each of Problems 27–32. Check using a calculator.

**27.** $\log N = 0.2345 - 2$ **28.** $\log N = 3.6200$ **29.** $\log N = 4.0362$

**30.** $\log N = 0.5293 - 3$ ★ **31.** $\log N = -2.6807$ ★ **32.** $\log N = -0.7921$

Evaluate in Problems 33–38 using logarithms. Find all logarithms by using Table C. Check without logarithms by using a calculator.

**33.** $(23)(58)$ **34.** $(593)(26.8)$ **35.** $\dfrac{0.0154}{0.0023}$

**36.** $\dfrac{0.254}{34.6}$ **37.** $\dfrac{(245)(30.1)}{(58.6)(24)}$ **38.** $\dfrac{(0.001593)(23.8)}{(0.014)(25.5)}$

Evaluate in Problems 39–46 using logarithms. Find all logarithms by using Table C. Check answers using a calculator with a power function (remember you may convert roots to fractional powers).

**39.** $(0.21)^4$ **40.** $(15.3)^5$ **41.** $\sqrt{0.0951}$ **42.** $\sqrt[5]{0.687}$

**43.** $\dfrac{(321)^5}{\sqrt{4.638}}$ **44.** $\dfrac{(28.6)^4}{\sqrt[3]{9.361}}$ ★**45.** $\dfrac{(8.613)\sqrt[3]{4.123}}{(1.364)^2}$ ★**46.** $\sqrt[3]{\dfrac{(3.621)(5.36)^2}{2.187}}$

Find the logarithms in Problems 47–50. Use only Table C, then check your answers with a calculator.

**47.** $\log_2 29$ **48.** $\log_5 35$ **49.** $\log_9 613$ **50.** $\log_3 0.035$

## 5-5 EXPONENTIAL AND LOGARITHMIC EQUATIONS

In certain equations the variable appears as an exponent or in a logarithmic expression. We will now look at the methods for solving such equations.

**Hint:** Many logarithmic equations may be solved by using the property if $\log M = \log N$ then $M = N$. We often seek the solution of logarithmic equations by trying to obtain the logarithm of one expression on each side of the equation. However, we must remember that *you cannot take the logarithm of zero or a negative number.*

EXAMPLE 1 Solve for $x$: $\log (x + 1) = \log x + \log 3$.

***Solution*** We use the First Law of Logarithms on the right-hand side.

$$\log (x + 1) = \log 3x$$

$$x + 1 = 3x \qquad \textit{If } \log M = \log N, \textit{ then } M = N.$$

$$1 = 2x$$

$$\frac{1}{2} = x$$

*Check* Can you verify that this is a solution to the original equation? ■

EXAMPLE 2 Solve for $x$: $\log (2x - 1) = \log (x + 2) + \log (x - 2)$.

***Solution*** Again we apply the First Law of Logarithms to the right-hand side.

$$\log (2x - 1) = \log (x + 2)(x - 2)$$

$$2x - 1 = (x + 2)(x - 2)$$

$$2x - 1 = x^2 - 4$$

$$0 = x^2 - 2x - 3$$

$$0 = (x - 3)(x + 1)$$

Thus we have two *possible* solutions.

$$x = 3 \qquad \text{or} \qquad x = -1$$

*Check* Checking our answers, we see that 3 is a solution, but if we substitute $x = -1$ in the last term of the equation $\log(x - 2)$, we obtain $\log(-3)$. But we cannot find the logarithm of a negative number. Thus $-1$ is not a solution of the original equation. Therefore the solution is 3. ■

**Note:** The major concern in checking answers for the apparent solutions to logarithmic equations is to be sure that *you do not obtain an expression* that involves the *logarithm of a nonpositive number*!

**EXAMPLE 3** Solve for $x$: $\log(2x - 1) = 1 - \log(x + 4)$. Express any approximate answers to the nearest hundredth.

***Solution***

$$\log(2x - 1) + \log(x + 4) = 1$$
$$\log(2x - 1)(x + 4) = 1$$
$$\log(2x^2 + 7x - 4) = 1$$

Now this logarithmic equation can be written as an equivalent exponential equation:

$$2x^2 + 7x - 4 = 10^1$$
$$2x^2 + 7x - 14 = 0$$

$$x = \frac{-7 \pm \sqrt{(7)^2 - 4(2)(-14)}}{2(2)}$$

*Since the quadratic will not factor, we use the quadratic formula.*

$$x = \frac{-7 \pm \sqrt{49 + 112}}{4}$$

$$x = \frac{-7 \pm \sqrt{161}}{4}$$

To the nearest hundredth, $x = 1.42$ and $x = -4.92$. We have two *possible* solutions.

*Check* We immediately discard $x = -4.92$, since $\log(x + 4)$ when $x = -4.92$ would yield the log of a negative number. If $x = 1.42$, we have

$$\log[2(1.42) - 1] = 1 - \log(1.42 + 4)$$

On a calculator we evaluate each side and round to nearest hundredth.

$$0.26 \doteq 1 - 0.73$$
$$0.26 \doteq 0.27$$

These are approximately equal. The fact that they disagree by one unit in the hundredths place is due to our roundoff. Thus $x = 1.42$ is a valid approximate solution. ■

The base of logarithms in a logarithmic equation can be any positive constant value. It is not necessarily 10 or $e$.

**EXAMPLE 4** Solve for $x$: $\log_8 4 = \log_8 (x + 1) - \log_8 (x)$.

***Solution*** Simplify the right-hand side by applying the Second Law of Logarithms:

$$\log_8 4 = \log_8 \left(\frac{x+1}{x}\right)$$

Thus we have $4 = \dfrac{x+1}{x}$.

Since $x \neq 0$ by definition of the original equation, we can multiply both sides by $x$:

$$4x = x + 1$$

$$3x = 1$$

$$x = \frac{1}{3}$$

*Check* See whether you can verify this answer. ■

In solving exponential equations we will often use the property that if $M, N > 0$ and $M = N$, then $\log M = \log N$. In other words, we take the logarithm of both sides of the equation.

**EXAMPLE 5** Solve for $x$: $3^x = 5$.

***Solution*** In this equation the variable occurs in the exponent. In such cases it is usually necessary to take the log of both sides of the equation.

$$3^x = 5$$

$$\log 3^x = \log 5$$

$$x \log 3 = \log 5 \qquad \textit{We use the Third Law of Logarithms.}$$

$$x = \frac{\log 5}{\log 3} \qquad \textit{Divide each side by } \log 3.$$

We will leave this solution in terms of logarithms. To find a numerical answer would require only that we use the table to find log 5 and log 3 and then divide, or else evaluate using a calculator.

On a calculator we would do this:

5 [log] [÷] 3 [log] [=] 1.464973521 ■

**EXAMPLE 6** Solve for $x$: $3^{x-4} = 2^{x+1}$. Leave your answer in terms of logarithms.

***Solution*** Taking the log of each side, we obtain

$$\log 3^{x-4} = \log 2^{x+1}$$

$$(x - 4) \log 3 = (x + 1) \log 2$$

$$x \log 3 - 4 \log 3 = x \log 2 + \log 2$$ *This is a critical step. Be sure you understand it.*

$$x \log 3 - x \log 2 = \log 2 + 4 \log 3$$ *We obtain all terms containing $x$ on left-hand side.*

$$x (\log 3 - \log 2) = \log 2 + 4 \log 3$$ *Then we factor out $x$.*

$$x = \frac{\log 2 + 4 \log 3}{\log 3 - \log 2}$$ *Divide both sides by $(\log 3 - \log 2)$.* ■

If the base $e$ is encountered in an exponential equation, it is advisable to take the natural logarithm of each side of the equation.

**EXAMPLE 7** Solve for $t$ in the equation $10 = 25e^{-t/250}$.

***Solution***

$$\frac{10}{25} = e^{-t/250}$$ *Divide each side by 25.*

$$0.4 = e^{-t/250}$$ *Express $\frac{10}{25}$ as a decimal.*

$$\ln 0.4 = \ln e^{-t/250}$$ *Take the natural log of each side.*

$$\ln 0.4 = (-t/250) \ln e$$ *Use the Third Law of Logarithms.*

$$\ln 0.4 = \frac{t}{-250}$$ *We know that $\ln e = 1$.*

$$-250 \ln 0.4 = t$$ *Multiply each side by $-250$.*

On a calculator we evaluate the left-hand expression as follows:

250 [+/−] [×] 0.4 [ln x] [=] 229.0726830

Rounded to nearest thousandth, $t = 229.073$. ■

## EXERCISE 5-5

Solve the equations in Problems 1–22. You may leave your answer in terms of logarithms.

1. $\log(x + 2) = \log x + \log 2$
2. $\log(5x - 3) = \log(2x + 1) + \log 2$
3. $\log(3x + 1) = \log(x + 2) + \log 4$
4. $\log(x + 3) = \log(2x - 1) - \log 3$
5. $\log(x + 5) - \log 2 = \log(5x + 2)$
6. $2 \log x = \log 9$
7. $2 \log(x + 3) = \log(x + 1) + \log(x + 4)$
8. $2 \log(x + 1) = \log(5x + 11)$
9. $\log(x - 1) + \log(x - 2) = \log 2 + \log(2x - 5)$
10. $\log(x + 6) = \log(x + 20) - \log(x + 2)$
11. $3^{x-2} = 9$
12. $5^{2x-1} = 125$
13. $2^x = 7$
14. $6^{x-1} = 11$
15. $2^{3x+4} = 9$
16. $7^{2x-3} = 1$
17. $2^{x+1} = 3^{2x+5}$
18. $5^{2x-1} = 11^{x+5}$
19. $3^{x-2} = 15^{2x-1}$
20. $\log(x + 3) + \log 5 = 2$
21. $\log 3 = 3 - \log(x - 2)$
22. $\log x + \log(x + 15) = 2$

### Calculator Problems

Solve the equations in Problems 23–42. If necessary, find an approximate solution. Round any approximations to the nearest thousandth.

23. $0.4^x = 2.3$
24. $0.6^x = 8.4$
25. $7^x = \frac{1}{37}$
26. $4^x = \frac{2}{23}$
27. $36 = 17e^{-t/360}$
28. $12 = 8e^{-t/700}$
29. $(1.7)^{x+2} = e^x$
30. $(4.7)^{x-1} = e^x$
31. $2 + \log x = 3 - \log(x - 3)$
32. $\log(x - 2) - \log(2x + 1) = \log\left(\frac{1}{x}\right)$
33. $\log_2(x - 1) + 2 = \log_2(2x + 1)$
34. $\log_5(3x + 1) = 1 - \log_5(x + 1)$
35. $\log_3(\log_4 x) = 0$
36. $2 \log_5 x = \log_5(2 - x)$
37. $\ln(20 - x) - \ln(15 - x) = 2$
38. $\ln(x + 4) - \ln(x - 3) = 5$
39. $\log_2 x + \log_2 5 = \log_2 x^2 - \log_2 14$
40. $\log_5(3x - 1) + \log_5(x + 1) = \log_5(4x + 7)$

(*Hint:* In Problems 41 and 42, first convert each logarithm to a common logarithm.)

★ 41. $\log_5 x = 2 - \log_7 x$

★ 42. $\log_2 x = 1 - \log_3 x$

★ 43. Given that $a, b > 0$, solve for $x$ in the equation $a(1.03)^x = b(1.07)^x$.

★ 44. Given $a, b, c > 0$, solve for $x$ in the equation $3a = a + b \log_c x^3$.

## 5-6 APPLICATIONS OF EXPONENTIAL AND LOGARITHMIC EQUATIONS

### Applications of Exponential Equations

In a number of applications in business and in the physical sciences it is necessary to solve for the variable in an exponential equation. Since the variable is an exponent or part of an exponent, we will need to employ logarithms.

The following example is one of interest in the field of biology.

**EXAMPLE 1** The number $N$ of bacteria present in a certain culture at the end of $t$ hours is given by $N = N_0 \times 10^{0.01t}$, where $N_0$ is the original number of bacteria. How many hours will it take for the number of bacteria to double?

***Solution*** When the number of bacteria doubles, the value of $N$ will be twice that of $N_0$. Thus, the formula will read

$$2N_0 = N_0 \times 10^{0.01t}$$

Dividing each side by $N_0$, we obtain

$$2 = 10^{0.01t}$$

We now take the logarithm of each side, obtaining

$$\log 2 = \log 10^{0.01t}$$

Using the Third Law of Logarithms, we have

$$\log 2 = (0.01t) \log 10$$

Since $\log 10 = 1$, we may write

$$\log 2 = 0.01t \qquad \text{or} \qquad t = \frac{\log 2}{0.01}$$

Evaluating $\frac{\log 2}{0.01}$ on a scientific calculator can be done as follows:

$$2 \boxed{\log x} \boxed{\div} 0.01 \boxed{=} \text{30.10299957}$$

Rounded to the nearest tenth of an hour, $t = 30.1$ hours. ■

The following example is of interest to students of business, economics, and finance.

**EXAMPLE 2** The total amount of money in an account that earns 16% interest compounded annually is given by $A = P(1.16)^t$ where $P$ is the principal invested and $t$ is the number of years. If \$150 is placed in such an account, how many years will it take to grow to a total of \$3000?

***Solution***

$$3000 = 150(1.16)^t \qquad \textit{Since } A = 3000 \textit{ and } P = 150.$$

$$20 = (1.16)^t \qquad \textit{Divide each side by } 150.$$

$$\log 20 = \log (1.16)^t \qquad \textit{Take the log of each side.}$$

$$\log 20 = t \log 1.16 \qquad \textit{Apply the Third Law of Logarithms.}$$

$$\frac{\log 20}{\log 1.16} = t$$

Evaluating on a calculator,

20 [log x] [÷] 1.16 [log x] [=] 20.18415423

Thus we see it would take just slightly over 20 years. ■

Similar techniques can be used in problems dealing with radioactive decay.

**EXAMPLE 3** Strontium 90 has a half-life of 28 years. The amount of strontium 90 left after $t$ years is given by $A = A_0 \left(\frac{1}{2}\right)^{t/28}$ where $A_0$ is the original amount of the element at time $t = 0$. How long does it take until only 30% of the original amount of strontium 90 is left?

***Solution*** We want to know what $t$ is when $A = 30\%$ of $A_0$. Thus we have the equation $0.3A_0 = A_0 \left(\frac{1}{2}\right)^{t/28}$. We divide each side by $A_0$:

$$0.3 = \left(\frac{1}{2}\right)^{t/28}$$

$$\log 0.3 = \log \left(\frac{1}{2}\right)^{t/28}$$ *Take the log of each side.*

$$\log 0.3 = \left(\frac{t}{28}\right)(\log 0.5)$$ *By the Third Law of Logarithms.*

$$\frac{28 \log 0.3}{\log 0.5} = t$$ *Divide each side by* log 0.5 *and multiply each side by* 28.

On a calculator this is evaluated thus:

28 [×] 0.3 [log x] [÷] 0.5 [log x] [=] 48.635037

Rounding to the nearest tenth, we have 48.6 years. ■

## Applications of Logarithmic Equations

In various branches of scientific study the magnitude of some measured quantity varies over such a wide range that it is not convenient to measure the largest and the smallest amounts in the same units. In such cases a logarithmic scale is often used so that the relative changes in measured quantity are more easily handled. We shall examine such areas.

In chemistry, when we wish to make a measurement to determine if a liquid is an acid, a base, or neutral, we employ the pH scale. The pH actually stands for hydrogen potential of the solution. The defining equation for this quantity is $pH = -\log [H^+]$ where $[H^+]$ is the concentration of hydrogen ions in moles per liter.

The pH of distilled water is 7.0. Any solution having a pH of 7.0 is considered neutral. If a solution has a pH greater than 7.0, it is considered a base. If a solution has a pH less than 7.0, it is considered an acid.

**EXAMPLE 4** Use the equation $pH = -\log [H^+]$ to find the desired value.

**a.** The concentration of hydrogen ions in fresh eggs is approximately $1.6 \times 10^{-8}$ moles per liter. What is the pH?

**b.** Find the concentration of hydrogen ions in cider if the pH value is approximately 3.1.

***Solution*** **a.** $pH = -\log (1.6 \times 10^{-8})$

On a calculator we have

1.6 [EXP] 8 [+/−] [log] [+/−] 7.79588002

Rounded to the nearest tenth, the pH of fresh eggs is 7.8.

**b.** If the pH of cider is 3.1, we have $3.1 = -\log [H^+]$.

The equation is easier to evaluate if we multiply both sides by $-1$ to obtain $-3.1 = \log [H^+]$. This is equivalent to the exponential equation

$$10^{-3.1} = H^+$$

On a calculator we have

3.1 [+/−] [$10^x$] 0.000794328

Rounded to two significant digits, the concentration of hydrogen ions is $7.9 \times 10^{-4}$ moles per liter. ■

The magnitude of an earthquake is measured on the Richter scale. This scale is logarithmic. If an earthquake has a magnitude of less than 2, it would probably not even be noticed. An earthquake of magnitude 5 will damage chimneys and plaster and knock pictures off walls. An earthquake of magnitude 6 or greater is considered dangerous. Earthquakes of magnitude 7 or greater cause tremendous destruction and loss of life if the earthquake occurs in populated areas.

An earthquake of magnitude 7 releases 729 times the intensity of an earthquake of magnitude 5!

The defining equation of the magnitude of an earthquake is given by

$M = \log\left(\frac{I}{I_0}\right)$ where $I$ is the intensity of the earthquake and $I_0$ is the minimum measurable intensity.

**EXAMPLE 5** The 1964 earthquake in Anchorage, Alaska, had a magnitude of 8.4. The 1975 earthquake in Turkey had a magnitude of 6.7. How much greater was the amount of energy released from the Alaska earthquake than from the Turkey earthquake?

***Solution*** For the Alaska earthquake,

$$8.4 = \log\left(\frac{I_A}{I_0}\right) = \log I_A - \log I_0 \tag{1}$$

where $I_A$ = intensity of Alaska earthquake.

For the Turkey earthquake,

$$6.7 = \log\left(\frac{I_T}{I_0}\right) = \log I_T - \log I_0 \tag{2}$$

where $I_T$ = intensity of Turkey earthquake.

If we solve for $\log I_0$ in equation (1), we obtain

$$\log I_0 = \log I_A - 8.4 \tag{3}$$

If we solve for $\log I_0$ in equation (2), we obtain

$$\log I_0 = \log I_T - 6.7 \tag{4}$$

Since $\log I_0$ is a constant value, we can set

$$\text{equation (3)} = \text{equation (4)}$$

$$\log I_A - 8.4 = \log I_T - 6.7$$

$$\log I_A - \log I_T = 8.4 - 6.7$$

$$\log \frac{I_A}{I_T} = 1.7 \qquad \textit{By the Second Law of Logarithms.}$$

This is equivalent to the exponential equation

$$10^{1.7} = \frac{I_A}{I_T}$$

On a calculator, $10^{1.7}$ is approximately **50.118723**. We can roughly state

$$50I_T = I_A$$

Thus the Alaska earthquake had roughly 50 times the intensity of the Turkey earthquake. ■

## EXERCISE 5-6

The use of a scientific calculator will be helpful in the solution of most of these applied problems.

### Applications of Exponential Equations

1. The number $N$ of bacteria present in a certain culture at the end of $t$ hours is given by $N = N_0 \times 10^{0.04t}$ where $N_0$ is the original number of bacteria. How many hours will it take for the number of bacteria to triple?
2. The number $N$ of bacteria present in a certain culture at the end of $t$ hours is given by $N = N_0 \times 10^{0.05t}$ where $N_0$ is the original number of bacteria. How many hours will it take for the number of bacteria to increase fourfold?
3. The number of bacteria present in a certain culture at the end of $t$ days is given by $N = N_0e^{0.0136t}$ How many days will it take for the number of bacteria to double?
4. The number of bacteria present in a certain culture at the end of $t$ days is given by $N = N_0e^{0.0216t}$ How many days will it take for the number of bacteria to be one and one half times the original number $N_0$?

The following information is needed in Problems 5–8: The total amount of money $A$ in an account that earns an interest rate $r$ compounded annually is given by $A = P(1 + r)^t$ where $P$ is the original principal and $t$ is the time in years.

5. Using the formula for $A$ above, determine approximately how long it will take the principal to triple if $r = 14\%$.
6. Using the formula for $A$ above, determine approximately how long it will take the principal to quadruple if $r = 12\%$.
7. Sam invested $5000 at 11% interest compounded annually. Alicia invested $4600 at 12% interest compounded annually. How many years will it take until the total amount of money each has invested is exactly the same?
8. Carlos invested $6000 at 18% interest compounded annually. Maria invested $6300 at 17% interest compounded annually. How many years will it take until the total amount of money each has invested is exactly the same?
9. Assume that the population of the world is growing at a rate of 2% per year and that there were 4 billion people on the earth in 1975. How many years will it take for the population to reach 7 billion people?
10. Assume that the population of a southwestern city is growing at the rate of 12% per year. How many years will it take for the city's population to double?
11. Americium 241 has a half-life of 433 years. The amount of the element left after $t$ years is given by $A = A_0\left(\frac{1}{2}\right)^{t/433}$. How long does it take until 90% of the original amount of americium 241 is left?
12. Einsteinium 253 has a half-life of 20.7 days. The amount of the element left after $t$ days is given by $A = A_0\left(\frac{1}{2}\right)^{t/20.7}$. How many days does it take until 36% of the original amount of einsteinium 253 is left?

★ **13.** Curium 242 has a half-life of 163 days. The amount of the element left after $t$ days is given by $A = A_0\left(\frac{1}{2}\right)^{t/163}$. The equation can also be expressed in the form $A = A_0e^{kt}$ where $k$ is a constant. Find the constant $k$ to the nearest six significant figures. (*Hint:* Set $\left(\frac{1}{2}\right)^{t/163} = e^{kt}$ and then take the natural logarithm of each side. Use properties of logarithms and solve the resulting equation for $t$.)

★ **14.** Radium 226 has a half-life of 1620 years. The amount of the element left after $t$ years is given by $A = A_0\left(\frac{1}{2}\right)^{t/1620}$. The equation can also be expressed in the form $A = A_0e^{kt}$ where $k$ is a constant. Find the constant to the nearest six significant figures. (*Hint:* Set $\left(\frac{1}{2}\right)^{t/1620} = e^{kt}$ and then take the natural logarithm of each side. Use properties of logarithms and solve the resulting equation for $t$.)

**15.** Many scientists theorize that the amount $A$ of radioactive carbon 14 remaining at any time $t$ (measured in years) is given by the formula $A = A_0(2)^{-\frac{t}{5750}}$ where $A_0$ is the amount that was originally present. An animal's tooth was tested and found to have lost 60% of its carbon 14 content. If this formula is valid, what is the approximate age of the animal?

**16.** The intensity $I$ (in lumens) of a beam of light passing through a substance is given by the formula $I = I_0e^{-at}$ where $I_0$ is the intensity of the light before passing through the substance, $a$ is the absorption coefficient of the substance, and $t$ is the thickness (in centimeters) of the substance. If the initial intensity is 1000 lumens and the absorption coefficient of the substance is 0.2, what thickness is necessary to reduce the illumination to 600 lumens?

### Applications of Logarithmic Equations

**17.** The concentration of hydrogen ions in grapefruit is approximately $6.3 \times 10^{-1}$ moles per liter. What is the pH?

**18.** An environmentalist was testing for "acid rain." The concentration of hydrogen ions in a local sample of rainwater is $2.3 \times 10^{-6}$ moles per liter. What is the pH of the sample? Is the sample acid or base?

**19.** The pH of lemon juice is approximately 2.3. Find the concentration of hydrogen ions to two significant digits.

**20.** The pH of milk is approximately 6.6. Find the concentration of hydrogen ions to two significant digits.

**21.** The 1906 earthquake in San Francisco had a magnitude of 8.3. The same year an earthquake was recorded in Taiwan with a magnitude of 7.1. How much greater was the amount of energy released from the San Francisco earthquake than from the Taiwan earthquake?

**22.** In 1933 an earthquake in Japan had a magnitude of 8.9. The previous year an earthquake was recorded in Mexico with a magnitude of 8.1. How much greater was the amount of energy released from the Japan earthquake than from the Mexico earthquake?

**23.** In 1960 an earthquake in Chile measured 8.4 in magnitude. What was the intensity of the earthquake? (Express as a multiple of $I_0$.)

**24.** In 1971 an earthquake in San Fernando, California, measured 6.8 in magnitude. What was the intensity of the earthquake? (Express as a multiple of $I_0$.)

The following discussion is needed to solve Problems 25 and 26. The brightness of a heavenly body is measured by a logarithmic equation. The measure of brightness is called the magnitude. The defining equation is $M = -2.5 \log\left(\frac{I}{I_0}\right)$ where $I$ is the intensity of light coming from the observed object and $I_0$ is the intensity of light coming from an object of magnitude zero. The full moon has a brightness $M = -12.5$. The sun has a brightness $M = -26.5$.

★ **25.** The planet Venus (at the brightest) has a magnitude of $-4.4$. The distant planet Pluto (at the brightest) has a magnitude of 15. How much more *intense* is the light from Venus than the light from Pluto?

★ **26.** Sirius is the brightest star seen from earth. The magnitude of Sirius is $-1.4$. The planet Neptune has a magnitude of 7.8. How much more intense is the light from Sirius than the light from Neptune?

## COMPUTER PROBLEMS FOR CHAPTER 5

See if you can apply your knowledge of computer programming and of Chapter 5 to do the following.

**1.** Write a computer program to indicate how much of any radioactive element will be left if the following three pieces of data are entered:

**a.** The amount of the radioactive element present initially measured in milligrams.

**b.** The half-life of the radioactive element measured in years.

**c.** The number of years from the initial time, for which the prediction is made.

Use your computer program to find how many milligrams of each element will be present in the given number of years if the given amount is initially present:

(i) Element A: half-life 36.4 years, 0.63 mg present initially. Find how much will be present after 100.5 years.

(ii) Element B: half-life 0.623 year, 1.46 mg present initially. Find how much will be present after 2.5 years.

(iii) Element C: half-life 18,624 years, 0.00667 mg present initially. Find how much will be present after 25,000 years.

(iv) Element D: half-life 0.002643 year, 0.18 mg present initially. Find how much will be present after 0.05 year.

**2.** Write a computer program to determine how many days a certain principal should be left invested at a particular *annual* interest rate that is compounded daily to accumulate to a given value. Round your answer to the nearest whole day. For this program assume the value of 365 days in a year. Use your computer program to evaluate the following:

**a.** How many days should $3200 be left at a 12% annual interest rate compounded daily in order to accumulate to a total of $4860?

**b.** How many days should $156,300 be left at a 17% annual interest rate compounded daily in order to accumulate to a total of $365,800?

**c.** How many days should $49,699 be left at a 15% annual interest rate compounded daily in order to accumulate to a total of $50,000?

## KEY TERMS AND CONCEPTS

Be sure you understand what is meant by these terms and can give an example of each.

Exponential function
Inverse of the exponential function
Compound interest
Population growth at a constant rate
Continuous compounding
Radioactive decay
Common logarithm
Natural logarithm
Characteristic
Mantissa
Scientific notation
Antilogarithm
Logarithmic scale
Exponential increase
Exponential decrease
Logarithmic function
Exponential form of an equation
Logarithmic form of an equation
Inverse of the logarithmic function

## SUMMARY OF PROCEDURES AND CONCEPTS

### Laws of Logarithms

1. $\log_b (xy) = \log_b x + \log_b y$
2. $\log_b \left(\frac{x}{y}\right) = \log_b x - \log_b y$
3. $\log_b (x)^n = n \log_b x$

### Properties of Logarithms

For all $a > 0$ where $a \neq 1$ and all $b > 0$ where $b \neq 1$:

1. $y = a^{\log_a y}$
2. $x = \log_a a^x$
3. $\log_a 1 = 0$
4. $\log_a a = 1$
5. $\log_a a^x = x$
6. The exponential function $y = a^x$ and the logarithmic function $y = \log_a x$ are inverse functions.
7. The *common logarithm* of a number $x$ is given by $\log x = \log_{10} x$ for all $x > 0$.
8. The *natural logarithm* of a number $x$ is given by $\ln x = \log_e x$ for all $x > 0$.
9. The base of a logarithm may be changed by $\log_b x = \dfrac{\log_a x}{\log_a b}$.

### Logarithmic Equations

1. Many logarithmic equations may be solved by transforming each side into one logarithm, then using the property that if $\log M = \log N$ then $M = N$.
2. Other logarithmic equations may be solved by changing from logarithmic to exponential form. For example, to solve $\log_4 (x + 2) = 3$, we would first write $x + 2 = 4^3$.
3. The solution of a logarithmic equation must be verified to determine that you do not obtain the logarithm of a negative number or zero in the original expression.

### Exponential Equations

1. Many exponential equations may be solved by first transforming each side into one exponential expression. If only one exponential expression is present in the equation, it should be isolated.
2. If $M = N$ for $M, N > 0$ and $M \neq 1$ and $N \neq 1$, then $\log M = \log N$. This is usually called "taking the logarithm of each side of the equation."
3. The resulting equation is solved for the desired variable.

## CHAPTER 5 REVIEW EXERCISE

Solve Problems 1–5 for the variable.

**1.** $\log_4 x = 3$ **2.** $\log_2 8 = y$ **3.** $\log_b 81 = 2$ **4.** $\log_2 x = -3$ **5.** $\log_3 \frac{1}{27} = y$

If $\log_b 2 = x$ and $\log_b 3 = y$, find the expressions in Problems 6–10.

**6.** $\log_b 6$ **7.** $\log_b 8$ **8.** $\log_b \frac{2}{9}$ **9.** $\log_b 72$ **10.** $\log_b \frac{4}{27}$

Find the common logarithms of the numbers in Problems 11–15 by using a calculator or Table C.

**11.** 8.04 **12.** 38.6 **13.** 49,000 **14.** 0.0062 **15.** 0.000156

Find $N$ for Problems 16–20. Use a calculator or Table C.

**16.** $\log N = 0.6405$ **17.** $\log N = 2.9581$ **18.** $\log 173.4 = N$
**19.** $\log 0.4507 = N$ **20.** $\log N = 0.8683 - 2$

(Optional Problems) Problems 21–30 are dependent on Section 5-4 in the text. Find the logarithms by using Table C.

Evaluate Problems 21–25 using logarithms.

**21.** $(2.7)(8.1)$ **22.** $(25)(470)$ **23.** $(0.013)(6.24)$ **24.** $\frac{381}{29.3}$ **25.** $\frac{0.034}{0.263}$

Evaluate Problems 26–30 using logarithms.

**26.** $(38.7)^3$ **27.** $(0.052)^3$ **28.** $\sqrt{529}$ **29.** $\sqrt[3]{71.4}$ **30.** $\sqrt[5]{0.0342}$

Find the logarithms in Problems 31–35 by using a calculator or Table C.

**31.** $\log_4 104$ **32.** $\log_2 33.6$ **33.** $\log_5 2.27$ **34.** $\ln 912$ **35.** $\ln 0.826$

Solve Problems 36–40.

**36.** $\log(3x - 1) = \log 2 + \log(x + 3)$ **37.** $\log(x - 6) = \log(x + 4) - \log 3$

**38.** $\log(x + 1) + \log(x + 2) = \log(7x + 23)$ **39.** $3^{x+1} = 7$ **40.** $2^{x-1} = 5^{2x+1}$

**41.** "The Law of Natural Growth" is given by the formula $A = A_0 e^{rn}$ where $A$ is the resulting amount, $A_0$ is the original amount, $r$ is the rate of growth, and $n$ is the time. If a city has a steady growth rate of 5%, in how many years will the population double?

**42.** How much money will Susan have in five years if she invests \$3500 at 18% annual interest rate which is compounded monthly?

**43.** Actinium 227 has a half-life of 21.8 years. If 1.5 mg are present now in a laboratory, how much will be present in 10 years?

**44.** This year an earthquake was recorded with a magnitude of 7.6. Ten years ago an earthquake was recorded at the same location with a magnitude of 5.8. How much greater was the amount of energy released from the recent earthquake than from the one that took place ten years ago?

**45.** A town with a population of 14,000 is growing at 6% per year. A town with a population of 12,000 is growing at 8% per year. If these rates continue, how many years will it be until both towns have the same population?

## PRACTICE TEST FOR CHAPTER 5

**1.** Write in logarithmic form: $3^2 = 9$. **2.** Write in exponential form: $\log_2 \frac{1}{4} = -2$.

Find $a$ in Problems 3–7.

**3.** $\log_4 16 = a$ **4.** $\log_3 a = 2$ **5.** $\log_a 27 = 3$ **6.** $\log_2 \frac{1}{16} = a$ **7.** $\log 10^5 = a$

If $\log_b 2 = x$ and $\log_b 3 = y$, find the following:

**8.** $\log_b 9$ **9.** $\log_b 12$

Find the logarithms in Problems 10–13 by using a calculator or Table C.

**10.** $\log 24.7$ **11.** $\log 0.000156$ **12.** $\log 7.143$ **13.** $\log_4 316$

Find $N$ in Problems 14–16 by using a calculator or Table C.

**14.** $\log N = 0.8331$ **15.** $\log N = 3.4742$ **16.** $\log N = 0.7364 - 2$

(Optional Problems) Problems 17 and 18 are dependent on Section 5-4 in the text.

Evaluate Problems 17 and 18 using logarithms.

**17.** $(23.6)^3$ **18.** $\sqrt{547}$

Solve Problems 19 and 20 for $x$.

**19.** $\log (x + 2) = \log 5 + \log x$ **20.** $3^{x-1} = 81$

**21.** The loudness of sound in decibels is defined by $L = 10 \log \left(\frac{I}{I_0}\right)$ where $I$ is the intensity of the sound and $I_0$ is the minimum intensity detectable. The loudness of student whispering was measured at 20 dB. How many times $I_0$ is the intensity of this whispering sound?

**22.** Einsteinium 253 has a half-life of 20.7 days.

**a.** How much of a 0.55-mg sample will be left after 90 days?

**b.** How long will it take until only 26% of the original amount of the element is left?

# 6 Trigonometric Functions: Right Triangle and General Angle

*Accurate reckoning of entering into things, knowledge of existing things all, mysteries . . .*

*Ahmes (An Egyptian scribe) circa 1650 B.C.*
*(A translation of part of the title page of the Rhind Papyrus)*

The early foundations of trigonometry were established by men who needed to make astronomical measurements and solve geometric construction problems (such as constructing a huge pyramid of stone blocks in Egypt). Trigonometry as a branch of mathematics had its origin in the study of the relationship between the sides and the angles of triangles.

## 6-1 ANGLES AND THEIR MEASURE

We shall begin our study of trigonometry by an overview of a few key geometric concepts.

A **circle** is the set of all points in a plane equidistant from a fixed point in that plane. The fixed point is the **center** of the circle. The distance from the center to the circle is the **radius** (plural radii).

A **central angle** of a circle is an angle with its vertex at the center of the circle, whereas an **arc** of a circle is a portion of the circumference of the circle. These concepts appear in Figure 6-1. $\pi$ is an irrational number (frequently approximated by 3.14) that is the ratio of the circumference to the diameter, $\pi = \frac{c}{d}$. (A more accurate approximation of $\pi$ obtained from a calculator is 3.14159265.)

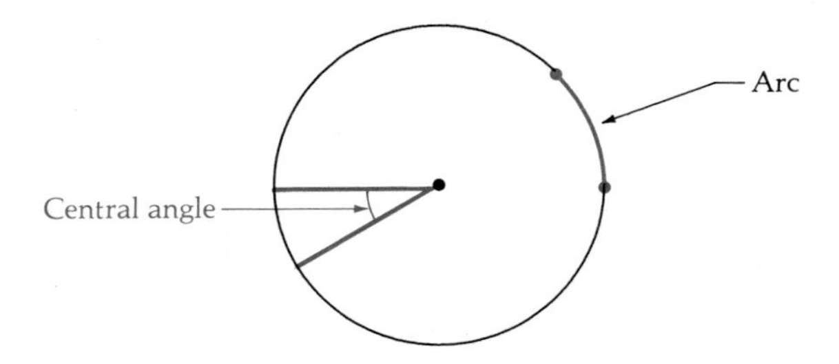


*Figure 6-1*

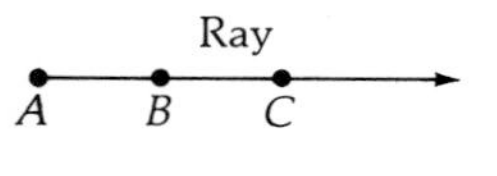


*Figure 6-2*

A **ray** is a portion of a straight line. The ray $AB$ is the point $A$, the point $B$, all points between $A$ and $B$, and all points $C$ such that $B$ is between $A$ and $C$. (See Figure 6-2.)

An **angle** is generated when a ray is rotated about its endpoint from some **initial** position to some **terminal** position. The amount of rotation is the **measure** of the angle. The endpoint of the rotated ray is called the **vertex** of the angle and is shown in Figure 6-3.

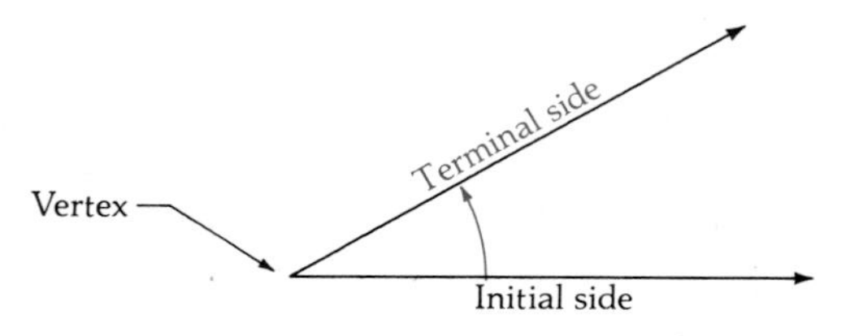


*Figure 6-3*

Figure 6-4 illustrates several ways to denote an angle. If the angle is clearly indicated, a single letter at the vertex is sufficient. In some cases a number is placed in the interior to denote the angle. In other cases a Greek letter such as $\alpha$, $\beta$, or $\theta$ is used. Most often, three letters are used—one represents a point on each ray (two rays) and one represents the vertex. The middle letter indicates the vertex.

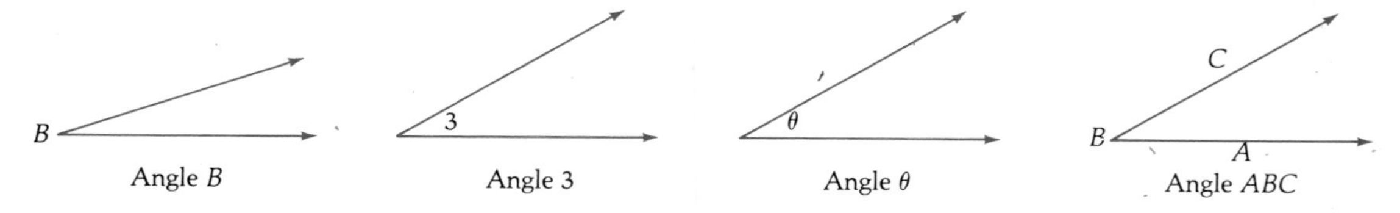


*Figure 6-4*

A **counterclockwise** rotation generates a **positive** angle; a **clockwise** rotation generates a **negative** angle, as shown in Figure 6-5.

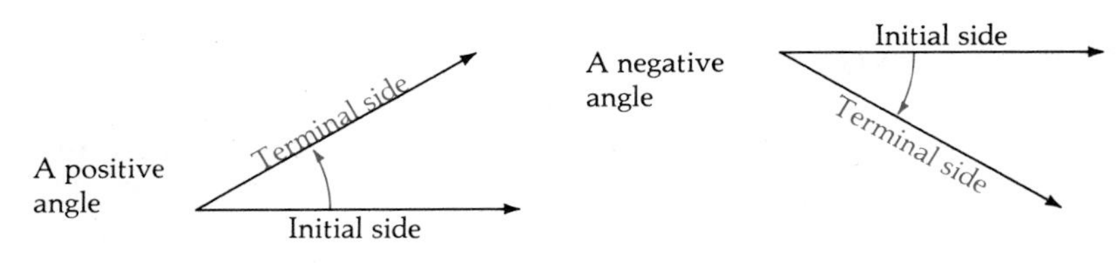


*Figure 6-5*

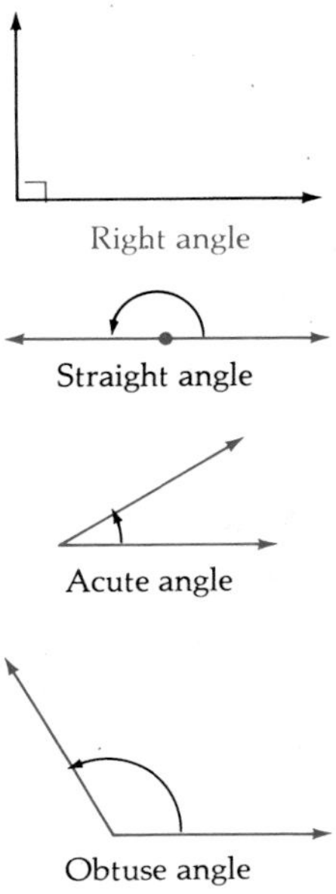

Right angle

Straight angle

Acute angle

Obtuse angle

*Figure 6-6*

A rotation of one fourth of a circle is called a **right** angle; a rotation of one half of a circle is called a **straight** angle; a rotation of less than one fourth of a circle is called an **acute** angle; a rotation of more than one fourth but less than one half of a circle is called an **obtuse** angle. These rotations are illustrated in Figure 6-6.

The measure of a rotation is always in reference to the circle. Angles are generally measured in units called **degrees**, or in units called **radians**.

***Definition***

A *degree* is a measurement of an angle that is $\frac{1}{360}$ of a complete revolution. It is written as $1^\circ$.

$1^\circ$ or $\frac{1}{360}$ of a rotation

There are $360^\circ$ in one rotation.

**EXAMPLE 1**

**a.** Referring to Figure 6-7, we can see that *a right angle* is one fourth of a rotation. Since $\frac{1}{4}$ of $360^\circ$ is $90^\circ$, we know that a right angle is $90^\circ$.

**b.** *A straight angle* is $180^\circ$. **c.** *One half of a right angle* is $45^\circ$. ■

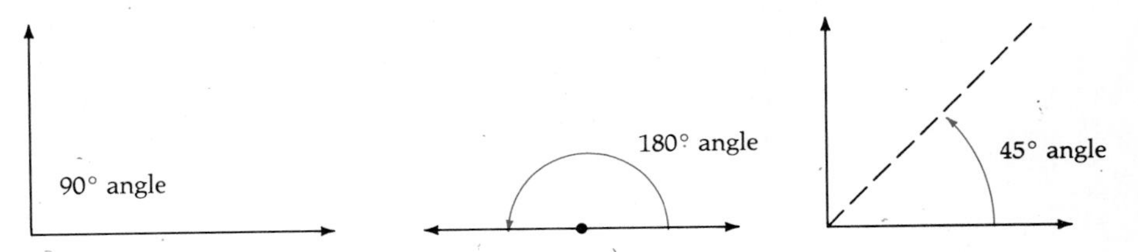


*Figure 6-7*

The measure of angles that is most frequently used in mathematics is the radian. The radian is a much larger unit of measure than the degree.

*Definition*

A *radian* is a measurement of an angle that is the amount of rotation used to make the arc of the circle equal in length to the radius.

In other words, we see from Figure 6-8 that an angle with a measure of one radian is the central angle of a circle that subtends an arc that is equal in length to the radius of the circle.

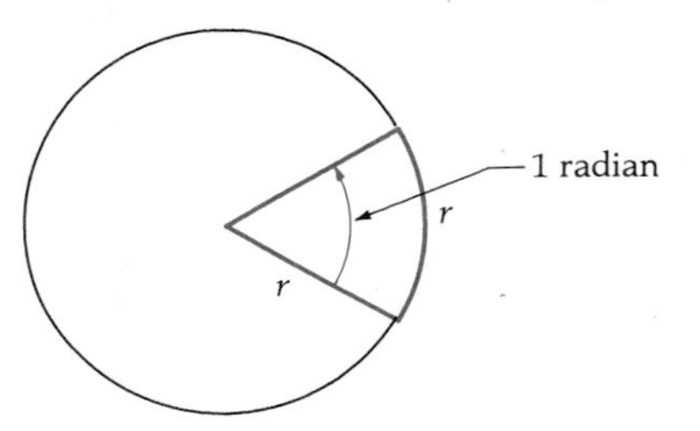


*Figure 6-8*

If one radian intercepts an arc of length $r$, then the number of radians in a complete revolution intercepts an arc that is equal in length to the circumference of the circle. Thus, the number of radians in a revolution is $\frac{2\pi r}{r} =$ $2\pi$ radians.

Some relationships are obtained immediately between the degree measurement and the radian measurement of an angle.

$$2\pi \text{ radians} = 360^\circ$$

If we divide both sides of this equation by 2, we obtain the most useful form of the equation. It is wise to commit the following formula to memory.

$$\pi \text{ radians} = 180^\circ \qquad \textbf{Formula (1)}$$

We can quickly use this formula to find an approximate measure of one radian in terms of degrees. If we divide both sides of formula (1) by $\pi$, we have

$$1 \text{ radian} = \frac{180^\circ}{\pi} \doteq 57.3^\circ$$

If we desire to find an approximate measure of one degree in terms of radians, we divide both sides of formula (1) by 180. We obtain

$$1^\circ = \frac{\pi}{180} \text{ radians} \doteq 0.0175 \text{ radian}$$

It is a wise idea to perform several of these conversion problems *without the use of a calculator* in order that you establish a good understanding of the relative comparisons between radians and degrees.

**EXAMPLE 2** Change $135^\circ$ to radians.

***Solution***

$$135^\circ = 135 \times \boxed{\frac{\pi}{180}} = \frac{135}{180}\pi \text{ radians} = \frac{3}{4}\pi \text{ radians}$$ ■

When converting from radians to degrees (when the radians are given in terms of $\pi$), remember that $\pi$ radians $= 180^\circ$.

**EXAMPLE 3** Change $\frac{9}{5}\pi$ radians to degrees.

***Solution***

$$\frac{9}{5}\pi \text{ radians} = \frac{9\pi}{5} \times \boxed{\frac{180}{\pi}} \text{ degrees}$$

$$\frac{9}{5}\pi \text{ radians} = \frac{9(180^\circ)}{5} = 324^\circ$$ ■

The previous examples involved $\pi$. In many instances the measure of the angles will not be expressed in terms of $\pi$. To obtain an approximate value when you do not have a scientific calculator available, two rules are helpful.

1. To change degrees to radians without using $\pi$, multiply the number of degrees by 0.0175 (since $1^\circ \doteq 0.0175$ radians).
2. To change radians to degrees without using $\pi$, multiply the radians by $57.3^\circ$ (since one radian $\doteq 57.3^\circ$). (A hand-held calculator is helpful for these multiplications.)

**EXAMPLE 4** Change $78^\circ$ to radians. (Express your approximation to three significant digits.)

***Solution***

$78^\circ = (78)(1^\circ) = (78)(0.0175 \text{ radians}) = 1.37$ radians ■

**Caution:** Note that since our radian approximation of $1^\circ$ has three significant digits, our final answer must have three significant digits.

**EXAMPLE 5** Change 5 radians to degrees. (Express your approximation to three significant digits.)

***Solution*** 5 radians $= 5(1 \text{ radian}) = 5(57.3^\circ) = 287^\circ$ ■

**Caution:** Note that the conversion units $57.3^\circ$ and 0.0175 radians are only approximate. A more accurate solution to Example 4 would be obtained on a calculator resulting in $78^\circ = 1.361356817$ radians. Similarly, the answer to Example 5 would be 5 radians $= 286.4788976^\circ$. Thus we see our answers to Examples 4 and 5 were off by one unit in the third significant digit.

### Calculator Conversions from Radians to Degrees and from Degrees to Radians

A scientific calculator is strongly recommended if more accurate conversions are to be made. The calculator approximation of $\pi$ is used in the conversion procedure.

**Note:** Some scientific calculators have a key that allows for direct conversion from radians to degrees and degrees to radians. Consult your owner's manual to see whether your calculator has this special feature.

**EXAMPLE 6** Change 2.63742 radians to degrees using a scientific calculator.

***Solution***

$$2.63742 \text{ radians} = \frac{2.63742 \times 180}{\pi} \text{ degrees}$$

2.63742 [×] 180 [÷] [π] [=] 151.11303° ■

**EXAMPLE 7** Change $37.62185^\circ$ to radians using a scientific calculator.

***Solution***

$$37.62185^\circ = \frac{37.62185 \times \pi}{180} \text{ radians}$$

37.62185 [×] [π] [÷] 180 [=] 0.6566252 radian ■

**Caution:** A word about accuracy. A student is warned to avoid making statements like $21^\circ = 0.3665191$ radian. A good rule of thumb in this type of problem is that the answer has as many significant digits as does the original problem. Thus we would in most cases use the value $21^\circ \doteq 0.37$ radians if the original angle $21^\circ$ was measured only to two significant digits.

## EXERCISE 6-1

Change to radians in terms of $\pi$ in Problems 1–10. Do *not* use a calculator.

**1.** $30^\circ$ **2.** $180^\circ$ **3.** $20^\circ$ **4.** $360^\circ$ **5.** $150^\circ$

**6.** $75^\circ$ **7.** $270^\circ$ **8.** $120^\circ$ **9.** $45^\circ$ **10.** $90^\circ$

Change to approximate radians in Problems 11–20. Use $1^\circ \doteq 0.0175$ radian. (A calculator may be helpful but is not necessary.)

**11.** $10^\circ$ **12.** $23^\circ$ **13.** $51^\circ$ **14.** $8^\circ$ **15.** $100^\circ$

**16.** $47^\circ$ **17.** $94^\circ$ **18.** $36^\circ$ **19.** $281^\circ$ **20.** $345^\circ$

Change to degrees in Problems 21–40. Use 1 radian $\doteq 57.3^\circ$ for those exercises not involving $\pi$. (A calculator may be helpful but is not necessary.)

**21.** $\frac{\pi}{4}$ radian **22.** $\frac{2}{3}\pi$ radians **23.** $\frac{\pi}{6}$ radian **24.** $\frac{5}{8}\pi$ radians

**25.** $\frac{3}{5}\pi$ radians **26.** 4 radians **27.** 15 radians **28.** $\frac{4}{9}\pi$ radians

**29.** 10 radians **30.** $\frac{5}{4}\pi$ radians **31.** 6 radians **32.** 2.5 radians

**33.** 16 radians **34.** 43 radians **35.** $\frac{7}{6}\pi$ radians **36.** 21.5 radians

**37.** $\frac{19}{20}\pi$ radians **38.** 0.4 radian **39.** $\frac{11}{8}\pi$ radians **40.** $\frac{1}{\pi}$ radian

The following problems require the use of a scientific calculator. Use the calculator approximation for $\pi$ in your calculations. Round your answer to the number of significant digits in the original problem.

Change to radians in Problems 41–46.

**41.** $27.38162^\circ$ **42.** $126.7513^\circ$ **43.** $0.5216782^\circ$

**44.** $0.001267317^\circ$ **45.** $193.64^\circ$ **46.** $1.368^\circ$

Change to degrees in Problems 47–54. (Do not leave $\pi$ in your answers.)

**47.** 126.3127 radians **48.** 99.62137 radians **49.** 0.026143 radian **50.** 1.6372 radians

**51.** $9.6342\sqrt{2}$ radians **52.** $5.3634\sqrt{3}$ radians **53.** $\left(\frac{1.4236}{\pi}\right)$ radian **54.** $\left(\frac{0.7261}{\pi}\right)$ radian

The following information will be needed to solve Problems 55–60:

Let $t$ be the radian measure of an angle with vertex at the center of a circle of radius $r$. This angle cuts off an arc of length $S$. The formula $S = rt$ can be used to find arc length as long as $t$ is an angle *measured in radians.*

**55.** Find the arc length $S$ for a circle of radius 2.0 meters with a central angle of 1.6 radians. (See Figure 6-9.)

**56.** Find the arc length $S$ for a circle of radius 3.5 meters with a central angle of 0.50 radian. (See Figure 6-10.)

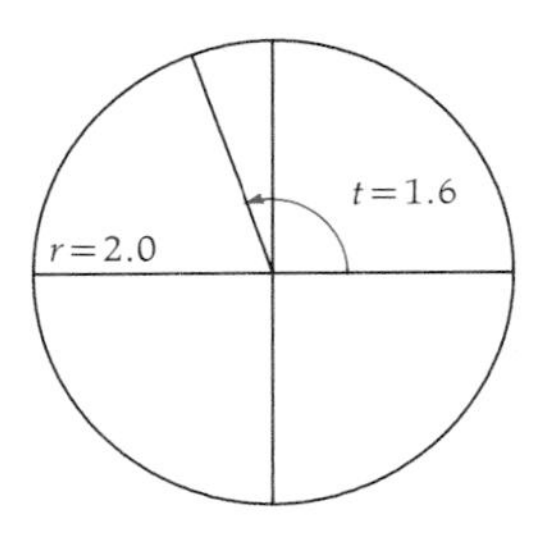


*Figure 6-9*

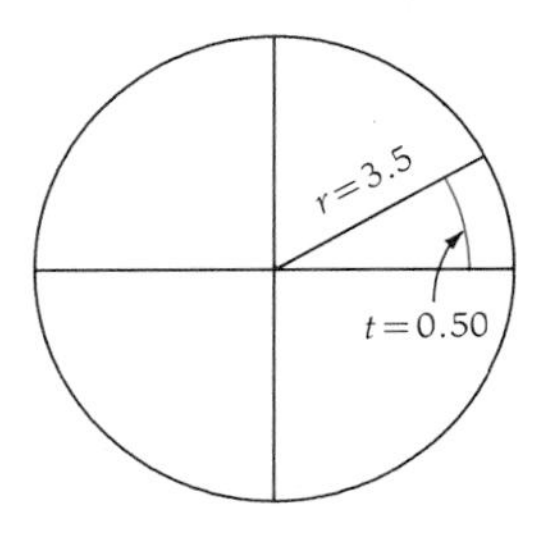


*Figure 6-10*

**57.** Find the radius of a circle when a central angle of 0.36 radian subtends an arc of 0.40 inch.

**58.** Find the central angle of a circle whose radius is 5.0 inches if the central angle subtends an arc of 0.25 inch.

**★59.** If a tire of radius 1.6 feet travels 0.7 foot on the roadway, through how much of a rotation has the wheel turned? (Express your answer in radians.)

**★60.** Through how many radians does the hour hand of a clock move in 20 minutes?

---

## 6-2 THE RIGHT TRIANGLE

Definitions of trigonometric functions can be approached in three different ways: (1) ratios of the sides of a right triangle, (2) ratios from an angle in a standard position, or (3) ratios from a representation of the real numbers on the coordinate plane. (We will establish a correspondence among all three approaches later.) These approaches can be taken in any order, but since the student is probably more familiar with triangles, we shall first establish the definitions from that direction.

A few previously learned concepts from geometry and elementary algebra are used in the discussion of this section. We shall quickly list them for convenience.

- Points are **collinear** if they lie on the same straight line.
- A **ratio** of two numbers $a{:}b$ may be written in fraction form as $\frac{a}{b}$.
- A **proportion** is a statement of the equality of two ratios.
  Two angles are **complementary** if their sum is equal to one right angle (90°).
- An **acute** angle is an angle whose measure is between 0° and 90°.

To rationalize the denominator of a fraction containing a radical such as $\frac{2}{\sqrt{10}}$, multiply both the numerator and the denominator by the radical. Thus,

$$\frac{2}{\sqrt{10}} = \frac{2}{\sqrt{10}} \cdot \frac{\sqrt{10}}{\sqrt{10}} = \frac{2\sqrt{10}}{10} = \frac{\sqrt{10}}{5}$$

***Definition***

A *triangle* is the figure formed by three noncollinear points and the three line segments that join them. The three points are the *vertices;* the line segments are the *sides* of the triangle.

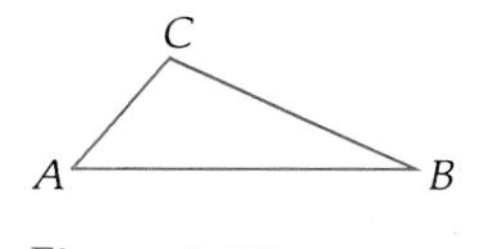


*Figure 6-11*

In general, a triangle is named by the uppercase letters that represent the three vertices, as shown in Figure 6-11.

***Definition***

A *right triangle* is a triangle with one right angle.

In a right triangle the middle letter designates the right angle. Thus we know from Figure 6-12 that in right triangle $ACB$, angle $C = 90°$.

We will be dealing extensively with right triangles, hence the need for some standard nomenclature as shown in Figure 6-13.

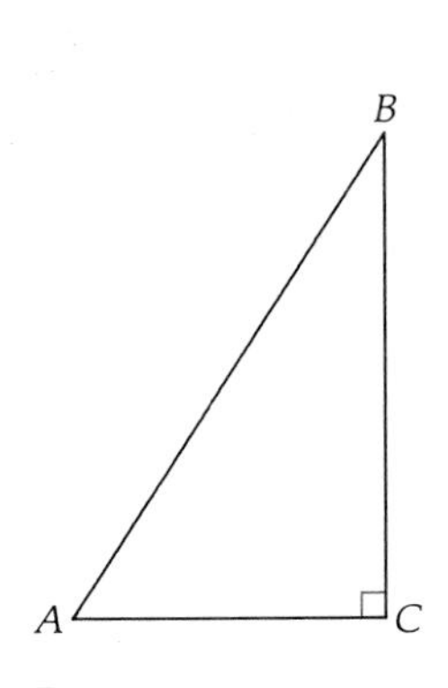


*Figure 6-12*

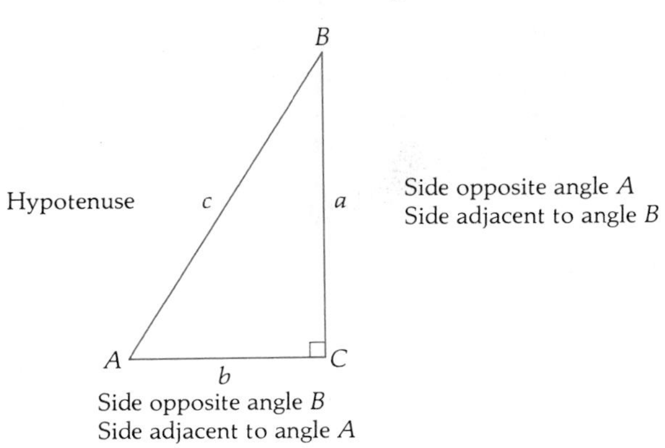


*Figure 6-13*

**Caution:** Note carefully the distinction between opposite and adjacent sides. The two terms *must* refer to an angle to have meaning.

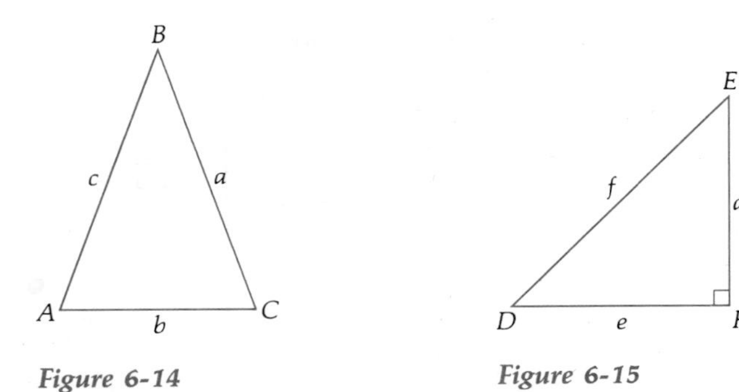


Figure 6-14

Figure 6-15

*Definition*

An *isosceles triangle* is a triangle that has two equal sides and two equal angles.

Triangle $ABC$ in Figure 6-14 is an isosceles triangle. Angle $A =$ angle $C$. Side $a =$ side $c$.

Triangle $DFE$ in Figure 6-15 is an isosceles right triangle. Angle $D =$ angle $E$. Side $d =$ side $e$.

In any right triangle the **hypotenuse** is the side opposite the right angle. The side opposite an angle is named by the lowercase of the letter that names the angle. In other words, side $a$ is opposite angle $A$, and side $b$ is opposite angle $B$.

*Pythagorean Theorem*

In a right triangle, the square of the hypotenuse is equal to the sum of the squares of the other two sides of the triangle.

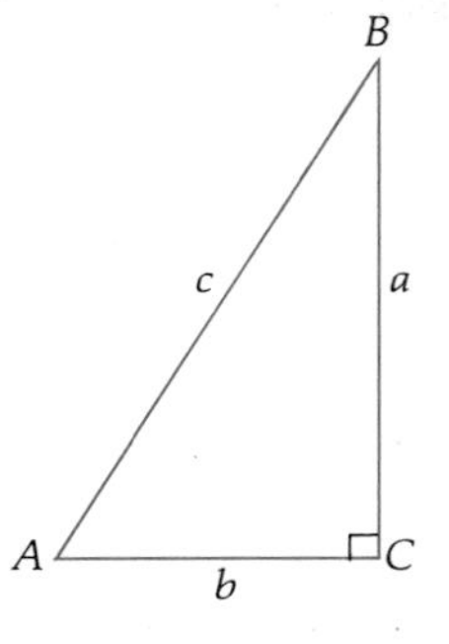


Figure 6-16

In a right triangle $ACB$, $c^2 = a^2 + b^2$. The general configuration appears in Figure 6-16.

**EXAMPLE 1** In a right triangle $ACB$, side $a = 6$, side $b = 10$. Find side $c$.

***Solution*** $c^2 = a^2 + b^2 = (6)^2 + (10)^2 = 136 \qquad c = \sqrt{136} = 2\sqrt{34}$ ■

**EXAMPLE 2** In a right triangle $ACB$, $a = 12$ and $c = 15$. Find $b$.

***Solution***

$$c^2 = a^2 + b^2$$

$$(15)^2 = (12)^2 + b^2$$

$$b^2 = 81$$

$$b = 9$$ ■

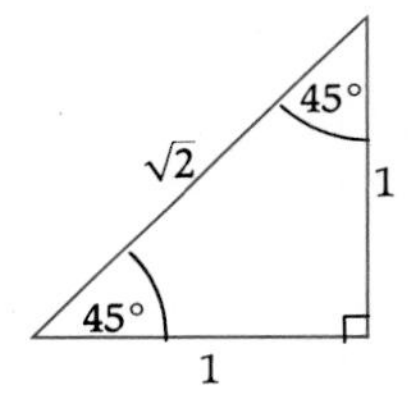


*Figure 6-17*

Special cases of right triangles that occur in plane geometry give rise to special angles in trigonometry. Two special right triangles are used very extensively in trigonometry.

### The 45°–45°–90° Triangle

In an isosceles right triangle (Figure 6-17) the sides are proportional to the numbers 1, 1, and $\sqrt{2}$, as easily shown by the Pythagorean Theorem.

### The 30°–60°–90° Triangle

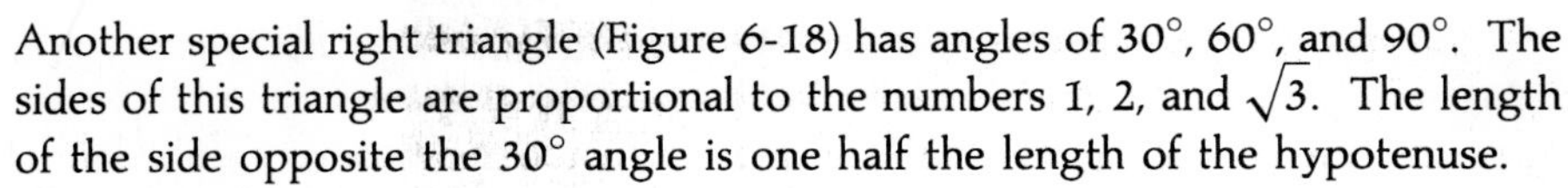

Another special right triangle (Figure 6-18) has angles of 30°, 60°, and 90°. The sides of this triangle are proportional to the numbers 1, 2, and $\sqrt{3}$. The length of the side opposite the 30° angle is one half the length of the hypotenuse.

**Note:** Do you see how this special triangle can be created by taking an *equilateral* triangle (a triangle with three equal sides) and dividing it in half?

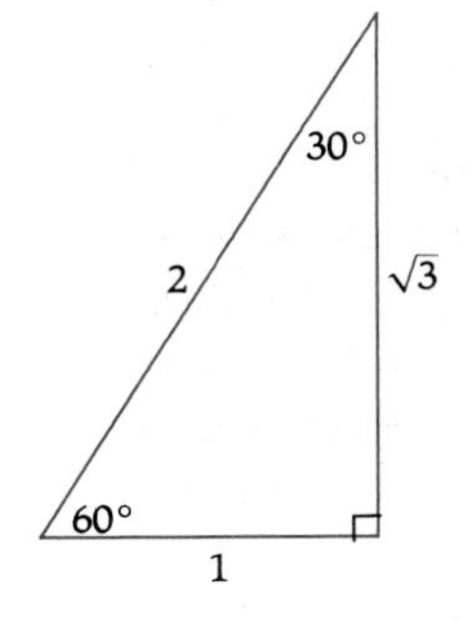


*Figure 6-18*

When we think of *similar* objects in geometry, we are considering objects of the same shape but not the same size. We now give a formal definition of similar triangles.

***Definition***

Two triangles are *similar* if the three angles of one are equal (respectively) to the three angles of the other.

Remember that the corresponding sides of similar triangles (Figure 6-19) are proportional. Similar triangles play an important part in the study of right-triangle trigonometry.

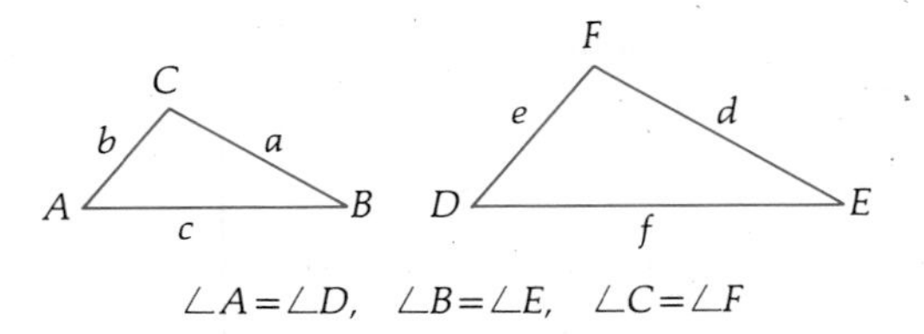


*Figure 6-19*

For example, we can write

$$\frac{b}{e} = \frac{a}{d}$$ *Since in similar triangles the corresponding sides are proportional.*

Do you see why we can write this equation? Some further work with corresponding sides is left as a homework exercise.

### The Six Trigonometric Functions

Note that with right triangles it is necessary to have only one other angle of the triangle equal a corresponding angle of another right triangle to give us similar triangles. In geometry this is stated as a theorem. We will state this theorem without proof.

***Theorem of Similar Triangles***

> If an acute angle of one right triangle is equal to an acute angle of another right triangle, the triangles are similar.

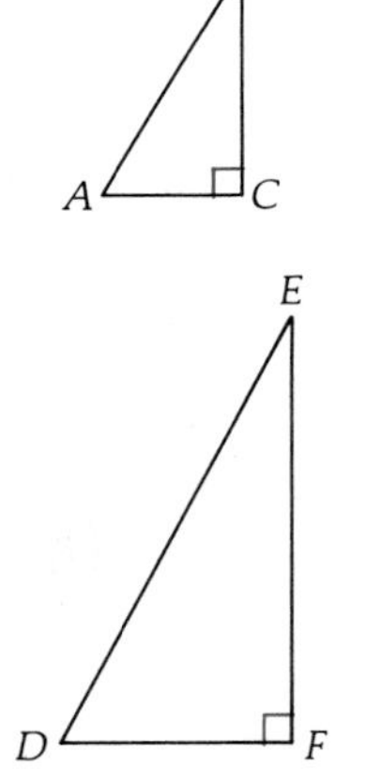


*Figure 6-20*

In right triangles $ACB$ and $DFE$ (Figure 6-20), if $\angle A = \angle D$, the triangles are similar. In other words, if an angle the size of $\angle A$ appears in any other right triangle, the ratios of the sides of the triangles are constant. These constant ratios for a given angle are the trigonometric functions.

Let us consider one specific instance, as shown in Figure 6-21.

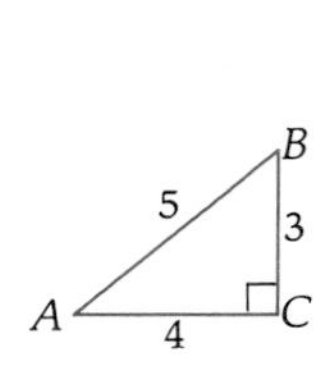


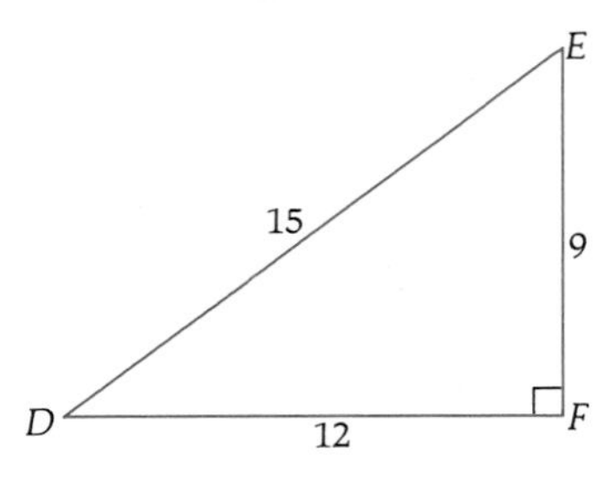


*Figure 6-21*

Suppose angle $A$ = angle $D$. Then triangle $ACB$ is similar to triangle $DFE$. The ratio of the side opposite angle $A$ to the hypotenuse is $\frac{3}{5}$.

The ratio of the side opposite angle $D$ to the hypotenuse is $\frac{9}{15}$.

Clearly,

$$\frac{3}{5} = \frac{9}{15}$$

This particular ratio is called the sine ratio (abbreviated sin). We have shown that $\sin A = \sin D$.

Given three sides of a right triangle, there are exactly six possible ratios. It is very important that you understand how they are formed. Consider the following right triangle $ACB$ shown in Figure 6-22.

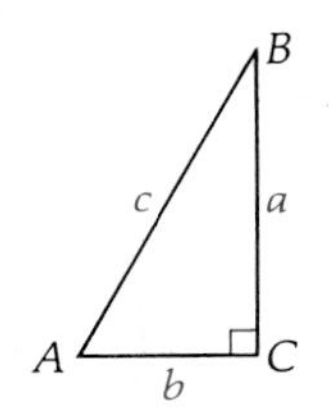


*Figure 6-22*

***Definition***

The six trigonometric functions of angle $A$ are defined as follows:

| *Function* | *Abbreviation* | *Definition* |
|---|---|---|
| sine of $\angle A$ | $\sin A$ | $\frac{\text{side opposite } \angle A}{\text{hypotenuse}} = \frac{a}{c}$ |
| cosine of $\angle A$ | $\cos A$ | $\frac{\text{side adjacent to } \angle A}{\text{hypotenuse}} = \frac{b}{c}$ |
| tangent of $\angle A$ | $\tan A$ | $\frac{\text{side opposite } \angle A}{\text{side adjacent to } \angle A} = \frac{a}{b}$ |
| cotangent of $\angle A$ | $\cot A$ | $\frac{\text{side adjacent to } \angle A}{\text{side opposite } \angle A} = \frac{b}{a}$ |
| secant of $\angle A$ | $\sec A$ | $\frac{\text{hypotenuse}}{\text{side adjacent to } \angle A} = \frac{c}{b}$ |
| cosecant of $\angle A$ | $\csc A$ | $\frac{\text{hypotenuse}}{\text{side opposite } \angle A} = \frac{c}{a}$ |

**This definition should be thoroughly memorized.**

**EXAMPLE 3** Referring to Figure 6-23, find the six trigonometric functions of angle $A$.

***Solution***

$$\sin A = \frac{\text{side opposite}}{\text{hypotenuse}} = \frac{5}{13} \qquad \csc A = \frac{\text{hypotenuse}}{\text{side opposite}} = \frac{13}{5}$$

$$\cos A = \frac{\text{side adjacent}}{\text{hypotenuse}} = \frac{12}{13} \qquad \sec A = \frac{\text{hypotenuse}}{\text{side adjacent}} = \frac{13}{12}$$

$$\tan A = \frac{\text{side opposite}}{\text{side adjacent}} = \frac{5}{12} \qquad \cot A = \frac{\text{side adjacent}}{\text{side opposite}} = \frac{12}{5}$$ ■

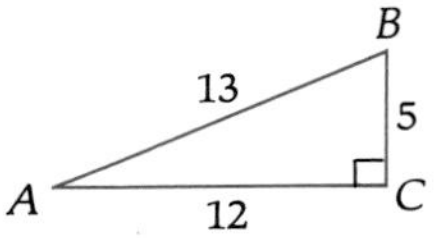

*Figure 6-23*

Often the angle is labelled by the Greek letter theta ($\theta$).

**EXAMPLE 4** Referring to Figure 6-24, find the six trigonometric functions of angle $\theta$.

***Solution***

$$\sin \theta = \frac{2}{4} = \frac{1}{2} \qquad \csc \theta = \frac{4}{2} = 2$$

$$\cos \theta = \frac{2\sqrt{3}}{4} = \frac{\sqrt{3}}{2} \qquad \sec \theta = \frac{4}{2\sqrt{3}} = \frac{2}{\sqrt{3}} = \frac{2\sqrt{3}}{3}$$

$$\tan \theta = \frac{2}{2\sqrt{3}} = \frac{1}{\sqrt{3}} = \frac{\sqrt{3}}{3} \qquad \cot \theta = \frac{2\sqrt{3}}{2} = \sqrt{3}$$ ■

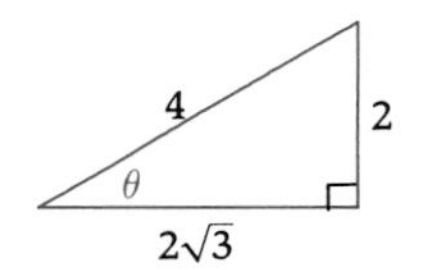


*Figure 6-24*

Take a minute to review Examples 3 and 4. Be sure you see how to obtain the trigonometric ratios.

We shall now make some close observations that reveal two important properties of trigonometric functions.

### Cofunctions of Angles

Referring to Figure 6-25, we can see that in right triangle $ACB$,

$$\sin A = \frac{7}{25} \quad \text{and} \quad \cos B = \frac{7}{25}$$

*Figure 6-25*

The side opposite angle $A$ is the side adjacent to angle $B$. Angles $A$ and $B$ are **complementary angles**. Note that $\sin A = \cos B$. The sine of any acute angle is equal to the cosine of its complementary angle. In this sense, the cosine is the "complementary sine." Hence, they are said to be **cofunctions**. The cofunctions are

sine — cosine

tangent — cotangent

secant — cosecant

---

A function of any angle is equal to the cofunction of its complementary angle.

---

In equation form we can state this general case:

Trigonometric function of acute angle $A$ = The cofunction of $(90° - A)$

The six specific cases of the equation are

$$\sin A = \cos (90° - A) \qquad \sec A = \csc (90° - A)$$
$$\cos A = \sin (90° - A) \qquad \csc A = \sec (90° - A)$$
$$\tan A = \cot (90° - A) \qquad \cot A = \tan (90° - A)$$

### The Reciprocal Functions

In right triangle $ACB$, $\sin A = \frac{24}{25}$ and $\csc A = \frac{25}{24}$. Hence $\sin A = \frac{1}{\csc A}$, since $\frac{24}{25} = \frac{1}{\frac{25}{24}}$. We could also state $\csc A = \frac{1}{\sin A}$, since $\frac{25}{24} = \frac{1}{\frac{24}{25}}$.

The sine and cosecant functions are called **reciprocal functions**. Other pairs of reciprocal functions are tangent and cotangent and also the cosine and secant. We will list all of these.

$$\tan A = \frac{1}{\cot A} \qquad \cot A = \frac{1}{\tan A}$$
$$\cos A = \frac{1}{\sec A} \qquad \sec A = \frac{1}{\cos A}$$
$$\sin A = \frac{1}{\csc A} \qquad \csc A = \frac{1}{\sin A}$$

### *A Special Word about Notation for Trigonometric Functions*

The abbreviations of the functions are written without the use of a period. When we write $\sin A$, $A$ is called the **argument**. It represents the measure of angle $A$. The abbreviations of any of the functions are not used without the arguments.

If we are given any trigonometric function of an acute angle, we can find the other five functions of that acute angle from the definitions and the Pythagorean Theorem.

EXAMPLE 5 If $\sin A = \frac{4}{5}$, find the other five functions.

***Solution*** Refer to Figure 6-26. Since $\sin A = \frac{4}{5}$, we know the ratio of $a$ to $c$ is 4 to 5. We can therefore let $a = 4$, $c = 5$, and find $b$ using the Pythagorean Theorem. (Remember that this is a special right triangle.)

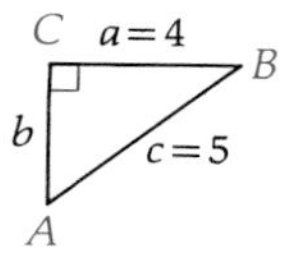

C a=4 B b c=5 A

*Figure 6-26*

$$c^2 = a^2 + b^2$$
$$(5)^2 = (4)^2 + b^2$$
$$b^2 = 9$$
$$b = \pm 3$$

Since a line segment cannot have a negative measurement, we know that $b = 3$. Therefore

$$\cos A = \frac{b}{c} = \frac{3}{5} \qquad \sec A = \frac{c}{b} = \frac{5}{3}$$
$$\tan A = \frac{a}{b} = \frac{4}{3} \qquad \csc A = \frac{c}{a} = \frac{5}{4} \quad \blacksquare$$
$$\cot A = \frac{b}{a} = \frac{3}{4}$$

If we are given a trigonometric function of one angle, we may be asked to find another trigonometric function of the other acute angle.

EXAMPLE 6 If $\cot B = \frac{1}{3}$, find $\sec A$.

***Solution*** Refer to Figure 6-27. Since $\cot B = \frac{a}{b}$, we may let $a = 1$, $b = 3$. We can now find $c$ by the Pythagorean Theorem.

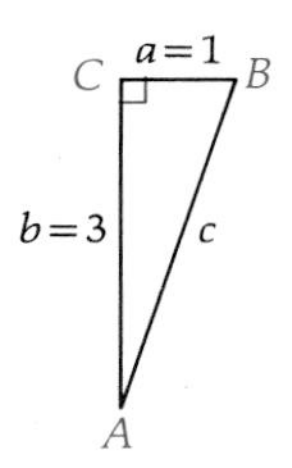

Figure 6-27

$$c^2 = a^2 + b^2$$
$$c^2 = (1)^2 + (3)^2$$
$$c^2 = 10$$
$$c = \sqrt{10}$$

Therefore $\sec A = \frac{c}{b} = \frac{\sqrt{10}}{3}$ ■

At this point it should be stressed that the six trigonometric ratios must be memorized and that every student should master the following skill level problems *without the aid of a calculator.* In each case write the exact answer unless otherwise directed.

## EXERCISE 6-2

**1.** In right triangle *ACB*, side $a = 2$ and side $b = 3$. Find side $c$.

**2.** In right triangle *ACB*, $a = 4$, $b = 4$. Find $c$.

**3.** In right triangle *ACB*, $a = 6$, $c = 10$. Find $b$.

**4.** In right triangle *ACB*, $b = 8$, $c = 11$. Find $a$.

**5.** In right triangle *ACB*, $a = 10$, $c = 26$. Find $b$.

**6.** Angle $A$ of right triangle *ACB* is 30°. Find side $a$ if the hypotenuse is 6.

***7.** The hypotenuse of an isosceles right triangle is 10. Find the other two sides.

**8.** The hypotenuse of a 30°–60°–90° triangle is 8. Find the length of the side opposite the 60° angle.

**9.** A guy wire is fastened to a telephone pole 20 meters above the ground. The other end of the wire is fastened to a stake in the ground that is 5 meters from the base of the pole. How long is the wire? (Approximate to the nearest tenth of a meter using Table A.)

**10.** Two people start from the same point. One walks 60 meters due north, the other walks 25 meters due east. How far apart are these two people?

**11.** Find the length of a diagonal of a square whose side is 14.

**12.** A rectangular room measures 16 meters by 12 meters. What is the distance between diagonally opposite corners?

**13.** The two right triangles in Figure 6-28 are similar, with angle $B$ = angle $B'$ and angle $A$ = angle $A'$:

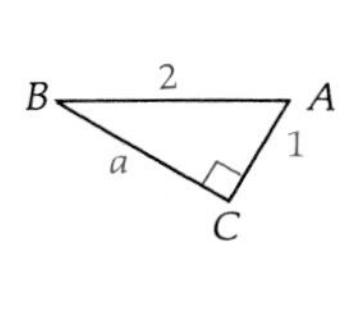

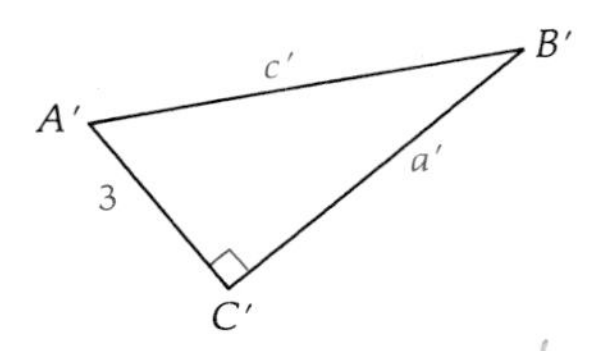


Figure 6-28

C = Right angle

**a.** Find $a$.

**b.** Find $a'$.

**c.** Find $c'$.

d. What can be said about angles $B$ and $B'$?

e. What is the ratio $\frac{b}{c}$?

f. What is the ratio $\frac{b'}{c'}$?

g. For any right triangle having an angle equal to $B$, what would be the ratio of the length of the side opposite $B$ to the length of the hypotenuse?

Referring to Figure 6-29, if a right triangle has sides 5, 12, and 13:

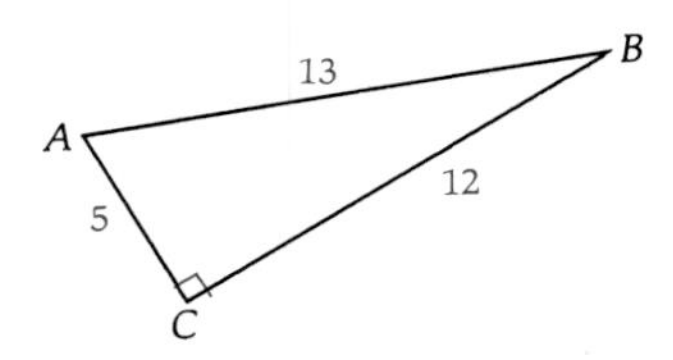


*Figure 6-29*

**14.** Find the six trigonometric functions of angle $A$.

**15.** Find the six trigonometric functions of angle $B$.

**16.** Find $\sin D$, $\tan D$, $\cos F$ for Figure 6-30.

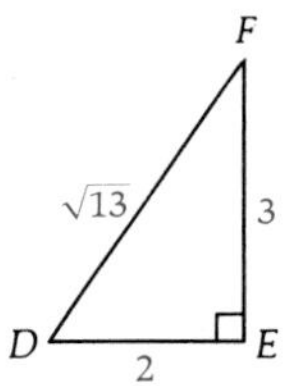


*Figure 6-30*

**17.** Find $\cos D$, $\sec F$, $\cot D$ for figure 6-31.

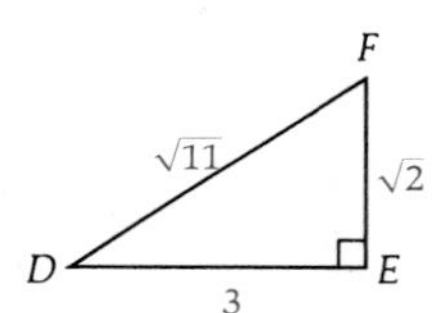


*Figure 6-31*

**18.** Find $\tan D$, $\tan F$, $\sin F$ for Figure 6-32.

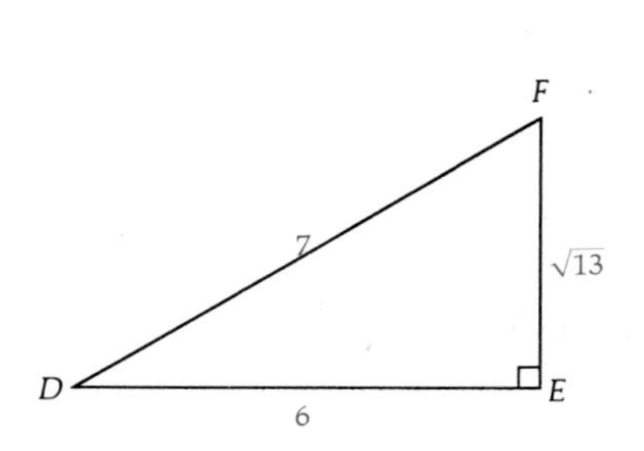


*Figure 6-32*

**19.** Find side $g$ for Figure 6-33. Then find $\cos F$, $\sec F$, $\cot F$.

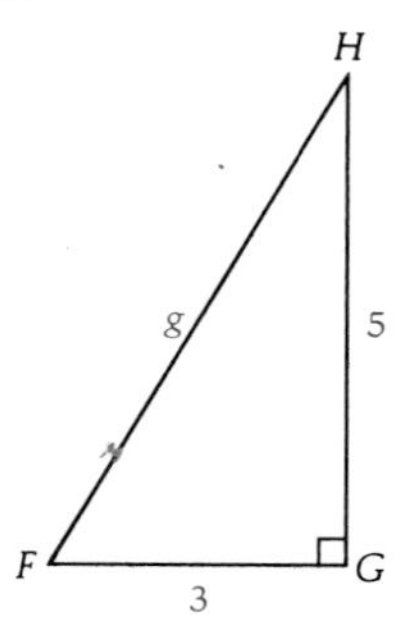


*Figure 6-33*

**20.** Find side $h$ for Figure 6-34. Then find sin $F$, csc $F$, tan $F$.

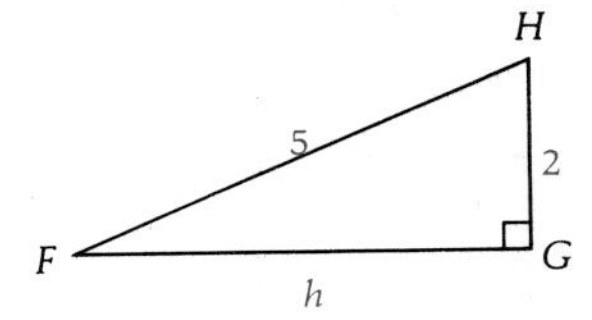


*Figure 6-34*

**21.** Two of the special right triangles mentioned in Section 6-2 are the isosceles right triangle (45°–45°–90° triangle) and the 30°–60°–90° triangle. Complete the following table using the triangles in Figure 6-35:

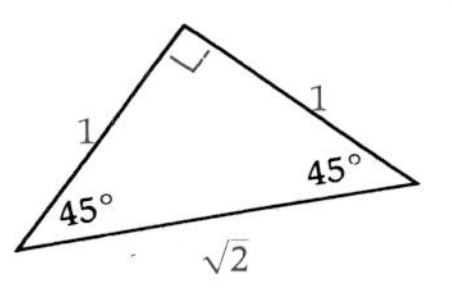


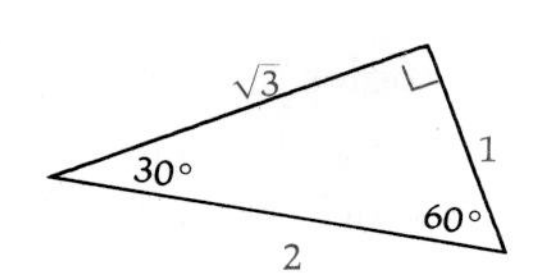


| *Angle* ($\theta$) | *30°* | *45°* | *60°* |
|---|---|---|---|
| sin $\theta$ | | | |
| cos $\theta$ | | | |
| tan $\theta$ | | | |
| cot $\theta$ | | | |
| sec $\theta$ | | | |
| csc $\theta$ | | | |

*Figure 6-35*

**22.** Is triangle $ACB$ similar to triangle $DGF$ if the lengths of the sides in Figure 6-36 are labelled correctly? Why or why not?

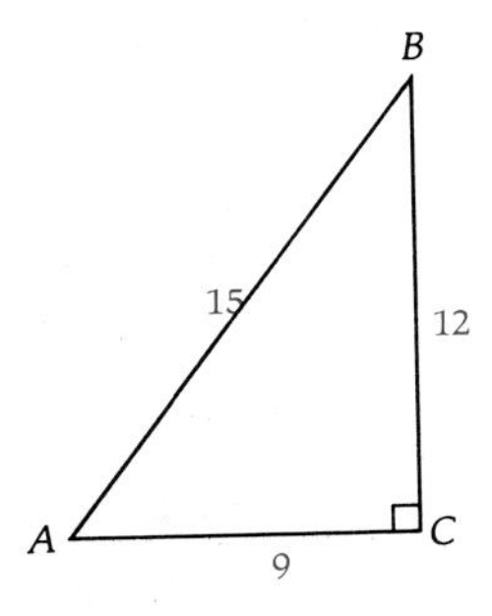


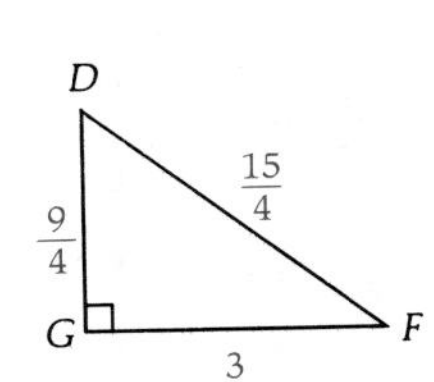


*Figure 6-36*

**23.** If $\tan A = \frac{2}{3}$, find the other five trigonometric functions of angle $A$.

**24.** If $\sin B = \frac{3}{7}$, find the other five trigonometric functions of angle $B$.

**25.** If $\sin A = \frac{1}{\sqrt{5}}$, find $\cot B$.

**26.** If $\sec B = \frac{5}{3}$, find $\cos A$.

**27.** If $\csc A = \frac{2}{\sqrt{3}}$, find $\tan B$.

**28.** If $\sin B = \frac{2}{3}$, find $\cot A$.

**29.** If $\cot B = \frac{5}{6}$, find $\csc B$.

**30.** If $\tan A = \frac{\sqrt{7}}{3}$, find $\sin B$.

**31.** If $\cos B = \frac{4}{7}$, find $\sec A$.

**32.** If $\tan A = 4$, find $\csc A$.

In Problems 33–40, express the value of each trigonometric function as the value of the cofunction of the complementary angle.

**33.** $\cos 45^\circ$ **34.** $\tan 56^\circ$ **35.** $\sin 18^\circ$ **36.** $\sec 84^\circ$

**37.** $\csc 44^\circ$ **38.** $\cot 35^\circ$ **39.** $\sin 1.5^\circ$ **40.** $\cos 3.8^\circ$

## 6-3 THE GENERAL ANGLE

Trigonometry literally means "measure of triangles." Surveyors, engineers, and others still make use of the indirect measure of triangles. However, trigonometry has uses in modern mathematics that are of a much different character. Accordingly, in this section we will define the six trigonometric functions as ratios of coordinates and distance on the Cartesian coordinate system.

*Definition*

An angle is in *standard position* on the Cartesian coordinate plane if (1) the vertex of the angle is at the origin and (2) the initial side lies along the positive $x$-axis. Refer to Figure 6-37.

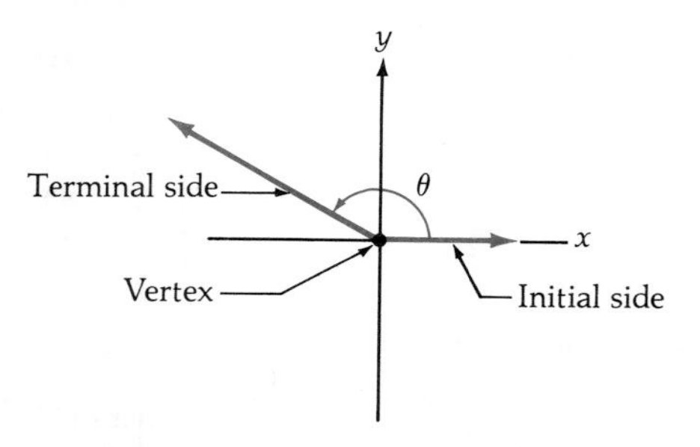


*Figure 6-37*

Angles in standard position may be positive or negative.

***Definition***

> The measure of an angle in standard position is *positive* if the rotation is *counterclockwise* and *negative* if the rotation is *clockwise*. (See Figure 6-38.)

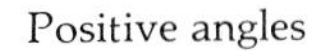


Negative angles

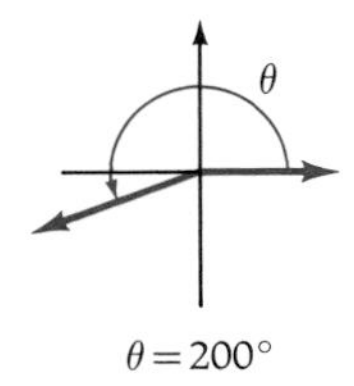


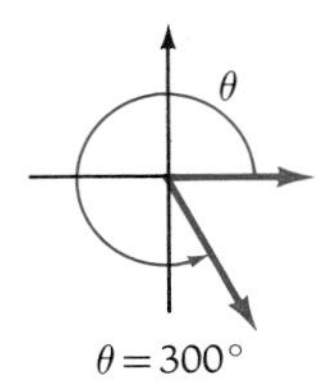


$\theta$

$\theta = -290°$

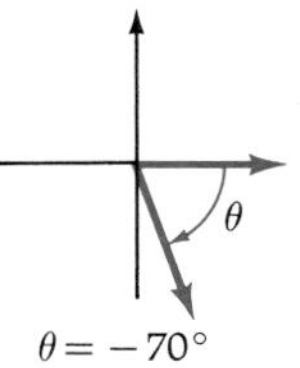


***Figure 6-38***

Many different angles in standard position have the same initial side and the same terminal side. All such angles are given the specific name of coterminal angles.

***Definition***

> Angles in standard position are *coterminal* if their terminal sides coincide.

For example, $30°$, $390°$, and $-330°$ are coterminal angles (Figure 6-39).

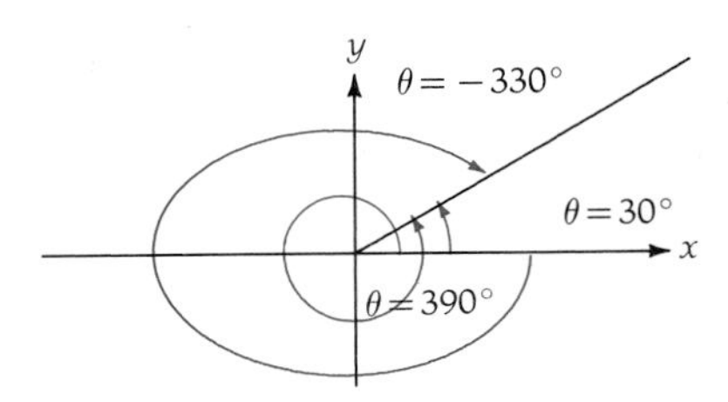


***Figure 6-39***

**EXAMPLE 1** Find the smallest positive and largest negative angles that are coterminal with an angle of $260°$.

***Solution*** Refer to Figure 6-40(a). If we add $260° + 360° = 620°$, we obtain the smallest positive angle coterminal with $260°$. If we subtract $260° - 360° = -100°$, we obtain the largest negative angle coterminal with $260°$. This is illustrated in Figure 6-40(b). ■

*Figure 6-40(a)* *Figure 6-40(b)*

In general, all angles coterminal with a given angle $\theta$ measured in degrees are angles of the form $\theta + n(360°)$ where $n$ is an integer. In Example 1 the two coterminal angles had $n = 1$ and $n = -1$.

Angles are classified according to the position of the terminal side. A first-quadrant angle has its terminal side in the first quadrant, etc.

*Definition*

If $\theta$ is an angle in standard position on the Cartesian coordinate plane, and if the point $(x, y)$ is on the terminal side of $\theta$, and if the point $(x, y)$ is $r$ units from the origin, then we define the following functions.

| *Function* | *Definition* | *Function* | *Definition* |
|---|---|---|---|
| sine $\theta$ | $\frac{y}{r}$ | cotangent $\theta$ | $\frac{x}{y}$ |
| cosine $\theta$ | $\frac{x}{r}$ | secant $\theta$ | $\frac{r}{x}$ |
| tangent $\theta$ | $\frac{y}{x}$ | cosecant $\theta$ | $\frac{r}{y}$ |

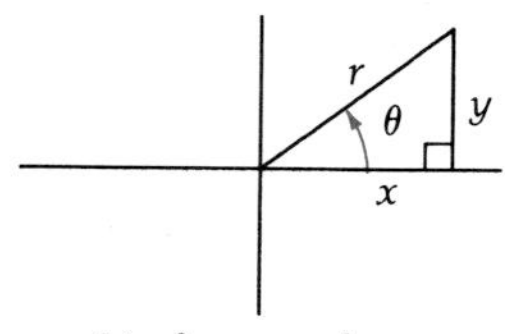


$\theta$ in first quadrant

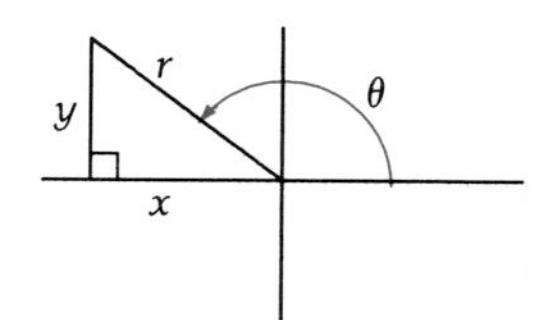


$\theta$ in second quadrant

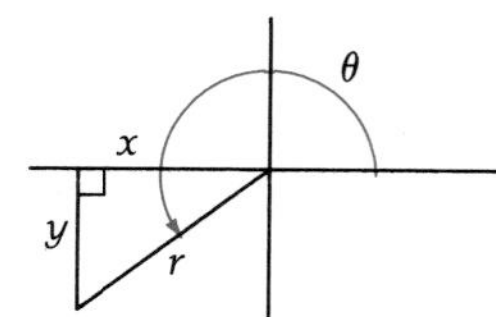


$\theta$ in third quadrant

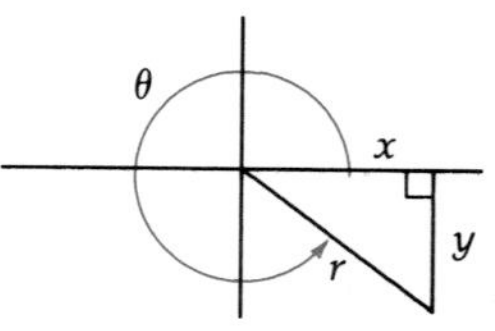


$\theta$ in fourth quadrant

*Figure 6-41*

Using the Pythagorean Theorem and referring to Figure 6-41, we can show that a point $(x, y)$ is $\sqrt{x^2 + y^2}$ units from the origin. Hence we have the relationship $r = \sqrt{x^2 + y^2}$. This enables us to find either $x$, $y$, or $r$ when two of the values are given.

EXAMPLE 2 Find the six trigonometric functions of $\theta$ if the point (5, 12) is on the terminal side of $\theta$.

***Solution*** Refer to Figure 6-42. We are given $x = 5$ and $y = 12$. Since

$$r = \sqrt{x^2 + y^2}$$

then

$$r = \sqrt{(5)^2 + (12)^2} = \sqrt{169} = 13$$

Therefore

$$\sin\theta = \frac{12}{13} \qquad \csc\theta = \frac{13}{12}$$

$$\cos\theta = \frac{5}{13} \qquad \sec\theta = \frac{13}{5}$$

$$\tan\theta = \frac{12}{5} \qquad \cot\theta = \frac{5}{12}$$ ■

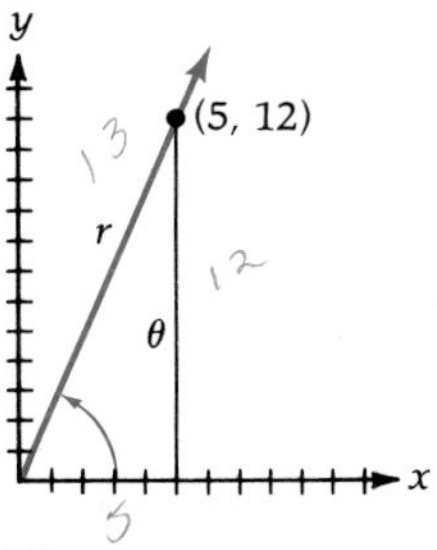

*Figure 6-42*

In Example 2 we used a sketch to aid us. However, a sketch, although helpful, is not actually necessary. If we know that the terminal side of $\theta$ is in a given **quadrant**, we can determine the sign (+ or −) of a trigonometric function.

Since $r$ is a distance (always positive), we can determine the following from the definitions and our knowledge of the coordinate system:

1. If the terminal side of $\theta$ is in the first quadrant, all functions are positive.
2. If the terminal side of $\theta$ is in the second quadrant, only the sine and cosecant are positive.
3. If the terminal side of $\theta$ is in the third quadrant, only the tangent and cotangent are positive.
4. If the terminal side of $\theta$ is in the fourth quadrant, only the cosine and secant are positive.

The accompanying table summarizes these facts.

| Quadrant II | Quadrant I |
|---|---|
| $\sin\theta$, $\csc\theta$ (+)<br>all others (−) | all functions (+) |
| Quadrant III | Quadrant IV |
| $\tan\theta$, $\cot\theta$ (+)<br>all others (−) | $\cos\theta$, $\sec\theta$ (+)<br>all others (−) |

Beginning in the first quadrant, the letters ASTC are the initials for the sentence: *A*ll *S*tudents *T*ake *C*lasses.

Of course, if sine is positive, its reciprocal, cosecant is positive. When a given ratio is positive, its reciprocal is positive. Thus when the cosine is positive, so also is the secant. When the tangent is positive, so also is the cotangent.

**EXAMPLE 3** Given $\cos\theta = \frac{3}{5}$ and $\theta$ in quadrant IV, find the other five trigonometric functions of $\theta$. See if you can do the problem without a sketch.

***Solution*** $\cos\theta = \frac{3}{5}$ gives us the ratio $\frac{x}{r} = \frac{3}{5}$. This does not necessarily mean that $x = 3$ and $r = 5$. However, we are dealing with ratios, so we can make this assumption without changing the ratio of $x$ to $y$ or $r$ to $y$.* Thus

$$r = \sqrt{x^2 + y^2}$$
$$5 = \sqrt{(3)^2 + y^2}$$
$$y^2 = 16$$
$$y = \pm 4$$

Since $\theta$ is in quadrant IV, we know that $y = -4$. Hence

$$\sin\theta = -\frac{4}{5} \qquad \tan\theta = -\frac{4}{3} \qquad \cot\theta = -\frac{3}{4}$$

$$\sec\theta = \frac{5}{3} \qquad \csc\theta = -\frac{5}{4}$$ ■

We can determine certain limits for the trigonometric functions from our knowledge of geometry. Referring to Figure 6-43, note that for all points $(x, y)$ on the plane, it must be true that neither $x$ nor $y$ can exceed $r$ in absolute value.

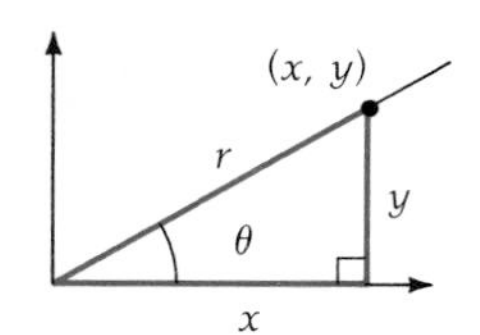

*Figure 6-43*

Also for all points $(x, y)$, $r \geq |x|$ and $r \geq |y|$. Hence the ratios $\frac{|x|}{r}$ and $\frac{|y|}{r}$ can never exceed 1, and $\frac{r}{|x|}$ and $\frac{r}{|y|}$ can never be less than 1.

Since $x$ and $y$ can be positive, negative, or zero and $r$ is always positive, this gives us the range for four trigonometric functions. The limiting values for

* The ratio $\frac{x}{r} = \frac{3}{5}$ actually tells us that $x = 3a$ and $r = 5a$, where $a$ is any nonzero number. Using this yields $y = 4a$. In any ratio involving $x$, $y$, and $r$, the factor $a$ will reduce.

the range of the four functions is as follows:

| | | |
|---|---|---|
| | $-1 \leq \sin\theta \leq 1$ | |
| | $-1 \leq \cos\theta \leq 1$ | |
| $\sec\theta \geq 1$ | or | $\sec\theta \leq -1$ |
| $\csc\theta \geq 1$ | or | $\csc\theta \leq -1$ |

Do you see how the above limiting values are determined? Do you see why it is *not* possible to place a limit for $\tan\theta$ or for $\cot\theta$ using a similar argument?

## EXERCISE 6-3

**1.** Which of the following angles are in standard position?

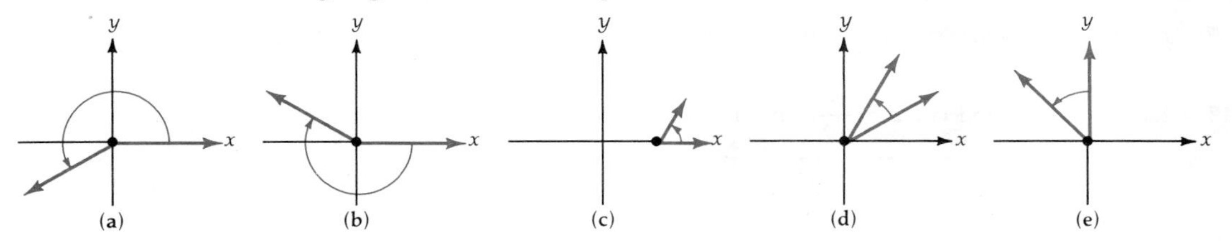


In which quadrant will we find each of the angles in Problems 2–13?

**2.** $32°$ **3.** $-32°$ **4.** $216°$ **5.** $154°$ **6.** $-385°$ **7.** $2{,}168°$

**8.** $-165°$ **9.** $330°$ **10.** $\frac{\pi}{3}$ **11.** $-\frac{8}{3}\pi$ **12.** $-\frac{5\pi}{6}$ **13.** $-\frac{11\pi}{4}$

Give the smallest positive and largest negative angles that are coterminal with each of Problems 14–24. Draw a sketch for each. Please express your answer in the same type of units as the original angle.

**14.** $32°$ **15.** $180°$ **16.** $0°$ **17.** $157°$ **18.** $45°$ **19.** $\frac{\pi}{3}$

**20.** $215°$ **21.** $\frac{5}{3}\pi$ **22.** $-\frac{\pi}{4}$ **23.** $-\frac{2\pi}{3}$ **24.** $\frac{7\pi}{3}$

In Problems 25–32 a point on the terminal side of an angle $\theta$ is given. Find the six trigonometric functions for each angle. Draw a sketch if needed.

**25.** $(3, 2)$ **26.** $(-4, 3)$ **27.** $(8, -5)$ **28.** $(4, -4)$

**29.** $(-2, -4)$ **30.** $(-6, 7)$ **31.** $(\sqrt{3}, -1)$ **32.** $(-5, -2)$

In Problems 33–46, find the other five trigonometric functions.

**33.** $\sin\theta = \frac{3}{5}$, $\theta$ in quadrant II

**34.** $\tan\theta = -\frac{3}{4}$, $\theta$ in quadrant IV

**35.** $\sec\theta = -\frac{13}{5}$, $\theta$ in quadrant III

**36.** $\cot\theta = -\frac{4}{7}$, $\theta$ in quadrant IV

**37.** $\csc\theta = \frac{5}{4}$, $\theta$ in quadrant II

**38.** $\tan\theta = 1$, $\theta$ in quadrant III

**39.** $\cos\theta = \frac{12}{13}$, $\sin\theta$ is negative

**40.** $\cot\theta = -\frac{5}{12}$, $\sin\theta$ is positive

**41.** $\sec\theta = -\frac{25}{24}$, $\cot\theta$ is positive

**42.** $\sin\theta = -\frac{4}{5}$, $\tan\theta < 0$

**43.** $\cot\theta = \frac{5}{8}$, $\sec\theta > 0$

**44.** $\cos\theta = -\frac{4}{9}$, $\tan\theta < 0$

**45.** $\csc\theta = -2$, $\cot\theta$ is positive

**46.** $\sec\theta = 4$, $\tan\theta$ is positive

★ **47.** $\cos A = -\frac{3}{5}$ and $\sin B = \frac{4}{5}$. In what possible ways is angle $A$ related to angle $B$?

★ **48.** $\tan A = -\frac{12}{13}$ and $\cot B = \frac{13}{12}$. In what possible ways is angle $A$ related to angle $B$?

---

## 6-4 TRIGONOMETRIC FUNCTIONS OF SPECIAL ANGLES

Throughout trigonometry and higher mathematics, one frequently encounters expressions involving trigonometric functions for 30°, 45°, and 60° angles. These angles are often called special angles. A wise student should be able to quickly evaluate trigonometric expressions of this type without referring to a calculator or to a table of trigonometric values. We will now closely examine these special angles.

Consider a triangle with dimensions as labelled in Figure 6-44. This gives us $x$, $y$, and $r$ in the ratios $\sqrt{3}:1:2$, respectively. Therefore, we can write

$$\sin 30° = \frac{1}{2} \qquad \csc 30° = 2$$

$$\cos 30° = \frac{\sqrt{3}}{2} \qquad \sec 30° = \frac{2\sqrt{3}}{3}$$

$$\tan 30° = \frac{\sqrt{3}}{3} \qquad \cot 30° = \sqrt{3}$$

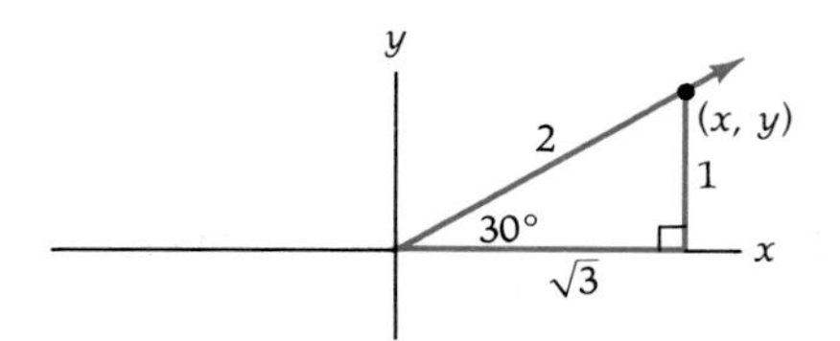


*Figure 6-44*

If a 45° angle is in standard position in such a right triangle, we see from Figure 6-45 that we can choose a triangle with $x = 1$, $y = 1$, and $r = \sqrt{2}$.

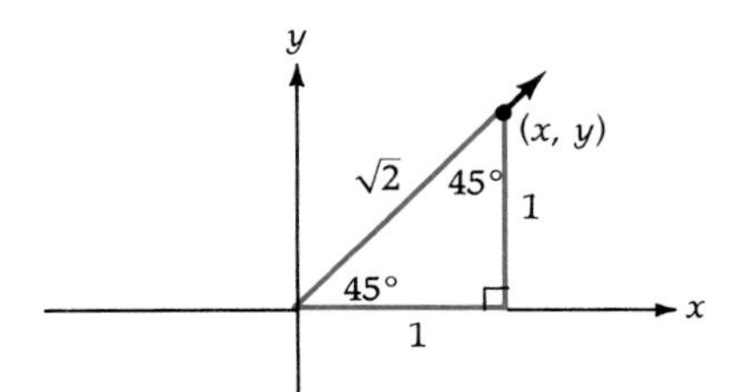


*Figure 6-45*

We can now write

$$\sin 45° = \frac{1}{\sqrt{2}} = \frac{\sqrt{2}}{2} \qquad \csc 45° = \frac{\sqrt{2}}{1} = \sqrt{2}$$

$$\cos 45° = \frac{\sqrt{2}}{2} \qquad \sec 45° = \sqrt{2}$$

$$\tan 45° = 1 \qquad \cot 45° = 1$$

Take a minute and verify that each of these values is correct.

***Definition***

> When the terminal side of an angle is on one of the coordinate axes, the angle is a *quadrantal angle.*

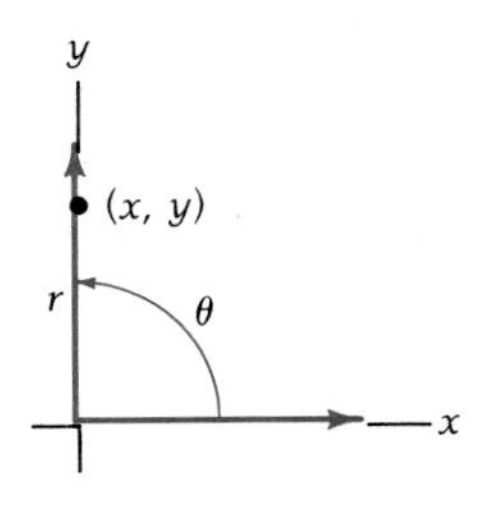


*Figure 6-46*

0°, 90°, 180°, 270°, 360°, and any positive or negative angle coterminal with them are the quadrantal angles.

We can determine the functions of a quadrantal angle from the definitions. First, choose a point on the terminal side and note that either $x = 0$ or $y = 0$. Also, note that the nonzero coordinate is equal to $\pm r$. For instance, if $\theta = 90°$, then the terminal side is on the positive $y$-axis and $x = 0$, $y = r$. (See Figure 6-46.)

$$\sin 90^\circ = \frac{y}{r} = \frac{r}{r} = 1 \qquad \cot 90^\circ = \frac{x}{y} = \frac{0}{y} = 0$$

$$\cos 90^\circ = \frac{x}{r} = \frac{0}{r} = 0 \qquad \sec 90^\circ = \frac{r}{x} = \frac{r}{0}$$ *Undefined. There is no value for* sec 90°.

$$\tan 90^\circ = \frac{y}{x} = \frac{y}{0}$$ *Undefined. There is no value for* tan 90°.

$$\csc 90^\circ = \frac{r}{y} = \frac{r}{r} = 1$$

**Caution:** Note that we can never divide by zero. Fractions with zero denominators have no meaning and are referred to as undefined.

In a similar fashion, if $\theta = 180^\circ$, we will have $y = 0$ and $x = -r$. Note that $r$ is always positive, but $x$ may be positive or negative. The terminal side of the angle is on the negative $x$-axis. (See Figure 6-47.)

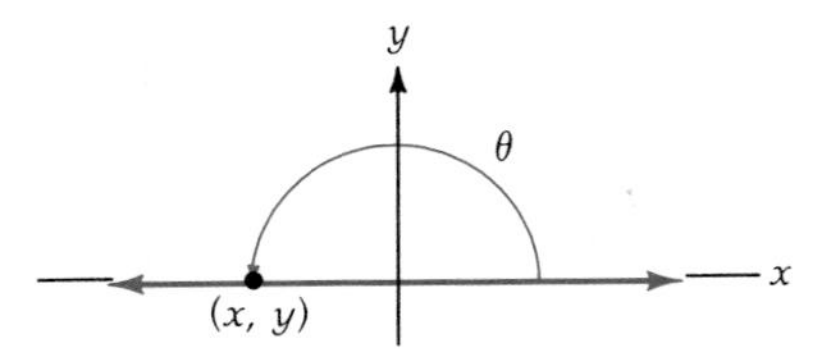


*Figure 6-47*

$$\sin 180^\circ = \frac{y}{r} = \frac{0}{r} = 0 \qquad \csc 180^\circ = \frac{r}{y} = \frac{r}{0}$$ *Undefined.*

$$\cos 180^\circ = \frac{x}{r} = \frac{-r}{r} = -1 \qquad \sec 180^\circ = \frac{r}{x} = \frac{r}{-r} = -1$$

$$\tan 180^\circ = \frac{y}{x} = \frac{0}{x} = 0 \qquad \cot 180^\circ = \frac{x}{y} = \frac{x}{0}$$ *Undefined.*

The evaluation of the trigonometric functions of 0° and 270° is left as an exercise.

Trigonometric functions are most easily obtained for angles from 0° to 90°. However, trigonometric functions of angles in quadrants II, III, and IV can be found readily on the basis of a related angle in quadrant I.

*Definition*

For each angle $\theta_1$ in quadrant I there are angles $\theta_2$ in quadrant II, $\theta_3$ in quadrant III, and $\theta_4$ in quadrant IV such that the trigonometric functions of $\theta_2$, $\theta_3$, and $\theta_4$ are equal in absolute value to the corresponding trigonometric functions of $\theta_1$. Also, $\theta_1 = 180^\circ - \theta_2 = \theta_3 - 180^\circ = 360^\circ - \theta_4$.

*Definition*

> The first-quadrant angle whose trigonometric functions are equal in absolute value to the functions of an angle in each of the other quadrants is called the *reference* angle of the other angles.

For $0^\circ \le \theta < 360^\circ$:

| *If $\theta$ Is in Quadrant* | *The Reference Angle Is* |
|---|---|
| I | $\theta$ |
| II | $180^\circ - \theta$ |
| III | $\theta - 180^\circ$ |
| IV | $360^\circ - \theta$ |

If $\theta \ge 360^\circ$ or if $\theta < 0^\circ$, we must first determine the positive angle less than $360^\circ$ that is coterminal with $\theta$ and then find the reference angle.

Be sure you understand the concept of reference angle. Remember that a reference angle is always the positive angle measured between the terminal side of the given angle and the nearest portion of the $x$-axis.

**EXAMPLE 1** Express the functions of $230^\circ$ in terms of the functions of its reference angle.

***Solution*** Since $230^\circ$ is in quadrant III, the reference angle is $230^\circ - 180^\circ = 50^\circ$. In quadrant III we know the tangent and cotangent are positive. However, the sine, cosine, secant, and cosecant are negative. Therefore

$$\begin{aligned}
\sin 230^\circ &= -\sin 50^\circ \\
\cos 230^\circ &= -\cos 50^\circ \\
\tan 230^\circ &= \tan 50^\circ \\
\cot 230^\circ &= \cot 50^\circ \\
\sec 230^\circ &= -\sec 50^\circ \\
\csc 230^\circ &= -\csc 50^\circ
\end{aligned}$$ ■

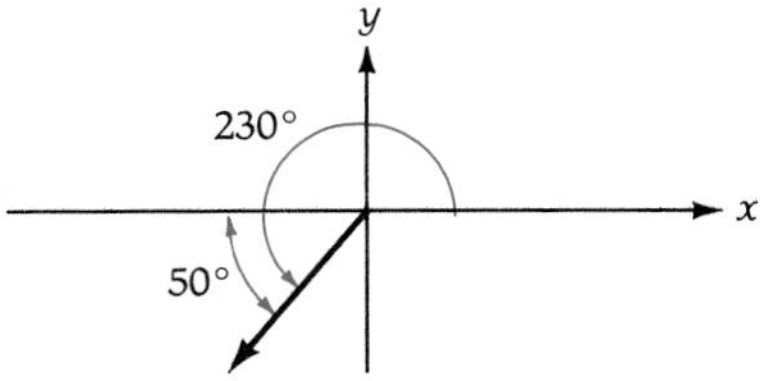

y
230°
x
50°

**Caution:** A common error is in the sign of the value of the function. Be sure to remember which quadrant the original angle is in.

**EXAMPLE 2** Find $\cos(-210^\circ)$.

***Solution*** $-210^\circ$ is coterminal with $360^\circ - 210^\circ = 150^\circ$. The reference angle for $150^\circ$ is $180^\circ - 150^\circ = 30^\circ$. $\cos 30^\circ = \frac{\sqrt{3}}{2}$ but $-210^\circ$ is a second-quadrant angle in which the cosine is negative. Hence $\cos(-210^\circ) = -\dfrac{\sqrt{3}}{2}$ ■

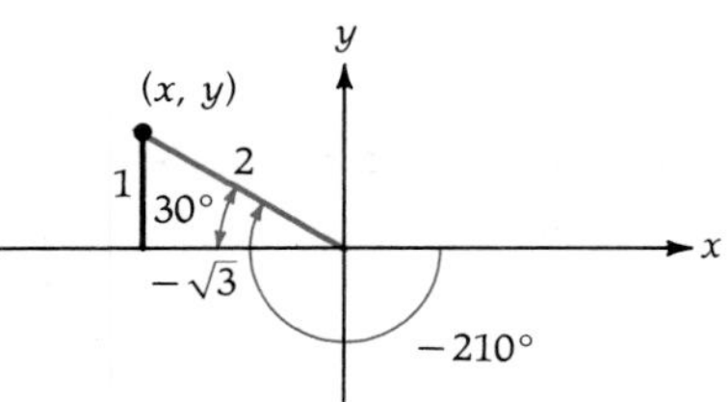


**EXAMPLE 3** Find $\tan 240^\circ$.

***Solution***

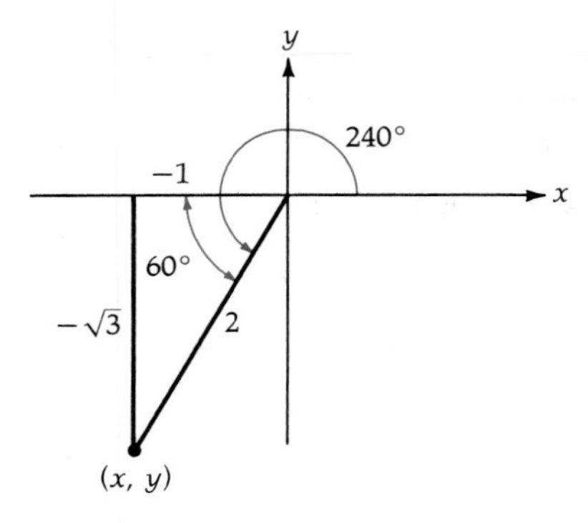


The reference angle for $240^\circ$ is $240^\circ - 180^\circ = 60^\circ$. We know that $\tan 60^\circ = \sqrt{3}$. Now $240^\circ$ is a third-quadrant angle in which the tangent is also positive. Hence $\tan 240^\circ = \sqrt{3}$. ■

Note that the list of special angles will now include 30°, 45°, 60°, and all angles for which these are reference angles. You are asked to find the functions of the special angles between 0° and 360° as an exercise.

A similar approach can be used for angles that exceed 360°.

**EXAMPLE 4** Find $\sec 495^\circ$.

***Solution***

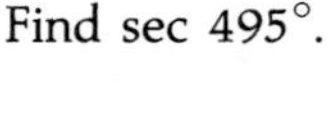

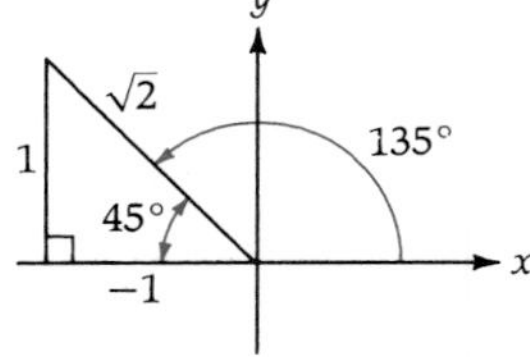


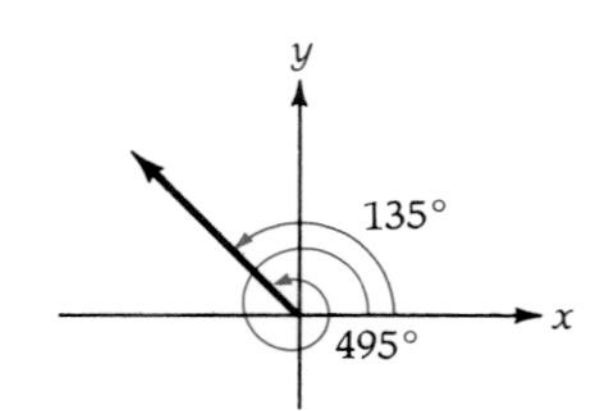


An angle of 495° is coterminal with 135°, since $(495^\circ = 360^\circ + 135^\circ)$. Therefore we have $\sec 495^\circ = \sec 135^\circ$, since the trigonometric functions of two coterminal angles are equal. The reference angle for 135° is 180° − 135° or 45°. The cosine and the secant of a second-quadrant angle are negative. Since $\sec 45° = \sqrt{2}$, we thus have $\sec 495° = \sec 135° = -\sqrt{2}$. ■

The use of the reference angle is very important in trigonometry. It becomes even more critical when we introduce inverse trigonometric functions. Be sure you understand the procedure to find reference angles in Examples 1–4.

## EXERCISE 6-4

Take special care to complete Problems 1 and 3 without referring back to the text. It is important to master the use of these special angles and quadrantal angles. (Do not use trigonometric tables or a calculator!)

**1.** Fill in the following table using exact values.

| $\theta$ | $\sin\theta$ | $\cos\theta$ | $\tan\theta$ | $\cot\theta$ | $\sec\theta$ | $\csc\theta$ |
|---|---|---|---|---|---|---|
| 0° | 0 | 1 | 0 | not def | 1 | not def |
| 30° | $\frac{1}{2}$ | $\frac{\sqrt{3}}{2}$ | $\frac{\sqrt{3}}{3}$ | $\sqrt{3}$ | $\frac{2\sqrt{3}}{3}$ | 2 |
| 45° | $\frac{\sqrt{2}}{2}$ | $\frac{\sqrt{2}}{2}$ | 1 | 1 | $\sqrt{2}$ | $\sqrt{2}$ |
| 60° | $\frac{\sqrt{3}}{2}$ | $\frac{1}{2}$ | $\sqrt{3}$ | $\frac{\sqrt{3}}{3}$ | 2 | $\frac{2\sqrt{3}}{3}$ |
| 90° | 1 | 0 | un def | 0 | not def | 1 |
| 180° | 0 | −1 | 0 | not def | −1 | not def |
| 270° | −1 | 0 | not def | 0 | not def | −1 |
| 360° | 0 | 1 | 0 | not def | 1 | not def |
| 450° | 1 | 0 | not def | 0 | not def | 1 |
| 540° | 0 | −1 | 0 | not def | −1 | not def |
| 630° | | | | | | |
| 720° | | | | | | |

**2.** Evaluate using the above table and show that each statement is true.

**a.** $(\sin 60°)^2 + (\cos 60°)^2 = 1$

**b.** $1 + (\tan 45°)^2 = (\sec 45°)^2$

**3.** Complete the following table using exact values.

| $\theta$ | $\sin\theta$ | $\cos\theta$ | $\tan\theta$ | $\cot\theta$ | $\sec\theta$ | $\csc\theta$ |
|---|---|---|---|---|---|---|
| 120° | $\frac{\sqrt{3}}{2}$ | $-\frac{1}{2}$ | $-\sqrt{3}$ | $-\frac{\sqrt{3}}{3}$ | $-2$ | $\frac{2\sqrt{3}}{3}$ |
| 135° | $\frac{\sqrt{2}}{2}$ | $-\frac{\sqrt{2}}{2}$ | $-1$ | $-1$ | $-\sqrt{2}$ | $\sqrt{2}$ |
| 150° | $\frac{1}{2}$ | $-\frac{\sqrt{3}}{2}$ | $-\frac{\sqrt{3}}{3}$ | $-\sqrt{3}$ | $-\frac{2\sqrt{3}}{3}$ | $2$ |
| 210° | $-\frac{1}{2}$ | $-\frac{\sqrt{3}}{2}$ | $\frac{\sqrt{3}}{3}$ | $\sqrt{3}$ | $-\frac{2\sqrt{3}}{3}$ | $-2$ |
| 225° | $-\frac{\sqrt{2}}{2}$ | $-\frac{\sqrt{2}}{2}$ | $1$ | $1$ | $-\sqrt{2}$ | $-\sqrt{2}$ |
| 240° | $-\frac{\sqrt{3}}{2}$ | $-\frac{1}{2}$ | $\sqrt{3}$ | $\frac{\sqrt{3}}{3}$ | $-2$ | $-\frac{2\sqrt{3}}{3}$ |
| 300° | $-\frac{\sqrt{3}}{2}$ | $\frac{1}{2}$ | $-\sqrt{3}$ | $-\frac{\sqrt{3}}{3}$ | $2$ | $-\frac{2\sqrt{3}}{3}$ |
| 315° | $-\frac{\sqrt{2}}{2}$ | $\frac{\sqrt{2}}{2}$ | $-1$ | $-1$ | $\sqrt{2}$ | $-\sqrt{2}$ |
| 330° | $-\frac{1}{2}$ | $\frac{\sqrt{3}}{2}$ | $-\frac{\sqrt{3}}{3}$ | $-\sqrt{3}$ | $\frac{2\sqrt{3}}{3}$ | $-2$ |

Evaluate Problems 4–15 by finding the exact values (without using a calculator).

**4.** $\tan(-225°)$ **5.** $\sin(-135°)$ **6.** $\cos(-330°)$ **7.** $\sin(-270°)$
**8.** $\cos(-45°)$ **9.** $\tan(-210°)$ **10.** $\cos(-180°)$ **11.** $\sin(-315°)$
**12.** $\csc(-150°)$ **13.** $\sec(-330°)$ **14.** $\cot(-120°)$ **15.** $\tan(-60°)$

Show that the statements in Problems 16–18 are always true. (*Hint:* If $(-\theta)$ is in quadrant IV, then $\theta$ is in quadrant I. If $(-\theta)$ is in quadrant III, then $\theta$ is in quadrant II, etc.)

**★16.** $\sin(-\theta) = -\sin\theta$ **★17.** $\cos(-\theta) = \cos\theta$ **★18.** $\tan(-\theta) = -\tan\theta$

Evaluate in Problems 19–27 by finding the exact values (without using a calculator).

**19.** $\cos 390°$ **20.** $\sin 480°$ **21.** $\tan 495°$ **22.** $\cos 600°$ **23.** $\sin 510°$
**24.** $\tan 405°$ **25.** $\sec 870°$ **26.** $\csc 765°$ **27.** $\cot 840°$

Verify that each of the statements in Problems 28–30 is true.

**28.** $\tan 45° = \dfrac{1 - \cos 90°}{\sin 90°}$ **29.** $\cos 60° = 1 - 2(\sin 30°)^2$ **30.** $(\sin 300°)^2 + (\cos 300°)^2 = 1$

Evaluate each of the expressions in Problems 31–34. Simplify your answer as much as possible. (Leave your answer as an exact value; do not use tables or a calculator.)

**31.** $(\sin 45^\circ)(\cos 315^\circ) + (\sin 225^\circ)(\cos 45^\circ)$

**32.** $1 + 2(\sin 210^\circ)(\cos 360^\circ)$

**33.** $1 - 2(\sin 495^\circ)$

**34.** $\dfrac{2(\tan 30^\circ)}{1 - (\tan 390^\circ)^2}$

## 6-5 TRIGONOMETRIC FUNCTION VALUES FROM A CALCULATOR

The evaluation of trigonometric functions of angles may be performed directly on any scientific calculator. There are three buttons labelled [sin], [cos], and [tan] that are used to evaluate sines, cosines, and tangents, respectively.

It should be noted that calculator values of trigonometric functions are *approximations* and not exact values. Most scientific calculators will display at least seven digits for trigonometric functions. Answers obtained on your calculator may vary slightly from the results shown in the text.

These problems involve angles measured in degrees. Be sure your calculator is in degree mode (and not radians or grads).

**EXAMPLE 1** Evaluate $\cos 36^\circ$.

***Solution*** First enter the 36 and then depress the button labelled [cos]. Your calculator should display **0.8090170** with perhaps a few more digits.

*Note:* Your calculator may display .8090170 without a zero to the left of the decimal point. Be sure to examine your own calculator display and become familiar with it. ■

**EXAMPLE 2** Evaluate $\tan 382.84^\circ$.

***Solution*** Enter 382.84 [tan], and your calculator should display **0.4211830.** ■

Be alert to read negative signs in your answer. Sometimes students forget to notice the negative sign before the number.

### Trigonometric Functions of Negative Angles

All scientific calculators will accept negative arguments. First enter the digits and then press the sign change key (usually denoted [+/−]) and finally the appropriate trigonometric key.

**EXAMPLE 3** Evaluate $\cos(-48.2653^\circ)$.

***Solution*** Enter 48.2653 [+/−] [cos], and your calculator will display **0.6656824.** ■

EXAMPLE 4 Evaluate $\tan(-556.7213°)$.

*Solution* Enter 556.7213 [+/−] [tan], and your calculator will display −0.3004196. ■

### Trigonometric Functions of Angles Measured in Degrees, Minutes, and Seconds

In higher mathematics the most common measurements of angles are made in degrees in decimal form. Sometimes, however, the angle is measured in parts of a degree called minutes.

*Definition*

> A degree is divided into 60 equal measures called minutes ($1° = 60'$).

Minutes are further subdivided into a smaller unit called a second.

*Definition*

> A minute is divided into 60 equal measures called seconds ($1' = 60''$).

EXAMPLE 5 Express an angle of $38°27'40''$ in decimal form.

*Solution* From the definitions we can write this as $38° + \left(\frac{27}{60}\right)° + \left(\frac{40}{60^2}\right)°$. (Do you see that 40 seconds is $\left(\frac{1}{60}\right)\left(\frac{40}{60}\right)$ of a degree and therefore $\frac{40}{60^2}$)? This can be quickly evaluated on a calculator with the following keystrokes:

38 [+] 27 [÷] 60 [+] 40 [÷] 60 [$x^2$] [=] 38.461111

Thus $38°27'40''$ is approximately $38.461111°$. ■

**Note:** Some scientific calculators will convert an angle measured in degrees, minutes, and seconds to decimal degrees directly. If your calculator has this feature, consult your owner's manual for specific directions.

### Finding Secant, Cosecant, and Cotangent on a Calculator

Almost all scientific calculators have buttons only for sine, cosine, and tangent functions. If the reciprocal functions are to be evaluated, we will need to use the following relationships:

$$\csc \theta = \frac{1}{\sin \theta}$$

$$\sec \theta = \frac{1}{\cos \theta}$$

$$\cot \theta = \frac{1}{\tan \theta}$$

**Caution:** When evaluating the reciprocal functions on a calculator, be sure to pause after depressing the [sin], [cos], or [tan] button and allow the screen to display an intermediate value. Then you may depress the [$\frac{1}{x}$] button. Students frequently obtain errors by depressing the buttons too rapidly to obtain reciprocal trigonometric functions.

**EXAMPLE 6** Evaluate cot 36°27′.

***Solution*** First we change 36°27′ to a decimal value:

36 [+] 27 [÷] 60 [=] 36.45

Now we use the property that $\cot 36.45° = \frac{1}{\tan 36.45°}$. To evaluate this on a calculator, we enter the following:

36.45 [tan] [$\frac{1}{x}$] 1.3538918 ■

### Finding Trigonometric Functions of Angles Measured in Radians on a Calculator

When you turn on most scientific calculators, they are automatically in degree mode. This means they will accept arguments measured in degrees for trigonometric functions. To find trigonometric functions for angles measured in radians, it is necessary to depress the proper button or move a switch to the radian mode.

There are various designations for this key. It is often denoted as [RAD] or [DRG]. Depressing the [RAD] key or the [DRG] key once will place the calculator in radian mode.

**Note:** Angles are assumed to be measured in radians unless they have a degree symbol (°) after the argument.

**EXAMPLE 7** Evaluate $\cos\left(\frac{\pi}{6}\right)$ on a calculator.

***Solution*** We are asked to find the cosine of $\frac{\pi}{6}$ radians. We will want to use the calculator value for $\pi$, so the following steps are used. (If you already have your calculator in radian mode, then you should *not* depress the [RAD] key again.)

[RAD] [$\pi$] [÷] 6 [=] [cos] 0.8660254 ■

**EXAMPLE 8** Evaluate sec 2.856.

***Solution*** We recall that $\sec 2.856 = \frac{1}{\cos 2.856}$. On a calculator we enter the following keystrokes (assuming your calculator is already in radian mode):

[RAD] 2.856 [cos] [$\frac{1}{x}$] = −1.0422151 ■

**Note:** Many calculators will also evaluate trigonometric functions of angles measured in grads. If your calculator has a [DRG] button, this capability is included on your calculator. A grad is a measure of an angle that is $\frac{1}{400}$ of a circle. Thus 400 grads = 360 degrees. The measure of angles in grads is not commonly used in mathematics and will not be used in this text.

In the algebraic hierarchy of a scientific calculator trigonometric functions take precedence over the operations [+], [−], [×], [÷]. This is helpful to remember when performing more involved calculations.

**EXAMPLE 9** **a.** Evaluate $\frac{(3.62)^5}{\sin 86.4^\circ}$. **b.** Evaluate $\frac{\sin 16.8^\circ}{8.69 + \cos 2.9^\circ}$.

***Solution*** **a.** 3.62 [$y^x$] 5 [÷] 86.4 [sin] [=] 622.87468

**b.** 16.8 [sin] [÷] [(] 8.69 + 2.9 [cos] [)] [=] 0.0298318

Note the need of parentheses here, since more than one operation is indicated in the denominator. ■

### Error Messages for Nondefined Trigonometric Values

Some trigonometric functions are not defined. For example tan 90° is not defined. Your calculator will display an error message if you attempt to find it.

Since $90° = \frac{\pi}{2}$ radians, we know that $\tan \frac{\pi}{2}$ is not defined. Some lower-priced scientific calculators do not display an error message when $\tan \frac{\pi}{2}$ is evaluated. It is best to become familiar with the capabilities and limitations of your own calculator.

EXAMPLE 10 Attempt to find on your calculator:

**a.** $\tan 90°$ **b.** $\tan \frac{\pi}{2}$

***Solution*** **a.** 90 [tan] *Error message such as flashing display, ERROR, or E.*

**b.** [RAD] [π] [÷] 2 [=] [tan] *If you obtain an error message, you are fortunate, since* $\tan \frac{\pi}{2}$ *is not defined.* ■

## EXERCISE 6-5

Evaluate on a scientific calculator in Problems 1–12. Round your answer to four decimal places.

**1.** $\sin 16°$ **2.** $\cos 26°$ **3.** $\tan 84°$ **4.** $\cos 127°$

**5.** $\sin 256°$ **6.** $\tan 344°$ **7.** $\sec 27°$ **8.** $\csc 18°$

**9.** $\cot 38°$ **10.** $\csc 146°$ **11.** $\cot 238°$ **12.** $\sec 296°$

Evaluate on a scientific calculator in Problems 13–24. Round your answer to six decimal places.

**13.** $\cos 38.42°$ **14.** $\sin 86.37°$ **15.** $\tan 47.76°$ **16.** $\cos(-126.31°)$

**17.** $\sin(-52.38°)$ **18.** $\tan(-188.63°)$ **19.** $\sec 387.04°$ **20.** $\csc(-251.38°)$

**21.** $\cot(-88.74°)$ **22.** $\sin(864.38°)$ **23.** $\csc(-744.72°)$ **24.** $\cos(1000.74°)$

**25.** Find tangent of 150° on your calculator. Show that it is equivalent to the previously measured value of $-\frac{\sqrt{3}}{3}$. (Use seven-decimal-place accuracy.)

**26.** Find the cos of 225° on your calculator. Show that it is equivalent to the previously learned value of $-\frac{\sqrt{2}}{2}$. (Use seven-decimal-place accuracy.)

Verify on your calculator that the statements in Problems 27 and 28 are true. (Use seven-decimal-place accuracy.)

**27.** $(\sin 36°)^2 + (\cos 36°)^2 = 1$ **28.** $1 + (\tan 54°)^2 = (\sec 54°)^2$

Evaluate on a scientific calculator in Problems 29–40. Round your answer to six decimal places. (Be sure your calculator is in radian mode.)

**29.** tan 1.326 **30.** cos 5.172 **31.** sin 0.768 **32.** tan 0.983

**33.** $\cos\left(\frac{5\pi}{12}\right)$ **34.** $\sin\left(\frac{7\pi}{8}\right)$ **35.** $\sec\left(\frac{\pi}{11}\right)$ **36.** $\csc\left(\frac{3\pi}{4}\right)$

**37.** $\cot\left(-\frac{3\pi}{8}\right)$ **38.** $\sec\left(\frac{-5\pi}{6}\right)$ **39.** $\sin(-127.341)$ **40.** $\cos(-866.356)$

Attempt to evaluate on your calculator in Problems 41–48. Then show (without using your calculator) why each one is not defined.

**41.** $\cot 180^\circ$ **42.** $\sec 90^\circ$ **43.** $\tan\frac{3\pi}{2}$ **44.** $\csc \pi$

**45.** $\sec\frac{5\pi}{2}$ **46.** $\cot 5\pi$ **47.** $\csc 900^\circ$ **48.** $\tan 1350^\circ$

Evaluate in Problems 49–52. Round your answer to six decimal places.

★ **49.** $\frac{\cos 18.64^\circ + \sin 36.38^\circ}{\tan 98.33^\circ}$ ★ **50.** $\cos 16.9^\circ + \sin 42.3^\circ - 2(\sin 18^\circ)(\cos 26^\circ)$

★ **51.** $\sqrt{\tan 18.94^\circ + \sec 16.98^\circ}$ ★ **52.** $\frac{\sqrt{158.62}}{\sin 88.6^\circ + \cos 397.2^\circ}$

## 6-6 TRIGONOMETRIC FUNCTION VALUES FROM A TABLE (OPTIONAL)

Trigonometric tables have been compiled to enable a person to find an approximation of the trigonometric values of angles. These tables are usually accurate to four decimal places. The tables in this book are compiled for angles measured to the nearest tenth of a degree from 0° to 90°. Table D, Four-Place Values of Trigonometric Functions, is referred to in this section. It is located on page A-10. Note the structure of this table. Verify your understanding of the table by finding the following.

**EXAMPLE 1** Find from the table an approximate value for:

**a.** sin 1.0° **b.** tan 1.4° **c.** csc 2.2°

***Solution*** By reading under the proper column we have the following:

**a.** sin 1.0° = 0.0175
**b.** tan 1.4° = 0.0244
**c.** csc 2.2° = 26.050 ■

For convenience we repeat the definition of degree from Section 6-5:

*Definition*

A degree is divided into 60 equal measures called minutes ($1^\circ = 60'$).

Sometimes angles are measured in degrees and minutes. In such cases the second column should be used.

EXAMPLE 2 Find from the table an approximate value for:

**a.** $\cos 1^\circ 12'$ **b.** $\cot 2^\circ 36'$

***Solution*** By reading under the proper column we have the following:

**a.** $\cos 1^\circ 12' = 0.9998$
**b.** $\cot 2^\circ 36' = 22.022$ ■

**Caution:** The most common error in using the table is looking in the wrong column. If the angle is between 45° and 90°, the angle is found in the *right* column, and the function value is read *from the bottom* upward to the correct value.

EXAMPLE 3 Find from the table an approximate value for:

**a.** $\sin 45.1^\circ$ **b.** $\tan 46.4^\circ$ **c.** $\csc 45^\circ 48'$

***Solution*** By reading the angle in the right-hand columns and reading from the bottom heading upward to the correct value we have

**a.** $\sin 45.1^\circ = 0.7083$
**b.** $\tan 46.4^\circ = 1.0501$
**c.** $\csc 45^\circ 48' = 1.3949$ ■

### Interpolation

If the angle we are seeking is not found in the table, we can find its functions by the method of **linear interpolation**. This process uses proportional parts.

EXAMPLE 4 Find $\sin 26.38^\circ$.

***Solution*** We can find in the table that $\sin 26.3^\circ = 0.4431$ and $\sin 26.4^\circ = 0.4446$. We know that $\sin 26.38^\circ$ lies between these two values. Linear interpolation assumes that $\sin 26.38^\circ$ lies at the same proportional distance between 0.4431 and 0.4446 as does 26.38° between 26.3° and 26.4°.

We can now set up the following pattern:

$$10\left[\begin{array}{ll} 8\left[\begin{array}{ll}\sin 26.30^\circ & 0.4431 \\ \sin 26.38^\circ & \_\_\_\_\end{array}\right] x & \\ \sin 26.40^\circ & 0.4446 \end{array}\right] 15$$

We can omit the decimal points when we write the proportion. (Do you see why?) The proportion is $\frac{8}{10} = \frac{x}{15}$. Solving for $x$, we have

$$x = \frac{(8)(15)}{10} = 12$$

Do you see that we need to add 0.0012 (and not 12) to the value 0.4431? Therefore we write

$$\sin 26.38^\circ = 0.4431 + 0.0012 = 0.4443 \qquad \textit{(To four decimal places.)}$$ ■

**Caution:** Take extra care to make sure that your final answer is between the two entries in the table. Some functions increase as the angle increases, whereas others can decrease as the angle increases.

EXAMPLE 5 Find $\cos 41.57^\circ$.

***Solution*** We can find in the table that $\cos 41.5^\circ = 0.7490$ and $\cos 41.6^\circ = 0.7478$. We now set up the following pattern:

$$10\left[\begin{array}{ll} 7\left[\begin{array}{ll}\cos 41.50^\circ & 0.7490 \\ \cos 41.57^\circ & \_\_\_\_\end{array}\right] x & \\ \cos 41.60^\circ & 0.7478 \end{array}\right] 12$$

The proportion is $\frac{7}{10} = \frac{x}{12}$. Solving for $x$, we have $x = 8.4$. Rounding, we have $x \doteq 8$.

Note that we must subtract since the function decreases from 41.5° to 41.6°.

$$\cos 41.57^\circ = 0.7490 - 0.0008 = 0.7482 \qquad \textit{(To four decimal places.)}$$ ■

### *How Accurate Is Interpolation?*

In the type of examples we have just done the answer is usually accurate to four decimal places. Sometimes the last decimal place may be slightly off.

A calculator value for $\cos 41.57^\circ$ is 0.7481456, which rounds to 0.7481. If we compare this to our answer for Example 5, we see that our approximation was off by one ten thousandth.

More accurate tables are available in many handbooks of math tables. However, the greatest degree of accuracy would obviously be obtained by using a computer or a scientific calculator.

**EXAMPLE 6** Find csc 220.9°.

***Solution*** We know that 220.9° is in the third quadrant and that the reference angle is 40.9°. (See Figure 6-48.) We know that the sine and its reciprocal are negative in the third quadrant. Thus we have

$$\csc 220.9^\circ = -\csc 40.9^\circ = -1.5273 \qquad \textit{From the table.}$$ ■

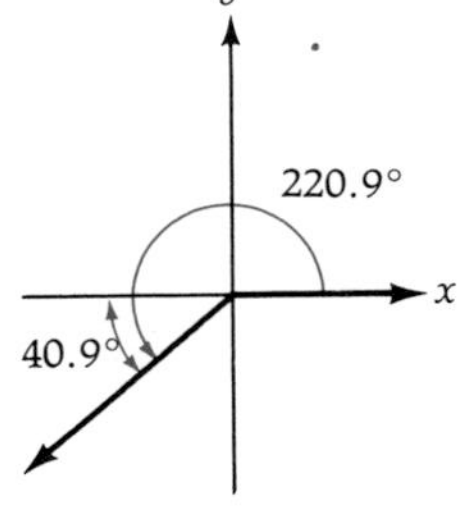


*Figure 6-48*

If the angle is negative, you first find the smallest positive angle that is coterminal with the negative angle.

## EXERCISE 6-6

Find the values in Problems 1–15 by using Table D.

| | | | | |
|---|---|---|---|---|
| **1.** sin 16.2° | **2.** cot 9.4° | **3.** tan 40.3° | **4.** sec 45.2° | **5.** cos 38.1° |
| **6.** csc 29.8° | **7.** cot 66.8° | **8.** sin 88.5° | **9.** sec 18.7° | **10.** cos 5.6° |
| **11.** csc 77°18′ | **12.** tan 12°18′ | **13.** sin 19°12′ | **14.** cot 87°42′ | **15.** cos 17°36′ |

Find the values in Problems 16–27 by using Table D. Draw a sketch if necessary to determine the reference angle.

| | | | |
|---|---|---|---|
| **16.** sin 130° | **17.** tan (−310°) | **18.** cos (−100°) | **19.** csc 200° |
| **20.** cot 100°12′ | **21.** sec 199°36′ | **22.** cos 1000.8° | **23.** tan 1826.2° |
| **24.** sin (−566.5°) | **25.** cos 286.3° | **26.** tan 167.8° | **27.** sin (−69.2°) |

Find the values in Problems 28–40 by using Table D. You will need to interpolate to obtain your answer.

| | | | |
|---|---|---|---|
| **28.** cos 88.42° | **29.** sin 33.66° | **30.** tan 71.63° | **31.** cot 12.03° |
| **32.** sec 16.54° | **33.** csc 56.07° | **34.** sin 127.33° | **35.** cos 211.42° |
| **36.** tan 288.07° | **37.** csc 266.48° | **38.** cot 153.25° | **39.** sin (−134°10′) |

**40.** cos (−211°40′)

★ **41.** If you interpolated to obtain an answer for tan 88.66°, how accurate do you feel your answer would be? Why?

★ **42.** If you interpolated to obtain an answer for csc 1.04°, how accurate do you feel your answer would be? Why?

## 6-7 DETERMINING AN ANGLE FROM A GIVEN TRIGONOMETRIC RATIO

### Determining an Angle by General Knowledge

Sometimes we are asked to find an angle $\theta$ if we are given the trigonometric ratio. For certain special angles and quadrantal angles these can be quickly identified by our general knowledge in trigonometry. In such cases the use of a calculator or a table is not necessary.

EXAMPLE 1 If $\sin\theta = \dfrac{\sqrt{2}}{2}$, find angle $\theta$ given that $0 \leq \theta < 360°$.

***Solution*** We know that in a 45°–45°–90° right triangle, $\sin 45° = \dfrac{1}{\sqrt{2}} = \dfrac{\sqrt{2}}{2}$. Therefore one answer is $\theta = 45°$. However, the sine is positive in both the first and second quadrants. If the reference angle of 45° is in the second quadrant, we can see that $\sin 135° = \sin 45° = \dfrac{\sqrt{2}}{2}$. So a second possibility is $\theta = 135°$. Thus the two answers as shown in Figure 6-49, are $\theta = 45°$ and $\theta = 135°$. ■

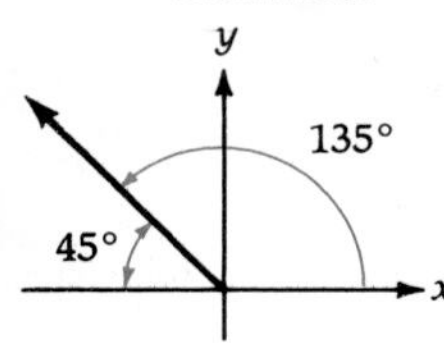


*Figure 6-49*

EXAMPLE 2 If $\cos\theta = -\dfrac{\sqrt{3}}{2}$, find $\theta$ given that $\sin\theta$ is negative and that $0 \leq \theta < 360°$.

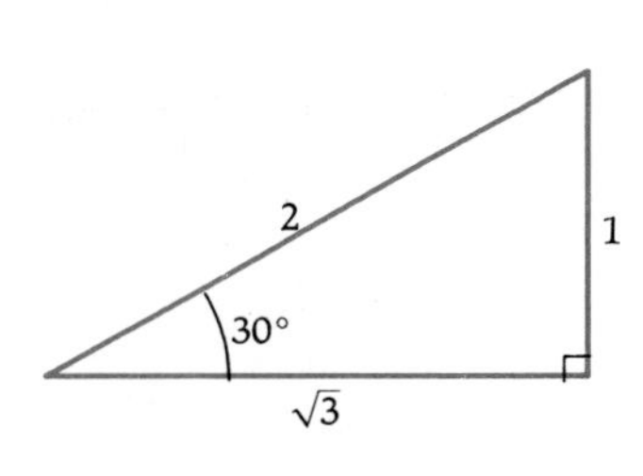


***Solution*** We know that $\cos 30° = \dfrac{\sqrt{3}}{2}$. We also know that the cosine and the sine are both negative in the third quadrant. If we place the reference angle of 30° in the third quadrant, we find that $\theta = 210°$. This is the only one value of $\theta$ that satisfies the conditions. ■

### Determining an Angle by a Calculator

If we are given the trigonometric function value, we can find the angle on most calculators by using an additional key before the [sin], [cos], or [tan] key. It is often labelled as [INV], [2nd], [Fn], or [Arc]. Some calculators have separate keys for $\boxed{\sin^{-1}}$, $\boxed{\cos^{-1}}$, and $\boxed{\tan^{-1}}$. On these calculators, only one key will need to be used.

**EXAMPLE 3** If $\sin \theta = 0.6293208$, find $\theta$ given that it is an acute angle.

***Solution*** Be sure your calculator is in the degree mode. Enter the number, the $\boxed{\text{INV}}$ key (or the corresponding key on your calculator), and finally the $\boxed{\sin}$ key.

$$0.6293208 \boxed{\text{INV}} \boxed{\sin} \; 39.000030$$

Thus $\theta = 39.000030°$. ■

**EXAMPLE 4** If $\tan \theta = \frac{3}{8}$, find $\theta$ given that it is an acute angle. Round your answer to the nearest degree.

***Solution*** We divide the two values to obtain $\tan \theta = 0.375$. If we evaluate $\theta$, we have $0.375 \boxed{\text{INV}} \boxed{\tan}$ 20.556045. Rounded to the nearest degree, $\theta = 21°$. ■

**EXAMPLE 5** If $\sin \theta = -0.26183$, find the values of $\theta$ to the nearest hundredth of a degree given that $0 \leq \theta < 360°$.

***Solution*** Clearly, the sine is negative in quadrants III and IV, so we will obtain both a third- and a fourth-quadrant angle. Using a calculator, we obtain

$$0.26183 \boxed{+/-} \boxed{\text{INV}} \boxed{\sin} \; -15.178676$$

If we round to the nearest hundredth of a degree, we have $-15.18°$. However, we want $0 \leq 0 < 360°$; so to obtain the positive fourth-quadrant angle, we calculate $360° - 15.18° = 344.82°$. Refer to Figure 6-50.

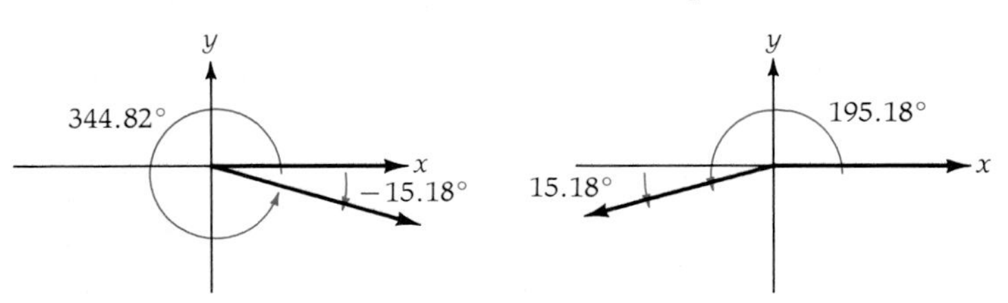


*Figure 6-50*

If the reference angle is in the third quadrant, we obtain $180° + 15.18° = 195.18°$. Thus the two values of $\theta$ satisfying all conditions of the problem are $\theta = 342.82°$ and $\theta = 195.18°$. ■

### Determining an Angle from a Trigonometric Table

One may read the trigonometric Table D to obtain the angle.

The trigonometric table may be used if the angle is in the second, third, or fourth quadrant. In such cases you find the acute reference angle from the table and then form the necessary steps to obtain the desired angle.

**EXAMPLE 6** Find $\theta$ if $\theta$ is in quadrant IV and if $\cos\theta = 0.6320$

***Solution*** From the table we can find $\theta$ if $\cos\theta = 0.6320$, and we obtain $\theta = 50.8°$. We use a reference angle of 50.8° in the fourth quadrant to obtain 309.2°. (See Figure 6-51.) Thus $\theta = 309.2°$. ■

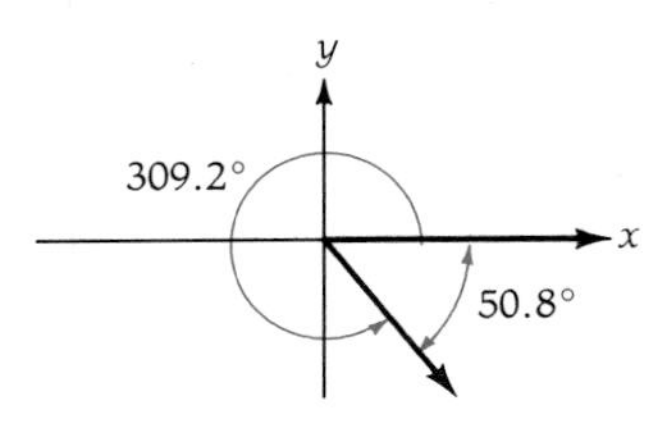


*Figure 6-51*

### Interpolation

Interpolation can be used if the given value does not appear in the table.

**EXAMPLE 7** If $\theta$ is an acute angle and $\sin\theta = 0.2345$, find $\theta$ to the nearest hundredth of a degree.

***Solution*** We observe from the table that $\sin 13.5° = 0.2334$ and $\sin 13.6° = 0.2351$. We therefore set up the following proportion:

$$10\left[\begin{array}{ll} x\left[\begin{array}{l}\sin 13.50° \\ \sin\theta\end{array}\right. & \left.\begin{array}{r}0.2334 \\ 0.2345\end{array}\right]11 \\ \quad\sin 13.60° & 0.2351\end{array}\right]17$$

$$\frac{x}{10} = \frac{11}{17}$$

$$x \doteq 6$$

Therefore $\theta = 13.50° + 0.06° = 13.56°$ *(To the nearest hundredth.)* ■

### Finding an Angle Given Two Sides of a Triangle

**EXAMPLE 8** In the triangle shown in Figure 6-52,

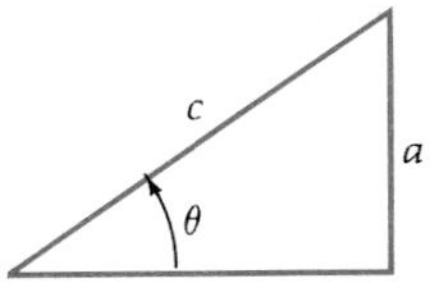

$$a = 7.237 \quad \text{and} \quad c = 9.643$$

*Figure 6-52*

**a.** Find $\sin\theta$ (to the nearest ten-thousandth).

**b.** Find $\theta$ measured to the nearest hundredth of a degree.

***Solution*** **a.** $\sin\theta = \dfrac{7.237}{9.643}$

$\sin\theta = 0.7504926$ *Calculator value.*

$\sin\theta = 0.7505$ *Rounded to nearest ten-thousandth.*

**b.** By calculator we obtain

0.7505 [INV] [sin] 48.633708

$\theta = 48.63°$ *Rounded to nearest hundredth of a degree.*

From the table we can use interpolation to obtain the same value. ■

## EXERCISE 6-7

In Problems 1–14, find the necessary value or values of $\theta$ without using a calculator or a table. Assume that $0 \le \theta < 360°$.

**1.** $\sin\theta = \dfrac{1}{2}$, $\theta$ is acute

**2.** $\cos\theta = \dfrac{\sqrt{3}}{2}$, $\theta$ is acute

**3.** $\tan\theta = -1$, $\theta$ is in quadrant II

**4.** $\sin\theta = -\dfrac{\sqrt{2}}{2}$, $\theta$ is in quadrant IV

**5.** $\cos\theta = \dfrac{\sqrt{2}}{2}$, $\sin\theta$ is negative

**6.** $\tan\theta = -\sqrt{3}$, $\cos\theta$ is positive

**7.** $\tan\theta = -\dfrac{\sqrt{3}}{3}$, $\cos\theta$ is negative

**8.** $\sin\theta = \dfrac{1}{2}$, $\tan\theta$ is negative

**9.** $\sin\theta = -1$

**10.** $\tan\theta = 1$

**11.** $\tan\theta = 0$

**12.** $\cos\theta = -1$

**13.** $\sec\theta = -2$

**14.** $\csc\theta = -\sqrt{2}$

In Problems 15–28, use a calculator or Table D. Express all answers $\theta$ to the nearest tenth of a degree. Assume that $0 \le \theta < 360°$.

**15.** $\tan\theta = 0.7239$

**16.** $\sin\theta = 0.6211$

**17.** $\cos\theta = 0.8290$

**18.** $\tan\theta = -0.5985$

**19.** $\sin\theta = -0.9293$

**20.** $\cos\theta = -0.3845$

**21.** $\tan\theta = -0.8788$

**22.** $\sin\theta = 0.6939$

**23.** $\sin\theta = \dfrac{1.20}{3.80}$ 1st quad

**24.** $\cos\theta = \dfrac{5.80}{9.20}$

**25.** $\cos\theta = \dfrac{7.36}{8.80}$

**26.** $\tan\theta = \dfrac{5.04}{3.10}$

**27.** $\cot\theta = 1.4932$

**28.** $\sec\theta = 1.2483$

In Problems 29–42, use a calculator or Table D. Express all answers $\theta$ to the nearest hundredth of a degree. Assume that $0 \le \theta < 360°$.

**29.** $\sin\theta = -0.6843$
$\cos\theta$ is positive

**30.** $\cos\theta = -0.7527$
$\tan\theta$ is positive

**31.** $\tan\theta = 0.8870$
$\sin\theta$ is negative

**32.** $\tan\theta = 0.6015$
$\cos\theta$ is positive

**33.** $\csc\theta = -1.1675$
$\tan\theta$ is positive

**34.** $\sec\theta = -1.0285$
$\sin\theta$ is positive

**35.** Find $\tan\theta$ and $\theta$.

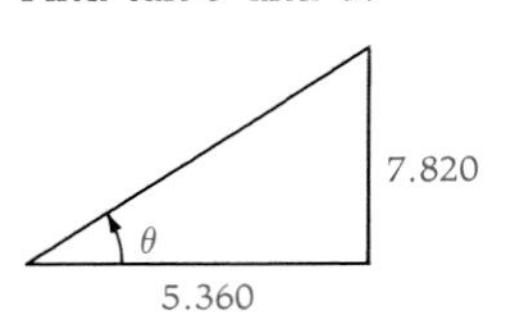


**36.** Find $\cos\theta$ and $\theta$.

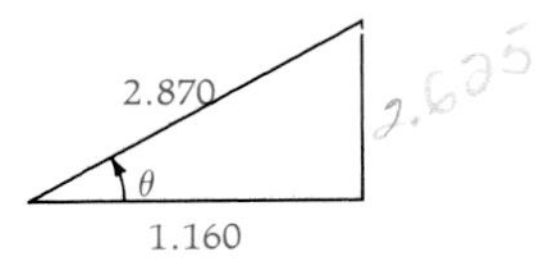


**37.** Find $\sin\theta$ and $\theta$.

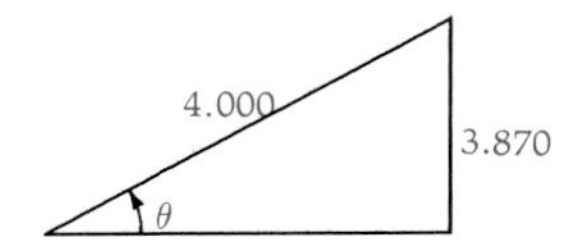


**38.** Find $\tan\theta$ and $\theta$.

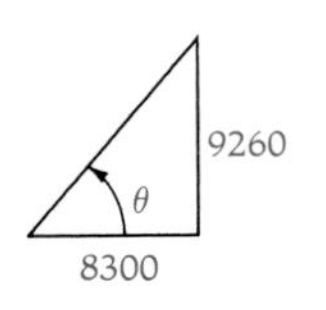


**39.** Find $\theta$.

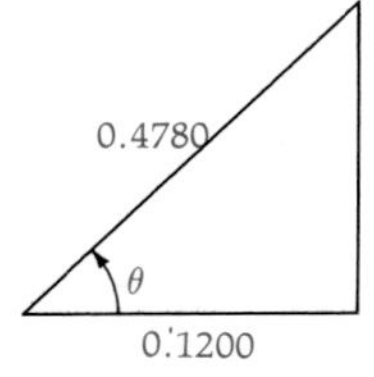


**40.** Find $\theta$.

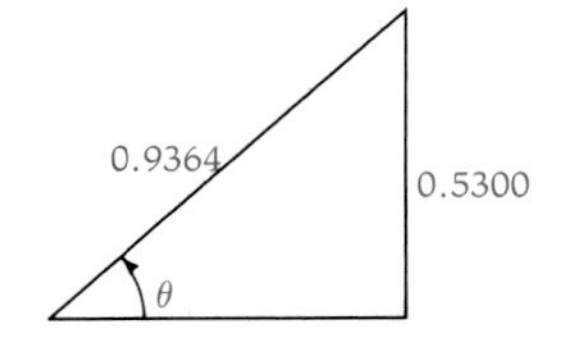


**41.** Find $\theta$.

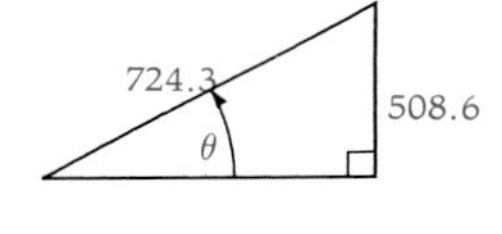


**42.** Find $\theta$.

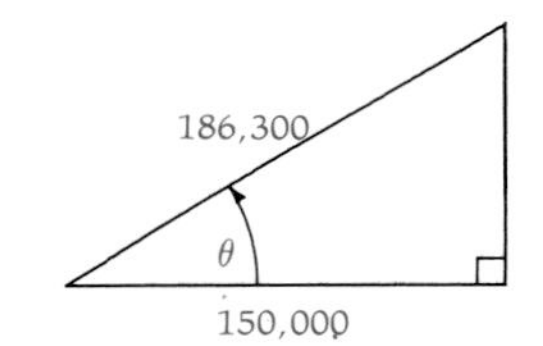


In Problems 43 and 44, you can first find $(x + y)$ by using the Pythagorean Theorem. Refer also to Figure 6-53.

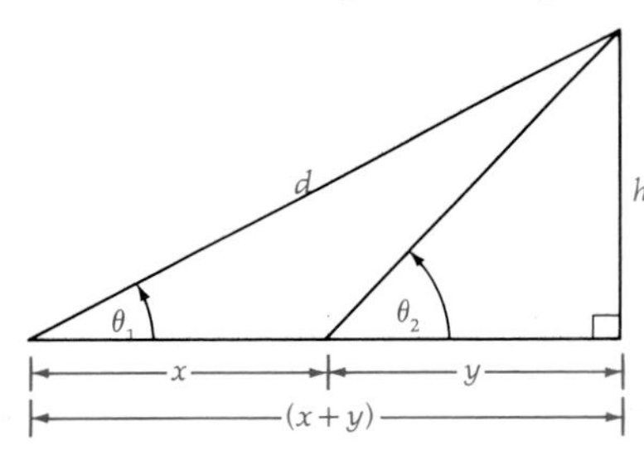


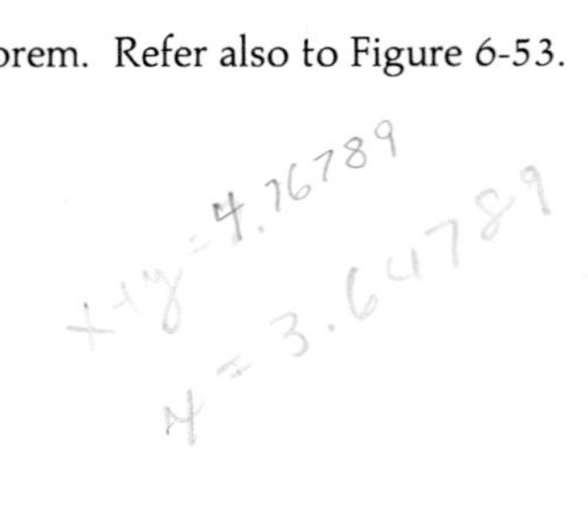

*Figure 6-53*

★ **43.** Using Figure 6-53 assume $d = 5.320$, $h = 2.360$, and $y = 1.120$.
- **a.** Find $x$
- **b.** Find $\theta_1$

★ **44.** Assume $d = 4.380$, $h = 3.200$, and $x = 1.240$.
- **a.** Find $y$
- **b.** Find $\theta_2$

## 6-8 SOLVING RIGHT TRIANGLES AND APPLICATIONS

If we know two sides or a side and an angle of a right triangle (other than the right angle), we can "solve" the triangle. "Solving" the triangle means finding the values of all of the unknown parts. This is accomplished by using the definitions of the trigonometric functions, and the tables or a calculator.

Use the following rules to determine how precise the measurements will be.

| *Number of Digits in the Length of a Side* | *Measure of an angle in Degrees, Minutes, Seconds* | *Measure of an angle in Decimal Degrees* |
|---|---|---|
| 2 | nearest degree | nearest degree |
| 3 | nearest ten minutes | nearest tenth of a degree |
| 4 | nearest minute | nearest hundredth of a degree |

1. If the sides of a triangle are precise to three digits, then the angles will be to the nearest ten minutes or the nearest tenth of a degree.
2. If you are using a trigonometric table, then use one whose precision is one digit greater than the number of digits in the given length of the sides of a triangle. If a side is precise to three digits, use a four-place table. If you are using a calculator, round all values to values of one digit more precision than the given data.
3. Perform all calculations, then round off the final result to the same number of digits in the given data.

**EXAMPLE 1** Solve right triangle $ACB$ if $a = 5.00$, $b = 12.00$. (Use trigonometric functions to find the solution.)

***Solution***

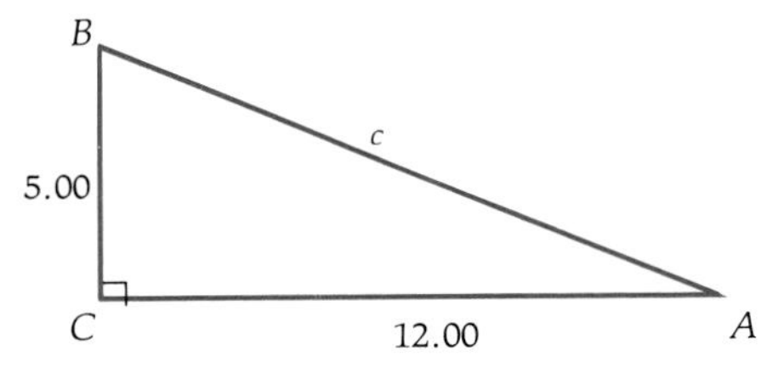


This problem could be solved by using the Pythagorean Theorem, but we want practice in using the trigonometric functions, so we note that

$$\tan A = \frac{5.00}{12.00} = 0.4167 \qquad \textit{Rounded to 4 places.}$$

Therefore $A = 22.6^\circ$ *Rounded to nearest tenth of a degree.*

Also $B = 90^\circ - A = 90 - 22.6 = 67.4^\circ$

To find side $c$, we may use either the sine, cosine, secant, or cosecant of either angle $A$ or angle $B$. Since the sine or cosine requires division in this case, if we do not have a calculator we will use

$$\sec A = \frac{c}{b}$$

$$\sec 22.6^\circ = \frac{c}{12.00}$$

$$c = (12.00)(1.0832) = 12.9984$$

$c = 13.0$ *Three significant digits, since $a = 5.00$ only has 3 significant digits.*

If we have a calculator, we find it more convenient to use

$$\cos A = \frac{b}{c}$$

$$\cos 22.6^\circ = \frac{12.00}{c}$$

Thus

$$c = \frac{12.00}{\cos 22.6^\circ}$$

We readily find this on a calculator with the following keystrokes:

12.00 [÷] 22.6 [cos] [=] 12.998123

$c = 13.0$ *Rounded to 3 significant digits.*

*Check:* You can use the Pythagorean Theorem to verify that these values of $a$, $b$, $c$ yield a right triangle:

$$a^2 + b^2 = c^2$$

$$(5.00)^2 + (12.00)^2 = (13.00)^2 \quad ■$$

The advantage of using a scientific calculator is clearly seen in the following example. (It can of course be done as well in a time-consuming manner by using the tables and interpolation.)

EXAMPLE 2 Solve right triangle $ABC$ if $c = 148.2$ and $A = 37.96^\circ$. Refer to Figure 6-54.

***Solution*** We can easily find $B = 90^\circ - 37.96^\circ = 52.04^\circ$. To find side $a$, we know that

$$\sin A = \frac{a}{c}$$

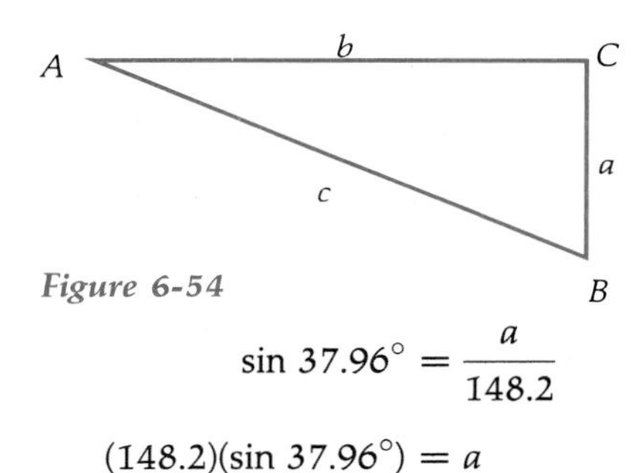


*Figure 6-54*

$$\sin 37.96^\circ = \frac{a}{148.2}$$

$$(148.2)(\sin 37.96^\circ) = a$$

On a calculator we enter

148.2 [×] 37.96 [sin] [=] 91.159478

$a = 91.16$ *Rounded to 4 significant digits.*

To find side $b$, we know that

$$\cos A = \frac{b}{c}$$

$$\cos 37.96^\circ = \frac{b}{148.2}$$

$$(148.2)(\cos 37.96^\circ) = b$$

148.2 [×] 37.96 [cos] [=] 116.84686

$b = 116.8$ *Rounded to 4 significant digits.* ■

A number of situations arise in which a given value cannot be measured directly, but it can be determined indirectly by the use of trigonometry. In order to have the necessary mathematical skills to determine necessary angles involved in these applications a few helpful properties from geometry are listed for reference.

## Geometric Properties for Review

**A.** Parallel lines are everywhere equidistant.

**B.** If two parallel lines are cut by a third line (*transversal*) as shown in Figure 6-55, then the following are true:

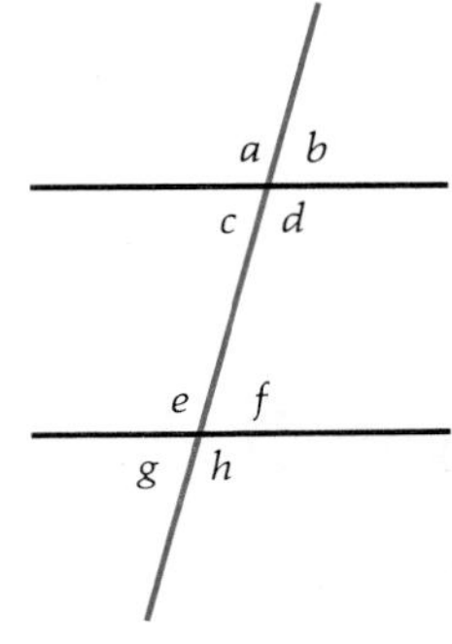


*Figure 6-55*

**1.** Alternate interior angles are equal.

$$\angle c = \angle f, \quad \angle d = \angle e$$

2. Corresponding angles are equal.

$$\angle a = \angle e, \quad \angle b = \angle f, \quad \angle d = \angle h, \quad \angle c = \angle g$$

3. Alternate exterior angles are equal.

$$\angle a = \angle h, \quad \angle b = \angle g$$

4. Interior angles on the same side of the transversal are supplementary.

$$\angle c + \angle e = 180^\circ, \quad \angle d + \angle f = 180^\circ$$

---

Finally, certain specialized terms are commonly used and must be clearly understood. Therefore the following definitions are necessary in order to solve indirect measurement problems.

*Definition*

The *line of sight* of an object is the straight line from the eye of the observer to the object.

*Definition*

If a horizontal plane in the eye of the observer is below the line of sight, the angle measured from the horizontal plane to the line of sight is called the *angle of elevation*. Refer to Figure 6-56.

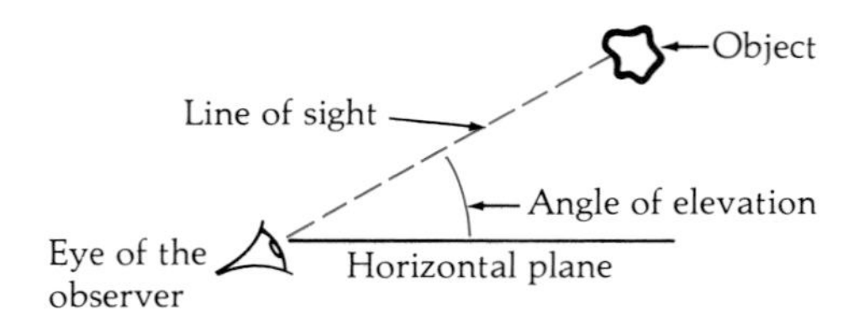


*Figure 6-56*

*Definition*

If a horizontal plane in the eye of the observer is above the line of sight, the angle measured from the horizontal plane to the line of sight is called the *angle of depression*. Refer to Figure 6-57.

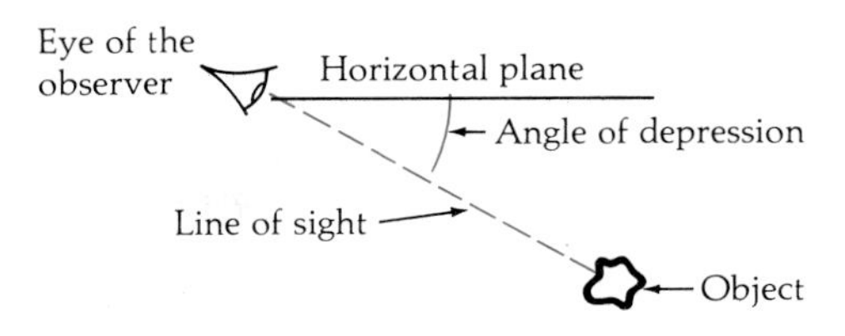


*Figure 6-57*

**EXAMPLE 3** A new railroad bed has been laid in a construction project. The angle of elevation of the rails is 5.6°. If the beginning of this stretch of rail is at sea level, and if the elevation of the rails at the end of the distance is 84.6 meters above sea level, how many meters of rail were laid?

***Solution*** We draw a sketch to represent the situation.

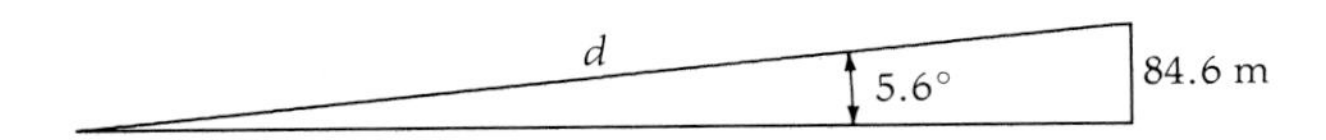


In order to determine the distance of rail, $d$, we use the trigonometric ratio

$$\sin 5.6^\circ = \frac{84.6}{d}$$

Thus we have

$$d = \frac{84.6}{\sin 5.6^\circ}$$

On a calculator we obtain

84.6 [÷] 5.6 [sin] [=] 866.95518

Rounding to three significant digits, we would estimate that 867 meters of track were laid. If we use Table D, we would obtain the answer more readily by finding $d$ using the equation $d = (84.6)(\csc 5.6^\circ)$. Either way we obtain an answer of 867 meters. ■

**EXAMPLE 4** Two buildings are 110 meters apart. From a window of one building the angle of depression to the base of the second building is 24.5°, and the angle of elevation to the top of the second building is 42.2°. How tall is the second building?

***Solution*** In solving a problem of this type we try to form a mathematical model that will represent the physical situation. Our first step is to use a geometric diagram. Refer to Figure 6-58(a).

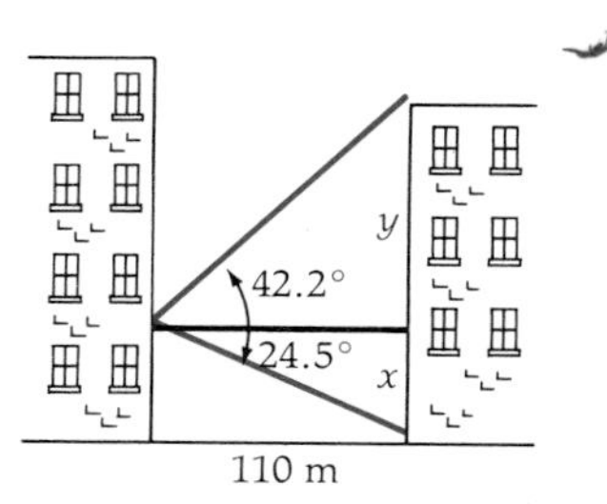


*Figure 6-58(a)*

Let $(x + y)$ be the height of the second building, where $x$ is the distance from the base of the building to a point on the horizontal plane even with the eye of the observer. Let $y$ represent the remaining height.

In essence, we are solving two separate problems. Refer to Figure 6-58(b).

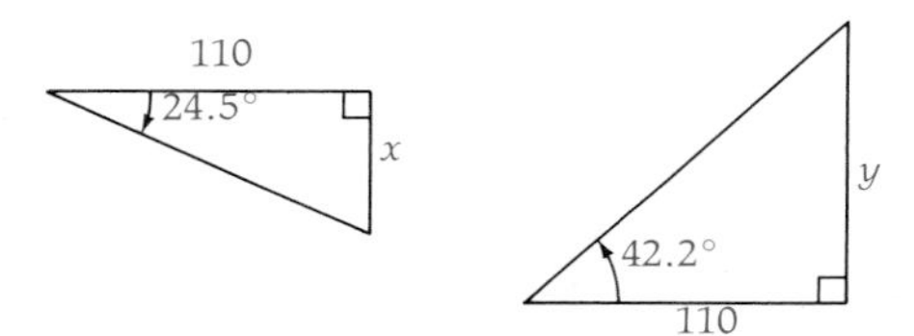


*Figure 6-58(b)*

$$\tan 24.5° = \frac{x}{110} \qquad \text{and} \qquad \tan 42.2° = \frac{y}{110}$$

$$x \doteq (110)(0.4557) \qquad\qquad y \doteq (110)(0.9067)$$

$$\doteq 50.127 \qquad\qquad \doteq 99.737$$

Rounding to three significant digits, we have $x + y = 150$ meters. ■

In navigation the north line of the compass is used as the reference line in measuring directions. Thus a positive angle is a clockwise rotation.

***Definition*** The *course* of an object is the angle that is measured from the north, clockwise, to the direction in which the object is moving. (See Figure 6-59.)

***Definition*** A *bearing* is a direction; it is described by the acute angle it makes with the north or south direction.

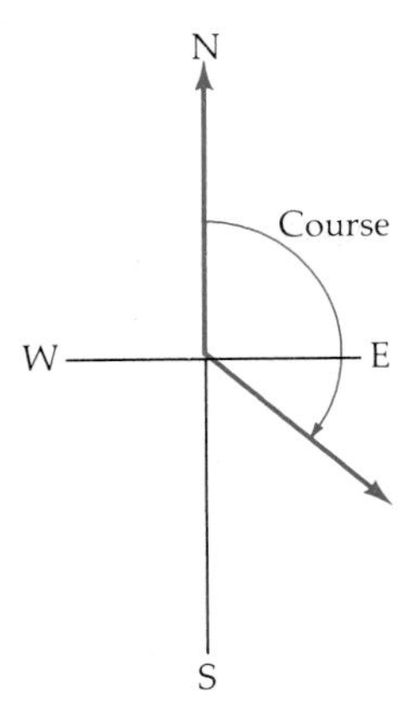


*Figure 6-59*

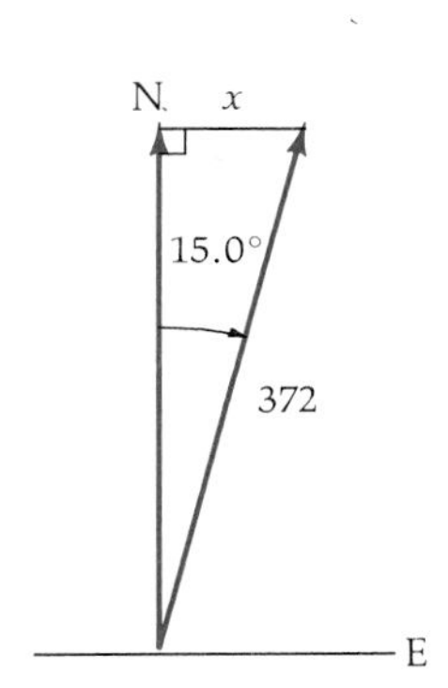


*Figure 6-60*

For example, a bearing of N 38°W is a direction that is 38° west of north and a bearing of S 50°E is a direction that is 50° east of south.

**EXAMPLE 5** A light plane traveled 372 miles on a bearing N 15.0°E. At the end of the trip, how far east of the original starting point was the plane? Refer to Figure 6-60.

***Solution*** To find the distance $x$ we use the trigonometric ratio

$$\sin 15.0^\circ = \frac{x}{372} \qquad (372)(\sin 15^\circ) = x$$

On a calculator we enter

372 [×] 15 [sin] [=] 96.280685

To the nearest tenth of a mile (three digits) the distance is 96.3 miles east of the starting point. ■

## EXERCISE 6-8

Use a calculator or Table D in the following problems. Referring to Figure 6-61, solve for the unknown parts of the general right triangle $ACB$ in Problems 1–20.

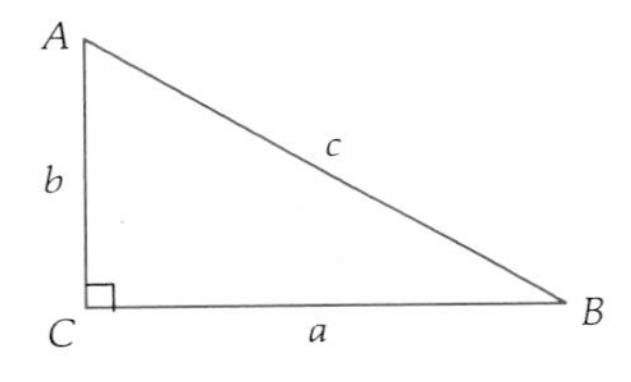


*Figure 6-61*

1. $A = 27^\circ, a = 3.5$
2. $B = 50^\circ, b = 62$
3. $A = 32^\circ, c = 20$
4. $A = 30.1^\circ, b = 2.34$
5. $A = 47.3^\circ, b = 41.3$
6. $B = 62.3^\circ, c = 12.6$
7. $B = 28.5^\circ, a = 19.4$
8. $A = 39.6^\circ, a = 22.3$
9. $B = 19.5^\circ, b = 3.15$
10. $B = 51.2^\circ, a = 20.0$
11. $A = 32.2^\circ, b = 21.7$
12. $B = 61.0^\circ, a = 23.5$
13. $A = 42.3^\circ, c = 24.9$
14. $A = 55.5^\circ, b = 13.7$
15. $a = 16.0, b = 14.0$
16. $a = 18.2, c = 25.0$
17. $b = 28.3, c = 36.2$
18. $A = 29.24^\circ, a = 143.1$
19. $B = 35.56^\circ, a = 16.65$
20. $a = 163.8, c = 187.5$
21. The angle of elevation from a point on the ground 20.0 meters from the base of a tree to the top of the tree is 32.2°. Find the height of the tree.
22. A searchlight is aimed directly upward to illuminate a spot on a cloud. From a station on the ground 40.0 meters from the searchlight, the angle of elevation to the spot on the cloud is 71.5°. Find the height of the cloud.
23. A person stands at the edge of a river directly opposite a rock on the other side of the river. She walks 15.0 meters along the edge of the river. She then measures the angle between her line of sight to the rock and her line of travel. She finds it to be 55.3°. What is the width of the river?
24. From the top of a lighthouse that measures 50.0 meters in height, the angle of depression of a ship on the water is 16.3°. How far is the ship from the base of the lighthouse? (Assume that the base of the lighthouse is at water level.)
25. From the top of a cliff, the angle of depression of a rock that is 81.4 meters from the base of the cliff is 22.2°. Find the height of the cliff.
26. A guy wire is attached to a vertical pole at a height of 7.00 meters from the ground to make an angle of 68.7° with the ground. Assuming that a total of one meter of wire is used in making the knots at both ends, what is the total length of the guy wire?
27. Find the angle of elevation of the sun when a 10.0-meter vertical flagpole casts a 7.20-meter shadow on level ground.
28. A 6.0-meter ladder rests against a vertical wall. The base of the ladder is 2.0 meters from the wall. Find the acute angle the ladder makes with the ground.
29. A ship sails due west a distance of 31.0 kilometers, then alters course and sails due south a distance of 45.0 kilometers. Find its bearing from the starting point.
30. A plane leaves Kennedy Airport and travels on a course of 247.7° at 640 kilometers per hour. How far south of the airport is the plane after 2 hours?
31. A boat is anchored in a lake with 16.5 meters of line. The angle of depression of the taut anchor line is 35.2°. If 1.20 meters of the line is above the water, how deep is the lake?
32. A taut string from a kite makes an angle 52.5° with the ground. If there are 43.5 meters of string between the ground and the kite, how high is the kite?
33. From a point on a level road to the base of a mountain, the angle of elevation to the top of the mountain is 17.6°. From a second point on the road 1 kilometer closer to the mountain, the angle of elevation is 21.2°. Find the height of the mountain to the nearest 10 meters.
34. A jet plane is headed north with an airspeed of 560 kilometers per hour. A west wind is blowing

with a speed of 128 kilometers per hour. Find the plane's speed with respect to the ground (groundspeed) and its course.

**35.** From the base of a 35.0-meter tall building the angle of elevation of a tower a distance away is 60.2°. From the top of the building the angle of elevation of the tower is 20.5°. Find the height of the tower.

**36.** A jet plane with an airspeed of 614 kilometers per hour has an angle of descent of 19.0°. Find the rate at which it is approaching the ground.

**37.** A ship sails 32.1 kilometers on course 260.0°. It then sails 10.3 kilometers on course 170.0°. Find the course and the distance it must sail to return to its starting point.

**38.** From the base of a tower the angle of elevation to the top of a 4.00-meter vertical pole a distance away is 18.8°. From the top of the tower the angle of depression of the base of the pole is 48.2°. Find the height of the tower.

**39.** A 36.0 Newton object is on a frictionless incline that makes an angle of 21°30′ with the horizontal. What force must be exerted parallel to the incline to prevent the weight from sliding down the incline?

**40.** A forest ranger at station *A* sights a fire at bearing N 24°40′E. A ranger at station *B*, 10.0 kilometers due east of station *A*, sights the fire at bearing N 65°20′W. How far is the fire from station *A*?

★ **41.** The angle of elevation from the ground to a helicopter flying at 2860 feet is 52.8°. The helicopter descends vertically 1000 feet. By how much will the angle of elevation change?

★ **42.** An astronaut measures the shadow in a photograph of a crater on the moon. He finds that the length of the shadow is 1.39 cm. The angle of elevation to the sun (from the moon's surface) is 49.1° when the photograph is taken. The scale on the photograph is 1 cm = 2500 m. How many meters deep is the crater?

★ **43.** In a recent briefing before the U.S. Congress an admiral commented that a renovated battleship with 12.0 inches of armor plate on its side has an effective protection of 17.0 inches of armor plate, since enemy fire can be predicted to come in at a specified angle. (See Figure 6-62.) What is the angle of elevation from the deck of the battleship to the predicted enemy fire? (Round your answer to the nearest tenth of a degree.)

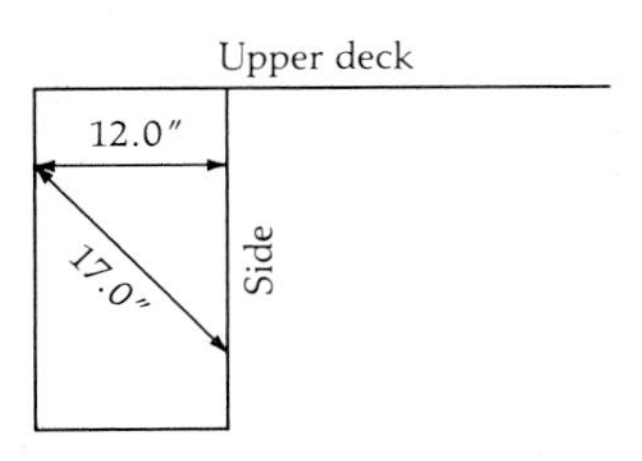


*Figure* **6-62**

★ **44.** Refer to the situation in Problem 43. In a simulated war exercise a simulated enemy missile reached the battleship. The angle of elevation from the deck of the battleship to the simulated missile was 34.5°. What would be the effective thickness of the armor plate under such a condition? (In other words, through how many inches of armor plate would the missile theoretically travel?)

## COMPUTER PROBLEMS FOR CHAPTER 6

See if you can apply your knowledge of computer programming to do Problems 1 and 2.

**1.** Using the library function in your computer for the sine, write a program to find the sine of 101 angles from sin 10° to sin 11° in measurements of one one-hundredth of a degree. (sin 10.00°, sin 10.01°, sin 10.02°, etc.).

**2.** Write a program to convert angles measured in decimal degrees to angles measured in degrees and minutes. Run the program for 40 angles from 20.01° to 20.40° (measurements of one one-hundredth of a degree).

## KEY TERMS AND CONCEPTS

Be sure you understand what is meant by these terms and can give an example of each.

Angle
Central angle
Arc of a circle
Vertex of an angle
Initial side of an angle
Terminal side of an angle
Positive angle
Negative angle
Right angle
Straight angle
Acute angle
Obtuse angle
Degrees
Radians
Triangle
Similar triangles
Complementary angles
Vertices of a triangle
Right triangle
Isosceles triangle
Pythagorean Theorem
Sine
Cosine
Tangent
Cosecant
Secant
Cotangent
Cofunction
Reciprocal functions
Argument of a trigonometric function
Standard position of an angle
Coterminal angles
Quadrants
30°–60°–90° triangle
45°–45°–90° triangle
Quadrantal angle
Reference angle
Minutes (measure of an angle)
Seconds (measure of an angle)
Solving a right triangle
Line of sight
Angle of elevation
Angle of depression
Course
Bearing

## SUMMARY OF PROCEDURES AND CONCEPTS

### Changing Radians to Degrees and Degrees to Radians

1. To change radians to degrees, multiply the angle by $\frac{180}{\pi}$.
2. To change degrees to radians, multiply the angle by $\frac{\pi}{180}$.

### Finding Coterminal Angles

All angles coterminal with $\theta$ are obtained by solving the equation $\theta + n(360^\circ)$ for integer values of $n$.

### Finding the Reference Angle

The first quadrant angle whose trigonometric functions are equal in absolute value to the functions of an angle in each of the other quadrants is called the *reference* angle of the other angles.

1. For $0^\circ \leq \theta < 360^\circ$:

| *If $\theta$ Is in Quadrant* | *The Reference Angle $\theta'$ Is* |
|---|---|
| I | $\theta$ |
| II | $180^\circ - \theta$ |
| III | $\theta - 180^\circ$ |
| IV | $360^\circ - \theta$ |

2. If $\theta \geq 360^\circ$ or if $\theta < 0^\circ$, we must first determine the positive angle less than $360^\circ$ that is coterminal with $\theta$ and then find the reference angle $\theta'$.

### Limits of Trigonometric Functions

1. $-1 \leq \sin\theta \leq 1$
2. $-1 \leq \cos\theta \leq 1$
3. $\sec\theta \geq 1$ or $\sec\theta \leq -1$
4. $\csc\theta \geq 1$ or $\csc\theta \leq -1$
5. There are no limits for $\tan\theta$ and $\cot\theta$.

## CHAPTER 6 REVIEW EXERCISE

Change to radians in terms of $\pi$ in Problems 1–5.

**1.** $10^\circ$ **2.** $210^\circ$ **3.** $300^\circ$ **4.** $25^\circ$ **5.** $225^\circ$

Change to approximate radians in Problems 6–10. (Use $1^\circ = 0.0175$ radian.)

**6.** $50^\circ$ **7.** $41^\circ$ **8.** $95^\circ$ **9.** $200^\circ$ **10.** $163^\circ$

Change to degrees in Problems 11–20. (Use 1 radian $= 57.3^\circ$ for those exercises that do not involve $\pi$.)

**11.** $\frac{\pi}{2}$ radians **12.** $\frac{3}{2}\pi$ radians **13.** 2 radians **14.** 0.1 radian **15.** $3\pi$ radians

**16.** 1.5 radians **17.** $\frac{8}{3}\pi$ radians **18.** $\frac{1}{3}$ radian **19.** $\frac{7}{4}\pi$ radians **20.** 3.14 radians

In Problems 21–26, use the Pythagorean Theorem to find the unknown side of the general right triangle $ACB$.

**21.** Find $c$ if $a = 9$, $b = 5$. **22.** Find $b$ if $a = 3$, $c = 8$. **23.** Find $a$ if $b = 7$, $c = 10$.

**24.** Find $c$ if $a = 6$, $b = 6$. **25.** Find $a$ if $b = 8$, $c = 10$. **26.** Find $b$ if $a = 4$, $c = 8$.

In Problems 27–30, find the exact value. Assume $A$ is an acute angle. (Do not use Table D or a calculator.)

**27.** Find the other five trigonometric functions of angle $A$ if $\tan A = \frac{4}{5}$.

**28.** Find $\csc A$ if $\cos A = \frac{3}{4}$. **29.** Find $\sin B$ if $\sec A = \frac{3}{2}$. **30.** Find $\cos A$ if $\cot B = \frac{3}{7}$.

In Problems 31–38, use a calculator or Table D when needed. Assume $\theta$ is an acute angle.

**31.** Find $\sin 23.4^\circ$. **32.** Find $\cot 56.3^\circ$. **33.** Find $\cos 72.2^\circ$.

**34.** Find $\theta$ if $\tan\theta = 0.7412$. **35.** Find $\theta$ if $\sin\theta = 0.3240$. **36.** Find $\theta$ if $\sec\theta = 2.329$.

**37.** Find $\theta$ if $\sin\theta = \frac{\sqrt{3}}{2}$. **38.** Find $\theta$ if $\cos\theta = \frac{\sqrt{2}}{2}$.

In Problems 39–44, give the quadrant of each angle.

**39.** $-100^\circ$ **40.** $250^\circ$ **41.** $490^\circ$ **42.** $-2{,}194^\circ$ **43.** $\frac{3}{4}\pi$ **44.** $-\frac{11}{6}\pi$

**45.** Which of the following angles are coterminal with $146^\circ$?

**a.** $34^\circ$ **b.** $506^\circ$ **c.** $-56^\circ$ **d.** $1{,}586^\circ$ **e.** $214^\circ$ **f.** $-934^\circ$

In Problems 46–49, a point on the terminal side of an angle $\theta$ is given. Find the six trigonometric functions for each. Express as exact values. (Do not use a calculator or Table D.)

**46.** $(-1, 4)$ **47.** $(3, 5)$ **48.** $(-2, -\sqrt{5})$ **49.** $(24, -7)$

In Problems 50–53, find the exact value. (Do not use a calculator or Table D.)

**50.** Find $\theta$ if $\sin\theta = \frac{1}{2}$, $\theta$ in quadrant II. **51.** Find $\theta$ if $\cot\theta = -1$, $\theta$ in quadrant IV.

**52.** Find $\theta$ if $\sec\theta = \sqrt{2}$, $\sin\theta < 0$.

**53.** Find $\theta$ if $\cos\theta = -\frac{\sqrt{3}}{2}$, $\tan\theta > 0$.

Use a calculator or Table D and your knowledge of the reference angle to solve Problems 54–61.

**54.** Find $\tan 120°$.

**55.** Find $\cot(-228°)$.

**56.** Find $\sin 540°$.

**57.** Find $\cos 100\pi$.

**58.** Find $\csc 990°$.

**59.** Find $\tan\left(-\frac{7}{3}\pi\right)$.

**60.** Find $\csc 1062.7°$.

**61.** Find $\cos(-2918.2°)$.

For Problems 62–68, find $\theta$ where $0° \le \theta \le 360°$. Use a calculator or Table D. (If no such angle exists, so state.)

**62.** $\sec\theta = 1.1707$, $\tan\theta < 0$

**63.** $\cot\theta = 0.8050$, $\sin\theta < 0$

**64.** $\cos\theta = -0.7333$, $\csc\theta > 0$

**65.** $\sin\theta = -0.9100$, $\tan\theta < 0$

**66.** $\csc\theta = 1.3022$, $\cot\theta < 0$

**67.** $\sin\theta = 2.146$, $\cot\theta < 0$

**68.** $\tan\theta = -2.2817$, $\sin\theta > 0$

**69.** Solve the general right triangle $ACB$ if $A = 33°10'$ and $b = 14.2$.

**70.** Solve the general right triangle $ACB$ if $a = 24.6$ and $c = 37.4$.

**71.** The top of a ladder rests against a vertical wall. The foot of the ladder is 3.10 meters from the base of the wall. It makes an angle of 72.8° with the ground. How long is the ladder?

**72.** A tree casts a shadow 20.5 meters long on the level ground when the angle of elevation of the sun is 34.1°. Find the height of the tree.

**73.** From the top of a 48.5 meter observation tower, a forest ranger spots a fire at an angle of depression of 11.2°. How far is the fire from the base of the tower? (Assume the ground is level from the base of the tower.)

**74.** A ship sails 21.4 kilometers due east, then 15.1 kilometers due north. What is the ship's bearing from the starting point?

**75.** From one bank of a river the angle of elevation to the top of a cliff on the opposite bank is 50.1°. The observer moves directly backwards away from the cliff a distance of 40.0 meters and finds the angle of elevation to be 33.3°. Find the width of the river. (Assume that the ground is level.)

## PRACTICE TEST FOR CHAPTER 6

In Problems 1–24, you may use a calculator or Table D unless otherwise directed.

**1.** Change 80° to radians in terms of $\pi$.

**2.** Approximate 92° in radians. (Use $1° = 0.0175$ radian.)

**3.** Change $\frac{2}{9}\pi$ radians to degrees.

**4.** Approximate 5.3 radians in degrees. (Use 1 radian $= 57.3°$.)

**5.** In right triangle $ACB$, $a = 10$, $c = 14$. Find $b$.

**6.** Find $\tan A$ if $\cos A = \frac{5}{9}$. (exact value)

7. Find $\sin B$ if $\sec A = \frac{6}{5}$. (exact value)

8. Find $\csc A$ if $\cot A = \frac{3}{8}$. (exact value)

9. Use interpolation or a calculator to find $\cot 58.24^\circ$.

10. Find $\theta$ if $\sin \theta = 0.7249$.

11. Name the quadrant of the angle $\theta = 1{,}985^\circ$.

12. Find an angle between $0^\circ$ and $-360^\circ$ that is coterminal with $500^\circ$.

13. The point $(-5, 8)$ lies on the terminal side of angle $\theta$. Find $\cos \theta$.

14. The point $(x, -4)$ lies on the terminal side of angle $\theta$ in the third quadrant a distance of five units from the origin. Find $\tan \theta$.

15. Find $\theta$ if $\sec \theta = -\sqrt{2}$ and $\tan \theta < 0$. (exact value)

16. Find $\cot 1{,}350^\circ$.

17. Find $\sin 575^\circ$.

18. Find $\cos(-771^\circ)$.

19. Find $\sec(-492.7^\circ)$.

20. Find $\theta$, if $\cos \theta = 0.6926$ and $\sin \theta < 0$.

21. Solve the general right triangle $ACB$ when $B = 34.3^\circ$ and $a = 19.6$.

22. In right triangle $ACB$, $a = 24.1$ and $b = 16.8$. Find angle $A$.

23. A building casts a shadow of 40.0 meters when the elevation of the sun is $40.3^\circ$. Find the height of the building.

24. The angle of depression of a boat from the top of a lighthouse is $6.8^\circ$. After the boat has traveled 50.0 meters toward the lighthouse, the angle of depression of the boat is $8.2^\circ$. Find the height of the lighthouse. (Assume that the base of the lighthouse is at water level.)

# 7 Trigonometric Functions and Their Graphs

■ *Mathematicians do not deal in objects, but in relations between objects; thus, they are free to replace some objects by others so long as the relations remain unchanged. Content to them is irrelevant: they are interested in form only.*

*Henry Poincaré*

Our work in Chapter 6 involved the development of the trigonometric functions of angles. This approach corresponds to the classical development of trigonometry.

However, much of the application of modern trigonometry in the higher levels of mathematics does not involve angles at all, but rather the set of real numbers. We thus turn our study to the definition of the trigonometric functions of all real numbers.

## 7-1 THE UNIT CIRCLE

We generally think of the real numbers as being represented by coordinates on a number line. In this section we will represent them as the length of an arc of a circle and define the trigonometric functions in this context.

***Definition***

A *unit circle* is a circle having a radius of one unit.

If we place a unit circle on the Cartesian coordinate system such that the center of the circle is at the origin, its equation is $x^2 + y^2 = 1$.

From the point (1, 0) on the circumference of the unit circle, measure the length of an arc of the circle in a counterclockwise direction and let the endpoint

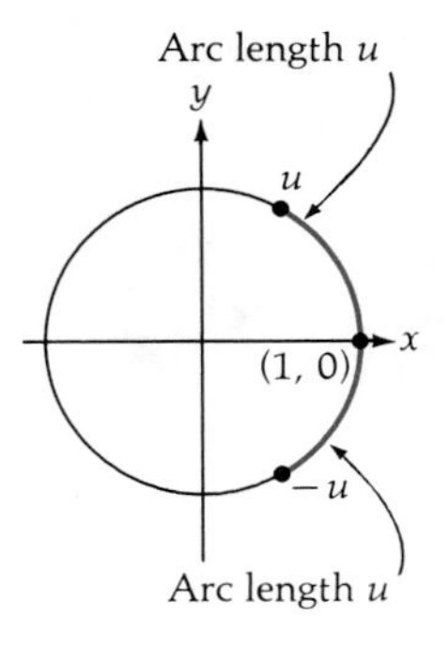


of the arc correspond to the positive real number that represents the length of the arc. If we measure the length of an arc from this point in a clockwise direction, then the endpoint of the arc corresponds to the negative of the real number that represents the length of the arc.

Since the circumference of a circle is $C = 2\pi r$ and $r = 1$, it should be recognized immediately that this pattern gives a unique representation of the real numbers 0 to $2\pi$. Other positive and negative numbers will correspond to points already represented by a number between 0 and $2\pi$.

The correspondence can be visualized in this way. Consider the real number line with an initial point at (1, 0) being wrapped around the unit circle.

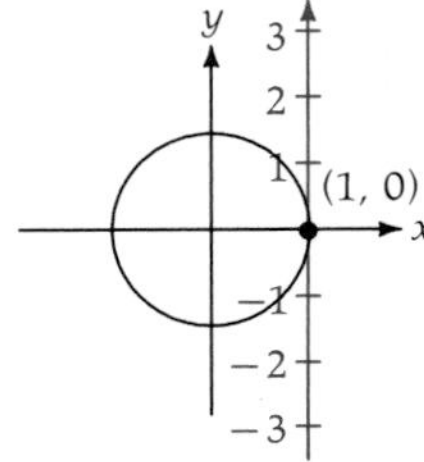


Number line with initial point at (1, 0) parallel to $y$-axis

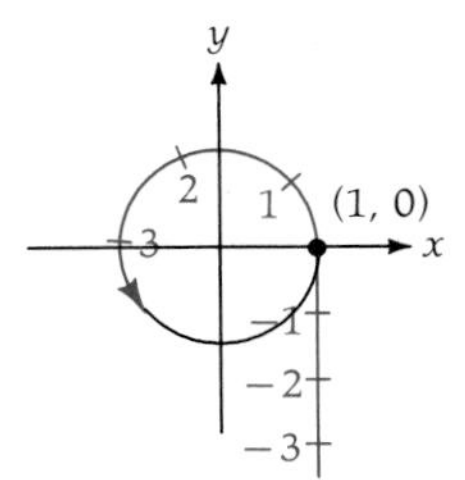


Number line with initial point at (1, 0) "wrapped around" the unit circle

If the number line is marked in terms of $\pi$, certain values are conveniently located on the $x$-axis or the $y$-axis.

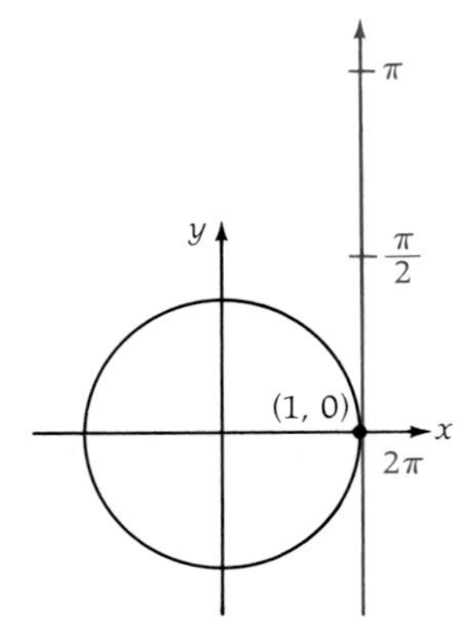


Number line marked in terms of units of $\frac{\pi}{2}$ parallel to the $y$-axis

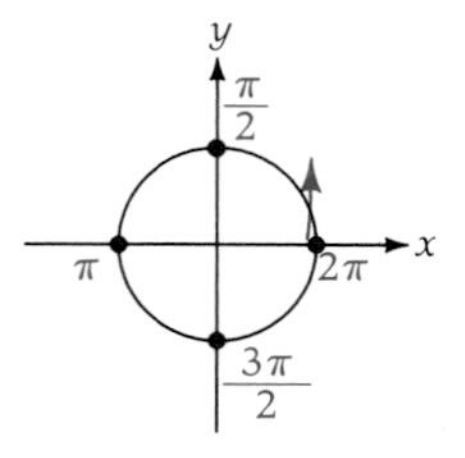


Number line marked in terms of units of $\frac{\pi}{2}$ "wrapped around" the unit circle

*Definition*

An arc is in *standard position* on the unit circle if its initial point is at the point (1, 0) and its endpoint (terminal point) is represented by a positive real number.

*Definition*

Arcs are *coterminal* if their terminal points coincide.

**EXAMPLE 1** Given an arc $A$ in standard position with length 2, find a negative real number and another positive real number that will represent arcs coterminal with $A$.

***Solution*** Since the length of arc $A$ is 2, then $(2 + 2\pi)$ is another positive real number that represents the length of an arc whose endpoint coincides with the endpoint of arc $A$, and $(2 - 2\pi)$ is a negative number that represents an arc coterminal with arc $A$. ■

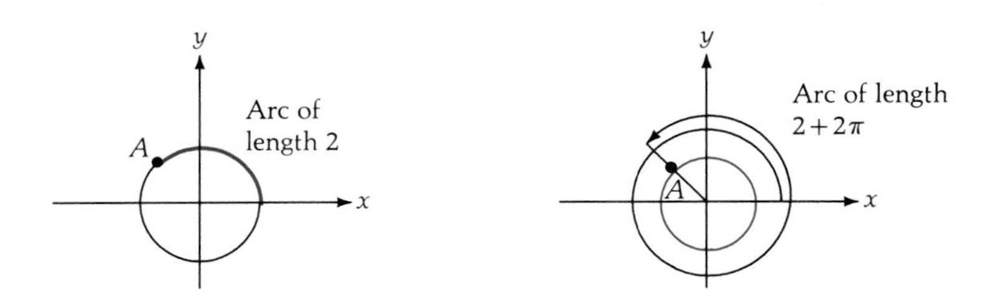


These examples lead us to a general conclusion.

*Property of Coterminal Arcs*

If $u$ is the length of an arc in standard position, then, in general, $(u \pm 2n\pi)$, where $n$ is any integer, is coterminal with $u$.

We see that a point on the unit circle can be represented by many different real numbers that are lengths of arcs in standard position. However, given any real number $u$ that represents the length of an arc in standard position, there is only *one* ordered pair $(x, y)$ on the coordinate system that represents the number

$u$. In other words, the ordered pair $(x, y)$ is a function of the real number $u$ under this representation.

Assume that $u$ is a real number and that it represents the length of an arc in standard position.

**EXAMPLE 2** Find the ordered pair $(x, y)$ that is the terminal point of the arc of length $u$ if

$$u = \frac{5\pi}{2}$$

***Solution*** If $u = \frac{5\pi}{2}$, we can rewrite $\frac{5\pi}{2} = \frac{\pi}{2} + \frac{4\pi}{2} = \frac{\pi}{2} + 2\pi$. Thus we will go around the circle one and one quarter times. The arc of length $\frac{5\pi}{2}$ is coterminal with an arc of length $\frac{\pi}{2}$. This value of $(x, y) = (0, 1)$.

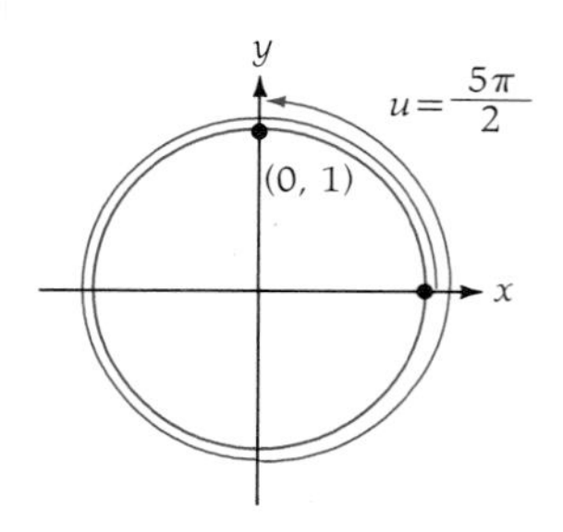


■

### Definitions of Trigonometric Functions of Real Numbers

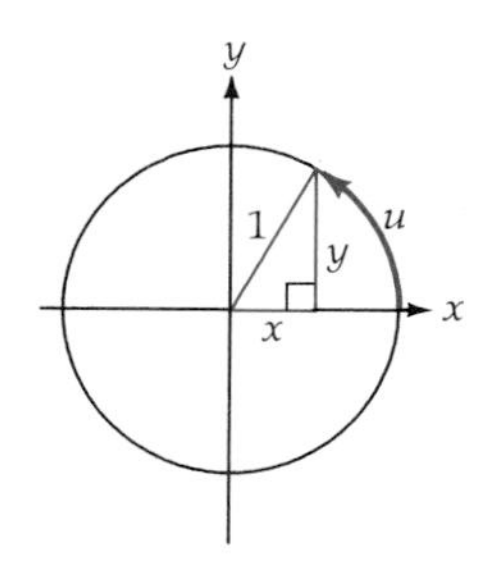


We have been investigating how any real number $u$ can be represented by a length of the arc of the unit circle. We also examined the terminal point of the arc and found that the ordered pair $(x, y)$ describing the terminal point of the arc is a function of the real number $u$.

This uniqueness of the pair $(x, y)$ for the real number $u$ allows us to define the trigonometric functions of $u$ in terms of $x$, $y$, and 1 (the radius of the unit circle).

*Definition*

If the real number $u$ is the endpoint of an arc in standard position on the unit circle and that point has coordinates $(x, y)$, then the following are true.

| *Function* | *Definition* | *Function* | *Definition* |
|---|---|---|---|
| $\sin u$ | $y$ | $\csc u$ | $\frac{1}{y}$ |
| $\cos u$ | $x$ | $\sec u$ | $\frac{1}{x}$ |
| $\tan u$ | $\frac{y}{x}$ | $\cot u$ | $\frac{x}{y}$ |

We observe that the general properties of the trigonometric functions observed in the previous chapter are also evident in this definition of these functions. Four important properties are the following:

1. The sine and cosecant, cosine and secant, tangent and cotangent are reciprocal functions.

$$\sin u = \frac{1}{\csc u} \qquad \cos u = \frac{1}{\sec u} \qquad \tan u = \frac{1}{\cot u}$$

2. Since neither $x$ nor $y$ can exceed 1 (the radius of the unit circle) in absolute value, the ranges of the sine, cosine, secant, and cosecant are limited.

$$\sin u = y \qquad -1 \leq \sin u \leq 1$$

$$\cos u = x \qquad -1 \leq \cos u \leq 1$$

$$\sec u = \frac{1}{x} \qquad \sec u \geq 1 \quad \text{or} \quad \sec u \leq -1$$

$$\csc u = \frac{1}{y} \qquad \csc u \geq 1 \quad \text{or} \quad \csc u \leq -1$$

3. The signs of the functions in each quadrant are determined by the signs of $x$ and $y$. These are left as an exercise.
4. Since $x^2 + y^2 = 1$, given the value of any function of a number $u$, the other functions can be found.

In some cases you may not know the quadrant of $u$, but you can determine it by the signs of the trigonometric functions.

**EXAMPLE 3** Given $\cos u = -\frac{1}{3}$ and $\csc u < 0$, find the other trigonometric function.

***Solution*** The cosine and the cosecant are negative in quadrant III. Do you see why? Thus we have an arc $u$ which terminates in the third quadrant.

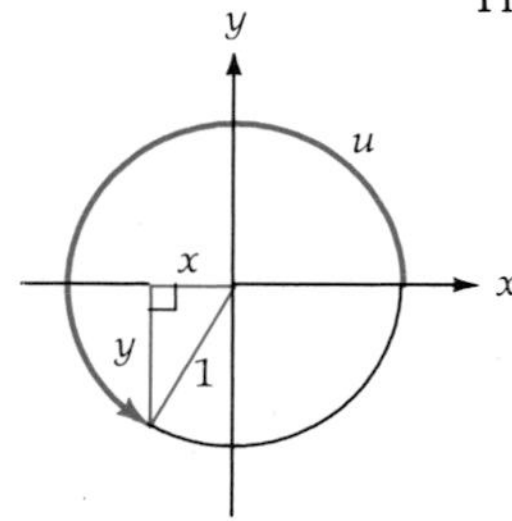


We know that $x = \cos u = -\frac{1}{3}$. Thus

$$x^2 + y^2 = 1$$

$$\left(-\frac{1}{3}\right)^2 + y^2 = 1$$

$$y^2 = \frac{8}{9}$$

Since $y$ is negative, we have $y = -\sqrt{\frac{8}{9}}$. After simplifying the radical we have $y = \frac{-2\sqrt{2}}{3}$. Therefore

$$\sin u = \frac{-2\sqrt{2}}{3} \qquad \tan u = \frac{y}{x} = \frac{\frac{-2\sqrt{2}}{3}}{\frac{1}{-3}} = 2\sqrt{2}$$

$$\csc u = \frac{1}{\sin u} = \frac{3}{-2\sqrt{2}} = \frac{3\sqrt{2}}{-4} \qquad \sec u = \frac{1}{\cos u} = \frac{1}{-\frac{1}{3}} = -3$$

$$\cot u = \frac{1}{\tan u} = \frac{1}{2\sqrt{2}} = \frac{\sqrt{2}}{4}$$

■

## Exercise 7-1

Find one positive and one negative real number that represents an arc coterminal with each of the values in Problems 1–14. (Use exact values.)

**1.** 1 **2.** 4 **3.** $-2$ **4.** 3.6 **5.** 0 **6.** $\pi$ **7.** $\frac{\pi}{2}$

**8.** $\frac{3}{2}\pi$ **9.** $2\pi$ **10.** $1 - \pi$ **11.** $\frac{3\pi}{4}$ **12.** $-\frac{3\pi}{4}$ **13.** $2 - \frac{\pi}{2}$ **14.** $3 + \frac{\pi}{4}$

By using a scientific calculator, find the smallest positive real number and the largest negative real number that represent an arc coterminal with each of the values in Problems 15–20. Find an approximation accurate to four decimal places.

**15.** $3 + \sqrt{2}$ **16.** $1 + \sqrt{5}$ **17.** $\frac{5\pi}{12}$ **18.** $\frac{7\pi}{8}$ **19.** $-\frac{3\pi}{5}$ **20.** $-\frac{2\pi}{9}$

In Problems 21–26, assume that $u$ is a real number that represents the length of an arc. Find the ordered pair $(x, y)$ that is the terminal point of $u$.

**21.** $\frac{\pi}{2}$ **22.** $\frac{3\pi}{2}$ **23.** $6\pi$ **24.** $7\pi$ **25.** $\frac{11\pi}{2}$ **26.** $\frac{9\pi}{2}$

In Problems 27–33, find the other five trigonometric functions. Do not use a calculator. Simplify your answers.

**27.** $\sin u = \frac{3}{5}$, $u$ in quadrant II **28.** $\sec u = -\frac{13}{5}$, $u$ in quadrant III

**29.** $\csc u = \frac{5}{4}$, $u$ in quadrant II **30.** $\cos u = \frac{12}{13}$, $\sin u$ is negative

**31.** $\sec u = -\frac{25}{24}$, $\cot u > 0$ **32.** $\sin u = -\frac{4}{5}$, $\tan u < 0$ **33.** $\cos u = -\frac{4}{9}$, $\tan u < 0$

**34.** If $\tan u = \frac{4}{3}$, $u$ in quadrant I, find the coordinates $(x, y)$ of the endpoint $u$ on the unit circle.

**35.** If $\cot u = -\frac{5}{12}$, $u$ in quadrant II, find the coordinates $(x, y)$ of $u$.

In Problems 36–40, find the other five trigonometric functions. Do not use a calculator. Simplify your answers.

**36.** $\tan u = -\frac{3}{4}$, $u$ in quadrant IV **37.** $\cot u = -\frac{4}{7}$, $u$ in quadrant II

**38.** $\tan u = 1$, $u$ in quadrant III **39.** $\cot u = -\frac{5}{12}$, $\sin u$ is positive **40.** $\tan u = \frac{8}{5}$, $\sec u < 0$

In Problems 41–44, find the other five trigonometric functions. Do not use a calculator or table. Simplify your answers.

**41.** $\tan u = -\frac{5}{3}$, $\sin u > 0$ **42.** $\sec u = -4$, $\sin u > 0$

**43.** $\csc u = -5$, $\tan u > 0$ **44.** $\sin u = -\frac{2}{3}$, $\tan u < 0$

★ **45.** Prove from the definition of the trigonometric functions of real numbers that $(\sin u)^2 + (\cos u)^2 = 1$.

★ 46. Prove from the definition of the trigonometric functions of real numbers that $1 + (\tan u)^2 = (\sec u)^2$.

★ 47. Prove from the definition of the trigonometric functions of real numbers that $1 + (\cot u)^2 = (\csc u)^2$.

## 7-2 TRIGONOMETRIC FUNCTIONS OF SPECIAL NUMBERS

### The Special Numbers $\frac{\pi}{2}$, $\pi$, $\frac{3\pi}{2}$, $2\pi$

We have noted that there is an ordered pair $(x, y)$ related to each arc length $u$. There is also a central angle $\theta$ related to each arc length $u$. We will use the angle $\theta$ and some prior knowledge from plane geometry to find the exact values of the trigonometric functions of some special numbers.

The formula $C = 2\pi r$ gives us the circumference of the unit circle to be an arc of length $2\pi$. Hence the central angle of 360° is related to the real number $2\pi$. From this relationship we obtain the following results.

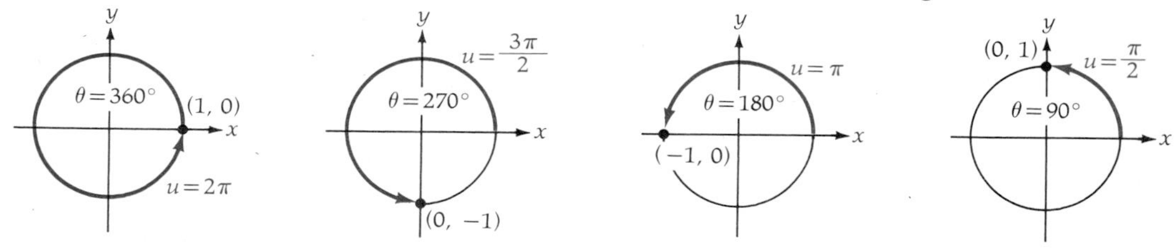


Finding the coordinates of the point $(x, y)$ on the unit circle related to each of these special numbers will give us the values of the trigonometric functions. [Remember that $(x, y)$ has been defined as $(\cos u, \sin u)$.]

The coordinates $(x, y)$ for the numbers $2\pi$ (or 0), $\frac{3}{2}\pi$, $\pi$, and $\frac{\pi}{2}$ are the points of the unit circle on the coordinate axes.

| *Value of u* | *Corresponding values of (x, y)* |
|---|---|
| $2\pi$ (or 0) | (1, 0) |
| $\frac{3}{2}\pi$ | (0, − 1) |
| $\pi$ | (− 1, 0) |
| $\frac{\pi}{2}$ | (0, 1) |

The functions of these special numbers are easily determined when we use the definitions. For example, to find $\sin u$ if $u = \pi$, we see that $x = -1$ and $y = 0$. By using these values we obtain

$$\sin \pi = y = 0 \qquad \tan \pi = \frac{y}{x} = 0 \qquad \csc \pi = \frac{1}{y} = \text{not defined}$$

$$\cos \pi = x = -1 \qquad \cot \pi = \frac{x}{y} = \text{not defined} \qquad \sec \pi = \frac{1}{x} = -1$$

The determination of the functions of these other numbers is left as an exercise.

### The Special Numbers $\frac{\pi}{6}, \frac{\pi}{4}, \frac{\pi}{3}$

The coordinates of the point on the unit circle for the numbers $\frac{\pi}{6}, \frac{\pi}{4}, \frac{\pi}{3}$ are found by using some theorems from plane geometry.

- For the number $\frac{\pi}{6}$, note that the central angle $\theta$ is 30°. [You can obtain this by writing $\frac{\pi}{6} = \frac{1}{3}\left(\frac{\pi}{2}\right) = \frac{1}{3}(90°) = 30°$.]
- For the number $\frac{\pi}{4}$, the central angle is 45°. (Do you see why?)
- For the number $\frac{\pi}{3}$, the central angle is 60°. (Can you verify this?)

These three arc lengths measuring less than $\frac{\pi}{2}$ are of special importance.

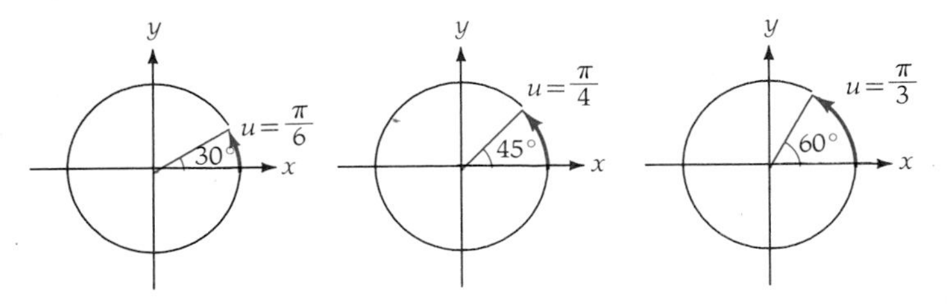


Let us consider in detail the case where $u = \frac{\pi}{6}$. If we drop a perpendicular from the endpoint of the arc to the $x$-axis, we form a triangle of 30°–60°–90°. We know from plane geometry that in this triangle the side opposite the 30° angle is one half the length of the hypotenuse. In our case we know the hypotenuse is 1, since it is the radius of the unit circle. Therefore, the coordinates

$(x, y)$ of the endpoint of the arc having a length $\frac{\pi}{6}$ are $\left(\frac{\sqrt{3}}{2}, \frac{1}{2}\right)$. We can now write the six trigonometric functions of $u = \frac{\pi}{6}$.

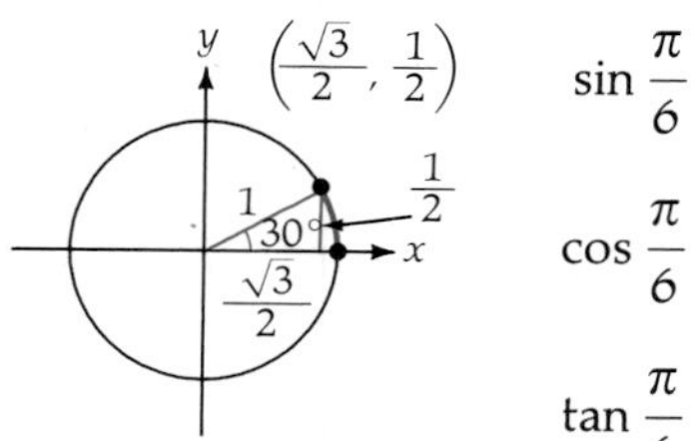


$$\sin\frac{\pi}{6} = \frac{1}{2} \qquad \cot\frac{\pi}{6} = \sqrt{3}$$

$$\cos\frac{\pi}{6} = \frac{\sqrt{3}}{2} \qquad \sec\frac{\pi}{6} = \frac{2\sqrt{3}}{3}$$

$$\tan\frac{\pi}{6} = \frac{\sqrt{3}}{3} \qquad \csc\frac{\pi}{6} = 2$$

For $u = \frac{\pi}{4}$, the related central angle is 45°. Dropping a perpendicular will give us an isosceles right triangle that has sides of $\frac{\sqrt{2}}{2}$, $\frac{\sqrt{2}}{2}$, 1. Thus the coordinates of the endpoint of the arc are $\left(\frac{\sqrt{2}}{2}, \frac{\sqrt{2}}{2}\right)$. Therefore we have

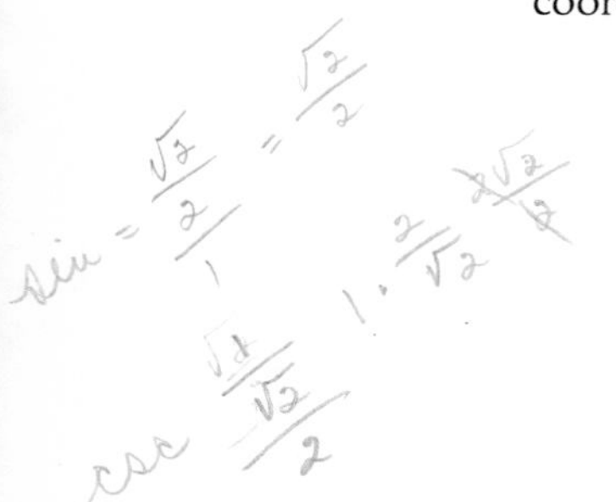

$$\sin\frac{\pi}{4} = \frac{\sqrt{2}}{2} \qquad \cot\frac{\pi}{4} = 1$$

$$\cos\frac{\pi}{4} = \frac{\sqrt{2}}{2} \qquad \sec\frac{\pi}{4} = \sqrt{2}$$

$$\tan\frac{\pi}{4} = 1 \qquad \csc\frac{\pi}{4} = \sqrt{2}$$

The exact values for the trigonometric functions of $\frac{\pi}{3}$ are included as an exercise.

### The Reference Arc

We have studied certain special numbers that are less than $\frac{\pi}{2}$. We will now develop the concept of the reference arc in order to deal with special numbers that do not satisfy that restriction.

For any point $P_1$ $(x_1, y_1)$ on the unit circle in quadrant I, there are points in quadrants II, III, and IV such that their coordinates are equal in absolute value to the coordinates of $P_1$. This is obvious from the equation of the unit circle $x^2 + y^2 = 1$.

*Theorem 7-1*

If $u_1$ is a first-quadrant number with endpoint $(x_1, y_1)$, then the second-quadrant number $u_2 = (\pi - u_1)$ has coordinates $(x_2, y_2)$ such that $x_2 = -x_1$ and $y_2 = y_1$. Hence, the functions of $u_2$ are equal in absolute value to the corresponding functions of $u_1$.

***Proof*** We choose $P_1 = (x_1, y_1)$ in quadrant I associated with the first-quadrant number $u_1$ and then locate $u_2 = (\pi - u_1)$ in quadrant II having endpoint $P_2$ $(x_2, y_2)$.

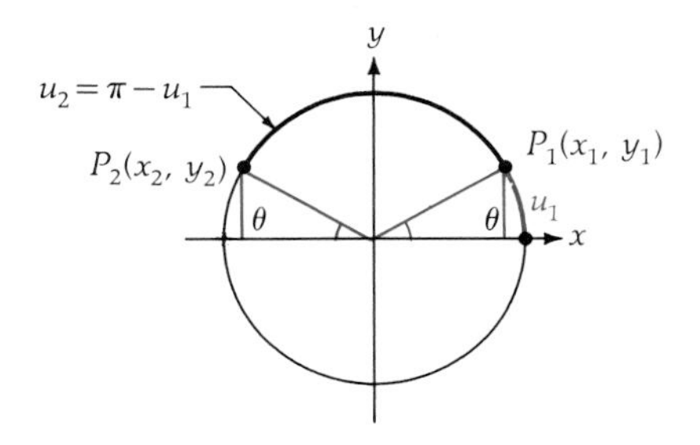


The length of arc $u_1$ is the same as the arc between $(\pi - u_1)$ and $\pi$. Hence the central angles related to these arcs are equal.

Dropping perpendiculars from $P_1$ and $P_2$ to the $x$-axis forms two right triangles that are congruent, since the hypotenuse and an acute angle of one are equal to the hypotenuse and an acute angle of the other, respectively. Hence corresponding sides are equal. This gives us

$$x_1 = |x_2| \qquad \text{and} \qquad y_1 = |y_2|$$

Since $x$ is negative in the second quadrant, then

$$x_2 = -x_1 \qquad \text{and} \qquad y_2 = y_1 \quad \blacksquare$$

The previous theorem proves that the functions of a second-quadrant number are equal in absolute value to the functions of a first-quadrant number and, further, that if the second-quadrant number is $u_2$, then the related first-quadrant number is $(\pi - u_2)$.

We will call the first-quadrant arc the **reference arc**. In a similar manner it can be proved that numbers in quadrants III and IV also have reference arcs in quadrant I. The following table shows the reference arcs for each quadrant.

**Table of Reference Arcs**

| *Given That u Is a Positive Number:* | |
|---|---|
| *If u Is in Quadrant* | *The Reference Arc Is* |
| I | $u$ |
| II | $\pi - u$ |
| III | $u - \pi$ |
| IV | $2\pi - u$ |

**Caution:** Note that the reference arc is *always* measured from the $x$-axis and *never* from the $y$-axis.

We will now combine our knowledge of reference arcs and special numbers to solve a few problems.

**EXAMPLE 1** Find the exact value of each of the six trigonometric functions for the real number $u$ if $u = \dfrac{5\pi}{6}$. Illustrate by drawing an appropriate sketch.

***Solution***

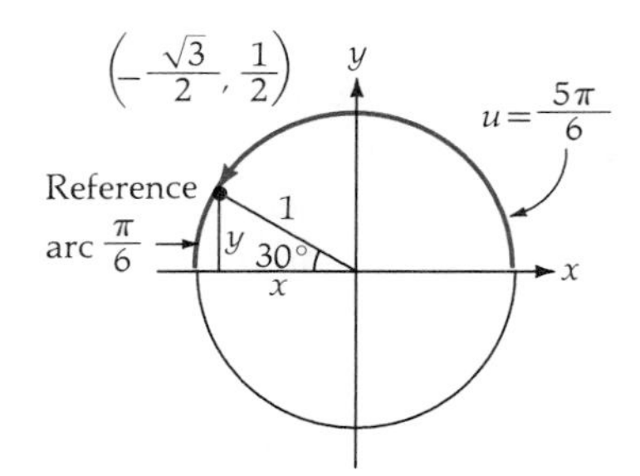


Since $u$ is in quadrant II, the reference arc is $\pi - u$. Thus we have $\pi - \dfrac{5\pi}{6} = \dfrac{\pi}{6}$. Therefore we may use a central angle of 30° in the reference triangle. Do you see that in this case $(x, y) = \left(-\dfrac{\sqrt{3}}{2}, \dfrac{1}{2}\right)$?

Thus from the definition,

$$\sin\frac{5\pi}{6} = y = \frac{1}{2}$$

$$\cos\frac{5\pi}{6} = x = \frac{-\sqrt{3}}{2}$$

$$\tan\frac{5\pi}{6} = \frac{y}{x} = \frac{\frac{1}{2}}{-\frac{\sqrt{3}}{2}} = -\frac{1}{\sqrt{3}} = -\frac{\sqrt{3}}{3}$$

By using the reciprocal functions we have

$$\csc\frac{5\pi}{6} = \frac{1}{\sin\frac{5\pi}{6}} = \frac{1}{\frac{1}{2}} = 2$$

$$\sec\frac{5\pi}{6} = \frac{1}{\cos\frac{5\pi}{6}} = \frac{1}{-\frac{\sqrt{3}}{2}} = \frac{-2}{\sqrt{3}} = \frac{-2\sqrt{3}}{3}$$

$$\cot\frac{5\pi}{6} = \frac{1}{\tan\frac{5\pi}{6}} = \frac{1}{-\frac{\sqrt{3}}{3}} = -\frac{3\sqrt{3}}{3} = -\sqrt{3} \quad ■$$

**EXAMPLE 2** Find the exact value of each of the six trigonometric functions for the real number $u$ if $u = \frac{-3\pi}{4}$. Illustrate by drawing an appropriate sketch.

***Solution*** We observe that $u$ is in quadrant III. The arc that measures $\frac{-3\pi}{4}$ is coterminal with an arc of $\frac{5\pi}{4}$. If we use the table of reference arcs, we must have a positive value of $u$; hence we use the coterminal value $u_1 = \frac{5\pi}{4}$. The reference arc is therefore $\frac{5\pi}{4} - \pi = \frac{\pi}{4}$.

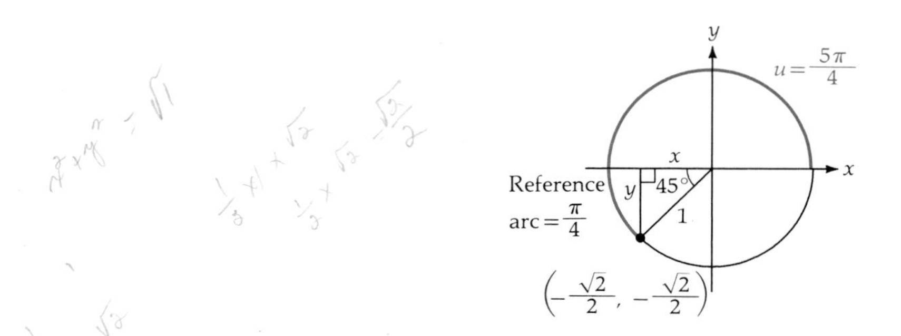


Therefore we have a central angle of 45° in the reference triangle. Can you verify that $(x, y) = \left(-\frac{\sqrt{2}}{2}, -\frac{\sqrt{2}}{2}\right)$?

We find therefore that

$$\sin\left(-\frac{3\pi}{4}\right) = \frac{-\sqrt{2}}{2} \qquad \csc\left(-\frac{3\pi}{4}\right) = -\sqrt{2}$$

$$\cos\left(-\frac{3\pi}{4}\right) = \frac{-\sqrt{2}}{2} \qquad \sec\left(-\frac{3\pi}{4}\right) = -\sqrt{2}$$

$$\tan\left(-\frac{3\pi}{4}\right) = 1 \qquad \cot\left(-\frac{3\pi}{4}\right) = 1 \quad \blacksquare$$

## EXERCISE 7-2

**1.** Review the derivation in the text that allowed us to find the six trigonometric functions of $u$ if $u = \frac{\pi}{6}$. In a similar manner, obtain exact value of the six trigonometric functions of $u$ if $u = \frac{\pi}{3}$.

Find the exact value of each of the six trigonometric functions for the real numbers in Problems 2–12. (Draw a sketch to illustrate each problem.)

**2.** $\frac{\pi}{2}$ **3.** $\frac{3}{2}\pi$ **4.** $2\pi$ **5.** $\frac{4\pi}{3}$ **6.** $\frac{2\pi}{3}$ **7.** $\frac{3}{4}\pi$

**8.** $\frac{7}{6}\pi$ **9.** $\frac{5}{4}\pi$ **10.** $\frac{5}{3}\pi$ **11.** $\frac{11}{6}\pi$ **12.** $\frac{7}{4}\pi$

Find the exact value of each of the six trigonometric functions for the real number $u$ in Problems 13–28. (Draw a sketch to illustrate each problem.)

**13.** $u = \frac{-3\pi}{2}$ **14.** $u = -3\pi$ **15.** $u = \frac{-4\pi}{3}$ **16.** $u = -\frac{5\pi}{6}$

**17.** $u = \frac{-7\pi}{4}$ **18.** $u = \frac{-5\pi}{4}$ **19.** $u = \frac{-\pi}{6}$ **20.** $u = -\frac{5\pi}{3}$

**21.** $u = \frac{-2\pi}{3}$ **22.** $u = \frac{-7\pi}{6}$ **23.** $u = \frac{17\pi}{6}$ **24.** $u = \frac{7\pi}{3}$

**25.** $u = \frac{15\pi}{4}$ **26.** $u = \frac{11\pi}{4}$ **27.** $u = \frac{-11\pi}{3}$ **28.** $u = \frac{-21\pi}{6}$

★ **29.** Study Theorem 7-1 in the text. In a similar fashion, prove the following theorem: If $u_1$ is a first-quadrant number with endpoint $(x_1, y_1)$ and the third quadrant number $u_3$ has the property $u_1 = (u_3 - \pi)$, then $u_3$ has the coordinates $(x_3, y_3)$ such that $x_3 = -x_1$ and $y_3 = -y_1$. Hence the functions of $u_3$ are equal in absolute value to the corresponding functions of $u_1$.

★ **30.** In a fashion similar to Problem 29, prove the following theorem: If $u_1$ is a first-quadrant number with endpoint $(x_1, y_1)$ and the fourth quadrant number $u_4$ has the property $u_1 = (2\pi - u_4)$, then $u_4$ has the coordinates $(x_4, y_4)$ such that $x_4 = x_1$ and $y_4 = -y_1$. Hence the functions of $u_4$ are equal in absolute value to the corresponding functions of $u_1$.

---

## 7-3 TRIGONOMETRIC FUNCTIONS OF ANY REAL NUMBER

### Introduction

There is a correspondence between finding the trigonometric function of an angle like $\sin \frac{\pi}{4}$ where $\frac{\pi}{4}$ is an angle measured in radians and the trigonometric function of a real number, such as finding $\sin u$ if $u$ represents the real number $\frac{\pi}{4}$. However this close correspondence does not mean that they are exactly the same.

The student should understand that it is not totally correct to say "$\frac{\pi}{2} = 90°$." There is no way that the real number $\frac{\pi}{2}$ is equal to the angle measure of 90°. Correct statements would be "the real number $\frac{\pi}{2}$ on the unit circle subtends an arc whose related central angle has a measure of 90°," or "if an angle is measured as $\frac{\pi}{2}$ radians, or 90°, it will subtend an arc of length $\frac{\pi}{2}$ on the unit circle." $\frac{\pi}{2}$ radians *corresponds* to the real number $\frac{\pi}{2}$, but they are not the same.

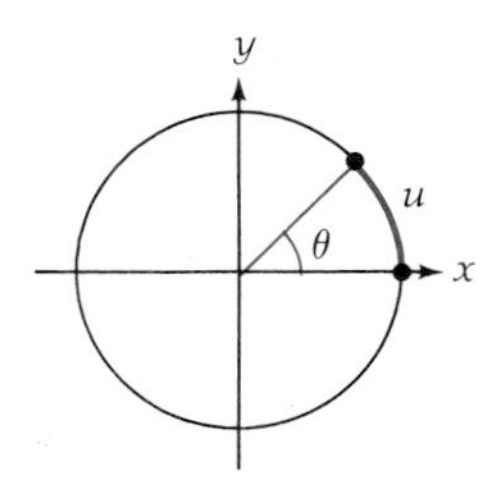


Of course, from the definition of a radian it should be clear that the radian measure of an angle is the same as the real number that gives the arc length on the unit circle.

If the arc length is $u$ and the angle measure is $\theta$ in radians, then $\theta = u$ radians. Hence trigonometric tables often list "real number $u$ or $\theta$ in radians" in the same column. These are technical points to be sure, but they are important to the future study of mathematics and the student should recognize the distinctions.

### Finding the Trigonometric Functions of Any Real Number on a Calculator

The evaluation of tan $\theta$ where $\theta = 1.2$ radians and the evaluation tan $u$ where $u =$ the real number 1.2 are done in the same way on the calculator. The calculator will solve either problem when it is placed in *radian mode*. When you wish to use the calculator to do such problems, first be sure that it is in the radian mode. Then the value of $\theta$ or $u$ should be entered. Finally, the [tan] button should be depressed. You will see displayed the calculator value 2.5721516. This approximation for tan 1.2 is accurate to seven decimal places.

**EXAMPLE 1** Evaluate on a calculator the following. **a.** sin 0.8 **b.** sin $(0.8 + 2\pi)$

***Solution***

**a.** [RAD] 0.8 [sin] 0.7173561

**b.** One approach is to evaluate the sum $0.8 + 2\pi$ by using the [=] button prior to depressing the [sin] button.

[RAD] 0.8 [+] 2 [×] [π] [=] [sin] 0.7173561

A second approach is to use the parentheses buttons to enclose the argument of the sine.

[RAD] [(] 0.8 [+] 2 [×] [π] [)] [sin] 0.7173561

*Note:* The student should try both of these methods on his or her calculator. A few scientific calculators will not accept the sequence of instructions including parentheses. Also, you *do not need* to place the calculator in the radian mode each time. Once your calculator is in the radian mode, it will stay in that mode until the mode key is pressed or (on some calculators) the calculator is shut off. ■

In the following three examples we will list all calculator answers to seven decimal places. However (as we will discuss in detail in a later chapter) this level of accuracy is not usually needed or warranted.

**EXAMPLE 2** Evaluate on a calculator: $\sec\left(\frac{3\pi}{7}\right)$.

***Solution*** We know that $\sec\left(\frac{3\pi}{7}\right) = \frac{1}{\cos\left(\frac{3\pi}{7}\right)}$. We enter the following steps:

$$\boxed{\text{RAD}}\ 3\ \boxed{\times}\ \boxed{\pi}\ \boxed{\div}\ 7\ \boxed{=}\ \boxed{\cos}\ \boxed{\tfrac{1}{x}}\ 4.4939592 \quad \blacksquare$$

**EXAMPLE 3** Find a real number $u$ such that $0 \leq u < \frac{\pi}{2}$ for which $\cos u = 0.6214873$.

***Solution*** We should confirm that the calculator is in radian mode. (We might obtain an angle in degrees if we forget this step.)

$$\boxed{\text{RAD}}\ 0.6214873\ \boxed{\text{INV}}\ \boxed{\cos}\ 0.9001566 \quad \blacksquare$$

**EXAMPLE 4** Find $u$ such that $0 \leq u < \frac{\pi}{2}$ if $\sec u = 2.955$.

***Solution*** We know that $\sec x = \frac{1}{\cos x}$ and $\cos x = \frac{1}{\sec x}$, since secant and cosine are reciprocal functions. Therefore

$$\sec u = \frac{1}{\cos u} = 2.955$$

Thus

$$\cos u = \frac{1}{2.955}$$

On a calculator we enter

$$2.955\ \boxed{\tfrac{1}{x}}\ 0.3384095$$

Thus we have $\cos u = 0.3384095$. To find $u$, we use

$$0.3384095\ \boxed{\text{INV}}\ \boxed{\cos}\ 1.2255702$$

Thus $u = 1.2255702$. $\blacksquare$

### Trigonometric Functions of Any Real Number from a Table (Optional)

If you do not have a calculator, you may use a table to find the trigonometric function of any number.

This section refers to Table E. An abbreviated portion of Table E appears on page 339. The table gives the values of the trigonometric functions of numbers (or radians) from 0 to $\frac{\pi}{2}$ in intervals of 0.01. Note the structure of the table. Columns read from the top of each page are

| *Real Number u* or $\theta$ *Radians* | $\sin u$ or $\sin\theta$ | $\cos u$ or $\cos\theta$ | $\tan u$ or $\tan\theta$ | $\cot u$ or $\cot\theta$ | $\sec u$ or $\sec\theta$ | $\csc u$ or $\csc\theta$ |
|---|---|---|---|---|---|---|

The real-number column starts at 0.00 and ends at 1.57 $\left(\text{approximately } \frac{\pi}{2}\right)$.

To find the value of a function of a real number, locate the real number in the left-hand column, and then look under the proper column for the value of the function.

**EXAMPLE 5** Use the portion of Table E to find $\cos 0.17$.

***Solution*** Under the "Real Number $u$ or Radians" column we find 0.17. Correspondingly, underneath the "$\cos u$ or $\cos\theta$" column we find 0.9856. Therefore we know that $\cos 0.17 = 0.9856$. This approximation is accurate to four decimal places. ■

**EXAMPLE 6** Evaluate by using Table E.

**a.** $\sin 0.23$ **b.** $\cos 1.30$ **c.** $\tan 1.48$

***Solution*** **a.** $\sin 0.23 = 0.2280$ **b.** $\cos 1.30 = 0.2675$ **c.** $\tan 1.48 = 10.983$ ■

Of course, finding the number when a function is given is another way of using the table.

**EXAMPLE 7** Use Table E. Given $\sin u = 0.7311$, find $u$ such that $0 \le u \le \frac{\pi}{2}$.

***Solution*** Looking under the sine column, we locate 0.7311. This corresponds to 0.82 in the real-number column. Thus $u = 0.82$. ■

TABLE E. *Trigonometric Functions of Real Numbers or Radians*

| Real Number $u$ or $\theta$ Radians | $\sin u$ or $\sin\theta$ | $\cos u$ or $\cos\theta$ | $\tan u$ or $\tan\theta$ | $\cot u$ or $\cot\theta$ | $\sec u$ or $\sec\theta$ | $\csc u$ or $\csc\theta$ |
|---|---|---|---|---|---|---|
| 0.00 | 0.0000 | 1.0000 | 0.0000 | No value | 1.000 | No value |
| 0.01 | 0.0100 | 1.0000 | 0.0100 | 99.997 | 1.000 | 100.0 |
| 0.02 | 0.0200 | 0.9998 | 0.0200 | 49.993 | 1.000 | 50.00 |
| 0.03 | 0.0300 | 0.9996 | 0.0300 | 33.323 | 1.000 | 33.34 |
| 0.04 | 0.0400 | 0.9992 | 0.0400 | 24.987 | 1.001 | 25.01 |
| 0.05 | 0.0500 | 0.9988 | 0.0500 | 19.983 | 1.001 | 20.01 |
| 0.06 | 0.0600 | 0.9982 | 0.0601 | 16.647 | 1.002 | 16.68 |
| 0.07 | 0.0699 | 0.9976 | 0.0701 | 14.262 | 1.002 | 14.30 |
| 0.08 | 0.0799 | 0.9968 | 0.0802 | 12.473 | 1.003 | 12.51 |
| 0.09 | 0.0899 | 0.9960 | 0.0902 | 11.081 | 1.004 | 11.13 |
| 0.10 | 0.0998 | 0.9950 | 0.1003 | 9.967 | 1.005 | 10.02 |
| 0.11 | 0.1098 | 0.9940 | 0.1104 | 9.054 | 1.006 | 9.109 |
| 0.12 | 0.1197 | 0.9928 | 0.1206 | 8.293 | 1.007 | 8.353 |
| 0.13 | 0.1296 | 0.9916 | 0.1307 | 7.649 | 1.009 | 7.714 |
| 0.14 | 0.1395 | 0.9902 | 0.1409 | 7.096 | 1.010 | 7.166 |
| 0.15 | 0.1494 | 0.9888 | 0.1511 | 6.617 | 1.011 | 6.692 |
| 0.16 | 0.1593 | 0.9872 | 0.1614 | 6.197 | 1.013 | 6.277 |
| 0.17 | 0.1692 | 0.9856 | 0.1717 | 5.826 | 1.015 | 5.911 |
| 0.18 | 0.1790 | 0.9838 | 0.1820 | 5.495 | 1.016 | 5.586 |
| 0.19 | 0.1889 | 0.9820 | 0.1923 | 5.200 | 1.018 | 5.295 |
| 0.20 | 0.1987 | 0.9801 | 0.2027 | 4.933 | 1.020 | 5.033 |
| 0.21 | 0.2085 | 0.9780 | 0.2131 | 4.692 | 1.022 | 4.797 |
| 0.22 | 0.2182 | 0.9759 | 0.2236 | 4.472 | 1.025 | 4.582 |
| 0.23 | 0.2280 | 0.9737 | 0.2341 | 4.271 | 1.027 | 4.386 |
| 0.24 | 0.2377 | 0.9713 | 0.2447 | 4.086 | 1.030 | 4.207 |
| 0.25 | 0.2474 | 0.9689 | 0.2553 | 3.916 | 1.032 | 4.042 |
| 0.26 | 0.2571 | 0.9664 | 0.2660 | 3.759 | 1.035 | 3.890 |

### Interpolation Using Table E

Table E is given in intervals of 0.01. When a function of a number with three decimal places is needed, we must find it by the method of *linear interpolation.* This process uses proportional parts.

**EXAMPLE 8** Find $u$ if $\sin u = 0.2345$. Use Table E.

***Solution*** From the table (under $\sin u$ column) we find that 0.2345 is between 0.2280 and 0.2377. Outline the problem as follows:

$$10\left[\begin{array}{ll} x\left[\begin{array}{ll}\sin 0.23 & 0.2280 \\ \sin u & 0.2345\end{array}\right]65 & \\ \sin 0.24 & 0.2377\end{array}\right]97$$

$$\frac{x}{10} = \frac{65}{97}$$

$$x = 7 \qquad \textit{To the nearest integer.}$$

Therefore $u = 0.23 + 0.007 = 0.237$ ■

**EXAMPLE 9** Find $u$ if $\tan u = -0.2447$ and $\cos u < 0$. Use Table E.

***Solution*** The table of course contains only positive values. From the table we find that $\tan 0.24 = 0.2447$. Then the reference arc is $u_1 = 0.24$. We now must determine the quadrant of $u$ if $\tan u < 0$ and $\cos u < 0$. Clearly, $u$ will terminate in quadrant II if both tangent $u$ and cosine $u$ are negative. If we approximate $\pi = 3.14$,

$$u = 3.14 - u_1 = 3.14 - 0.24$$

So we have $u = 2.90$.

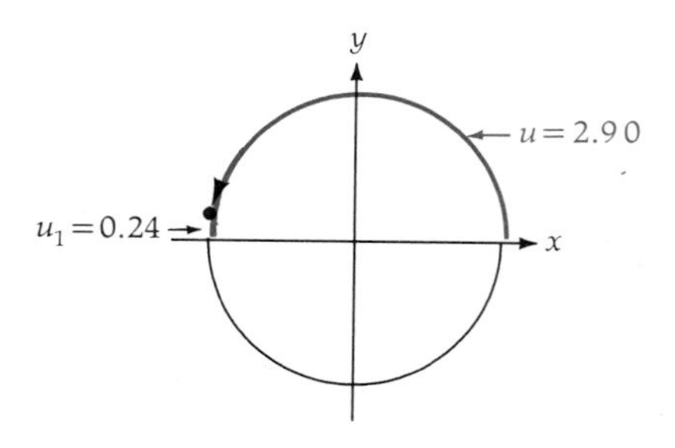


■

If you do not have a calculator, it will take somewhat longer to use the tables to find the trigonometric functions of $u$ if $u$ is not restricted to $0 \leq u < \frac{\pi}{2}$. However, it can be readily done by using reference arcs.

**Note:** In all the calculations that follow, we will approximate $\pi$ by the value 3.14. The trigonometric functions can thus be evaluated quite quickly, but the accuracy is greatly diminished. If you find it necessary to solve problems like Examples 10 and 11 with four-digit accuracy, the use of a scientific calculator is strongly encouraged.

**EXAMPLE 10** Find cos 4. Use Table E and the approximation $\pi \doteq 3.14$.

***Solution*** Since 4 is a third-quadrant number, the reference arc is approximately $(4 - 3.14)$ or 0.86. From Table E we obtain $\cos 0.86 = 0.6524$. We know that $\cos u < 0$ in quadrant III, so $\cos 4 = -0.6524$. ■

**Negative Real Numbers Using Table E**

For negative real numbers we must find a positive number between 0 and $2\pi$ whose endpoint is coincident with the endpoint of the negative number and then find the reference arc.

**EXAMPLE 11** Find $\tan(-1.68)$ using Table E and the approximation $\pi \doteq 3.14$.

***Solution***

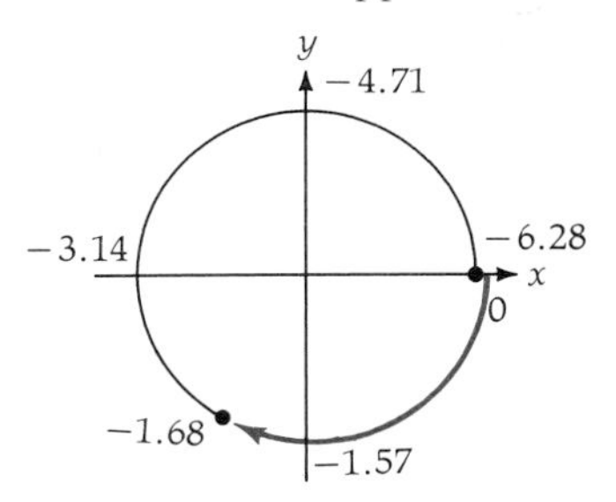


$-1.68$ is coincident with $(2\pi - 1.68)$ or $(6.28 - 1.68) = 4.6$. 4.6 is a third-quadrant number, so the reference arc is $(4.6 - 3.14)$, or 1.46, and $\tan 1.46 = 8.989$. Since $\tan u > 0$ in quadrant III, we have $\tan(-1.68) = 8.989$. ■

## EXERCISE 7-3

Problems 1–35 require the use of a scientific calculator. If you do not have a calculator start with Problem 36.

| | | | |
|---|---|---|---|
| **1.** sin 1.365 | **2.** cos 0.721 | **3.** tan 1.873 | **4.** sin 2.982 |
| **5.** cos 3.116 | **6.** tan 6.145 | **7.** sec 3.241 | **8.** cot 1.867 |
| **9.** cot 3.578 | **10.** csc 4.813 | **11.** cos (−2.361) | **12.** sin (−4.072) |
| **13.** $\tan\left(\frac{5\pi}{12}\right)$ | **14.** $\cos\left(\frac{3\pi}{5}\right)$ | **15.** $\sin\left(\frac{-2\pi}{9}\right)$ | **16.** $\tan\left(\frac{-\pi}{8}\right)$ |

**17.** On your calculator, verify that $\tan(1.6) = \tan(1.6 + 2\pi)$.

**18.** On your calculator, verify that $\cos(3.7) = \cos(3.7 + 2\pi)$.

**19.** On your calculator, verify that $\csc(-5.2) = -\csc(5.2)$.

**20.** On your calculator, verify that $\cot(-0.83) = -\cot(0.83)$.

**21.** On your calculator, verify that $\sin(2\pi - 1.3) = -\sin(1.3)$.

**22.** On your calculator, verify that $\cos(2\pi - 1.5) = \cos(1.5)$.

Without a calculator, show why each of the statements in Problems 23–25 is true; then give a numerical example and verify it on your calculator. (*Hint:* If $(-u)$ is in quadrant IV, then $u$ is in quadrant I; if $(-u)$ is in quadrant III, then $u$ is in quadrant II; etc.)

★ **23.** $\sin(-u) = -\sin u$ ★ **24.** $\cos(-u) = \cos u$ ★ **25.** $\tan(-u) = -\tan u$

Find a real number $u$ such that $0 \le u < \pi$ in Problems 26–35. Round your approximation to four decimal places.

**26.** $\cos u = 0.863142$ **27.** $\cos u = 0.731726$

**28.** $\tan u = 1.437126$ **29.** $\cos u = -0.716483$

**30.** $\cos u = -0.277631$ **31.** $\tan u = -0.771344$

**32.** $\sec u = 1.341127$ **33.** $\csc u = 1.762314$

**34.** $\cot u = -1.763218$ **35.** $\cot u = -0.778613$

Optional Problems 36–56 are written to develop some facility in using Table E to determine the value of trigonometric functions.

Evaluate in Problem 36–40 using Table E.

**36.** tan 0.27 **37.** sec 1.18 **38.** cot 1.51 **39.** csc 1.07 **40.** cos 1.49

Find $u$ such that $0 \le u \le \frac{\pi}{2}$ in Problems 41–42 using Table E.

**41.** $\cos u = 0.7838$ **42.** $\tan u = 0.3425$

In Problems 43–46, use Table E and the method of interpolation to find the expression.

**43.** sin 0.134 **44.** cot 0.613 **45.** tan 0.367 **46.** sec 0.921

Find $u$ such that $0 \le u \le \frac{\pi}{2}$ in Problems 47–50 using Table E.

**47.** $\cos u = 0.6142$ **48.** $\tan u = 0.4232$ **49.** $\csc u = 1.209$ **50.** $\cot u = 1.348$

In Problems 51–54, use Table E. Use interpolation when necessary to find the expression. (Use $\pi = 3.14$.)

**51.** sin 2 **52.** cos 1.9 **53.** tan 3 **54.** csc 4.12

In Problems 55–56, find $u$ where $0 \le u < 2\pi$ using Table E.

**55.** $\sin u = 0.4706$, $\tan u < 0$ **56.** $\sec u = 1.380$, $\sin u > 0$

## 7-4 THE GRAPH OF THE SINE AND COSINE FUNCTIONS

### Graphing the Sine Function

To sketch the graph of the function $y = \sin x$, proceed as follows.

First, regard the argument $x$ as a real number, keeping in mind the correspondence that exists between real numbers, radian measure, and the degree measure of an angle. The coordinate axes will thus be the Cartesian coordinate system with two perpendicular real number lines.

Second, construct a table of values by assigning values to $x$ and finding the corresponding values of $y$. To be very accurate, we need to supply values for $x$ in very small intervals. However, by using only some of the special numbers, we can obtain a fairly accurate representation.

| $x$ | $0$ | $\frac{\pi}{6}$ | $\frac{\pi}{3}$ | $\frac{\pi}{2}$ | $\frac{2\pi}{3}$ | $\frac{5\pi}{6}$ | $\pi$ | $\frac{7\pi}{6}$ | $\frac{4\pi}{3}$ | $\frac{3\pi}{2}$ | $\frac{5\pi}{3}$ | $\frac{11\pi}{6}$ | $2\pi$ |
|---|---|---|---|---|---|---|---|---|---|---|---|---|---|
| Exact value of $\sin x$ | $0$ | $\frac{1}{2}$ | $\frac{\sqrt{3}}{2}$ | $1$ | $\frac{\sqrt{3}}{2}$ | $\frac{1}{2}$ | $0$ | $-\frac{1}{2}$ | $-\frac{\sqrt{3}}{2}$ | $-1$ | $-\frac{\sqrt{3}}{2}$ | $-\frac{1}{2}$ | $0$ |
| Decimal Approximation of $\sin x$ | 0 | 0.50 | 0.87 | 1 | 0.87 | 0.50 | 0 | −0.50 | −0.87 | −1 | −0.87 | −0.50 | 0 |

Now plot the points on the coordinate system and draw the "curve of best fit."

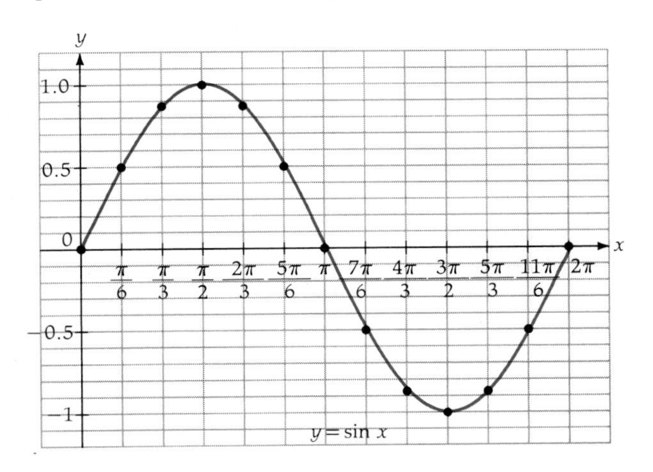


This graph gives us a pictorial representation of a portion of the function $y = \sin x$. If we continue with numbers greater than $2\pi$ or less than 0, we will find that the graph continuously repeats itself.

Many specific quantities in economics, in mathematics, and in the physical sciences are exactly described by the sine function and precisely visualized by the graph of a sine function. For example, the voltage induced by a rotating armature in a generator may be graphed by a sine curve. In this case the value of the induced voltage is $y$ and varies directly with the sine of $x$, where $x$ is a measure of the angle of rotation of the armature measured in radians.

Graphs like that of the sine function that have a regular repeating pattern are very important in mathematics. Functions that have such a graph are called periodic functions.

***Definition***

A function is said to be a *periodic function* if there exists a smallest value $k$ such that for any integer $n$, $f(x) = f(x + nk)$ for all values of $x$ in the domain of $f$. The *period* of the function $f$ is $k$.

From this definition we know the sine function to be periodic because $\sin x = \sin (x + 2n\pi)$. Therefore, the period of the sine function is $2\pi$. Graphically, this means that the sine curve will repeat every $2\pi$ units.

There are many examples of periodic functions in the physical world. An electrocardiograph makes a chart of the heart's electrical impulses. The regular patterns thus generated can be approximated by a periodic function.

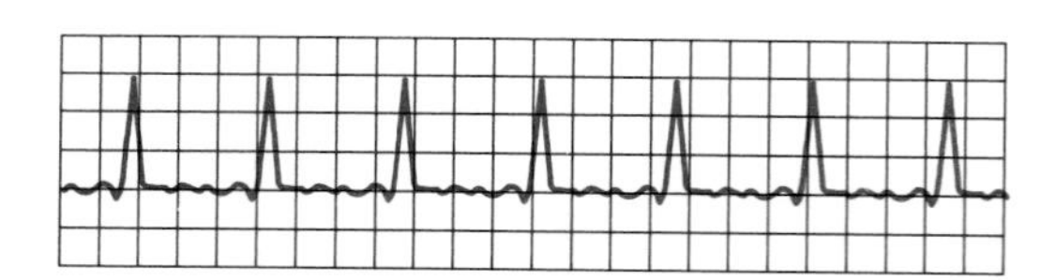

***Definition***

The *amplitude* of a periodic function is one half the difference of the maximum and minimum values of the function.

Note that this value will always be positive. The amplitude is measured as a positive number.

Now we will graph a function of the form $y = a \sin bx$ where $a$ and $b$ are real numbers. Such functions are best thought of as variations of $y = \sin x$. Note that the real number $a$ affects the amplitude. The amplitude of $y = \sin x$ is 1. Thus, $y = a \sin x$ has an amplitude of $|a|$, since each value of $\sin x$ is multiplied by $a$ to obtain the value for $y$.

**EXAMPLE 1** Sketch the graph of $y = 3 \sin x$.

***Solution*** If we regard this function as a variation of $y = \sin x$, we should note that for each value of $\sin x$, $y$ is three times as large. Hence the amplitude is 3. The period of $2\pi$ is not affected by the coefficient 3. (Usually, we will sketch the graph of only one period because this is sufficient to cover all values of the function.) ■

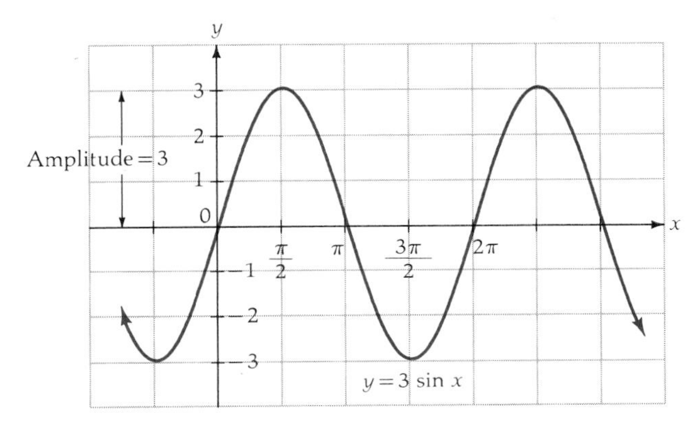


Notice the change in the period when a function of the form $y = a \sin bx$ has a value of $b$ that is not equal to 1.

**EXAMPLE 2** Sketch three periods of the graph of $y = \sin 3x$.

***Solution*** Again, think of this function as a variation of $y = \sin x$. Note that the amplitude is not affected by the 3; as $3x$ varies from 0 to $2\pi$, $x$ varies from 0 to $\frac{2\pi}{3}$.

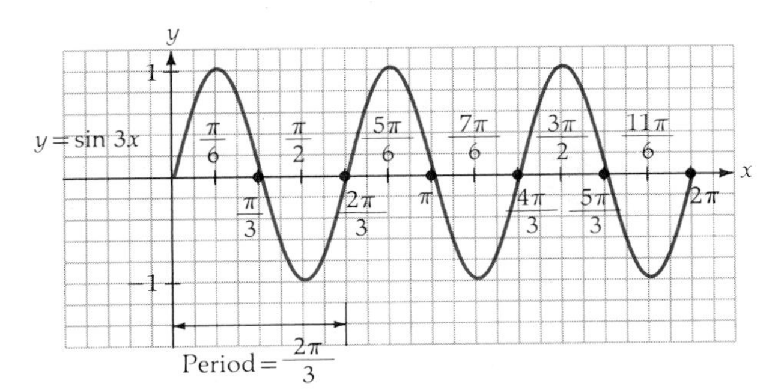


We see from the above graph that the graph of $y = \sin 3x$ has a period of $\frac{2\pi}{3}$. It goes through one cycle of the sine wave three times as fast as the graph of $y = \sin x$. ■

In general, the graph of $y = a \sin bx$ has an amplitude of $|a|$ and a period of $\frac{2\pi}{|b|}$.

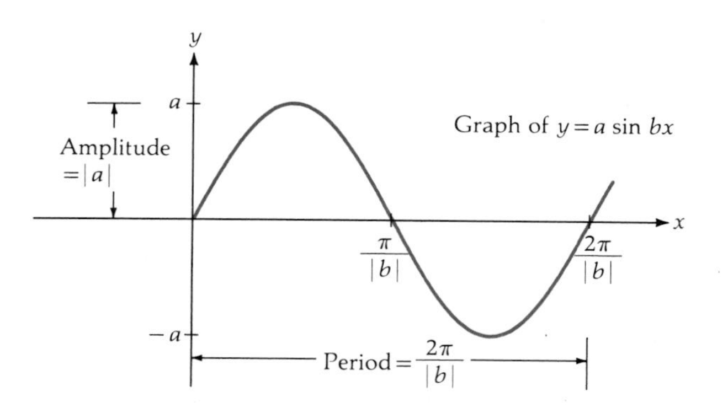


One must take particular care in graphing the sine function if the value of $a$ is negative. A negative value of $a$ in the curve $y = a \sin x$ has the effect of reflecting the curve about the $x$-axis.

**EXAMPLE 3** Give the amplitude and period of the graph of $y = -0.5 \sin 2x$; graph one period of the function.

***Solution*** Since $|a| = |-0.5| = 0.5$, the amplitude is 0.5. However, since $a$ is negative, the curve will start downward and obtain negative values of $y$ first. Since $b = 2$, the period of this sine function is $\frac{2\pi}{2} = \pi$.

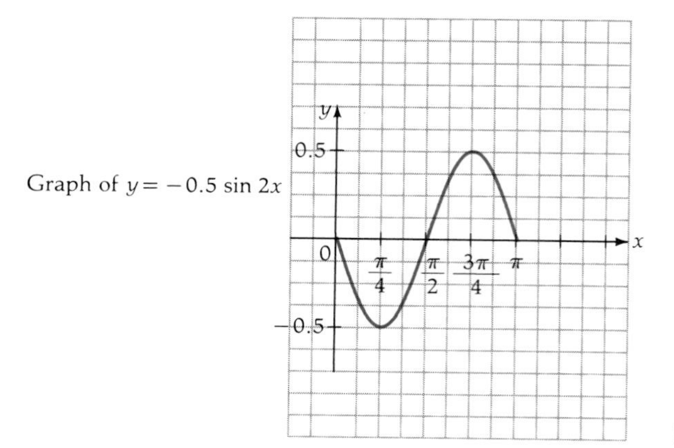


■

### The Phase Shift

The sine function does not always start at zero. In the graph of the function $y = a \sin (bx + c)$, if $c \neq 0$, the sine curve is shifted to the right or left, depending on the value of $c$. If $c$ is *positive*, the shift is to the *left*. If $c$ is *negative*, the shift is to the *right*. This shift is commonly called the *phase shift*.

Let us examine three simple sine functions. We will graph $y = \sin\left(x + \frac{\pi}{4}\right)$, $y = \sin x$, and $y = \sin\left(x - \frac{\pi}{4}\right)$ and compare the three graphs.

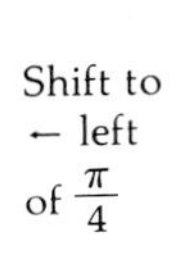


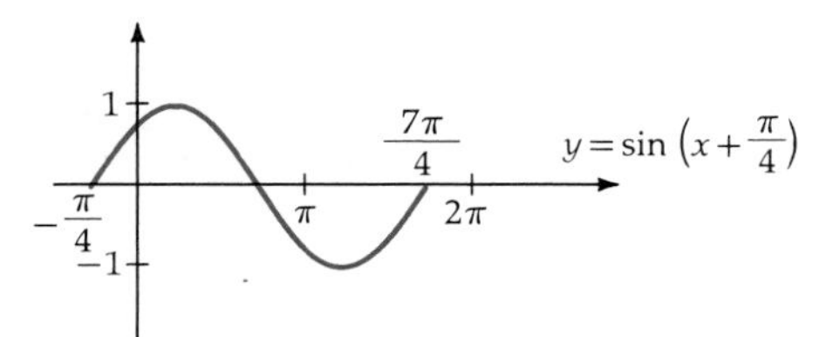


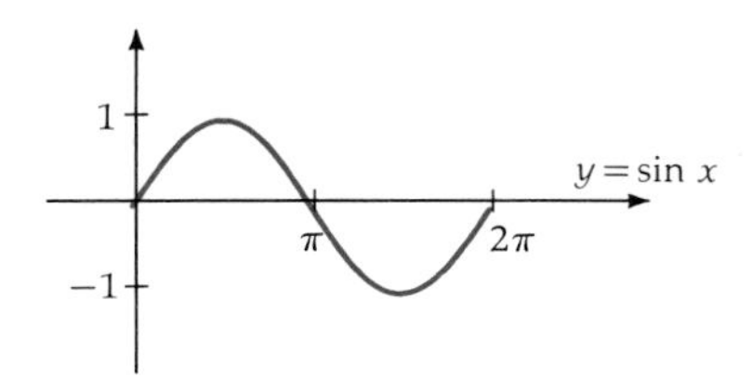


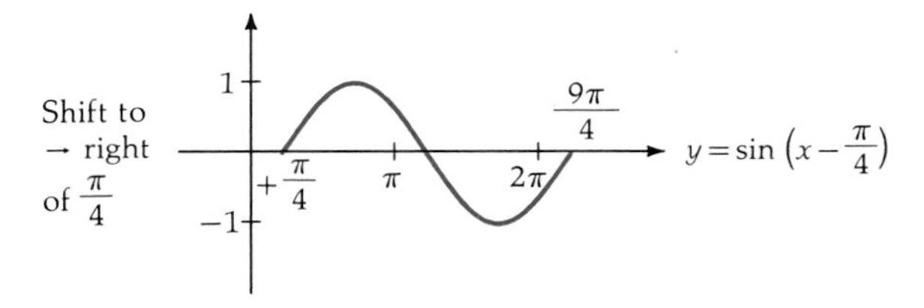


Observe that the graph of $y = \sin\left(x + \frac{\pi}{4}\right)$ is shifted to the *left* by $\frac{\pi}{4}$ units from the graph of $y = \sin x$. Likewise the graph of $y = \sin\left(x - \frac{\pi}{4}\right)$ is shifted to the *right* by $\frac{\pi}{4}$ units from the graph of $y = \sin x$.

---

**Summary of Graphing the Sine Function**

---

For the function $y = a \sin (bx + c)$, the following points will aid in quickly sketching the graph.

**1.** The *amplitude* is $|a|$.

**2.** The *period* is $\frac{2\pi}{|b|}$.

3. The *phase shift* is $-\frac{c}{|b|}$. $\left[\begin{array}{l}\text{If } c \text{ is positive, the shift is to the left.}\\ \text{If } c \text{ is negative, the shift is to the right.}\end{array}\right]$
4. If we start a period where the graph crosses the axis, then it will also cross the axis at the middle and the end of the period.
5. The maximum and minimum points are halfway between the points where the graph crosses the $x$-axis.

---

Points 4 and 5 in the summary are very helpful. Without this, many students find that they have difficulties in obtaining the places where one period of the function starts and stops and where the function takes on maximum and minimum values.

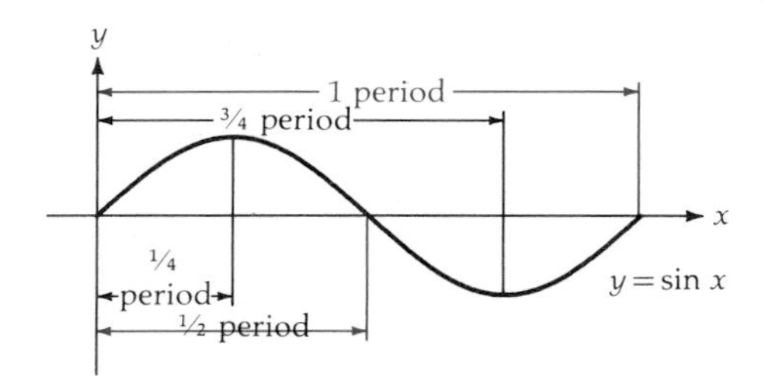


Let us examine this carefully in the next example.

**EXAMPLE 4** Sketch one period of the graph of

$$y = 1.5 \sin (2x - \pi)$$

by finding the amplitude, period, and phase shift of the function.

***Solution*** The amplitude is 1.5. The period is $\frac{2\pi}{2} = \pi$. The value of $c = -\pi$. So the phase shift is

$$-\frac{-\pi}{|b|} = -\frac{-\pi}{2} = +\frac{\pi}{2}$$

We shift to the right $\frac{\pi}{2}$ units. One period of the function starts at $\frac{\pi}{2}$. To obtain the endpoint, we add the period $\pi$ to the starting point. One period of the function therefore ends at $\frac{\pi}{2} + \pi = \frac{3\pi}{2}$. The function is zero at $\frac{1}{2}$ period, which is $\frac{\pi}{2} + \frac{\pi}{2} = \pi$. At $\frac{1}{4}$ of the period we have the maximum point. It is at:

$$\frac{\pi}{2} + \frac{1}{4}(\pi) = \frac{\pi}{2} + \frac{\pi}{4} = \frac{3\pi}{4}$$

The minimum point occurs at $\frac{3}{4}$ of the period:

$$\frac{\pi}{2} + \frac{3}{4}(\pi) = \frac{\pi}{2} + \frac{3\pi}{4} = \frac{5\pi}{4}$$

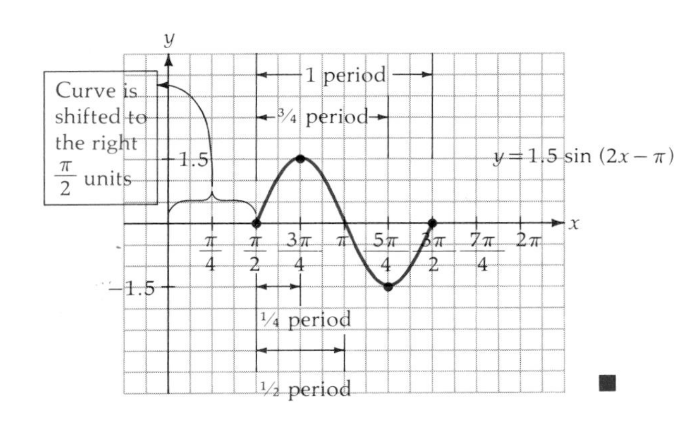


### Graphing the Cosine Function

The cosine function has many similarities to the sine function. We shall begin by sketching the graph. We will first set up a table using special numbers as was done in the previous section.

For $y = \cos x$ we obtain the following table of values using special numbers, as was done in the previous section.

| $x$ | $0$ | $\frac{\pi}{6}$ | $\frac{\pi}{3}$ | $\frac{\pi}{2}$ | $\frac{2\pi}{3}$ | $\frac{5\pi}{6}$ | $\pi$ | $\frac{7\pi}{6}$ | $\frac{4\pi}{3}$ | $\frac{3\pi}{2}$ | $\frac{5\pi}{3}$ | $\frac{11\pi}{6}$ | $2\pi$ |
|---|---|---|---|---|---|---|---|---|---|---|---|---|---|
| Exact values of $\cos x$ | $1$ | $\frac{\sqrt{3}}{2}$ | $\frac{1}{2}$ | $0$ | $-\frac{1}{2}$ | $-\frac{\sqrt{3}}{2}$ | $-1$ | $-\frac{\sqrt{3}}{2}$ | $-\frac{1}{2}$ | $0$ | $\frac{1}{2}$ | $\frac{\sqrt{3}}{2}$ | $1$ |
| Decimal approximation of $\cos x$ | 1 | 0.87 | 0.50 | 0 | −0.50 | −0.87 | −1 | −0.87 | −0.50 | 0 | 0.50 | 0.87 | 1 |

Plotting the points and sketching the curve of best fit give the following graph of a portion of $y = \cos x$.

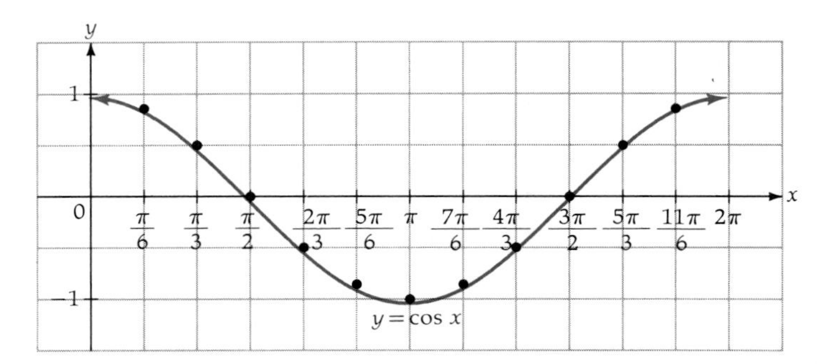


If we sketch $y = \cos x$ for other values of $x$, we see immediately that the cosine function is a periodic function with a period of $2\pi$ and an amplitude of 1.

In the next chapter we will show $\sin\left(x + \dfrac{\pi}{2}\right) = \cos x$ in a more formal proof.

Discussions dealing with the amplitude, period, and phase shift for $y = a \sin(bx + c)$ in the previous section are obviously valid for $y = a \cos(bx + c)$.

The function $y = a \cos(bx + c)$ can then be considered a variation of $y = \cos x$ where

$$\text{amplitude} = |a| \qquad \text{period} = \frac{2\pi}{|b|} \qquad \text{phase shift} = -\frac{c}{|b|}$$

We can state a summary as follows:

---

**Summary of Graphing the Cosine Function**

---

For the function $y = a \cos(bx + c)$ the following points will aid in quickly sketching the graph.

1. The *amplitude* is $|a|$.
2. The *period* is $\dfrac{2\pi}{|b|}$.
3. The *phase shift* is $-\dfrac{c}{|b|}$. [If $c$ is positive, the shift is to the left. If $c$ is negative, the shift is to the right.]
4. Consider $y = a \cos(bx + c)$ where $a > 0$. At the start of one period the function value is $a$. The graph will cross the axis at $\dfrac{1}{4}$ of the period and at $\dfrac{3}{4}$ of the period. At $\dfrac{1}{2}$ of the period the graph will reach a minimum at $-a$.

5. Consider $y = a \cos (bx + c)$ where $a < 0$. At the start and at the end of one period the function value is $a$. In each case these points are the minimum value of the function. The maximum value of the function is $-a$, which will take place at $\frac{1}{2}$ of the period.

---

**EXAMPLE 5** Sketch one period of the graph of $y = 2 \cos \frac{1}{2}x$.

***Solution*** Amplitude = 2 Period = $4\pi$ Phase shift = 0

In one period from $x = 0$ to $x = 4\pi$ the function $y = 2 \cos \frac{1}{2}x$ will have a maximum value of 2 when $x = 0$ and $x = 4\pi$. It will have a minimum value of $-2$ when $x = 2\pi$. ■

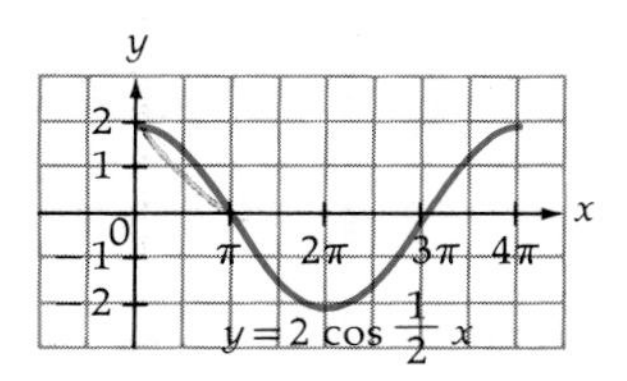


**EXAMPLE 6** Sketch one period of the graph of $y = -1.5 \cos (\pi x)$.

***Solution*** We see that $|a| = 1.5$ and that $a$ is negative, so the cosine curve will begin and end at the minimum value $-1.5$.

$$\text{Period} = \frac{2\pi}{\pi} = 2$$

Thus the real number 2 is the period.

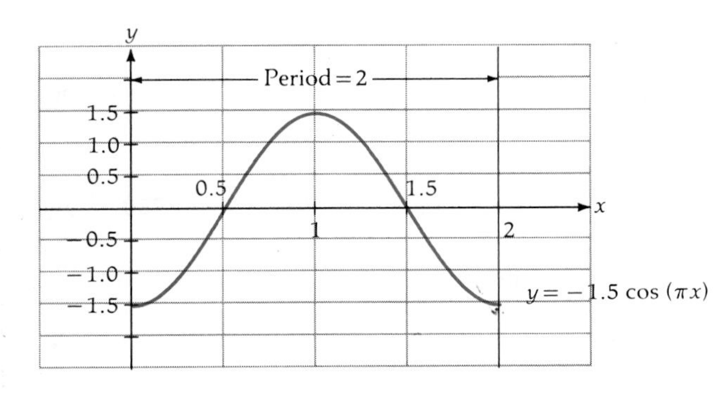


Note that the maximum value occurs halfway through the period. ■

### Applications of Sine and Cosine Functions

You have probably seen an oscilloscope at some time. An oscilloscope is an electronic device. It displays a graph on a scaled cathode-ray tube. A number of the outputs studied on the screen are periodic functions that can be precisely given or at least approximated by sine and/or cosine functions. By using the proper controls, the operator of an oscilloscope can manually change the amplitude, period and phase shift of the displayed graph. The voltage $e$ in an alternating circuit can be obtained by the equation

$e = E \cos (bt + c)$ where

$E =$ the maximum voltage attainable in the circuit

$t =$ the time measured in seconds

$b =$ the frequency measured in radians/seconds

$c =$ the "phase angle" *Note: This is not the same as the phase shift.*

**EXAMPLE 7** Sketch one period of the graph of $e = E \cos (bt + c)$ when $E = 110$ volts, $b = 180$ radians/second, and $c = -\frac{\pi}{4}$.

***Solution*** The independent variable is $t$, which is displayed on the horizontal axis. The dependent variable is $e$, which is displayed on the vertical axis.

$$\text{Amplitude} = 110 \qquad \text{Period} = \frac{2\pi}{180} = \frac{\pi}{90}$$

The phase shift will be to the right, since $c < 0$. We find that

$$-\frac{c}{|b|} \quad \text{is} \quad -\frac{\frac{-\pi}{4}}{180} = \frac{\pi}{4} \cdot \frac{1}{180} = \frac{\pi}{720}$$

The phase shift of $\frac{\pi}{720}$ provides the starting point. The endpoint of the cycle is

$$\frac{\pi}{720} + \frac{\pi}{90} = \frac{\pi}{720} + \frac{8\pi}{720} = \frac{9\pi}{720}$$

The minimum value is at $\frac{1}{2}$ of the period or

$$\frac{\pi}{720} + \frac{\pi}{180} = \frac{\pi}{720} + \frac{4\pi}{720} = \frac{5\pi}{720}$$

It is left to the student to show that $\frac{1}{4}$ of the period would be a value of $t = \frac{3\pi}{720}$ and that $\frac{3}{4}$ of the period would be a value of $t = \frac{7\pi}{720}$.

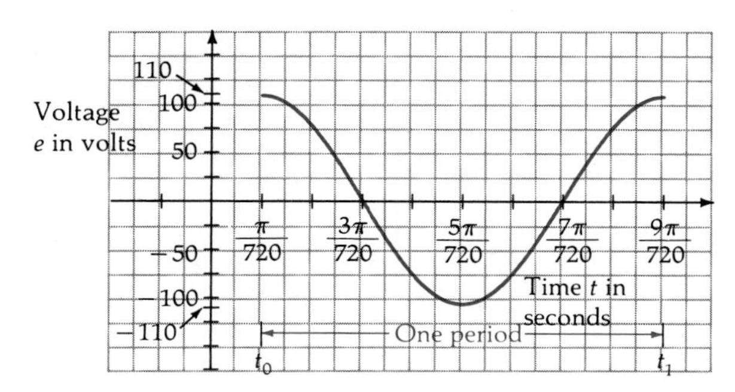


Graph of $e = 110 \cos\left(180t - \frac{\pi}{4}\right)$

The start of the cycle is at $t_0 = \frac{\pi}{720}$ seconds:

[π] [÷] 720 [=] 0.0043633

The end of the cycle is at $t_1 = \frac{9\pi}{720}$ seconds:

9 [×] [π] [÷] 720 [=] 0.0392699

$$t_1 - t_0 = 0.0349066$$

Thus the time for one cycle in the alternating current is about 35 thousandths of a second!

## EXERCISE 7-4

State the amplitude, period, and phase shift and sketch one period of the graph of each of the functions in Problems 1–28.

**1.** $y = 2 \sin x$

**2.** $y = \sin 2x$

**3.** $y = \frac{1}{2} \sin x$

**4.** $y = \sin \frac{1}{2}x$

**5.** $y = -\sin x$

**6.** $y = \sin (x + \pi)$

**7.** $y = 4 \sin 2x$

**8.** $y = 2 \sin 6x$

**9.** $y = -0.8 \sin\left(\frac{x}{3}\right)$

**10.** $y = -0.6 \sin\left(\frac{x}{2}\right)$

**11.** $y = \sin\left(x + \frac{\pi}{2}\right)$

**12.** $y = 3 \sin(x - \pi)$

**13.** $y = \sin(2x + \pi)$

**14.** $y = 2 \sin\left(\frac{1}{2}x - \frac{2}{3}\pi\right)$

★**15.** $y = -3 + 2 \sin\left(x + \frac{3\pi}{2}\right)$

★**16.** $y = 4 - 3 \sin\left(x - \frac{\pi}{2}\right)$

**17.** $y = 2 \cos x$

**18.** $y = \cos 2x$

**19.** $y = -\cos x$

**20.** $y = 3 \cos \frac{1}{2} x$

**21.** $y = 1.5 \cos\left(\frac{2}{3} x\right)$

**22.** $y = -4 \cos\left(\frac{3}{4} x\right)$

**23.** $y = 2 \cos\left(\frac{1}{3} x + 2\pi\right)$

**24.** $y = \frac{1}{3} \cos\left(2x - \frac{1}{3}\pi\right)$

**25.** $y = 3 \cos(2x - \pi)$

**26.** $y = -2 \cos(0.25x + \pi)$

**27.** $y = 0.8 \cos\left(\frac{\pi x}{3} + \pi\right)$

**28.** $y = 0.6 \cos\left(\frac{\pi x}{2} - \frac{\pi}{4}\right)$

### Application Problems

★**29.** Use the procedure of Example 7 to graph one period of $e = E \cos(bt + c)$ when $E = 100$ volts, $b = 150$ radians/second, and $c = \frac{\pi}{4}$.

★**30.** Use the procedure of Example 7 to graph one period of $e = E \cos(bt + c)$ when $E = 150$ volts, $b = 200$ radians/second, and $c = \frac{\pi}{2}$.

### Calculator Problems

★**31.** Draw the graph of the normal average temperature of a city where the temperature $T$ in degrees Fahrenheit is predicted by

$$T = 30 + 35 \sin\left[\frac{2\pi(d - 120)}{365}\right]$$

where $d$ is the number of days from January 1. Plot the values $d = 0, 20, 40, 60, \ldots, 380$ days.

(*Hint:* To simplify computation write the equation in the form $T = 30 + 35 \sin\left[\left(\frac{2\pi}{365}\right)(d - 120)\right]$ and store the value $\frac{2\pi}{365}$ in memory.)

★**32.** Follow the directions of Problem 31 if the equation that predicts the normal average temperature is $T = 20 + 40 \cos\left[\frac{2\pi(d - 100)}{365}\right]$.

State the amplitude, period, and phase shift and sketch one period of the graph of each of the functions in Problems 33 and 34.

★**33.** $y = 2 + 4\cos\left[3\left(x - \frac{\pi}{6}\right)\right]$ ★**34.** $y = 3 - 2\sin\left[2\left(x + \frac{\pi}{3}\right)\right]$

---

## 7-5 GRAPHING THE TANGENT, COTANGENT, SECANT, COSECANT FUNCTIONS

The graphs of the four remaining trigonometric functions do not have as many physical world and scientific applications. They are quite different in character, since the curves are not continuously connected as were the sine and cosine.

### The Graph of the Tangent Function

Suppose we wanted to sketch the graph of $y = \tan x$. When we discussed the range of the tangent function, we found it to be all real numbers. However, the domain does not include odd-integral multiples of $\frac{\pi}{2}$ because at these points $x$ is zero and $\frac{y}{x}$ is undefined. We also found the tangent to be positive from 0 to $\frac{\pi}{2}$ and from $\pi$ to $\frac{3\pi}{2}$ (first and third quadrants) and negative elsewhere. Also, $\tan x = 0$ when $x = n\pi$ where $n$ is an even integer. Some numerical substitutions into $\frac{y}{x}$ indicate that $y = \tan x$ becomes very large in the positive direction as $x$ approaches $\frac{\pi}{2}$ and very large in the negative direction as $x$ approaches $\frac{3\pi}{2}$.

We will now make a table of values for $y = \tan x$ using some special numbers and sketch the graph.

| $x$ | $0$ | $\frac{\pi}{6}$ | $\frac{\pi}{3}$ | $\frac{\pi}{2}$ | $\frac{2\pi}{3}$ | $\frac{5\pi}{6}$ | $\pi$ | $\frac{7\pi}{6}$ | $\frac{4\pi}{3}$ | $\frac{3\pi}{2}$ | $\frac{5\pi}{3}$ | $\frac{11\pi}{6}$ | $2\pi$ |
|---|---|---|---|---|---|---|---|---|---|---|---|---|---|
| Exact value of $\tan x$ | 0 | $\frac{\sqrt{3}}{3}$ | $\sqrt{3}$ | No value | $-\sqrt{3}$ | $-\frac{\sqrt{3}}{3}$ | 0 | $\frac{\sqrt{3}}{3}$ | $\sqrt{3}$ | No value | $-\sqrt{3}$ | $-\frac{\sqrt{3}}{3}$ | 0 |
| Decimal approximation of $\tan x$ | 0 | 0.58 | 1.73 | No value | $-1.73$ | $-0.58$ | 0 | 0.58 | 1.78 | No value | 1.78 | $-0.58$ | 0 |

We will also make use of the fact that $\tan x = 1$ when $x = \frac{\pi}{4}, \frac{3\pi}{4}, \frac{5\pi}{4}, \frac{7\pi}{4}, \ldots$. This property is most helpful when graphing the tangent and cotangent.

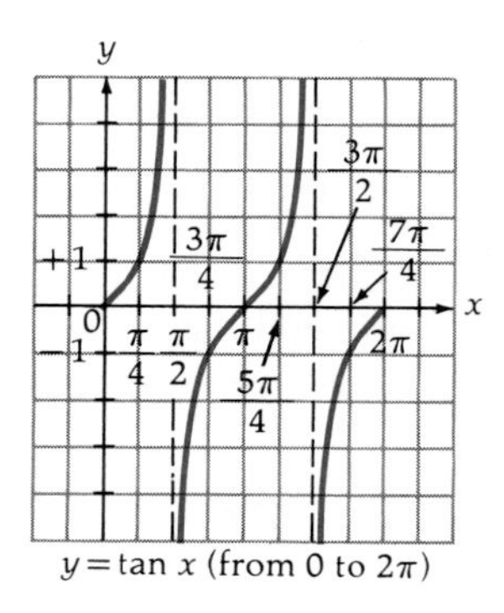


$y = \tan x$ (from 0 to $2\pi$)

---

Note the following observations regarding the tangent function.

1. Amplitude has no meaning for $y = \tan x$, since there is no maximum or minimum value for $y = \tan x$.
2. $y = \tan x$ is repeated for $y = \tan (x + \pi)$. *Hence the period of the tangent function is* $\pi$.
3. The lines $x = \frac{\pi}{2} + \pi n$, where $n$ is any integer, are *asymptotes* to the graph of $y = \tan x$.

---

You may find it helpful to try a few additional calculator values as $x$ approaches $\frac{\pi}{2}$. A calculator approximation for $\frac{\pi}{2}$ is 1.5707963. Let us find a few values for $\tan x$, where $x$ is slightly less than this value.

**EXAMPLE 1** Evaluate by using a scientific calculator.

**a.** $\tan (1.5)$ **b.** $\tan (1.57)$ **c.** $\tan (1.570796)$ **d.** $\tan \left(\frac{\pi}{2}\right)$

***Solution***

**a.** [RAD] 1.5 [tan] 14.10142
**b.** [RAD] 1.57 [tan] 1255.7656
**c.** [RAD] 1.570796 [tan] 3059975.5

We observe as $x$ approaches $\frac{\pi}{2}$ that the value of tan $x$ is becoming very large.

**d.** [RAD] [$\pi$] [$\div$] 2 [=] [tan] error message

On most calculators you will obtain E, a flashing display, or "error" when you evaluate $\tan\left(\frac{\pi}{2}\right)$. Since the value of tan $x$ is increasing without bound as $x$ approaches $\frac{\pi}{2}$, we see that the calculator is encountering a number too large for its memory. ■

**Note:** If you do not obtain an error message, this is due to some roundoff errors in your calculator. You should be aware of this limitation on your calculator.

**EXAMPLE 2** Sketch one period of the graph $y = 3 \tan x$.

***Solution*** The coefficient 3 will make each value of $y = \tan x$ three times as large. This will have an effect on the graph for small values in the range, but for values that are increasing without bound, multiplying by three makes no significant change in the appearance of the graph. Do you see that in any curve of the form $y = a \tan x$ if $a > 0$, the curve is "pulled" or "stretched" from top to bottom?

You may find it helpful to note just a few easily obtained values.

| $x$ | 0 | $\frac{\pi}{4}$ | $\frac{\pi}{2}$ | $\frac{3\pi}{4}$ | $\pi$ |
|---|---|---|---|---|---|
| tan $x$ | 0 | 1 | No value | $-1$ | 0 |
| 3 tan $x$ | 0 | 3 | No value | $-3$ | 0 |

**Note:** Do you see that all curves of the form $y = a \tan x$ will have the value $+a$ or $-a$ when $x = \frac{\pi}{4}$ and when $x = \frac{3\pi}{4}$? In our equation since $a = 3$, we know that $y = 3$ when $x = \frac{\pi}{4}$ and $\frac{3\pi}{4}$. Now examine the graph of $y = 3 \tan x$ on the next page.

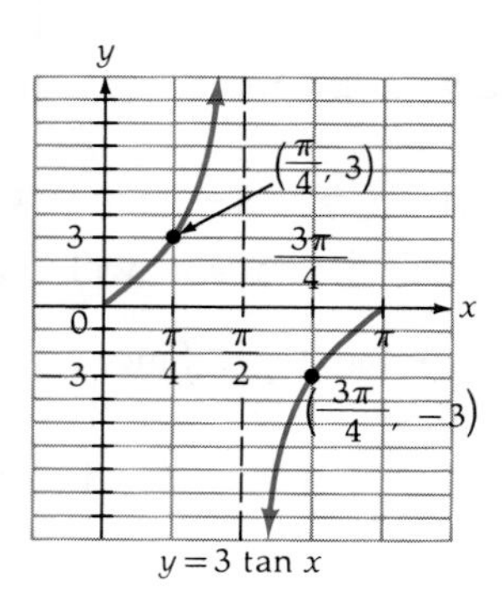


$y = 3 \tan x$ ■

Since the tangent and cotangent functions are reciprocals, when one is zero, the other is undefined. The graph of $y = \cot x$ then has the same period as $y = \tan x$. We now look at the graph of $y = \cot x$.

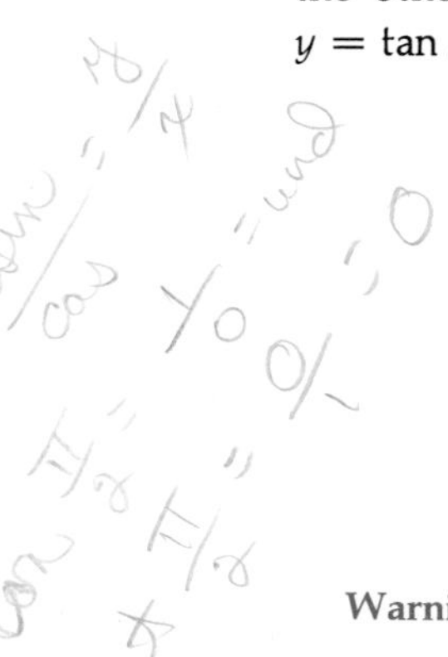

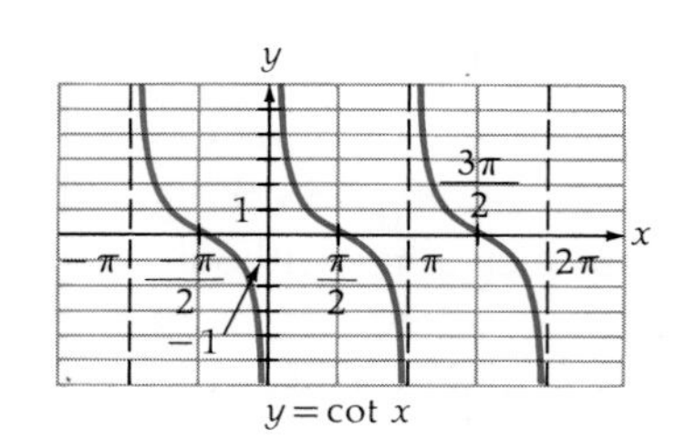


$y = \cot x$

**Warning:** Be sure you understand why $\cot x$ is undefined when $x = 0$ and $x = \pi$ but $\cot x = 0$ when $x = \frac{\pi}{2}$. If you depend only on your calculator here without a good understanding of the definition of cotangent, you will get into trouble!

To illustrate, to evaluate $\cot x$ on your calculator, you use the property that $\cot x = \frac{1}{\tan x}$. But to evaluate $\cot \frac{\pi}{2} = \frac{1}{\tan \frac{\pi}{2}}$ on the calculator is not possible.

[RAD] [π] [÷] 2 [=] [tan] [1/x] *You will usually obtain an error message.*

**Note:** If your calculator gives you an answer like $4 \times 10^{-8}$, it is due to a rounding error in the calculator itself. Your calculator has actually given you for $x$ an approximation quite close to $\frac{\pi}{2}$ and for $y$ an approximation fairly close to zero!

### Properties of Cotangent Function

1. Amplitude has no meaning for $y = \cot x$, since there is no maximum or minimum value for $y = \cot x$.
2. $y = \cot x$ has a *period of* $\pi$.
3. The lines $x = \pi n$ where $n$ is any integer are *asymptotes* to the graph of $y = \cot x$.

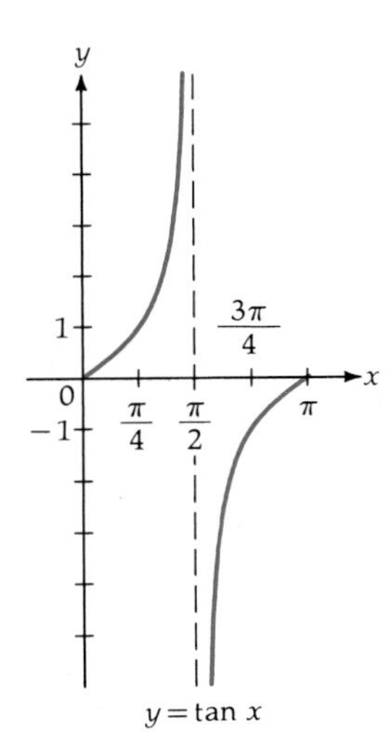

$y = \tan x$

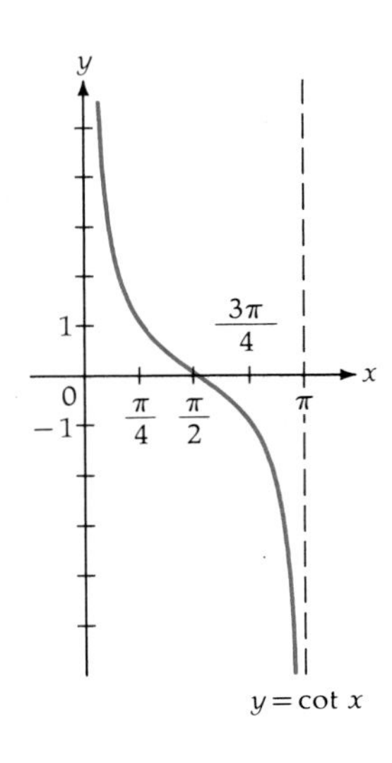

$y = \cot x$

Note the difference between the graph of $y = \tan x$ and $y = \cot x$. The cotangent is asymptotic at integral multiples of $\pi$, whereas the tangent is asymptotic at odd-integral multiples of $\frac{\pi}{2}$.

An interesting and helpful property that is observable from the above graphs is that $\cot x = \tan x$ when $x = \frac{\pi}{4}$ and $x = \frac{3\pi}{4}$.

**EXAMPLE 3** Graph one period of the function $y = \cot\left(\frac{\pi}{2} x\right)$.

***Solution*** The period for $y = \cot x$ is $\pi$. The period for $y = \cot(bx)$ is $\frac{\pi}{|b|}$. Thus the period is $\frac{\pi}{\frac{\pi}{2}} = \pi \cdot \frac{2}{\pi} = 2$. If the entire period of the function is 2, then at $x = 1$ the function value will be 0. If $x = \frac{1}{2}$, the function value will be 1, while at $x = \frac{3}{2}$ the function will be $-1$. Can you verify these statements?

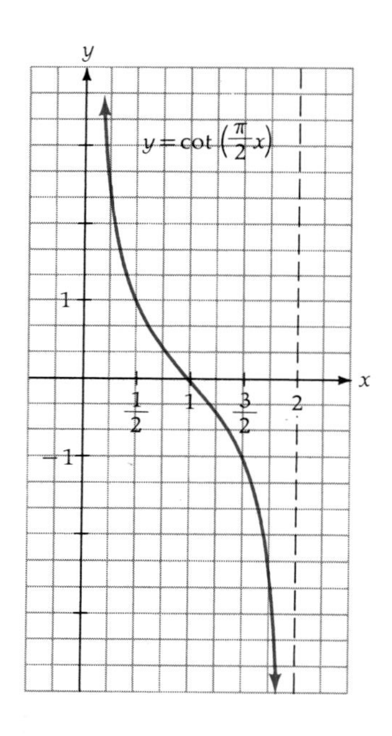


■

## GRAPHING THE SECANT AND THE COSECANT FUNCTIONS

The graphs of the secant and the cosecant functions are quite different from the graphs of the tangent and cotangent functions. The graph of each consists of disjoint branches that are separated by vertical asymptotes. Both functions have a period of $2\pi$.

We will now make a table of values for $y = \csc x$ using some special numbers and then sketch the graph.

| $x$ | $0$ | $\frac{\pi}{6}$ | $\frac{\pi}{3}$ | $\frac{\pi}{2}$ | $\frac{2\pi}{3}$ | $\frac{5\pi}{6}$ | $\pi$ | $\frac{7\pi}{6}$ | $\frac{4\pi}{3}$ | $\frac{3\pi}{2}$ | $\frac{5\pi}{3}$ | $\frac{11\pi}{6}$ | $2\pi$ |
|---|---|---|---|---|---|---|---|---|---|---|---|---|---|
| Exact value of $\csc x$ | No value | $2$ | $\frac{2\sqrt{3}}{3}$ | $1$ | $\frac{2\sqrt{3}}{3}$ | $2$ | No value | $-2$ | $\frac{-2\sqrt{3}}{3}$ | $-1$ | $\frac{-2\sqrt{3}}{3}$ | $-2$ | No value |
| Decimal approximation of $\csc x$ | No value | $2$ | $1.15$ | $1$ | $1.15$ | $2$ | No value | $-2$ | $-1.15$ | $-1$ | $-1.15$ | $-2$ | No value |

In our graph we will continue the same pattern to show two complete periods.

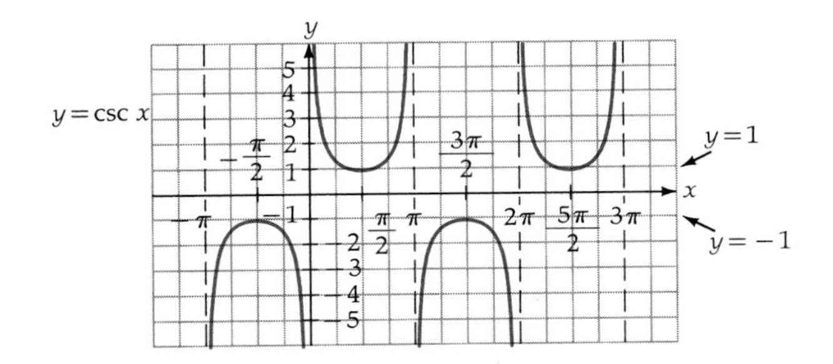


## Properties of the Cosecant Function

1. The graph of $y = \csc x$ has a minimum positive value at $y = 1$ and a negative maximum value at $y = -1$.
2. The period of $y = \csc x$ is $2\pi$.
3. The lines $x = \pi n$ where $n$ is any integer are asymptotes to the graph of $y = \csc x$.

Normally, we *do not* sketch the cosecant function by plotting points. We instead use our understanding of the properties of the cosecant.

Since $\sec x = \dfrac{1}{\cos x}$ and $\csc x = \dfrac{1}{\sin x}$, these functions can be sketched as reciprocals of the sine and cosine curves. Observe that the period of these functions would be $2\pi$, since the reciprocals have a period of $2\pi$.

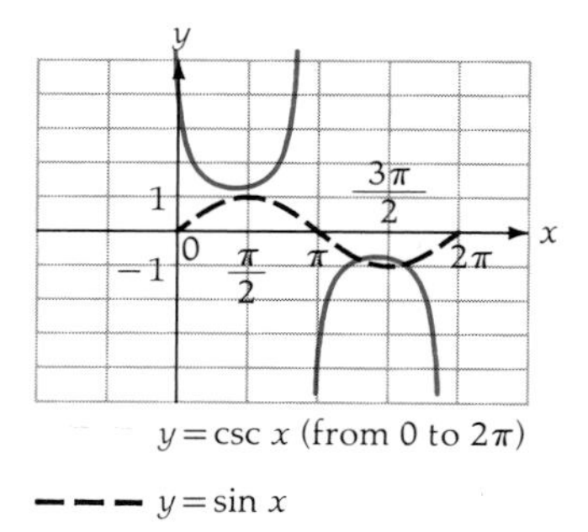


$y = \csc x$ (from 0 to $2\pi$)

– – – $y = \sin x$

The secant function is similar to the cosecant function. The vertical asymptotes, however, occur at different values.

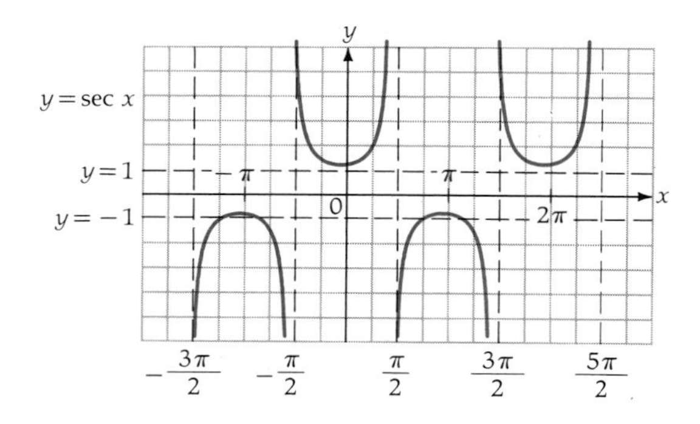


### Properties of the Secant Function

1. The graph of $y = \sec x$ has a minimum positive value at $y = 1$ and a maximum negative value at $y = -1$.
2. The period of $y = \sec x$ is $2\pi$.
3. The lines $x = \dfrac{\pi}{2} + \pi n$ where $n$ is any integer are asymptotes to the graph of $y = \sec x$.

When asked to sketch the secant we do not usually plot points, just as with the cosecant.

To sketch $y = \sec x$, first sketch $y = \cos x$ and then sketch its reciprocal. (Sketching the asymptotes is an aid in sketching any asymptotic curve.) This same method can be applied to $y = a \csc x\,(bx + c)$ and to $y = a \sec\,(bx + c)$.

**EXAMPLE 4** Use the fact that $\sec x = \dfrac{1}{\cos x}$ to sketch one period of $y = \sec x$.

***Solution*** We first sketch $y = \cos x$ from $x = 0$ to $x = 2\pi$. Since $\cos x = 0$ at $x = \dfrac{\pi}{2}$ and $x = \dfrac{3\pi}{2}$, we know that the secant is undefined for those values of $x$.

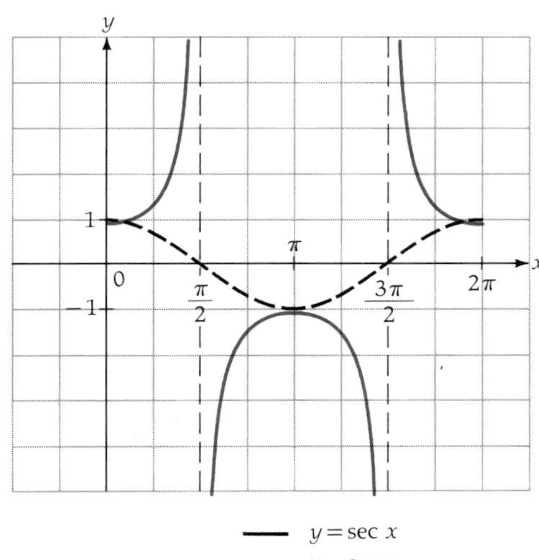


If there is a phase shift in the argument, we merely use the properties for graphing $y = a \sin(bx + c)$ and $y = a \cos(bx + c)$ that were previously developed.

**EXAMPLE 5** Sketch one period of the function $y = \sec\left(2x + \dfrac{\pi}{2}\right)$.

***Solution*** First we note that the equation can be written as $y = \dfrac{1}{\cos\left(2x + \dfrac{\pi}{2}\right)}$. To graph $\cos\left(2x + \dfrac{\pi}{2}\right)$, we note that the period $= 2\pi \div 2 = \pi$. The phase shift is to the left $\dfrac{\pi}{2} \div 2 = \dfrac{\pi}{2} \cdot \dfrac{1}{2} = \dfrac{\pi}{4}$. First we will sketch one period of $y = \cos\left(2x + \dfrac{\pi}{2}\right)$ and then show the reciprocal. If the graph of one period of $y = \cos\left(2x + \dfrac{\pi}{2}\right)$

starts at $= -\frac{\pi}{4}$, it will end at $\frac{3\pi}{4}$. Do you see why? Now we sketch the cosine curve with a dashed line and the secant curve with a solid line.

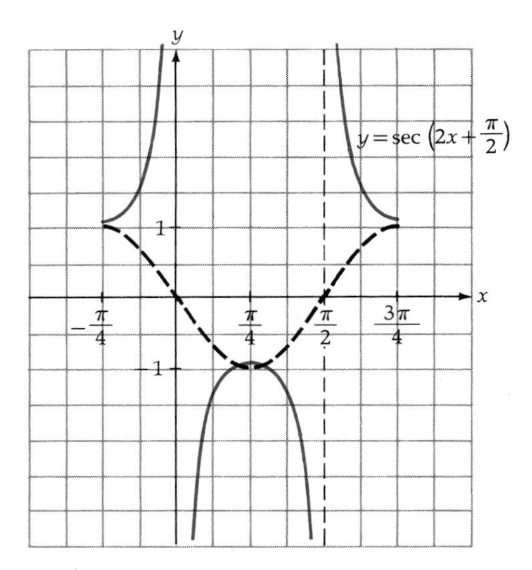


We see that the asymptotes are at $x = 0$ and $x = \frac{\pi}{2}$. ■

## EXERCISE 7-5

1. Sketch the graph of $y = \tan x$ from $-\frac{3\pi}{2}$ to $\frac{5\pi}{2}$.
2. Is $y = \tan x$ a periodic function? Why?
3. Give the period for the function $y = a \tan (bx + c)$. Give the period for the function $y = a \tan \left(\frac{x}{d} + c\right)$.
4. Sketch the graph of $y = \tan 2x$ from 0 to $2\pi$.
5. Sketch one period of the graph of $y = \cot \frac{1}{4}x$.
6. State the phase shift for the function $y = a \tan (bx + c)$.
7. Sketch the graph of $y = \tan \left(x + \frac{\pi}{2}\right)$ from 0 to $2\pi$.
8. Sketch the graph of $y = -\cot x$ from 0 to $2\pi$.
9. Sketch the graph of $y = \tan \left(4x + \frac{\pi}{2}\right)$ from $-\frac{\pi}{2}$ to $\frac{\pi}{2}$.
10. State a function, in terms of the tangent, that has the same graph as $y = \cot (2x - \pi)$.

**11.** Graph one period of the function $y = \cot(2x) + 2$.

**12.** Graph one period of the function $y = \tan\left(\frac{1}{4}x\right) - 2$.

**13.** Graph one period of the function $y = \frac{1}{2}\tan(\pi x)$.

**14.** Graph one period of the function $y = 2\cot(2\pi x)$.

**15.** Graph on one axis $y = \cot x$ and $y = \tan x$ from $x = \pi$ to $x = 2\pi$.

**16.** Graph on one axis $y = \tan x$ and $y = \tan(x + \pi)$ from $x = \pi$ to $x = 2\pi$.

**17.** Sketch $y = \csc x$ from $-2\pi$ to $4\pi$. **18.** Sketch $y = \sec x$ from $-\frac{3\pi}{2}$ to $\frac{9\pi}{2}$.

**19.** Does amplitude have any meaning regarding the secant or cosecant functions? Explain.

**20.** Sketch one period of $y = \sec\left(x - \frac{\pi}{2}\right)$. **21.** Sketch one period of $y = \csc\left(x + \frac{\pi}{2}\right)$.

**22.** State the phase shift of $y = a\csc(bx + c)$. **23.** State the period of $y = a\sec(bx + c)$.

**24.** State a function in terms of cosecant that would have the same graph as $y = \sec(x + \pi)$.

**25.** Sketch one period of the graph of $y = 4\csc 2x$.

**26.** Sketch one period of the graph of $y = 2\sec(3x - \pi)$.

**27.** Sketch one period of the graph of $y = 2\sec(\pi x)$.

**28.** Sketch one period of the graph of $y = 3\csc\left(\frac{\pi}{2}x\right)$.

**29.** Sketch one period of the graph $y = 3\sec\left(x - \frac{\pi}{4}\right)$.

**30.** Sketch one period of the graph $y = 2\csc\left(x + \frac{\pi}{4}\right)$.

★ **31.** Using the properties of the secant and the cosecant, find a value $c$ such that $y = -\sec x$ and $y = \csc(x + c)$ have the same graph.

★ **32.** Using the properties of the secant and the cosecant, find a value $c$ such that $y = -\csc x$ and $y = \sec(x + c)$ have the same graph.

---

## 7-6 GRAPHING COMPOSITE TRIGONOMETRIC FUNCTIONS

The vibration of a simple musical tone can be displayed as a sine wave. The display could be shown on an oscilloscope or on a computer screen. If more than one tone is involved, it can be expressed mathematically as a combination of sine and cosine waves. In this section we will consider how such waves can be combined to form one composite wave.

Many phenomena from the physical sciences can be represented mathematically as sums or differences of functions. We will now establish a "mechanical" method of graphing such functions.

**EXAMPLE 1** Sketch the graph of $y = \sin x + \cos x$.

***Solution*** Sketch $y = \sin x$ and $y = \cos x$ separately, then obtain the final graph by the addition of ordinates ($y$ values) of each function.

First, observe that this function is the sum of two functions. Represent these two functions as

$$y_1 = \sin x \qquad \text{and} \qquad y_2 = \cos x$$

where

$$y = y_1 + y_2 \qquad \text{for each } x$$

Now, sketch each of the two functions on the same coordinate system. For several values of $x$, add the two corresponding $y$ values to obtain the value of the final function. Connect each of these final values to obtain a smooth curve. To observe the actual process, take a portion of the graph on a larger scale.

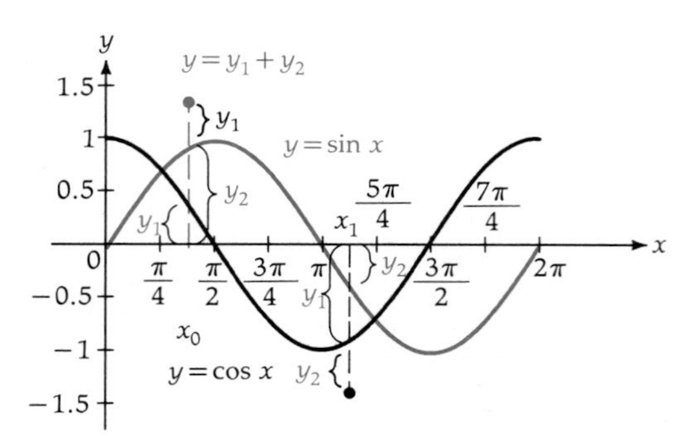


For a value of $x$ such as $x_0$ or $x_1$ (as shown), we measure $y_1$ and $y_2$, then locate $y = y_1 + y_2$.

**A helpful hint:** If one function value is zero (at $x = \frac{3\pi}{2}$, for example), then the desired point is on the other curve. Also, when the curves cross, you simply double the value (such as at $x = \frac{\pi}{4}$).

By continuing this process for a number of points we obtain a number of points of the composite curve. Finally, we connect these points by a smooth curve.

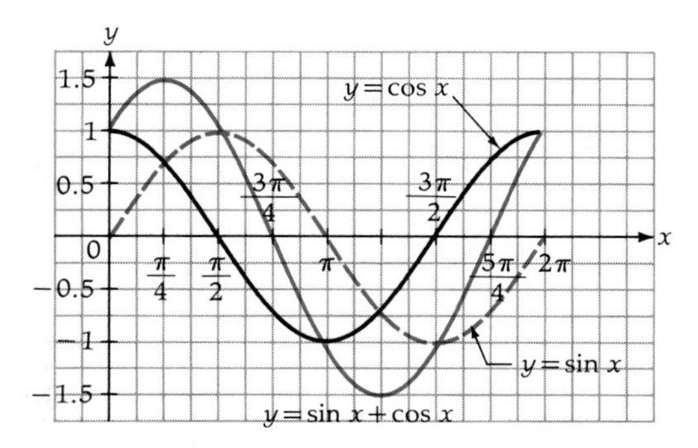


**EXAMPLE 2** Graph $y = 2 \cos x + \cos 2x$ from $x = 0$ to $x = 2\pi$.

***Solution*** First we draw each curve separately, noting that $2 \cos x$ has an amplitude of 2 and a period of $2\pi$ while $\cos 2x$ has an amplitude of 1 and a period of $\pi$. We draw each graph separately. Then we graphically add some ordinates to obtain points on the graph we seek. (It is helpful to use a compass to measure an ordinate on one curve and add it to the ordinate on the other curve.) For convenience we have marked the resulting point with a small x. Now we connect together the points marked with small x's to form a smooth curve.

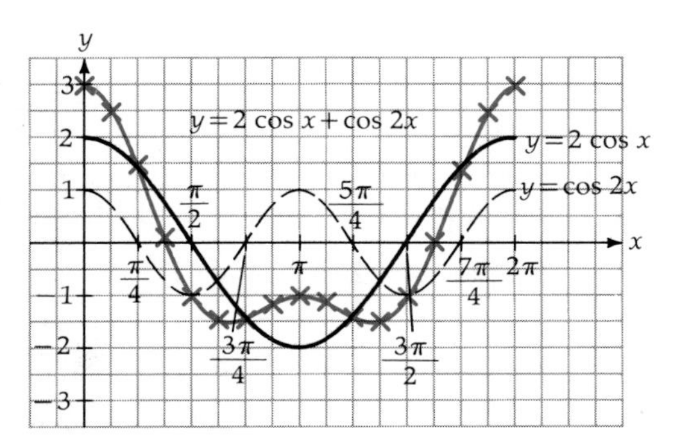


**EXAMPLE 3** Sketch the graph of $y = \sin x + 2x$. Sketch the graph from $x = 0$ to $x = 2\pi$.

***Solution*** (*Hint:* It is most helpful to use the type of trigonometric graph paper that labels the horizontal axis in terms of multiples of $\pi$ and the vertical axis in terms of units 1, 2, 3, etc.) In graphing the straight line $y = 2x$, you will note that if $x = \pi$, then $y = 2\pi$. It may be convenient to approximate the point

$(\pi, 2\pi)$ as $(3.14, 6.28)$ so as to obtain the correct height on the $y$-axis. This will ensure that the straight line is drawn accurately. Graphing each of these functions and adding the $y$-components gives us the following.

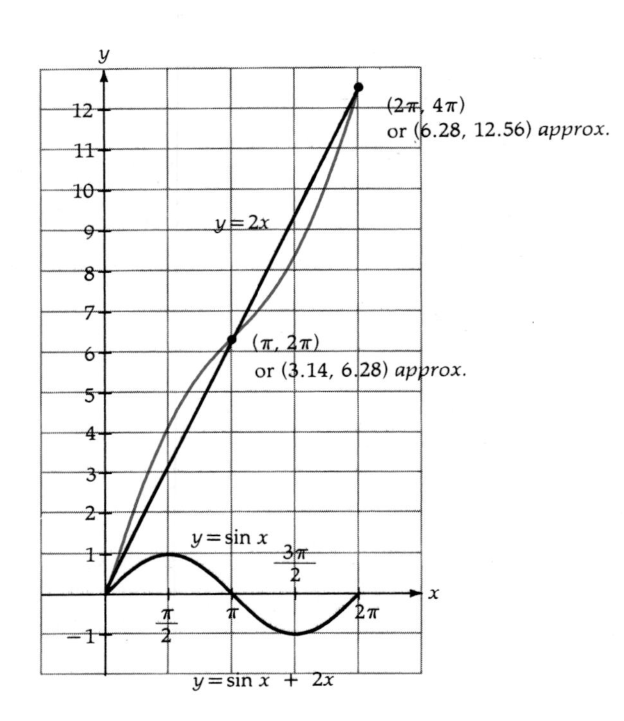


■

## EXERCISE 7-6

Sketch graphs for the equations in Problems 1–18, using the techniques discussed in this section.

**1.** $y = 2 \sin x + \cos x$ **2.** $y = 2 \cos x + \sin x$ **3.** $y = \cos x + \cos 2x$

**4.** $y = \sin x + \sin 2x$ **5.** $y = 3 \sin x + \cos 2x$ **6.** $y = 2 \sin x + \cos 2x$

**7.** $y = \sin x - \cos x$ (*Hint:* Write it as $y = \sin x + (-\cos x)$.)

**8.** $y = 2 \cos x - \sin x$ (*Hint:* Write it as $y = 2 \cos x + (-\sin x)$.)

**9.** $y = 1 + \cos x$ **10.** $y = 2 + \sin 3x$ **11.** $y = x + \sin x$

**12.** $y = x - \sin x$ **13.** $y = \frac{1}{2}x - \cos 2x$ **14.** $y = \frac{1}{2}x + \cos 2x$

**15.** $y = \cos x + \frac{1}{2} \sin 2x$ **16.** $y = 2 \cos \frac{x}{2} + 3 \sin x$ **★17.** $y = \tan x - \sin x$

**★18.** $y = \cos x + \sec x$

## 7-7 INVERSE TRIGONOMETRIC FUNCTIONS

In this section we will discuss the concept of inverse functions and the inverse of a function. If those ideas are not clear to you, perhaps you should take a few minutes and review Section 3-6.

### The Inverse Sine Function

Consider the inverse of the function $y = \sin x$. We know that the range of the function $y = \sin x$ is $-1 \le y \le 1$. If we choose an element in this range and look for the $x$ from which it came, we can see that the inverse of $y = \sin x$ is not a function. For instance, if $y$ is taken as $\frac{1}{2}$, then $x$ can be either $\frac{\pi}{6}$ or $\frac{5}{6}\pi$, or in fact $\frac{\pi}{6} + 2n\pi$, and $\frac{5\pi}{6} + 2n\pi$, where $n$ is an integer. Therefore if we wish the inverse in this case to be a function, we must place some limits on the range. To distinguish between the inverse of the sine function and the inverse sine function, use the following notation and definition.

*Definition*

The *inverse sine function* will be denoted by $y = \text{Arcsin } x$ or $y = \text{Sin}^{-1} x$. This is the inverse of the sine function with the range limited to the values $-\frac{\pi}{2} \le y \le \frac{\pi}{2}$ $\left(\text{i.e., } y = \text{Sin}^{-1} x \text{ is } y = \sin^{-1} x \text{ if } -\frac{\pi}{2} \le y \le \frac{\pi}{2}\right)$.

This definition is represented by the following pair of graphs.

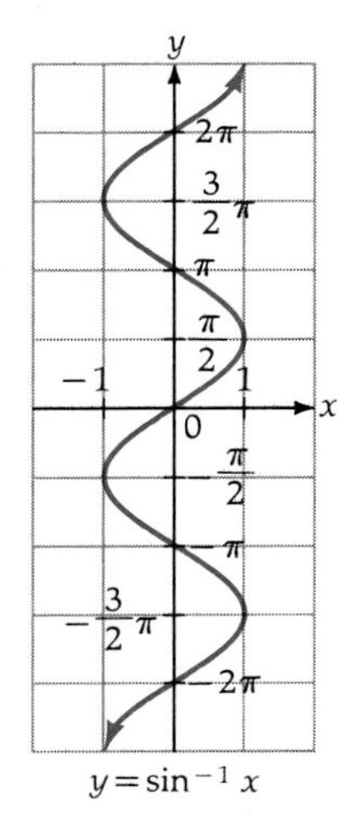


$y = \sin^{-1} x$

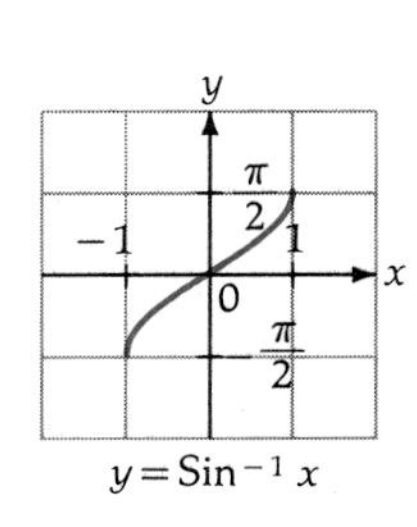


$y = \text{Sin}^{-1} x$

The limited range of the inverse of the sine function is necessary if we are to have a function. This particular limitation of $-\frac{\pi}{2} \leq y \leq \frac{\pi}{2}$ is chosen by agreement among mathematicians. (Actually, many possible arbitrary range restrictions could have been chosen.) *We must always be aware of this limited range when working problems. The uppercase letters must always be noted as specifying a function and therefore indicating the limited range.*

The uppercase letters are also used to denote a trigonometric function with limited domain. $y = \text{Sin } x$ denotes that part of $y = \sin x$ such that the domain is limited to $-\frac{\pi}{2} \leq x \leq \frac{\pi}{2}$. This then is the inverse of $y = \text{Sin}^{-1} x$.

Be sure you see the difference between the two parts of Example 1.

EXAMPLE 1

**a.** Find $x$ if $\sin x = \frac{\sqrt{3}}{2}$.

**b.** Solve $y = \text{Sin}^{-1} \frac{\sqrt{3}}{2}$.

***Solution***

**a.** The difference in these two problems is in the range. For part a we have infinitely many solutions: $\frac{\pi}{3}, \ldots, \left(\frac{\pi}{3} \pm 2n\pi\right)$ or $\frac{2}{3}\pi, \ldots, \left(\frac{2\pi}{3} \pm 2n\pi\right)$.

**b.** By our definition of $y = \text{Sin}^{-1} x$ we are limited to $-\frac{\pi}{2} \leq y \leq \frac{\pi}{2}$. Thus we have the single solution $y = \frac{\pi}{3}$. ■

In our definition of $y = \text{Sin } x$ the values of $x$ may be real numbers or angles. If they are angles, they may be measured in degrees. Therefore it is clear that for $y = \text{Sin}^{-1} x$ the range of the function will sometimes be expressed as $-90° \leq y \leq 90°$.

EXAMPLE 2

Find $\cos\left[\text{Arcsin}\left(-\frac{3}{5}\right)\right]$.

***Solution***

Let us be sure we understand the problem. The problem says, in other words, find the cosine of a fourth-quadrant angle or real number whose sine is $\left(-\frac{3}{5}\right)$.

It may help you to visualize this:

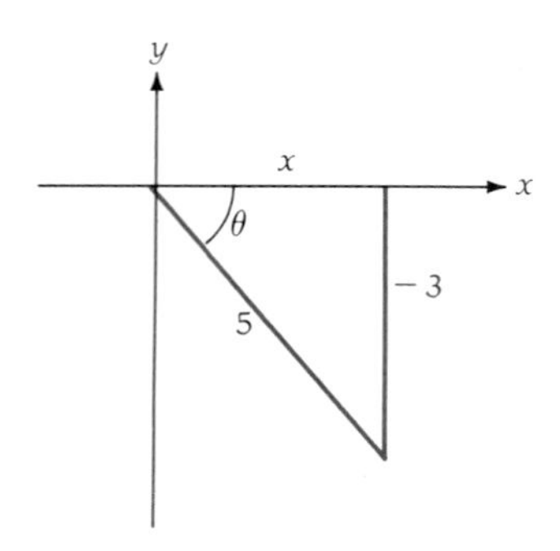


By the Pythagorean Theorem we find that $x = 4$. Since the ratios involved are in a 3–4–5 right triangle, we have $\cos\left[\text{Arcsin}\left(-\frac{3}{5}\right)\right] = \frac{4}{5}$. ■

### The Inverse Cosine Function

The Arccos $x$ or $\text{Cos}^{-1}\, x$ is similarly defined. The important thing is to understand and remember the limited range.

*Definition*

> The inverse cosine function will be denoted either by $y = \text{Arccos}\, x$ or else by $y = \text{Cos}^{-1}\, x$ and is the inverse of the cosine function with the range limited to $0 \leq y \leq \pi$ (i.e., $y = \text{Cos}^{-1}\, x$ is $y = \cos^{-1} x$ if $0 \leq y \leq \pi$).

Note the accompanying graphs.

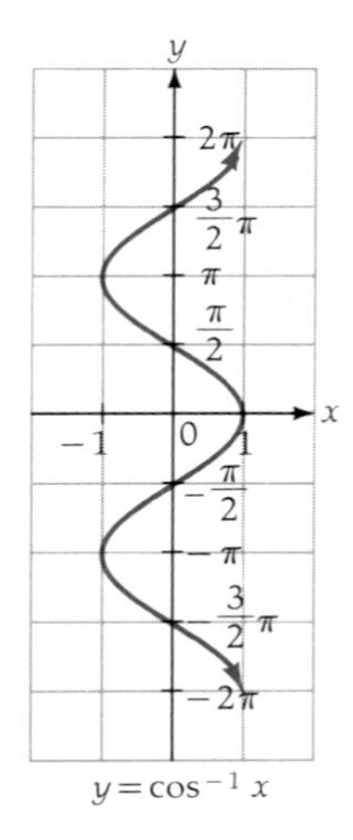


$y = \cos^{-1} x$

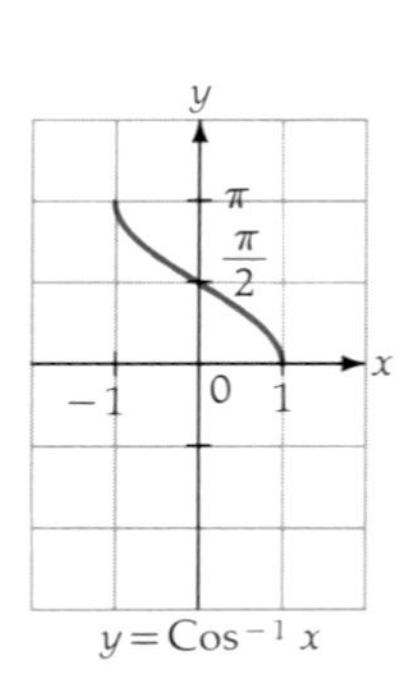


$y = \text{Cos}^{-1}\, x$

EXAMPLE 3 Find Arccos [Sin $(-30°)$].

***Solution*** $\text{Sin}(-30°) = -\frac{1}{2}$. Therefore $\text{Arccos}\left(-\frac{1}{2}\right) = 120°$. ■

EXAMPLE 4 Find an expression for $\text{Cos}\left(\text{Arcsin}\frac{u}{5}\right)$. Assume $u > 0$.

***Solution*** We can picture an angle $\theta$ where $\theta = \text{Arcsin}\frac{u}{5}$ by the following sketch.

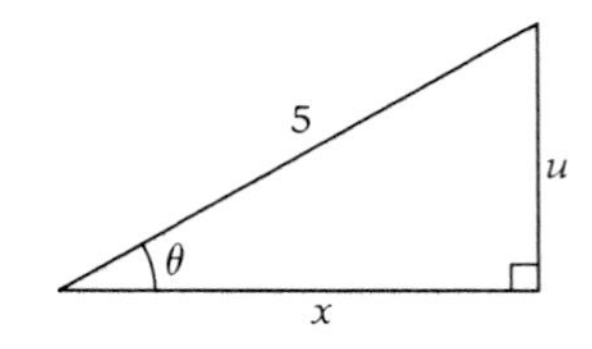


We can find $x$ by the Pythagorean Theorem:

$$5^2 = u^2 + x^2$$
$$25 - u^2 = x^2$$

Since $x$ is positive, we have $x = \sqrt{25 - u^2}$. Thus $\text{Cos}\left(\text{Arcsin}\frac{u}{5}\right) = \text{Cos}\,\theta = \frac{x}{5} = \frac{\sqrt{25 - u^2}}{5}$. ■

The following table summarizes our discussions of the domain and range of inverse sine and cosine functions.

| *Function* | *Domain* | *Range* |
|---|---|---|
| $y = \text{Sin}\,x$ | $-\frac{\pi}{2} \leq x \leq \frac{\pi}{2}$ | $-1 \leq y \leq 1$ |
| $y = \text{Cos}\,x$ | $0 \leq x \leq \pi$ | $-1 \leq y \leq 1$ |
| $y = \text{Sin}^{-1}\,x$ | $-1 \leq x \leq 1$ | $-\frac{\pi}{2} \leq y \leq \frac{\pi}{2}$ |
| $y = \text{Cos}^{-1}\,x$ | $-1 \leq x \leq 1$ | $0 \leq y \leq \pi$ |

It is wise to memorize the values of the domain and range of the inverse sine and cosine functions. A knowledge of them is frequently needed in a variety of advanced mathematics courses.

### Inverse Tangent, Cotangent, Secant, and Cosecant Functions

In a similar fashion we can identify the inverse trigonometric functions corresponding to the tangent, cotangent, secant, and cosecant.

The following are graphs of the tangent, cotangent, secant, and cosecant functions that we have previously studied in this chapter.

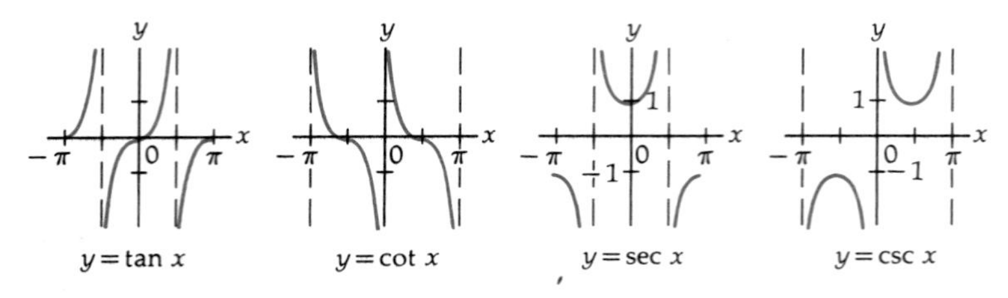


$y = \tan x$ $y = \cot x$ $y = \sec x$ $y = \csc x$

Interchanging the domain and the range in each case gives us the equations of the inverses of these functions.

Obviously, the range of these inverses must be limited if they are to be functions. There are many ways this can be done. The following graphs indicate the limitations that are often accepted by mathematicians.

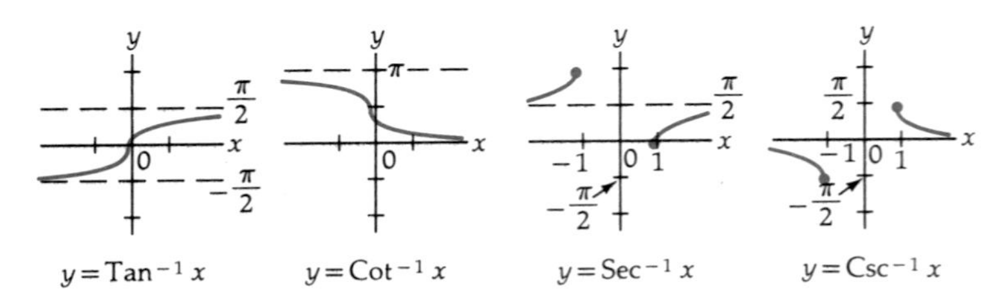


$y = \text{Tan}^{-1} x$ $y = \text{Cot}^{-1} x$ $y = \text{Sec}^{-1} x$ $y = \text{Csc}^{-1} x$

The Arctangent function is commonly used, and it is wise to memorize the limitations.

Not all mathematicians agree with the limitations for the Arccotangent, Arcsecant, and Arccosecant functions. It is not worthwhile to memorize these limitations. These three inverse functions are not used extensively in higher mathematics.

The following table summarizes the domain and range for the inverse tangent, cotangent, secant, and cosecant functions.

| *Function* | *Domain* | *Range* | |
|---|---|---|---|
| $y = \text{Tan } x$ | $-\frac{\pi}{2} < x < \frac{\pi}{2}$ | all reals | **Note** |
| $y = \text{Cot } x$ | $0 < x < \pi$ | all reals | |
| $y = \text{Sec } x$ | $0 \le x < \pi$; $x \ne \frac{\pi}{2}$ | $y \ge 1$ or $y \le -1$ | |
| $y = \text{Csc } x$ | $-\frac{\pi}{2} \le x \le \frac{\pi}{2}$; $x \ne 0$ | $y \ge 1$ or $y \le -1$ | |
| $y = \text{Tan}^{-1} x$ | all reals | $-\frac{\pi}{2} < y < \frac{\pi}{2}$ | **Note** |
| $y = \text{Cot}^{-1} x$ | all reals | $0 < y < \pi$ | |
| $y = \text{Sec}^{-1} x$ | $x \ge 1$ or $x \le -1$ | $0 \le y \le \pi$; $y \ne \frac{\pi}{2}$ | |
| $y = \text{Csc}^{-1} x$ | $x \ge 1$ or $x \le -1$ | $-\frac{\pi}{2} \le y \le \frac{\pi}{2}$; $y \ne 0$ | |

**Note:** You are asked to take special care to study $y = \text{Tan } x$ and $y = \text{Tan}^{-1} x$, which are more commonly used than the others.

We will now do a few examples.

**EXAMPLE 5** Find $\text{Tan}\left[\text{Arcsin}\left(-\frac{1}{2}\right)\right]$.

***Solution*** There are two distinct steps. First we find $\text{Arcsin}\left(-\frac{1}{2}\right) = -\frac{\pi}{6}$. Do you see that because of the capital A there is only *one possible value* and it is $-\frac{\pi}{6}$? Then we obtain $\text{Tan}\left(-\frac{\pi}{6}\right) = -\frac{\sqrt{3}}{3}$. Therefore $\text{Tan}\left[\text{Arcsin}\left(-\frac{1}{2}\right)\right] = -\frac{\sqrt{3}}{3}$. ■

Notice that because of the restrictions stated with the definitions, not all stated problems have an answer.

**EXAMPLE 6** Find $\text{Tan}^{-1}\left(\cos\frac{\pi}{2}\right)$.

***Solution*** We know that $\cos\frac{\pi}{2} = 0$. So we now evaluate $\text{Tan}^{-1}(0)$. This will occur when $x = 0$. Thus $\text{Tan}^{-1}\left(\cos\frac{\pi}{2}\right) = 0$. ■

The limitations in domain and range we have discussed for the trigonometric values are often referred to as the principal values of the function. For example, the *principal values* of the sine function $y = \sin x$ would be all $x$ such that $-\frac{\pi}{2} < x < \frac{\pi}{2}$. This corresponds to the domain we defined for $y = \text{Sin } x$.

When you are using a calculator to evaluate inverse trigonometric functions, you will need to keep the principal values in mind. To find the inverse sine on a calculator, you usually need to press two keys: [INV] [sin] or [2nd Fn] [sin] or [ARC] [sin]. (A few calculators have a special button [sin⁻¹].)

**EXAMPLE 7** Evaluate the following on a calculator. Express your answer in radians to four decimal places.

**a.** $\text{Sin}^{-1}(-0.2684)$
**b.** $\text{Cos}^{-1}(-0.3618)$

***Solution*** **a.** [RAD] .2684 [+/−] [INV] [sin] −0.2717317

Rounded to four decimal places,

$$\text{Sin}^{-1}(-0.2684) = -0.2717$$

**b.** [RAD] .3618 [+/−] [INV] [cos] 1.9409943

Rounded to four decimal places,

$$\text{Cos}^{-1}(-0.3618) = 1.9410$$ ■

Notice that the results are consistent with our convention of principal values. The function $y = \text{Sin}^{-1} x$ has negative $x$ values when $y$ is in the fourth quadrant $\left(-\frac{\pi}{2} < x < 0\right)$. The function $y = \text{Cos}^{-1} x$ has negative $x$ values when $y$ is in the second quadrant $\left(\frac{\pi}{2} < x < \pi\right)$.

**Note:** If your calculator did not obtain the correct results, it probably does not follow the standard procedure for principal values of inverse trigonometric functions. You will need to make some mental adjustments to obtain the correct values.

## EXERCISE 7-7

1. Why are we not able to define the inverse sine function to have a range of $0 \le y \le \pi$?
2. Can we define the inverse sine function over a range other than $-\frac{\pi}{2} \le y \le \frac{\pi}{2}$?
3. Can the inverse cosine function be defined over the range $-\frac{\pi}{2} \le y \le \frac{\pi}{2}$? Explain.
4. Find another range that could have been used to define the inverse cosine function.

In Problems 5–14, find the values in terms of $\pi$.

5. $\text{Arcsin}\ \frac{1}{2}$
6. $\text{Sin}^{-1} \frac{\sqrt{2}}{2}$
7. $\text{Arccos}\ \frac{1}{2}$
8. $\text{arccos}\ \frac{\sqrt{3}}{2}$
9. $\sin^{-1} 0$
10. $\text{Cos}^{-1}(-1)$
11. $\text{Sin}^{-1}\left(-\frac{\sqrt{3}}{2}\right)$
12. $\text{Arcsin}\left(-\frac{1}{2}\right)$
13. $\text{Arccos}\ 0$
14. $\text{Cos}^{-1}\left(-\frac{1}{2}\right)$

5–53 odds

Do not use a calculator in Problems 15–44. You should understand trigonometric functions well enough to find the values directly. Evaluate each of the expressions in Problems 15–26.

15. $\sin\left(\text{Arcsin}\ \frac{\sqrt{3}}{2}\right)$
16. $\cos\left[\text{Cos}^{-1}\left(-\frac{1}{2}\right)\right]$
17. $\sin\left(\text{Sin}^{-1} \frac{\sqrt{2}}{2}\right)$
18. $\cos(\text{Sin}^{-1} 1)$
19. $\cos\left[\text{Arcsin}\left(-\frac{5}{13}\right)\right]$
20. $\text{Arcsin}(\sin 60^\circ)$
21. $\text{Arcsin}(\sin 120^\circ)$
22. $\text{Cos}^{-1}(\cos 200^\circ)$
23. $\text{Sin}^{-1}[\cos(-60^\circ)]$
24. $\text{Cos}^{-1}[\sin(-45^\circ)]$
25. $\text{Sin}^{-1} \frac{\sqrt{3}}{2} + \text{Sin}^{-1}\left(-\frac{\sqrt{3}}{2}\right)$
26. $\text{Cos}^{-1} \frac{\sqrt{3}}{2} + \text{Cos}^{-1}\left(-\frac{\sqrt{3}}{2}\right)$

Find the value of each of the expressions in Problems 27–44.

27. $\text{Arctan}\ \sqrt{3}$
28. $\text{Arccot}(-1)$
29. $\text{Sec}^{-1} 2$

**30.** $\text{Csc}^{-1}(-1)$
**31.** $\text{Tan}^{-1} 0$
**32.** $\text{Arcsec}(-\sqrt{2})$
**33.** $\text{Tan}\left(\text{Cos}^{-1}\dfrac{\sqrt{3}}{2}\right)$
**34.** $\tan\left[\text{Arccos}\left(-\dfrac{4}{5}\right)\right]$
**35.** $\text{Tan}\left[\text{Arccos}\left(-\dfrac{4}{5}\right)\right]$
**36.** $\cos(\text{Sec}^{-1} 2)$
**37.** $\text{Sin}[\text{Arccsc}(-2)]$
**38.** $\text{Tan}\left(\text{Sin}^{-1}\dfrac{\sqrt{3}}{2}\right)$
**39.** $\cos\left[\text{Sec}^{-1}\left(-\dfrac{13}{12}\right)\right]$
**40.** $\csc\left[\text{Arcsin}\left(-\dfrac{1}{3}\right)\right]$
**41.** $\sec[\text{Arctan}(-1)]$
**42.** $\text{Cot}\left[\text{Cos}^{-1}\left(-\dfrac{1}{2}\right)\right]$
**43.** $\text{Sin}\left(\text{Cot}^{-1}\dfrac{3}{4}\right)$
**44.** $\text{Tan}\left[\text{Arcsin}\left(-\dfrac{24}{25}\right)\right]$

### Calculator Problems

Evaluate each of the expressions in Problems 45–60 on a calculator. Round your answer to four decimal places.

**45.** $\text{Sin}^{-1} 0.3628$
**46.** $\text{Cos}^{-1} 0.7214$
**47.** $\text{Tan}^{-1}(-0.3615)$
**48.** $\text{Sin}^{-1}(-0.7283)$
**49.** $\text{Csc}^{-1} 1.3624$
**50.** $\text{Sec}^{-1} 2.0089$
**51.** $\text{Sin}(\text{Cos}^{-1} 0.6317)$
**52.** $\text{Tan}(\text{Sin}^{-1} 0.0783)$
**53.** $\text{Tan}^{-1}(\cos 0.7318)$
**54.** $\text{Cos}^{-1}(\sin 1.523)$
**55.** $\text{Sin}^{-1}\left(\text{Sec}\dfrac{\pi}{12}\right)$
**56.** $\text{Cos}^{-1}\left(\text{Cot}\dfrac{\pi}{8}\right)$
**57.** $\sin[\text{Cot}^{-1}(-1.6231)]$
**58.** $\text{Tan}[\text{Sec}^{-1}(-1.7431)]$
**59.** $\text{Sec}[\text{Cos}^{-1}(-1.832)]$
**60.** $\text{Cot}[\text{Sin}^{-1}(-2.6314)]$

★ **61.** Find an expression for $\sin\left(\text{Arccos}\dfrac{3}{u}\right)$. Assume $u > 0$.

★ **62.** Find an expression for $\cos\left(\text{Arctan}\dfrac{u}{7}\right)$. Assume $u > 0$.

## COMPUTER PROBLEMS FOR CHAPTER 7

See if you can apply your knowledge of computer programming and of Chapter 7 to do the following.

**1.** Write a computer program to print out two periods of the graph of the function $y = \cos(bx)$ where $x$ is measured in radians. Write the program so that you can input the value of $b$. Run the program and graph the following:

**a.** $y = \cos(2x)$
**b.** $y = \cos(0.5x)$
**c.** $y = \cos(-3.61x)$
**d.** $y = \cos(0.076x)$

**2.** Write a computer program to print out two periods of the graph of the function $y = \sin(x + c)$ where $x$ is measured in radians. Write the program so that you can input the value of $c$. Run the program and graph the following:

**a.** $y = \sin(x + 1.57)$
**b.** $y = \sin(x - 3.14)$
**c.** $y = \sin(x - 5.6231)$
**d.** $y = \sin(x + 0.6734)$

## KEY TERMS AND CONCEPTS

Be sure you understand what is meant by these terms and can give an example of each.

Arc
Unit circle
Standard position of an arc
Coterminal arcs
Reciprocal trigonometric functions
Reference arc
Trigonometric function of a real number
Graph of a trigonometric function
Periodic function
Period of a function
Amplitude of a periodic function
Phase shift of a periodic function
Asymptote of the graph of a trigonometric function
Composite trigonometric function
Inverse trigonometric function
Principal values of a trigonometric function

## SUMMARY OF PROCEDURES AND CONCEPTS

### Finding Arcs Coterminal with a Given Arc

If $u$ is the length of an arc and it is in standard position on a unit circle, then $u \pm 2n\pi$ where $n$ is any integer is coterminal with $u$.

### Graphing Sine and Cosine Functions

1. For functions of the form $y = a \sin(bx + c)$ and $y = a \cos(bx + c)$:
   **a.** The amplitude is $|a|$.
   **b.** The period is $\frac{2\pi}{|b|}$.
   **c.** The phase shift is $-\frac{c}{|b|}$. If $c > 0$, the shift is to the left. If $c < 0$, the shift is to the right.
2. One period of any sine or cosine function can be graphed by making appropriate adjustments to the basic sine and cosine curves using the information in part 1 above.

### Graphing the Tangent and Cotangent Functions

1. The curves $y = \tan x$ and $y = \cot x$ have a period of $\pi$ and have no maximum or minimum value.
2. The graph of $y = \tan x$ is asymptotic to the lines $x = \frac{\pi}{2} + \pi n$ where $n$ is any integer.
3. The graph of $y = \cot x$ is asymptotic to the lines $x = \pi n$ where $n$ is any integer.
4. Graphs of the form $y = a \tan (bx + c)$ and $y = a \cot (bx + c)$ can be made by making appropriate adjustments for the values $a$, $b$, $c$ to the graphs in 2 or 3 above. For the two curves tangent and cotangent:

$$\text{function value for a quarter of a period} = a \text{ or } -a$$

$$\text{the length of one period} = \frac{\pi}{|b|}$$

$$\text{the right or left displacement} = -\frac{c}{|b|}$$

### Graphing the Secant and Cosecant Functions

1. The graphs of $y = \sec x$ and $y = \csc x$ each have a period of $2\pi$.
2. The secant function
   a. The graph of $y = \sec x$ has a minimum positive value of $y = 1$ and a maximum negative value of $y = -1$.
   b. The lines $x = \frac{\pi}{2} + \pi n$ where $n$ is any integer are asymptotes to the graph of $y = \sec x$.
3. The cosecant function
   a. The graph of $y = \csc x$ has a minimum positive value of $y = 1$ and a maximum negative value of $y = -1$.
   b. The lines $x = \pi n$ where $n$ is any integer are asymptotes to the graph of $y = \csc x$.
4. The graphs of $y = a \sec (bx + c)$ and $y = a \csc (bx + c)$ can be made by making appropriate adjustments to the curves of $y = \sec x$ and $y = \csc x$.

## CHAPTER 7 REVIEW EXERCISE

In Problems 1–4, an arc length is given. Find the coordinates $(x, y)$ for the endpoint of the arc in standard position on the unit circle. (Do not use a calculator or Table E.)

**1.** $\frac{\pi}{4}$ **2.** $\frac{3}{2}\pi$ **3.** $3\pi$ **4.** $\frac{11}{6}\pi$

In Problems 5–9, find the exact value. (Do not use a calculator or Table E.)

**5.** Find $\cos u$ if $\csc u = \frac{13}{5}$ and $\tan u < 0$.

**6.** Find $\sec u$ if $\tan u = -\frac{3}{4}$ and $\sin u < 0$.

**7.** Find $\cot u$ if $\sin u = \frac{2}{3}$ and $\tan u < 0$.

**8.** Find the coordinates of the endpoint of $u$ if $u$ is in quadrant IV and $\sec u = \frac{13}{5}$.

**9.** Find the exact length of an arc $u$ on the unit circle if $\sin u = \frac{1}{2}$ and $\tan u < 0$.

Find the expressions in Problems 10–13. (Use Table E or a calculator.) Round your answer to 4 decimal places.

**10.** $\sin 1.253$ **11.** $\tan 0.816$ **12.** $\cot 0.730$ **13.** $\cos 0.560$

Find $u$ in Problems 14–17. (Use Table E or a calculator.) Assume that $0 \le u \le \frac{\pi}{2}$. Round your answer to 3 decimal places.

**14.** $\cos u = 0.3333$ **15.** $\csc u = 4.5018$ **16.** $\sec u = 1.4658$ **17.** $\sin u = 0.9608$

In Problems 18–24, find the value of the given function. (Use $\pi = 3.14$ as an approximation if using Table E. If using a calculator, use calculator values, but round your answer to 4 decimal places.)

**18.** $\cot(-5)$ **19.** $\csc(-0.92)$ **20.** $\tan(-2.35)$ **21.** $\cos(9.42)$

**22.** $\sin\left(\frac{\pi}{20}\right)$ **23.** $\cot\left(-\frac{\pi}{20}\right)$ **24.** $\csc\left(\frac{3\pi}{20}\right)$

In Problems 25–30, find $u$ where $0 \le u < 2\pi$. (Use Table E or a calculator.) Round your answer to 3 decimal places.

**25.** $\sec u = 1.1730$, $\tan u < 0$

**26.** $\cot u = 0.8100$, $\sin u < 0$

**27.** $\cos u = -0.7317$, $\csc u > 0$

**28.** $\sin u = -0.9168$, $\tan u < 0$

**29.** $\csc u = 1.3083$, $\cot u < 0$

**30.** $\tan u = -2.2958$, $\sin u > 0$

In Problems 31–36, sketch one period of the indicated function.

**31.** $y = \frac{1}{3}\sin x$ **32.** $y = 4\cos x$ **33.** $y = \sin\left(x - \frac{\pi}{3}\right)$

**34.** $y = \tan 3x$ **35.** $y = 3\sin\left(x + \frac{3\pi}{2}\right)$ **36.** $y = \frac{1}{3}\cos\left(x - \frac{\pi}{8}\right)$

**37.** Sketch the graph of $y = \frac{1}{2}x + \cos x$ from 0 to $2\pi$.

**38.** State a function in terms of the cosine that has the same graph as $y = \sin(x - 3\pi)$.

**39.** Sketch the graph of $y = -\cos\left(x + \dfrac{\pi}{2}\right)$ from 0 to $2\pi$.

In Problems 40–43, sketch two periods of the indicated function.

**40.** $y = -\sin 2x$ **41.** $y = \sin(x + 1)$ **42.** $y = \dfrac{1}{2}\sin(4x - 2\pi)$

**43.** $y = \dfrac{1}{2}\cos\left(\dfrac{2}{3}x - \pi\right)$ **44.** Sketch $y = x - 2\sin x$ from 0 to $2\pi$.

**45.** Sketch two periods of $y = \cos 2x - 2\sin x$.

In Problems 46–53, find the value of each expression.

**46.** $\text{Arccos}\,\dfrac{\sqrt{2}}{2}$ **47.** $\text{Tan}^{-1}(-1)$ **48.** $\text{Sin}^{-1}\left(-\dfrac{1}{2}\right)$

**49.** $\cos\left(\text{Arccsc}\,\dfrac{13}{5}\right)$ **50.** $\sin\left[\text{Arccos}\left(-\dfrac{5}{13}\right)\right]$ **51.** $\tan\left(\text{Cos}^{-1}\dfrac{\sqrt{2}}{3}\right)$

**52.** $\sec\left(\text{Sin}^{-1}\dfrac{3}{5}\right)$ **53.** $\cos\left(\text{Arcsin}\,\dfrac{7}{25}\right)$

## PRACTICE TEST FOR CHAPTER 7

In Problems 1–5, do not use a calculator or Table E.

**1.** $\dfrac{19}{6}\pi$ is an arc in standard position on the unit circle. Find the coordinates $(x, y)$ of its endpoint.

**2.** Give the smallest positive real number that is coterminal with 14.19. (Use $\pi = 3.14$.)

**3.** Find $\cos u$ if $\cot u = -3$ and $\sin u > 0$.

**4.** Find the coordinates of the endpoint of $u$ if $u$ is in quadrant IV and $\cos u = \dfrac{1}{3}$.

**5.** Find the length of an arc $u$ on the unit circle if $\cos u = \dfrac{1}{2}$ and $\tan u < 0$.

In Problems 6–11, use a calculator or Table E as needed.

**6.** Find $u$ if $\tan u = 4.1990$. $(0 \le u < 1.57)$ **7.** Find csc 8.21. **8.** Find $\tan(-5.14)$.

**9.** Find $u$ $(0 \le u < 2\pi)$ when $\sin u = -0.3146$ and $\tan u > 0$.

**10.** Find a positive value for $u$ less than $2\pi$ when $\sec u = -16.9007$ and $\cot u < 0$.

**11.** Find $\cos\left(\dfrac{\pi}{15}\right)$.

**12.** Sketch the graph of $y = 2\cos\left(2x + \dfrac{\pi}{2}\right)$. **13.** State the period of $y = \dfrac{1}{2}\cos(3x + \pi)$.

**14.** Sketch one period of the graph of $y = 2 \sin (3x - \pi)$.

**15.** Sketch the graph of $y = \frac{1}{2}x - \cos x$ from 0 to $2\pi$.

Find the exact value of each expression in Problems 16–18. (Do not use a calculator or Table E.)

**16.** $\csc\left(\text{Arccos}\,\frac{12}{13}\right)$ **17.** $\text{Arccsc}\,(-2)$ **18.** $\text{Arcsin}\left[\cos\left(-\frac{5\pi}{6}\right)\right]$

# 8 Trigonometric Relationships and Applications

*In mathematics, I can report no deficiency, except it may be that men do not sufficiently understand the excellent use of the Pure Mathematics.*

*Francis Bacon*

The quote from Francis Bacon emphasizes the importance of theoretical mathematics. It is often important for a mathematician to study properties and relationships of mathematical expressions without actually considering any numerical values. The topics in this chapter are appropriate for this type of study.

## 8-1 BASIC TRIGONOMETRIC IDENTITIES

It is evident from the definitions themselves that trigonometric functions are interrelated. Other relationships are a result of these and are very important to the study and use of trigonometry. In this chapter we will develop relationships, some of which will result in formulas that will be used throughout the remainder of the text.

There are two types of equations—those that are **conditional** and those that are **identical**. An algebraic example of a conditional equation is $x + 2 = 5$, which is true only for $x = 3$. An identical equation, or an identity, is $2(x + 3) = 2x + 6$, which is true for every value of the variable $x$. The equation $\sin\theta = \cos\theta$ is true for $\theta = 45°$, but certainly not true for all angles $\theta$, and is therefore a conditional equation. We are presently interested in trigonometric equations that are identities, that is, statements that are true for all permissible values of the variable.

There are eight basic identities that are evident from the definitions and the Pythagorean Theorem. We have noted some of these in previous chapters. They are listed here for easy reference and should be memorized. We will use the symbol $\equiv$ to distinguish an identity from a conditional equation. It should be understood that an identity can only hold for permissible values of $\theta$.

Trigonometric expressions are often squared or raised to a higher power. A special notation is used by mathematicians to express this. We write $(\sin\theta)^2$ as $\sin^2\theta$. We write $(\cos\theta)^3$ as $\cos^3\theta$. This abbreviated notation will be used throughout this chapter.

**Reciprocal Identities**

$$\sin\theta\csc\theta \equiv 1 \tag{8-1}$$

$$\cos\theta\sec\theta \equiv 1 \tag{8-2}$$

$$\tan\theta\cot\theta \equiv 1 \tag{8-3}$$

**Ratio Identities**

$$\tan\theta \equiv \frac{\sin\theta}{\cos\theta} \tag{8-4}$$

$$\cot\theta \equiv \frac{\cos\theta}{\sin\theta} \tag{8-5}$$

**Pythagorean Identities**

$$\sin^2\theta + \cos^2\theta \equiv 1 \tag{8-6}$$

$$1 + \tan^2\theta \equiv \sec^2\theta \tag{8-7}$$

$$1 + \cot^2\theta \equiv \csc^2\theta \tag{8-8}$$

These basic identities can occur in more than one form; you should learn to recognize them in each of their various forms. Study the three columns below for some of the different ways in which each identity can occur.

| | | | |
|---|---|---|---|
| $\sin\theta\csc\theta \equiv 1$ | $\sin\theta \equiv \frac{1}{\csc\theta}$ | $\csc\theta \equiv \frac{1}{\sin\theta}$ | **(8-1)** |
| $\cos\theta\sec\theta \equiv 1$ | $\cos\theta \equiv \frac{1}{\sec\theta}$ | $\sec\theta \equiv \frac{1}{\cos\theta}$ | **(8-2)** |
| $\tan\theta\cot\theta \equiv 1$ | $\tan\theta \equiv \frac{1}{\cot\theta}$ | $\cot\theta \equiv \frac{1}{\tan\theta}$ | **(8-3)** |
| $\tan\theta \equiv \frac{\sin\theta}{\cos\theta}$ | $\cos\theta\tan\theta \equiv \sin\theta$ | $\cos\theta \equiv \frac{\sin\theta}{\tan\theta}$ | **(8-4)** |
| $\cot\theta \equiv \frac{\cos\theta}{\sin\theta}$ | $\sin\theta\cot\theta \equiv \cos\theta$ | $\sin\theta \equiv \frac{\cos\theta}{\cot\theta}$ | **(8-5)** |
| $\sin^2\theta + \cos^2\theta \equiv 1$ | $\sin^2\theta \equiv 1 - \cos^2\theta$ | $\cos^2\theta \equiv 1 - \sin^2\theta$ | **(8-6)** |
| $1 + \tan^2\theta \equiv \sec^2\theta$ | $\sec^2\theta - \tan^2\theta \equiv 1$ | $\tan^2\theta \equiv \sec^2\theta - 1$ | **(8-7)** |
| $1 + \cot^2\theta \equiv \csc^2\theta$ | $\csc^2\theta - \cot^2\theta \equiv 1$ | $\cot^2\theta \equiv \csc^2\theta - 1$ | **(8-8)** |

To verify an identity, we must reduce the right-hand member to the left-hand member, or reduce the left-hand member to the right-hand member, or reduce each member to the same form.

**Caution:** We cannot perform an algebraic operation on both members, such as multiplying both sides by a quantity. Performing such an operation would be assuming the identity to be true. Our task is to *prove* it true.

There is no set method of verifying an identity—no right or wrong way, no best method. We must simply use the basic identities and the algebraic manipulations to accomplish our task. One basic rule that is valuable is that it is generally more desirable to change a complicated form to a simpler form than to do the reverse.

While there are no absolute rules, here are some suggestions that will provide us with a possible starting point.

---

1. Perform an indicated operation.
2. Use the list of memorized formulas to replace an expression by the other half of the identity.
3. Change all trigonometric functions into sines and cosines.
4. Factor one side.
5. If there are two or more fractions on one side, find the LCD and then combine the fractions.

---

EXAMPLE 1 Verify $\tan \theta \equiv \dfrac{\sec \theta}{\csc \theta}$.

***Solution*** Here the right-hand side is the more complicated, so we will attempt to reduce $\dfrac{\sec \theta}{\csc \theta}$ to $\tan \theta$ from the basic identities.

$$\frac{\sec \theta}{\csc \theta} \equiv \frac{\dfrac{1}{\cos \theta}}{\dfrac{1}{\sin \theta}}$$ *Formulas (8-1) and (8-2).*

$$\frac{\sec \theta}{\csc \theta} \equiv \frac{\sin \theta}{\cos \theta}$$ *Definition of division of two fractions.*

$$\frac{\sec \theta}{\csc \theta} \equiv \tan \theta$$ *Formula (8-4).* ■

**EXAMPLE 2** Verify $\sec^2 \theta + \csc^2 \theta \equiv \sec^2 \theta \csc^2 \theta$.

***Solution*** We start by working with the left-hand side.

$$\sec^2 \theta + \csc^2 \theta \equiv \frac{1}{\cos^2 \theta} + \frac{1}{\sin^2 \theta}$$ *Formulas (8-1) and (8-2).*

$$\sec^2 \theta + \csc^2 \theta \equiv \frac{\sin^2 \theta + \cos^2 \theta}{\cos^2 \theta \sin^2 \theta}$$ *Addition of fractions.*

$$\sec^2 \theta + \csc^2 \theta \equiv \frac{1}{\cos^2 \theta \sin^2 \theta}$$ *Why?*

$$\sec^2 \theta + \csc^2 \theta \equiv \sec^2 \theta \csc^2 \theta$$ *Formulas (8-1) and (8-2).*

Note that we converted to sine and cosine and then back to secant and cosecant. ■

**EXAMPLE 3** Verify $(\tan u + \cot u)^2 \equiv \sec^2 u + \csc^2 u$.

***Solution*** If one side of an identity contains parentheses and the other does not, it is usually helpful to remove the parentheses.

$$(\tan u + \cot u)^2 \equiv \tan^2 u + 2 \tan u \cot u + \cot^2 u$$ *Square the binomial.*

$$(\tan u + \cot u)^2 \equiv \tan^2 u + 2 + \cot^2 u$$ *Formula (8-3).*

$$(\tan u + \cot u)^2 \equiv \tan^2 u + 1 + 1 + \cot^2 u$$ $1 + 1 = 2$.

$$(\tan u + \cot u)^2 \equiv \sec^2 u + \csc^2 u$$ *Formulas (8-7) and (8-8).* ■

In the preceding example, the only "trick" is $1 + 1 = 2$. Looking ahead to what is needed is usually the key to verifying an identity.

**EXAMPLE 4** Verify $\dfrac{\sin \theta}{1 - \cos \theta} \equiv \dfrac{1 + \cos \theta}{\sin \theta}$.

***Solution*** The first question here is "Which side is more complicated?" Since the left-hand member has two terms in the denominator, we will work with this side and try to simplify the denominator.

$$\frac{\sin \theta}{1 - \cos \theta} \equiv \frac{\sin \theta\,(1 + \cos \theta)}{(1 - \cos \theta)(1 + \cos \theta)}$$ *Multiply the numerator and denominator by the expression* $(1 + \cos \theta)$.

$$\frac{\sin \theta}{1 - \cos \theta} \equiv \frac{\sin \theta\,(1 + \cos \theta)}{1 - \cos^2 \theta}$$ *Simplify denominator.*

$$\frac{\sin\theta}{1-\cos\theta} \equiv \frac{\sin\theta\,(1+\cos\theta)}{\sin^2\theta}$$ *Formula (8-6).*

$$\frac{\sin\theta}{1-\cos\theta} \equiv \frac{1+\cos\theta}{\sin\theta}$$ *Fundamental principle of fractions.* ■

Of course, some identities are more difficult to prove than others because they require more substitutions or more algebraic manipulations. However, if an identity is true, one side can be reduced to the other.

Remember that each of the preceding examples represents only one way of verifying the identity. The many different ways by which an identity can be proved provide the student with an opportunity to develop his or her ingenuity and originality.

Certain identities are more easily verified if all expressions are written in terms of $\sin\theta$ and $\cos\theta$.

**EXAMPLE 5** Verify $\dfrac{\tan\theta\cos\theta-\sin^2\theta}{1-\sin\theta} \equiv \sin\theta$.

***Solution*** We will start with the most complicated expression and replace the $\tan\theta$ expression.

$$\frac{\tan\theta\cos\theta-\sin^2\theta}{1-\sin\theta} \equiv \frac{\left(\dfrac{\sin\theta}{\cos\theta}\right)\cos\theta-\sin^2\theta}{1-\sin\theta}$$ *Formula (8-4).*

$$\frac{\tan\theta\cos\theta-\sin^2\theta}{1-\sin\theta} \equiv \frac{\sin\theta-\sin^2\theta}{1-\sin\theta}$$ *Basic principle of fractions.*

$$\frac{\tan\theta\cos\theta-\sin^2\theta}{1-\sin\theta} \equiv \frac{\sin\theta\,(1-\sin\theta)}{1-\sin\theta}$$ *Factoring.*

$$\frac{\tan\theta\cos\theta-\sin^2\theta}{1-\sin\theta} \equiv \sin\theta$$ *Basic principle of fractions.*

■

**EXAMPLE 6** Verify $\dfrac{\sec^4\theta-1}{\tan^2\theta} \equiv 2+\tan^2\theta$.

***Solution*** Sometimes factoring a more complicated expression will assist us in the process of simplification.

$$\frac{\sec^4 \theta - 1}{\tan^2 \theta} \equiv \frac{(\sec^2 \theta + 1)(\sec^2 \theta - 1)}{\tan^2 \theta}$$ *Factoring the numerator.*

$$\frac{\sec^4 \theta - 1}{\tan^2 \theta} \equiv \frac{(\sec^2 \theta + 1)(\tan^2 \theta)}{\tan^2 \theta}$$ *Formula (8-7).*

$$\frac{\sec^4 \theta - 1}{\tan^2 \theta} \equiv \sec^2 \theta + 1$$ *Basic principle of fractions.*

$$\frac{\sec^4 \theta - 1}{\tan^2 \theta} \equiv (1 + \tan^2 \theta) + 1$$ *Formula (8-7).*

$$\frac{\sec^4 \theta - 1}{\tan^2 \theta} \equiv 2 + \tan^2 \theta$$ *Simplification.* ■

Before doing the exercises it is best to remember that learning to verify identities is a matter of practice as well as "trial and error." If the first approach you take is leading to a progressively more complicated expression that does not resemble the desired result, it is probably best to abandon that method and try another approach. There are often many different ways to approach the same identity. Do not be concerned if you find you have done it a different way from the answer key or from what your instructor explained, as long as all the steps used in your method are correct.

## EXERCISE 8-1

Verify the identities in Problems 1–30.

1-19 odd

**1.** $\sin \theta \equiv \dfrac{\cos \theta}{\cot \theta}$

**2.** $\tan u + \cot u \equiv \sec u \csc u$

**3.** $\tan^2 x \equiv \dfrac{\sin x \tan x}{\cos x}$

**4.** $\cos^2 \theta - \sin^2 \theta \equiv 1 - 2 \sin^2 \theta$

**5.** $\dfrac{1 - \sin^2 \theta}{\sin \theta} \equiv \dfrac{\cos \theta}{\tan \theta}$

**6.** $\dfrac{1 - \cos \alpha}{\sin \alpha} \equiv \dfrac{\sin \alpha}{1 + \cos \alpha}$

**7.** $1 - 2 \sin^2 t \equiv 2 \cos^2 t - 1$

**8.** $\sin^2 u \sec^2 u \equiv \sec^2 u - 1$

**9.** $\sin \theta \cot \theta + \sin \theta \csc \theta \equiv \cos \theta + 1$

**10.** $\dfrac{\tan^2 x + 1}{\cot^2 x + 1} \equiv \tan^2 x$

**11.** $(\tan u + \sec u)^2 \equiv \dfrac{1 + \sin u}{1 - \sin u}$

**12.** $\cos \beta + \tan \beta \sin \beta \equiv \sec \beta$

**13.** $\dfrac{1}{\tan^2\theta} - \dfrac{1}{\sec^2\theta} \equiv \dfrac{\cos^4\theta}{\sin^2\theta}$

**14.** $\dfrac{\cos^2\theta}{\sin\theta} + \dfrac{1}{\csc\theta} \equiv \csc\theta$

**15.** $\dfrac{1+\sec\theta}{\tan\theta} - \dfrac{\tan\theta}{\sec\theta} \equiv \dfrac{1+\sec\theta}{\sec\theta\tan\theta}$

**16.** $\tan\theta - \sin^2\theta \equiv \dfrac{\sin\theta + \cos^3\theta - \cos\theta}{\cos\theta}$

**17.** $(\tan\theta + \cot\theta)(\cos\theta + \sin\theta) \equiv \csc\theta + \sec\theta$

**18.** $(\sec\theta - \tan\theta)(1 + \csc\theta) \equiv \cot\theta$

**19.** $\dfrac{\cot^2\theta - 1}{1 + \cot^2\theta} \equiv 1 - 2\sin^2\theta$

**20.** $\dfrac{1-\cos\theta}{1+\cos\theta} \equiv (\cot\theta - \csc\theta)^2$

**21.** $\dfrac{1-\cos\theta}{\csc\theta} \equiv \dfrac{\sin^3\theta}{1+\cos\theta}$

**22.** $\dfrac{\cot\theta\cos\theta}{1-\sin\theta} \equiv 1 + \csc\theta$

**23.** $\cot\theta + \tan\theta \equiv \cot\theta\sec^2\theta$

**24.** $\dfrac{\cos u}{1-\sin u} \equiv \sec u + \tan u$

**25.** $\cos^2 t - \sin^2 t \equiv \dfrac{1-\tan^2 t}{1+\tan^2 t}$

**26.** $\cos^4 t - \sin^4 t \equiv 2\cos^2 t - 1$

★ **27.** $\dfrac{\sin\alpha}{\csc\alpha - \cot\alpha} \equiv 1 + \cos\alpha$

★ **28.** $\dfrac{1}{\csc\beta - \cot\beta} - \dfrac{1}{\csc\beta + \cot\beta} \equiv \dfrac{2}{\tan\beta}$

★ **29.** $\sec u + \cos u + \sin^2 u \sec u \equiv \dfrac{2\sin u}{\cos^2 u \tan u}$

★ **30.** $\sec\theta - \tan\theta \equiv \pm\sqrt{\dfrac{1-\sin\theta}{1+\sin\theta}}$

## 8-2 TRIGONOMETRIC FUNCTIONS OF THE SUM AND THE DIFFERENCE OF TWO ARGUMENTS

The trigonometric identities discussed in the previous section involved various trigonometric functions. However, each trigonometric function had an argument of only one real number or one angle. Sometimes we will require a trigonometric function for the sum or difference of two real numbers or two angles. In these next two sections we will establish the identities of the sum and difference of two arguments. (The arguments may be expressed as real numbers or as angles.)

### The Cosine of the Sum and Difference of Two Arguments

First, we will prove a formula for cosine of the difference of two angles. This proof will require some geometry. The proof depends on the following three basic concepts:

1. In the same circle or equal circles, equal central angles subtend equal arcs.
2. In the same circle or equal circles, equal arcs subtend equal chords.

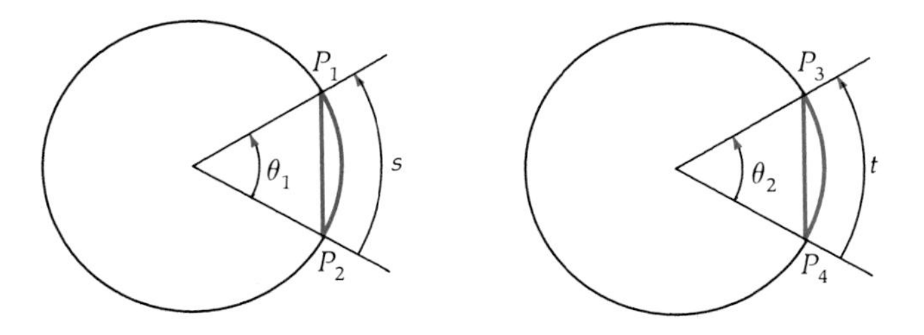


In other words, in the above two equal circles if $\theta_1 = \theta_2$, then arc length $s$ = arc length $t$; and if arc length $s$ = arc length $t$, then

$$\overline{P_1P_2} = \overline{P_3P_4}$$

**3.** The distance from $P_1$ $(x_1, y_1)$ to $P_2$ $(x_2, y_2)$, on the coordinate plane is given by the formula

$$\overline{P_1P_2} = \sqrt{(x_1 - x_2)^2 + (y_1 - y_2)^2}$$

We will now proceed to prove the formula for the cosine of the difference of two angles.

*Prove*

$$\cos(A - B) \equiv \cos A \cos B + \sin A \sin B. \qquad \textbf{(8-9)}$$

***Proof*** Consider the unit circle in which $A$ and $B$ are any two angles in standard position.

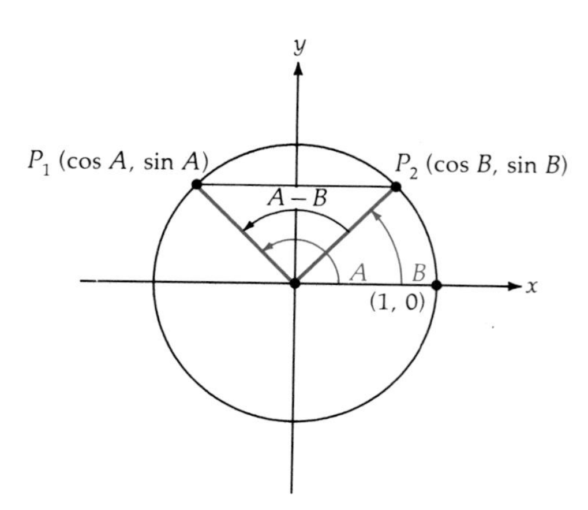


The proof does not depend on the size of the angles. $P_1$ on the terminal side of angle $A$ has coordinates $(\cos A, \sin A)$; $P_2$ on the terminal side of angle $B$

has coordinates (cos $B$, sin $B$). From the distance formula we have

$$\overline{P_1P_2} = \sqrt{(\cos B - \cos A)^2 + (\sin B - \sin A)^2}$$

$$(\overline{P_1P_2})^2 = (\cos B - \cos A)^2 + (\sin B - \sin A)^2$$

$$(\overline{P_1P_2})^2 = \cos^2 B - 2 \cos A \cos B + \cos^2 A + \sin^2 B - 2 \sin A \sin B + \sin^2 A$$

$$(\overline{P_1P_2})^2 = 2 - 2 \cos A \cos B - 2 \sin A \sin B \qquad \textbf{(Equation A)}$$

Set this information aside for the moment and turn once more to the diagram. Now place angle $(A - B)$ in standard position on the unit circle. Now reflect on the properties of geometry we discussed at the beginning of this section. Do you see that $P_2$ will now have coordinates (1, 0), and $P_1$ will have coordinates [cos $(A - B)$, sin $(A - B)$]?

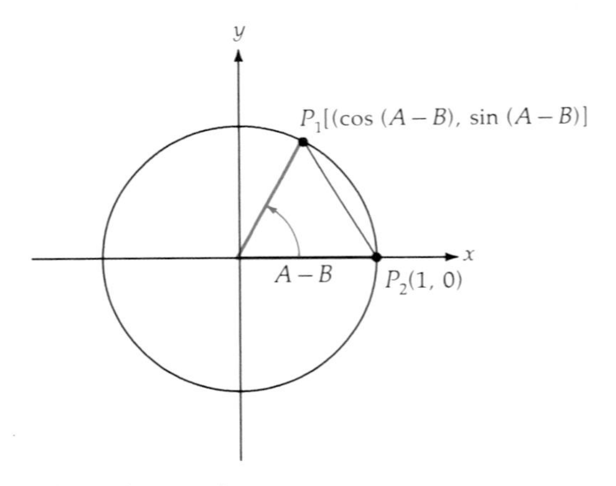


With the distance formula we obtain

$$\overline{P_1P_2} = \sqrt{[1 - \cos (A - B)]^2 + [0 - \sin (A - B)]^2}$$

$$(\overline{P_1P_2})^2 = [1 - \cos (A - B)]^2 + [0 - \sin (A - B)]^2$$

$$(\overline{P_1P_2})^2 = 1 - 2 \cos (A - B) + \cos^2 (A - B) + \sin^2 (A - B)$$

$$(\overline{P_1P_2})^2 = 2 - 2 \cos (A - B) \qquad \textbf{(Equation B)}$$

We now use equation A and equation B together to equate the two expressions derived for $(\overline{P_1P_2})^2$ to obtain

$$2 - 2 \cos (A - B) \equiv 2 - 2 \cos A \cos B - 2 \sin A \sin B$$

or

$$\cos (A - B) \equiv \cos A \cos B + \sin A \sin B \qquad \textbf{(8-9)}$$

This identity is known as the "cosine of the difference of two arguments," and is one of the most basic formulas for composite arguments. It should be *memorized.* ■

We can now use algebraic methods to prove a similar formula for the cosine of the sum of two angles. The key element of the proof is the property that for all $x$ it is true that $x = -(-x)$.

*Prove*

$$\cos (A + B) \equiv \cos A \cos B - \sin A \sin B. \qquad \textbf{(8-10)}$$

***Proof*** Let $(A + B) \equiv [A - (-B)]$. Then

$$\cos (A + B) \equiv \cos [A - (-B)]$$
$$\equiv \cos A \cos (-B) + \sin A \sin (-B)$$

We have seen earlier that

$$\cos (-B) \equiv \cos B \qquad \text{and} \qquad \sin (-B) \equiv -\sin B$$

Thus we may write

$$\cos (A + B) \equiv \cos A \cos B - \sin A \sin B \quad ■ \qquad \textbf{(8-10)}$$

It is wise to memorize this formula as well.

We will now apply these formulas in a variety of situations.

EXAMPLE 1 Without the use of calculators or tables, find the exact value of $\cos 15°$. (*Hint:* Use the fact that $15° = 45° - 30°$.)

***Solution*** $\cos (A - B) = \cos A \cos B + \sin A \sin B$ *Formula (8-9).*

$$\cos 15° = \cos (45° - 30°) = \cos 45° \cos 30° + \sin 45° \sin 30°$$
$$= \left(\frac{\sqrt{2}}{2}\right)\left(\frac{\sqrt{3}}{2}\right) + \left(\frac{\sqrt{2}}{2}\right)\left(\frac{1}{2}\right) = \frac{\sqrt{6} + \sqrt{2}}{4} \quad ■$$

The formulas are helpful in cases where we do not directly know the measure of the angles $A$ or $B$.

EXAMPLE 2 If $\sin A = \frac{3}{5}$ ($A$ in quadrant II) and $\tan B = \frac{5}{12}$ ($B$ in quadrant I), find an exact value for $\cos (A + B)$.

***Solution*** Since $\tan B = \frac{y}{x}$, we have $y = 5$ and $x = 12$. By the Pythagorean Theorem we obtain $r = 13$. Also, $\sin A = \frac{y}{r}$, so $y = 3$, $r = 5$, and $x = -4$. A sketch is not necessary, but it may be helpful to ensure that you understand the steps to solve for these values.

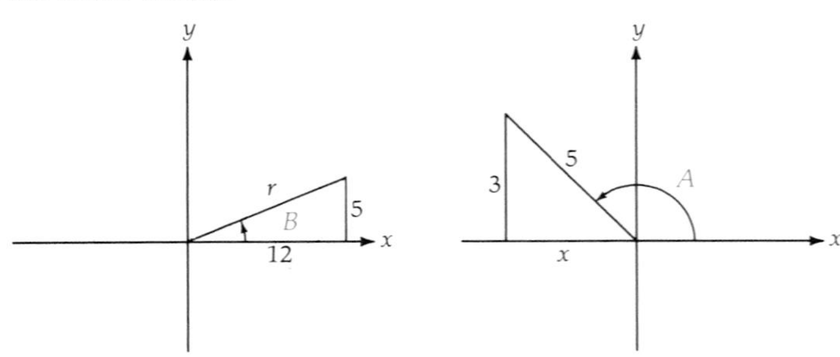


Now that we know $x, y, r$ in each case, we can evaluate $\cos A$, $\cos B$, $\sin A$, and $\sin B$ in the formula.

$$\cos (A + B) \equiv \cos A \cos B - \sin A \sin B$$ *Formula (8-10).*

we have

$$\cos (A + B) = \left(-\frac{4}{5}\right)\left(\frac{12}{13}\right) - \left(\frac{3}{5}\right)\left(\frac{5}{13}\right)$$

$$= -\frac{48}{65} - \frac{15}{65} = -\frac{63}{65} \quad ■$$

*Note that cos A is negative, since A is in quadrant II.*

In some cases a formula may be used to simplify a more involved trigonometric expression.

EXAMPLE 3 Prove that $\cos (180° + A) \equiv -\cos A$

***Solution***

$$\cos (180° + A) \equiv \cos 180° \cos A - \sin 180° \sin A$$ *Formula (8-10).*

$$\equiv (-1) \cos A - (0) \sin A$$

$$\equiv -\cos A \quad ■$$

### The Sine and Tangent of the Sum and Difference of Two Arguments

We now turn to the discussion of formulas that can be derived for the sine and tangent functions. We will state the formulas without proof.

$$\sin (A + B) \equiv \sin A \cos B + \cos A \sin B. \qquad \textbf{(8-11)}$$

$$\sin (A - B) \equiv \sin A \cos B - \cos A \sin B. \qquad \textbf{(8-12)}$$

We will now use Formulas (8-11) and (8-12) in a few different areas of application.

EXAMPLE 4 Find the exact value of $\sin 105^\circ$. (Use the fact that $105^\circ = 60^\circ + 45^\circ$.)

*Solution* Using $\sin(A + B) \equiv \sin A \cos B + \cos A \sin B$, we have

$$\sin 105^\circ = \sin(60^\circ + 45^\circ) = \sin 60^\circ \cos 45^\circ + \cos 60^\circ \sin 45^\circ$$
$$= \left(\frac{\sqrt{3}}{2}\right)\left(\frac{\sqrt{2}}{2}\right) + \left(\frac{1}{2}\right)\left(\frac{\sqrt{2}}{2}\right) = \frac{\sqrt{6} + \sqrt{2}}{4}$$

■

EXAMPLE 5 Derive a formula for $\sin\left(A - \frac{\pi}{6}\right)$ using formula (8-12).

*Solution* Using $\sin(A - B) \equiv \sin A \cos B - \cos A \sin B$, we have

$$\sin\left(A - \frac{\pi}{6}\right) = \sin A \cos\frac{\pi}{6} - \cos A \sin\frac{\pi}{6}$$
$$= (\sin A)\left(\frac{\sqrt{3}}{2}\right) - (\cos A)\left(\frac{1}{2}\right)$$
$$= \frac{\sqrt{3}}{2}\sin A - \frac{1}{2}\cos A \quad \text{or} \quad \frac{1}{2}(\sqrt{3}\sin A - \cos A)$$

■

### The Tangent of the Sum and Difference of Two Arguments

Next, we turn to the formulas for the tangent of the sum and difference of two angles. We will state the formulas without proof.

$$\tan(A + B) \equiv \frac{\tan A + \tan B}{1 - \tan A \tan B} \qquad \textbf{(8-13)}$$

$$\tan(A - B) \equiv \frac{\tan A - \tan B}{1 + \tan A \tan B} \qquad \textbf{(8-14)}$$

We will apply the two tangent formulas thus derived to some different problem situations.

EXAMPLE 6 If $\tan A = -\frac{3}{4}$ ($A$ in quadrant II) and $\sin B = \frac{5}{13}$ ($B$ in quadrant I), find an exact value for $\tan(A - B)$.

***Solution*** See if you can follow the following reasoning without resorting to a sketch.

Since $\tan A = -\frac{3}{4}$ in quadrant II, is a given fact, we do not need to find any additional information regarding angle $A$.

Since $\sin B = \frac{5}{13}$ in quadrant I, we have $y = 5$, $r = 13$. Thus $x = 12$. Then $\tan B = \frac{5}{12}$. Thus we have

$$\tan(A - B) \equiv \frac{\tan A - \tan B}{1 + \tan A \tan B} \qquad \textit{Formula (8-14).}$$

$$= \frac{\left(-\frac{3}{4}\right) - \left(\frac{5}{12}\right)}{1 + \left(-\frac{3}{4}\right)\left(\frac{5}{12}\right)}$$

$$= \frac{\frac{-9}{12} - \frac{5}{12}}{1 - \frac{5}{16}} = \frac{\frac{-14}{12}}{\frac{16}{16} - \frac{5}{16}} = \frac{\frac{-7}{6}}{\frac{11}{16}} = -\frac{56}{33}$$

■

**EXAMPLE 7** Derive the formula $\tan(A + \pi) \equiv \tan A$.

***Solution***

$$\tan(A + \pi) \equiv \frac{\tan A + \tan \pi}{1 - \tan A \tan \pi} \qquad \textit{Formula (8-13).}$$

$$\equiv \frac{\tan A + 0}{1 - (\tan A)(0)} \equiv \tan A$$

■

The student should take a few minutes to be sure that Formulas (8-9)–(8-14) are firmly committed to memory. The six formulas can be listed as three if you consider the sign differences as follows:

$$\cos(A \pm B) \equiv \cos A \cos B \mp \sin A \sin B$$

$$\sin(A \pm B) \equiv \sin A \cos B \pm \cos A \sin B$$

$$\tan(A \pm B) \equiv \frac{\tan A \pm \tan B}{1 \mp \tan A \tan B}$$

(When using the formulas in this form, if you use the top sign on the left-hand side of an equation, you must also use the top sign on the right-hand side.)

## EXERCISE 8-2

In Problems 1–6, find the exact value of the function by using the sum or the difference of two arguments.

**1.** $\cos 75^\circ$ **2.** $\cos 105^\circ$ **3.** $\cos \frac{3}{4}\pi$

**4.** $\cos 165^\circ$ **5.** $\cos \frac{13}{12}\pi$ **6.** $\cos(-15^\circ)$

**7.** Use the cosine of the difference of two angles to prove that $\cos(-\theta) \equiv \cos\theta$. (*Hint:* Use $A = 0^\circ$ and $B = \theta$.)

Use formulas (8-9) and (8-10) to prove the identities in Problems 8–12.

**8.** $\cos(90^\circ + \theta) \equiv -\sin\theta$ **9.** $\cos(360^\circ - \theta) \equiv \cos\theta$ **10.** $\cos(180^\circ - \theta) \equiv -\cos\theta$

**11.** $\cos(270^\circ + \theta) \equiv \sin\theta$ **12.** $\cos(270^\circ - \theta) \equiv -\sin\theta$

In Problems 13–16 find the exact value by using the sum or difference of two arguments.

**13.** If $\cos A = -\frac{5}{13}$ ($A$ in quadrant III) and $\sin B = \frac{4}{5}$ ($B$ in quadrant II), find:

**a.** $\cos(A - B)$ **b.** $\cos(A + B)$

**14.** If $\cos A = -\frac{3}{5}$ ($A$ in quadrant II) and $\tan B = \frac{5}{12}$ ($B$ in quadrant I), find:

**a.** $\cos(A - B)$ **b.** $\cos(A + B)$

**15.** If $\tan A = \frac{1}{2}$ ($A$ in quadrant III) and $\csc B = \frac{3}{2}$ ($B$ in quadrant II), find:

**a.** $\cos(A - B)$ **b.** $\cos(A + B)$

**16.** If $\cos A = \frac{12}{13}$ ($A$ is in quadrant IV) and $\sin B = \frac{3}{5}$ ($B$ is in quadrant I), find:

**a.** $\cos(A - B)$ **b.** $\cos(A + B)$

**17.** Reduce $\cos 3A \cos 2A - \sin 3A \sin 2A$ to a single term.

**18.** Reduce $\cos 6A \cos(-2A) + \sin 6A \sin(-2A)$ to a single term.

In Problems 19–37, find the exact value of the function by using the sum or the difference of two arguments. (Do not use a calculator or table.)

**19.** $\sin 15^\circ$ **20.** $\sin \frac{5}{12}\pi$ **21.** $\sin 75^\circ$ **22.** $\sin \frac{11}{12}\pi$

**23.** $\sin 255^\circ$ **24.** $\sin 345^\circ$ **25.** $\tan 15^\circ$ **26.** $\tan \frac{7}{12}\pi$

**27.** $\tan 105^\circ$ **28.** $\tan 165^\circ$ **29.** $\tan \frac{17}{12}\pi$ **30.** $\tan(-75^\circ)$

**31.** If $\cos A = \frac{12}{13}$ ($A$ in quadrant IV) and $\sin B = \frac{3}{5}$ ($B$ in quadrant II), find:

**a.** $\sin(A + B)$ **b.** $\sin(A - B)$

**32.** If $\csc A = -\frac{5}{4}$ ($A$ in quadrant III) and $\tan B = \frac{1}{4}$ ($B$ in quadrant III), find:

**a.** $\sin (A + B)$ **b.** $\sin (A - B)$

**33.** If $\tan A = -\frac{2}{3}$ ($A$ in quadrant IV) and $\cot B = -\frac{1}{2}$ ($B$ in quadrant II), find:

**a.** $\sin (A + B)$ **b.** $\sin (A - B)$

**34.** If $\cos A = \frac{1}{3}$ ($A$ in quadrant IV) and $\cos B = -\frac{1}{3}$ ($B$ in quadrant III), find:

**a.** $\sin (A + B)$ **b.** $\sin (A - B)$

**35.** If $\tan A = \frac{3}{4}$ ($A$ in quadrant III) and $\tan B = -\frac{12}{5}$ ($B$ in quadrant II), find:

**a.** $\tan (A + B)$ **b.** $\tan (A - B)$

**36.** If $\sin A = \frac{4}{5}$ ($A$ in quadrant II) and $\sec B = \frac{13}{12}$ ($B$ in quadrant I), find:

**a.** $\tan (A + B)$ **b.** $\tan (A - B)$

**37.** If $\cot A = -\frac{1}{2}$ ($A$ in quadrant IV) and $\sin B = -\frac{3}{4}$ ($B$ in quadrant III), find:

**a.** $\tan (A + B)$ **b.** $\tan (A - B)$

**38.** Reduce the expression $\sin 5x \cos 3x - \cos 5x \sin 3x$ to a single term.

**39.** Verify $\sin 2A \equiv 2 \sin A \cos A$.

**40.** Verify $\sin (90^\circ - B) \equiv \cos B$ using the sine of the difference of two arguments.

**41.** Verify that $\sin (270^\circ + A) \equiv -\cos A$.

**42.** Reduce the expression $\dfrac{\tan 3u - \tan u}{1 + \tan 3u \tan u}$ to a single expression.

**43.** Verify $\tan 2\theta \equiv \dfrac{2 \tan \theta}{1 - \tan^2 \theta}$.

**44.** Verify $\tan (90^\circ - \theta) \equiv \cot \theta$. $\left[\textit{Hint: } \text{Use } \tan (A - B) \equiv \dfrac{\sin (A - B)}{\cos (A - B)}.\right.$ Why can't you use $\left.\tan (A - B) \equiv \dfrac{\tan A - \tan B}{1 + \tan A \tan B}?\right]$

In Problems 45–50, simplify each expression as a function of $\theta$.

**45.** $\sin \left(\theta - \frac{\pi}{4}\right)$

**46.** $\sin \left(\theta + \frac{3\pi}{2}\right)$

**47.** $\tan (\pi - \theta)$

**48.** $\tan \left(\frac{\pi}{6} + \theta\right)$

★ **49.** $\tan \left(\theta + \frac{\pi}{4}\right) - \tan \left(\theta + \frac{\pi}{2}\right)$

★ **50.** $\sin \left(\theta + \frac{\pi}{6}\right) + \sin (\theta - \pi)$

## 8-3 TRIGONOMETRIC FUNCTIONS OF DOUBLE ARGUMENTS AND HALF-ARGUMENTS

### Double Arguments

Formulas for the functions of double arguments can be found directly from the formulas for the functions of the sum of two arguments. In the expression $(A + B)$ we replace $B$ with $A$ and obtain $(A + B) = (A + A) = 2A$. Therefore

$$\sin 2A \equiv \sin (A + A) \equiv \sin A \cos A + \cos A \sin A$$

or

$$\sin 2A \equiv 2 \sin A \cos A \tag{8-15}$$

This formula for the sine of a double argument should be memorized.

The formula for the cosine of a double argument can be expressed in three different forms. Each form is useful in certain situations, so we shall develop all three forms.

$$\cos 2A \equiv \cos (A + A) \equiv \cos A \cos A - \sin A \sin A$$

or

$$\cos 2A \equiv \cos^2 A - \sin^2 A \tag{8-16}$$

Other forms of this formula may be obtained by using the identity

$$\sin^2 A + \cos^2 A \equiv 1$$

First

$$\begin{aligned} \cos 2A &\equiv \cos^2 A - \sin^2 A \\ &\equiv \cos^2 A - (1 - \cos^2 A) \end{aligned}$$

or

$$\cos 2A \equiv 2 \cos^2 A - 1 \tag{8-17}$$

Also

$$\begin{aligned} \cos 2A &\equiv \cos^2 A - \sin^2 A \\ &\equiv (1 - \sin^2 A) - \sin^2 A \end{aligned}$$

or

$$\cos 2A \equiv 1 - 2 \sin^2 A \tag{8-18}$$

The tangent of a double argument may also be found as follows.

$$\tan 2A \equiv \tan(A + A) \equiv \frac{\tan A + \tan A}{1 - \tan A \tan A}$$

or

$$\tan 2A \equiv \frac{2 \tan A}{1 - \tan^2 A} \tag{8-19}$$

The five formulas (8-15) to (8-19) should be thoroughly memorized. We shall now apply these formulas to a numerical situation.

EXAMPLE 1 When $\csc A = \frac{5}{4}$ ($A$ in quadrant II):

**a.** Find $\sin 2A$. **b.** Find $\cos 2A$. **c.** Find $\tan 2A$.

***Solution*** **a.** Since $\csc A = \frac{5}{4}$, we have $r = 5$, $y = 4$, and by the Pythagorean Theorem and the fact that $A$ is in quadrant II, we obtain $x = -3$. This enables us to state

$$\sin A = \frac{4}{5} \qquad \cos A = -\frac{3}{5} \qquad \text{and} \qquad \tan A = -\frac{4}{3}$$

Thus

$$\sin 2A \equiv 2 \sin A \cos A$$

$$\sin 2A = 2\left(\frac{4}{5}\right)\left(-\frac{3}{5}\right) = -\frac{24}{25}$$

**b.** We may use Formula (8-16), (8-17), or (8-18). The choice here is somewhat arbitrary. This time, let us use Formula (8-18).

$$\cos 2A \equiv 1 - 2 \sin^2 A$$

$$\cos 2A = 1 - 2\left(\frac{4}{5}\right)^2 = 1 - \frac{32}{25} = -\frac{7}{25}$$

**c.**

$$\tan 2A \equiv \frac{2 \tan A}{1 - \tan^2 A} = \frac{2\left(-\frac{4}{3}\right)}{1 - \left(-\frac{4}{3}\right)^2} = \frac{24}{7}$$

*Warning:* It is easy to make a careless error when evaluating fractions such as this. You should always obtain a single fraction in the numerator and denominator before dividing. Take a minute to work the problem on your own and verify the above solution to part c. Did you obtain the same value? ■

EXAMPLE 2 Verify $\dfrac{\sin 2x}{1 + \cos 2x} \equiv \tan x$.

***Solution***

$$\frac{\sin 2x}{1 + \cos 2x} \equiv \frac{2 \sin x \cos x}{1 + 2 \cos^2 x - 1} \equiv \frac{2 \sin x \cos x}{2 \cos^2 x} \equiv \frac{\sin x}{\cos x} \equiv \tan x \qquad ■$$

In addition to Formulas (8-15)–(8-19), other formulas for multiple angles such as $3A$, $4A$, etc., can be derived. We shall develop one formula here. Other problems of this type are left as exercises.

In addition to the five suggestions we made in Section 8-1 as to beginning points in the verification of trigonometric identities, we provide one other basic suggestion.

---

If multiple angles or half-angles occur, use appropriate formulas to transform the expressions to equivalent ones involving only single angles.

---

EXAMPLE 3 Verify that $\sin 3\theta \equiv 3 \sin \theta \cos^2 \theta - \sin^3 \theta$

*Solution* We write $3\theta \equiv \theta + 2\theta$. Thus

$$\begin{aligned} \sin 3\theta &\equiv \sin(\theta + 2\theta) \equiv \sin\theta \cos 2\theta + \cos\theta \sin 2\theta \\ &\equiv \sin\theta\,[\cos^2\theta - \sin^2\theta] + \cos\theta\,[2\sin\theta\cos\theta] \\ &\equiv \sin\theta\cos^2\theta - \sin^3\theta + 2\sin\theta\cos^2\theta \\ &\equiv 3\sin\theta\cos^2\theta - \sin^3\theta \end{aligned}$$

*Using formulas (8-16) and (8-15).* ■

### Trigonometric Functions of Half-Arguments

Early in the history of trigonometry, formulas were developed to find a trigonometric function of half of a given angle. These were used repeatedly to construct the earliest trigonometric table of values.

Formulas for the functions of half-arguments may be obtained directly from the formulas for the functions of double arguments.

From the identity $\cos 2\theta \equiv 2\cos^2\theta - 1$, we solve for $\cos\theta$ and obtain

$$\cos\theta \equiv \pm\sqrt{\frac{1 + \cos 2\theta}{2}}$$

Now we let $2\theta = A$ and obtain

$$\cos\frac{A}{2} \equiv \pm\sqrt{\frac{1 + \cos A}{2}} \qquad \textbf{(8-20)}$$

The sign of this function is determined by the quadrant of $\frac{A}{2}$.

In the same manner we can use $\cos 2\theta \equiv 1 - 2\sin^2\theta$ and solve for $\sin\theta$ to obtain

$$\sin\theta \equiv \pm\sqrt{\frac{1 - \cos 2\theta}{2}}$$

If we let $2\theta = A$, we obtain

$$\sin\frac{A}{2} \equiv \pm\sqrt{\frac{1 - \cos A}{2}} \qquad \textbf{(8-21)}$$

The sign is determined by the quadrant of $\frac{A}{2}$.

We will apply these formulas to a couple of different situations.

EXAMPLE 4 Find $\cos\frac{5\pi}{8}$ using a half-angle formula.

***Solution*** Since $\frac{5\pi}{8}$ is greater than $\frac{\pi}{2}$ but less than $\pi$, we know that the angle is in the second quadrant. The cosine of a second quadrant angle is negative. Thus we will use the negative sign with formula (8-20):

$$\cos\frac{A}{2} = -\sqrt{\frac{1+\cos A}{2}}$$

Note that $A = \frac{5\pi}{4}$ and $\frac{A}{2} = \frac{5\pi}{8}$ in this case.

$$\cos\frac{5\pi}{8} = -\sqrt{\frac{1+\cos\left(\frac{5\pi}{4}\right)}{2}} = -\sqrt{\frac{1+\left(\frac{-\sqrt{2}}{2}\right)}{2}} = -\sqrt{\frac{\frac{2}{2}-\frac{\sqrt{2}}{2}}{2}}$$

$$= -\sqrt{\frac{2-\sqrt{2}}{4}} = -\frac{\sqrt{2-\sqrt{2}}}{2}$$

It is left to the student to verify that the formula is correct by evaluation of $\cos\frac{5\pi}{8}$ directly on the calculator and then evaluate the expressions $\frac{-\sqrt{2-\sqrt{2}}}{2}$ on the calculator. Your results should agree completely to at least seven decimal places. ■

We will now list several forms of the formula of the tangent of a half-angle.

$$\tan\frac{A}{2} \equiv \pm\sqrt{\frac{1-\cos A}{1+\cos A}} \tag{8-22}$$

The sign is determined by the quadrant of $\frac{A}{2}$ in Formula (8-22).

It is possible to obtain more convenient forms for the tangent of a half-argument. These forms do not involve a radical in the formula.

$$\tan\frac{A}{2} \equiv \frac{\sin A}{1+\cos A} \tag{8-23}$$

$$\tan\frac{A}{2} \equiv \frac{1-\cos A}{\sin A} \tag{8-24}$$

In most cases where it is necessary to find the tangent of a half-angle it is best to employ Formula (8-23) or (8-24).

We now will apply the tangent of a half-angle formula to various problems.

**EXAMPLE 5** Find $\tan\frac{A}{2}$ if $\sin A = \frac{3}{5}$ ($A$ in quadrant II). Do not use a calculator or tables.

***Solution*** Since $\sin A = \frac{3}{5}$ and $A$ is in quadrant II, then $y = 3$, $r = 5$, and $x = -4$. Therefore $\cos A = -\frac{4}{5}$.

$$\tan\frac{A}{2} \equiv \frac{\sin A}{1 + \cos A} = \frac{\frac{3}{5}}{1 - \frac{4}{5}} = \frac{\frac{3}{5}}{\frac{1}{5}} = 3$$ ■

Before doing the homework exercises, you should memorize formulas (8-15) to (8-24). In addition to being printed in this section of the text, they are also summarized on page 436.

## EXERCISE 8-3

In Problems 1–7, find the exact value using the appropriate formula. Do not use a calculator or a trigonometric table.

**1.** If $\sin A = \frac{4}{5}$ ($A$ in quadrant II), find:

**a.** $\sin 2A$ **b.** $\cos 2A$ **c.** $\tan 2A$

**2.** If $\sec\theta = -\frac{13}{12}$ ($\theta$ in quadrant III), find:

**a.** $\sin 2\theta$ **b.** $\cos 2\theta$ **c.** $\tan 2\theta$

**3.** If $\tan\theta = \frac{5}{12}$ and $\sin\theta < 0$, find:

**a.** $\sin 2\theta$ **b.** $\cos 2\theta$ **c.** $\tan 2\theta$

**4.** If $\sin\theta = \frac{8}{17}$ and $\cos\theta < 0$, find:

**a.** $\sin 2\theta$ **b.** $\cos 2\theta$ **c.** $\tan 2\theta$

**5.** Find $\cot 2\beta$ if $\cos\beta = \frac{3}{5}$ ($\beta$ in quadrant IV).

**6.** Find $\tan A$ if $\tan\frac{A}{2} = -\frac{3}{4}\left(\frac{A}{2}\text{ in quadrant II}\right)$.

**7.** Find $\sin A$ if $\cos\frac{A}{2} = \frac{5}{13}\left(\frac{A}{2}\text{ in quadrant IV}\right)$.

**8.** Verify the formula that $\sin 3\theta \equiv 3\sin\theta - 4\sin^3\theta$.

**9.** Express $\cos 3\theta$ in terms of $\cos\theta$.

**10.** Verify $\tan\theta \equiv \frac{1 - \cos 2\theta}{\sin 2\theta}$.

**11.** Verify $\sin 2x \equiv \frac{2\tan x}{1 + \tan^2 x}$.

**12.** Verify $\sec 2\theta \equiv \frac{1}{1 - 2\sin^2\theta}$.

**13.** Verify $\tan 2A \equiv \frac{2}{\cot A - \tan A}$.

**14.** Verify that $\sin 4A \equiv 4\cos A\sin A\,(1 - 2\sin^2 A)$.

**15.** Verify that $\cos 4A \equiv 8\cos^4 A - 8\cos^2 A + 1$.

**16.** Verify that $\tan 3A \equiv \frac{3\tan A - \tan^3 A}{1 - 3\tan^2 A}$.

**17.** Verify that $\cos 2A \equiv \dfrac{\dfrac{\cos^3 A}{\sin A} - \cos A \sin A}{\cot A}$.

**18.** Verify that $2 \csc 2A \equiv \sec A \csc A$.

Use the half-argument formulas to find the exact value of the expressions in Problems 19–34.

**19.** $\tan 15°$ **20.** $\sin 22\frac{1}{2}°$ **21.** $\cos 75°$ **22.** $\tan 75°$ **23.** $\sin \dfrac{7\pi}{8}$

**24.** $\cos \dfrac{5\pi}{12}$ **25.** $\cos 195°$ **26.** $\sin 112.5°$ **27.** $\sec 30°$ **28.** $\cot \dfrac{3}{8}\pi$

**29.** Find $\cos \dfrac{A}{2}$ if $\tan A = -\dfrac{3}{4}$ ($A$ in quadrant IV).

**30.** Find $\sin \dfrac{A}{2}$ if $\sec A = -\dfrac{13}{5}$ ($A$ in quadrant III).

**31.** Find $\tan \dfrac{\theta}{2}$ if $\sin \theta = \dfrac{5}{13}$ ($\theta$ in quadrant II).

**32.** Find $\csc \dfrac{\alpha}{2}$ if $\cot \alpha = \dfrac{3}{4}$ ($\alpha$ in quadrant III).

**33.** Find $\sin \dfrac{\theta}{2}$ if $\tan \theta = -\dfrac{8}{15}$ and $\csc \theta < 0$.

**34.** Find $\tan \dfrac{\theta}{2}$ if $\cos \theta = -\dfrac{12}{13}$ and $\sin \theta < 0$.

**35.** Derive the identity $\tan \dfrac{A}{2} \equiv \dfrac{1 - \cos A}{\sin A}$ by using the cosine of the difference of two arguments $\left[\text{i.e., } \cos \dfrac{A}{2} \equiv \cos\left(A - \dfrac{A}{2}\right)\right]$.

**36.** Verify $\left(\cos \dfrac{\theta}{2} - \sin \dfrac{\theta}{2}\right)^2 \equiv 1 - \sin \theta$.

**37.** Verify $\dfrac{1 - \tan \dfrac{x}{2}}{1 + \tan \dfrac{x}{2}} \equiv \dfrac{1 - \sin x}{\cos x}$.

**38.** Verify $\sin^2 \dfrac{\alpha}{4} + \cos \dfrac{\alpha}{2} \equiv \cos^2 \dfrac{\alpha}{4}$.

**39.** Verify $\sin \dfrac{A}{2} \cos \dfrac{A}{2} \equiv \dfrac{\sin A}{2}$.

**40.** Verify $\tan \dfrac{A}{2} \equiv \csc A - \cot A$.

★ **41.** Verify $\tan\left(\dfrac{\pi}{4} + \dfrac{\theta}{2}\right) \equiv \tan \theta + \sec \theta$.

★ **42.** Verify $\tan\left(\dfrac{3\pi}{4} + \dfrac{\theta}{2}\right) \equiv \tan \theta - \sec \theta$.

## 8-4 PRODUCT, SUM, AND DIFFERENCE IDENTITIES (OPTIONAL)

The formulas in this section are used in more advanced science and mathematics courses where it is sometimes desirable to express the sum or difference of two trigonometric functions as a product, or vice versa.

If we add the formulas for $\sin(A+B)$ and $\sin(A-B)$,

$$\sin(A+B) \equiv \sin A \cos B + \cos A \sin B$$

$$\sin(A-B) \equiv \sin A \cos B - \cos A \sin B$$

we obtain

$$\sin(A+B) + \sin(A-B) \equiv 2 \sin A \cos B$$

or

$$\sin A \cos B \equiv \frac{1}{2}[\sin(A+B) + \sin(A-B)] \qquad \textbf{(8-25)}$$

If we subtract $\sin(A-B)$ from $\sin(A+B)$ and simplify, we obtain

$$\cos A \sin B \equiv \frac{1}{2}[\sin(A+B) - \sin(A-B)] \qquad \textbf{(8-26)}$$

Using formulas for $\cos(A-B)$ and $\cos(A+B)$, we can obtain the following identities

$$\cos A \cos B \equiv \frac{1}{2}[\cos(A+B) + \cos(A-B)] \qquad \textbf{(8-27)}$$

$$\sin A \sin B \equiv -\frac{1}{2}[\cos(A+B) - \cos(A-B)] \qquad \textbf{(8-28)}$$

If we let $u = (A+B)$ and $v = (A-B)$ and substitute these values into the preceding formulas and rewrite them, we obtain the formulas for the sum of two trigonometric functions.

$$\sin u + \sin v \equiv 2 \sin \frac{1}{2}(u+v) \cos \frac{1}{2}(u-v) \qquad \textbf{(8-29)}$$

$$\sin u - \sin v \equiv 2 \cos \frac{1}{2}(u+v) \sin \frac{1}{2}(u-v) \qquad \textbf{(8-30)}$$

$$\cos u + \cos v \equiv 2 \cos \frac{1}{2}(u+v) \cos \frac{1}{2}(u-v) \qquad \textbf{(8-31)}$$

$$\cos u - \cos v \equiv -2 \sin \frac{1}{2}(u+v) \sin \frac{1}{2}(u-v) \qquad \textbf{(8-32)}$$

The formulas thus developed are usually used to verify other identities or to simplify expressions. First we will practice the general use of the formulas.

EXAMPLE 1 Rewrite $\sin 3x - \sin 5x$ as a product of two functions.

***Solution*** By using Formula (8-30), $\sin u - \sin v \equiv 2 \cos \frac{1}{2}(u+v) \sin \frac{1}{2}(u-v)$, when

$u = 3x$ and $u = 5x$, we have

$$\sin 3x - \sin 5x \equiv 2 \cos \frac{1}{2}(3x + 5x) \sin \frac{1}{2}(3x - 5x)$$

$$\equiv 2 \cos 4x \sin(-x)$$

$$\equiv -2 \cos 4x \sin x \qquad \textit{Since } \sin(-x) = -\sin x. \quad ■$$

We will now use the formulas to verify identities.

EXAMPLE 2 Verify $\sin 5x \cos 3x \equiv \frac{1}{2}(\sin 8x + \sin 2x)$.

***Solution*** We will use formula (8-25):

$$\sin A \cos B \equiv \frac{1}{2}[\sin(A + B) + \sin(A - B)]$$

Letting $A = 5x$ and $B = 3x$, we have $(A + B) = 8x$ and $(A - B) = 2x$. Thus the identity for $\sin A \cos B$ gives

$$\sin 5x \cos 3x \equiv \frac{1}{2}(\sin 8x + \sin 2x)$$

■

EXAMPLE 3 Prove $\sin 3\theta + \sin \theta \equiv 2 \sin 2\theta \cos \theta$.

***Solution*** By using Formula (8-29), $\sin u + \sin v \equiv 2 \sin \frac{1}{2}(u + v) \cos \frac{1}{2}(u - v)$, and letting $u = 3\theta$, $v = \theta$ we have

$$\sin 3\theta + \sin \theta \equiv 2 \sin \frac{1}{2}(3\theta + \theta) \cos \frac{1}{2}(3\theta - \theta)$$

so

$$\sin 3\theta + \sin \theta \equiv 2 \sin 2\theta \cos \theta \quad ■$$

EXAMPLE 4 Prove $\dfrac{\sin u + \sin v}{\cos u + \cos v} \equiv \tan \frac{1}{2}(u + v)$.

***Solution*** Working on the left-hand side, we will use Formula (8-29) in the numerator and Formula (8-31) in the denominator to obtain

$$\frac{\sin u + \sin v}{\cos u + \cos v} \equiv \frac{2 \sin \frac{1}{2}(u + v) \cos \frac{1}{2}(u - v)}{2 \cos \frac{1}{2}(u + v) \cos \frac{1}{2}(u - v)}$$

and by cancelling common factors we obtain

$$\equiv \frac{\sin \frac{1}{2}(u + v)}{\cos \frac{1}{2}(u + v)} \equiv \tan \frac{1}{2}(u + v)$$ ■

The formulas (8-25) to (8-32) discussed in this section of the text are summarized on page 437 for reference purposes.

## EXERCISE 8-4

Convert each of the expressions in Problems 1–10 to a product. Do not evaluate the expressions.

**1.** $\sin 60^\circ + \sin 20^\circ$ **2.** $\sin 24^\circ - \sin 18^\circ$  **3.** $\cos 55^\circ + \cos 22^\circ$

**4.** $\cos 38^\circ - \cos 70^\circ$ **5.** $\sin 2x + \sin 4x$ **6.** $\cos 6x - \cos 2x$

**7.** $\sin 12x - \sin 3x$ **8.** $\cos 2x + \cos 8x$ **9.** $\cos 7x - \cos 3x$

**10.** $\sin 5x + \sin 7x$

Convert each of the expressions in Problems 11–20 to a sum. Do not evaluate the expressions.

**11.** $\sin 30^\circ \cos 20^\circ$ **12.** $\sin 40^\circ \sin 28^\circ$ **13.** $\cos 43^\circ \cos 35^\circ$

**14.** $\cos 5x \cos 3x$ **15.** $2 \sin 5\theta \cos \theta$ **16.** $\sin 2\alpha \sin 9\alpha$

**17.** $\sin 4x \sin 2x$ **18.** $\sin 2x \cos 2x$ **19.** $\sin\left(\frac{A}{2}\right) \cos\left(\frac{3A}{2}\right)$

**20.** $\cos\left(\frac{3A}{2}\right) \sin\left(\frac{5A}{2}\right)$

**21.** Using the identity $\sin A \cos B \equiv \frac{1}{2}[\sin(A + B) + \sin(A - B)]$, let $(u + v) = 2A$ and $(u - v) = 2B$ to derive the formula for $\sin u + \sin v$.

**22.** Verify $\frac{\sin x - \sin 3x}{\cos 3x - \cos x} \equiv \cot 2x$.

**23.** Verify $\frac{\cos A - \cos B}{\sin A + \sin B} \equiv -\tan \frac{1}{2}(A - B)$.

★ **24.** Verify $\sec 2\theta \equiv \frac{1}{2}[\tan(45^\circ + \theta) + \tan(45^\circ - \theta)]$.

**25.** Using the formulas for $\cos(A - B)$ and $\cos(A + B)$, derive Formula (8-27).

**26.** Using the formulas for $\cos(A - B)$ and $\cos(A + B)$, derive Formula (8-28).

**27.** Verify $\frac{\sin A + \sin B}{\cos A + \cos B} \equiv \tan \frac{1}{2}(A + B)$.

**28.** Verify $\frac{\sin 9x - \sin 5x}{\sin 14x} \equiv \frac{\sin 2x}{\sin 7x}$.

★ **29.** Verify $\frac{\cot x - \tan x}{\cot x + \tan x} \equiv \cos 2x$.

★ **30.** Verify $\cos 6x \cos 2x + \sin^2 4x \equiv \cos^2 2x$.

## 8-5 THE LAW OF SINES

In Chapter 6 we studied a section on solving the right triangle. At that point in our development of trigonometry, it was necessary to have a right angle and use the definitions of the trigonometric functions to solve a triangle. Since we have now studied the general angle, we can develop laws for solving any triangle. The arithmetic computations involved are usually long and the use of a scientific calculator is strongly encouraged.

We will sometimes refer to the "obtuse angle" and "obtuse triangle." Make sure that you understand those words precisely so that there is no misunderstanding.

- An *obtuse angle* is an angle whose measure is between $90^\circ \left(\frac{\pi}{2} \text{ radians}\right)$ and $180^\circ$ ($\pi$ radians).
- An *obtuse triangle* is a triangle having one obtuse angle.

We will construct an acute triangle (a triangle in which all three angles are acute) and an obtuse triangle and label sides $a$, $b$, $c$ and the angles $A$, $B$, $C$.

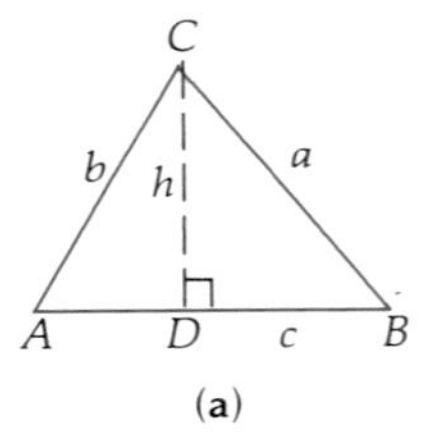


(a)

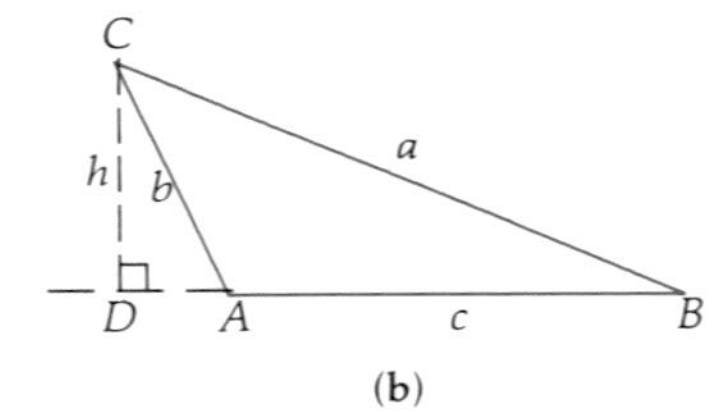


(b)

*Figure 8-1*

We will now prove the following theorem.

*Theorem 8-1*

> **The Law of Sines:** In any triangle $ABC$ with sides $a$, $b$, $c$ the following ratios hold true:
>
> $$\frac{a}{\sin A} = \frac{b}{\sin B} = \frac{c}{\sin C}$$

***Proof*** Consider the acute triangle and the obtuse triangle (see Figures 8-1a and 8-1b). In each, drop a perpendicular from vertex $C$ to side $AB$. In both figures, we have $\sin B = \frac{h}{a}$ by definition. Also by definition in the first figure, $\sin A = \frac{h}{b}$. In the second figure, angle $A$ is a second-quadrant angle, and its reference angle is its supplementary angle $DAC$. Therefore in the second figure, $\sin A = \frac{h}{b}$ also. This gives $h = a \sin B$ and $h = b \sin A$. Equating these and dividing by $\sin A \sin B$, we obtain

$$\frac{a}{\sin A} = \frac{b}{\sin B}$$

(Note that we do not have to be concerned about dividing by 0. We know that $\sin A \neq 0$ and $\sin B \neq 0$ because $A$ and $B$ cannot be $0°$; neither can they be $180°$. If either angle were $0°$ or $180°$, we would not have a triangle.)

In a similar fashion we could drop a perpendicular from vertex $A$ and establish the following relationship $\frac{b}{\sin B} = \frac{c}{\sin C}$. Therefore by combining our two results we have

$$\frac{a}{\sin A} = \frac{b}{\sin B} = \frac{c}{\sin C} \tag{8-33}$$

This theorem is referred to as the *law of sines.* It should be memorized. ■

This theorem is sometimes stated in words as follows: "The sides of a triangle are proportional to the sines of the opposite angles."

We can use the law of sines in certain cases to find all remaining parts of the triangle if only three parts of information are given. This is called "solving a triangle." In determining the degree of accuracy of our answers, we will follow the rules of precision of measurement developed in Section 6-8.

### Solving Triangles When Given Two Angles and One Side

The law of sines lends itself readily to solving a triangle if we are given two angles and a side. If we are given two angles, it is equivalent to being given all three angles, since the measure of the three angles of a triangle totals $180°$. Thus a proportion containing the given side would contain only one unknown value and it would be a side. (This situation is sometimes referred to from geometry as the AAS case, where the letters represent Angle, Angle, Side.)

EXAMPLE 1 In triangle $ABC$, $A = 20°$, $B = 47°$, and $b = 12$. Solve the triangle.

***Solution*** We see immediately that

$$C = 180° - (A + B) = 180° - (20° + 47°) = 113°$$

Next let us find the length of side $a$. Since we know that $A = 20°$, $B = 47°$, and $b = 12$ and we want to find $a$, we use that portion of the law of sines that relates the three known parts ($b$, $\sin B$, $\sin A$) to the unknown side $a$.

$$\frac{a}{\sin 20°} = \frac{12}{\sin 47°}$$

$$(a)(\sin 47°) = (12)(\sin 20°) \qquad \textit{Cross-multiply.}$$

$$a = \frac{(12)(\sin 20°)}{\sin 47°} \qquad \textit{Solve for a.}$$

We may now find sin 20° and sin 47° by using the trigonometric table or a calculator. We will round our answer to two significant digits. Using Table D, we would obtain

$$a = \frac{(12)(0.3420)}{(0.7314)} = 5.6$$

Next, we need to find side $c$. The portion of the law of sines we will use now is

$$\frac{b}{\sin B} = \frac{c}{\sin C}$$

$$\frac{12}{\sin 47°} = \frac{c}{\sin 113°} \qquad \textit{Substitute in known values.}$$

$$(12)(\sin 113°) = c(\sin 47°)$$

$$\frac{(12)(\sin 113°)}{\sin 47°} = c$$

*Note:* We do not use $\frac{a}{\sin A} = \frac{c}{\sin C}$. The value of $a$ is calculated and therefore is less accurate than the given value of $b$.

We can evaluate this directly on a calculator. If you need to use Table D, you will need the property that $\sin 113° = \sin(180° - 113°) = \sin(67°)$. Performing the calculation, we have

$$15 = c \qquad \textit{Rounded to two digits.}$$

We have now found the unknown parts of the triangle:

$$C = 113°, \qquad a = 5.6, \qquad \text{and} \qquad c = 15 \quad ■$$

**Note:** It is always a good idea to see whether your answer is reasonable. The longest side should be opposite the largest angle, the shortest side should be opposite the smallest angle, etc.

**EXAMPLE 2** A ship is heading due north. Its initial bearing to a radio signal tower is N35.62°E. After traveling 114.0 kilometers, the bearing to the station is N64.35°E. To three significant digits, how far is the signal tower from the ship at the time of the second observation?

***Solution*** First we draw a sketch. Notice that at the second observation point $B$ we have a bearing N64.35°E. In order to find the angle $CBA$ we will need to calculate $180° - 64.35° = 115.65°$. For convenience we will denote this as angle $B$.

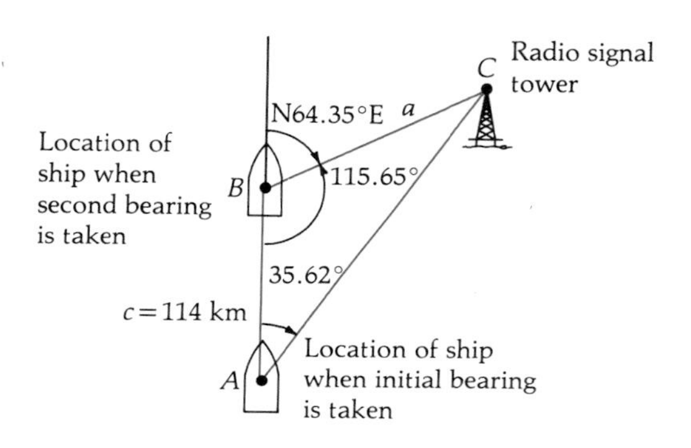


We need to find angle $C$:

$$C = 180° - (B + A) = 180° - (115.65° + 35.62°) = 28.73°$$

The distance we desire is labeled side $a$ on the triangle. Thus we will use the formula

$$\frac{a}{\sin A} = \frac{c}{\sin C}$$

$$\frac{a}{\sin 35.62°} = \frac{114.0}{\sin 28.73°}$$

$$a = \frac{(114.0)(\sin 35.62°)}{\sin 28.73°}$$

If you do not have a calculator, use Table D and interpolate to obtain the necessary values. On the calculator this can be performed directly:

114 [×] 35.62 [sin] [÷] 28.73 [sin] [=] 138.12515

Rounding to four digits, we have the distance 138.1 km. ■

**Note:** On many scientific calculators it is important to wait and allow the display to appear after pushing the [sin] button. Then you enter the next keystroke. If you enter values too rapidly, you will obtain an incorrect value.

### Solving Triangles When Given Two Sides and an Angle Opposite One of the Sides

This situation is sometimes referred to as SSA.

When we are given two sides and an angle opposite one of them, there will be three possibilities:

1. A unique solution.
2. Two possible solutions. (This is known as the ambiguous case.)
3. No solution.

Let us examine these three situations by observing a sketch. Assume that sides $a$ and $b$ and acute angle $A$ are known values. What types of triangles could occur?

#### *Case I*

One triangle. There is a unique solution when we solve the triangle.

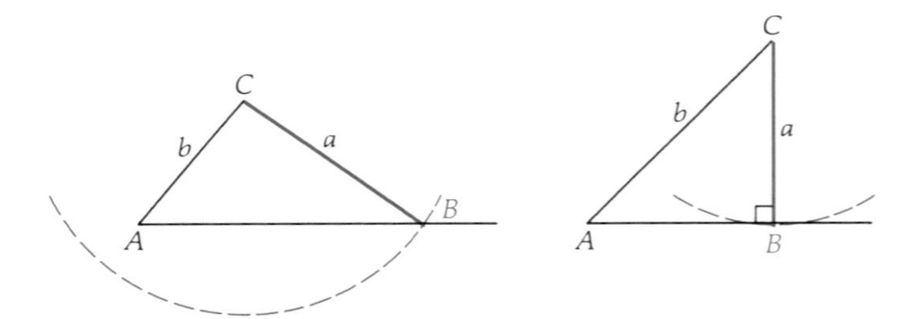


If we start at point $C$ and draw an arc of radius $a$, it will intersect the base of the triangle at only one point in this case.

#### *Case II*

Two possible triangles. The given facts describe either of two triangles.

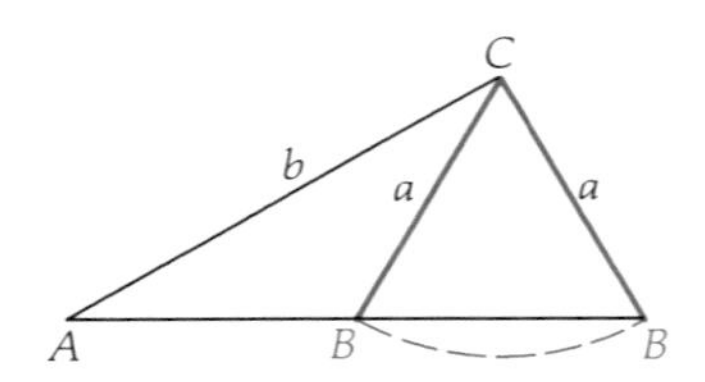

If we start at point $C$ and draw an arc of radius $a$, it will intersect the base of the triangle at two different points in this case.

### *Case III*

No triangle is possible. There is no solution.

The no-solution case arises when the given sides and the angle cannot form a triangle. The figure below represents this case.

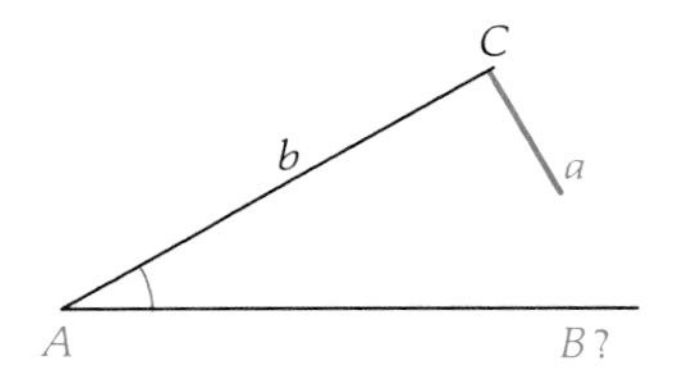


In this example, we are given side $a$, side $b$, and angle $A$, but side $a$ is not long enough to complete the triangle with the given angle $A$. We will recognize this case when our computation gives $\sin B > 1$. (We know that the sine of an angle cannot exceed 1.)

We can recognize this situation graphically if we start at point $C$ and draw an arc of radius $a$. The arc will not intersect the base of the triangle.

We will now proceed to solve some triangles that illustrate each possible case.

### *Case I Example*

EXAMPLE 3 In triangle $ABC$, side $b = 9.0$, side $a = 18$, angle $A = 27°$. Find angle $B$.

***Solution*** Here we expect angle $B$ to be smaller than angle $A$, since side $b$ is shorter than side $a$. If we draw a sketch of the triangle, this same conclusion seems reasonable. (Do you see why two different triangles are *not* possible? Think carefully and study the sketch.) We wish a formula that relates sides $a$, $b$ and angles $A$

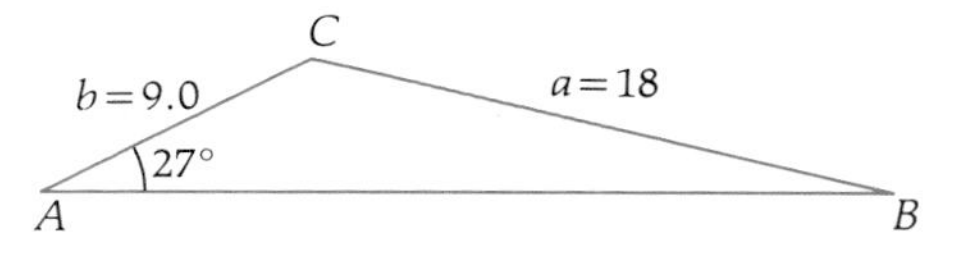


and $B$. Therefore we will use

$$\frac{a}{\sin A} = \frac{b}{\sin B}$$

$$\frac{18}{\sin 27°} = \frac{9}{\sin B}$$

Using Table D we have

$$\sin B = \frac{(9)(\sin 27^\circ)}{18} = \frac{(9)(0.4540)}{18}$$

$$\sin B = 0.2270$$

$$B = 13^\circ \qquad \textit{To the nearest degree.}$$

On a calculator, angle $B$ can be quickly found by the following keystrokes:

9 [×] 27 [sin] [÷] 18 [=] [INV] [sin] 13.120234

which we round to the nearest degree to obtain $B = 13^\circ$. ■

***Case II Example***

**EXAMPLE 4** In triangle $ABC$, side $b = 15$, side $a = 12$, angle $A = 27^\circ$. Find angle $B$ to the nearest degree.

***Solution*** We should first note that we expect angle $B$ to be larger than $27^\circ$, since side $b$ is longer than side $a$.

$$\frac{a}{\sin A} = \frac{b}{\sin B}$$

$$\frac{12}{\sin 27^\circ} = \frac{15}{\sin B}$$

$$\sin B = \frac{(15)(\sin 27^\circ)}{12}$$

$$\sin B = 0.5675$$

therefore $B = 35^\circ$ *To the nearest degree.*

But this is not the only possible answer. Do you see why? We also have

$$B = 145^\circ \qquad \textit{To the nearest degree.}$$

Since both angles are larger than the given $27^\circ$, we must conclude that there are two correct solutions. The figures given represent both solutions.

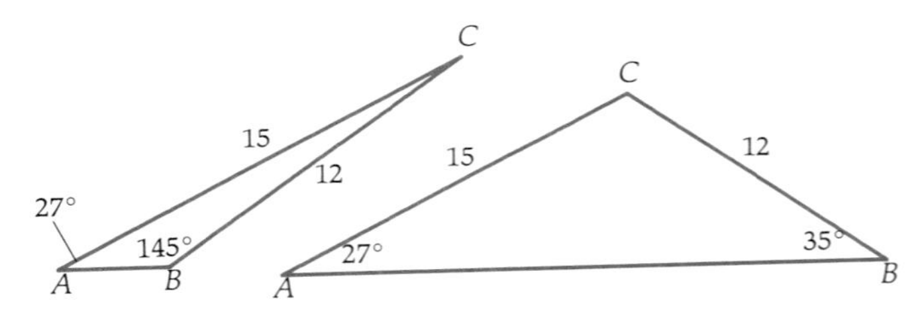

Examine the two diagrams on the previous page. Can you visualize how a radius of length 12 drawn from point $C$ would intersect the base of the triangle in two locations?

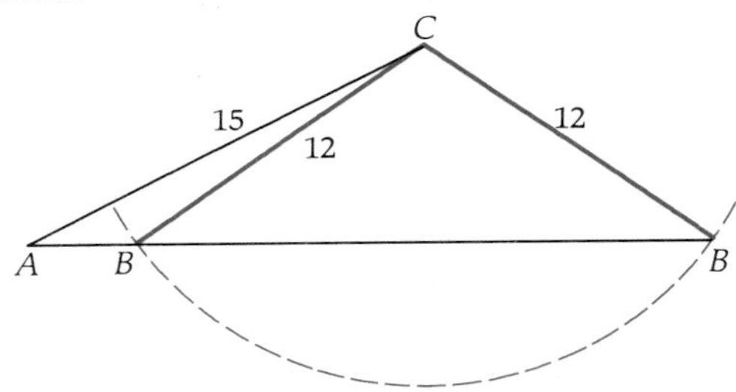


■

**Case III Example**

**EXAMPLE 5** In triangle $ABC$, $a = 5.0$, $c = 12$, $A = 27°$. Find $C$.

***Solution***

$$\frac{a}{\sin A} = \frac{c}{\sin C}$$

$$\frac{5}{\sin 27°} = \frac{12}{\sin C}$$

$$\sin C = \frac{(12)(\sin 27°)}{5} = \frac{(12)(0.4540)}{5}$$

$$\sin C = 1.0896 \qquad \textit{To the nearest ten thousandth.}$$

This gives *no solution*, since $\sin C > 1$ is impossible. ■

When you are solving a problem, it is important that you can recognize that if $\sin C > 1$, there is no solution. You should not depend on a calculator or a sketch to help you out. You must depend on your knowledge of the important property that $-1 \leq \sin \theta \leq 1$ and $-1 \leq \cos \theta \leq 1$ for all values of $\theta$.

Some students get confused in determining when there are two answers, one answer, or no answers to a problem. It is wise therefore to study carefully each of the examples and the description of each of the three cases presented in this section.

## EXERCISE 8-5

Problems 1–5 refer to the general triangle $ABC$. Use the law of sines to find the indicated parts.

**1.** Find $a$ if $A = 42°$, $B = 63°$, $b = 15$.

**2.** Find $c$ if $B = 25°$, $C = 68°$, $b = 16$.

**3.** Find $b$ if $A = 100°$, $C = 34°$, $a = 24$.

**4.** Find $B$ if $A = 55°$, $a = 21$, $b = 14$.

**5.** Find $B$ if $C = 30°$, $b = 2.4$, $c = 1.2$.

In Problems 6–23, solve for all unknown parts of the triangle. If no solution is possible, so state.

**6.** $A = 42°, B = 73°, a = 11$

**7.** $B = 69°, C = 64°, b = 26$

**8.** $A = 103°, C = 48°, a = 25$

**9.** $A = 24°, B = 95°, c = 18$

**10.** $C = 75°, c = 95, b = 75$

**11.** $A = 110°, a = 75, b = 95$

**12.** $A = 35°, b = 10, a = 8.0$

**13.** $A = 34°, a = 22, b = 32$

**14.** $C = 42°, c = 15, b = 24$

**15.** $C = 62°, c = 21, b = 23$

**16.** $A = 51.2°, C = 65.7°, a = 8.34$

**17.** $A = 23.8°, C = 13.3°, b = 10.56$

**18.** $A = 76.2°, B = 27.6°, c = 13.9$

**19.** $B = 27.4°, C = 61.2°, c = 2.63$

**20.** $B = 43.8°, b = 17.5, c = 10.2$

**21.** $C = 22.5°, c = 4.87, a = 5.43$

**22.** $A = 108.5°, a = 112, c = 73.4$

**23.** $A = 102.6°, a = 327, b = 56.2$

**24.** To find the distance from an observer at point $A$ to a point $B$ on the other side of a river, the line $AC$ was measured to be 2530 feet. Angle $A$ was measured to be 47.0°. Angle $C$ was measured to be 78.0°. Find the distance from $A$ to $B$ to the nearest foot.

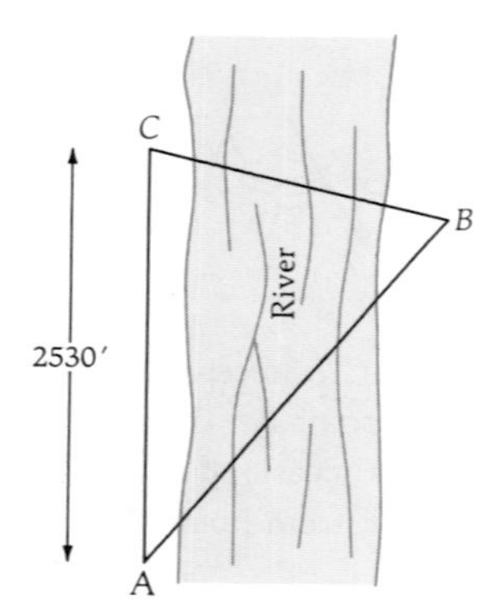


**25.** Two ranger towers ($A$ and $B$) are located 2.0 kilometers apart. A ranger at tower $A$ spots a fire ($F$) and measures angle $FAB$ to be 60°. A ranger at tower $B$ measures angle $FBA$ to be 55°. How far is tower $A$ from the fire?

**26.** A ship takes a bearing on a lighthouse of N20°W. After sailing 55 nautical miles due north, the bearing is N74°W. How far is the ship from the lighthouse?

**27.** A vertical pole is supported by two guy wires on opposite sides. One wire is 13.5 meters long. The wire makes an angle of 42.2° with the ground. The other wire is 10.5 meters long. Find the angle between the two guy wires.

**28.** A rock ($R$) lies on the side of a hill 321 meters above the base ($B$) of the hill. From a point ($P$) on the horizontal ground 75.0 meters from $B$, the angle of elevation of the rock is 13.6°. What angle does the hill make with the horizontal? (*Hint:* Find angle $RBD$.)

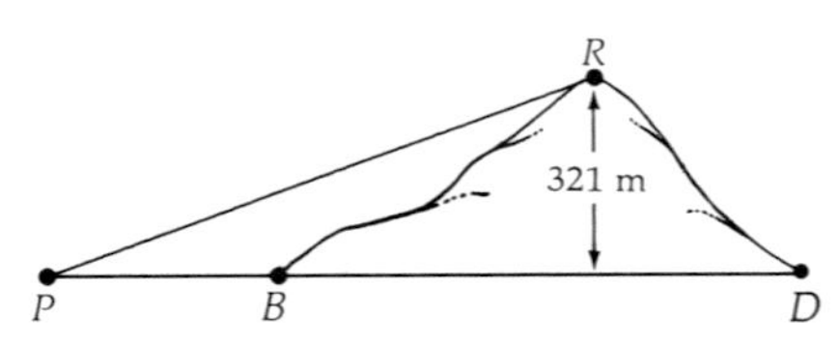

**29.** Two people ($A$ and $B$) stand on one side of a river 50.0 meters apart and look at a tree ($T$) on the opposite shore. How wide is the river if angle $ABT = 38.2°$ and angle $BAT = 62.3°$?

**30.** A radio tower is secured by two primary guy wires and several secondary holding wires. The wire from point $B$ to point $C$ is 1648 feet long. The wire from point $A$ to point $C$ is 1536 feet long. The angle of elevation of the wire at point $B$ is 46.23°. Find the angle of elevation of the wire at point $A$.

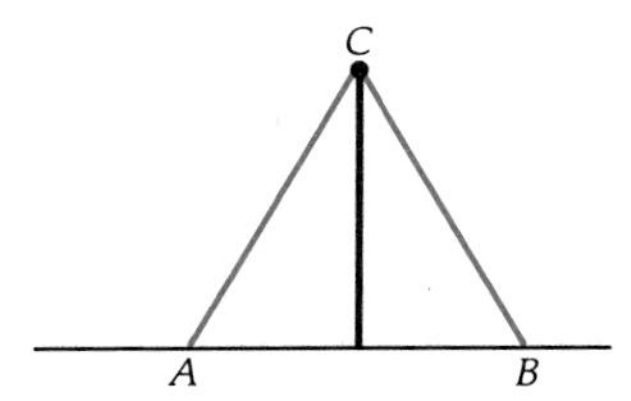


**31.** The diagonal of a parallelogram meets the sides at angles of 30° and 75°. The length of the diagonal line shown in the sketch is 38 cm. Find the perimeter of the parallelogram. Round your answer to the nearest centimeter.

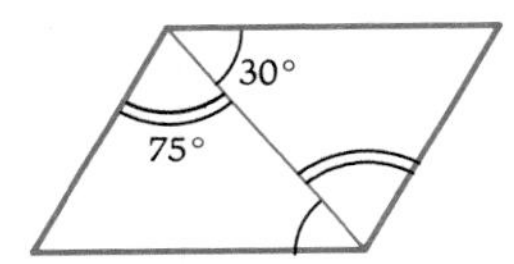


★ **32.** A radio signal from an unknown point $C$ is received by two listening stations at points $A$ and $B$. Point $A$ is 8.42 kilometers due north of point $B$. Town $D$ is due north of point $A$. Angle $DAC$ was measured to be 128.4°. Angle $CBA$ was measured to be 43.14°. How far is it from point $C$ to the listening station at point $A$? How far is it from point $C$ to the listening station at point $B$?

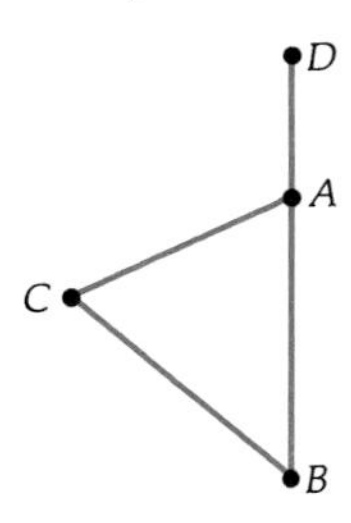


★ **33.** A ship leaves a harbor and travels west. It then changes course to a course N61.5°W. It then travels 44.3 kilometers on the new course to a point 64.5 kilometers from the harbor. How far from the harbor is the point where the ship turned?

Use the law of sines to prove that the statements in Problems 34 and 35 are true for any triangle $ABC$ with sides $a$, $b$, $c$.

**34.** $\dfrac{a-b}{a+b} = \dfrac{\sin A - \sin B}{\sin A + \sin B}$ **35.** $\dfrac{a+b}{b} = \dfrac{\sin A + \sin B}{\sin B}$

## 8-6 THE LAW OF COSINES

Triangles can be classified as to the information given about them in the following ways:

1. two angles and a side
2. two sides and an angle opposite one of the sides
3. two sides and the included angle
4. three sides

We have used the law of sines to solve triangles in the first two cases listed. However, the law of sines cannot be used in all of the preceding cases. For instance, if we are given two sides and the included angle—side $a$, side $b$, and angle $C$—then the proportions $\frac{a}{\sin A} = \frac{b}{\sin B}$, $\frac{a}{\sin A} = \frac{c}{\sin C}$, and $\frac{b}{\sin B} = \frac{c}{\sin C}$ will each have two unknown quantities, and no solution is possible. The same is true if we are given three sides and no angle.

We will now develop the law of cosines in order to solve triangles in the last two cases listed.

*Theorem 8-2*

**The Law of Cosines:** In any triangle $ABC$, the square of a side is equal to the sum of the squares of the other two sides *minus* twice the product of those sides and the cosine of the angle between them.

There are three statements that can be derived from this theorem:

$$a^2 = b^2 + c^2 - 2bc \cos A \qquad \textbf{(8-34)}$$

$$b^2 = a^2 + c^2 - 2ac \cos B \qquad \textbf{(8-35)}$$

$$c^2 = a^2 + b^2 - 2ab \cos C \qquad \textbf{(8-36)}$$

***Proof*** Place triangle $ABC$ on the coordinate axes with angle $B$ in standard position. We shall derive Formula (8-35).

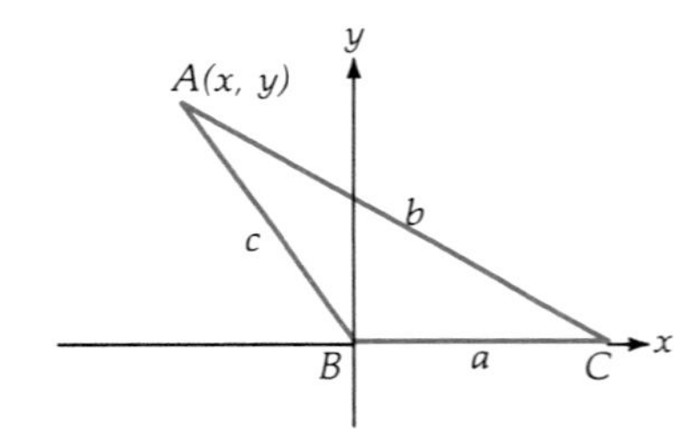

Using $(x, y)$ as the coordinates of point $A$, we know from our definitions that

$$\cos B = \frac{x}{c} \qquad \text{or} \qquad x = c \cos B$$

and

$$\sin B = \frac{y}{c} \qquad \text{or} \qquad y = c \sin B$$

giving us the following coordinates for $A$ and $C$:

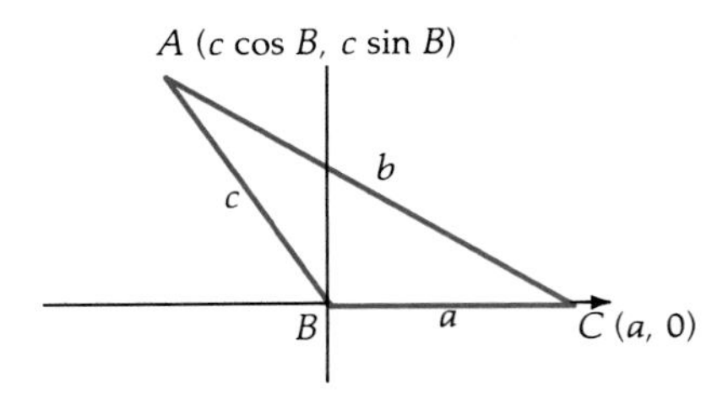


If we compute the distance between $A$ and $C$, we have

$$b^2 = (a - c \cos B)^2 + (0 - c \sin B)^2$$
$$b^2 = a^2 - 2ac \cos B + c^2 \cos^2 B + c^2 \sin^2 B$$
$$b^2 = a^2 - 2ac \cos B + c^2(\cos^2 B + \sin^2 B)$$
$$b^2 = a^2 + c^2 - 2ac \cos B$$

Since the placement of the letters $ABC$ is arbitrary, the other two forms of the law of cosines are easily established. [This law of cosines should be memorized in all three of its forms, (8-34), (8-35), and (8-36). This can be easily done by observing the pattern of the letters.

$$a^2 = b^2 + c^2 - 2bc \cos A.] \quad ■$$

This law is used to solve triangles when three sides, or two sides and the included angle are given. There is no allowance for an ambiguous case, because if $\cos \theta < 0$, we have a second-quadrant angle, and if $\cos \theta > 0$, we have a first-quadrant angle. However, when three sides are given, we must be sure that the sum of the lengths of any two of the given sides is greater than the length of the third side. If this is not the case, then no triangle can be formed.

### Solving Triangles When All Sides Are Known (SSS)

EXAMPLE 1 In triangle $ABC$, $a = 5$, $b = 7$, $c = 10$. Find $A$ to the nearest degree.

***Solution*** Since we want to find angle $A$, we will use the following form of the law of cosines:

$$a^2 = b^2 + c^2 - 2bc \cos A \qquad \textit{Formula (8-34).}$$

$$(5)^2 = (7)^2 + (10)^2 - 2(7)(10) \cos A$$

$$(2)(7)(10) \cos A = (7)^2 + (10)^2 - (5)^2$$

$$\cos A = \frac{(7)^2 + (10)^2 - (5)^2}{(2)(7)(10)}$$

The use of a calculator is helpful at this point to obtain

$$\cos A = 0.8857 \qquad \textit{Rounded to the nearest ten-thousandth.}$$

$$A = 28^\circ \qquad \textit{To the nearest degree.} \quad ■$$

When solving a triangle in the case in which all three sides of the triangle are known, it is best to solve for the largest angle first.

### Solving Triangles When Two Sides and an Included Angle Are Known (SAS)

**EXAMPLE 2** In triangle $ABC$, $c = 8.0$, $b = 14$, $A = 100^\circ$. Find $C$.

***Solution***

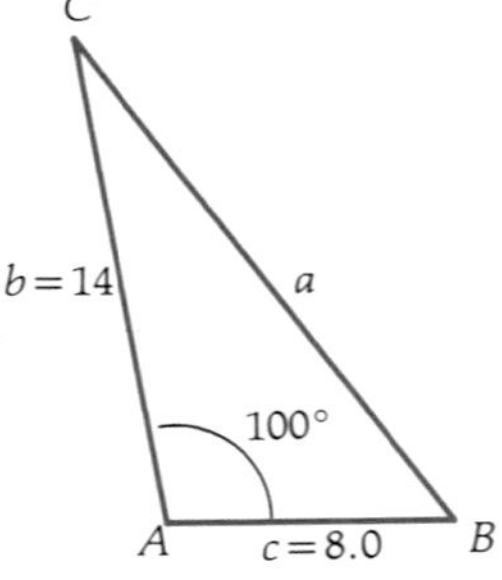


To find angle $C$, we must first know all three sides. Therefore use the given information to find side $a$:

$$a^2 = b^2 + c^2 - 2bc \cos A$$

$$a^2 = (14)^2 + (8)^2 - 2(14)(8) \cos 100^\circ$$

$$a^2 = 196 + 64 - 224 \cos 100^\circ$$

Note that $100^\circ$ is in the second quadrant. Therefore the cosine is negative. If we use Table D, we must use the fact that

$$\cos 100^\circ = -\cos 80^\circ = -0.1736$$

Hence $$a^2 = 196 + 64 + 224\ (0.1736)$$

Thus we obtain

$$a^2 = 298.9 \qquad \textit{Rounded to the nearest tenth.}$$

$$a = 17 \qquad \textit{Rounded to two significant digits.}$$

Now using

$$c^2 = a^2 + b^2 - 2ab \cos C$$

$$2ab \cos C = a^2 + b^2 - c^2$$

$$\cos C = \frac{a^2 + b^2 - c^2}{2ab}$$

$$\cos C = \frac{(17)^2 + (14)^2 - (8)^2}{2(17)(14)}$$

$$\cos C = 0.8845 \qquad \textit{Rounded to the nearest ten-thousandth.}$$

$$C = 28^\circ \qquad \textit{To the nearest degree.}$$ ■

## EXERCISE 8-6

**1.** Simplify the law $c^2 = a^2 + b^2 - 2ab \cos C$ when angle $C = 90°$. Where have you seen this expression before?

Problems 2–8 refer to the general triangle $ABC$. Solve for the indicated parts. If no solution is possible, so state.

**2.** Find $c$ if $C = 39°$, $a = 4.0$, and $b = 6.0$.

**3.** Find $b$ if $B = 65°$, $a = 4.0$, and $c = 8.0$.

**4.** Find $a$ if $A = 106°$, $b = 2.4$, and $c = 3.2$.

**5.** Find $A$ if $a = 3.0$, $b = 5.0$, and $c = 7.0$.

**6.** Find $B$ if $a = 2.0$, $b = 6.1$, and $c = 8.4$.

**7.** Find $C$ if $a = 3.9$, $b = 5.2$, and $c = 6.5$.

**8.** Find $A$ if $a = 17$, $b = 12$, and $c = 10$.

In Problems 9–19, solve for all unknown parts of the triangle.

**9.** $C = 45°$, $a = 11$, $b = 17$

**10.** $a = 2.7$, $b = 3.8$, $c = 3.2$

**11.** $A = 120°$, $b = 27$, $c = 10$

**12.** $A = 135°$, $b = 15$, $c = 10$

**13.** $a = 14$, $b = 16$, $c = 20$

**14.** $a = 9.26$, $b = 7.31$, $c = 5.23$

**15.** $a = 2.12$, $b = 3.22$, $c = 4.06$

**16.** $B = 103.5°$, $a = 234$, $c = 162$

**17.** $C = 67.4°$, $a = 35.2$, $b = 42.3$

**18.** $A = 54.1°$, $b = 7.56$, $c = 6.12$

**19.** $a = 2.56$, $b = 1.72$, $c = 3.44$

**20.** A ship sails 40 nautical miles due east from a starting point. It then sails on course 326° for a distance of 28 nautical miles. How far is the ship from its starting point?

**21.** Two guy wires on opposite sides of a TV antenna reach the ground at points that are 21 meters apart. One wire is 30 meters long; the other is 25 meters long. Find the angle that the 30-meter wire makes with the ground.

22. Two forces act at the same point. One component is 51.0 newtons. If the resultant force is 107 newtons and makes an angle of 38.3° with the known force, find the magnitude of the other component.
23. Points $A$ and $B$ are on opposite sides of a lake. From a third point $C$ on the same side of the lake as $A$ an angle is measured. Angle $C$ is 46.1°. The distance $AC$ is 342 meters long, and the distance $BC$ is 283 meters long. How far is point $A$ from point $B$?
24. A surveyor desires to measure a distance between points $A$ and $B$. He knows that point $C$ is 420 meters from point $A$. He measures the distance from point $C$ to point $B$ to be 564 meters. Finally, he measures angle $CAB$ and finds that it is 36.8°. What is the distance between points $A$ and $B$?
25. A small vertical flagpole is 580 centimeters tall. The top of the pole is 984 centimeters from an observer's eyes. The bottom of the pole is 763 centimeters from the observer's eyes. What angle does the flagpole subtend at the observer's eyes?

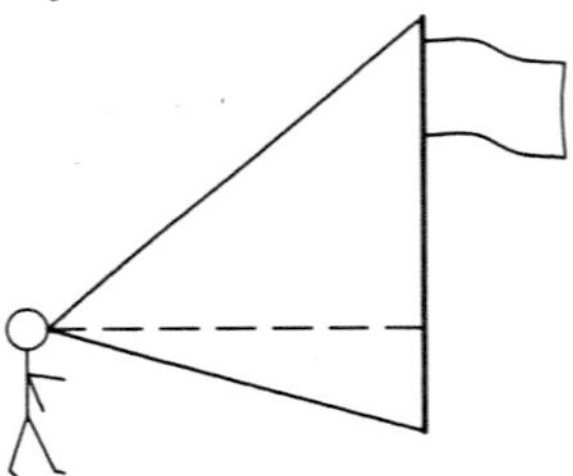

26. The city has created a triangular park at the edge of the beach. It is 264 feet along the ocean, 436 feet along Spruce Street, and 374 feet along Western Avenue. Find the angle of the intersection when Western Avenue meets Spruce Street (assume that both roads and the beach are exactly straight).
27. The distance from the student union to the gymnasium is 1640 meters. If you then turn and go to the science building, it is a distance of 1320 meters. The angle between these two paths is 115°. There is a direct path from the student union to the science building. How long is this path?
28. Two boats leave the dock. One is traveling at 18 km/hour. The other travels at 29 km/hour. They leave on straight paths. Their courses are 74° from each other. How far apart are they after 6.5 minutes?

## Calculator Problems

A more advanced application of the law of cosines is required in modern technology. Dr. Wayne Hoover of Cape Cod Community College has developed Problems 29 and 30 to introduce students to some of the situations encountered in studying radar systems and satellite communication. (See figure on next page.)

★ 29. A satellite (labeled $C$) is orbiting the earth. At 1:00 P.M. it is 154.3 miles above a point $B$ on the earth's surface. A radar station is tracking the satellite. The distance from the radar station (labeled $A$) to the satellite at 1:00 P.M. is 251.7 miles. At 1:00 P.M., how far is the curved distance (dashed line) on the earth's surface from point $A$ to point $B$? (Assume that the earth is a sphere of radius $r = 3960$ miles. Round your answer to the nearest thousandth of a mile.)

★ 30. Refer to Problem 29. On another day at 1:00 P.M. the following measurements were made. The distance from $A$ to $C$ is 238.4 miles, and the distance from $B$ to $C$ is 146.8 miles. Find the measure of angle $C$ to the nearest thousandth of a degree.

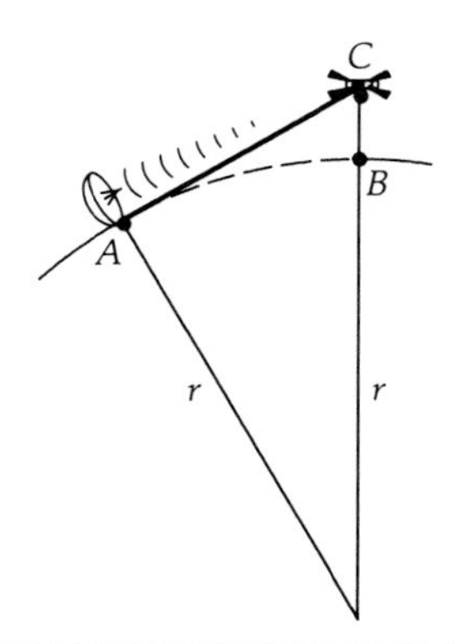


## 8-7 SOLUTION OF TRIGONOMETRIC EQUATIONS

In the earlier part of this chapter, we dealt with a special type of equation—the identity. An identity is valid for all values of the argument for which the functions exist. In this section we will solve conditional equations—those that are true for only certain values of the argument.

We have solved some simple equations in previous chapters without specifically calling attention to them as equations. For instance, if we are given $\sin \theta = 0.7653$ and asked to find $\theta$, we are solving the given equation for $\theta$.

In Sections 8-5 and 8-6 we solved specific types of trigonometric equations to obtain a side or an angle in a triangle using the law of sines or the law of cosines. The equations in this section are a type that can be solved by using only algebraic methods, such as factoring, adding, or subtracting on both sides of the equation, and so on.

**EXAMPLE 1** Solve $\sin^2 A - \sin A = 0$ for all values of $A$ such that $0° \leq A < 360°$.

***Solution***

$$\sin^2 A - \sin A = 0$$

$$\sin A\,(\sin A - 1) = 0$$

Since the product is zero, one or both of the factors must be zero. Thus

$$\sin A = 0 \qquad \text{or} \qquad \sin A - 1 = 0$$

$$A = 0°,\ 180° \qquad\qquad \sin A = 1$$

$$A = 90°$$

Our solution set is $\{0°, 90°, 180°\}$. ■

The argument of the trigonometric functions may be expressed in radians as in the next example.

EXAMPLE 2 Solve for $x$: $\cos x + 2 \cos x \sin x = 0$. Express your answer $x$ in radians where $0 \le x < 2\pi$.

***Solution***

$$\cos x + 2 \cos x \sin x = 0$$

$$\cos x(1 + 2 \sin x) = 0$$

$$\cos x = 0 \quad \text{or} \quad 1 + 2 \sin x = 0$$

$$x = \frac{\pi}{2}, \frac{3}{2}\pi \qquad 2 \sin x = -1$$

$$\sin x = -\frac{1}{2}$$

*Since the sine is negative, we want quadrants III and IV.*

$$x = \frac{7}{6}\pi, \frac{11}{6}\pi$$

The solution set is $\left\{\frac{\pi}{2}, \frac{7}{6}\pi, \frac{3}{2}\pi, \frac{11}{6}\pi\right\}$. ■

Some trigonometric equations do not yield special values for the angles in the solution. In those cases a calculator or a table must be used.

EXAMPLE 3 Solve for $\alpha$: $\tan^2 \alpha - 2 \tan \alpha - 3 = 0$, $0^\circ \le \alpha < 360^\circ$. Round your answer to the nearest tenth of a degree.

***Solution***

$$\tan^2\alpha - 2 \tan \alpha - 3 = 0$$

$$(\tan \alpha - 3)(\tan \alpha + 1) = 0$$

$\tan \alpha = 3$ or $\tan \alpha = -1$

Using a calculator or a table, we obtain $\alpha = 71.6^\circ$. Since the tangent is positive, we want quadrants I and III, so we also have $\alpha = 251.6^\circ$.

$\alpha = 135^\circ, 315^\circ$

Since the tangent is negative, we want quadrants II and IV.

The solution set is $\{71.6^\circ, 135^\circ, 251.6^\circ, 315^\circ\}$. ■

EXAMPLE 4 Solve for $\theta$: $2 \cos 2\theta - \sqrt{3} = 0$, $0^\circ \le \theta < 360^\circ$.

***Solution***

$$2 \cos 2\theta - \sqrt{3} = 0$$

$$2 \cos 2\theta = \sqrt{3}$$

gives

$$\cos 2\theta = \frac{\sqrt{3}}{2}$$

Now we must find the value of $2\theta$. Do you see that since $0^\circ \leq \theta < 360^\circ$ that we must have $0^\circ \leq 2\theta < 720^\circ$?

$$2\theta = 30^\circ, 330^\circ, 390^\circ, 690^\circ$$

$$\theta = 15^\circ, 165^\circ, 195^\circ, 345^\circ$$

Thus the solution set is $\{15^\circ, 165^\circ, 195^\circ, 345^\circ\}$. ■

**Caution:** If an equation is given in $n\theta$, keep in mind that if $0^\circ \leq \theta < 360^\circ$ or $0 \leq \theta < 2\pi$, then $0^\circ \leq n\theta < n(360^\circ)$ or $0 \leq n\theta < 2n\pi$. Failure to note this could result in an incomplete solution set.

### Solution of Trigonometric Equations Through the Use of Identities

If an equation contains more than one function such that factors of the equation also contain more than one function, we must use identities to reduce the expression to a single function.

EXAMPLE 5 Solve $\sin^2\theta + 2\cos\theta + 2 = 0$, $0^\circ \leq \theta < 360^\circ$.

***Solution*** Note here that factoring is not possible in the present form. We need to use identities to express the equation in a single function:

$$\sin^2\theta + 2\cos\theta + 2 = 0$$

$$(1 - \cos^2\theta) + 2\cos\theta + 2 = 0$$

$$-\cos^2\theta + 2\cos\theta + 3 = 0$$

$$\cos^2\theta - 2\cos\theta - 3 = 0$$

$$(\cos\theta - 3)(\cos\theta + 1) = 0$$

$$\cos\theta = 3 \quad \text{or} \quad \cos\theta = -1$$

$$\text{(no solution)} \qquad \theta = 180^\circ$$

Note that $\cos\theta = 3$ does not give a solution, since $-1 \leq \cos\theta \leq 1$. Therefore the only solution is $\{180^\circ\}$. ■

If one trigonometric function is expressed in terms of $\theta$ and another a multiple of $\theta$, it is usually wise to use the multiple angle formulas to express everything in terms of $\theta$.

EXAMPLE 6 Solve $\sin 2\theta\cos\theta - \sin\theta = 0$, $0^\circ \leq \theta < 360^\circ$.

***Solution*** Here the function of the double angle prevents us from factoring. Hence we will use the identity for $\sin 2\theta$:

$$\sin 2\theta \cos \theta - \sin \theta = 0$$

$$(2 \sin \theta \cos \theta) \cos \theta - \sin \theta = 0$$

$$2 \sin \theta \cos^2 \theta - \sin \theta = 0$$

$$\sin \theta\,(2 \cos^2 \theta - 1) = 0$$

$$\sin \theta = 0 \quad \text{or} \quad 2 \cos^2 \theta - 1 = 0$$

$$\theta = 0°, 180° \qquad \cos^2 \theta = \frac{1}{2}$$

*Note that both sign values are needed.*

$$\cos \theta = \pm\sqrt{\frac{1}{2}}$$

$$\cos \theta = \pm\frac{\sqrt{2}}{2}$$

$$\theta = 45°, 135°, 225°, 315°$$

Thus the solution set is $\{0°, 45°, 135°, 180°, 225°, 315°\}$. ■

## EXERCISE 8-7

Solve for $\theta$ or $x$ in degrees when $0° \le \theta < 360°$ or $0° \le x < 360°$ in Problems 1–25.

**1.** $\cos^2 \theta - \cos \theta = 0$

**2.** $\sin \theta - \sin \theta \cos \theta = 0$

**3.** $2 \cos^2 x - \cos x = 0$

**4.** $\tan^2 x - 1 = 0$

**5.** $2 \sin^2 \theta + \sin \theta = 1$

**6.** $\tan \theta \cos \theta - 2 \cos \theta - \tan \theta + 2 = 0$

**7.** $2 \sin x + \sqrt{3} = 0$

**8.** $2 \cos^2 x + 3 \cos x + 1 = 0$

**9.** $2 \sec x - \sqrt{3} = 0$

**10.** $\tan 2x - 1 = 0$

**11.** $2 \sec^2 \theta - 3 \sec \theta = 2$

**12.** $\sin 3x + 1 = 0$

**13.** $2 \cos 2\theta \sin \theta + 2 \cos 2\theta - \sin \theta = 1$

**14.** $5 \tan^2 2x - 3 \tan 2x - 36 = 0$

**15.** $2 \sin 2\theta \sin \theta - 4 \sin 2\theta - \sin \theta + 2 = 0$

**16.** $|\tan x| = 1$

**17.** $|\cos x| = \frac{1}{2}$

**18.** $2 \cos^2 x - \sqrt{3} \cos x = 0$

**19.** $4 - \cos x = 4.5664$

**20.** $\tan^2 x - 3 = 0$

**21.** $4 \cos^2 x - 3 = 0$

**22.** $\sqrt{3} + 2 \cos 2\theta = 0$

**23.** $\frac{2}{\sqrt{3}} \sin 2\theta = 1$

**24.** $2 \sin^2 \theta - 5 \sin \theta + 2 = 0$

**25.** $6 \cos^2 \theta + 5 \cos \theta + 1 = 0$

all odd 1-25

Solve for $\theta$ or $x$ in radians where $0 \leq \theta < 2\pi$ or $0 \leq x < 2\pi$ in Problems 26–35. If no solution is possible, so state.

**26.** $\sec^2 x + 1 = 0$

**27.** $\csc^2 2x = 1$

**28.** $2 \cos^2 x + \cos x - 1 = 0$

**29.** $\tan^2 x + 2 \tan x + 1 = 0$

**30.** $2 \cos^2 x \tan x - \tan x = 0$

**31.** $2 \sin^2 x - 7 \sin x + 3 = 0$

**32.** $2 \cos 3x + 1 = 0$

**33.** $2 \sin 3x + \sqrt{3} = 0$

★**34.** $3 \tan^2 \theta + \tan \theta = 1$

★**35.** $4 \cos^2 \theta - 1 = \cos \theta$

Solve for $\theta$ or $x$ in degrees where $0° \leq \theta < 360°$ or $0° \leq x < 360°$ in Problems 36–46. Use trigonometric identities wherever necessary. If no solution is possible, so state.

**36.** $\sin^2 \theta - \cos^2 \theta = 0$

**37.** $\sin^2 \theta - \cos^2 \theta + 1 = 0$

**38.** $3 \cos x + 2 \sin^2 x - 3 = 0$

**39.** $\cot^2 x = \cot x$

**40.** $\cos^2 x - \sin^2 x = \sin x$

**41.** $2 \sin^2 \theta - 2 \cos^2 \theta = 3$

**42.** $\cos 2\theta + \sin \theta - 1 = 0$

**43.** $\cot 2x - \tan x = 0$

**44.** $\cos \theta + 1 = \sin \theta$

**45.** $3 \csc \theta + 2 = \sin \theta$

**46.** $2 \cos x - 5 + 2 \sec x = 0$

## 8-8 VECTORS: DEFINITIONS AND BASIC OPERATIONS

Vectors are very important quantities in the physical sciences. A vector may be defined mathematically as a directed line segment. Physically, vectors represent quantities that have both **magnitude** and **direction**. Common examples are force, acceleration, velocity, and displacement. We will limit our discussion to vectors in two dimensions, although vector notation and vector operations are more powerful in three or more dimensions.

Consider Figure 8-2 as an example of a displacement vector. A particle has moved from point $A$ to point $B$ along the curved path $S$, but the **displacement** is represented by the directed line segment $\overrightarrow{AB}$. The length of the segment represents the magnitude of the displacement, written as $|\overrightarrow{AB}|$, and the direction is indicated by placing an arrowhead at point $B$. Point $A$ is then defined as the **tail** of the vector and point $B$ as the **head**.

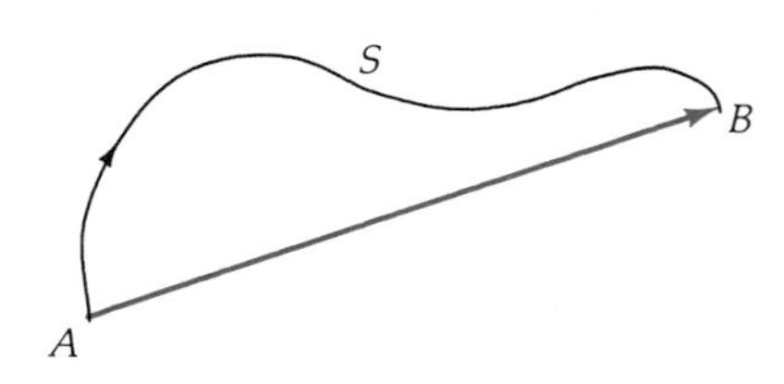


*Figure 8-2*

## Vector Operations (Geometrically)

### *Vector Addition*

In Figure 8-3a, point $P$ on the path $S$ represents an intermediate position of the particle. Notice that we can now identify two displacements: $\overrightarrow{AP}$ and $\overrightarrow{PB}$. The vector sum of $\overrightarrow{AP}$ and $\overrightarrow{PB}$ must equal $\overrightarrow{AB}$ (see Figure 8-3b).

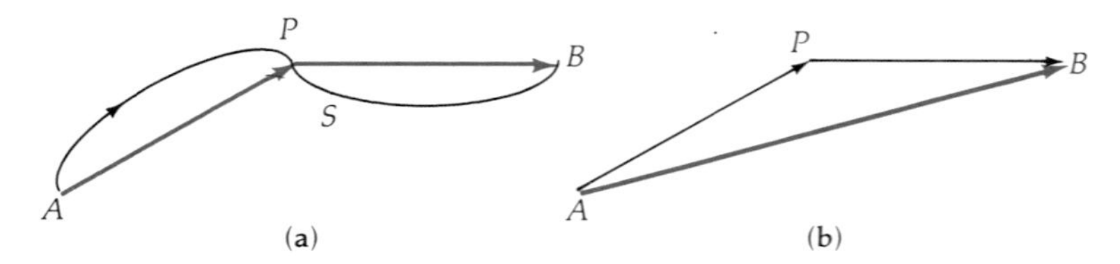


*Figure 8-3*

The vector $\overrightarrow{AB}$ is called the **vector sum** or **resultant**.

Notice that the sum is not an algebraic sum and a number is not sufficient to identify it. This example illustrates a basic rule for the geometrical addition of vectors.

*Rule I*

Given any number of vectors, representing the same physical quantity, we may find their *sum (or resultant)* geometrically by placing them head to tail, and then a vector drawn from the tail of the first to the head of the last will be the vector sum.

For the vectors in Figure 8-3b we may state the rule by the vector equation:

$$\overrightarrow{AP} + \overrightarrow{PB} = \overrightarrow{AB}$$

$\overrightarrow{AP}$ and $\overrightarrow{PB}$ are then defined as the **vector components** of $\overrightarrow{AB}$. Vector notation is greatly simplified if we write the directed line segment $\overrightarrow{AB}$ as a single entity; i.e., let $\overrightarrow{AB} = \mathbf{a}$ and $|\overrightarrow{AB}| = |\mathbf{a}|$. This notation will be used throughout the remainder of this section.

*Definition*

*Equal vectors* are vectors that have the same magnitude and direction.

Combining Rule I and the definition of equal vectors gives a geometrical method of obtaining the sum of any number of vectors regardless of their direction.

EXAMPLE 1 Referring to Figure 8-4, sketch $\mathbf{a} + \mathbf{b} + \mathbf{c}$.

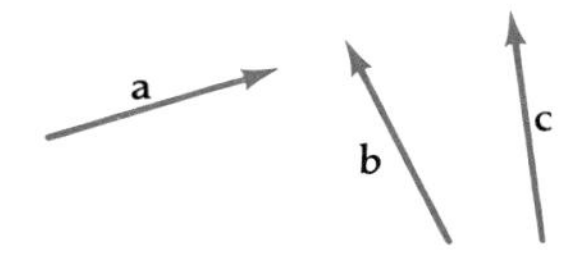


*Figure 8-4*

***Solution*** We reposition vector **b** with its tail beginning at the head of vector **a**. We reposition vector **c** with its tail beginning at the head of vector **b**. When vectors are repositioned, we are careful to preserve the original magnitude and direction. The resultant is **r**, which is $\mathbf{r} = \mathbf{a} + \mathbf{b} + \mathbf{c}$. ■

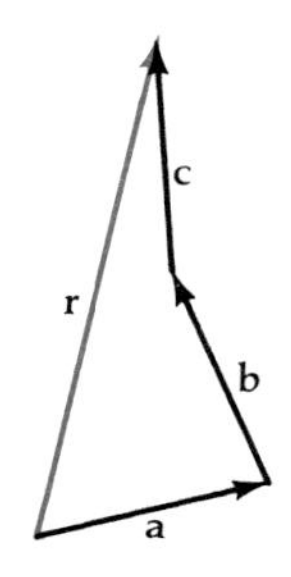


*Rule II*

For all vectors **a**, **b**, **c**,

1. Vector addition is commutative: $\mathbf{a} + \mathbf{b} = \mathbf{b} + \mathbf{a}$
2. Vector addition is associative: $(\mathbf{a} + \mathbf{b}) + \mathbf{c} = \mathbf{a} + (\mathbf{b} + \mathbf{c})$

It makes no difference as to the order or grouping of the individual vectors in vector addition.

### Scalar Multiplication

*Definition*

If $k$ is any real number (sometimes called a scalar) and **a** is the vector, then $k\mathbf{a}$ is defined as the scalar multiplication of $k$ and **a**.

If $k$ is positive, the operation changes the magnitude of the vector only. Moreover, if $k$ is negative, the resulting vector is $k$ times as long and oppositely directed to the original. These ideas are shown in the following example.

EXAMPLE 2 Given the vector **a** represented graphically as

a

draw a graphical representation of $2\mathbf{a}$, $-2\mathbf{a}$, $\frac{1}{2}\mathbf{a}$, and $-\mathbf{a}$.

***Solution***

2a  −2a  $\frac{1}{2}$a  −a

*Note:* In examining the scalar multiplication for $-\mathbf{a}$ in the form $k\mathbf{a}$, we have $k = -1$. ■

### *Vector Subtraction*

Scalar multiplication of a vector when $k = -1$ allows us to define vector subtraction.

*Definition*

> Vector subtraction is merely vector addition where the vectors to be subtracted have been negated; i.e., $\mathbf{a} - \mathbf{b} = \mathbf{a} + (-\mathbf{b})$.

The following example illustrates the concept of vector subtraction.

EXAMPLE 3 Referring to Figure 8-5, sketch $\mathbf{a} - \mathbf{b}$.

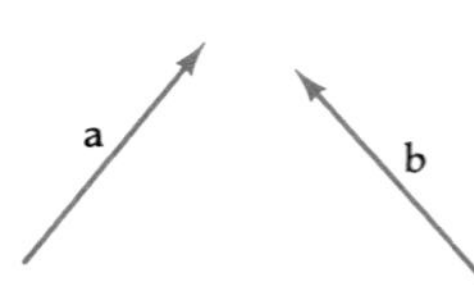


*Figure 8-5*

***Solution***

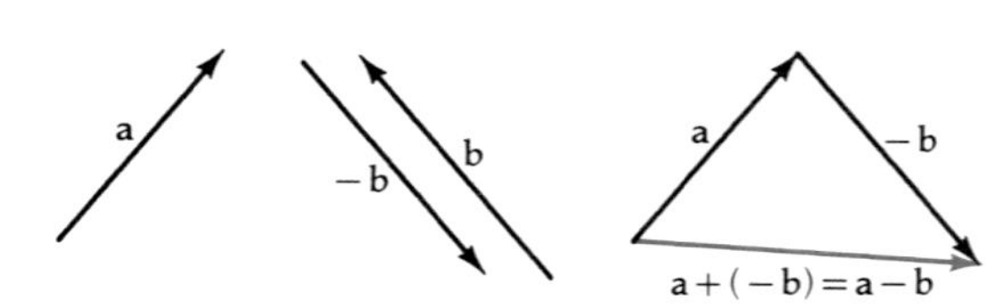

We add the vectors **a** and ($-$**b**), since $\mathbf{a} - \mathbf{b} = \mathbf{a} + (-\mathbf{b})$. A word of caution here. Vector subtraction is *not commutative*. That is, $\mathbf{a} - \mathbf{b} \neq \mathbf{b} - \mathbf{a}$. ■

The geometrical addition and subtraction of vectors is important because it gives insight and a visual perspective of the process. However, from a quantitative point of view it is generally inadequate, and the analytical or algebraic approach is preferred. This requires the introduction of a coordinate plane and trigonometry.

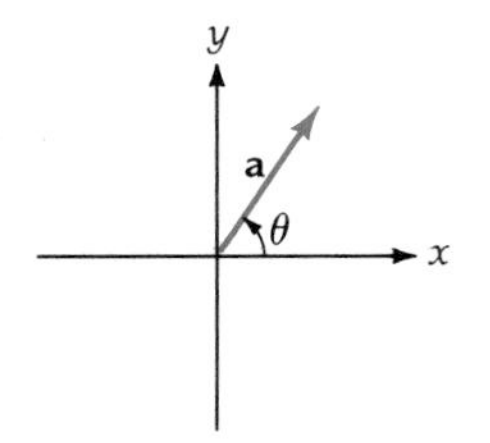

*Figure 8-6*

## Algebraic Properties of Vectors

### Polar Form of a Vector

Figure 8-6 shows the vector **a** whose origin or tail is located at the origin of the coordinate axis and whose direction is specified by the angle $\theta$ in standard position.

The vector **a** is now uniquely defined by its magnitude $|\mathbf{a}|$ and the angle $\theta$. We define the polar form of **a** to be $\mathbf{a} = |\mathbf{a}| \angle \theta$.

EXAMPLE 4 Write the polar form of vectors **b**, **c**, and **d** shown in the sketches below.

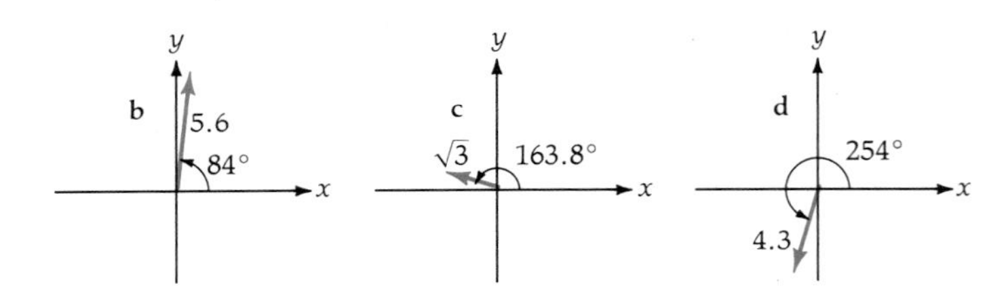


*Solution* $\mathbf{b} = 5.6 \angle 84^\circ$ $\quad \mathbf{c} = \sqrt{3} \angle 163.8^\circ$ $\quad \mathbf{d} = 4.3 \angle 254^\circ$ ■

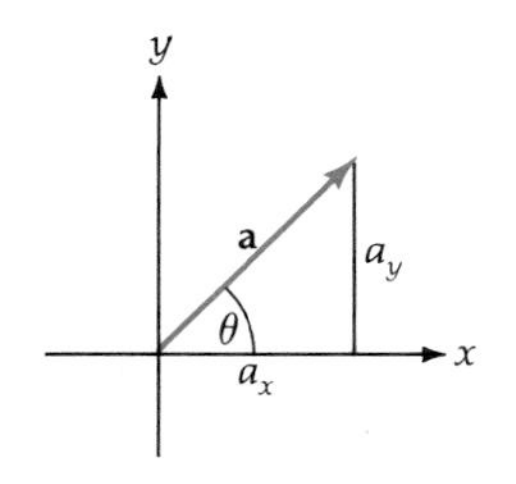

*Figure 8-7*

### Resolution of a Vector into Rectangular Coordinates

We may now resolve **a** into its **rectangular** (or $x$ and $y$) **components** by the trigonometric functions sine and cosine. Figure 8-7 illustrates these quantities. $a_x$ is defined as the **scalar component** of **a** in the $x$-direction, and $a_y$ is defined as the **scalar component** of **a** in the $y$-direction where $a_x = |\mathbf{a}| \cos \theta$ and $a_y = |\mathbf{a}| \sin \theta$. When $a_x$ and $a_y$ have been calculated, **a** is said to have been resolved into its rectangular components. These components will be a necessary requirement for the algebraic additions of vectors. They may be referred to as *scalar components* or as *rectangular components.*

EXAMPLE 5 Resolve vector $\mathbf{a} = 26 \angle 30^\circ$ into its components $a_x$ and $a_y$.

***Solution***

$$a_x = |\mathbf{a}| \cos \theta = 26 \cos 30^\circ = (26)\left(\frac{\sqrt{3}}{2}\right) = \frac{26\sqrt{3}}{2} = 13\sqrt{3}$$

$$a_y = |\mathbf{a}| \sin \theta = 26 \sin 30^\circ = 26\left(\frac{1}{2}\right) = 13$$

A convenient notation for writing the components of a vector is to enclose them within the symbol $\langle\ \rangle$. Thus we have

$$\mathbf{a} = \langle 13\sqrt{3}, 13 \rangle \quad ■$$

### *Algebraic Operations on Vectors Expressed in Terms of Their Rectangular Components*

*Definition*

> If vectors are written in terms of their scalar components, the components of the sum of two vectors are the sum of their corresponding components. If $\mathbf{A} = \langle x, y \rangle$ and $\mathbf{B} = \langle w, z \rangle$, then $\mathbf{C} = \mathbf{A} + \mathbf{B}$ is obtained by the operation $\langle x, y \rangle + \langle w, z \rangle = \langle x + w, y + z \rangle$.

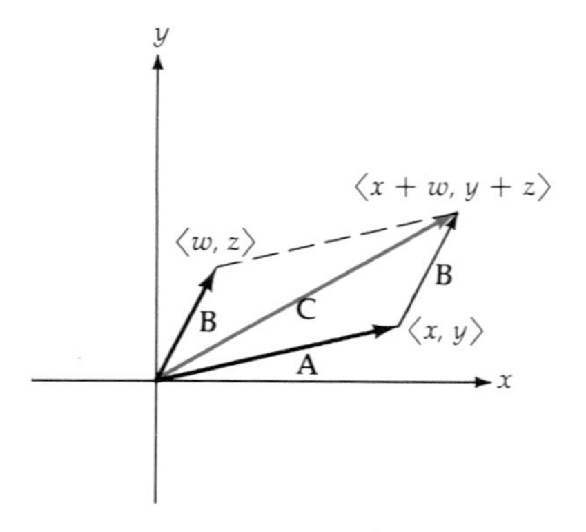


*Figure 8-8*

By examining Figure 8-8 and recalling the geometric approach to addition of two vectors, the student will see that the definition of addition of the components of two vectors seems reasonable.

In similar fashion, we state the following definitions.

*Definition*

For all vectors $\mathbf{A} = \langle x, y\rangle$, $\mathbf{B} = \langle w, z\rangle$ and all real scalars $c$, the following operations are defined:

1. When a vector is multiplied by a scalar, the components of the vector are multiplied by the scalar, that is,
$$c\langle x, y\rangle = \langle cx, cy\rangle$$
2. The vector $-\mathbf{A}$ can be written as $(-1)\mathbf{A}$, that is,
$$-\langle x, y\rangle = \langle -x, -y\rangle$$
3. To subtract vector **B** from vector **A**, subtract the components of **B** from the components of **A**, that is,
$$\langle x, y\rangle - \langle w, z\rangle = \langle x - w, y - z\rangle$$

EXAMPLE 6 Perform the following calculations.

**a.** $\langle 3, 2\rangle + \langle -8, 7\rangle + \langle 2, -1\rangle$ **b.** $\frac{5}{3}\langle 2, -6\rangle$

**c.** $-\langle -3, 5\rangle$ **d.** $\langle 1.6, -2.5\rangle - \langle 8.6, 3.7\rangle$

*Solution*

**a.** $\langle 3, 2\rangle + \langle -8, 7\rangle + \langle 2, -1\rangle = \langle -3, 8\rangle$

**b.** $\frac{5}{3}\langle 2, -6\rangle = \left\langle \frac{10}{3}, -10\right\rangle$

**c.** $-\langle -3, 5\rangle = \langle 3, -5\rangle$

**d.** $\langle 1.6, -2.5\rangle - \langle 8.6, 3.7\rangle = \langle 1.6 - 8.6, -2.5 - 3.7\rangle = \langle -7.0, -6.2\rangle$ ■

The ability to rapidly perform these algebraic operations using vector components is most useful in a variety of vector applications.

#### Determining Magnitude and Director of a Vector Expressed in Terms of Its Rectangular Components

The Pythagorean Theorem enables us to find the magnitude of any vector if we know its components.

*Definition*

The *magnitude* of any vector $\mathbf{A} = \langle x, y\rangle$ is given by
$$|\mathbf{A}| = \sqrt{x^2 + y^2}$$

EXAMPLE 7 Find the magnitude of each vector.

a. $\mathbf{A} = \langle 7, -5 \rangle$ b. $\mathbf{B} = \langle -6, -8 \rangle$

***Solution*** a. $|\mathbf{A}| = \sqrt{7^2 + (-5)^2} = \sqrt{49 + 25} = \sqrt{74} = 8.60$ (nearest hundredth)

b. $|\mathbf{B}| = \sqrt{(-6)^2 + (-8)^2} = \sqrt{36 + 64} = \sqrt{100} = 10$ ■

Our knowledge of trigonometry enables us to find the direction $\theta$ of a vector if we know the components.

***Definition***

> The positive *direction angle* $\theta$ measured counterclockwise from the positive $x$-axis to any vector $\mathbf{A} = \langle x, y \rangle$ is given by $\tan \theta = \dfrac{y}{x}$ if $x \neq 0$.

EXAMPLE 8 Find the smallest positive direction angle $\theta$ for $\mathbf{A} = \langle -1, -\sqrt{3} \rangle$.

***Solution*** $\mathbf{A} = \langle -1, -\sqrt{3} \rangle$

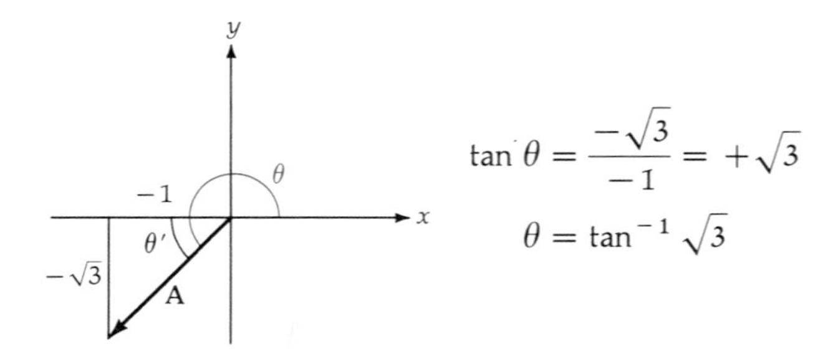


$$\tan \theta = \frac{-\sqrt{3}}{-1} = +\sqrt{3}$$

$$\theta = \tan^{-1} \sqrt{3}$$

■

Now the reference angle $\theta'$ is $60°$, but since $\theta$ is in quadrant III, we have $\theta = 240°$.

## EXERCISE 8-8

Given vectors **a**, **b**, **c**, **d**, **e** as shown below. Perform the indicated operations graphically in Problems 1–12.

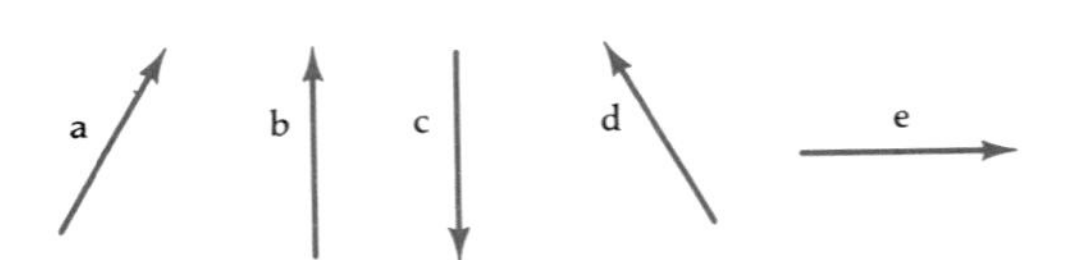


1. **a** + **b** 2. **c** + **e** 3. **a** − **b** 4. **c** − **e** 5. **e** + **d** + **b** 6. **a** + **c** + **b**
7. 3**d** 8. 2**c** 9. −4**a** 10. −3**d** 11. 2**a** − 3**d** 12. 4**b** − 3**a**

In Problems 13 and 14 assume that the angle $\theta$ is measured counterclockwise from the positive $x$-axis.

**13.** A vector **a** with a magnitude of 160 is inclined upward 28° from the horizontal. Find the horizontal and vertical components of **a**. Round your answer to the nearest hundredth.

**14.** A vector **b** has a magnitude of 88. It is inclined upward 75° from the horizontal. Find the horizontal and vertical components of **b**. Round your answer to the nearest hundredth.

In Problems 15–22, find 2**A**, **A** + **B**, **A** − **B**, 3**A** − 2**B**.

**15.** $\mathbf{A} = \langle 3, 7 \rangle$ $\mathbf{B} = \langle 2, 1 \rangle$

**16.** $\mathbf{A} = \langle 4, 2 \rangle$ $\mathbf{B} = \langle 3, 5 \rangle$

**17.** $\mathbf{A} = \langle -6, 2 \rangle$ $\mathbf{B} = \langle 2, -1 \rangle$

**18.** $\mathbf{A} = \langle -7, -3 \rangle$ $\mathbf{B} = \langle -4, 2 \rangle$

**19.** $\mathbf{A} = \left\langle 0, \frac{1}{2} \right\rangle$ $\mathbf{B} = \langle -3, 4 \rangle$

**20.** $\mathbf{A} = \langle -1, 2 \rangle$ $\mathbf{B} = \left\langle \frac{1}{3}, 0 \right\rangle$

**21.** $\mathbf{A} = \langle 1.6, -2.3 \rangle$ $\mathbf{B} = \langle 5.1, 2.6 \rangle$

**22.** $\mathbf{A} = \langle 1.1, 2.8 \rangle$ $\mathbf{B} = \langle -7.6, 0.5 \rangle$

**23.** If $\mathbf{A} = \langle -6, -1 \rangle$, $\mathbf{B} = \langle 2, 3 \rangle$, and $\mathbf{C} = \langle -5, 4 \rangle$, find $\mathbf{A} + 3\mathbf{B} - \mathbf{C}$.

**24.** If $\mathbf{A} = \langle 0, -7 \rangle$, $\mathbf{B} = \langle -3, -2 \rangle$, and $\mathbf{C} = \langle 6, -8 \rangle$, find $\mathbf{A} - \mathbf{B} + 2\mathbf{C}$.

In Problems 25–32, write each vector in polar form with $\theta$ measured from the positive $x$-axis such that $0 \le \theta < 360°$.

**25.**

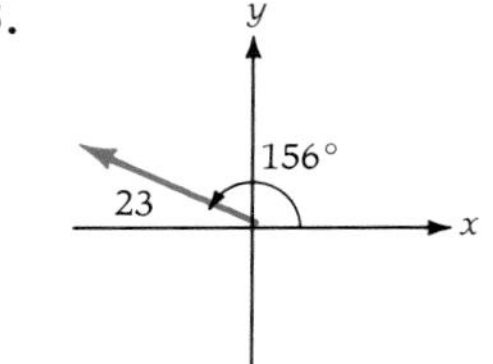


**26.**

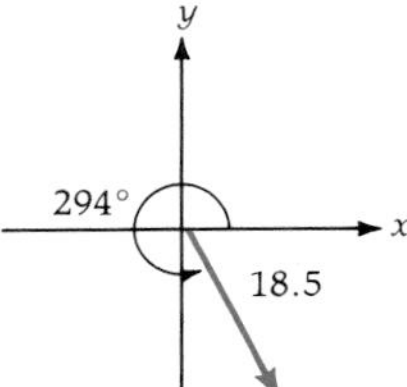


**27.**

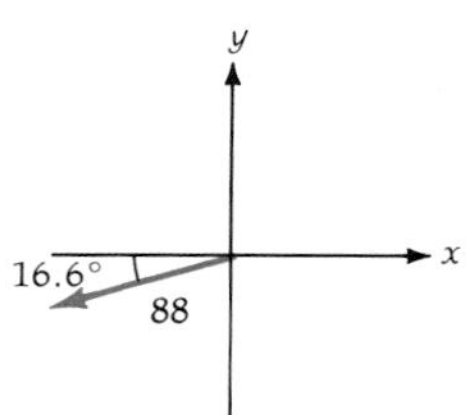


**28.**

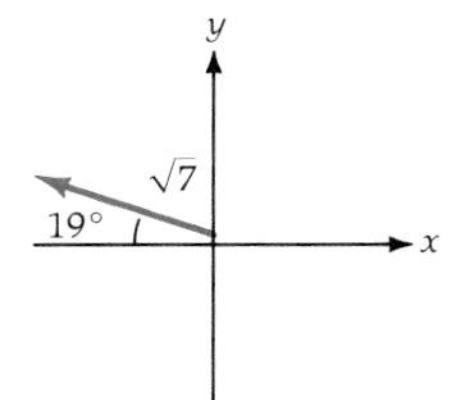


**29.**

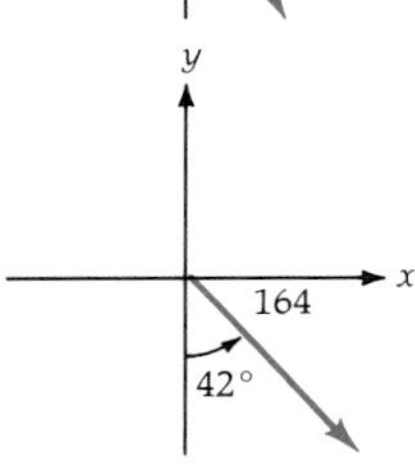


**30.**

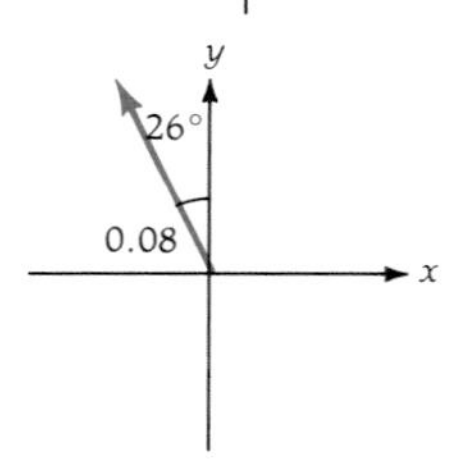


**31.**

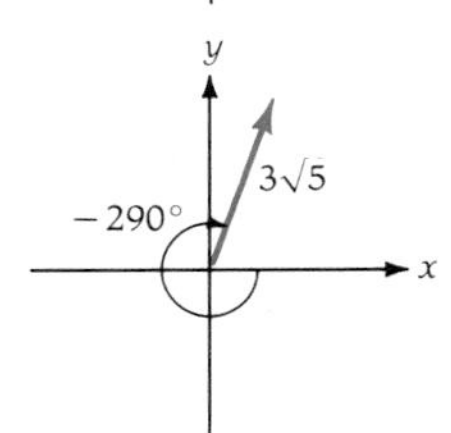


**32.**

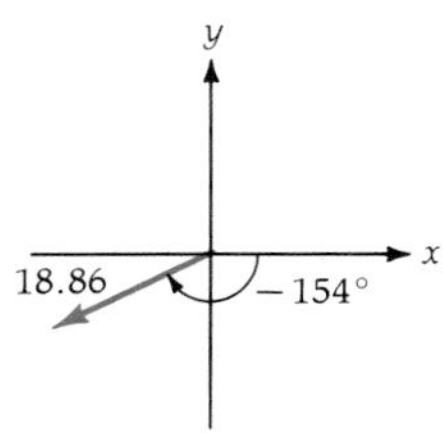


In Problems 33–44, find the magnitude of each vector to the nearest hundredth. Find the positive direction of $\theta$ where $0 \le \theta < 360°$ and $\theta$ is measured from the positive $x$-axis. Express your value for $\theta$ to the nearest tenth of a degree.

**33.** $\mathbf{A} = \langle -3, 3 \rangle$

**34.** $\mathbf{B} = \langle -4, -4 \rangle$

**35.** $\mathbf{C} = \langle \sqrt{3}, -1 \rangle$

**36.** $D = \left\langle -\frac{\sqrt{3}}{2}, -\frac{1}{2} \right\rangle$ **37.** $E = \langle -2, 5 \rangle$ **38.** $F = \langle 6, -2 \rangle$

**39.** $G = \langle 0, -1.8 \rangle$ **40.** $H = \langle -10.2, 5.1 \rangle$ **41.** $J = \left\langle \frac{1}{4}, \frac{1}{6} \right\rangle$

**42.** $K = \left\langle \frac{2}{3}, -\frac{1}{5} \right\rangle$ **43.** $L = \langle -1.62, 0 \rangle$ **44.** $M = \langle -8.62, -8.62 \rangle$

In Problems 45–52, find the $x$-component and the $y$-component of each vector. Round your answer to the nearest hundredth.

**45.** $1.6 \angle 30°$ **46.** $2.5 \angle 120°$ **47.** $8.0 \angle 315°$ **48.** $12.5 \angle 225°$

**49.** $|\mathbf{A}| = 15, \theta = 126°$ **50.** $|\mathbf{B}| = 9.5, \theta = 224°$

**51.** $|\mathbf{C}| = 0.62, \theta = 314.6°$ **52.** $|\mathbf{D}| = 0.06, \theta = 156.3°$

★ **53.** Draw three vectors **A**, **B**, and **C** and show graphically that $(\mathbf{A} + \mathbf{B}) + \mathbf{C} = \mathbf{A} + (\mathbf{B} + \mathbf{C})$.

★ **54.** Draw two vectors **A** and **B** where $|\mathbf{A}| \neq |\mathbf{B}|$ and graphically show that, in general, $\mathbf{A} - \mathbf{B} \neq \mathbf{B} - \mathbf{A}$.

## COMPUTER PROBLEMS FOR CHAPTER 8

See if you can apply your knowledge of computer programming to do Problems 1 and 2.

**1.** Assume that in a uniquely defined triangle you are given two sides and an angle opposite one of them. Write a computer program that will print out the measure in degrees of the other two angles and the one unknown side. Test your program by doing Problem 10 of Exercise 8-5.

**2.** Modify your program in Problem 1 above so that it will print out "no solution" if the triangle is not determined and it will print out "a second solution is" if there are two possible triangles. Test your program by doing Problems 11 and 20–23 of Exercise 8-5.

## KEY TERMS AND CONCEPTS

Be sure you can explain what is meant by each of the following.

Conditional equation
Identity
Acute triangle
Obtuse triangle
Law of sines
Bearing
Law of cosines
Conditional trigonometric equations
Vector
Scalar
Rectangular components of a vector
Vector components of a vector
Magnitude of a vector
Direction of a vector
Equal vectors
Vector addition
Multiplication of a vector by a scalar
Polar form of a vector
Algebraic representation of a vector

## SUMMARY OF PROCEDURES AND CONCEPTS

The important formulas established in this chapter are listed below for your convenience.

### Basic Identities

$$\sin A \csc A \equiv 1 \qquad \cot A \equiv \frac{\cos A}{\sin A}$$

$$\cos A \sec A \equiv 1$$

$$\tan A \cot A \equiv 1 \qquad \sin^2 A + \cos^2 A \equiv 1$$

$$\tan A \equiv \frac{\sin A}{\cos A} \qquad 1 + \tan^2 A \equiv \sec^2 A$$

$$1 + \cot^2 A \equiv \csc^2 A$$

### Sum and Difference of Two Arguments

$$\cos (A + B) \equiv \cos A \cos B - \sin A \sin B$$

$$\cos (A - B) \equiv \cos A \cos B + \sin A \sin B$$

$$\sin (A + B) \equiv \sin A \cos B + \cos A \sin B$$

$$\sin (A - B) \equiv \sin A \cos B - \cos A \sin B$$

$$\tan (A + B) \equiv \frac{\tan A + \tan B}{1 - \tan A \tan B}$$

$$\tan (A - B) \equiv \frac{\tan A - \tan B}{1 + \tan A \tan B}$$

### Double Argument

$$\sin 2A \equiv 2 \sin A \cos A$$

$$\cos 2A \equiv \cos^2 A - \sin^2 A \equiv 1 - 2 \sin^2 A \equiv 2 \cos^2 A - 1$$

$$\tan 2A \equiv \frac{2 \tan A}{1 - \tan^2 A}$$

### Half-Arguments

$$\cos \frac{A}{2} \equiv \pm\sqrt{\frac{1 + \cos A}{2}}$$

$$\sin \frac{A}{2} \equiv \pm\sqrt{\frac{1 - \cos A}{2}}$$

$$\tan \frac{A}{2} \equiv \pm\sqrt{\frac{1 - \cos A}{1 + \cos A}} \equiv \frac{\sin A}{1 + \cos A} \equiv \frac{1 - \cos A}{\sin A}$$

**Product**

$$\sin A \cos B \equiv \frac{1}{2}[\sin (A + B) + \sin (A - B)]$$

$$\cos A \sin B \equiv \frac{1}{2}[\sin (A + B) - \sin (A - B)]$$

$$\cos A \cos B \equiv \frac{1}{2}[\cos (A + B) + \cos (A - B)]$$

$$\sin A \sin B \equiv -\frac{1}{2}[\cos (A + B) - \cos (A - B)]$$

**Sum and Difference**

$$\sin A + \sin B \equiv 2 \sin \frac{1}{2}(A + B) \cos \frac{1}{2}(A - B)$$

$$\sin A - \sin B \equiv 2 \cos \frac{1}{2}(A + B) \sin \frac{1}{2}(A - B)$$

$$\cos A + \cos B \equiv 2 \cos \frac{1}{2}(A + B) \cos \frac{1}{2}(A - B)$$

$$\cos A - \cos B \equiv -2 \sin \frac{1}{2}(A + B) \sin \frac{1}{2}(A - B)$$

**Law of Sines**

In any triangle $ABC$ with sides $a$, $b$, $c$,

$$\frac{a}{\sin A} = \frac{b}{\sin B} = \frac{c}{\sin C}$$

**Law of Cosines**

In any triangle $ABC$ with sides $a$, $b$, $c$ and angle $C$ as the angle between sides $a$ and $b$,

$$c^2 = a^2 + b^2 - 2ab \cos C$$

## CHAPTER 8 REVIEW EXERCISE

Complete these problems without the use of a calculator or a trigonometric table.

**1.** Find $\sin 30°$ by computing:

**a.** $\sin (90° - 60°)$ **b.** $\sin \frac{60°}{2}$

**2.** Find $\sin 60°$ by computing:

**a.** $\sin (90° - 30°)$ **b.** $\sin 2(30°)$

3. Find $\cos 75°$ by using the difference of two angles.
4. Find $\tan 75°$ by using the sum of two angles.

In Problems 5–14, assume $\sin B = \frac{5}{13}$ ($B$ in quadrant II) and $\cos A = \frac{4}{5}$ ($A$ in quadrant IV). Find the value of each function.

5. $\cos 2A$
6. $\tan 2A$
7. $\sin \frac{A}{2}$
8. $\sec \frac{A}{2}$
9. $\tan \frac{B}{2}$
10. $\sin 2B$
11. $\sin (A + B)$
12. $\tan (A - B)$
13. $\sec (A - B)$
14. $\sin (2A + B)$
15. Find $\sin \theta$ if $\tan 2\theta = \frac{4}{3}$ ($2\theta$ in quadrant III).
16. Express $\tan 3\theta$ in terms of $\tan \theta$.
17. Express $\cos 25° - \cos 55°$ as a product.
18. Express $\sin 7x + \sin 3x$ as a product.
19. Express $2 \cos 38° \cos 61°$ as a sum.
20. Express $\sin 3x \cos x$ as a sum.

Verify the identities in Problems 21–40.

21. $\dfrac{1 + \sin \theta}{\cot^2 \theta} \equiv \dfrac{\sin \theta}{\csc \theta - 1}$
22. $(1 - \sin^2 \alpha)(1 + \tan^2 \alpha) \equiv 1$
23. $\dfrac{\tan^2 x + 1}{\cot^2 x + 1} \equiv \tan^2 x$
24. $\cos A - \sin A \equiv \dfrac{\csc A - \sec A}{\sec A \csc A}$
25. $\dfrac{\tan x - 1}{\tan x + 1} \equiv \dfrac{1 - \cot x}{1 + \cot x}$
26. $\dfrac{\sec \theta}{\cot \theta + \tan \theta} \equiv \sin \theta$
27. $\cot^2 \beta \equiv \dfrac{\cos^2 \beta - \cot \beta}{\sin^2 \beta - \tan \beta}$
28. $\sec 2x \equiv \tan 2x \tan x + 1$
29. $\cos 2\alpha \equiv \dfrac{\cot^2 \alpha - 1}{\csc^2 \alpha}$
30. $\dfrac{\cos A - \cos B}{\cos A + \cos B} \equiv -\dfrac{\tan \frac{1}{2}(A + B)}{\cot \frac{1}{2}(A - B)}$
31. $\dfrac{\cos 2\alpha - \cos 2\beta}{\cos \alpha - \cos \beta} \equiv 2(\cos \alpha + \cos \beta)$
32. $\dfrac{\sec^2 \theta}{\sec^2 \theta - 1} \equiv \csc^2 \theta$
33. $\cos \alpha + \cot \alpha \equiv \dfrac{1 + \sin \alpha}{\tan \alpha}$
34. $\sin (A + B) \sin (A - B) \equiv \cos^2 B - \cos^2 A$
35. $\sin x \sin 5x \equiv \sin^2 3x - \sin^2 2x$
36. $\sin \theta \equiv \cot \frac{\theta}{2} (1 - \cos \theta)$
37. $\dfrac{1 + \cos \beta}{\cos \beta + \sin \beta + 1} \equiv \dfrac{1 + \sec \beta}{\sec \beta + \tan \beta + 1}$
38. $\sin x \cos x \equiv \dfrac{\sin x - \cos x}{\tan x \csc x - \sec x \cot x}$

**39.** $\sin^2 \frac{A}{2} \equiv \frac{\sin A \tan \frac{A}{2}}{2}$

**40.** $\frac{1}{\csc 3\theta} \equiv \frac{\sin \theta}{\sec 2\theta} + \frac{\cos \theta}{\csc 2\theta}$

In Problems 41–48, find the indicated part of the general triangle $ABC$. If no solution is possible, so state.

**41.** $B = 51^\circ$, $C = 28^\circ$, $c = 23$. Find $b$.

**42.** $B = 38^\circ$, $b = 13$, $c = 15$. Find $C$.

**43.** $B = 120^\circ$, $a = 31$, $c = 46$. Find $b$.

**44.** $A = 30^\circ$, $a = 8.0$, $b = 10$. Find $B$.

**45.** $C = 28^\circ$, $a = 43$, $b = 29$. Find $c$.

**46.** $a = 23$, $b = 17$, $c = 19$. Find $A$.

**47.** $C = 41^\circ$, $a = 18$, $c = 10$. Find $A$.

**48.** $a = 22$, $b = 45$, $c = 34$. Find $B$.

In Problems 49–55, solve for all unknown parts of the general triangle $ABC$.

**49.** $A = 30^\circ$, $a = 12$, $b = 10$

**50.** $C = 27^\circ$, $a = 16$, $b = 20$

**51.** $A = 59^\circ$, $B = 43^\circ$, $b = 24$

**52.** $A = 57^\circ 20'$, $a = 16.7$, $b = 19.2$

**53.** $A = 118^\circ$, $b = 1.3$, $c = 1.6$

**54.** $B = 76^\circ 40'$, $C = 68^\circ 30'$, $b = 67.3$

**55.** $a = 2.0$, $b = 4.2$, $c = 2.5$

**56.** Person $A$ flies a kite using 150 meters of taut string. Person $B$, directly under the string a distance of 38 meters from $A$, finds the angle of elevation of the kite to be $54^\circ$. Find the angle the kite string makes with the ground.

**57.** Two trees, $A$ and $B$, are on opposite sides of a pond. A third tree $C$ is measured to be 35 meters from tree $A$ and 54 meters from tree $B$. Angle $BAC$ is found to be $48^\circ$. Find the distance across the pond.

**58.** Tower $A$ is 1.9 kilometers north of tower $B$. Tower $C$, to the east of line $AB$, is 4.3 kilometers from tower $A$ and 3.1 kilometers from tower $B$. Find the bearing of tower $C$ from tower $B$.

**59.** Two lighthouses, $A$ and $B$, are 3500 meters apart along the north shoreline of a lake. Lighthouse $A$ is west of lighthouse $B$. A boat leaves lighthouse $B$ and sails on course $237.6^\circ$ for 2840 meters. The boat then finds the angle between the lines of sight to the two lighthouses to be $98.8^\circ$. How far is the boat from lighthouse $A$ (to the nearest ten meters)?

**60.** A ship sails on a course $132^\circ$ for 46 nautical miles. It then changes course to $62^\circ$ and sails 28 nautical miles. What should the course be in order to return to the starting point?

Express all answers in degrees $\theta$ or $x$ where $0^\circ \leq \theta < 360^\circ$ or $0^\circ \leq x < 360^\circ$ in Problems 61–84. If there is no solution, so state.

**61.** $\sin^2 \theta + \sin \theta = 0$

**62.** $2 \sin^2 x - 5 \sin x + 2 = 0$

**63.** $2 \tan 2x - 3 = 0$

**64.** $2 \sin^2 \theta + 7 \sin \theta - 4 = 0$

**65.** $2 \cos^2 \theta + 5 \sin \theta + 1 = 0$

**66.** $2 \sin^2 x - \cos 2x = 0$

**67.** $\tan^2 \theta - \sec \theta = 1$

**68.** $\sqrt{3} \sec x - 2 = 0$

**69.** $\cos^2 x + \sin x + 1 = 0$

**70.** $2 \sin^2 \theta + \sin \theta = 0$

**71.** $\sin 2x + \sin x = 0$

**72.** $\sin 2x \sin x + \sin 2x + \sin x + 1 = 0$

**73.** $2 \cos^2 \theta - \sqrt{3} \cos \theta = 0$

**74.** $\cos x + 3 \sin x = 2$

**75.** $\sec \theta = \cos \theta - \tan \theta$

**76.** $\sec 3x + 2 = 0$

**77.** $2\sin^4 x - 3\sin^2 x + 1 = 0$ **78.** $\tan\frac{x}{2} + \cot x = 0$

**79.** $\sin\theta = 2\cos\frac{1}{2}\theta$ **80.** $\sin 3x - \sin x = 0$

**81.** $\mathbf{a} = \langle 2, 1.5\rangle$ and $\mathbf{b} = \langle -7, 3.8\rangle$, find $\mathbf{a} + \mathbf{b}$, $\mathbf{a} - 2\mathbf{b}$.

**82.** $\mathbf{a} = \left\langle \frac{1}{3}, \frac{1}{4} \right\rangle$ and $\mathbf{b} = \langle 2, -5\rangle$, find $3\mathbf{a} + \mathbf{b}$, $\frac{1}{2}\mathbf{a} - \mathbf{b}$.

Express the vector **a** in terms of its rectangular components.

**83.** If $|\mathbf{a}| = 17$ and $\theta = 224°$. **84.** If $|\mathbf{a}| = 2.6$ and $\theta = 175°$.

## PRACTICE TEST FOR CHAPTER 8

Complete the following problems without the use of a calculator or trigonometric table.

In Problems 1–2, find the indicated function if $\tan A = \frac{1}{2}$ ($A$ in quadrant III) and $\sin B = \frac{12}{13}$ ($B$ in quadrant II).

**1.** $\tan 2A$ **2.** $\tan(A - B)$

**3.** Find $\tan\theta$ if $\sin 2\theta = -\frac{12}{13}$ ($2\theta$ in quadrant III).

**4.** Verify $\cot A + \tan A \equiv \cot A \sec^2 A$.

**5.** Find $\tan 4\theta$ if $\tan\theta = \frac{1}{2}$ ($\theta$ in quadrant I).

**6.** Verify $\dfrac{1 + \tan A}{1 + \cot A} \equiv \dfrac{\sec A}{\csc A}$.

**7.** Verify $\cot^2\theta - \cos^2\theta \equiv \cot^2\theta\cos^2\theta$.

**8.** In triangle $ABC$, $B = 73°$, $C = 24°$, and $b = 13$. Find side $a$.

**9.** In triangle $ABC$, $a = 31$, $b = 37$, and $c = 28$. Find angle $C$.

**10.** A ship takes a bearing on a buoy of N28°E. The ship travels 5.2 nautical miles due north and takes another bearing of N81°E. How far is the ship from the buoy?

Express your answer in degrees, where $0° \le \theta < 360°$ or $0° \le x < 2 < 360°$ in Problems 11–15.

**11.** $\sec^2\theta - 3\sec\theta + 2 = 0$ **12.** $2\cos^2 x - 5\sin x + 1 = 0$

**13.** $\sin x \sec^2 x - 2\sin x = 0$

**14.** If $\mathbf{a} = \left\langle \frac{1}{2}, 7 \right\rangle$ and $\mathbf{b} = \left\langle \frac{1}{3}, -5 \right\rangle$, find $\mathbf{a} - 3\mathbf{b}$.

**15.** Express the vector **a** in terms of its rectangular components if $|\mathbf{a}| = 15.3$ and $\theta = 165°$.

# 9 Complex Numbers

■ *Civilization advances by extending the number of operations which we can perform without thinking about them.*

*Alfred North Whitehead*

The concept of a complex number took quite a long time to become acceptable in the mathematical community. Most early mathematicians could not make any sense out of the expression $\sqrt{-1}$. The thought was considered either ridiculous or beneath the dignity of an educated person. The fact that we retain the word "imaginary number" in our vocabulary is testimony to the fact that early mathematicians did not think these numbers reflected phenomena in the "real" world.

## 9-1 BASIC PROPERTIES OF COMPLEX NUMBERS

*Definition 9-1*

The imaginary number unit $i$ is defined as the square root of $-1$ so that $i = \sqrt{-1}$ and $i^2 = -1$.

Simplification of imaginary expressions is often done by using the next definition.

*Definition 9-2*

For all real numbers $a > 0$ we have

$$\sqrt{-a} = \sqrt{-1}\sqrt{a} = i\sqrt{a}$$

Using Definition 9-2, we can write $\sqrt{-25}$ as $5i$ or $\sqrt{-7}$ as $i\sqrt{7}$.

**Warning:** It is necessary to use Definition 9-2 *before* performing algebraic operations on square roots of negative numbers. For example, to multiply $\sqrt{-3}\sqrt{-4}$, we would proceed as follows:

$$
\begin{aligned}
(\sqrt{-3})(\sqrt{-4}) &= (i\sqrt{3})(i\sqrt{4}) && \textit{By Definition 9-2.}\\
&= (i\sqrt{3})(2i) && \textit{Simplify } \sqrt{4}.\\
&= 2i^2\sqrt{3} && \textit{Multiplication.}\\
&= -2\sqrt{3} && \textit{By Definition 9-1.}
\end{aligned}
$$

Do you see *why* it is *incorrect* for a student to say $\sqrt{-3}\sqrt{-4} = \sqrt{12} = 2\sqrt{3}$?

From these definitions it follows any number of the form $bi$ where $b$ is a real number is called an **imaginary number**.

Finally, we have numbers that have real parts added to imaginary parts. These are the complex numbers.

*Definition 9-3*

A number that can be written as the sum of a real number and an imaginary number of the form $(a + bi)$ where $a$ and $b$ are real is called a *complex number.*

It can be seen from this definition that all real numbers are complex numbers. This is true because all real numbers can be written in the form $a + bi$ with $b = 0$.

We will now define the equality of complex numbers.

*Definition 9-4*

Two complex numbers $(a + bi)$ and $(c + di)$ are equal if and only if $a = c$ and $b = d$.

**EXAMPLE 1** Find real numbers $g$ and $h$ if $g + i\sqrt{7} = -3 + hi$.

***Solution*** By Definition 9-4 we have $g = -3$ and $h = \sqrt{7}$. ■

Complex numbers can be added, subtracted, multiplied, divided, raised to powers, etc. Rules for these operations follow.

*Definition 9-5*

For all real numbers $a$, $b$, $c$, $d$,

$$(a + bi) + (c + di) = (a + c) + (b + d)i$$

*Definition 9-6*

For all real numbers $a$, $b$, $c$, $d$,

$$(a + bi) - (c + di) = (a - c) + (b - d)i$$

The strict application of this definition is not actually required.

In actual practice, when a student subtracts one complex number from another, it is usually somewhat more convenient to remove parentheses and add real terms and imaginary terms separately.

EXAMPLE 2 Subtract:

**a.** $(3 + 4i) - (2 - 6i)$ **b.** $(2 - 5i) - (-3 + 7i)$

***Solution***

**a.** $(3 + 4i) - (2 - 6i) = 3 + 4i - 2 + 6i = 1 + 10i$

**b.** $(2 - 5i) - (-3 + 7i) = 2 - 5i + 3 - 7i = 5 - 12i$ ■

Rather than memorize Definitions 9-5 and 9-6, students often verbalize the rules together as follows:

---

To add or subtract complex numbers, combine the real parts and combine the imaginary parts.

---

Note that this verbal rule is quite consistent with our concept of combining like terms. Obviously, a real number and an imaginary number would not be considered like terms.

*Definition 9-7*

For all real numbers $a$, $b$, $c$, $d$,

$$(a + bi)(c + di) = (ac - bd) + (ad + bc)i$$

Definition 9-7 is not a useful form to memorize; rather it is more helpful to understand the following:

---

To multiply complex numbers, consider them as binomials and use the distributive property.

---

Let us examine two different types of multiplication problems involving complex numbers.

**EXAMPLE 3** Multiply: $(5 + 2i)(4 + 6i)$.

***Solution***

$$\begin{aligned}(5 + 2i)(4 + 6i) &= 20 + 30i + 8i + 12i^2 && \textit{Multiply the two binomials.}\\ &= 20 + 38i + 12i^2 && \textit{Collect like terms.}\\ &= 20 + 38i + 12(-1) && \textit{Use Definition 9-1.}\\ &= 20 - 12 + 38i\\ &= 8 + 38i \quad \blacksquare\end{aligned}$$

**EXAMPLE 4** Multiply: $5(3 + 4i)$.

***Solution*** $5(3 + 4i) = 15 + 20i$ *Here we need to apply the distributive property only once.* ■

One other principle of multiplication is important to notice. How would you evaluate $i^n$ where $n$ is any positive integer?

We have defined $i^2 = -1$

We could write $i^3 = i^2 \cdot i = (-1)i = -i$

and likewise $i^4 = i^2 \cdot i^2 = (-1)(-1) = +1$

$i^5 = i \cdot i^4 = (+1)i = +i$

Using the same technique, we can obtain the following interesting pattern:

| | | | |
|---|---|---|---|
| $i = i$ | $i^2 = -1$ | $i^3 = -i$ | $i^4 = 1$ |
| $i^5 = i$ | $i^6 = -1$ | $i^7 = -i$ | $i^8 = 1$ |
| $i^9 = i$ | $i^{10} = -1$ | $i^{11} = -i$ | $i^{12} = 1$ |

A continuation of this pattern is left as an exercise.

**EXAMPLE 5** Evaluate:

**a.** $i^{36}$ **b.** $i^{27}$

***Solution***

**a.** $i^{36} = (i^4)^9 = (1)^9 = 1$

**b.** $i^{27} = (i^{26})(i) = (i^2)^{13}i = (-1)^{13}i = -i$ ■

The solutions to Example 5 can be obtained by using other approaches as well. Can you think of another way to obtain the answers?

---

To divide two complex numbers, multiply the numerator and the denominator by the *conjugate* of the denominator.

---

Be sure you understand the idea of a **conjugate**. The conjugate of $a + bi$ is $a - bi$. The conjugate of $a - bi$ is $a + bi$.

EXAMPLE 6 Divide $(2 - 5i) \div (1 - 3i)$ by using the conjugate of the denominator.

***Solution***

$$(2 - 5i) \div (1 - 3i) = \frac{2 - 5i}{1 - 3i}$$ *The conjugate of $1 - 3i$ is $1 + 3i$.*

$$= \frac{(2 - 5i)}{(1 - 3i)} \cdot \frac{(1 + 3i)}{(1 + 3i)}$$ *Multiplication of both numerator and denominator by the conjugate.*

$$= \frac{2 + 6i - 5i - 15i^2}{1 - 9i^2} = \frac{2 + i - 15(-1)}{1 - 9(-1)} = \frac{17 + i}{10}$$

If necessary, we may now express the answer in the form $a + bi$:

$$= \frac{17}{10} + \frac{1}{10}i$$ ■

## EXERCISE 9-1

Perform the following operations in Problems 1–32. Simplify your answers.

**1.** $(2 + 3i) + (5 + 4i)$ **2.** $(6 + 5i) + (1 - 6i)$ **3.** $(5 - 2i) - (8 - i)$

**4.** $(11 + 7i) - (3 + 4i)$ **5.** $(3 - 2i) + (6 - 7i) - (4 - 5i)$

**6.** $(3 + 8i) - (2 - i) + (-1 - 7i)$ **7.** $5(2 - 3i)$ **8.** $-7(-4 + i)$

**9.** $-3i(-i + 6)$ **10.** $i\sqrt{2}(1 + i\sqrt{3})$ **11.** $(11 + i) - 3(2 - 5i)$

**12.** $7 - i - 4(2 + 3i)$ **13.** $(2 + 3i)(1 + 4i)$ **14.** $(2 - 5i)(1 + 2i)$

**15.** $(-4 + 3i)(2 - i)$ **16.** $(5 - i)(2 - 3i)$ **17.** $(6 - 5i)^2$

**18.** $(5 + 4i)^2$ **19.** $(x + iy)(x - iy)$ **20.** $(x - iy)^2$

**21.** $\frac{2}{3 - 4i}$ **22.** $\frac{5}{1 + 2i}$ **23.** $5 \div (4 + 3i)$

**24.** $2i \div (6 + i)$ **25.** $\frac{1 + 2i}{3 - i}$ **26.** $\frac{2 - i}{1 + 2i}$

**27.** $(2 - 3i) \div (1 - 4i)$ **28.** $(2 + 3i) \div (1 - 2i)$ **29.** $\frac{7 + 2i}{3 - 5i}$

**30.** $\dfrac{-1+3i}{5-2i}$ **31.** $\dfrac{3-2i}{3i}$ **32.** $\dfrac{4+2i}{-4i}$

Simplify in Problems 33–38 and write in the standard form $a + bi$.

**33.** $\sqrt{-16}$ **34.** $\sqrt{-36}$ **35.** $\sqrt{12} + \sqrt{-50}$

**36.** $\sqrt{48} + \sqrt{-8}$ **37.** $\sqrt{-5}\sqrt{-6}$ **38.** $\sqrt{-3}\sqrt{-6}$

Find each of the powers of $i$ in Problems 39–44.

**39.** $i^{33}$ **40.** $i^{56}$ **41.** $i^{70}$ **42.** $i^{43}$ **43.** $(i^5)^3$ **44.** $(i^7)^5$

★ **45.** Write a formula or rule to find the value of $i^n$ where $n$ is any odd positive integer.

★ **46.** Write a formula or rule to find the value of $i^n$ where $n$ is any even positive integer.

★ **47.** Evaluate $(4 + 3i)^3$ ★ **48.** Evaluate $(3 - 2i)^3$

Simplify in Problems 49 and 50 and express your answer in the form $a + bi$.

★ **49.** $\dfrac{1}{3-2i} + 3i$ ★ **50.** $7 + \dfrac{2-i}{3+i}$

---

## 9-2 COMPLEX ROOTS OF EQUATIONS

In Section 2-4 we solved some quadratic equations that had complex roots. We will now study in detail various types of equations with complex roots.

### Linear Equations with Complex Roots

We can use the same basic operations to solve linear equations with complex coefficients since complex numbers obey the field properties.

**EXAMPLE 1**

**a.** Solve for $x$: $3ix - 4 = -2x - 4i$.

**b.** Express your answer in the form $a + bi$.

***Solution***

**a.**

$3ix - 4 = -2x - 4i$

$3ix + 2x = -4i + 4$ *Obtain all the $x$ terms on one side and all the other terms on the other side.*

$x(3i + 2) = -4i + 4$ *Factor out the $x$ variable.*

$x = \dfrac{-4i + 4}{3i + 2}$ *Divide by the coefficient of $x$.*

or $x = \dfrac{4 - 4i}{2 + 3i}$

**b.** To simplify the answer to the standard form $a + bi$, we will multiply both numerator and denominator by the conjugate of the denominator:

$$\frac{4 - 4i}{2 + 3i} \cdot \frac{2 - 3i}{2 - 3i} = \frac{8 - 12i - 8i + 12i^2}{4 - 9i^2}$$

$$= \frac{8 - 20i - 12}{4 - 9(-1)}$$

$$= \frac{-4 - 20i}{4 + 9}$$

$$= \frac{-4 - 20i}{13}$$

$$= -\frac{4}{13} - \frac{20}{13}i$$

The complex root of the equation is

$$x = -\frac{4}{13} - \frac{20}{13}i \quad ■$$

### Quadratic Equations with Complex Roots

From our discussion in Section 2-4 it is clear that all quadratic equations $ax^2 + bx + c = 0$, where $a$, $b$, $c$ are real, will yield complex roots when $b^2 - 4ac < 0$.

EXAMPLE 2 Solve for $x$: $x^2 - 4x + 8 = 0$.

***Solution*** We will use quadratic formula where $a = 1$, $b = -4$, and $c = 8$:

$$x = \frac{-(-4) \pm \sqrt{(-4)^2 - 4(1)(8)}}{2(1)}$$

$$x = \frac{4 \pm \sqrt{16 - 32}}{2}$$

$$x = \frac{4 \pm \sqrt{-16}}{2}$$

$$x = \frac{4 \pm 4i}{2} \qquad x = 2 \pm 2i$$

Therefore the two roots of the equation are $x = 2 + 2i$ and $x = 2 - 2i$. ■

The procedures that were developed in Section 9-1 allow us to check the complex roots in linear or quadratic equations.

**EXAMPLE 3** Verify that $x = 2 + 2i$ is a root of $x^2 - 4x + 8 = 0$.

***Solution***

$(2 + 2i)^2 - 4(2 + 2i) + 8 \stackrel{?}{=} 0$ *Substitute $2 + 2i$ into the original equation.*

$4 + 8i + 4i^2 - 8 - 8i + 8 \stackrel{?}{=} 0$ *Perform the indicated multiplications.*

$4 + 4i^2 \stackrel{?}{=} 0$ *Since $-8i + 8i = 0$ and $-8 + 8 = 0$.*

$4 + 4(-1) \stackrel{?}{=} 0$ *Replace $i^2$ by $-1$.*

$4 - 4 = 0$ ✓ ■

We may also solve some quadratic equations that have complex coefficients.

**EXAMPLE 4** Solve for $x$: $3x^2 + (2 + 3i)x + i = 0$.

***Solution*** Here we use the quadratic formula with $a = 3$, $b = 2 + 3i$, and $c = i$:

$$x = \frac{-(2 + 3i) \pm \sqrt{(2 + 3i)^2 - 4(3)(i)}}{2(3)}$$

$$= \frac{-2 - 3i \pm \sqrt{4 + 12i + 9i^2 - 12i}}{6}$$

$$= \frac{-2 - 3i \pm \sqrt{-5}}{6} = \frac{-2 - 3i \pm i\sqrt{5}}{6}$$

If we write these roots in the form $a + bi$, we obtain

$$x = \frac{-2 - 3i + i\sqrt{5}}{6} \quad \text{and} \quad x = \frac{-2 - 3i - i\sqrt{5}}{6}$$

$$x = -\frac{1}{3} + \frac{(-3 + \sqrt{5})}{6}i \qquad x = -\frac{1}{3} + \frac{(-3 - \sqrt{5})}{6}i$$

■

In general, the solutions to the quadratic equation $ax^2 + bx + c = 0$ where $a$, $b$, or $c$ are complex numbers may be quite involved. Sometimes the roots are of the form where the radicand contains an imaginary number. In these cases the solutions may take a form such as

$$x = \frac{3 + 2i \pm \sqrt{6i}}{5}$$

Using more advanced techniques, expressions such as $\sqrt{6i}$ could be further evaluated. However, these will be left for a more advanced course in complex variables. A few problems illustrating this type of solution are left as exercises for the student.

### Determining Equations with Given Roots

It can be shown that if $x = a$ is a root of a polynomial equation in standard form, then $x - a$ is a factor of that polynomial.

EXAMPLE 5 Find an equation that has roots $x = 2i$, $x = -2i$, and $x = 1$.

***Solution*** We use the factors $(x - 2i)$, $(x + 2i)$, and $(x - 1)$ and form the equation

$$(x - 2i)(x + 2i)(x - 1) = 0$$

$$(x^2 - 4i^2)(x - 1) = 0 \quad \textit{Multiplying the first two binomials.}$$

$$(x^2 + 4)(x - 1) = 0 \quad \textit{Simplifying.}$$

$$x^3 - x^2 + 4x - 4 = 0 \quad \textit{Multiplying.}$$

■

A general discussion of the nature of the roots of polynomial equations of degree greater than 2 will be given in Chapter 11.

The multiplication to obtain the equation from the roots may involve one or more trinomial factors.

EXAMPLE 6 Find an equation that has roots $x = 1 + 3i$ and $x = 1 - 3i$.

***Solution***

$$(x - 1 - 3i)(x - 1 + 3i) = 0$$

$$[(x - 1) - 3i][(x - 1) + 3i] = 0$$

*The multiplication can be done more easily by considering the form $(a - b)(a + b) = a^2 - b^2$ where $a = x - 1$ and $b = 3i$.*

$$(x - 1)^2 - (3i)^2 = 0$$

$$x^2 - 2x + 1 - 9i^2 = 0$$

$$x^2 - 2x + 10 = 0$$

■

## EXERCISE 9-2

1–14

Solve the given equation in Problems 1–20. Express your answers in the form $a + bi$ if possible. If you obtain an $i$ term inside the radical, you do not need to express your answer as $a + bi$.

**1.** $2ix - 3 = 8$

**2.** $5ix + 2 = -4$

**3.** $5x + 2i = 7ix$

**4.** $-8x + 3i = 2ix$

**5.** $(2 - 3i)x + 4 = 2i$

**6.** $(8 - i)x - 3 = 4i$

**7.** $(2 + i)x + (4 + 7i) = (1 - 2i)x$

**8.** $(3 - i)x + (2 - 2i) = (4 + 5i)x$

**9.** $3 + 4ix - 7 - 2x = 3 + 8i$

**10.** $-5 - 2ix - 3i + x = 2 - ix$

**11.** $x^2 - 4x + 5 = 0$

**12.** $x^2 - 2x + 5 = 0$

**13.** $3x^2 + x + 2 = 0$

**14.** $2x^2 + 3x + 4 = 0$

★ **15.** $x^2 + (1 + i)x - i = 0$

**16.** $x^2 + (1 - 2i)x - i = 0$

★ **17.** $2x^2 + 3ix + 4i = 3i - 2 + i$

**18.** $2x^2 + 2 = i + 2ix$

**19.** $3ix + 2x - i = 0$

**20.** $2ix - 3 + 2i = 0$

**21.** Verify by substitution that $x = \frac{1}{2} + \frac{1}{2}i$ is a root of the equation $2x^2 - 2x + 1 = 0$.

**22.** Verify by substitution that $x = 1 + i\sqrt{3}$ is a root of the equation $x^2 - 2x + 4 = 0$.

Find an equation having the given roots in Problems 23–34.

**23.** $3i, -3i$

**24.** $4i, -4i$

**25.** $i\sqrt{2}, -i\sqrt{2}$

**26.** $i\sqrt{3}, -i\sqrt{3}$

**27.** $3, \frac{1}{2}i, -\frac{1}{2}i$

**28.** $2, 4i, -4i$

**29.** $2 + 3i, 2 - 3i$

**30.** $5 + i, 5 - i$

★ **31.** $2, i, 1 + i$

★ **32.** $3, i, 1 - i$

★ **33.** $i, 3 + i, 3 - i$

★ **34.** $i, 2 - i, 2 + i$

### Calculator Problems

Find the roots of the equations in Problems 35 and 36. Express your answers in the form $a \pm bi$ where $a$ and $b$ are real numbers approximated to the nearest hundredth.

**35.** $1.21x^2 + 4.88x + 8.08 = 0$

**36.** $5.36x^2 - 2.35x + 7.49 = 0$

## 9-3 POLAR FORM OF COMPLEX NUMBERS

To use trigonometric functions with complex numbers, we express them in **polar form** in the following manner.

First, take the rectangular coordinate system and designate the horizontal axis as the **real axis** and the vertical axis as the **imaginary axis**. The plane so specified is called the **complex plane**. Sometimes it is also referred to as the Gaussian plane in honor of the great mathematician Carl Friedrich Gauss (1777–1855).

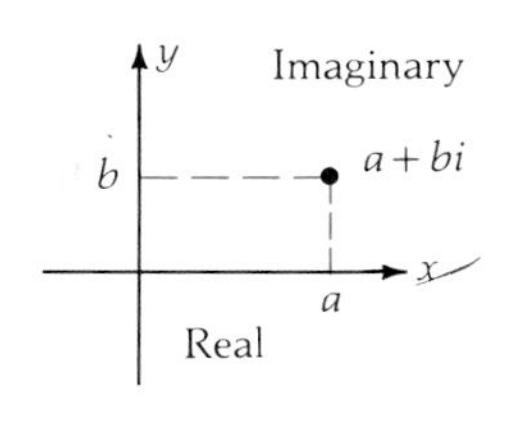


We plot the complex number $a + bi$ on this system by finding the point $(a, b)$. The real axis is labeled as we have previously done, but the $y$-axis is labeled so that its points correspond to multiples of $i$.

**EXAMPLE 1** Graph the following on the complex plane.

**a.** $2 - 3i$ **b.** $-4 - 5i$ **c.** $-3 + \frac{5}{2}i$ **d.** $4i$

***Solution***

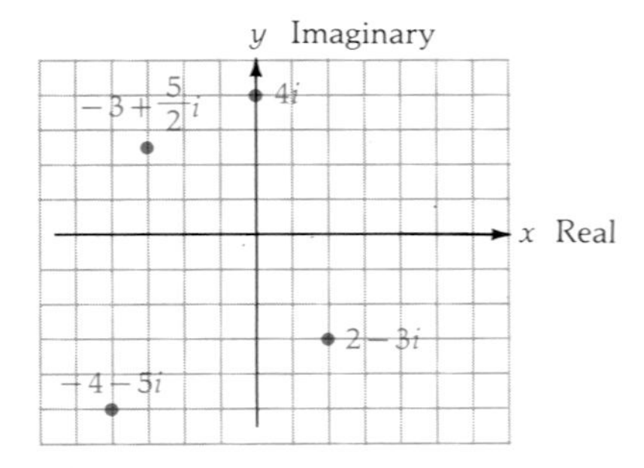


■

Plotting an arbitrary number $a + bi$ on the plane, we apply some prior knowledge of trigonometry and obtain the following.

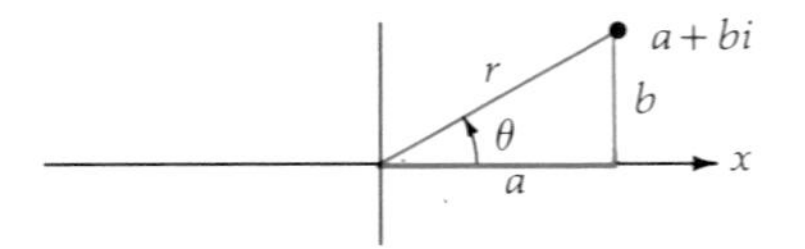


Note the following important relationships:

**1.** $r = \sqrt{a^2 + b^2}$ **2.** $a = r\cos\theta$ **3.** $b = r\sin\theta$

Thus, the complex number $a + bi$ can be expressed as

$$a + bi = r\cos\theta + ir\sin\theta = r(\cos\theta + i\sin\theta)$$

where $\theta$ is the angle in standard position formed by the positive horizontal axis (*polar axis*) and the line from the origin (*pole*) to the point $(a, b)$. The **polar form** of a complex number is $r(\cos\theta + i\sin\theta)$.

$$a + bi = r(\cos\theta + i\sin\theta)$$

We will use these properties to transform complex numbers in rectangular form to polar form.

*Definition*

> In the polar form of $a + bi = r(\cos\theta + i\sin\theta)$, $r$ is called the *absolute value* or *modulus* of the complex number, and $\theta$ is called the *argument* or *amplitude* of the complex number. The modulus of the complex number is denoted by the notation $|a + bi|$.

**EXAMPLE 2** Write $-2 + 2i\sqrt{3}$ in polar form.

***Solution***

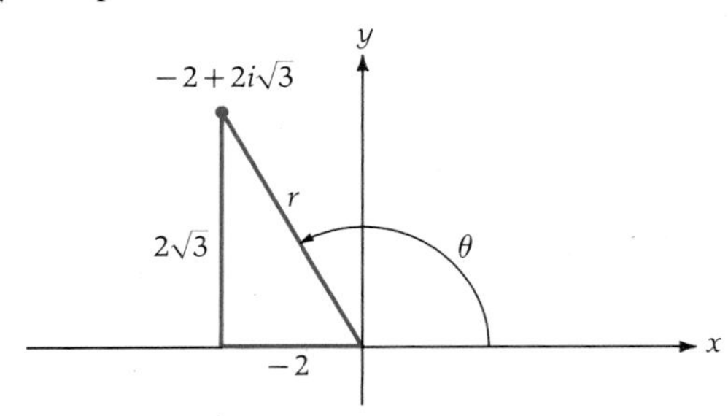


If we plot the point, we find that it is in the second quadrant. The *modulus* of the complex number $|a + bi| = r = \sqrt{4 + 12} = \sqrt{16} = 4$. The *argument* of the complex number can be obtained by using

$$\cos\theta = \frac{-2}{4} = -\frac{1}{2}$$

(Do you see why we used $\cos\theta$ instead of $\sin\theta$? It is very important to determine the correct quadrant for the angle $\theta$.) From this we obtain

$$\theta = 120^\circ \qquad \textit{$\theta$ is in the second quadrant.}$$

Hence $$-2 + 2i\sqrt{3} = 4(\cos 120^\circ + i\sin 120^\circ)$$ ■

Note that the answer to Example 2 is exact and does not involve an approximation.

The notation $r(\cos\theta + i\sin\theta)$ is somewhat cumbersome to work with. The following abbreviated notation is often used to save time.

*Definition*

> The expression $r(\cos\theta + i\sin\theta)$ is abbreviated as $r \text{ cis } \theta$.

EXAMPLE 3 Write 4 cis 45° in rectangular form.

***Solution*** $4 \text{ cis } 45° = 4(\cos 45° + i \sin 45°) = 4\left(\frac{\sqrt{2}}{2} + i\frac{\sqrt{2}}{2}\right)$

$= 2\sqrt{2} + 2i\sqrt{2}$ *Note that this is an exact answer.* ■

## EXERCISE 9-3

Plot the complex numbers in Problems 1–6 on the complex plane.

**1.** $2 - 3i$ **2.** $-3 + 5i$ **3.** $-7 - \frac{1}{2}i$ **4.** $3 - \frac{7}{2}i$ **5.** $-2i$ **6.** $\frac{5}{3}i$

Find the *modulus* of the complex numbers in Problems 7–12. (Leave your answer in radical form when necessary. Do not approximate.)

**7.** $-4-5i$ **8.** $2-7i$ **9.** $-12+5i$ **10.** $-6-4i$ **11.** $\frac{1}{2}-3i$ **12.** $-2+\frac{3}{2}i$

Find the *argument* of the complex numbers in Problems 13–18. (When necessary approximate your answer to the nearest tenth of a degree.)

**13.** $5 - 5i$ **14.** $-3 + 3i\sqrt{3}$ **15.** $-5 - 2i$

**16.** $-7 - i$ **17.** $-\frac{1}{2} + i\sqrt{2}$ **18.** $\sqrt{3} - i\sqrt{2}$

In Problems 19–28, change each to rectangular form. (When necessary, approximate your answer to three decimal places.)

**19.** 5 cis 45° **20.** 12 cis 120° **21.** $3\sqrt{2}$ cis 225° **22.** 3 cis 300°

**23.** 4 cis 27°38′ **24.** 2 cis 100° **25.** 2 cis 158° **26.** 5 cis 261.5°

**27.** $10(\cos 312.7° + i \sin 312.7°)$ **28.** $7(\cos 166.2° + i \sin 166.2°)$

In Problems 29–38, change each to polar form. (When necessary, approximate the argument to the nearest tenth of a degree.)

**29.** $1 - i\sqrt{3}$ **30.** $\sqrt{3} + i$ **31.** $2 + i$ **32.** $-2 - i$ **33.** $2 + 3i$

**34.** $-2 + 3i$ **35.** $-10\sqrt{3} + 10i$ **36.** $5 - 5i\sqrt{3}$ **37.** $6i$ **38.** $-4.5$

★ **39.** Find the product (3 cis 120°)(2 cis 210°) by first changing each factor to rectangular form, multiplying, and then changing the product to polar form. Compare the polar form of the answer with the polar form of the original problem. Can you see a relationship?

★**40.** Divide the numbers $\dfrac{\sqrt{3}+i}{\sqrt{3}-i}$ and simplify your answer. Express the original problem and the final answer in polar form. Compare these two polar expressions. Can you see a relationship?

---

## 9-4 DE MOIVRE'S THEOREM

You can see that multiplication and division of complex numbers in rectangular form is not a simple operation. It is, however, quite simple if the complex numbers are written in polar form. We will first prove a useful theorem for multiplication of complex numbers.

*Theorem 9-1* For any complex numbers $r_1 \text{ cis } A$ and $r_2 \text{ cis } B$, the product $(r_1 \text{ cis } A)(r_2 \text{ cis } B)$ is equal to $r_1 r_2 \text{ cis } (A + B)$.

***Proof***

$$(r_1 \text{ cis } A)(r_2 \text{ cis } B) = r_1 r_2 (\cos A + i \sin A)(\cos B + i \sin B)$$ *Notation.*

$$= r_1 r_2 (\cos A \cos B + i \sin A \cos B + i \sin B \cos A - \sin A \sin B)$$ *Multiplication of two binomials.*

$$= r_1 r_2 [(\cos A \cos B - \sin A \sin B) + i(\sin A \cos B + \sin B \cos A)]$$ *Rearranging order of terms and factoring.*

$$= r_1 r_2 [\cos (A + B) + i \sin (A + B)]$$ *Why?*

$$= r_1 r_2 \text{ cis } (A + B)$$ *Notation.* ■

This theorem can be extended to give the product of any number of complex numbers. For example,

$$(r_1 \text{ cis } A)(r_2 \text{ cis } B)(r_3 \text{ cis } C) = r_1 r_2 r_3 \text{ cis } (A + B + C)$$

EXAMPLE 1 Find the products of the following complex numbers:

**a.** $(5 \text{ cis } 30^\circ)(3 \text{ cis } 40^\circ)$
**b.** $(2 \text{ cis } 50^\circ)(5 \text{ cis } 80^\circ)(4 \text{ cis } 20^\circ)$

***Solution***

**a.** $(5 \text{ cis } 30^\circ)(3 \text{ cis } 40^\circ) = 15 \text{ cis } 70^\circ$

**b.** $(2 \text{ cis } 50^\circ)(5 \text{ cis } 80^\circ)(4 \text{ cis } 20^\circ) = 40 \text{ cis } 150^\circ$ ■

**EXAMPLE 2** Change $3 + 2i$ and $4 + 7i$ to polar form and find their product. Approximate angles to the nearest tenth of a degree. Leave your answer in polar form.

***Solution*** For $3 + 2i$, $a = 3$, $b = 2$, and $r = \sqrt{13}$. If we use $\tan\theta = \frac{2}{3}$, we get

$$\theta \doteq 33.7^\circ \qquad \textit{Rounding to the nearest tenth.}$$

For $4 + 7i$, $a = 4$, $b = 7$, and $r = \sqrt{65}$. If we use $\tan\theta = \frac{7}{4}$, then

$$\theta \doteq 60.3^\circ \qquad \textit{Again, rounding to the nearest tenth.}$$

Therefore we have

$$3 + 2i \doteq \sqrt{13}\text{ cis }33.7^\circ$$

and

$$4 + 7i \doteq \sqrt{65}\text{ cis }60.3^\circ$$

Thus

$$\begin{aligned}&(\sqrt{13}\text{ cis }33.7^\circ)(\sqrt{65}\text{ cis }60.3^\circ)\\ &= \sqrt{(13)(65)}\text{ cis }(33.7^\circ + 60.3^\circ)\\ &= 13\sqrt{5}\text{ cis }94^\circ \qquad \textit{Do you see why?}\end{aligned}$$ ■

Now consider the following:

$$(r\text{ cis }A)^2 = (r\text{ cis }A)(r\text{ cis }A) = r^2\text{ cis }(A + A) = r^2\text{ cis }2A$$

In like fashion we can find

$$\begin{aligned}(r\text{ cis }A)^3 &= (r\text{ cis }A)(r\text{ cis }A)(r\text{ cis }A)\\ &= r^3\text{ cis }(A + A + A) = r^3\text{ cis }3A\end{aligned}$$

The general case of these products is named after the French mathematician Abraham De Moivre (1667–1754).

***Theorem 9-2 (De Moivre's Theorem)***

For any complex number $r$ cis $\theta$ and any positive integer $n$,

$$(r\text{ cis }\theta)^n = r^n\text{ cis }n\theta$$

The proof of this theorem involves mathematical induction, which will be studied in a later chapter. We will not prove the theorem now.

**EXAMPLE 3** Find $(4\text{ cis }126^\circ)^4$ and express your answer as $r$ cis $\theta$ where $0^\circ \le \theta < 360^\circ$.

***Solution***

$$(4 \text{ cis } 126^\circ)^4 = 4^4 \text{ cis } (4 \cdot 126^\circ) = 256 \text{ cis } 504^\circ$$
$$= 256 \text{ cis } 144^\circ \qquad \textit{Since } 144^\circ \textit{ is coterminal with } 504^\circ. \quad ■$$

Finally, we turn to the case of division of two complex numbers.

***Theorem 9-3***

For any two complex numbers $r_1$ cis $A$ and $r_2$ cis $B$ where $r_2 \neq 0$,

$$\frac{r_1 \text{ cis } A}{r_2 \text{ cis } B} = \frac{r_1}{r_2} \text{ cis } (A - B)$$

EXAMPLE 4 Find $\dfrac{28 (\cos 126^\circ + i \sin 126^\circ)}{7 (\cos 200^\circ + i \sin 200^\circ)}$ and express your answer in the form $r$ cis $\theta$ where $0^\circ \leq \theta < 360^\circ$.

***Solution*** We can write the problem as $\dfrac{28 \text{ cis } 126^\circ}{7 \text{ cis } 200^\circ}$

$$= 4 \text{ cis } (126^\circ - 200^\circ)$$
$$= 4 \text{ cis } (-74^\circ) = 4 \text{ cis } 286^\circ$$

*Since 286° is coterminal with −74°.* ■

More than one theorem may be necessary to evaluate some expressions.

EXAMPLE 5 Find $\dfrac{(2 \text{ cis } 29^\circ)^8}{(4 \text{ cis } 231^\circ)^3}$ and express your answer as $r$ cis $\theta$ where $0^\circ \leq \theta < 360^\circ$.

***Solution***

$$(2 \text{ cis } 29^\circ)^8 = 2^8 \text{ cis } (8 \cdot 29^\circ) = 256 \text{ cis } 232^\circ$$
$$(4 \text{ cis } 231^\circ)^3 = 4^3 \text{ cis } (3 \cdot 231^\circ) = 64 \text{ cis } 693^\circ$$

The original problem can then be written as

$$\frac{(2 \text{ cis } 29^\circ)^8}{(4 \text{ cis } 231^\circ)^3} = \frac{256 \text{ cis } 232^\circ}{64 \text{ cis } 693^\circ} = 4 \text{ cis } (-461^\circ) = 4 \text{ cis } 259^\circ \quad ■$$

## EXERCISE 9-4

Find the products in Problems 1–8. Express your answers in the form $r$ cis $\theta$ where $0^\circ \leq \theta < 360^\circ$.

**1.** $(3 \text{ cis } 35^\circ)(2 \text{ cis } 18^\circ)$

**2.** $(10 \text{ cis } 129^\circ)(4 \text{ cis } 200^\circ)$

**3.** $5(\cos 228^\circ + i \sin 228^\circ) \cdot 8(\cos 230^\circ + i \sin 230^\circ)$

4. $6(\cos 77° + i \sin 77°) \cdot 9(\cos 344° + i \sin 344°)$

5. $(\sqrt{3} \text{ cis } 20.3°)(\sqrt{2} \text{ cis } 48.9°)$

6. $(8 \text{ cis } 140.3°)(\sqrt{3} \text{ cis } 296.4°)$

7. $(7 \text{ cis } 36°)(5 \text{ cis } 124°)(3 \text{ cis } 229°)$

8. $(1.5 \text{ cis } 200°)(3 \text{ cis } 58°)(2 \text{ cis } 154°)$

9. a. Find the product of $(\sqrt{3} + i)(1 - i\sqrt{3})$ in rectangular form.
   b. Change the answer found in part a to polar form.
   c. Change $\sqrt{3} + i$ and $1 - i\sqrt{3}$ to polar form and then find their product.

10. a. Find the product $(1 + i)(2 - 2i)$ in rectangular form.
   b. Change the answer found in part a to polar form.
   c. Change $(1 + i)$ and $(2 - 2i)$ to polar form and then find their product.

Use De Moivre's Theorem to find the expressions in Problems 11–14. Express your answer in the form $r \text{ cis } \theta$ where $0 \le \theta < 360°$.

11. $(2 \text{ cis } 45°)^3$  12. $(3 \text{ cis } 210°)^5$  13. $(1 \text{ cis } 100°)^4$  14. $[6(\cos 122° + i \sin 122°)]^3$

Evaluate in Problems 15–18. Change the answers to rectangular form.

15. $(3 \text{ cis } 30°)^2$  16. $(2 \text{ cis } 120°)^3$  17. $(5 \text{ cis } 18°)^3$  18. $(2 \text{ cis } 35°)^6$

Change to polar form in Problems 19–22. Use De Moivre's Theorem to evaluate, then express the answers in rectangular form.

19. $(\sqrt{3} - i)^5$  20. $(2 - 2i\sqrt{3})^4$  21. $\dfrac{(3 + 3i)^2}{\sqrt{3} - i}$  22. $\dfrac{\left(\frac{\sqrt{2}}{2} + \frac{\sqrt{2}}{2}i\right)^3}{\frac{\sqrt{3}}{2} - \frac{1}{2}i}$

Find the expressions in Problems 23–28. Express the answers in polar form $r \text{ cis } \theta$ where $0 \le \theta < 360°$.

23. $\dfrac{2 \text{ cis } 130°}{5 \text{ cis } 70°}$  24. $\dfrac{258 \text{ cis } 37°}{129 \text{ cis } 250°}$  25. $\dfrac{6 \text{ cis } 36°}{(2 \text{ cis } 20°)^3}$

26. $\dfrac{5 \text{ cis } 188°}{(5 \text{ cis } 37°)^2}$  ★ 27. $\dfrac{(3 \text{ cis } 28°)^3}{(2 \text{ cis } 19.5°)^2}$  ★ 28. $\dfrac{(5 \text{ cis } 220°)^3}{(10 \text{ cis } 186°)^2}$

## Calculator Problems

Change each of Problems 29 and 30 to polar form before evaluating. Express your answer in both polar form and rectangular form. Round the angles to the nearest thousandth of a degree. Round other values to six significant digits.

★ 29. $(5 - 2i)^8$  ★ 30. $\dfrac{4.6 - 2.3i}{0.7 + 2.8i}$

## 9-5 ROOTS OF COMPLEX NUMBERS

We will now discuss the concept of finding all roots of a complex number. First we will discuss the general concept, and then we will investigate how we can develop a powerful procedure to find the $n$ roots of any complex number based on our knowledge of De Moivre's Theorem.

**EXAMPLE 1** By algebraic methods, find the roots to the equation $x^2 + 4 = 0$.

***Solution***

$$x^2 = -4$$
$$x = \pm\sqrt{-4} = \pm 2i$$

In solving the equation we found that the two square roots of $-4$ are $+2i$ and $-2i$. ■

Now consider how our knowledge of De Moivre's Theorem can be applied if we want to find any root of any complex number. (The following discussion is not at all obvious. You may find it necessary to study some numerical examples like Examples 2–3 before it becomes clear to you.) First we can write any complex number in the form $r(\cos\theta + i\sin\theta)$. Now we wish to find all roots of the complex number $r(\cos\theta + i\sin\theta)$. Assume that the complex number $r_1(\cos\theta_1 + i\sin\theta_1)$ is an $n$th root of $r(\cos\theta + i\sin\theta)$. Then, by De Moivre's Theorem,

$$r_1{}^n(\cos n\theta_1 + i\sin n\theta_1) = r(\cos\theta + i\sin\theta)$$

Hence $r_1{}^n = r$, giving $r_1 = \sqrt[n]{r}$. Therefore $r_1$ is the principal $n$th root of $r$. Also, since $\cos n\theta_1 = \cos\theta$ and $\sin n\theta_1 = \sin\theta$, it follows that $n\theta_1 = \theta + k\cdot 360°$ where $k$ is an integer or

$$\theta_1 = \frac{\theta + k\cdot 360°}{n}$$

Note that if $k = 0, 1, 2, \ldots, n-1$, the angles given by $\dfrac{\theta + k\cdot 360°}{n}$ will be $n$ non-coterminal angles. However, if $k = n, n+1, n+2, \ldots$, the expression $\dfrac{\theta + k\cdot 360°}{n}$ will give angles that differ by 360° from those obtained when $k = 0, 1, 2, \ldots, n-1$. For example, if $k = n$, $\dfrac{\theta + k\cdot 360°}{n}$ becomes $\dfrac{\theta + n\cdot 360°}{n} = \dfrac{\theta}{n} + 360°$, which is coterminal with the angle obtained when $k = 0$.

*Theorem 9-4*

> The formula $\sqrt[n]{r}\left[\cos\left(\frac{\theta + k \cdot 360^\circ}{n}\right) + i \sin\left(\frac{\theta + k \cdot 360^\circ}{n}\right)\right]$ where the values of $k$ are $k = 0, 1, 2, \ldots, n - 1$ gives all $n$ roots of $r(\cos\theta + i \sin\theta)$.

EXAMPLE 2 Find all fourth roots of $-1 + i\sqrt{3}$.

***Solution*** To express $-1 + i\sqrt{3}$ in polar form, we note that $a = -1$, $b = \sqrt{3}$, and $r = 2$. Also, $\tan\theta = -\frac{\sqrt{3}}{1}$, so $\theta = 120^\circ$. Hence $-1 + i\sqrt{3} = 2(\cos 120^\circ + i \sin 120^\circ)$. Accordingly, the four fourth roots of $-1 + i\sqrt{3}$ are given by

$$\sqrt[4]{2}\left[\cos\left(\frac{120^\circ + k \cdot 360^\circ}{4}\right) + i \sin\left(\frac{120^\circ + k \cdot 360^\circ}{4}\right)\right]$$

We now assign the values $k = 0, 1, 2, 3$.

$$k = 0 \text{ gives } \sqrt[4]{2}(\cos 30^\circ + i \sin 30^\circ) = \frac{\sqrt[4]{2}}{2}(\sqrt{3} + i)$$

$$k = 1 \text{ gives } \sqrt[4]{2}(\cos 120^\circ + i \sin 120^\circ) = \frac{\sqrt[4]{2}}{2}(-1 + i\sqrt{3})$$

$$k = 2 \text{ gives } \sqrt[4]{2}(\cos 210^\circ + i \sin 210^\circ) = \frac{\sqrt[4]{2}}{2}(-\sqrt{3} - i)$$

$$k = 3 \text{ gives } \sqrt[4]{2}(\cos 300^\circ + i \sin 300^\circ) = \frac{\sqrt[4]{2}}{2}(1 - i\sqrt{3})$$ ■

One interesting property can be observed by plotting all the roots $r_0, r_1, r_2, r_3, \ldots, r_{n-1}$ of any complex number on a complex plane. We will illustrate this property in the solution to Example 3.

EXAMPLE 3 Find the three cube roots of 8.

***Solution*** First, write 8 in complex form as $8 = 8 + 0i$. Since $a = 8$, $b = 0$, $r = 8$, and $\theta = 0^\circ$, we may express 8 in polar form as $8 = 8 \text{ cis } 0^\circ$. Hence the three cube roots are given by

$$\sqrt[3]{8} \text{ cis}\left(\frac{0^\circ + k \cdot 360^\circ}{3}\right) \qquad \text{as } k = 0, 1, 2.$$

$$k = 0 \text{ gives } 2 \text{ cis } 0^\circ = 2$$

$$k = 1 \text{ gives } 2 \text{ cis } 120^\circ = -1 + i\sqrt{3}$$

$$k = 2 \text{ gives } 2 \text{ cis } 240^\circ = -1 - i\sqrt{3}$$

Do you see that you can check these answers by showing $2^3 = 8$ as well as by showing $(+1 + i\sqrt{3})^3 = 8$; and $(-1 - i\sqrt{3})^3 = 8$? ■

Let us graph the roots on the complex plane. All three roots lie on a circle of radius 2 and are equally spaced along the circle.

$$r_0 = 2 \text{ cis } 0^\circ$$

$$r_1 = 2 \text{ cis } 120^\circ$$

$$r_2 = 2 \text{ cis } 240^\circ$$

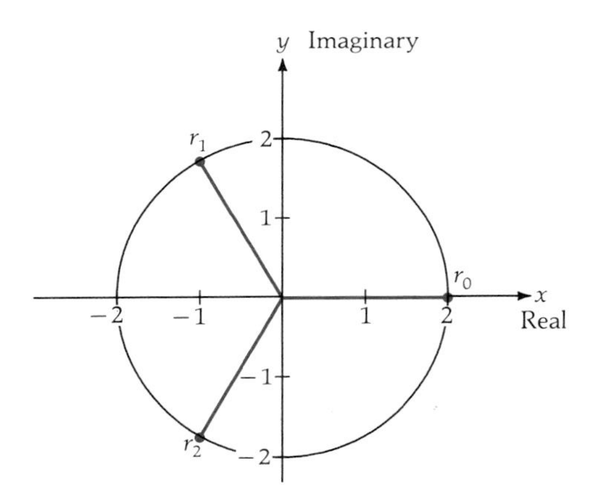


In general if the $n$th roots of any complex number $r$ cis $\theta$ are graphed on the complex plane, they will lie on the circle of radius $\sqrt[n]{r}$ and will be equally spaced along the circle. The roots $r_0, r_1, r_2, r_3, \ldots, r_{n-1}$ will divide the circle into $n$ equal arcs, each subtending an angle of $\frac{360^\circ}{n}$. These properties are useful in providing a check to see whether the answer you obtain to this type of problem seems reasonable.

In finding the roots of some complex numbers we will find it necessary to use approximations, particularly if we do not obtain a trigonometric function of a special angle.

**EXAMPLE 4** Find all roots to the equation $x^6 = 3 + 2i$. Round the argument to the nearest tenth of a degree. Round your final answers to three decimal places. *Leave your answer in rectangular form.*

***Solution*** We want to obtain the six roots of $3 + 2i$. Since $a = 3$ and $b = 2$, we have $r = \sqrt{9+4} = \sqrt{13}$. Since $\cos\theta = \dfrac{a}{r} = \dfrac{3}{\sqrt{13}}$, we will need to use a calculator or Table D to obtain $\theta \doteq 33.7^\circ$ (to the nearest tenth of a degree). We can simplify $\sqrt[6]{r} = \sqrt[6]{\sqrt{13}}$ to the form $\sqrt[12]{13}$. Thus the six roots of $3 + 2i$ will be obtained by the formula $\sqrt[12]{13}\ \text{cis}\left(\dfrac{33.7^\circ + k\cdot 360^\circ}{6}\right)$, where $k = 0, 1, 2, 3, 4, 5$. When $k = 0$,

$$r_0 \doteq \sqrt[12]{13}\ \text{cis}\left(\frac{33.7^\circ}{6}\right) \doteq \sqrt[12]{13}\ \text{cis}\ 5.6^\circ$$

Evaluating each of these expressions, we have

$$r_0 \doteq 1.238(\cos 5.6^\circ + i \sin 5.6^\circ) \doteq 1.232 + 0.121i$$

In similar fashion we can find

$$\begin{aligned}
r_1 &\doteq 1.238(\cos 65.6^\circ + i\sin 65.6^\circ) \doteq 0.511 + 1.127i\\
r_2 &\doteq 1.238(\cos 125.6^\circ + i\sin 125.6^\circ) \doteq -0.721 + 1.007i\\
r_3 &\doteq 1.238(\cos 185.6^\circ + i\sin 185.6^\circ) \doteq -1.232 - 0.121i\\
r_4 &\doteq 1.238(\cos 245.6^\circ + i\sin 245.6^\circ) \doteq -0.511 - 1.127i\\
r_5 &\doteq 1.238(\cos 305.6^\circ + i\sin 305.6^\circ) \doteq 0.721 - 1.007i
\end{aligned}$$

If we represent the roots on the complex plane, we observe that the six roots of the equation are equally spaced on a circle of radius $r = \sqrt[12]{13}$.

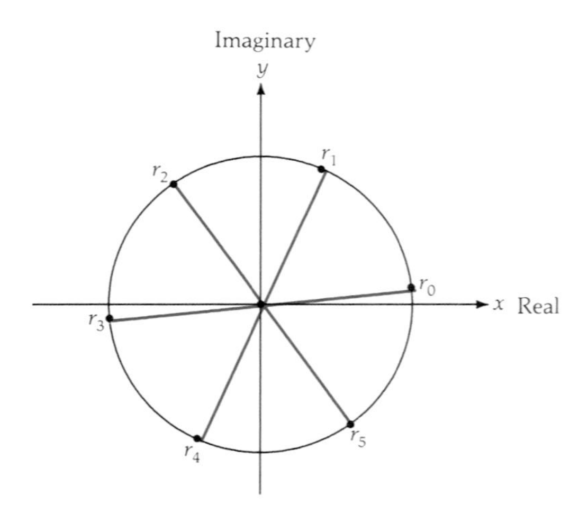


Observe that each arc subtends of a central angle of $\dfrac{360}{6} = 60^\circ$. ■

## EXERCISE 9-5

Find the specified complex $n$th roots in Problems 1–6. Use exact values. Do not approximate with table or calculator. Leave your answer in rectangular form.

**1.** The square roots of $i$
**2.** The square roots of $-7i$
**3.** The square roots of $-1 - i\sqrt{3}$
**4.** The square roots of $-\sqrt{2} + i\sqrt{2}$
**5.** The cube roots of $-8$
**6.** The cube roots of $27i$

In Problems 7–10, find the specified complex $n$th roots. Use a calculator or Table D to evaluate trigonometric functions. Expresss your answer in both polar form and rectangular form. Round answers in rectangular form to the nearest thousandth.

**7.** The cube roots of $-1 + i$
**8.** The cube roots of $-2 - 2i$
**9.** The fourth roots of $-5 + 5i\sqrt{3}$
**10.** The fourth roots of $+8 + 8i\sqrt{3}$

Find the specified complex $n$th roots in Problems 11–14. You may leave your answers in polar form.

**11.** The fifth roots of $-16\sqrt{2} - 16i\sqrt{2}$
**12.** The fifth roots of $7\sqrt{3} - 7i$
**13.** The sixth roots of $-8$
**14.** The sixth roots of $-64i$

Find all complex roots to the equations in Problems 15–18. Leave your answer in polar form.

**15.** $x^3 - 27 = 0$ **16.** $x^3 = 1 - i$ **17.** $x^5 + 32 = 0$ **18.** $x^7 + 1 = 0$

Find the specified complex $n$th roots in Problems 19–22. Approximate all arguments to the nearest tenth of a degree. Express your final answer to the nearest thousandth. Use appropriate tables or a calculator. Leave your answer in polar form.

**19.** The fifth roots of 32 cis 200°
**20.** The fourth roots of 16 cis 300°
**21.** The cube roots of $3 - 4i$
**22.** The cube roots of $-2 + 5i$

Find all the complex roots to the equations in Problems 23–26. Leave your answer in polar form. Approximate all arguments to the nearest tenth of a degree. Express the modulus to the nearest thousandth.

★ **23.** $x^3 = -3 + 5i$ ★ **24.** $x^3 - 5 + i = 0$
★ **25.** $x^4 + 1 + 3i = 0$ ★ **26.** $x^4 = 1 - 5i$

### COMPUTER PROBLEM FOR CHAPTER 9

See if you can apply your knowledge of computer programming and of Chapter 9 to do the following.

Write a computer program to print out the two imaginary roots of a quadratic equation of the form $ax^2 + bx + c = 0$ when $b^2 - 4ac < 0$ and the real roots when $b^2 - 4ac \geq 0$. Run your program to find roots for the following quadratic equations.

a. $3x^2 + 5x + 1 = 0$
b. $2.6x^2 - 9.2x + 2.3 = 0$
c. $5.4x^2 + 2.2x - 1.9 = 0$

## KEY TERMS AND CONCEPTS

Be sure you understand what is meant by these terms and can give an example of each.

Real number
Imaginary number
Complex number
Conjugate of a complex number
Equality of complex numbers
Complex root of an equation
Polar form of a complex number
Complex plane
Real axis
Imaginary axis
Polar form of a complex number
Rectangular form of a complex number
Absolute value of a complex number
Modulus of a complex number
Argument of a complex number
De Moivre's Theorem
Root of a complex number

## SUMMARY OF PROCEDURES AND CONCEPTS

### Polar Form of Complex Numbers

1. To change a complex number of the form $a + bi$ to the polar form $r(\cos\theta + i\sin\theta)$, use the relationships

$$r = \sqrt{a^2 + b^2} \qquad a = r\cos\theta \qquad b = r\sin\theta$$

The polar form $r(\cos\theta + i\sin\theta)$ is abbreviated as $r \text{ cis } \theta$.

2. To change a complex number from polar form to rectangular form, evaluate each trigonometric expression and multiply each one by $r$.

### Operations on Complex Numbers

For any two complex numbers $r_1 \text{ cis } A$ and $r_2 \text{ cis } B$ and any positive integer $n$:

1. To multiply two numbers, use $(r_1 \text{ cis } A)(r_2 \text{ cis } B) = r_1r_2 \text{ cis } (A + B)$.
2. To divide two numbers, use

$$\frac{r_1 \text{ cis } A}{r_2 \text{ cis } B} = \frac{r_1}{r_2} \text{ cis } (A - B) \qquad (\text{when } r_2 \neq 0)$$

3. To raise a number to a power, use $(r \text{ cis } \theta)^n = r^n \text{ cis } n\theta$

**Roots of Complex Numbers**

1. The $n$ roots of $r(\cos\theta + i\sin\theta)$ are given by

$$\sqrt[n]{r}\left[\cos\left(\frac{\theta + k\cdot 360^\circ}{n}\right) + i\sin\left(\frac{\theta + k\cdot 360^\circ}{n}\right)\right]$$

where $k = 0, 1, 2, 3, \ldots, n-1$

2. The $n$ roots of $r(\cos\theta + i\sin\theta)$ can be graphed on the complex plane. The roots will be equally spaced on a circle with the center at the origin and a radius of $\sqrt[n]{r}$.

## CHAPTER 9 REVIEW EXERCISE

Perform the specified operations in Problems 1–12. Simplify your answers.

**1.** $(5 + 2i) + (3 - 4i)$ **2.** $(8 - 5i) + (2 - 7i)$ **3.** $(4 - 5i) - (-2 + i)$

**4.** $(6 - i) - (3 - 5i)$ **5.** $2i(3 - i)$ **6.** $-7(4 - 5i)$

**7.** $(4 - 5i)(3 - 2i)$ **8.** $(-5 - 3i)(2 + 3i)$ **9.** $\dfrac{5}{2 + i}$

**10.** $\dfrac{3i}{1 - i}$ **11.** $\dfrac{6 + i}{3 - 2i}$ **12.** $\dfrac{5 - i}{2 + 5i}$

Solve to find the complex root(s) of each equation in Problems 13–18. Simplify all answers.

**13.** $5x + 3i = 2ix + 5 - 7i$ **14.** $3(x + 5i) = 2x - 3i + 4$

**15.** $4x^2 + 2x + 1 = 0$ **16.** $5x^2 + 4x + 2 = 0$

**17.** $3x^2 - 7x + 5 = 0$ **18.** $x^2 - 7x + 14 = 0$

In Problems 19–26, express each complex number in polar form. Determine the angles to the nearest tenth of a degree.

**19.** $3i$ **20.** $-5$ **21.** $2 + 2i$ **22.** $-4\sqrt{2} + 4i\sqrt{2}$

**23.** $4 + 3i$ **24.** $-12 + 5i$ **25.** $-5 - 4i$ **26.** $2 - 5i$

In Problems 27–34, express each complex number in rectangular form.

**27.** $4(\cos 240^\circ + i\sin 240^\circ)$ **28.** $7(\cos 0^\circ + i\sin 0^\circ)$ **29.** $2(\cos 360^\circ + i\sin 360^\circ)$

**30.** $4(\cos 225^\circ + i\sin 225^\circ)$ **31.** $3 \text{ cis } 30^\circ$ **32.** $3 \text{ cis } 300^\circ$

**33.** $10 \text{ cis } 120^\circ$ **34.** $4 \text{ cis } 123^\circ$

In Problems 35–50, perform the indicated operations. Leave your answer in polar form.

**35.** $(3 \text{ cis } 33^\circ)(2 \text{ cis } 27^\circ)$ **36.** $(3 \text{ cis } 18^\circ)(5 \text{ cis } 43^\circ)$ **37.** $(\sqrt{5} \text{ cis } 80^\circ)(2 \text{ cis } 72^\circ)$

**38.** $\dfrac{6 \text{ cis } 140^\circ}{2 \text{ cis } 30^\circ}$ **39.** $\dfrac{3 \text{ cis } 75^\circ}{6 \text{ cis } 125^\circ}$ **40.** $(4 \text{ cis } 23^\circ)^3$

**41.** $(2 \text{ cis } 45°)^5$

**42.** $\dfrac{8 \text{ cis } 120°}{2 \text{ cis } 40°}$

**43.** $\dfrac{3\sqrt{2} \text{ cis } 185°}{1 + i}$

**44.** $(2\sqrt{3} \text{ cis } 200°)^6$

**45.** $(3 \text{ cis } 25°)^6$

**46.** $(2 + 2i)^4$

**47.** $(-1 + i\sqrt{3})^5$

**48.** $(-\sqrt{2} + i\sqrt{2})^{10}$

**49.** $\dfrac{-5}{15 \text{ cis } 300°}$

**50.** $\left(\dfrac{\sqrt{2}}{2} + i\dfrac{\sqrt{2}}{2}\right)^{200}$

**51.** Find the three cube roots of unity in rectangular form.

**52.** Find the four fourth roots of 81 cis 120° in rectangular form.

**53.** Find the four fourth roots of $25(\cos 48° + i \sin 48°)$ in rectangular form.

**54.** Find the five fifth roots of $i$ in polar form. Graph resulting roots on the complex plane.

**55.** Find the four fourth roots of $-8 + 8i\sqrt{3}$ in rectangular form. Graph resulting roots on the complex plane.

**56.** Find the five fifth roots of 32 cis 300° in rectangular form. (Round off the answers to three decimal places.)

**57.** Find all solutions of $x^4 - 16 = 0$ in rectangular form.

**58.** Find all solutions of $x^5 - 32 = 0$ in rectangular form. (Round off the answers to three decimal places.)

## PRACTICE TEST FOR CHAPTER 9

Simplify your answers. Use a table or calculator as needed in this Practice Test.

Perform the specified operations in Problems 1–4.

**1.** $7 + 2i - 3(4 - 2i)$

**2.** $(6 + 4i) + (2 - 3i) - (5 + i)$

**3.** $(5 - 7i)(2 + 3i)$

**4.** $(5 - 3i) \div (4 + 3i)$

Solve to find the complex root(s) of each equation in Problems 5–8. Simplify all answers.

**5.** $3x + 5i - 2ix = 5(x + 3i) - 4$

**6.** $(3 + i)x + (2 - 4i) = 3ix$

**7.** $3x^2 = 10x - 12$

**8.** $4x^2 - 3x + 2 = 0$

**9.** Change $-3 + 2i$ to polar form.

**10.** Change 6 cis 135° to rectangular form.

**11.** Multiply (7 cis 46°)(4 cis 227°).

**12.** Simplify $(\sqrt{2} \text{ cis } 78°)^5$.

**13.** Simplify $\left(-\dfrac{\sqrt{3}}{2} + \dfrac{1}{2}i\right)^5$. Express your answer in rectangular form.

**14.** Divide $\dfrac{6(\cos 215° + i \sin 215°)}{15(\cos 329° + i \sin 329°)}$.

**15.** Find the three cube roots of 64 cis 180°. Express your answer in rectangular form.

# 10 Systems of Equations and Inequalities and the Use of Matrices

■ *A theory is the more impressive the greater the simplicity of its premises, the more diverse the things it relates, and the more extended its area of applicability.*

*Albert Einstein*

The concept of dealing with two or more equations at one time and finding the roots that satisfy all equations simultaneously is not a difficult one. An elementary understanding of systems of equations is apparent in various early civilizations. Several historical documents showing the solution of such mathematical problems have been uncovered that are dated in the period 300–200 B.C.

In modern times, through the development of matrices and determinants and aided by the technology of the computer, systems of equations are used to obtain the solution of a wide variety of applied problems.

## 10-1 SOLUTION OF A SYSTEM OF TWO LINEAR EQUATIONS IN TWO VARIABLES

A collection of two or more equations is called a **system** of equations. We often enclose such a system with a brace to indicate that the two equations are being considered together.

To illustrate,

$$\begin{cases} 3x + y = 5 \\ 2x - 3y = 7 \end{cases}$$

is a system of two linear equations in the variables $x$ and $y$. Usually, the system is written in this fashion, with the variables on the left and the constant terms on the right.

A pair of numbers $(a, b)$ is a **solution to a system** of two equations in $x$ and $y$ if a true statement is obtained when $a$ and $b$ are substituted for $x$ and $y$, respectively, in each of the two equations.

A linear system of two equations in two variables can be represented graphically by two straight lines. If a linear system of two equations in two variables has a solution, we can obtain the solution $(a, b)$ by graphing each line and obtaining the coordinates of the point of intersection. (Do you see why?) The graphs of each line must be extremely accurate in order to arrive at the solution.

**EXAMPLE 1** Solve by graphing the system

$$\begin{cases} x + y = 5 \\ 2x + y = 8 \end{cases}$$

***Solution*** Set up a table of values for each equation and sketch their graphs. (We should recognize each of these equations as graphically representing a straight line.)

$x + y = 5$

| $x$ | $-2$ | 0 | 5 |
|---|---|---|---|
| $y$ | 7 | 5 | 0 |

$2x + y = 8$

| $x$ | 0 | 2 | 4 |
|---|---|---|---|
| $y$ | 8 | 4 | 0 |

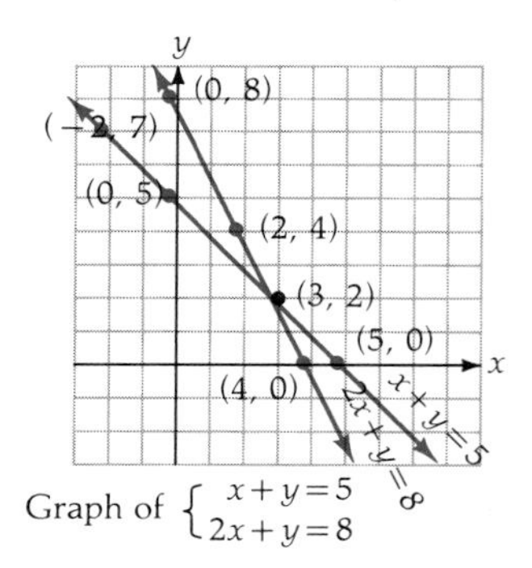


Graph of $\begin{cases} x+y=5 \\ 2x+y=8 \end{cases}$

The point of intersection is (3, 2) and should be labeled on the graph as shown. We can check the correctness of our solution by substituting (3, 2) into both of the equations to see that it is actually a solution. (Can you verify that $x = 3$, $y = 2$ is a solution?) ■

The use of graphs to find solutions for a system of two linear equations in two variables is time consuming and not very accurate unless the solutions are integer values. We now turn to the discussion of algebraic methods to find the solution.

### The Method of Substitution

The first method to be discussed is the method of *substitution*. This method involves solving for one unknown (variable) in terms of the other in one of the two equations and then substituting this expression into the other equation.

In general, the process is as listed below:

---

#### Solution of a Linear System by Substitution

---

1. Solve one of the equations for one variable.
2. Substitute the expression obtained in step 1 into the other equation. This will yield an equation with one variable.
3. Solve this resulting equation for the variable.
4. Substitute the value obtained in step 3 into an equation that contains each variable. Solve that resulting equation.
5. Check by substituting each value into *both* original equations.

---

EXAMPLE 2 Solve by the substitution method the system

$$\begin{cases} 2x + 3y = 1 \\ x - 2y = 4 \end{cases}$$

***Solution*** *Step 1*

$$x = 4 + 2y$$

*Step 2* Substituting $(4 + 2y)$ for $x$, we obtain

$$2(4 + 2y) + 3y = 1$$

*Step 3* Solve this equation for the variable $y$.

$$8 + 4y + 3y = 1$$
$$7y = -7$$
$$y = -1$$

*Step 4* Substitute $y = -1$ into an equation containing both $x$ and $y$ to find the corresponding value for $x$.

$$x = 4 + 2y = 4 + 2(-1) = 4 - 2 = 2$$

Thus we have the solution $(2, -1)$.

*Step 5* Check the solution in both equations. Remember that the solution for a system must be a solution of each equation in the system. Since

$$2x + 3y = 2(2) + 3(-1) = 4 - 3 = 1$$

and

$$x - 2y = 2 - 2(-1) = 2 + 2 = 4$$

we see that the solution $(2, -1)$ does check. ■

### The Method of Elimination

We will next study the *method of elimination.* The method of elimination is based on the idea of equivalent systems of equations. Two systems of equations are said to be **equivalent** if they have exactly the same solutions.

---

A system of equations can be transformed to an equivalent system by the following three operations:

1. Interchanging the position of any two equations.
2. Multiplying or dividing each side of any equation by a nonzero constant.
3. Multiplying each side of one equation by a constant and adding the result to another equation.

---

The primary approach of the *elimination method* (or the *addition method* as it is often called) is to add a multiple of one equation to another in such a way as to eliminate one variable.

EXAMPLE 3 Solve the following system by the elimination method:

$$\begin{cases} 4x + 5y = -6 \\ 3x - 2y = -16 \end{cases}$$

***Solution*** *Step 1* Let us eliminate the variable $x$. We will multiply each side of the first equation by 3 and each side of the second equation by $-4$.

$$\begin{cases} 12x + 15y = -18 \\ -12x + 8y = 64 \end{cases}$$ *We obtain this equivalent system.*

*Step 2* Add the equations to obtain

$$23y = 46$$

*Step 3* Solve the resulting equation for $y$:

$$y = 2$$

*Step 4* Find the other variable by substituting $y = 2$ into one of the original equations. We will choose the second equation:

$$3x - 2(2) = -16$$

$$3x = -12$$

$$x = -4$$

*Step 5* *Check:* We must verify the solution $x = -4$, $y = 2$ into *both* of the original equations.

$$4x + 5y = -6 \qquad\qquad 3x - 2y = -16$$

$$4(-4) + 5(2) \stackrel{?}{=} -6 \qquad\qquad 3(-4) - 2(2) \stackrel{?}{=} -16$$

$$-16 + 10 \stackrel{?}{=} -6 \qquad\qquad -12 - 4 \stackrel{?}{=} -16$$

$$-6 = -6 \checkmark \qquad\qquad -16 = -16 \checkmark \quad \blacksquare$$

### Situations in Which There Is No Unique Solution

A system of two linear equations does not always have a unique solution, and we should be aware of the other possibilities. An **inconsistent** system of equations has no solution. A **consistent** system of equations has one or more solutions.

Since we are dealing with linear equations, we may think of the lines represented by the equations.

---

Two linear equations in two variables represent one of three possible situations.

1. *Inconsistent system of equations:* The two lines are parallel. In this case we have no solution.
2. *Independent equations:* The two lines intersect in a point. In this case we have a unique solution. (These equations form a consistent system.)
3. *Dependent equations:* The two equations give the same line. In this case any solution of one equation is a solution of the other. (These equations form a consistent system.)

---

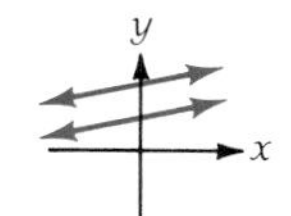


1. Inconsistent system of equations

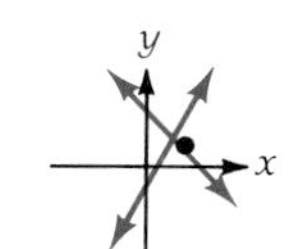


2. Independent equations

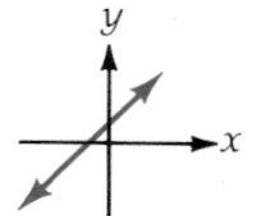


3. Dependent equations

We can recognize any of the three situations by proceeding to work the problem either by the substitution or elimination method.

1. If the system of equations is *inconsistent,* we will obtain a *contradiction.*
2. If the equations are *independent* and *consistent,* we will obtain a *unique solution.*
3. If the equations are *dependent,* we will obtain an *identity.*

**EXAMPLE 4** Solve the system

$$\begin{cases} x + y = 6 \\ x + y = 8 \end{cases}$$

***Solution*** If we multiply both sides of the second equation by $-1$ and add, we obtain

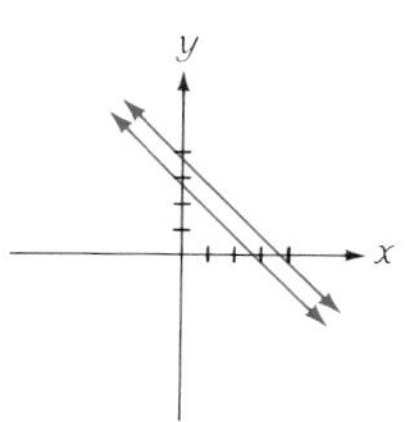


$$0 = -2$$

Since this is a *contradiction,* the system of equations is *inconsistent.* The graphs of these equations would be parallel lines. This system has **no solution.** ■

**EXAMPLE 5** Solve the system

$$\begin{cases} x + y \ \ = 6 \\ 2x + 2y = 12 \end{cases}$$

***Solution*** Multiplying both sides of the first equation by $-2$ and adding yields

$$0 = 0$$

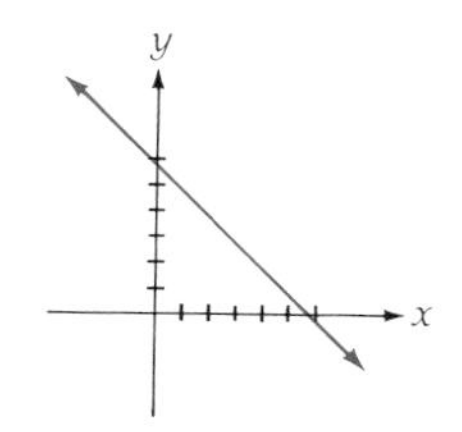


Since this is an *identity,* the equations are *dependent* and represent the same line on the coordinate plane. Therefore any ordered pair that satisfies one equation will also satisfy the other. There are infinitely many solutions.

The graph of $x + y = 6$ and $2x + 2y = 12$ yield exactly the same line. ■

In Example 5 there are infinitely many solutions. It would therefore be impossible to list all the solutions; however, we would like to have some way of indicating what these solutions are. A convenient tool which is used for this purpose is *set notation.*

Set notation is a precise shorthand technique of stating in a concise way what might otherwise take several written statements to explain. The solution to this example may be written as

$$\{(x, y) \mid x + y = 6\}$$

and at first glance might seem quite complex. Remember, however, that the symbols used are a shorthand technique and therefore need translation. The braces $\{ \}$ mean "the set," and the vertical bar $\mid$ means "such that." Thus the solution would be read as "the set of all ordered pairs $(x, y)$ such that $x + y = 6$."

### Applications

We often use systems of equations to solve verbal problems.

**EXAMPLE 6** It took a boat 3 hours to travel 75 miles downstream and 5 hours to make the return trip upstream. Find the average speed of the boat in still water and the average rate of the flow of the stream.

***Solution*** We must use a formula from physics to solve this problem. The formula states

$$\text{distance} = \text{rate} \times \text{time}$$

If we first consider the trip downstream, we recognize that the rate of travel will be the sum of the speed of the boat and the speed of the stream. If we let

$$b = \text{speed of the boat}$$

and

$$s = \text{speed of the stream}$$

$$(\text{rate of boat going downstream})(3 \text{ hours}) = 75 \text{ miles}$$

then

$$(b + s)(3) = 75$$

The rate of travel for the trip upstream will be the speed of the boat decreased by the speed of the stream. Thus

$$(\text{rate of boat going upstream})(5 \text{ hours}) = 75 \text{ miles}$$

Thus we have

$$(b - s)(5) = 75$$

We now have the system

$$\begin{cases} 3b + 3s = 75 \\ 5b - 5s = 75 \end{cases}$$

Solving the system yields $b = 20$ and $s = 5$. (Can you verify this?) Thus the average speed of the boat in still water is 20 mph and the average rate of the stream is 5 mph. ■

## EXERCISE 10-1

Solve the systems in Problems 1 and 2 by graphing.

**1.** $\begin{cases} x + y = 2 \\ 2x - y = 1 \end{cases}$ **2.** $\begin{cases} 3x + y = 0 \\ 2x - y = -5 \end{cases}$

Solve by the substitution method in each of the systems in Problems 3–10.

**3.** $\begin{cases} 2x + y = 4 \\ 3x - 2y = -1 \end{cases}$ **4.** $\begin{cases} x + 5y = 2 \\ 2x + 3y = -3 \end{cases}$ **5.** $\begin{cases} x + y = 5 \\ 2x - y = -5 \end{cases}$

**6.** $\begin{cases} 2x + y = -2 \\ x - 2y = 5 \end{cases}$ **7.** $\begin{cases} 5x - y = 11 \\ 2x - 3y = 7 \end{cases}$ **8.** $\begin{cases} x + 3y = -1 \\ 2x + 5y = 1 \end{cases}$

**9.** $\begin{cases} 3x - 2y = -6 \\ x + 2y = 22 \end{cases}$ **10.** $\begin{cases} 2x + 3y = 2 \\ 3x - y = -19 \end{cases}$

Solve the systems in Problems 11–18 by the elimination (or addition) method.

**11.** $\begin{cases} 2x + y = 1 \\ x - y = 5 \end{cases}$ **12.** $\begin{cases} 2x - 3y = 5 \\ 2x + y = 9 \end{cases}$ **13.** $\begin{cases} 2x + y = 1 \\ 3x - 2y = 5 \end{cases}$

**14.** $\begin{cases} 2x + 3y = 21 \\ 3x + 2y = 19 \end{cases}$ **15.** $\begin{cases} 5x - 3y = 10 \\ 2x + y = 4 \end{cases}$ **16.** $\begin{cases} 2x - y = 13 \\ 3x + 5y = 13 \end{cases}$

**17.** $\begin{cases} 5x + 2y = 2 \\ 2x + 3y = 14 \end{cases}$ **18.** $\begin{cases} 2x + 5y = -6 \\ 3x + 4y = 5 \end{cases}$

Classify each of the systems in Problems 19–33 as containing independent equations, dependent equations, or else as an inconsistent system of equations. If the system contains independent equations, find its solution by a convenient algebraic method.

**19.** $\begin{cases} 2x + y = 1 \\ 3x - y = 9 \end{cases}$ **20.** $\begin{cases} x + y = 4 \\ x - y = 6 \end{cases}$ **21.** $\begin{cases} x + 2y = 9 \\ 3x - y = -1 \end{cases}$

**22.** $\begin{cases} 2x - y = 1 \\ 6x - 3y = 3 \end{cases}$ **23.** $\begin{cases} 3x - y = 2 \\ 2x + 3y = -6 \end{cases}$ **24.** $\begin{cases} 3x - 2y = 3 \\ 6x - 4y = 1 \end{cases}$

**25.** $\begin{cases} 6x - 4y = 2 \\ 3x - 2y = 1 \end{cases}$ **26.** $\begin{cases} x - 2y = 15 \\ 3x + y = 3 \end{cases}$ **27.** $\begin{cases} 3x + 2y = 5 \\ 2x + y = 1 \end{cases}$

**28.** $\begin{cases} 2x + y = 5 \\ 10x + 5y = 10 \end{cases}$ **29.** $\begin{cases} 2x + 3y = 10 \\ 6x - 2y = -3 \end{cases}$ **30.** $\begin{cases} x + 3y = 1 \\ 2x - 9y = -8 \end{cases}$

**31.** $\begin{cases} 2x - 4y = 6 \\ 3x - 6y = 9 \end{cases}$ **32.** $\begin{cases} 6x + 3y = 3 \\ 10x + 5y = 15 \end{cases}$ **33.** $\begin{cases} 5x + 2y = -1 \\ 4x + 3y = -12 \end{cases}$

Solve Problems 34–40 by using two linear equations in two variables.

**34.** The sum of two numbers is 147 and their difference is 19. Find the numbers.

**35.** An airliner took 6 hours to travel 2400 miles from New York to Las Vegas. The return trip took 4 hours. Find the average speed of the plane in still air and the average speed of the wind.

**36.** A professor has 85 students enrolled in two classes; 49 of these students are freshmen. If two thirds of the first class and one half of the second class are freshmen, how many students are in each class?

**37.** A boat travels 15 miles per hour downstream and 9 miles per hour upstream. Find the speed of the current and the speed of the boat in still water.

**38.** How much of a 40% solution of alcohol and how much of an 80% solution should be mixed to give 40 liters of a 50% solution?

★ **39.** A dietician must supplement the food of a hospital patient with 37 units of vitamin A and 51 units of vitamin B each day. Additive $X$ contains five units of vitamin A and seven units of vitamin B per milligram. Additive $Y$ contains three units of vitamin A and four units of vitamin B per milligram. How many milligrams of each additive should be used each day?

★ **40.** A chemist must add some preservatives to a machine that is used to process ketchup for a food company each day. It is necessary to have 33 units of Chemical A and 85 units of Chemical B added to the machine each day in order for the ketchup to have the correct minimum shelf life. The chemist has available Additive $X$, which has three units of Chemical A and eight units of Chemical B in each gram. She also has available Additive $Y$, which has six units of Chemical A and 15 units of Chemical B in each gram. How many grams of each additive should be placed in the machine each day?

## 10-2 SOLUTION OF A SYSTEM OF THREE LINEAR EQUATIONS IN THREE VARIABLES

An ordered triple of numbers $(a, b, c)$ is a **solution** to a system of three equations in $x, y, z$ if a true statement is obtained when $a$ is substituted for $x$, $b$ is substituted for $y$, and $c$ is substituted for $z$ in all three equations.

The approach we will use reduces a system of three equations in three variables to two equations in two variables by elimination. This method is best illustrated by example.

EXAMPLE 1 Solve the system

$$\begin{cases} x - y + 2z = 6 & (1) \\ 2x + 3y - z = -3 & (2) \\ 3x + 2y + 2z = 5 & (3) \end{cases}$$

***Solution*** *Step 1* Eliminate $y$ by multiplying both sides of equation (1) by 3 and adding the result to equation (2). Three times equation (1) added to equation (2) gives

$$5x + 5z = 15$$

*Step 2* Eliminate $y$ by multiplying both sides of equation (1) by 2 and adding the result to equation (3). This yields

$$5x + 6z = 17$$

*Step 3* Solve the system

$$\begin{cases} 5x + 5z = 15 \\ 5x + 6z = 17 \end{cases}$$

We find the solution to be

$$x = 1 \qquad \text{and} \qquad z = 2$$

*Step 4* Substitute $x = 1$ and $z = 2$ in equation (1) to solve for $y$. We find the solution to be

$$y = -1$$

*Step 5* The ordered triple $(1, -1, 2)$ checks in all three original equations. Can you verify that $x = 1$, $y = -1$, $z = 2$ is the solution to this system? ■

If one of the variables is missing from one or more of the equations, finding the solution becomes easier.

**EXAMPLE 2** Solve the system

$$\begin{cases} x + z = 3 & \textbf{(1)} \\ 2x + y + z = 3 & \textbf{(2)} \\ 3x - y + 2z = 8 & \textbf{(3)} \end{cases}$$

***Solution*** Note that equation (1) contains only two of the variables. We can eliminate $y$ using equations (2) and (3), and then we will have two equations in two variables ($x$ and $z$). Equation (2) added to equation (3) gives

$$5x + 3z = 11$$

Now we solve the system

$$\begin{cases} x + z = 3 \\ 5x + 3z = 11 \end{cases}$$

We obtain as our solution

$$x = 1 \qquad \text{and} \qquad z = 2$$

Substituting these values into equation (2) yields

$$y = -1$$

The solution $(1, -1, 2)$ checks in all three original equations. Can you verify this? ■

### Special Cases in Which There Is No Unique Solution to the System

A system of linear equations that has a unique solution is said to have **independent equations**. A system of linear equations that has solutions is said to be **consistent**. But we cannot always obtain a unique solution to a system of three linear equations in three variables. A system of linear equations that has no solutions is said to be **inconsistent**.

EXAMPLE 3 Solve the following system if possible.

$$\begin{cases} x - y + 5z = 2 & \textbf{(1)} \\ 4x + 3y - z = 3 & \textbf{(2)} \\ 8x + 6y - 2z = 7 & \textbf{(3)} \end{cases}$$

***Solution*** Let us eliminate the variable $z$. Multiply equation (2) by 5 and add to equation (1). We obtain $21x + 14y = 17$. Now multiply equation (2) by $-2$ and prepare to add to equation (3). At this point we have

$$-8x - 6y + 2z = -6$$

$$8x + 6y - 2z = 7$$

When we add, we obtain $0 = 1$. Thus the system is *inconsistent*, and there is **no solution** to the system. ■

EXAMPLE 4 Solve the following system if possible.

$$\begin{cases} 2x + y + z = 9 & \textbf{(1)} \\ x - y + z = 2 & \textbf{(2)} \\ x - 4y + 2z = -3 & \textbf{(3)} \end{cases}$$

If there are an infinite number of solutions, show the form that they will take and list two examples.

***Solution*** Let us add equation (1) and (2) to eliminate the variable $y$. We obtain

$$3x + 2z = 11 \qquad \textbf{(4)}$$

Now multiply equation (1) by 4 and add to equation (3) to again eliminate $y$. We obtain

$$9x + 6z = 33 \qquad \textbf{(5)}$$

These equations (4) and (5) are dependent. If we multiply equation (4) by $-3$ and add to equation (5), we obtain $0 = 0$.

*There are an infinite number of solutions to the original system of equations.* So that we can describe these solutions, we will solve (4) for $z$:

$$2z = 11 - 3x$$

$$z = \frac{11 - 3x}{2} \tag{6}$$

We now replace this value for $z$ in equation (1) and solve for $y$:

$$2x + y + \left(\frac{11 - 3x}{2}\right) = 9$$

$$4x + 2y + 11 - 3x = 18$$

$$2y + 11 + x = 18$$

$$2y = 7 - x$$

Now isolate the variable $y$:

$$y = \frac{7 - x}{2} \tag{7}$$

By using equations (6) and (7) we can state that *all the solutions to the system are of the form*

$$\left(x, \frac{7 - x}{2}, \frac{11 - 3x}{2}\right)$$

Let us find two solutions by letting $x$ be an arbitrary value. If $x = 0$, we have $\left(0, \frac{7 - (0)}{2}, \frac{11 - 3(0)}{2}\right)$. So one solution is $\left(0, \frac{7}{2}, \frac{11}{2}\right)$. If $x = 3$, we have $\left(3, \frac{7 - (3)}{2}, \frac{11 - 3(3)}{2}\right)$. So another solution is (3, 2, 1).

*Check:* Can you verify that $\left(0, \frac{7}{2}, \frac{11}{2}\right)$ and (3, 2, 1) are solutions to the system? ■

### Applications

**EXAMPLE 5** The equation of a vertical parabola can be put in the form $y = ax^2 + bx + c$. Find the equation of the vertical parabola that passes through the points $(-3, 10)$, $(-2, 3)$, and $(1, 6)$.

***Solution*** We must find the values of $a$, $b$, and $c$. Since each point must satisfy the equation, we have for any point $(x, y)$

$$y = ax^2 + bx + c$$

Now we investigate the three values given

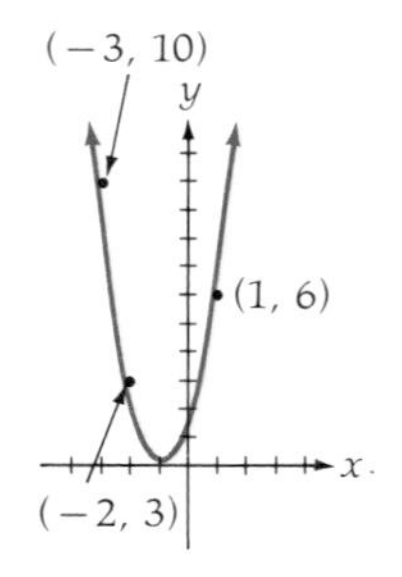


$$\begin{aligned} \text{for } (-3, 10)&: \quad 10 = a(-3)^2 + b(-3) + c \\ \text{for } (-2, 3)&: \quad 3 = a(-2)^2 + b(-2) + c \\ \text{and for } (1, 6)&: \quad 6 = a(1)^2 + b(1) + c \end{aligned}$$

These simplify to the system

$$\begin{cases} 9a - 3b + c = 10 \\ 4a - 2b + c = 3 \\ a + b + c = 6 \end{cases}$$

This system can be readily solved by eliminating the variable $c$. The solution yields $a = 2$, $b = 3$, $c = 1$. Can you verify this? Thus the desired equation is $y = 2x^2 + 3x + 1$.

*Check:* Do the points $(-3, 10)$, $(-2, 3)$, and $(1, 6)$ satisfy the quadratic equation $y = 2x^2 + 3x + 1$?

$$\begin{aligned} \text{for } (-3, 10)?& \quad 2(-3)^2 + 3(-3) + 1 = 18 - 9 + 1 = 10 \checkmark \\ \text{for } (-2, 3)?& \quad 2(-2)^2 + 3(-2) + 1 = 8 - 6 + 1 = 3 \checkmark \\ \text{for } (1, 6)?& \quad 2(1)^2 + 3(1) + 1 = 2 + 3 + 1 = 6 \checkmark \end{aligned}$$

■

## EXERCISE 10-2

Solve the systems in Problems 1–12.

**1.** $\begin{cases} x + y + z = 6 \\ 2x - y + z = 3 \\ x - y + 2z = 5 \end{cases}$

**2.** $\begin{cases} x + 2y + z = 0 \\ x - 3y - z = -2 \\ x + y - z = -2 \end{cases}$

**3.** $\begin{cases} 2x + y + z = 0 \\ 3x - 2y - z = -11 \\ x - y + 2z = 3 \end{cases}$

**4.** $\begin{cases} x - y + z = 8 \\ 5x + 4y - z = 7 \\ 2x + y - 3z = -7 \end{cases}$

**5.** $\begin{cases} x + 5y - 2z = 13 \\ 6x + y + 3z = 4 \\ x - y + 2z = -5 \end{cases}$

**6.** $\begin{cases} x + y = 6 \\ 2x - y + z = 7 \\ x + y - 3z = 12 \end{cases}$

**7.** $\begin{cases} 3x + 4y + z = -2 \\ y + z = 1 \\ 2x - y - z = -5 \end{cases}$

**8.** $\begin{cases} y - z = -3 \\ x + y = 1 \\ 2x + 3y + z = 1 \end{cases}$

**9.** $\begin{cases} 2x + 3y + z = 2 \\ -x + 2y + 3z = -1 \\ -3x - 3y + z = 0 \end{cases}$

**10.** $\begin{cases} 3x - y + 2z = 4 \\ 2x - 4y + 2z = -8 \\ 2x + y - z = 10 \end{cases}$ **11.** $\begin{cases} x - y + 3z = 12 \\ 2x + 5y + 2z = -2 \\ x + y + z = 2 \end{cases}$ **12.** $\begin{cases} 3x - y - 2z = 1 \\ 2x - 3y + z = 10 \\ x + 2y + 4z = 12 \end{cases}$

In Problems 13 and 14, solve the systems with four equations in four unknowns. (*Hint:* Seek to eliminate one variable in three ways. Remember that each equation must be used at least once. Then solve the resulting three equations in three unknowns.)

★ **13.** $\begin{cases} 3x + 2y - z + w = 8 \\ x + 2y + z - w = 4 \\ 2x + y + 2z - 3w = -2 \\ x + y - 3z + 2w = 2 \end{cases}$ ★ **14.** $\begin{cases} x + y + 2z + w = 5 \\ 2x + y - z + w = -2 \\ 3x + 2y + z + w = 5 \\ x - y + 2z - w = 11 \end{cases}$

Solve the systems *if possible,* in Problems 15–20. If there are an infinite number of solutions, describe the general case and give two examples of solutions.

**15.** $\begin{cases} 2x + 3y - z = 1 \\ 3x - 2y + z = 13 \\ x + y + 2z = 6 \end{cases}$ **16.** $\begin{cases} 2x + y - 3z = -12 \\ 3x + 4y + z = 1 \\ x + y + z = 2 \end{cases}$

**17.** $\begin{cases} 2x + y - z = 3 \\ -4x - 2y + 2z = 4 \\ x + y + z = 6 \end{cases}$ **18.** $\begin{cases} 4x - 3y - 2z = 4 \\ 3x + 2y + z = -2 \\ -2x - 7y - 4z = 1 \end{cases}$

**19.** $\begin{cases} 4x + 3y + 5z = 4 \\ 2x - y + 3z = -2 \\ x + 2y + z = 3 \end{cases}$ **20.** $\begin{cases} -3x + y + 3z = -4 \\ x - y + z = 2 \\ 2x - y - z = 3 \end{cases}$

**21.** A child has $4.50 in nickels, dimes, and quarters. He has a total of 28 coins, and the number of dimes is twice the number of nickels. How many of each coin does the child have?

**22.** The equation of a circle can be put in the form $x^2 + y^2 + ax + by + c = 0$. If the circle contains the points $(-1, 0)$, $(0, 1)$, and $(1, 3)$, find the values of $a$, $b$, and $c$ and write the equation of the circle.

**23.** A chemist wishes to make 9 liters of a 30% acid solution by mixing three solutions of 5%, 20%, and 50%. How much of each must she use if she uses twice as much 50% solution as she does 5% solution?

**24.** A swimming pool can be filled by two pipes $A$ and $B$ and drained by a third pipe $C$. If the drain is closed and both pipes $A$ and $B$ are open, it takes 6 hours to fill the pool. If $A$ and $B$ are open and the drain is also open, it takes 10 hours to fill the pool. If only pipes $A$ and $C$ are open, it takes 15 hours to fill the pool. How long would it take to drain the pool if $A$ and $B$ were closed and $C$ were open?

★ **25.** The perimeter of a triangle is 155 centimeters. The side $z$ is 20 centimeters shorter than the side $y$, and the side $y$ is 5 centimeters longer than the side $x$. Find the lengths of each of the sides of this triangle.

★**26.** Susan Dawson invested $30,000. Part of the investment was made at 12% interest, part at 10% interest, and part at 8% interest. Interest is compounded annually. Susan invested twice as much money at 10% interest as she did at 8% interest. At the end of the year she earned a total of $2960 from the three investments. How much did she invest at each amount?

## 10-3 MATRICES

Systems of linear equations are encountered in many different fields. In addition to the methods we have studied in Sections 10-1 and 10-2, a variety of other methods are available to solve linear systems. A number of efficient methods have been developed for use by a computer. A procedure used by a computer usually depends on the concept of a matrix. In this section we will develop a method for solving linear systems using matrices.

A *matrix* is a rectangular array of numbers called elements arranged in horizontal rows and vertical columns.

*Definition*

If $m$ and $n$ are positive integers, an *$m$ by $n$ matrix* is an array of the form

$$\begin{bmatrix} a_{11} & a_{12} & a_{13} & \cdots & a_{1n} \\ a_{21} & a_{22} & a_{23} & \cdots & a_{2n} \\ a_{31} & a_{32} & a_{33} & \cdots & a_{3n} \\ \vdots & & & & \vdots \\ a_{m1} & a_{m2} & a_{m3} & \cdots & a_{mn} \end{bmatrix}$$

where $a_{ij}$ represents an element in row $i$ and column $j$.

For example,

$$\begin{bmatrix} 1 & 2 & 3 & 6 \\ 2 & 1 & 4 & 0 \\ -1 & 3 & 2 & 1 \end{bmatrix}$$

is a 3 by 4 matrix, since it has three rows and four columns.

Now for any matrix, $a_{ij}$ represents the element in row $i$ and column $j$. The notation $a_{23}$ would indicate the element in the second row and third column. In the matrix just mentioned, the element $a_{23}$ is the number 4.

With any system of linear equations there are associated two matrices—the **matrix of coefficients** and the **augmented matrix**.

For example, given the system

$$\begin{cases} 2x + 3y - z = 11 \\ x + 2y + z = 12 \\ 3x - y + 2z = 5 \end{cases}$$

the matrix of coefficients is

$$\begin{bmatrix} 2 & 3 & -1 \\ 1 & 2 & 1 \\ 3 & -1 & 2 \end{bmatrix}$$

Notice that the first row corresponds to the coefficients of the first equation, the second row to the second equation, and the third row to the third equation. Also, the first column contains the $x$-coefficients, the second column the $y$-coefficients, and the third column the $z$-coefficients.

The augmented matrix of this system is

$$\begin{bmatrix} 2 & 3 & -1 & 11 \\ 1 & 2 & 1 & 12 \\ 3 & -1 & 2 & 5 \end{bmatrix}$$

This matrix is the coefficient matrix to which has been added a fourth column, which is composed of the constant terms in the equation.

**Augmented Matrix of the System**

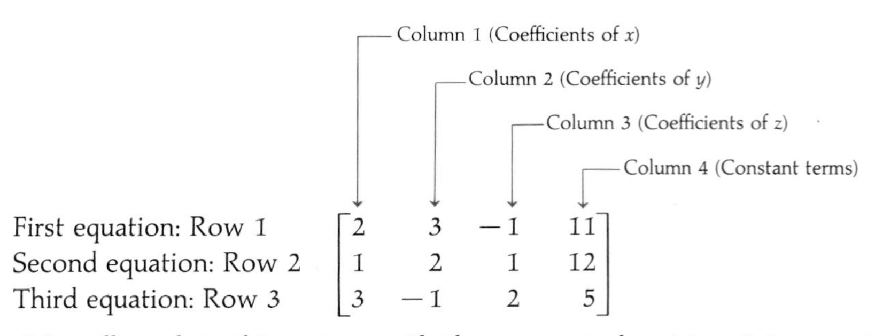


We will work in this section with the augmented matrix. It is important to remember that systems in three variables must have their equations in the form $ax + by + cz = d$ in order to form the augmented matrix of the system.

Any two systems of equations that have the same solution set are said to be **equivalent**. It can be shown that certain operations on the augmented matrix will result in a matrix of an equivalent system and therefore not change the solution of the system. The operations we will use are referred to as **elementary row operations**.

*Theorem 10-1*

The following operations on an augmented matrix will result in a matrix of an equivalent system:

1. Interchanging any two rows.
2. Multiplying all elements of a row by a constant $k$ ($k \neq 0$).
3. Multiplying the elements of a row by a constant and adding them to the corresponding elements of any other row.

You will notice that the operations listed in Theorem 10-1 are quite similar to the operations that will preserve equivalent systems of equations listed in Section 10-1.

We often refer to the main diagonal of a matrix.

*Definition*

The *main diagonal* of a matrix consists of all elements $a_{ij}$ where $i = j$.

In the matrix

$$\begin{bmatrix} 1 & 7 & 2 & 4 \\ 6 & 3 & 1 & 8 \\ -2 & -5 & -4 & -1 \end{bmatrix}$$

the main diagonal consists of the elements $a_{11}$, $a_{22}$, $a_{33}$ or 1, 3, $-4$.

We will work with an augmented matrix and seek to transform it to a form known as triangular form.

*Definition*

A matrix is in *triangular form* when all elements below the main diagonal are zero:

$$\begin{bmatrix} a_{11} & a_{12} & a_{13} & a_{14} & \cdots & a_{1n} \\ 0 & a_{22} & a_{23} & a_{24} & \cdots & a_{2n} \\ 0 & 0 & a_{33} & a_{34} & \cdots & a_{3n} \\ 0 & 0 & 0 & a_{44} & \cdots & a_{4n} \\ \vdots & & & & & \vdots \\ 0 & 0 & 0 & 0 & \cdots & a_{mn} \end{bmatrix}$$

We will now use the elementary row operations to reduce a matrix to an equivalent matrix in triangular form. This is known as the **matrix method** (or Gaussian elimination method) of solving a system of equations.

**EXAMPLE 1** Solve the following system using the matrix method:

$$\begin{cases} 3x + 4y = 1 \\ x - 3y = 9 \end{cases}$$

***Solution*** The augmented matrix of the system is

$$\begin{bmatrix} 3 & 4 & 1 \\ 1 & -3 & 9 \end{bmatrix}$$

Our goal is to use elementary row operations to obtain an augmented matrix in triangular form. (Remember that all elements *below* the main diagonal should be 0 if the matrix is in triangular form.) First we multiply all elements of row two by $-3$:

$$\begin{bmatrix} 3 & 4 & 1 \\ -3 & 9 & -27 \end{bmatrix}$$

Now we add row one to row two and replace row two with the result.

$$\begin{bmatrix} 3 & 4 & 1 \\ 0 & 13 & -26 \end{bmatrix}$$

This augmented matrix represents the system

$$\begin{cases} 3x + 4y = 1 \\ \phantom{3x + }13y = -26 \end{cases}$$

We first find $y$ and then use **back-substitution** in the first equation to find $x$:

$$13y = -26$$

$$y = -2$$

Substituting $y = -2$ in the first equation, we have

$$3x + 4(-2) = 1$$

$$3x = 9$$

$$x = 3$$

Our solution is $(3, -2)$ for the original system.

*Check:* $3(3) + 4(-2) = 9 - 8 = 1$ ✓

$3 - 3(-2) = 3 + 6 = 9$ ✓ ■

**EXAMPLE 2** Solve the following system using the matrix method:

$$\begin{cases} 2x + 3y - z = 11 \\ x + 2y + z = 12 \\ 3x - y + 2z = 5 \end{cases}$$

***Solution*** The augmented matrix is

$$\begin{bmatrix} 2 & 3 & -1 & 11 \\ 1 & 2 & 1 & 12 \\ 3 & -1 & 2 & 5 \end{bmatrix}$$

If any variables were missing, we would place a zero in that position. Our goal is to use the elementary row operations to obtain a matrix of an equivalent system with zero entries in the $a_{21}$, $a_{31}$, and $a_{32}$ positions. This might be accomplished by many different combinations of the elementary row operations. We will proceed as follows.

First we wish to obtain a zero in all elements in the first column beneath the element $a_{11}$ (which is 2). Multiply row two by $-3$, add it to row three, and replace row three with the result.

$$\begin{bmatrix} 2 & 3 & -1 & 11 \\ 1 & 2 & 1 & 12 \\ 0 & -7 & -1 & -31 \end{bmatrix}$$

Multiply row two by $-2$:

$$\begin{bmatrix} 2 & 3 & -1 & 11 \\ -2 & -4 & -2 & -24 \\ 0 & -7 & -1 & -31 \end{bmatrix}$$

Add row one to row two and replace row two with the result.

$$\begin{bmatrix} 2 & 3 & -1 & 11 \\ 0 & -1 & -3 & -13 \\ 0 & -7 & -1 & -31 \end{bmatrix}$$

Next we want to obtain a zero in the place of element $a_{32}$. Multiply row two by $-7$:

$$\begin{bmatrix} 2 & 3 & -1 & 11 \\ 0 & 7 & 21 & 91 \\ 0 & -7 & -1 & -31 \end{bmatrix}$$

Add row two to row three and replace row three with the result.

$$\begin{bmatrix} 2 & 3 & -1 & 11 \\ 0 & 7 & 21 & 91 \\ 0 & 0 & 20 & 60 \end{bmatrix}$$

Since we have used only elementary row operations, the final matrix is the augmented matrix of a system that is equivalent to our original system. This equivalent system is

$$\begin{cases} 2x + 3y - z = 11 \\ \quad 7y + 21z = 91 \\ \qquad 20z = 60 \end{cases}$$

Solving the third equation for $z$ gives $z = 3$. Now we use *back-substitution*. Substituting $z = 3$ in the second equation gives $y = 4$. Substituting $z = 3$ and $y = 4$ in the first equation gives $x = 1$. Our solution is (1, 4, 3), and this solution checks in the original system. (Can you verify that this solution satisfies all three original equations?) ■

A triangular matrix could be reached by a different combination of the elementary row operations, but the solution would not be changed. Looking ahead carefully toward the desired form can result in arriving at the triangular form in a minimum number of steps.

### Special Cases

If at any stage of using this method to solve a system of $n$ equations in $n$ unknowns you obtain an *entire row of zeros*, the system contains *dependent* equations.

**EXAMPLE 3** If possible, solve the following system using the matrix method:

$$\begin{cases} \quad 4x + y = 12 \\ x + 2y - z = 4 \\ 3x - y + z = 8 \end{cases}$$

***Solution*** Form the augmented matrix. Note the need for $a_{13}$ to be zero, since the first equation has no $z$ term.

$$\begin{bmatrix} 4 & 1 & 0 & 12 \\ 1 & 2 & -1 & 4 \\ 3 & -1 & 1 & 8 \end{bmatrix}$$

Our first goal is to obtain a first column whose elements $a_{21}$ and $a_{31}$ are zero. Multiply row two by $-3$ and add to row three. Replace row three with the result:

$$\begin{bmatrix} 4 & 1 & 0 & 12 \\ 1 & 2 & -1 & 4 \\ 0 & -7 & 4 & -4 \end{bmatrix}$$

Multiply row two by $-4$:

$$\begin{bmatrix} 4 & 1 & 0 & 12 \\ -4 & -8 & 4 & -16 \\ 0 & -7 & 4 & -4 \end{bmatrix}$$

Add row one to row two and replace row two with the result:

$$\begin{bmatrix} 4 & 1 & 0 & 12 \\ 0 & -7 & 4 & -4 \\ 0 & -7 & 4 & -4 \end{bmatrix}$$

We now wish to obtain a zero in the second column in the position designated by $a_{32}$. We multiply row two by $-1$, add to row three and replace row three with the result. We obtain

$$\begin{bmatrix} 4 & 1 & 0 & 12 \\ 0 & -7 & 4 & -4 \\ 0 & 0 & 0 & 0 \end{bmatrix}$$

*Thus the equations are dependent. There are an infinite number of solutions.* The augmented matrix represents the system

$$\begin{cases} 4x + y = 12 & \textbf{(1)} \\ -7y + 4z = -4 & \textbf{(2)} \end{cases}$$

If we solve equation (2) for $y$, we have

$$4z + 4 = 7y$$

$$\frac{4z + 4}{7} = y$$

By substituting that expression in equation (1) we have

$$4x + \left(\frac{4z + 4}{7}\right) = 12$$

$$28x + (4z + 4) = 84$$

$$28x = 84 - 4 - 4z$$

$$28x = 80 - 4z$$

$$x = \frac{80 - 4z}{28}$$

$$x = \frac{20 - z}{7}$$

Thus the solutions to the original system of equations are all of the form $(x, y, z) = \left(\frac{20 - z}{7}, \frac{4z + 4}{7}, z\right)$. ■

If at any stage of using this method to solve a system of $n$ equations in $n$ unknowns you obtain a row in which *all elements are zero except the right-hand element*, the system is *inconsistent*. In such a case, there is *no solution*.

**EXAMPLE 4** Solve if possible the system

$$\begin{cases} 2x - 4y = -6 \\ -3x + 6y = 8 \end{cases}$$

***Solution*** The augmented matrix is

$$\begin{bmatrix} 2 & -4 & -6 \\ -3 & 6 & 8 \end{bmatrix}$$

Multiply row two by 2:

$$\begin{bmatrix} 2 & -4 & -6 \\ -6 & 12 & 16 \end{bmatrix}$$

Multiply row one by 3 and add to row two and replace row two with the result:

$$\begin{bmatrix} 2 & -4 & -6 \\ 0 & 0 & -2 \end{bmatrix}$$

This matrix represents the system

$$2x - 4y = -6$$
$$0 = -2$$

Since the statement $0 = -2$ is false, we know that the *system of equations is inconsistent*. There is *no solution*. ■

## EXERCISE 10-3

**1.** In matrix $\begin{bmatrix} 3 & 6 & 5 \\ 2 & -1 & 8 \end{bmatrix}$

**a.** What are the elements in the main diagonal?

**b.** What is the value of element $a_{21}$?

**c.** What element is in the second row and the third column?

**2.** In matrix $\begin{bmatrix} 1 & 4 & -3 \\ 3 & 7 & -2 \\ 6 & 2 & 8 \end{bmatrix}$

**a.** What are the elements in the main diagonal?
**b.** What is the value of element $a_{32}$?
**c.** What element is in the third row and first column?

In Problems 3–6, write the augmented matrix for each system of equations. (Be sure to place the equations in the necessary form prior to forming the matrix.) *Do not solve the system of equations.*

**3.** $7x - 8y + 12 = 0$, $3x + 5 = 2y$

**4.** $8x + 2(y - 3) = 0$, $2y - 6x = 3$

**5.** $3x + 2y + z = 5$, $9x - y = 8$, $6x - 3y + 2z = 5$

**6.** $-5x + 2y + z = -8$, $3x + 2y - 3z = 5$, $-4x + 2z = 0$

Solve the systems of equations in Problems 7–20 by using the matrix method.

**7.** $\begin{cases} 2x + 3y = 5 \\ 5x + y = 19 \end{cases}$

**8.** $\begin{cases} 3x + 5y = -15 \\ 2x + 7y = -10 \end{cases}$

**9.** $\begin{cases} 2x + y = -3 \\ 5x - y = 24 \end{cases}$

**10.** $\begin{cases} x + 5y = -9 \\ 4x - 3y = -13 \end{cases}$

**11.** $\begin{cases} 5x + 2y = 6 \\ 3x + 4y = 12 \end{cases}$

**12.** $\begin{cases} -5x + y = 24 \\ x + 5y = 10 \end{cases}$

**13.** $\begin{cases} 3x - 2y + 3 = 5 \\ x + 4y - 1 = 9 \end{cases}$

**14.** $\begin{cases} 3x + y - 4 = 12 \\ -2x + 3y + 2 = -5 \end{cases}$

**15.** $\begin{cases} x - 2y - 3z = 4 \\ 2x + 3y + z = 1 \\ -3x + y - 2z = 5 \end{cases}$

**16.** $\begin{cases} x + y - z = -2 \\ 2x - y + 3z = 19 \\ 4x + 3y - z = 5 \end{cases}$

**17.** $\begin{cases} 5x - y + 4z = 5 \\ 6x + y - 5z = 17 \\ 2x - 3y + z = -11 \end{cases}$

**18.** $\begin{cases} 4x + 3y + 5z = 2 \\ 2y + 7z = 16 \\ 2x - y = 6 \end{cases}$

**19.** $\begin{cases} 6x - y + z = 9 \\ 2x + 3z = 16 \\ 4x + 7y + 5z = 20 \end{cases}$

**20.** $\begin{cases} 3x + 2y = 44 \\ 4y + 3z = 19 \\ 2x + 3z = -5 \end{cases}$

**21.** A woman made two investments, one at 5% interest and the other at 8%. Her total interest for the year was \$1700. The interest at 8% was \$200 more than twice the interest at 5%. How much did she have invested at each rate? (Assume that the interest discussed is simple interest.)

**22.** The sum of three numbers is 58. Twice the first number added to the sum of the second and third is 71. If the first number is added to four times the second number and the sum is decreased by three times the third number, the result is 18. Find the numbers.

If possible, solve the systems of equations in Problems 23–28 by using the matrix method. If there are an infinite number of solutions, describe the general form of the solutions.

**23.** $\begin{cases} 2x - y + z = 3 \\ x + 2y - z = 3 \\ 3x - 4y + 2z = -1 \end{cases}$

**24.** $\begin{cases} 2x + 3y + z = 1 \\ -3x + y - 2z = 5 \\ x - 2y - 3z = 4 \end{cases}$

**25.** $\begin{cases} x + y - z = 8 \\ -x + 5y - 5z = 10 \\ 2x - y + z = -2 \end{cases}$

**26.** $\begin{cases} 3x - 5y + 2z = 1 \\ x - z = 2 \\ -y + z = -3 \end{cases}$ **27.** $\begin{cases} x + 2y - z = 6 \\ 5x + 3y = 21 \\ 3x - y + 2z = 9 \end{cases}$ **28.** $\begin{cases} 3x + 2y - z = 5 \\ 2x + 5y + z = 11 \\ -x + 3y + 2z = 6 \end{cases}$

Find the unique solution of each of Problems 29 and 30 by using the matrix method.

**★ 29.** $\begin{cases} x + 7y - 3z - w = 5 \\ -2x + 5y - z + w = 5 \\ 3x - 8y + 6z + 3w = 6 \\ x + 4y - z - 2w = -7 \end{cases}$ **★ 30.** $\begin{cases} x + y + 2z + 3w = 1 \\ 3x - y - 2z - w = 3 \\ 2x - y - 3z - w = -2 \\ x + 4y + z + 2w = -3 \end{cases}$

## 10-4 DETERMINANTS

A square matrix is a matrix with the same number of rows and columns. Every square matrix has associated with it a unique real number called the **determinant of the matrix**. In the following sections we will discuss the determinant of two by two and three by three matrices and their uses in solving systems of linear equations.

*Definition*

The *determinant* of the second-order matrix

$$\begin{bmatrix} a & c \\ b & d \end{bmatrix} \quad \text{represented by} \quad \begin{vmatrix} a & c \\ b & d \end{vmatrix}$$

is the real number $ad - bc$.

Often a capital letter is used to denote a matrix. The determinant of a matrix $A$ is indicated by the notation $|A|$ or det $A$.

**EXAMPLE 1** Given $A = \begin{bmatrix} 4 & 7 \\ 3 & 2 \end{bmatrix}$ and $B = \begin{bmatrix} -5 & -6 \\ -2 & 1 \end{bmatrix}$:

**a.** Find $|A|$. **b.** Find $|B|$.

*Solution* **a.** $|A| = \begin{vmatrix} 4 & 7 \\ 3 & 2 \end{vmatrix} = (4)(2) - (3)(7) = 8 - 21 = -13$

**b.** $|B| = \begin{vmatrix} -5 & -6 \\ -2 & 1 \end{vmatrix} = (-5)(1) - (-2)(-6) = -5 - 12 = -17$ ■

The determinant of a third-order matrix can be found in more than one way, but we will use a general method that can be used to evaluate the determinant of a square matrix of any order.

***Definition***

> The *minor* of an element of a matrix is the determinant of the matrix that results when the row and column in which the element appears are eliminated.

**EXAMPLE 2** Given matrix $A = \begin{bmatrix} 3 & 1 & 4 \\ 2 & 6 & 5 \\ 3 & -1 & 2 \end{bmatrix}$:

Find the minor of 1 (the element $a_{12}$).

***Solution*** In the matrix

$$\begin{bmatrix} 3 & 1 & 4 \\ 2 & 6 & 5 \\ 3 & -1 & 2 \end{bmatrix}$$

the minor of 1 (first row, second column) is obtained by eliminating the first row and second column

$$\begin{bmatrix} \not{3} & \textcircled{1} & \not{4} \\ 2 & \not{6} & 5 \\ 3 & -\not{1} & 2 \end{bmatrix}$$

giving

$$\begin{vmatrix} 2 & 5 \\ 3 & 2 \end{vmatrix}$$ ■

Now we will need to define and use a new term called the **cofactor**.

***Definition***

> The *cofactor* of an element of a matrix is its minor multiplied by $(-1)^{i+j}$, where $i$ is the number of the row and $j$ is the number of the column of the element.

Notice that $(-1)^{i+j}$ will be positive if $(i + j)$ is an even number and negative if $(i + j)$ is an odd number.

Be sure you understand that the cofactor of an element consists of the product of $(-1)^{i+j}$ *and* the minor of that element. Thus the cofactor of any element in a third-order matrix consists of the product of $-1$ raised to a power and a second-order determinant.

**EXAMPLE 3** Given matrix $B = \begin{bmatrix} 6 & 5 & 2 \\ 9 & 3 & 4 \\ -7 & 8 & 1 \end{bmatrix}$:

Find the cofactor of $a_{21}$.

***Solution*** In this matrix the element $a_{21} = 9$. The cofactor of the element 9 is

$$(-1)^{2+1}\begin{vmatrix} 5 & 2 \\ 8 & 1 \end{vmatrix} = (-1)\begin{vmatrix} 5 & 2 \\ 8 & 1 \end{vmatrix}$$

since $(-1)^3 = -1$. ■

We can now use these definitions to write the definition of a third-order determinant.

***Definition***

The *determinant of a third-order matrix* is the sum of the products of each element of a row or column and its cofactor.

For instance, in

$$\begin{bmatrix} a_{11} & a_{12} & a_{13} \\ a_{21} & a_{22} & a_{23} \\ a_{31} & a_{32} & a_{33} \end{bmatrix}$$

the determinant is

$$(a_{11})(\text{cofactor of } a_{11}) + (a_{21})(\text{cofactor of } a_{21}) + (a_{31})(\text{cofactor of } a_{31})$$

**EXAMPLE 4** Find the following determinant: $\begin{vmatrix} 2 & 1 & 3 \\ 3 & -5 & 4 \\ 5 & 0 & 2 \end{vmatrix}$.

Evaluate the determinant by using the third row.

***Solution*** If we apply the definition to row three, we get

$$\begin{vmatrix} 2 & 1 & 3 \\ 3 & -5 & 4 \\ 5 & 0 & 2 \end{vmatrix}$$

$$= 5(-1)^{3+1}\begin{vmatrix} 1 & 3 \\ -5 & 4 \end{vmatrix} + 0(-1)^{3+2}\begin{vmatrix} 2 & 3 \\ 3 & 4 \end{vmatrix} + 2(-1)^{3+3}\begin{vmatrix} 2 & 1 \\ 3 & -5 \end{vmatrix}$$

$$= 5(1)(19) + 0(-1)(-1) + 2(1)(-13)$$

$$= 95 + 0 - 26 = 69 \quad \blacksquare$$

The value of a determinant is a constant and can be found by applying the definition to any row or column. Notice that choosing a row or column containing one or more zeros cuts the amount of computation necessary.

EXAMPLE 5 Evaluate the following: $\begin{vmatrix} 3 & 0 & 5 \\ 2 & 1 & 6 \\ -4 & 3 & -2 \end{vmatrix}$.

***Solution*** It is wise to choose either the first row or the second column, since the zero will simplify the work. Using the second column gives

$$\begin{vmatrix} 3 & 0 & 5 \\ 2 & 1 & 6 \\ -4 & 3 & -2 \end{vmatrix} = 1(-1)^{2+2}\begin{vmatrix} 3 & 5 \\ -4 & -2 \end{vmatrix} + 3(-1)^{3+2}\begin{vmatrix} 3 & 5 \\ 2 & 6 \end{vmatrix}$$

$$= 1(1)(14) + 3(-1)(8) = 14 - 24 = -10 \quad \blacksquare$$

If you use determinants extensively, several properties can be observed.

*Theorem 10-2*

For any square matrices $A$ and $B$ with exactly the same number of rows:

1. If any row or column of $A$ contains only zeros, then $|A| = 0$.
2. If any two rows of $A$ or any two columns of $A$ are the same, then $|A| = 0$.
3. If a new matrix $B$ is obtained from $A$ by *exchanging rows for columns*, then $|B| = |A|$.
4. If a new matrix $B$ is obtained by adding a multiple of one row (or column) of $A$ to another row (or column) of $A$, then $|B| = |A|$.
5. If a new matrix $B$ is obtained by *interchanging any two rows* (or columns) of $A$, then $|B| = -|A|$.
6. If a new matrix $B$ is obtained by multiplying one row (or column) of matrix $A$ by a constant $c$, then $|B| = c|A|$.

The definition for a third order determinant can be expanded for fourth and higher order determinants. The determinant of any square matrix is the sum of the products of each element of a row (or column) and its cofactor.

## EXERCISE 10-4

Evaluate in Problems 1–15.

**1.** $\begin{vmatrix} 2 & 1 \\ 3 & 5 \end{vmatrix}$ **2.** $\begin{vmatrix} 3 & -2 \\ 1 & 2 \end{vmatrix}$ **3.** $\begin{vmatrix} 1 & 2 \\ 4 & 5 \end{vmatrix}$ **4.** $\begin{vmatrix} 2 & 4 \\ 3 & 6 \end{vmatrix}$ **5.** $\begin{vmatrix} 5 & -2 \\ 3 & 1 \end{vmatrix}$

**6.** $\begin{vmatrix} -2 & -3 \\ 5 & 4 \end{vmatrix}$ **7.** $\begin{vmatrix} 4 & 2 \\ -3 & -2 \end{vmatrix}$ **8.** $\begin{vmatrix} 4 & 0 \\ 6 & -3 \end{vmatrix}$ **9.** $\begin{vmatrix} -15 & -5 \\ 3 & 1 \end{vmatrix}$ **10.** $\begin{vmatrix} 3 & 7 \\ 0 & -2 \end{vmatrix}$

**11.** $\begin{vmatrix} 5 & 9 \\ 4 & -13 \end{vmatrix}$ **12.** $\begin{vmatrix} 0 & 0 \\ 3 & -6 \end{vmatrix}$ **13.** $\begin{vmatrix} 4 & 0 \\ 9 & 0 \end{vmatrix}$ **14.** $\begin{vmatrix} 8 & -4 \\ 6 & -3 \end{vmatrix}$ **15.** $\begin{vmatrix} 0 & -6 \\ 9 & 0 \end{vmatrix}$

Given $A = \begin{vmatrix} 3 & 6 & 5 \\ 2 & -1 & 8 \\ 7 & 4 & -3 \end{vmatrix}$:

**16.** What is element $a_{32}$?

**17.** What is element $a_{22}$?

**18.** What is element $a_{13}$?

**19.** What is element $a_{21}$?

**20.** What is the cofactor of element $a_{13}$?

**21.** What is the cofactor of element $a_{23}$?

**22.** What is the cofactor of element $a_{33}$?

**23.** What is the cofactor of element $a_{12}$?

**24.** What is the cofactor of element $a_{21}$?

Evaluate in Problems 25–34.

**25.** $\begin{vmatrix} 1 & 2 & 5 \\ 3 & 1 & 4 \\ 2 & 0 & 3 \end{vmatrix}$ **26.** $\begin{vmatrix} 3 & 4 & 1 \\ 1 & 0 & -2 \\ 5 & -3 & 1 \end{vmatrix}$ **27.** $\begin{vmatrix} 4 & -2 & 1 \\ -2 & 3 & 5 \\ 1 & 0 & 2 \end{vmatrix}$

**28.** $\begin{vmatrix} 6 & -1 & 5 \\ 1 & 0 & 4 \\ -5 & -2 & 3 \end{vmatrix}$ **29.** $\begin{vmatrix} -1 & -2 & 4 \\ 3 & 1 & -1 \\ 5 & 6 & 5 \end{vmatrix}$ **30.** $\begin{vmatrix} 4 & 0 & 1 \\ -2 & 1 & 0 \\ 7 & 3 & 5 \end{vmatrix}$

**31.** $\begin{vmatrix} 3 & 5 & -3 \\ 4 & 1 & -2 \\ -2 & 6 & -4 \end{vmatrix}$ **32.** $\begin{vmatrix} 0 & 0 & 0 \\ -2 & 3 & -1 \\ 4 & 8 & -2 \end{vmatrix}$ **33.** $\begin{vmatrix} 4 & -2 & 0 \\ 3 & 4 & 0 \\ 1 & 3 & 0 \end{vmatrix}$

**34.** $\begin{vmatrix} -1 & 3 & 5 \\ -2 & 1 & -1 \\ 4 & 6 & 5 \end{vmatrix}$

★ **35.** Show that the equation of a line containing the points $(x_1, y_1)$ and $(x_2, y_2)$ may be expressed as

$$\begin{vmatrix} x & y & 1 \\ x_1 & y_1 & 1 \\ x_2 & y_2 & 1 \end{vmatrix} = 0$$

(*Hint:* Evaluate the determinant on the left. Then show that the two-point form of the equation of a line is equivalent to this form.)

★ **36.** Use the determinant form discussed in Problem 35 to find the equation of a line through the points (3, 1) and (2, 3).

Use $A = \begin{bmatrix} a & b \\ c & d \end{bmatrix}$ to represent any $2 \times 2$ square matrix in Problems 37–40. Prove the specified portion of Theorem 10-2 for any $2 \times 2$ matrix.

★ **37.** Prove part 1 of Theorem 10-2. ★ **38.** Prove part 2 of Theorem 10-2.

★ **39.** Prove part 3 of Theorem 10-2. ★ **40.** Prove part 4 of Theorem 10-2.

★ **41.** Use the cofactor method to evaluate determinants to prove for any $3 \times 3$ matrix $A$ written as

$$A = \begin{bmatrix} a & b & c \\ d & e & f \\ g & h & i \end{bmatrix}$$

that $|A| = aei + bfg + cdh - gec - hfa - idb$. (*Hint:* Expand by cofactors along the top row.)

Use some of the properties of Theorem 10-2 to evaluate in Problems 42 and 43.

★ **42.** Find $|B|$ when

$$B = \begin{bmatrix} 1 & 2 & 3 & -1 \\ 2 & 0 & 5 & 1 \\ 3 & 1 & 2 & 1 \\ -3 & -1 & 5 & 2 \end{bmatrix}$$

★ **43.** Find $|C|$ when

$$C = \begin{bmatrix} 5 & 6 & 8 & 9 \\ 11 & 3 & -2 & 1 \\ 0 & 1 & 0 & 3 \\ 0 & -4 & 0 & 6 \end{bmatrix}$$

---

## 10-5 CRAMER'S RULE

The **standard form** of a system of two linear equations is

$$\begin{cases} ax + cy = e & (1) \\ bx + dy = f & (2) \end{cases}$$

Let us solve this system for the variable $y$ using the method of elimination. Multiply equation (1) by the constant $d$ and equation (2) by the constant $-c$. We obtain the equivalent system

$$\begin{cases} adx + cdy = de & (3) \\ -bcx - cdy = -cf & (4) \end{cases}$$

Now add equation (3) to equation (4). We now have

$$adx - bcx = de - cf$$

$$x(ad - bc) = de - cf$$

$$x = \frac{de - cf}{ad - bc}$$

In a similar fashion we could eliminate the variable $x$ to obtain the corresponding expression for $y$:

$$y = \frac{af - be}{ad - bc}$$

From the definition of the value of the determinant of a second-order matrix, we see that these values for $x$ and $y$ could be expressed as

$$x = \frac{\begin{vmatrix} e & c \\ f & d \end{vmatrix}}{\begin{vmatrix} a & c \\ b & d \end{vmatrix}} \qquad y = \frac{\begin{vmatrix} a & e \\ b & f \end{vmatrix}}{\begin{vmatrix} a & c \\ b & d \end{vmatrix}}$$

This gives us a method of solving a system of two linear equations by determinants. This method is often referred to as *Cramer's Rule.* The rule is named after the Swiss mathematician Gabriel Cramer (1704–1752).

**Caution:** In a sense this is a formula and, like all formulas, must come from the standard form. If an equation is not in standard form, then it must first be put in standard form before this method will apply.

We will now state the formula as it is most often encountered.

---

### Cramer's Rule for Two Equations in Two Variables

---

Let

$$\begin{cases} ax + by = c \\ dx + ey = f \end{cases}$$

be any system of two linear equations in standard form. The solution to the system is given by

$$x = \frac{D_x}{D} \quad \text{and} \quad y = \frac{D_y}{D} \quad \text{if } D \neq 0$$

where $D = \begin{vmatrix} a & b \\ d & e \end{vmatrix}$, $D_x = \begin{vmatrix} c & b \\ f & e \end{vmatrix}$, and $D_y = \begin{vmatrix} a & c \\ d & f \end{vmatrix}$.

---

**Note the special notation here:** The symbol $D$ represents a determinant that contains the coefficients of the $x$ and $y$ terms in each equation. $D_x$ is a determinant where the constant values $c, f$ replace the coefficients of the $x$ terms. $D_y$ is a determinant where the constant values $c, f$ replace the coefficients of the $y$ terms.

**EXAMPLE 1** Solve by Cramer's Rule:

$$\begin{cases} 3x - y = 5 \\ x + y = 3 \end{cases}$$

***Solution*** The equations are already in standard form so

$$x = \frac{D_x}{D} = \frac{\begin{vmatrix} 5 & -1 \\ 3 & 1 \end{vmatrix}}{\begin{vmatrix} 3 & -1 \\ 1 & 1 \end{vmatrix}} = \frac{8}{4} = 2 \qquad y = \frac{D_y}{D} = \frac{\begin{vmatrix} 3 & 5 \\ 1 & 3 \end{vmatrix}}{\begin{vmatrix} 3 & -1 \\ 1 & 1 \end{vmatrix}} = \frac{4}{4} = 1$$

The solution is (2, 1).

*Check:* We substitute $x = 2$, $y = 1$ into each original equation:

$$3(2) - 1 = 6 - 1 = 5 \quad \checkmark$$

$$2 + 1 = 3 \quad \checkmark \quad \blacksquare$$

**Warning:** Be sure you copy the correct elements in the correct order when you form $D$, $D_x$, $D_y$.

---

Some special patterns can be noted that make it easy to set up the determinants involved.

When the equations are in standard form:

1. The denominator for each variable is the determinant of the matrix of coefficients.
2. The numerator is the same as the denominator *except* that the column of the variable being found is replaced by the column of constants.

---

### Special Cases

We have defined Cramer's Rule only for the case $D \neq 0$. This is a necessary condition for there to be a unique solution to the system

$$\begin{cases} ax + by = c \\ dx + ey = f \end{cases}$$

Now let us consider what would happen if

$$D = \begin{vmatrix} a & b \\ d & e \end{vmatrix} = 0$$

1. If $D = 0$ and also $D_x = D_y = 0$, then the two equations are **dependent**. There are an infinite number of solutions to the system.
2. If $D = 0$ but either one or both of $D_x$ and $D_y$ are nonzero, then the two equations are **inconsistent**. There is no solution to the system.

Suppose we tried to use Cramer's Rule when the system contains dependent equations.

EXAMPLE 2 Try to solve by determinants:

$$\begin{cases} x - 2y = 3 \\ 3x - 6y = 9 \end{cases}$$

***Solution***

$$x = \frac{\begin{vmatrix} 3 & -2 \\ 9 & -6 \end{vmatrix}}{\begin{vmatrix} 1 & -2 \\ 3 & -6 \end{vmatrix}} = \frac{-18 + 18}{-6 + 6} = \frac{0}{0} \qquad y = \frac{\begin{vmatrix} 1 & 3 \\ 3 & 9 \end{vmatrix}}{\begin{vmatrix} 1 & -2 \\ 3 & -6 \end{vmatrix}} = \frac{9 - 9}{-6 + 6} = \frac{0}{0}$$

Notice that *both* numerator and denominator are zero. These, of course, are undefined quantities, and we are at an impasse in our attempts to find a solution by using Cramer's Rule.

Try solving the system by the addition method. Do you see that the equations are dependent?

Since the equations are dependent, there are an infinite number of solutions. ■

We have thus verified a characteristic of a system of dependent equations when solving by determinants: *If the determinants in both numerator and denominator are zero, the equations are dependent.*

In a similar approach we will now try to solve an inconsistent system of equations.

EXAMPLE 3 Try to solve by determinants

$$\begin{cases} 2x + 4y = 5 \\ x + 2y = 6 \end{cases}$$

***Solution***

$$x = \frac{\begin{vmatrix} 5 & 4 \\ 6 & 2 \end{vmatrix}}{\begin{vmatrix} 2 & 4 \\ 1 & 2 \end{vmatrix}} = \frac{10 - 24}{4 - 4} = \frac{-14}{0} \qquad y = \frac{\begin{vmatrix} 2 & 5 \\ 1 & 6 \end{vmatrix}}{\begin{vmatrix} 2 & 4 \\ 1 & 2 \end{vmatrix}} = \frac{12 - 5}{4 - 4} = \frac{7}{0}$$

Here again we see that we cannot obtain values for $x$ and $y$, since the denominators are zero.

Try to solve the system by the elimination method. Do you see that the equations are inconsistent? Therefore, there is *no solution*. ■

Thus when solving a system by determinants: *If the determinant of the matrix of coefficients is zero but the other determinants (numerators) are not, the equations are inconsistent. There is no solution.*

*Cramer's Rule can be extended for any size of linear system that has the same number of variables as equations.* We will specifically treat the case for three equations in three variables.

---

### Cramer's Rule for Three Equations in Three Variables

---

Let

$$\begin{cases} ax + by + cz = d \\ ex + fy + gz = h \\ kx + ly + mz = n \end{cases}$$

be any system of three linear equations in standard form. The solution to the system is given by $x = \dfrac{D_x}{D}$, $y = \dfrac{D_y}{D}$, $z = \dfrac{D_z}{D}$ if $D \neq 0$, where

$$D = \begin{vmatrix} a & b & c \\ e & f & g \\ k & l & m \end{vmatrix} \qquad D_x = \begin{vmatrix} d & b & c \\ h & f & g \\ n & l & m \end{vmatrix} \qquad D_y = \begin{vmatrix} a & d & c \\ e & h & g \\ k & n & m \end{vmatrix}$$

and

$$D_z = \begin{vmatrix} a & b & d \\ e & f & h \\ k & l & n \end{vmatrix}$$

---

**EXAMPLE 4** Solve by Cramer's Rule:

$$\begin{cases} x - 2y + z = -2 \\ 2x + z = 5 \\ x + y - z = 6 \end{cases}$$

***Solution*** The unique solution is given by

$$x = \frac{D_x}{D}, \qquad y = \frac{D_y}{D}, \qquad z = \frac{D_z}{D} \qquad \text{if } D \neq 0$$

We formulate each determinant and evaluate it:

$$D = \begin{vmatrix} 1 & -2 & 1 \\ 2 & 0 & 1 \\ 1 & 1 & -1 \end{vmatrix} = 2(-1)^{2+1}\begin{vmatrix} -2 & 1 \\ 1 & -1 \end{vmatrix} \qquad \textit{Expanding by the second row.}$$

$$+ 0 + (1)(-1)^{2+3}\begin{vmatrix} 1 & -2 \\ 1 & 1 \end{vmatrix}$$

$$= (2)(-1)(1) + (1)(-1)(3) = -2 - 3 = -5$$

Since $D = -5 \neq 0$, we know we will have a unique solution.

In similar fashion we find

$$D_x = \begin{vmatrix} -2 & -2 & 1 \\ 5 & 0 & 1 \\ 6 & 1 & -1 \end{vmatrix} = -15$$

$$D_y = \begin{vmatrix} 1 & -2 & 1 \\ 2 & 5 & 1 \\ 1 & 6 & -1 \end{vmatrix} = -10$$

$$D_z = \begin{vmatrix} 1 & -2 & -2 \\ 2 & 0 & 5 \\ 1 & 1 & 6 \end{vmatrix} = 5$$

Thus

$$x = \frac{D_x}{D} = \frac{-15}{-5} = 3, \qquad y = \frac{D_y}{D} = \frac{-10}{-5} = 2, \qquad z = \frac{D_z}{D} = \frac{5}{-5} = -1$$

The solution to the system is $(3, 2, -1)$.

Solving a system of $n$ equations in $n$ unknowns by Cramer's Rule may be solved by setting up quotients of the determinants and using the value of the $D$ determinant in the denominator each time.

Should the value of the denominator be zero, two possibilities exist.

1. If the numerators of all the variables are zero, the system is dependent.
2. If at the least one of the variables has a nonzero numerator, the system is inconsistent.

## EXERCISE 10-5

Attempt to find a unique solution for each system in Problems 1–15 by using determinants. If there is no unique solution to the system, determine whether the equations are dependent or if the system is inconsistent.

**1.** $\begin{cases} x + y = 3 \\ 2x - y = 0 \end{cases}$ **2.** $\begin{cases} 2x + y = 7 \\ x + 2y = 11 \end{cases}$ **3.** $\begin{cases} x - y = -1 \\ 2x - y = 1 \end{cases}$

**4.** $\begin{cases} 2x + 3y = 10 \\ x - y = -5 \end{cases}$ **5.** $\begin{cases} 3x + y = 2 \\ 6x + 2y = 4 \end{cases}$ **6.** $\begin{cases} 3x + 2y = -8 \\ 2x - 5y = 1 \end{cases}$

**7.** $\begin{cases} 2x - 3y = 6 \\ x + 2y = 3 \end{cases}$ **8.** $\begin{cases} 2x + 8y = 1 \\ x + 4y = 3 \end{cases}$ **9.** $\begin{cases} x + 4y = 1 \\ 2x - 3y = 13 \end{cases}$

**10.** $\begin{cases} 2x + y = 0 \\ 4x + 3y = 6 \end{cases}$ **11.** $\begin{cases} x = 2y - 4 \\ 3y = 5 + x \end{cases}$ **12.** $\begin{cases} 2x - 3y = 5 \\ \frac{4}{3}x - 2y = \frac{10}{3} \end{cases}$

**13.** $\begin{cases} x - 2y = 6 \\ \dfrac{x-1}{3} + \dfrac{y}{2} = \dfrac{1}{2} \end{cases}$ **14.** $\begin{cases} y = x + 3 \\ x = \dfrac{2y}{3} - 2 \end{cases}$ **15.** $\begin{cases} \dfrac{x+1}{5} + \dfrac{y-3}{2} = 1 \\ y = -5x \end{cases}$

**16.** The sum of Tom's age and twice Kathy's age is 57. Three times Kathy's age, decreased by twice Tom's age, is 12. Find their ages. (First form two equations in two variables, then solve by using determinants.)

Solve the systems in Problems 17–24 by determinants.

**17.** $\begin{cases} x + y + z = 6 \\ 2x - y + 2z = 6 \\ 3x + 2y - z = 4 \end{cases}$ **18.** $\begin{cases} x - y + 2z = 3 \\ 3x + y + z = -1 \\ 2x - 3y + 5z = 8 \end{cases}$ **19.** $\begin{cases} x + 2y - z = 13 \\ 2x - y - 2z = 11 \\ 3x + y + z = 4 \end{cases}$

**20.** $\begin{cases} 2x + y - z = -2 \\ 3x + 2y + z = -4 \\ x - y + 3z = 13 \end{cases}$ **21.** $\begin{cases} 3x - y + z = -10 \\ 2x + 3y - 2z = -3 \\ x - 5y + 3z = -8 \end{cases}$ **22.** $\begin{cases} x + y = -2 \\ 3x - z = 1 \\ 2x + y + z = 1 \end{cases}$

**23.** $\begin{cases} x + 2y - 3z = -15 \\ 3x - y + z = 9 \\ 2x + 3y - z = -8 \end{cases}$ **24.** $\begin{cases} x - 3z = -15 \\ x - 2y = 2 \\ y + z = 4 \end{cases}$

First form three equations in three variables, then solve by determinants in Problems 25 and 26.

**25.** The sum of three numbers is zero. Twice the first number, added to the second, is 11 less than the third. The third number is 17 more than the second. Find the numbers.

**26.** A couple buys 11 gallons of three different kinds of paint, some at \$8.00 a gallon, some at \$7.00 a gallon, and some at \$5.00 a gallon. They have twice as many gallons of \$5.00 paint as they do of \$7.00 paint. If the total bill for the paint is \$74.00, how many gallons of \$5.00 paint did they buy?

A system of linear equations whose right-hand constants are all zero is called a **homogeneous system**. If the system has three equations in three variables, all homogeneous systems of this type can be placed in the form

$$\begin{aligned} ax + by + cz &= 0 \\ dx + ey + fz &= 0 \\ gx + hy + jz &= 0 \end{aligned}$$

All homogeneous systems of this type have the obvious solution (0, 0, 0). Some have other solutions as well.

If

$$D = \begin{vmatrix} a & b & c \\ d & e & f \\ g & h & j \end{vmatrix} \neq 0,$$

then the solution (0, 0, 0) is unique. If $D = 0$, then there are other solutions, which are best found by using augmented matrix operations. Use determinants to determine whether (0, 0, 0) is a unique solution to the systems in Problems 27 and 28.

**★27.** $$\begin{aligned} x + 2y - z &= 0 \\ 5x - z &= 0 \\ 3x - 4y + z &= 0 \end{aligned}$$

**★28.** $$\begin{aligned} 3x - y + 2z &= 0 \\ x + y - 3z &= 0 \\ 2x + y + 4z &= 0 \end{aligned}$$

---

## 10-6 SYSTEMS OF INEQUALITIES AND LINEAR PROGRAMMING

### Graphing an Inequality in Two Variables

In previous sections we solved inequalities in one variable. In this section we will graph inequalities in two variables and solve systems of inequalities in two variables by graphical methods.

There is a definite relationship between the graphs of equations and inequalities. For instance, the graph of $2x + 3y = 7$ is a straight line, while the graph of $2x + 3y > 7$ is the region above the line, and the graph of $2x + 3y < 7$ is the region below the line. Thus we see that the graph of a linear inequality in two variables corresponds to a half-plane composing the region on one side of a straight line.

### Procedures for Graphing an Inequality in Two Variables

1. Graph the equation found by replacing the inequality sign with an equal sign.

   If the inequality is $<$ or $>$, draw a dashed line.
   If the inequality is $\leq$ or $\geq$, draw a solid line.

2. Pick a convenient test point on one side of the line. If the coordinates of the test point satisfy the inequality, shade in that side of the line. If they do not satisfy the inequality, shade in the other side of the line.

**EXAMPLE 1** Graph $4x + 3y \geq 0$.

***Solution*** *Step 1* We first graph the line $4x + 3y = 0$. We can write this as $3y = -4x$ or $y = -\frac{4}{3}x$. This is a line going through the origin with a slope of $-\frac{4}{3}$. We graph the line as a solid line, since we encounter the $\geq$ symbol.

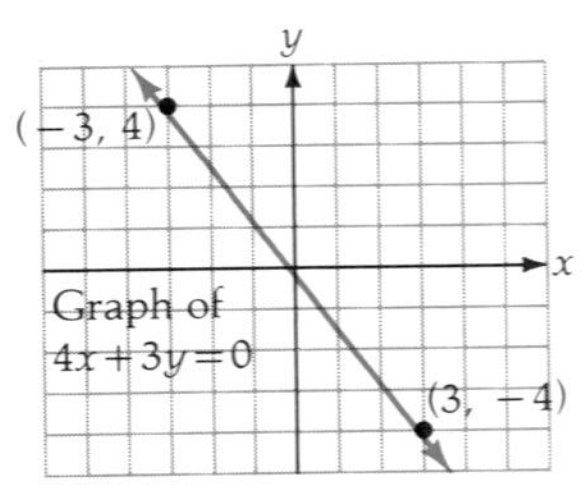


*Step 2* Since the graph passes through (0, 0), we pick a different point that is clearly not on the line. Arbitrarily, we select $(-2, -2)$ as a test point. It lies below the line. We substitute $(-2, -2)$ into the inequality

$$4(-2) + 3(-2) \geq 0 \quad ?$$

$$-8 - 6 \geq 0 \quad ?$$

$$-14 \geq 0 \quad \textbf{Not true}$$

A false statement is obtained. Therefore we do *not* shade in the half plane that includes our test point. We shade in the opposite side of the line. The answer is the line $4x + 3y = 0$ itself *and* the entire region above the line.

The graph is shown on the next page.

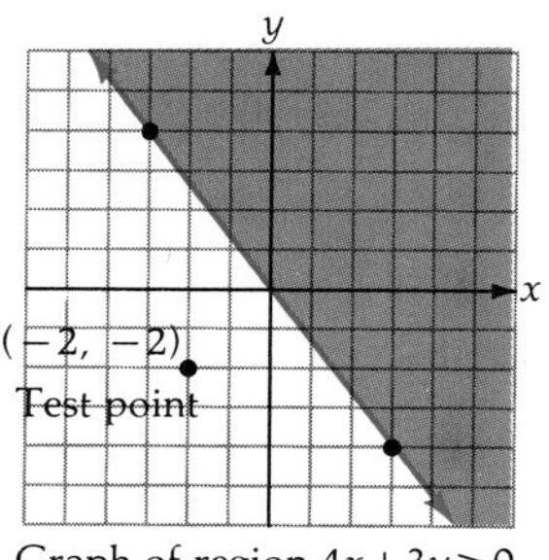


Graph of region $4x + 3y \geq 0$ ■

If the graph of an inequality is not linear, we use our knowledge of the graph of a function to complete the graph.

Consider the equation of a circle. The graph of $x^2 + y^2 = 36$ is a circle, and the graph of $x^2 + y^2 < 36$ is the interior of the circle, while the graph of $x^2 + y^2 > 36$ is the exterior of the circle.

Take a minute to examine these figures. Do you see why we shaded the interior of the circle $x^2 + y^2 < 36$? Do you see why we used a dashed line in the second and third figures below?

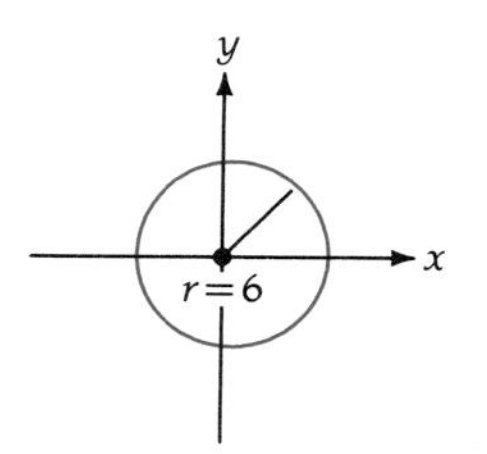


(a) Graph of $x^2 + y^2 = 36$

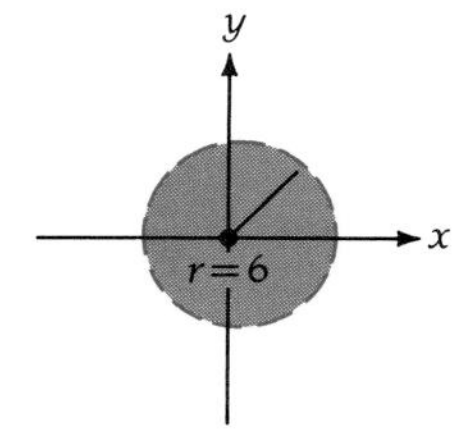


(b) Graph of $x^2 + y^2 < 36$

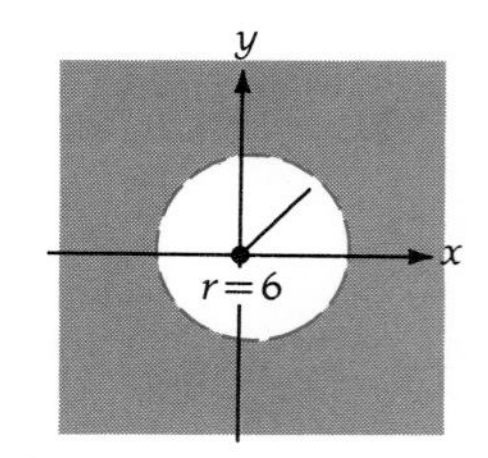


(c) Graph of $x^2 + y^2 > 36$

We will often need to take extra steps to put the equation into standard form before we graph an inequality in two variables.

**EXAMPLE 2** Graph $x^2 + y^2 - 6x - 10y \geq 2$.

***Solution*** *Step 1* *Graphing the equation of the curve.* First we will complete the square and get the equation $x^2 + y^2 - 6x - 10y = 2$ into standard form.

$$x^2 - 6x + \underline{9} + y^2 - 10y + \underline{25} = 2 + \underline{9} + \underline{25}$$

$$(x - 3)^2 + (y - 5)^2 = 36$$

We now see that we have the equation of a circle with its center at (3, 5) and a radius of 6. Since the inequality is not strict ($\geq$), we will draw the circle with a solid line. The graph of our original inequality

$x^2 + y^2 - 6x - 10y \geq 2$ is either the interior or exterior of the circle together with the circle itself.

*Step 2* *We use a test point.* We may use the center (3, 5) as a test point. Substituting (3, 5) into the inequality, we have

$$(3)^2 + (5)^2 - 6(3) - 10(5) \geq 2 \quad ?$$

$$9 + 25 - 18 - 50 \geq 2 \quad ?$$

$$-34 \geq 2 \quad \textbf{Not true}$$

This is a false statement. Therefore the region containing (3, 5) is not in the solution set. As mentioned, since we have $\geq$ as our relation, the circle itself is also a part of the solution set and is drawn as an unbroken curve to indicate this fact. The solution is the circle and all the region outside the circle.

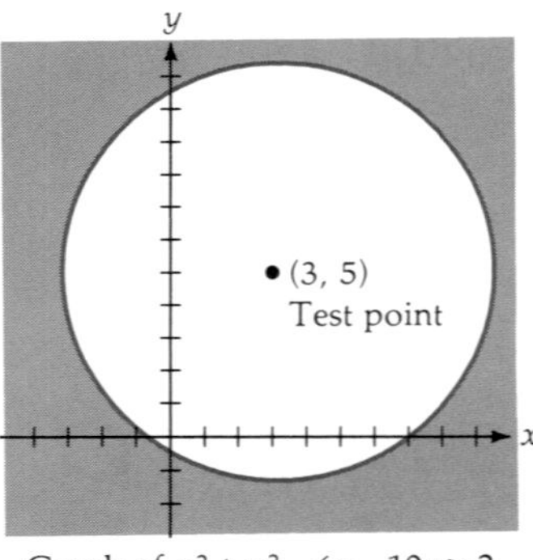


Graph of $x^2 + y^2 - 6x - 10y \geq 2$ ■

## Graphing a System of Two or More Inequalities

When we graph a system of two or more inequalities on the same coordinate axes, the intersection of the solution sets will be the solution set of the system.

**EXAMPLE 3** Graph the solution set of the system $\begin{cases} x - y \geq 6 \\ 2x + 3y \leq 5 \end{cases}$

***Solution*** *Step 1* *We graph the region $x - y \geq 6$.* First we will graph the straight line $x - y = 6$ using a solid line, since we do not have a strict inequality ($\geq$). As a test point we use (0, 0). Since $0 - 0 \geq 6$ is a false statement, we shade below the line $x - y = 6$.

*Step 2* *We graph the region $2x + 3y \leq 5$.* Again we draw the straight line determined by $2x + 3y = 5$ using a solid line. As a test point we use (0, 0). Substituting (0, 0) into $2x + 3y \leq 5$ gives $0 \leq 5$, which implies that (0, 0) is in the solution set of $2x + 3y \leq 5$. This is true. Thus we shade in the region that includes (0, 0), which is the region below the line $2x + 3y = 5$.

*Step 3* The solution set of the system is the intersection of the solution sets of the two inequalities. It is that region that the two inequalities have in common.

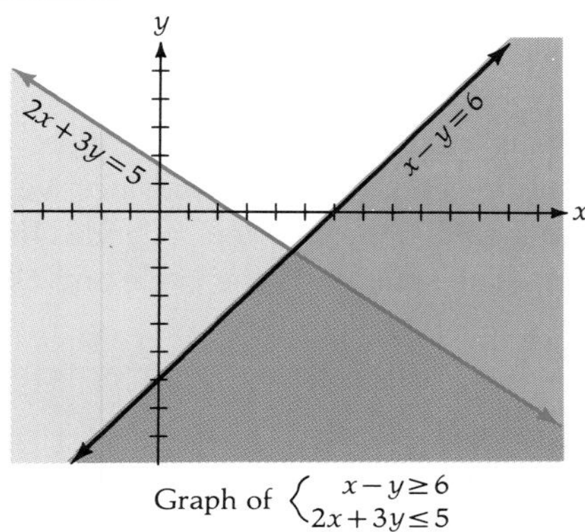


Graph of $\begin{cases} x - y \geq 6 \\ 2x + 3y \leq 5 \end{cases}$

The double-shaded portion of the plane is the solution set of the system of the two inequalities. It includes both the double-shaded region and the borders of the double-shaded region. ■

EXAMPLE 4 Graph the system

$$\begin{cases} x + y \leq -1 \\ y \geq x^2 + x - 1 \end{cases}$$

***Solution*** The line $x + y = -1$ passes through $(-3, 2)$ and $(2, -3)$. The region $x + y \leq -1$ lies *below the line,* since $0 + 0 \leq -1$ is a false statement and we want to shade the side that does not contain (0, 0).

We observe that the equation $y = x^2 + x - 1$ is a parabola opening upward. We can graph the parabola more quickly if we locate the vertex, which is at $x = \frac{-b}{2a} = -\frac{1}{2}$. If $x = -\frac{1}{2}$, $y = \frac{1}{4} - \frac{1}{2} - 1 = -\frac{5}{4}$.

Obtaining a few other values, we have

| $x$ | $-3$ | $-2$ | $-1$ | $-\frac{1}{2}$ | $0$ | $1$ | $2$ |
|---|---|---|---|---|---|---|---|
| $y$ | $5$ | $1$ | $-1$ | $-\frac{5}{4}$ | $-1$ | $1$ | $5$ |

If we test (0, 0) in the inequality $y \geq x^2 + x - 1$, we have $0 \geq -1$, which is true. So we shade *above the parabola.*

Thus the graph of the system is given by the shaded region and the boundary lines of the shaded region.

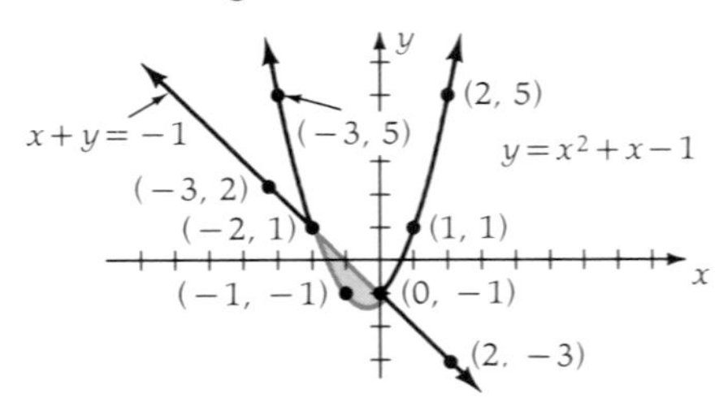


### Linear Programming

Sometimes a linear equation is used in conjunction with inequalities involving the same variables. This occurs in the branch of mathematics known as **linear programming**. The following example illustrates the process.

EXAMPLE 5 A small manufacturing company produces two models of fishing reels. Model $A$ sells for \$25, and model $B$ sells for \$40. The company cannot sell more than 100 reels a week. It takes 1 hour to produce a model $A$ reel and 3 hours to produce a model $B$ reel. The total number of labor hours available in a week is 150. How many reels of each model should the company produce in a week to maximize income?

***Solution*** First we shall let

$$x = \text{number of model } A \text{ reels to be produced}$$

$$y = \text{number of model } B \text{ reels to be produced}$$

Then, since the total number of reels produced cannot exceed 100, we have

$$x + y \leq 100$$

It would take a total of $x$ hours to produce the model $A$ reels and $3y$ hours to produce the model $B$ reels. Thus

$$x + 3y \leq 150$$

The number of reels for each model cannot be negative, so $x \geq 0$ and $y \geq 0$. We therefore have the following system of inequalities:

$$\begin{cases} x + y \leq 100 \\ x + 3y \leq 150 \\ x \geq 0 \qquad y \geq 0 \end{cases}$$

We graph our lines using a scale in which each grid mark equals 10. The use of a scale such as this is often necessary in linear programming problems.

We graph the line $x + y = 100$. Since (0, 0) satisfies $x + y \leq 100$, we shade the region *below* that line.

Next we graph the line $x + 3y = 150$. Since (0, 0) satisfies $x + 3y \leq 150$, we shade *below* that line also.

As we continue, we determine the graph of the system. The graph of $x \geq 0$ and $y \geq 0$ is the region in the first quadrant.

Thus we can now determine the graph of the system. It is the shaded region below, including the boundary lines. (Note that the line $x + y = 100$ and the line $x + 3y = 150$ intersect at (75, 25). Do you see that this can be determined either from the graph or from algebraic methods?)

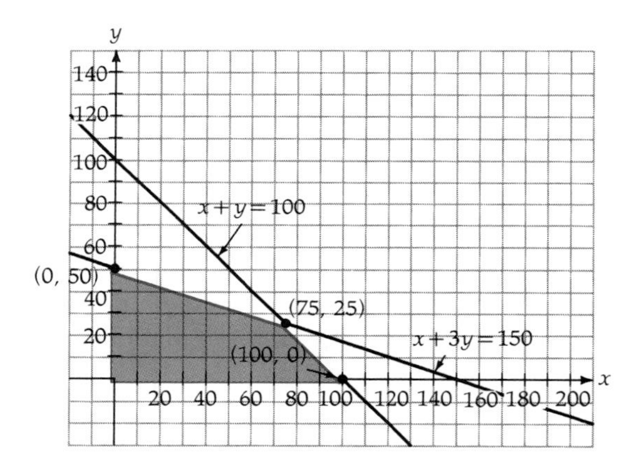


Now the income for the company can be expressed as

$$I = 25x + 40y$$

This expression has certain restrictions or **constraints** on the values of $x$ and $y$ as provided for by the given system of inequalities. We wish to find which values for $x$ and $y$ under the given constraints will produce a maximum income.

At this point we must introduce an important theorem from linear programming. It states that when we have a region, such as in this example: *The maximum and minimum values (if they exist) for the linear expression will be found at the vertices of the region.*

There are four vertices, which are determined by the intersection of the boundary lines. The four vertices are (0, 0), (0, 50), (75, 25), and (100, 0). Thus if there is a maximum income, it must occur at one of the vertices. We therefore try the coordinates for each vertex in the expression for income.

$(x, y)$: $I = 25x + 40y =$ Income for the company for given values of $x$ and $y$

$(0, 0)$: $I = 25(0) + 40(0) = \$0$

$(0, 50)$: $I = 25(0) + 40(50) = \$2000$

$(75, 25)$: $I = 25(75) + 40(25) = \boxed{\$2875}$ *Maximum value.*

$(100, 0)$: $I = 25(100) + 40(0) = \$2500$

We see that the vertex (75, 25) produces the maximum income. Therefore the company should produce 75 model $A$ reels and 25 model $B$ reels. ■

---

### Procedure to Solve Elementary Linear Programming Problems

---

1. Express each limitation in the original problem as a linear inequality with one or two variables.
2. Graph the system of linear inequalities obtained in Step 1. Be sure to determine the coordinates of the vertices where the boundary lines intersect.
3. Write a function in two variables that describes the quantity you desire to maximize or minimize.
4. Evaluate that function at each vertex to determine at what point the maximum or minimum value is obtained.

---

## EXERCISE 10-6

Graph the solution set for each of the inequalities in Problems 1–10.

**1.** $x + y > 3$ **2.** $2x - 3y \le 0$ **3.** $4x - 3y - 12 \le 0$

**4.** $7x + 2y + 14 > 0$ **5.** $x^2 + y^2 - 2x - 6y < -6$ **6.** $x^2 + y^2 + 8y \ge 0$

**7.** $9x^2 + 18x + 16y^2 - 64y \le 71$ **8.** $x^2 - 8x - 8y \ge -40$ **9.** $y^2 - 4x^2 > 16$

**10.** $y^2 - 12x > 12$

Graph the solution sets of the systems in Problems 11–24.

**11.** $\begin{cases} x + y \le 2 \\ 2x - y \ge 1 \end{cases}$ **12.** $\begin{cases} 3x + y > -9 \\ x - 2y < 4 \end{cases}$ **13.** $\begin{cases} 3x + 4y > -12 \\ 2x - 3y > 6 \end{cases}$

**14.** $\begin{cases} 2x + y \le 2 \\ x - y \le -2 \end{cases}$ **15.** $\begin{cases} 2x - y \le 2 \\ 2x + 3y \ge 6 \end{cases}$ **16.** $\begin{cases} 3x + 2y > 4 \\ x - y < 3 \end{cases}$

**17.** $\begin{cases} x - y \geq -1 \\ y \leq x^2 - 2x + 1 \end{cases}$

**18.** $\begin{cases} 2x + y > 8 \\ x^2 + y^2 - 6x - 2y \leq -1 \end{cases}$

**19.** $\begin{cases} y \geq 0 \\ x \geq -1 \\ x + y \geq 1 \end{cases}$

**20.** $\begin{cases} 3x + y \leq 0 \\ 2x - y \geq -5 \\ y \geq 0 \end{cases}$

**21.** $\begin{cases} x^2 + y^2 < 25 \\ x \geq -2 \\ y \geq 3 \end{cases}$

**22.** $\begin{cases} y \leq (x-2)^2 \\ x \geq -2 \\ y \geq -3 \end{cases}$

**23.** $\begin{cases} 3x - 4y \geq -12 \\ x + y \geq 1 \\ 3x + 4y \leq 12 \\ 2x - y \leq 2 \end{cases}$

**24.** $\begin{cases} 2x + y \leq 10 \\ y \geq \frac{1}{2}x \\ x + 2 \geq 0 \\ \frac{2}{3}y - 4 \leq 0 \end{cases}$

**25.** Find the minimum value of the expression $c = 2x + 3y$ subject to the restrictions

$$\begin{cases} x + 2y \geq 12 \\ x - 6 \geq 0 \\ y \geq 0 \end{cases}$$

**26.** Find the maximum value of the expression $c = 3x + 4y$ subject to the restrictions

$$\begin{cases} 2x + y \leq 7 \\ x + 3y \leq 6 \\ x \geq 0, \quad y \geq 0 \end{cases}$$

**27.** Find the maximum value of the expression $p = 3x + 2y$ subject to the restrictions

$$\begin{cases} x + 2y \geq 4 \\ x - 2y \geq 1 \\ x - 6 \leq 0 \\ y \geq 0 \end{cases}$$

**28.** Find the maximum value of the expression $p = 4x + 3y$ subject to the restrictions

$$\begin{cases} y - 2x \leq 0 \\ 3x + 2y \geq 18 \\ x - 6 \leq 0 \\ y \geq 0 \end{cases}$$

**★ 29.** Two numbers are such that the first number is not less than 10 and the sum of twice the first number and three times the second number is not more than 60. Also twice the first number decreased by the second number is not greater than 40. Find the two numbers satisfying these constraints such that their sum will be minimum. Find the two numbers whose sum will be maximum.

**★ 30.** A company wishes to purchase advertising time on a television network. The network offers a package containing half-minute commercials and one-minute commercials. The company must purchase a total of at least 100 minutes of air time. The number of one-minute commercials must be at least one and one-half that of the half-minute commercials. The number of one-minute commercials cannot exceed 125. If each half-minute commercial costs $30,000 and each one-minute commercial costs $50,000, how many of each should the company buy to minimize its expenses?

★ **31.** A farmer is trying to grow wheat and corn for a maximum profit. He has 300 acres available for these crops. He has promised a local grain company that he will sell them what he produces and that he will have at least 150 acres of wheat and at least 50 acres of corn. Because of the time period when he will harvest the crops, he wants to plant at least twice as many acres in wheat as in corn. He predicts that he will get $100 per acre profit from the wheat and $80 per acre profit from the corn. How many acres of each should he plant for maximum profit?

★ **32.** It takes a company two hours on a molding machine and four hours on a welding machine to make a customized trailer hitch. It takes the company four hours on a molding machine and two hours on a welding machine to make a customized set of stabilizer bars. With overtime production the company can use the molding machine up to 20 hours per day. However, the welding machine cannot be used more than 16 hours per day owing to daily maintenance. The profit on a customized trailer hitch is $64.00. The profit on a customized set of stabilizer bars is $58.00. How many of each unit should be made each day in order to achieve a maximum profit?

---

## 10-7 SYSTEMS INVOLVING NONLINEAR EQUATIONS

The methods of substitution and elimination can sometimes be used to find the solution to a system of equations where one or more equations is not linear.

EXAMPLE 1 Solve the following system by algebraic methods.

$$\begin{cases} x - y = -1 \\ y = x^2 - 2x + 1 \end{cases}$$

***Solution*** From the first equation we have

$$-y = -1 - x$$

and therefore

$$y = 1 + x.$$

Now we substitute this expression for $y$ into the second equation.

$$1 + x = x^2 - 2x + 1$$

$$0 = x^2 - 2x - x + 1 - 1$$

$$0 = x^2 - 3x$$

$$0 = x(x - 3)$$

$$x = 0 \qquad x = 3$$

For each value of $x$ we find the corresponding value of $y$.

If $x = 0$, we have

$$y = 1 + x = 1 + 0 = 1$$

If $x = 3$, we have

$$y = 1 + 3 = 4$$

Thus the solutions to the system are (0, 1) and (3, 4). ■

**EXAMPLE 2** Solve the system algebraically and verify your answer with a sketch.

$$\begin{cases} x^2 + y^2 - 2y - 3 = 0 \\ x + 2y = 4 \end{cases}$$

***Solution*** Solve the second equation for $x$:

$$x = 4 - 2y$$

Now we replace $x$ by this expression in the first equation:

$$(4 - 2y)^2 + y^2 - 2y - 3 = 0$$
$$16 - 16y + 4y^2 + y^2 - 2y - 3 = 0$$
$$5y^2 - 18y + 13 = 0$$
$$(5y - 13)(y - 1) = 0$$
$$5y = 13 \qquad y - 1 = 0$$
$$y = \frac{13}{5} \qquad y = 1$$

For each value of $y$ we need the corresponding value of $x$:

If $y = \frac{13}{5}$, $\quad x = 4 - 2y = 4 - 2\left(\frac{13}{5}\right) = \frac{20}{5} - \frac{26}{5} = -\frac{6}{5}$

If $y = 1$, $\quad x = 4 - 2 = 2$

Thus two solutions to the system are $\left(-\frac{6}{5}, \frac{13}{5}\right)$ and (2, 1). To draw a rough sketch of the circle $x^2 + y^2 - 2y - 3 = 0$, we will want to place it in standard form:

$$x^2 + y^2 - 2y + \_\_\_\_ = 3 + \_\_\_\_$$
$$x^2 + y^2 - 2y + 1 = 3 + 1$$
$$x^2 + (y - 1)^2 = 4$$

This is a circle with radius 2 and center at (0, 1). The straight line $x + 2y = 4$ passes through (0, 2) and (4, 0). The graphs are shown on the next page.

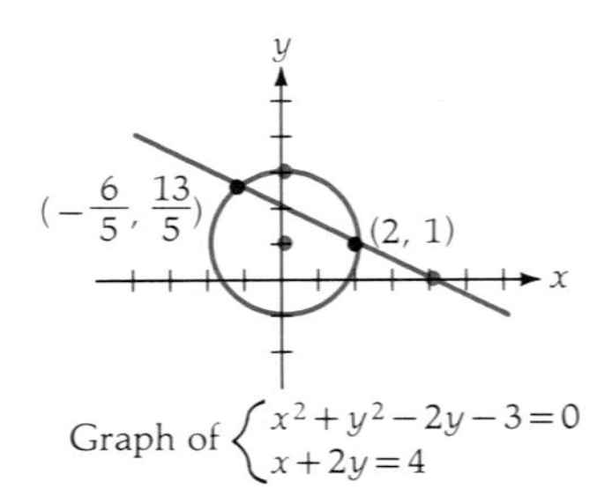


Graph of $\begin{cases} x^2+y^2-2y-3=0 \\ x+2y=4 \end{cases}$

The solution obtained algebraically seems consistent with our graphical approach. It is possible in some systems to have more than two solutions. ■

If both equations are of the second degree, you may be able to use the method of elimination.

**EXAMPLE 3** Solve the system

$$\begin{cases} 9x^2 - 4y^2 = 36 \\ 25x^2 - 16y^2 = 400 \end{cases}$$

***Solution*** Multiply the first equation by $-4$ and then add the equations together to eliminate the $y^2$-variable.

$$\begin{array}{r} -36x^2 + 16y^2 = -144 \\ 25x^2 - 16y^2 = 400 \\ \hline -11x^2 \qquad\quad = 256 \end{array}$$

$$x^2 = -\frac{256}{11}$$

$$x = \pm\sqrt{-\frac{256}{11}}$$

These values are imaginary. Thus there are *no real solutions* to this system. Can you verify by a sketch that these two hyperbolas do not intersect? ■

## EXERCISE 10-7

Solve each of the systems of equations in Problems 1–12 by algebraic methods and then verify that your answer is reasonable by drawing a rough sketch of the system of equations.

**1.** $\begin{cases} y = 2x^2 \\ y - 2x = 4 \end{cases}$

**2.** $\begin{cases} x - y = -2 \\ y = x^2 \end{cases}$

**3.** $\begin{cases} x + y = -1 \\ y = x^2 + x - 1 \end{cases}$

4. $\begin{cases} y = x^2 - 2x + 3 \\ x + y = 5 \end{cases}$

5. $\begin{cases} \dfrac{x^2}{4} + \dfrac{y^2}{9} = 1 \\ y = 2x \end{cases}$

6. $\begin{cases} \dfrac{x^2}{16} + \dfrac{y^2}{9} = 1 \\ y = x \end{cases}$

7. $\begin{cases} (x - 3)^2 + (y - 2)^2 = 16 \\ y + x = 2 \end{cases}$

8. $\begin{cases} (x + 4)^2 + (y - 1)^2 = 9 \\ y - x = 2 \end{cases}$

★9. $\begin{cases} 9x^2 + 2y^2 = 18 \\ y = -2x + 3 \end{cases}$

★10. $\begin{cases} x^2 + 16y^2 = 16 \\ y = 3x + 1 \end{cases}$

★11. $\begin{cases} \dfrac{x^2}{16} - \dfrac{y^2}{9} = 1 \\ y = x - 4 \end{cases}$

★12. $\begin{cases} \dfrac{x^2}{25} - y^2 = 1 \\ y = 5 - x \end{cases}$

Solve each of the systems of equations in Problems 13–22 entirely by algebraic methods. If there is no real solution, so state.

13. $\begin{cases} x + 2y = 8 \\ xy = 4 \end{cases}$

14. $\begin{cases} 3x + y = 6 \\ xy = 1 \end{cases}$

15. $\begin{cases} x - y = -2 \\ 2x^2 + 3y = 2 \end{cases}$

16. $\begin{cases} x + y = 1 \\ 3x^2 + 2y = 2 \end{cases}$

17. $\begin{cases} x^2 + 5y^2 = 1 \\ 5x^2 + y^2 = 1 \end{cases}$

18. $\begin{cases} x^2 + y^2 = 4 \\ x^2 + 8y^2 = 16 \end{cases}$

19. $\begin{cases} 3x^2 + 4y^2 = 24 \\ -2x^2 + 3y^2 = 35 \end{cases}$

20. $\begin{cases} 2x^2 + y^2 = 16 \\ 2x^2 - 3y^2 = 4 \end{cases}$

★21. $\begin{cases} x^2 - xy + y^2 = 20 \\ x^2 + 2xy - 3y^2 = 0 \end{cases}$

★22. $\begin{cases} 3x^2 + 2xy + y^2 = 6 \\ x^2 + 4xy - 2y^2 = 0 \end{cases}$

★23. Given a system of one straight line and one parabola, discuss the possibilities for the solution set.

★24. Given a system of one straight line and one ellipse, discuss the possibilities for the solution set.

★25. Given a system of one ellipse and one hyperbola, discuss the possibilities for the solution set.

★26. Given a system of two hyperbolas, discuss the possibilities for the solution set.

## COMPUTER PROBLEMS FOR CHAPTER 10

See if you can apply your knowledge of computer programming and of Chapter 10 to do the following.

1. Write a computer program to find the value of any $3 \times 3$ determinant in the form

$$\begin{vmatrix} a & b & c \\ d & e & f \\ g & h & i \end{vmatrix}$$

(*Hint:* Use the concept discussed in Problem 41 of Exercise 10-4.) Run the program and evaluate:

a. $\begin{vmatrix} 2 & -6 & 8 \\ 58 & -27 & 11 \\ 12 & 16 & -13 \end{vmatrix}$ b. $\begin{vmatrix} 0.6 & 1.23 & 5.8 \\ -7.3 & 1.71 & -6.1 \\ -3.8 & 2.74 & 1.7 \end{vmatrix}$

**2.** Modify your program in Problem 1 to find the solution of any three equations in three unknowns in the form

$$ax + by + cz = d$$
$$ex + fy + gz = h$$
$$jk + ky + lz = m$$

If $\begin{vmatrix} a & b & c \\ e & f & g \\ j & k & l \end{vmatrix} = 0$, have your program print out "This system of equations is either inconsistent or else it contains dependent equations. No unique solution is determined." Run your program to find the solution of the following systems:

**a.** $\begin{cases} 2x + 4y + 3z = -2 \\ x - 3y + 4z = 9 \\ 5x + 7y - z = -9 \end{cases}$ **b.** $\begin{cases} 2.16x + 3.08y + 2.98z = 8.93 \\ 5.07x - 1.98y + 7.78z = 5.78 \\ 3.97x - 1.02y + 5.13z = -1.08 \end{cases}$

## KEY TERMS AND CONCEPTS

Be sure you can explain and give an example for each of the following terms.

Systems of equations
Solution to a system of equations
Method of substitution
Method of elimination
Equivalent systems of equations
Independent equations
Inconsistent system of equations
Dependent equations
Matrix
Matrix of coefficients
Augmented matrix
Main diagonal of a matrix
Triangular form of a matrix
Determinant of a matrix
Minor of an element of a matrix
Cofactor of an element of a matrix
Cramer's Rule
Systems of inequalities
Linear programming

## SUMMARY OF PROCEDURES AND CONCEPTS

### Matrix Solution to a System of Equations

1. Form the augmented matrix of the system.
2. Using elementary row operations, obtain an equivalent matrix that is in triangular form.
3. Use the corresponding linear equation for the last row of the matrix in triangular form to find the value of one variable.
4. Using back-substitution, proceed to the row directly above it to find the value of one more variable.
5. Repeat this process till all variables are found.

### Evaluation of Determinants

1. The value of $\begin{vmatrix} a & c \\ b & d \end{vmatrix}$ is defined to be the real number $ad - bc$.
2. The value of any determinant larger than $2 \times 2$ is obtained by finding the sum of the products of each element of a row or column and its cofactor.

### Properties of Determinants

1. If any row or column of $A$ contains only zeros, then $|A| = 0$.
2. If any two rows of $A$ or any two columns of $A$ are the same, then $|A| = 0$.
3. If a new matrix $B$ is obtained from $A$ by *interchanging rows and columns,* then $|B| = |A|$.
4. If a new matrix $B$ is obtained by adding a multiple of one row (or column) of $A$ to another row (or column) of $A$, then $|B| = |A|$.
5. If a new matrix $B$ is obtained by *interchanging any two rows* (or columns) of $A$, then $|B| = -|A|$.
6. If a new matrix $B$ is obtained by multiplying one row (or column) of matrix $A$ by a constant $c$, then $|B| = c|A|$.

### General Form of Cramer's Rule

Consider a system of $n$ equations in $n$ unknowns of the form $ax + by + cz + \cdots = p$. Let $D$ be the *determinant* of the coefficient matrix of the system. Let $D_x$ represent the determinant of the matrix obtained by replacing the *first column* of $D$ by the constants of the system. Let $D_y$ represent the determinant of the matrix obtained by replacing the *second column* of $D$ by the constants of the system. Continue in similar fashion (if necessary) to find $D_z$, etc.

Then if $D \neq 0$, the unique solution to the system is

$$x = \frac{D_x}{D}, \qquad y = \frac{D_y}{D}, \qquad z = \frac{D_z}{D}, \ldots$$

## CHAPTER 10 REVIEW EXERCISE

Classify each of the systems in Problems 1–10 as containing independent equations, dependent equations, or else as an inconsistent system of equations. If the system contains independent equations, find the solution to the system.

1. $\begin{cases} x + y = -1 \\ 2x - y = 4 \end{cases}$

2. $\begin{cases} x - y = 3 \\ x + 2y = -6 \end{cases}$

3. $\begin{cases} 2x + y = -3 \\ x - 2y = -4 \end{cases}$

4. $\begin{cases} 3x - y = 7 \\ 2x + y = 8 \end{cases}$

5. $\begin{cases} 5x + 2y = -1 \\ 4x + 3y = -5 \end{cases}$

6. $\begin{cases} 3x - 5y = 1 \\ 4x - 3y = -6 \end{cases}$

7. $\begin{cases} 2x - y = 4 \\ 6x - 3y = 8 \end{cases}$

8. $\begin{cases} 2x + y = -5 \\ 3x + 2y = -10 \end{cases}$

**9.** $\begin{cases} 3x + 6y = 15 \\ 2x + 4y = 10 \end{cases}$ **10.** $\begin{cases} 2x + 3y = -5 \\ 5x - 2y = -22 \end{cases}$

Solve the systems in Problems 11–15 algebraically.

**11.** $\begin{cases} x + y - z = 4 \\ 2x + y + z = 3 \\ 2x + 2y + z = 5 \end{cases}$ **12.** $\begin{cases} x + y - z = -3 \\ x + y + z = 3 \\ 3x - y + z = 7 \end{cases}$ **13.** $\begin{cases} 2x - y + z = 5 \\ x + 2y - z = -2 \\ x + y - 2z = -5 \end{cases}$

**14.** $\begin{cases} 2x - 3y + z = 11 \\ x + y + 2z = 8 \\ x + 3y - z = -11 \end{cases}$ **15.** $\begin{cases} x + y = -2 \\ x - z = -1 \\ x + 2y - z = 1 \end{cases}$

Solve the systems of equations in Problems 16–20 using the matrix method.

**16.** $\begin{cases} x + 2y = 3 \\ 2x + 5y = 1 \end{cases}$ **17.** $\begin{cases} 2x + 3y = 6 \\ 3x - 4y = -8 \end{cases}$ **18.** $\begin{cases} x + y - z = 0 \\ 3x - y + z = 12 \\ 2x + 3y + 2z = 7 \end{cases}$

**19.** $\begin{cases} 3x - 2y + z = 4 \\ 4x + 3z = 12 \\ x - 5y + 2z = 8 \end{cases}$ **20.** $\begin{cases} 3x + z = -2 \\ 5y - 3z = 23 \\ 6x + 7y = 26 \end{cases}$

Evaluate the determinants in Problems 21–25.

**21.** $\begin{vmatrix} 3 & -6 \\ 1 & 2 \end{vmatrix}$ **22.** $\begin{vmatrix} -4 & 16 \\ 0 & 3 \end{vmatrix}$ **23.** $\begin{vmatrix} 2 & -3 \\ 8 & -12 \end{vmatrix}$

**24.** $\begin{vmatrix} 2 & 1 & 4 \\ 1 & 0 & 2 \\ -3 & 1 & -3 \end{vmatrix}$ **25.** $\begin{vmatrix} -2 & 1 & 0 \\ 0 & 6 & -3 \\ 4 & 2 & -5 \end{vmatrix}$

Solve the systems in Problems 26–30 by determinants.

**26.** $\begin{cases} 2x + y = 5 \\ 3x - 2y = 11 \end{cases}$ **27.** $\begin{cases} x - y = 6 \\ 3x + 2y = -7 \end{cases}$ **28.** $\begin{cases} 2x + 3y = 2 \\ x - 5y = 14 \end{cases}$

**29.** $\begin{cases} x + 2y = -6 \\ 2x - 3y = 9 \end{cases}$ **30.** $\begin{cases} 2x + 5y = 0 \\ 3x - 2y = 19 \end{cases}$

Solve the equations in Problems 31–35 by determinants.

**31.** $\begin{cases} x + y + z = 2 \\ 2x - y + z = -1 \\ x - y + z = -2 \end{cases}$ **32.** $\begin{cases} x + y - z = -3 \\ x + y + z = 3 \\ 3x - y + z = 7 \end{cases}$ **33.** $\begin{cases} 2x - y + z = 5 \\ x + 2y - z = -2 \\ x + y - 2z = -5 \end{cases}$

**34.** $\begin{cases} 2x - 3y + z = 11 \\ x + y + 2z = 8 \\ x + 3y - z = -11 \end{cases}$ **35.** $\begin{cases} x + y = -2 \\ x - z = -1 \\ x + 2y - z = 1 \end{cases}$

Use systems of linear equations to solve Problems 36 and 37.

**36.** A boat went a certain distance upstream and back to the starting point in 5 hours. The speed of the boat in still water is 15 miles per hour. The current of the stream is 3 miles per hour. How long did it take the boat to go upstream?

**37.** A person has 15 coins consisting of nickels, dimes, and quarters. The value of the coins is \$2.20. If there are two more quarters than nickels, how many dimes are there?

Graph the solution sets of the systems in Problems 38–42.

**38.** $\begin{cases} 3x + y \le 9 \\ 2x - 3y \ge -6 \end{cases}$ **39.** $\begin{cases} x - y \le -1 \\ 3x + 2y \le 6 \end{cases}$ **40.** $\begin{cases} x + y < -1 \\ y \ge x^2 - 2x - 3 \end{cases}$

**41.** $\begin{cases} y \ge x^2 + 1 \\ x^2 + y^2 \le 16 \\ y \ge 3 \end{cases}$ **42.** $\begin{cases} x - y \ge -4 \\ x + y \ge -2 \\ 2x + 3y \le 6 \\ x \le 0 \end{cases}$

**43.** A broker has a maximum of \$12,000 to invest in two types of bonds. Bond $A$ returns 8% and bond $B$ returns 10% per year. The investment in bond $B$ must not exceed 50% of the investment in bond $A$. How much should be invested at each rate in order to obtain a maximum profit?

Solve the systems in Problems 44 and 45 and check your solutions by making a rough sketch of the system.

**44.** $\begin{cases} y = 9 - 2x^2 \\ y = x^2 + 5x + 1 \end{cases}$ **45.** $\begin{cases} x^2 + 9y^2 = 9 \\ y = 2x + 1 \end{cases}$

## PRACTICE TEST FOR CHAPTER 10

Solve the systems in Problems 1 and 2 by algebraic methods.

**1.** $\begin{cases} x + y = 4 \\ 2x - y = 5 \end{cases}$ **2.** $\begin{cases} 2x + y = 2 \\ x - 3y = 15 \end{cases}$

**3.** Solve the following system of equations using the matrix method. Show your work.

$$\begin{cases} 2x - y + z = 14 \\ 3x + 2y + 2z = 17 \\ x + 5y + 3z = 11 \end{cases}$$

Evaluate the determinants in Problems 4–6.

**4.** $\begin{vmatrix} 6 & 3 \\ 5 & -1 \end{vmatrix}$ **5.** $\begin{vmatrix} 0 & 0 & -6 \\ 0 & 1 & 2 \\ 1 & 3 & 4 \end{vmatrix}$ **6.** $\begin{vmatrix} 4 & 0 & -3 \\ 1 & 2 & 5 \\ 3 & 1 & 6 \end{vmatrix}$

**7.** Solve the following system by determinants. Show your work.

$$\begin{cases} 2x + 3y = -2 \\ 3x + 2y = 7 \end{cases}$$

**8.** Solve the following system by any method. Show your work.

$$\begin{cases} x + 2y + z = 1 \\ 2x + 3y - z = 0 \\ x - 2y + 3z = 7 \end{cases}$$

**9.** Graph the solution set of the following system:

$$\begin{cases} y \geq x^2 + 2x - 3 \\ x - y \leq 2 \\ y \leq 0 \\ x \leq -1 \end{cases}$$

**10.** Find the maximum value of $p = 3x + 2y$ subject to the constraints

$$\begin{cases} y \leq 6 - x \\ 2x \leq 10 - y \\ x + 3y \leq 12 \\ x \geq 0 \\ y \geq 0 \end{cases}$$

**11.** Find the real solution(s), if any, to the system. Verify the accuracy of your solution by a rough sketch of the system.

$$\begin{cases} (x - 2)^2 + y^2 = 25 \\ y = 2x - 1 \end{cases}$$

# 11 Polynomials and Zeros of Polynomials

■ *There is no branch of mathematics, however abstract, which may not some day be applied to the phenomena of the real world.*

*Nicholas Lobachevsky*

Some of the more theoretical aspects of the properties of polynomials and their roots were studied by mathematicians for many years before practical applications of the properties were discovered. In other cases, simple properties of graphing polynomial functions were observed by mathematicians, but many years elapsed before the properties were formally stated and actually proved.

Thus the mathematical knowledge of polynomials has advanced sometimes in the realm of concepts and formal mathematical properties and at other times in more practical intuitive areas.

## 11-1 PROPERTIES OF POLYNOMIALS

The functional notation discussed in Chapters 3 and 4 can be useful in the study of polynomials. In general we will denote a polynomial in $x$ as $P(x)$.

The student should convince himself or herself that every polynomial $P$ does define a function. In other words, for a given polynomial $P$ in $x$, does each value $x_0$ yield one and only one value $P(x_0)$?

A general polynomial function is usually written as

$$P(x) = a_0x^n + a_1x^{n-1} + a_2x^{n-2} + \cdots + a_{n-1}x + a_n$$

The coefficients $a_i$ and the variable $x$ may be either real or complex.

*Definition*

> The *degree* of a polynomial in $x$ is the highest exponent of $x$ that appears in the polynomial.

It is generally agreed that no degree is assigned to $P(x) = 0$.

*Definition*

> If for a polynomial $P$ in $x$, $P(a) = 0$, then $a$ is a *zero* of the polynomial.

A **zero** of a polynomial is also called a **root** of the polynomial equation of the form $P(x) = 0$. The student should be familiar with the use of each term.

EXAMPLE 1 For $P(x) = x^2 + 5x + 6$, is $-2$ a zero of the polynomial $P(x)$?

***Solution*** We are given

$$P(x) = x^2 + 5x + 6$$

Since $$P(-2) = (-2)^2 + 5(-2) + 6 = 4 - 10 + 6 = 0$$

we conclude that $-2$ is a zero of the polynomial. ■

### Division of Polynomials

In previous chapters we added, subtracted, and multiplied polynomials. In this section we wish to discuss the other fundamental operation—division.

From arithmetic we are familiar with a process called "long division." For instance, to divide 30 by 8 using long division we write

$$\begin{array}{r|l} & 3 \\ \hline 8 & 30 \\ & \underline{24} \\ & \phantom{0}6 \end{array}$$

and say that 30 divided by 8 gives a quotient of 3 and a remainder of 6. We can express this as

$$30 = 8(3) + 6$$

In fact, for integers $a$ and $b$ where $b < a$ we can write

$$a = bq + r$$

where $q$ is the quotient and $r$ is the remainder such that $r$ is less than $b$. This is the **division algorithm for integers**.

A similar process can be used to divide one polynomial by another.

**EXAMPLE 2** Divide $6x^3 - 17x^2 - x + 15$ by $(2x - 5)$.

***Solution***

*Step 1* Make sure that both polynomials are arranged in descending powers of $x$. Supply zero coefficients for any missing terms and arrange as follows:

$$2x - 5 \,\overline{)\, 6x^3 - 17x^2 - x + 15}$$

*Step 2* To obtain the first term of the quotient, divide the first term of the dividend by the first term of the divisor. In this case, $6x^3 \div 2x$ gives $3x^2$. Write as follows:

$$\begin{array}{r|l} & 3x^2 \\ \hline 2x - 5 & 6x^3 - 17x^2 - x + 15 \end{array}$$

*Step 3* Multiply the *entire* divisor by the term obtained in Step 2 and subtract the result from the dividend as follows:

$$\begin{array}{r|l} & 3x^2 \\ \hline 2x - 5 & 6x^3 - 17x^2 - x + 15 \\ & 6x^3 - 15x^2 \\ \cline{2-2} & \quad\; - 2x^2 - x + 15 \end{array}$$

*Step 4* Divide the first term of the remainder by the first term of the divisor to obtain the next term of the quotient. Then multiply the entire divisor by the term thus obtained and subtract again as follows:

$$\begin{array}{r|l} & 3x^2 - x \\ \hline 2x - 5 & 6x^3 - 17x^2 - x + 15 \\ & 6x^3 - 15x^2 \\ \cline{2-2} & \quad\; - 2x^2 - x + 15 \\ & \quad\; - 2x^2 + 5x \\ \cline{2-2} & \qquad\qquad - 6x + 15 \end{array}$$

This process is repeated until the remainder is zero or a polynomial of degree *less* than the divisor.

$$\begin{array}{r|l} & 3x^2 - \phantom{17}x - 3 \\ \hline 2x - 5 & 6x^3 - 17x^2 - x + 15 \\ & \underline{6x^3 - 15x^2} \\ & \phantom{6x^3} - 2x^2 - x + 15 \\ & \phantom{6x^3} \underline{- 2x^2 + 5x} \\ & \phantom{6x^3 - 2x^2} - 6x + 15 \\ & \phantom{6x^3 - 2x^2} \underline{- 6x + 15} \\ & \phantom{6x^3 - 2x^2 - 6x + 1} 0 \end{array}$$

Therefore our example shows that

$$(6x^3 - 17x^2 - x + 15) \div (2x - 5) = 3x^2 - x - 3$$

or $$6x^3 - 17x^2 - x + 15 = (2x - 5)(3x^2 - x - 3) + 0 \quad \blacksquare$$

**EXAMPLE 3** Divide $5x^3 - 24x^2 + 7$ by $(5x + 1)$.

***Solution*** Notice that the dividend is missing an $x$ term. This is because the coefficient of that term is zero. We must remember to enter the $x$ term as $0x$ as follows:

$$\begin{array}{r|l} & x^2 - 5x + 1 \\ \hline 5x + 1 & 5x^3 - 24x^2 + 0x + 7 \\ & \underline{5x^3 + x^2} \\ & \phantom{5x^3} - 25x^2 + 0x + 7 \\ & \phantom{5x^3} \underline{- 25x^2 - 5x} \\ & \phantom{5x^3 - 25x^2} 5x + 7 \\ & \phantom{5x^3 - 25x^2} \underline{5x + 1} \\ & \phantom{5x^3 - 25x^2 5x + } 6 \end{array}$$

so $$5x^3 - 24x^2 + 7 = (5x + 1)(x^2 - 5x + 1) + 6 \quad \blacksquare$$

Examples 2 and 3 have illustrated the **division algorithm for polynomials**. We will state it here without proof.

***Theorem 11-1*** If $f(x)$ and $g(x)$ are polynomials and if $g(x) \neq 0$, then there exist unique polynomials $q(x)$ and $r(x)$ such that $f(x) = g(x) \cdot q(x) + r(x)$ where either $r(x) = 0$ or $r(x)$ has degree less than the degree of $g(x)$.

In this theorem, $q(x)$ is the quotient and $r(x)$ is the remainder.

*Theorem 11-2 (Remainder Theorem)*

> If a polynomial $f(x)$ is divided by $(x - c)$, the remainder is $f(c)$.

EXAMPLE 4 Find the remainder, using the remainder theorem, if $5x^2 - 7x + 12$ is divided by $(x + 4)$.

***Solution*** Here we must note that the theorem states that we divide by $(x - c)$. To use the theorem, we must express $(x + 4)$ in terms of $(x - c)$:

$$(x + 4) = [x - (-4)]$$

so that $c = -4$.

$$\begin{aligned} f(x) &= 5x^2 - 7x + 12 \\ f(-4) &= 5(-4)^2 - 7(-4) + 12 \\ &= 5(16) + 28 + 12 = 120 \end{aligned}$$

Thus if $5x^2 - 7x + 12$ is divided by $(x + 4)$, the remainder is 120. ■

Another theorem closely related to the remainder theorem is the factor theorem.

*Theorem 11-3 (Factor Theorem)*

> A polynomial $f(x)$ has a factor $(x - c)$ if and only if $f(c) = 0$.

The words "if and only if" in this theorem mean that the statement and its converse are both true. That is, the statements:

**1.** If $f(c) = 0$, then $f(x)$ has a factor $(x - c)$
**2.** If $f(x)$ has a factor $(x - c)$, then $f(c) = 0$

are both true.

EXAMPLE 5 Is $(x + 1)$ a factor of $x^3 - 6x^2 + 11x - 6$?

***Solution***

$$f(x) = x^3 - 6x^2 + 11x - 6$$

$$f(-1) = (-1)^3 - 6(-1)^2 + 11(-1) - 6 = -24$$

Since $f(-1) \neq 0$, $(x + 1)$ is not a factor of $x^3 - 6x^2 + 11x - 6$. ■

## EXERCISE 11-1

**1.** If $P(x) = x^4 + x^3 - 13x^2 - x + 12$, find
**a.** $P(2)$ **b.** $P(-1)$ **c.** $P(0)$ **d.** $P(3)$ **e.** $P(-4)$ **f.** $P(1)$
**g.** Did you find any values for $x$ in parts **a–f** that are zeros of $P(x)$? If so, which values are zeros?

**2.** If $P(x) = x^4 + 2x^3 - 2x^2 + 2x - 3$, find
**a.** $P(3)$ **b.** $P(2)$ **c.** $P(1)$ **d.** $P(0)$ **e.** $P(-3)$ **f.** $P(-1)$
**g.** Did you find any values for $x$ in parts **a–f** that are zeros of $P(x)$? If so, which values are zeros?

**3.** Is $-2$ a zero of $P(x) = x^5 + 2x^3 - 27x^2 - 54$?

**4.** Is $-4$ a zero of $P(x) = x^4 + 2x^3 + 2x^2 - 4x - 8$?

Two polynomials in $x$ are said to be *equal* if and only if they are of the same degree and their respective coefficients of terms of the same degree are equal. Use this definition to solve Problems 5–6.

**5.** If $P(x) = 5x^2 - 2x + 1$ and $Q(x) = (a + 4)x^2 + (b - 9)x + 1$, find $a$ and $b$ so that $P(x) = Q(x)$.

**6.** If $P(x) = 3x^3 + 2x^2 - 5x + 1$ and $Q(x) = (c - 8)x^3 + (d + 2)x^2 + ex + 1$, find $c$, $d$, $e$ so that $P(x) = Q(x)$.

Find the quotient and remainder for the divisions in Problems 7–15 by long division.

**7.** $(3x^3 - x^2 + 4x - 2) \div (x + 1)$

**8.** $(2x^4 + 13x^3 + 15x^2 - 4x - 9) \div (2x + 3)$

**9.** $(x^5 - 13x^3 + 11x^2 + 15) \div (x + 4)$

**10.** $(3x^5 - 4x^3 + x + 1) \div (x^2 - 2)$

**11.** $(x^5 - x^4 + 2x^3 - x + 10) \div (x^2 + 1)$

**12.** $(6x^4 - 51x^2 - 27) \div (x - 3)$

**13.** $(x^4 + 3x^2 - 340) \div (x - 4)$

**14.** $(6x^3 + 7x^2 - 18x + 15) \div (2x^2 + 3x - 5)$

**15.** $(2x^4 + 5x^3 - 11x^2 - 20x + 12) \div (x^2 + x - 6)$

In each of the divisions in Problems 16–24, find the remainder by (1) long division and (2) the remainder theorem.

**16.** $(2x^3 - x^2 + 1) \div (x - 1)$

**17.** $(5x^2 + x - 4) \div (x - 4)$

**18.** $(x^5 + 3x^2 - x) \div (x + 1)$

**19.** $(3x^4 + 5x^3 + 2) \div (x + 3)$

**20.** $(2x^4 + x^3 - 16x^2 + 18) \div (x + 2)$

**21.** $(x^5 - 32) \div (x - 2)$

**22.** $(x^5 + x^3 - 64) \div (x + 3)$

**23.** $(2x^4 + x^3 + x^2 + 10x - 8) \div (x + 2)$

**24.** $(x^4 - 3x^3 - 5x^2 + 2x - 16) \div (x + 2)$

**25.** Is $(x + 4)$ a factor of $3x^3 + 5x^2 - 2x - 1$?

**26.** Is $(x + 4)$ a factor of $5x^3 + 16x^2 - 18x - 8$?

**27.** Is $(x - 5)$ a factor of $3x^3 - 17x^2 + 11x - 5$?

**28.** Is $(x + 1)$ a factor of $5x^4 + 10x^3 - 15x^2 + 24x + 16$?

**29.** Is $\left(x - \frac{1}{2}\right)$ a factor of $4x^4 - 8x^3 + 3x^2 + 8x - 4$?

**30.** Is $\left(x - \frac{1}{3}\right)$ a factor of $3x^4 + 5x^3 + x^2 + 5x - 2$?

**31.** If $P(x) = x^3 - x^2 - 16x - 12$, is $(x + 3)$ a factor of $P(x)$? Can you determine a zero for this polynomial?

**32.** If $P(x) = x^3 - x^2 - 8x + 12$, is $(x - 4)$ a factor of $P(x)$? Can you determine a zero for this polynomial?

**★ 33.** The polynomial $P(x) = x^3 + 3x^2 - 10x + b$ has a factor of $(x - 3)$. Determine the value of $b$.

**★ 34.** The polynomial $P(x) = x^3 - x^2 + bx - 20$ has a factor of $(x + 2)$. Determine the value of $b$.

---

## 11-2 SYNTHETIC DIVISION

Polynomial division is a process often encountered in mathematics. Whenever a process is used often in algebra, it is highly desirable to shorten or simplify that process when possible. The long division process can be simplified by a procedure known as **synthetic division** when the divisor is of first degree.

Note that in the process of long division, only the coefficients of the terms are actually used in finding a quotient.

We will show the development of synthetic division by the following example:

Divide $3x^4 - 5x^3 + 9x + 5$ by $(x - 2)$. First, by long division we obtain

$$\begin{array}{r|l} & 3x^3 + x^2 + 2x + 13 \\ \hline x - 2 & 3x^4 - 5x^3 + 0x^2 + 9x + 5 \\ & \underline{3x^4 - 6x^3} \\ & \qquad x^3 + 0x^2 + 9x + 5 \\ & \qquad \underline{x^3 - 2x^2} \\ & \qquad\qquad 2x^2 + 9x + 5 \\ & \qquad\qquad \underline{2x^2 - 4x} \\ & \qquad\qquad\qquad 13x + 5 \\ & \qquad\qquad\qquad \underline{13x - 26} \\ & \qquad\qquad\qquad\qquad 31 \end{array}$$

The quotient is $3x^3 + x^2 + 2x + 13$, and the remainder is 31.

If we are careful to keep like powers of the variable under one another and replace missing terms by 0, we can eliminate some of the unnecessary detail in the division.

First, we note that the repeated terms $3x^4$, $x^3$, $2x^2$, and $13x$ could be eliminated. Also it is not necessary to keep bringing down terms from the dividend. Thus we could write the division as

$$
\begin{array}{r|rrrrr}
 & 3x^3 & +\ x^2 & +\ 2x & +\ 13 & \\ \cline{2-6}
x-2 & 3x^4 & -\ 5x^3 & +\ 0x^2 & +\ 9x & +\ 5 \\
 & & -\ 6x^3 & & & \\ \cline{2-3}
 & & x^3 & & & \\
 & & & -\ 2x^2 & & \\ \cline{3-4}
 & & & 2x^2 & & \\
 & & & & -\ 4x & \\ \cline{4-5}
 & & & & 13x & \\
 & & & & & -\ 26 \\ \cline{5-6}
 & & & & & 31
\end{array}
$$

Next, since we are always dividing by $(x - c)$, we could eliminate the $x$ in $(x - 2)$. We could also eliminate the other powers of $x$. Doing this, we obtain

$$
\begin{array}{r|rrrrr}
 & 3 & 1 & 2 & 13 & \\ \cline{2-6}
-2 & 3 & -5 & 0 & 9 & 5 \\
 & & -6 & & & \\ \cline{2-3}
 & & 1 & & & \\
 & & & -2 & & \\ \cline{3-4}
 & & & 2 & & \\
 & & & & -4 & \\ \cline{4-5}
 & & & & 13 & \\
 & & & & & -26 \\ \cline{5-6}
 & & & & & 31
\end{array}
$$

We may write this in a more compact form as

$$
\begin{array}{r|rrrrr}
 & 3 & 1 & 2 & 13 & \\ \cline{2-6}
-2 & 3 & -5 & 0 & 9 & 5 \\
 & & -6 & -2 & -4 & -26 \\ \cline{3-6}
 & & 1 & 2 & 13 & 31
\end{array}
$$

Now note that the bottom line is the same as the quotient (top line) except for the first entry on the top line and the last entry (remainder) on the bottom line. By eliminating the top line and entering 3 in the bottom line, we further reduce the problem to the form

$$
\begin{array}{r|rrrrr}
\underline{-2} & 3 & -5 & 0 & 9 & 5 \\
 & & -6 & -2 & -4 & -26 \\ \cline{2-6}
 & 3 & 1 & 2 & 13 & 31
\end{array}
$$

If we now notice that each entry in line two is obtained by multiplying the preceding entry of line three by $-2$, we can construct the form just given without referring to the intervening steps.

One final improvement may be made on this form. It is possible to avoid the subtraction in this form by using $c$ instead of $-c$ when dividing by $(x - c)$. This, in effect, changes all the signs of the entries in the second row.

### Method of Synthetic Division

---

Now, if we wish to divide $(5x^2 - 2x + 3)$ by $(x - 3)$, we write

$$\begin{array}{r|rrr} 3 & 5 & -2 & 3 \\ & & 15 & 39 \\ \hline & 5 & 13 & 42 \end{array}$$

Notice that we add rather than subtract. Therefore the quotient is $5x + 13$ and the remainder is 42. This process is called *synthetic division.*

---

**Caution:** Be careful of the sign of the number you divide by. For example, if you are dividing by $(x - 3)$, use 3 in your synthetic division. Likewise, if you are dividing by $(x + 5)$, use $-5$ in your synthetic division.

EXAMPLE 1 Using synthetic division, find the quotient and remainder if $2x^4 + 5x^3 - 2x^2 + x - 8$ is divided by $(x + 3)$.

***Solution*** If we divide by $(x + 3)$, we will want to use $c = -3$. Thus we perform the synthetic division with $-3$.

$$\begin{array}{r|rrrrr} -3 & 2 & 5 & -2 & 1 & -8 \\ & & -6 & 3 & -3 & 6 \\ \hline & 2 & -1 & 1 & -2 & -2 \end{array}$$

So $\quad 2x^4 + 5x^3 - 2x^2 + x - 8 = (x + 3)(2x^3 - x^2 + x - 2) - 2$

The quotient is $2x^3 - x^2 + x - 2$, and the remainder is $-2$. ■

The statement "A polynomial $f(x)$ has a factor $(x - c)$ if and only if $f(c) = 0$" does not restrict $c$ to be an integer. It is sometimes necessary to perform synthetic division when $c$ is a fraction (rational number).

EXAMPLE 2 Using synthetic division, find the quotient and remainder if $(2x^4 - x^3 + 4x - 2)$ is divided by $\left(x - \frac{1}{2}\right)$.

***Solution*** Note that the $x^2$ term is missing, and hence we will need to express the dividend $2x^4 - x^3 + 4x - 2$ as $2x^4 - x^3 + 0x^2 + 4x - 2$ to allow for that missing place.

$$\begin{array}{c|ccccc} \frac{1}{2} & 2 & -1 & 0 & 4 & -2 \\ & & 1 & 0 & 0 & 2 \\ \hline & 2 & 0 & 0 & 4 & 0 \end{array}$$

Thus we see that

$$2x^4 - x^3 + 4x - 2 = \left(x - \frac{1}{2}\right)(2x^3 + 0x^2 + 0x + 4)$$

There is no remainder. The quotient is $2x^3 + 4$. ■

### Finding a Function Value by Synthetic Division

We can use the remainder theorem to find a function value fairly quickly for a higher-order polynomial.

EXAMPLE 3 If $f(x) = 3x^5 - 38x^3 + 5x^2 + 4$, find $f(4)$ by synthetic division.

***Solution*** By the remainder theorem, $f(4)$ is the remainder when $f(x)$ is divided by $(x - 4)$. We divide by synthetic division, obtaining

$$\begin{array}{c|cccccc} 4 & 3 & 0 & -38 & 5 & 0 & 4 \\ & & 12 & 48 & 40 & 180 & 720 \\ \hline & 3 & 12 & 10 & 45 & 180 & 724 \end{array}$$

Therefore $f(4) = 724$. ■

## EXERCISE 11-2

In Problems 1–5, find the quotient and remainder using synthetic division.

**1.** $(x^3 - 4x^2 + 2x + 4) \div (x - 3)$

**2.** $(x^3 - 8) \div (x - 2)$

**3.** $(2x^4 + x^3 - 1) \div (x + 2)$

**4.** $(x^6 + x^4 - x) \div (x - 1)$

**5.** $(x^6 + 4x^5 + 2x^3 + 7x^2 - 4x + 1) \div (x + 4)$

Use synthetic division to solve Problems 6–19.

**6.** Find $f(2)$ if $f(x) = 2x^3 - x^2 + x + 4$.

**7.** Find $f(4)$ if $f(x) = x^3 - 4x^2 + x - 13$.

8. Find $g(-1)$ if $g(x) = 3x^3 + 5x^2 + x - 1$.
9. Find $h(-3)$ if $h(x) = 2x^5 + 5x^4 + 10x^2 + 2$.
10. Find $g(5)$ if $g(x) = x^4 - 2x^3 - 17x^2 + 8x + 10$.
11. Is 2 a zero of $f(x) = 3x^4 - 5x^3 + x + 10$?
12. Is 7 a zero of $g(x) = x^3 - 6x^2 - 11x + 28$?
13. Is $-1$ a zero of $h(x) = 4x^5 - 3x^3 + 1$?
14. Is $\frac{1}{3}$ a zero of $f(x) = 3x^5 + 5x^4 - 2x^3 + 3x^2 - 4x + 1$?
15. Is $\frac{1}{2}$ a zero of $g(x) = 2x^5 - 3x^3 + 2x^2 - 5x - 4$?
16. Find $h(-3)$ if $h(x) = 2x^4 - 3x^3 + 5x^2 - 2x + 4$.
17. Find $f(-2)$ if $f(x) = 3x^4 + 4x^3 - 6x^2 - 5x - 2$.
★ 18. Find $g\left(\frac{1}{2}\right)$ if $g(x) = 5x^5 + 3x^3 - 2x^2 - x - 4$.
★ 19. Find $h\left(\frac{1}{4}\right)$ if $h(x) = -4x^5 - 3x^4 + 2x^2 + 5x - 1$.

Although we have been applying the properties of polynomials to real numbers, we should not forget that these principles also hold for the field of complex numbers. To illustrate this, use the process of synthetic division to solve Problems 20 and 21.

★ 20. Is $(1 - i)$ a zero of $f(x) = 3x^3 - 5x^2 + 4x + 2$?
★ 21. Is $(1 + i)$ a zero of $f(x) = 3x^3 - 5x^2 + 4x + 2$?

**Calculator Problems**

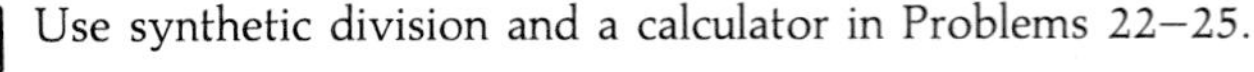

Use synthetic division and a calculator in Problems 22–25.

22. Find $f(1.4)$ if $f(x) = 4x^5 - 3x^3 + 5x - 6$.
23. Find $g(2.2)$ if $g(x) = 3x^6 - 5x^4 - 2x^3 - x + 4$.
24. Is $-2.8$ a zero of $h(x) = -2x^6 + 3x^5 - 2x^3 + 6x^2 - 5$?
25. Is $-1.9$ a zero of $f(x) = 3x^5 + 2x^4 - 7x^3 - 6x^2 + 1$?

## 11-3 ZEROS OF POLYNOMIALS WITH COMPLEX COEFFICIENTS

The factor theorem in Section 11-1 shows that there is a definite connection between the zeros of a polynomial and the factors of a polynomial. In fact, if we can find a number $c$ such that $f(c) = 0$, then we have a factor $(x - c)$. However, zeros of a polynomial are usually very difficult to find. In special cases, such as linear or quadratic polynomials, we have developed methods of finding zeros. But in other cases, such as $f(x) = x^5 - 3x^4 + 5x^3 - 2x + 10$, there are no obvious zeros and there are no special formulas, such as the quadratic formula, that will yield zeros. Even though it is difficult to locate zeros of polynomials, we can make some headway concerning the theory of such zeros. The following theorem is basic to the development of the theory.

***Theorem 11-4 (Fundamental Theorem of Algebra)***

If $f(x)$ is a polynomial of degree greater than zero, with complex coefficients, then $f(x)$ has at least one complex zero.

The proof of this theorem is beyond the scope of this text, but we will accept it without proof and proceed to use it in the development of our theory.

The student should take a minute to realize that we are talking about polynomial equations with complex coefficients. Since real numbers are a subset of the complex numbers, the coefficients may be real or imaginary. The three polynomials $f(x) = 3x^3 + 5x^2 - 6x + 1$, $g(x) = (5 + 2i)x^2 + (4 - i)x + 3$, and $h(x) = 2x^2 + 3ix - 4$ are all polynomials with complex coefficients. Do you see why this is so?

***Theorem 11-5***

If $f(x)$ is a polynomial of degree $n > 0$ with complex coefficients, then $f(x)$ has exactly $n$ zeros. (Note that this does not imply $n$ different zeros. If a zero occurs more than once, it is counted each time.)

The main purpose we will have for Theorem 11-5 is to help us determine the number of zeros any polynomial will have. Finding the zeros can be easier when we know exactly how many we must find. We know that $f(x) = 2x^3 - 7x^2 + x - 3$ has three zeros, for instance.

EXAMPLE 1 A polynomial $g(x)$ is of degree 3. It has zeros 2, $-3$, and 1.

**a.** Write an equation for $g(x)$.
**b.** Is this answer unique?

***Solution***

**a.** We know that if some polynomial $g(x)$ of degree 3 has zeros 2, $-3$, 1, then it has factors $(x - 2)$, $(x + 3)$, and $(x - 1)$ and that $g(x) = (x - 2)(x + 3)(x - 1)$ or $g(x) = x^3 - 7x + 6$.
**b.** Please note that this answer is not unique. Any polynomial of the general form $a_n(x - 2)(x + 3)(x - 1)$, where $a_n$ is a nonzero number, would also have roots 2, $-3$, and 1. Usually, we just let $a_n = 1$ to obtain the simplest polynomial. ■

**EXAMPLE 2** The polynomial $p(x) = x^3 - 4x^2 + x + 6$ has a zero of $-1$. Find the other two zeros.

***Solution*** If $p(x)$ has a zero of $-1$, then $x + 1$ is a factor. We use synthetic division to find the other factor:

$$\begin{array}{r|rrrr} -1 & 1 & -4 & 1 & 6 \\ & & -1 & +5 & -6 \\ \hline & 1 & -5 & 6 & 0 \end{array}$$

Thus we see that

$$x^3 - 4x^2 + x + 6 = (x + 1)(x^2 - 5x + 6)$$

By continuing to factor we have

$$(x + 1)(x - 2)(x - 3)$$

Thus in addition to $-1$, the other two zeros of $p(x)$ are 2 and 3. ■

**EXAMPLE 3** The polynomial $f(x) = x^3 - x^2 - 11x - 4$ has a zero of 4. Find the other two zeros.

***Solution*** We use synthetic division:

$$\begin{array}{r|rrrr} 4 & 1 & -1 & -11 & -4 \\ & & 4 & 12 & 4 \\ \hline & 1 & 3 & 1 & 0 \end{array}$$

Thus we see that

$$x^3 - x^2 - 11x - 4 = (x - 4)(x^2 + 3x + 1)$$

The quadratic expression $x^2 + 3x + 1$ cannot be factored, so we will use the quadratic formula

$$x^2 + 3x + 1 = 0$$

$$x = \frac{-3 \pm \sqrt{9 - 4(1)(1)}}{2(1)} = \frac{-3 \pm \sqrt{5}}{2}$$

Thus in addition to 4, the other two zeros of $f(x)$ are $-\frac{3}{2} + \frac{\sqrt{5}}{2}$ and $-\frac{3}{2} - \frac{\sqrt{5}}{2}$. ■

In some cases it may be necessary to use synthetic division with imaginary or complex numbers to obtain all the zeros.

**EXAMPLE 4** The polynomial $h(x) = 2x^3 + (-7 + 4i)x^2 + (-4 - 14i)x - 8i$ has a zero of $-2i$. Find the other two zeros.

***Solution***

$$\begin{array}{r|rrrr} -2i & 2 & -7+4i & -4-14i & -8i \\ & & -4i & +14i & +8i \\ \hline & 2 & -7 & -4 & 0 \end{array}$$

Thus we have $h(x) = (x + 2i)(2x^2 - 7x - 4)$. Since we can factor the quadratic, $h(x)$ can be written as

$$h(x) = (x + 2i)(2x + 1)(x - 4).$$

Thus in addition to $-2i$, the other two roots of $h(x)$ are $-\frac{1}{2}$ and 4. ■

## EXERCISE 11-3

How many zeros does each of the polynomials have in Problems 1–6?

**1.** $f(x) = 4x^2 - 2x + 1$

**2.** $f(x) = x^2$

**3.** $f(x) = x^3 - 1$

**4.** $f(x) = 3x^5 - (2 + i)x^3 + ix - 2$

**5.** $f(x) = 3x + 3 + i$

**6.** $f(x) = (x - 1)(x - 1)(x - 1)(x - 1)$

**7.** Find a third-degree polynomial that has zeros $\frac{1}{2}, \frac{1}{3}, -2$.

**8.** Find a third-degree polynomial that has zeros $-1, 1, 2$.

**9.** Find a third-degree polynomial that has zeros $1, \sqrt{2}, 3\sqrt{2}$.

**10.** Find a third-degree polynomial that has zeros $4, \sqrt{5}, -\sqrt{5}$.

**11.** Find a third-degree polynomial that has zeros $-3, \sqrt{7}, -2\sqrt{7}$.

**12.** Find a third-degree polynomial that has zeros $-2, 5, i$.

**13.** Find a third-degree polynomial that has zeros $i, -i, 2i$.

**14.** Find a third-degree polynomial that has zeros $1 + i, 1 - i, 3$.

**15.** Find a fourth-degree polynomial that has zeros $2, -2, 5, -5$.

**16.** Find a fourth-degree polynomial that has zeros $1 + i, 1 - i, i, -i$.

**17.** Find the zeros of $f(x) = x^2 - x - 30$.

**18.** Find the zeros of $f(x) = 2x^2 + x - 6$.

**19.** Find the zeros of $g(x) = x^4 - 1$.

**20.** Find the zeros of $h(x) = x^4 - 16$.

**21.** Find the zeros of $g(x) = 16x^4 - 81$.

Use synthetic division in Problems 22–27.

**22.** If $f(x) = x^3 - 3x^2 - 6x + 8$ has 1 as one of its zeros, find the other zeros.

**23.** If $f(x) = x^3 + 4x^2 - 36x - 144$ has $-4$ as one of its zeros, find the other zeros.

**24.** If $f(x) = x^3 - ix^2 - x + i$ has $i$ as one of its zeros, find the other zeros.

**25.** If $h(x) = x^3 + ix^2 - 3x - 3i$ has $-i$ as one of its zeros, find the other zeros.

**26.** If $f(x) = 2x^3 - x^2 - 4x + 2$ has $\frac{1}{2}$ as one of its zeros, find the other zeros.

**27.** If $g(x) = x^3 + 3x^2 + x + 3$ has $-3$ as one of its zeros, find the other zeros.

**28.** Use the polynomial $h(x) = x^3 - 8$ to find the three cube roots of 8. [*Hint:* Inspection should give you one zero. Then use synthetic division to factor $h(x)$.]

**29.** Use the polynomial $f(x) = x^3 - 1$ to find the three cube roots of 1. [*Hint:* Inspection should give you one zero. Then use synthetic division to factor $f(x)$.]

The following facts are helpful in solving Problems 30–33: If the same zero appears in a polynomial in $n$ factors, then we say that the zero is of multiplicity $n$. For example, if $f(x) = (x - 5)(x - 5)(x + 1)$, we say that the zero 5 is of multiplicity 2.

★ **30.** Find a polynomial of degree 3 that has 1 as a zero of multiplicity 3.

★ **31.** Find a polynomial of degree 4 that has 3 as a zero of multiplicity 4.

★ **32.** Find a polynomial of degree 4 that has 2 as a zero of multiplicity 2 and has a zero of $-1$ of multiplicity 2.

★ **33.** Find a polynomial of degree 3 that has $\frac{1}{2}$ as a zero of multiplicity 2 and $-3$ as a zero.

## 11-4 ZEROS OF POLYNOMIALS WITH REAL COEFFICIENTS

Since the set of real numbers is a subset of the set of complex numbers, any theorem concerning complex numbers will also hold true in the real numbers. In this section we wish to concentrate on zeros of polynomials with real coefficients.

***Theorem 11-6*** If a polynomial $f(x)$ of degree $n > 0$ with real coefficients has as a zero $(a + bi)$, then $(a - bi)$ is also a zero.

**EXAMPLE 1** Find all the zeros of $x^3 - 3x^2 + 4x - 12$ if one zero is $2i$.

***Solution*** Since we have real coefficients if $a + bi$ is a zero, $a - bi$ is also a zero. Thus if $0 + 2i$ is a zero, we also have $0 - 2i$ or $-2i$. Thus two factors of $f(x)$ are $(x + 2i)(x - 2i)$. If we write this as

$$(x + 2i)(x - 2i) = x^2 - 4i^2 = x^2 + 4$$

we can perform division as follows:

$$\begin{array}{r|l} & x \quad - 3 \\ \hline x^2 + 0x + 4 & x^3 - 3x^2 + 4x - 12 \\ & x^3 + 0x^2 + 4x \\ \hline & \quad - 3x^2 + 0x - 12 \\ & \quad - 3x^2 + 0x - 12 \\ \hline & \qquad\qquad\qquad 0 \end{array}$$

Thus $f(x) = (x + 2i)(x - 2i)(x - 3)$. The three zeros of $f(x)$ are $2i$, $-2i$, and 3. ■

*Theorem 11-7*

Every polynomial $f(x)$ with real coefficients and of odd degree has at least one real zero.

***Proof*** We are given that $f(x)$ is of degree $n$ and $n$ is an odd number. We know from the previous section that $f(x)$ has $n$ zeros. Since complex zeros that are not real must occur in pairs, $(a + bi)$ and $(a - bi)$, it follows that $f(x)$ must have at least one real zero. ■

Note that if the polynomial $f(x)$ has degree $n$ where $n$ is even, then all zeros of that polynomial could be nonreal.

A polynomial $f(x)$ with real coefficients is said to have a **sign variation** if, when arranged in descending powers of the variable $x$, the sign of a term is different from the sign of the preceding term. (We ignore missing terms in counting the sign variations.) For instance,

$$f(x) = x^5 + 3x^4 - 2x^3 - x + 5$$

has two sign variations—one between the second and third terms and one between the fourth and fifth terms.

We now state, without proof, a useful theorem called **Descartes' Rule of Signs**.

*Theorem 11-8 (Descartes' Rule of Signs)*

If $f(x)$ is a polynomial with real coefficients, the number of positive real zeros of $f(x)$ is either equal to the number of sign variations of $f(x)$ or less than the number of sign variations by a positive even integer. Also, the number of negative real zeros of $f(x)$ is either equal to the number of sign variations of $f(-x)$ or less than the number of sign variations of $f(-x)$ by a positive even integer.

**EXAMPLE 2** Determine the possible number of positive and negative real zeros of $f(x) = 3x^5 - x^4 + 3x^3 - 2x^2 - x - 1$.

***Solution*** $f(x)$ has three sign variations. Therefore the number of positive real zeros is either three or one.

$$f(-x) = -3x^5 - x^4 - 3x^3 - 2x^2 + x - 1$$

has two sign variations, and therefore the number of negative real zeros is 2 or 0. ■

**EXAMPLE 3** Show by Descartes' Rule of Signs that $f(x) = x^3 + 2x + 3$ has only one real zero.

***Solution*** Descartes' Rule of Signs can be used to check for positive and negative real zeros. However, since 0 is neither positive nor negative, we must check to see whether 0 is a zero of $f(x)$. Checking this, we find

$$f(0) = 3$$

which indicates that 0 is not a zero. Now continuing with Descartes' Rule of Signs, we see that $f(x)$ has no sign variation and therefore no positive real zeros.

$$f(-x) = -x^3 - 2x + 3$$

has one sign variation, and therefore $f(x)$ has one negative real zero. Since we know that $f(x)$ is of degree 3 and therefore has three zeros, the other two cannot be real.

Hence $f(x)$ has one negative real zero and two nonreal zeros. ■

## EXERCISE 11-4

Find a polynomial with real coefficients that has the zeros given in Problems 1–4. In each case the degree of the desired polynomial is specified.

1. Degree 3, zeros $3 - i$ and $-2$.
2. Degree 3, zeros $1 - 2i$, and 4.
3. Degree 4, zeros $2 + 2i$, $-1$, and 2.
4. Degree 4, zeros $1 + 3i$, $-2$, and 2.

Find all the zeros of the polynomials in Problems 5–8.

5. $p(x) = 2x^3 + 3x^2 + 2x + 3$ if one zero is $-i$.
6. $h(x) = x^3 - 5x^2 + 9x - 45$ if one zero is $+3i$.
7. $f(x) = x^3 + x - 10$ if one zero is $-1 + 2i$.
8. $j(x) = 2x^3 - 3x^2 + 18x + 10$ if one zero is $1 + 3i$.
9. Given $f(x) = 3x^4 + 2x^3 + x^2 - 2x - 4$:
   a. What are the possible numbers of positive real zeros of $f(x)$?
   b. What are the possible numbers of negative real zeros of $f(x)$?
   c. Is 0 a zero of $f(x)$? Explain.

**10.** Given $f(x) = 2x^4 - 5x^3 + x^2$:
  **a.** How many positive real zeros could $f(x)$ have?
  **b.** How many negative real zeros could $f(x)$ have?
  **c.** Is 0 a zero of $f(x)$? Explain.

**11.** What condition is necessary and sufficient for any polynomial $f(x)$ to have 0 as a zero?

**12.** Show that $f(x) = x^5 + 2x^3 + x - 3$ has only one real zero.

**13.** Show that $g(x) = x^4 + x^2 + 2x - 3$ has one positive real zero and one negative real zero.

**14.** Show that $h(x) = x^4 - 5x^3 + 2x^2 - 3x + 1$ has no negative zeros.

**15.** Show that $f(x) = x^6 + 3x^4 + 5x^2 + 1$ has no real zeros.

**16.** How many real zeros does $f(x) = x^6 - 5$ have?

**17.** How many zeros of $f(x) = x^4 + x^2 + x$ are not real?

★ **18.** Given $f(x) = x^n + 1$
  **a.** If $n$ is odd, how many real zeros does $f(x)$ have?
  **b.** If $n$ is even, how many real zeros does $f(x)$ have?

---

## 11-5 ZEROS OF POLYNOMIALS WITH RATIONAL OR INTEGRAL COEFFICIENTS

There is a theorem similar to Theorem 11-6 that deals with polynomials with rational coefficients. We will state the theorems without proof.

*Theorem 11-9*

> If a polynomial $p(x)$ of degree $n > 0$ with rational coefficients has a zero of the form $a + b\sqrt{c}$ where $a$, $b$ are rational numbers and $c$ is not a perfect square, then $a - b\sqrt{c}$ is also a zero.

**EXAMPLE 1** Find a polynomial of degree three with rational coefficients that has zeros of $3 - 2\sqrt{5}$ and $-1$.

*Solution* By Theorem 11-9 if $3 - 2\sqrt{5}$ is a zero, so is $3 + 2\sqrt{5}$. A polynomial with zeros of $-1$, $3 + 2\sqrt{5}$, and $3 - 2\sqrt{5}$ can be written in the form $p(x) = (x + 1)(x - 3 - 2\sqrt{5})(x - 3 + 2\sqrt{5})$. The last two expressions can be multiplied quickly:

$$[(x - 3) - 2\sqrt{5}][(x - 3) + 2\sqrt{5}] = [(x - 3)^2 - 4 \cdot 5]$$
$$= [x^2 - 6x + 9 - 20] = (x^2 - 6x - 11)$$

Thus

$$p(x) = (x + 1)(x^2 - 6x - 11)$$
$$= x^3 - 5x^2 - 17x - 11 \quad \blacksquare$$

If $f(x)$ is a polynomial having some nonintegral rational coefficients (that is, fractions), then the polynomial can easily be changed to one having integral coefficients by multiplying by a common denominator. This would yield a different polynomial but would not affect the zeros as the original polynomial. For instance,

$$f(x) = \frac{2}{3}x^3 + x^2 - \frac{1}{2}x + \frac{1}{3}$$

would have the same zeros as

$$g(x) = 4x^3 + 6x^2 - 3x + 2$$

obtained by multiplying all terms of $f(x)$ by 6.

In the remainder of this chapter we will develop some theorems that deal particularly with polynomials with integral coefficients.

*Theorem 11-10*

Let the polynomial $f(x) = a_0x^n + a_1x^{n-1} + a_2x^{n-2} + \cdots + a_n$ have only integer coefficients with $a_0 \neq 0$. If $f(x)$ has a rational zero $\frac{p}{q}$ $\left(\text{where } \frac{p}{q} \text{ is in reduced form}\right)$, then $p$ is an exact divisor of $a_n$ and $q$ is an exact divisor of $a_0$.

**EXAMPLE 2** List all possible rational zeros of $f(x) = 5x^3 + 3x^2 + 2x - 6$.

***Solution*** The divisors of 5 are $\pm 1$ and $\pm 5$, and the divisors of 6 are $\pm 1$, $\pm 2$, $\pm 3$, and $\pm 6$. By Theorem 11-10, if $\frac{p}{q}$ is a zero, then $p$ must be an exact divisor of 6, and $q$ must be an exact divisor of 5. Therefore the possible rational zeros are

$$\left\{\pm\frac{1}{5}, \pm\frac{2}{5}, \pm\frac{3}{5}, \pm 1, \pm\frac{6}{5}, \pm 2, \pm 3, \pm 6\right\} \quad \blacksquare$$

**Caution:** Note that the theorem does not guarantee that such zeros exist, but simply limits the possible rational zeros should they exist.

Two obvious corollaries of the theorem are as follows.

*Corollary 11-11*

An integral zero of the polynomial $f(x)$ must be an exact divisor of the constant term $a_n$.

*Corollary 11-12*

If the polynomial $f(x)$ has the leading coefficient $a_0 = 1$, then any rational zero is an integer.

**EXAMPLE 3** Find all zeros of $f(x) = 2x^3 - 5x^2 + x + 2$.

***Solution*** From the work in this chapter we know that

1. $f(x)$ has at least one real zero. (Why?)
2. The number of positive real zeros is 2 or 0. (Why?)
3. The number of negative real zeros is 1. (Why?)
4. The possible rational zeros are in the set

$$\left\{\pm\frac{1}{2}, \pm 1, \pm 2\right\}$$

Since we have this information, it is logical to check to see whether $-1$, $-2$, or $-\frac{1}{2}$ is a zero. (Why?) We may do this by substituting each of these values for $x$ in $f(x)$ or by using synthetic division. The advantage of using synthetic division is that when we find a zero, we will also have the coefficients of the other factor of the polynomial.

First we try $-\frac{1}{2}$:

$$\begin{array}{r|rrrr} -\frac{1}{2} & 2 & -5 & 1 & 2 \\ & & -1 & 3 & -2 \\ \hline & 2 & -6 & 4 & 0 \end{array}$$

We see here that $-\frac{1}{2}$ is a zero, and we may write

$$f(x) = \left(x + \frac{1}{2}\right)(2x^2 - 6x + 4)$$

Since $2x^2 - 6x + 4 = 2(x - 1)(x - 2)$, we now have

$$f(x) = 2\left(x + \frac{1}{2}\right)(x - 1)(x - 2)$$

Thus the zeros of $f(x)$ are $\left\{-\frac{1}{2}, 1, 2\right\}$. ■

**Caution:** Note that if no zero of $f(x)$ had been rational, this method would not have given us the answers. Of course, if enough of the zeros are rational to allow us to get to a quadratic factor, we may solve the quadratic to give us the remaining two zeros whether they are rational or not.

EXAMPLE 4 Find all the zeros of $p(x) = x^4 - 2x^3 - 7x^2 + 8x + 12$.

***Solution***

**a.** Since $p(x)$ is of degree 4, we *cannot* use Theorem 11-7 to conclude anything about the existence of real zeros.

**b.** There are two sign variations. Thus there will be 2 or 0 positive real zeros.

**c.** $p(-x) = x^4 + 2x^3 - 7x^2 - 8x + 12$. Here we have two sign variations. There will be 2 or 0 negative real zeros.

**d.** The possible rational zeros are in the set $\{\pm 1, \pm 2, \pm 3, \pm 4, \pm 6, \pm 12\}$

At this point, some trial and error steps will be needed to see whether any of these values is in fact a root. Perhaps the first zero we would discover is $-1$:

$$\begin{array}{r|rrrrr} -1 & 1 & -2 & -7 & 8 & 12 \\ & & -1 & 3 & 4 & -12 \\ \hline & 1 & -3 & -4 & 12 & 0 \end{array}$$

Thus $p(x) = (x + 1)(x^3 - 3x^2 - 4x + 12)$. Now we try to factor the polynomial $x^3 - 3x^2 - 4x + 12$ further. A value that we might try would be $+2$:

$$\begin{array}{r|rrrr} 2 & 1 & -3 & -4 & 12 \\ & & 2 & -2 & -12 \\ \hline & 1 & -1 & -6 & 0 \end{array}$$

*Observe that we are dividing $x^3 - 3x^2 - 4x + 12$ by $x - 2$. We do not use the original polynomial $x^4 - 2x^3 - 7x^2 + 8x + 12$ again.*

Thus we have

$$p(x) = (x + 1)(x - 2)(x^2 - x - 6)$$

The last polynomial can be quickly factored, so we have

$$p(x) = (x + 1)(x - 2)(x - 3)(x + 2)$$

Thus the zeros of $p(x)$ are $\{-1, 2, -2, 3\}$. ■

## EXERCISE 11-5

Find a polynomial with rational coefficients that has the zeros given in Problems 1–4.

**1.** Degree 3, zeros $2 + \sqrt{3}$, 2.

**2.** Degree 3, zeros $3 - \sqrt{2}$, $-1$.

**3.** Degree 4, zeros $1 + \sqrt{5}$, $+3$, $-3$.

**4.** Degree 4, zeros $2 - \sqrt{3}$, $-1$, $+2$.

Find all the zeros of the polynomials in Problems 5–8.

**5.** $f(x) = 2x^3 + x^2 - 10x - 5$ if one zero is $\sqrt{5}$.

**6.** $p(x) = 3x^3 - 2x^2 - 9x + 6$ if one zero is $-\sqrt{3}$.

**7.** $h(x) = 3x^3 - 13x^2 + x + 1$ if one zero is $2 - \sqrt{5}$.

**8.** $g(x) = x^3 - 11x^2 + 33x - 23$ if one zero is $5 + \sqrt{2}$.

Using Theorem 11-10, list all the *possible* rational zeros of the polynomials in Problems 9–12. (Do not attempt to actually find the zeros; merely list the possible zeros.)

**9.** $f(x) = 5x^3 - 3x^2 + 7x - 25$

**10.** $f(x) = 4x^3 - 12x^2 + 7x - 8$

**11.** $p(x) = 3x^3 - 7x^2 + 5x - 24$

**12.** $p(x) = 4x^3 + 6x^2 - 7x + 81$

Using the theory developed in this chapter, find all zeros of the polynomials in Problems 13–25.

**13.** $f(x) = x^3 - 2x^2 - x + 2$

**14.** $f(x) = x^3 - 4x^2 + x + 6$

**15.** $f(x) = 2x^3 - 3x^2 - 11x + 6$

**16.** $f(x) = 2x^3 + 3x^2 - 14x - 21$

**17.** $f(x) = 18x^3 - 9x^2 - 5x + 2$

**18.** $f(x) = 16x^3 + 4x^2 - 4x - 1$

**19.** $f(x) = 2x^3 + 7x^2 + 2x - 6$

**20.** $f(x) = x^3 + 5x^2 + x + 5$

**21.** $f(x) = 2x^4 + 5x^3 - 5x^2 - 5x + 3$

**22.** $f(x) = x^4 - 2x^3 - 13x^2 + 38x - 24$

**★23.** $f(x) = 3x^4 - 11x^3 + 9x^2 + 13x - 10$

**★24.** $f(x) = x^4 + x^3 - 5x^2 - 15x - 18$

**★25.** $f(x) = x^4 + 4x^3 - x^2 - 20x - 20$

## 11-6 APPROXIMATIONS AND BOUNDS FOR THE ZEROS OF POLYNOMIALS

It was previously stated that zeros of polynomials are difficult to find except in special cases. No general method exists for finding the zeros of a fifth- or higher-degree polynomial. The theorems on zeros from the previous sections help us to find zeros in many situations. In this section we wish to investigate some other theorems that will aid in our search for zeros and also discuss some methods of estimating irrational zeros.

### Upper Bounds of Zeros

*Theorem 11-13*

If $f(x) = a_0x^n + a_1x^{n-1} + \cdots + a_n$ and if $a_k$ is the coefficient of greatest absolute value, then no positive zero can exceed $\left|\frac{a_k}{a_0}\right| + 1$.

This theorem is stated without proof.

EXAMPLE 1 Determine the upper limit for the positive zeros of $f(x)$:

$$f(x) = 8x^7 + 3x^6 - 15x^3 + x - 12$$

***Solution*** The coefficient of greatest absolute value is $-15$, so we have $a_k = -15$. Using Theorem 11-13, we know that we cannot have a positive zero greater than

$$\left|\frac{-15}{8}\right| + 1 = 2\frac{7}{8}$$

Thus if we were searching for zeros for $f(x)$, we would not bother to check the possibilities 3, 4, 6, or 12. ■

### Lower Bounds of Zeros

To find **bounds** for negative zeros (i.e., points between which negative zeros occur), it is necessary only to find $f(-x)$ and use the bounds for positive zeros with the opposite sign. For the theorem just given, since coefficients of $f(-x)$ would have the same absolute value, the lower bounds of negative zeros are represented by $-\left(\left|\frac{a_k}{a_0}\right| + 1\right)$.

EXAMPLE 2 Determine the lower limit for the negative zeros of $p(x)$:

$$p(x) = 4x^3 + 4x^2 - 5x - 6$$

***Solution*** If we did not have Theorem 11-13, we would have to consider $-6, -3, -2, -\frac{3}{2}, -1, -\frac{3}{4}, -\frac{1}{2}, -\frac{1}{4}$ as possible negative zeros. Using Theorem 11-13 to find a lower limit, we find

$$p(-x) = -4x^3 + 4x^2 + 5x - 6$$

Then the coefficient of greatest absolute value is $-6$; hence we will have $a_k = -6$. Thus the lower bound is

$$-\left(\left|\frac{a_k}{a_0}\right| + 1\right) = -\left(\left|\frac{-6}{-4}\right| + 1\right) = -\left(\frac{3}{2} + 1\right) = -\frac{5}{2}$$

The lower bound is $-\frac{5}{2}$ or $-2\frac{1}{2}$. (Thus we know that the possible zeros $-6$ and $-3$ could be eliminated. No negative zero of $p(x)$ could be less than $-2\frac{1}{2}$.) ■

There are three additional theorems that will help us to restrict the region in which real zeros occur for any given polynomial. We will state each theorem without proof.

***Theorem 11-14***

If, in the synthetic division of a polynomial $f(x)$ with real coefficients by $(x - a)$ where $a > 0$, all the numbers in the third row are positive, then $f(x)$ has no real zero larger than $a$.

***Theorem 11-15***

If, in the synthetic division of a polynomial $f(x)$ with real coefficients by $(x - a)$ where $a < 0$, the numbers in the third row are alternately positive and negative, then $f(x)$ has no real zero less than $a$.

***Theorem 11-16 (Intermediate Value Theorem)***

If for the polynomial $f(x)$ with real coefficients, $f(a)$ and $f(b)$ are opposite in sign, then $f(x)$ has a real zero between $a$ and $b$.

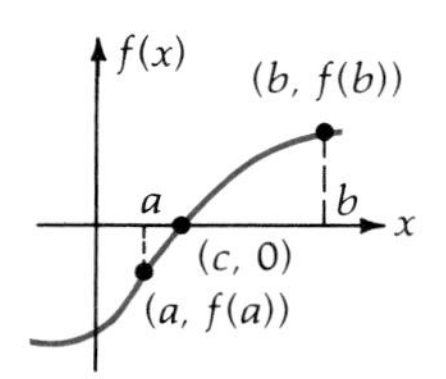


A graphical representation, assuming that a polynomial forms a continuous curve, illustrates Theorem 11-16.

A restatement of the theorem in terms of the graph would be: "If $f(a)$ is below the $x$-axis and $f(b)$ is above the $x$-axis, then the curve must cross the $x$-axis between $a$ and $b$." Of course, a zero of $f(x)$ is where $f(x) = 0$ or where the curve crosses the $x$-axis. It is noted that if $c$ is a zero of $f(x)$, then the point on the polynomial curve would be $(c, 0)$.

### Locating the Interval in Which Irrational Zeros Occur

EXAMPLE 3 Discuss the zeros of $f(x) = x^3 + 4x^2 + x - 3$. Find an interval in which each irrational zero occurs.

***Solution*** From the theorems in the chapter we know that

1. $f(x)$ has three zeros. (Why?)
2. the possible rational zeros are $\{\pm 1, \pm 3\}$. (Why?)
3. $f(x)$ has one positive real zero. (Why?)
4. $f(x)$ has two or no negative real zeros. (Why?)
5. an upper bound of the zeros is 5 and a lower bound is $-5$. (Why?)

Since we know there is one positive real zero, we will first try to locate it. We first note that

$$f(0) = -3$$

We now find $f(1)$ by using synthetic division.

$$\begin{array}{r|rrrr} 1 & 1 & 4 & 1 & -3 \\ & & 1 & 5 & 6 \\ \hline & 1 & 5 & 6 & 3 \end{array}$$

Thus $f(1) = 3$.

Now since $f(0) < 0$ and $f(1) > 0$, there must be a real zero between 0 and 1. Also, since any rational zero must be an integer (see 2 above), we know that this zero must be irrational. (We should note here that all the numbers in the third row are positive, and therefore there are no real zeros greater than 1. However, we know that there is only one positive real zero, and we have already located it.)

Next we check for negative real zeros. We know that if there are any, there will be two of them. First we check $-1$ by synthetic division:

$$\begin{array}{r|rrrr} -1 & 1 & 4 & 1 & -3 \\ & & -1 & -3 & 2 \\ \hline & 1 & 3 & -2 & -1 \end{array}$$

Thus $f(-1) = -1$. Similarly, we check $-2$:

$$\begin{array}{r|rrrr} -2 & 1 & 4 & 1 & -3 \\ & & -2 & -4 & 6 \\ \hline & 1 & 2 & -3 & 3 \end{array}$$

Thus $f(-2) = 3$. We note that $f(-1) < 0$ and $f(-2) > 0$. Hence there is a real zero between $-2$ and $-1$, and it must also be irrational. We next check $-3$:

$$\begin{array}{r|rrrr} -3 & 1 & 4 & 1 & -3 \\ & & -3 & -3 & 6 \\ \hline & 1 & 1 & -2 & 3 \end{array}$$

Thus $f(-3) = 3$. Next, checking $-4$, we obtain

$$\begin{array}{r|rrrr} -4 & 1 & 4 & 1 & -3 \\ & & -4 & 0 & -4 \\ \hline & 1 & 0 & 1 & -7 \end{array}$$

Hence $f(-4) = -7$. We see that $f(-3) > 0$ and $f(-4) < 0$; thus there is a real zero between $-4$ and $-3$ and it is also irrational.

We thus conclude that there are three irrational zeros. One of them is in the interval $(-4, -3)$, another is in the interval $(-2, -1)$, and the third is in the interval $(0, 1)$. ■

### Approximating Irrational Zeros

Attempting to estimate irrational zeros seems to be the next logical step. Let us use the example just discussed and approximate the zero in the interval $(-2, -1)$.

We found in Example 3 that for $f(x) = x^3 + 4x^2 + x - 3$,

$$f(-1) = -1 \qquad \text{and} \qquad f(-2) = 3$$

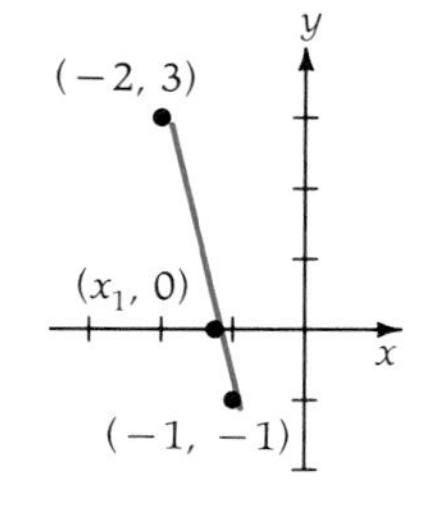


The curve representing $f(x)$ must intersect the $x$-axis somewhere between $-1$ and $-2$. If we approximate the curve between the points $(-2, 3)$ and $(-1, -1)$ by a line segment, we obtain the figure at left.

A line segment containing the points $(-2, 3)$ and $(-1, -1)$ will intersect the $x$-axis close (in some sense) to the point where $f(x)$ intersects the $x$-axis, $(x_1, 0)$. Using the two-point form of a straight line,

$$\frac{y - y_1}{x - x_1} = \frac{y_2 - y_1}{x_2 - x_1}$$

with $\qquad (x_1, y_1) = (-1, -1)$

and $\qquad (x_2, y_2) = (-2, 3)$

gives $$\frac{y + 1}{x + 1} = \frac{3 + 1}{-2 + 1} = \frac{4}{-1}$$

The first approximation for $x$ we will call $x_1$. Now if we substitute $(x_1, 0)$ for $(x, y)$, we obtain

$$\frac{1}{x_1 + 1} = \frac{4}{-1}$$

or

$$x_1 = -\frac{5}{4} = -1.25$$

This is an approximation of the zero between $-2$ and $-1$. If we wished a better approximation, we would continue by finding $f(-1.25)$, using synthetic division:

| $-1.25$ | 1 | 4 | 1 | $-3$ |
|---|---|---|---|---|
| | | $-1.25$ | $-3.4375$ | 3.0469 |
| | 1 | 2.75 | $-2.4375$ | 0.0469 |

*This value is rounded to the nearest ten-thousandth.*

Thus $f(-1.25) \doteq 0.0469$. Since $f(-1) < 0$ and $f(-1.25) > 0$, the irrational zero must be between $-1$ and $-1.25$.

If we use the two points $(-1, -1)$ and $(-1.25, 0.0469)$ and join them with a line segment, that line segment will intersect the $x$-axis close (in some sense) to the point where $f(x)$ intersects the $x$-axis. This time we have $(x_1, y_1) = (-1, -1)$ and $(x_2, y_2) = (-1.25, 0.0469)$.

$$\frac{y + 1}{x + 1} = \frac{0.0469 + 1}{-1.25 + 1}$$

or

$$\frac{y + 1}{x + 1} = \frac{1.0469}{-0.25}$$

We are now finding the second approximation for $x$ denoted by $x_2$. If we substitute $(x_2, 0)$ for $(x, y)$ and simplify, we obtain

$$x_2 \doteq -1.2388$$ *Rounded to the nearest ten-thousandth.*

which is our second approximation of the zero. If we round this to the nearest hundredth, we have $x_2 = -1.24$. This answer is correct to the nearest hundredth.

This method can be continued to obtain the approximation of the zero to any degree of accuracy desired. A calculator would aid greatly in the computation.

This method of approximation is rapid and efficient only if the graph of the equation of the polynomial is close to that of a straight line. In some cases it may take several repetitions before you can obtain a root to a desired level of accuracy. More advanced methods can be developed by using calculus that are more efficient in approximating the root.

**EXAMPLE 4** Find the zero between 2 and 3 in the polynomial $f(x) = x^3 + x^2 - 8x - 6$. Approximate your answer to the nearest hundredth.

***Solution*** *Step 1* We can easily find $f(2) = -10$ and $f(3) = 6$ by synthetic division. Now approximate the curve between $(2, -10)$ and $(3, 6)$ by a straight line. Let $(2, -10) = (x_1, y_1)$ and $(3, 6) = (x_2, y_2)$.

$$\frac{y - y_1}{x - x_1} = \frac{y_2 - y_1}{x_2 - x_1}$$

$$\frac{y + 10}{x - 2} = \frac{6 + 10}{3 - 2} \qquad \text{thus} \qquad \frac{y + 10}{x - 2} = 16$$

The first approximation we will call $x_1$. Substitute $(x_1, 0)$ for $(x, y)$:

$$\frac{10}{x_1 - 2} = 16$$

$$x_1 = \frac{42}{16} = 2.625$$

$$x_1 \doteq 2.6 \qquad \textit{Rounded to nearest tenth.}$$

$f(2.6)$ by synthetic division $= -2.464$. If we try a slightly larger value 2.7, we obtain $f(2.7)$ by synthetic division $= -0.627$. Once again a slightly larger value $f(2.8)$ by synthetic division $= 1.392$. Since $f(2.7) < 0$ and $f(2.8) > 0$, we know that the irrational root lies between 2.7 and 2.8.

*Step 2* We know that $f(2.8) = 1.392$ and $f(2.7) = -0.627$. Now we find the $x$-intercept between the line joining $(2.7, -0.627)$ and $(2.8, 1.392)$:

$$\frac{y + 0.627}{x - 2.7} = \frac{1.392 + 0.627}{2.8 - 2.7}$$

$$\frac{y + 0.627}{x - 2.7} = \frac{2.019}{0.1} = 20.19$$

If we substitute $(x_2, 0)$ for $(x, y)$, we obtain a second approximation:

$$\frac{0.627}{x_2 - 2.7} = 20.19$$

$$55.14 = 20.19x_2$$

$$2.73 \doteq x_2 \qquad \textit{Rounded to nearest hundredth.}$$

This zero is accurate to the nearest hundredth.

*How can this accuracy be verified?* To verify this, we can find by synthetic division

$$f(2.72) = -0.237952$$
$$f(2.73) = -0.040683$$
$$f(2.74) = +0.158424$$

Clearly, $f(2.73)$ is closer to 0 than the next highest and lowest values. Thus 2.73 is the approximate zero of $f(x)$ in the interval (2, 3) and is correct to the nearest hundredth. ■

## EXERCISE 11-6

Locate the intervals for the real zeros of the polynomials in Problems 1–8.

**1.** $f(x) = x^3 + x^2 - 2x - 1$
**2.** $f(x) = x^3 - 7x + 5$
**3.** $f(x) = x^3 - 33x + 20$
**4.** $f(x) = x^3 - x^2 - 2$
**5.** $f(x) = x^3 - x^2 + x - 4$
**6.** $f(x) = x^3 - x^2 + 2x - 3$
**7.** $f(x) = x^3 + 2x - 5$
**8.** $f(x) = x^3 + 3x^2 - 2x - 6$

Find, to one decimal place, the indicated real zero in each of the polynomials in Problems 9–12.

**9.** $p(x) = x^3 - 2x - 7$, the zero between 2 and 3.
**10.** $p(x) = x^3 + x^2 - 5x - 5$, the zero between 2 and 3.
**11.** $p(x) = 2x^3 - 9x^2 + 6x - 1$, the zero between 3 and 4.
**12.** $p(x) = x^3 - 8x^2 + 14x - 4$, the zero between 5 and 6.

### Calculator Problems

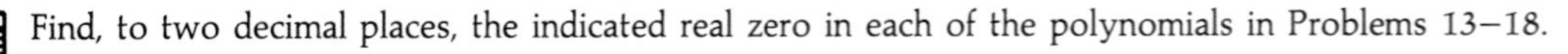

Find, to two decimal places, the indicated real zero in each of the polynomials in Problems 13–18.

**13.** $f(x) = x^3 - 5x^2 + 3$, the zero between 4 and 5.
**14.** $f(x) = x^3 + x^2 - 10x + 4$, the zero between 0 and 1.
**15.** $f(x) = x^3 + 3x^2 - 9x - 3$, the zero between 2 and 3.
**16.** $f(x) = x^4 - 5x^3 + 2x^2 + x + 7$, the zero between 1 and 2.
**17.** $f(x) = 4x^3 + 6x^2 + 3x + 5$, the zero between $-2$ and $-1$.
**18.** $f(x) = 2x^3 - 7x^2 - 16x + 33$, the zero between $-3$ and $-2$.

Find all real zeros of the polynomials in Problems 19–22 to two decimal places.

**19.** $f(x) = 2x^3 - 4x^2 - 10x - 3$
**20.** $f(x) = 2x^3 - 5x^2 - 8x + 11$
★**21.** $p(x) = 2x^4 + x^3 + 2x^2 - x - 1$
★**22.** $p(x) = x^4 - 6x^3 + 8x^2 + 2x - 1$

## 11-7 PARTIAL FRACTIONS (OPTIONAL)

In Section 1-5 we reviewed the process of adding or subtracting two or more fractions to obtain a single reduced fraction. In integral calculus it is sometimes desirable to reverse this process, that is, to find the individual fractions that sum to a given fraction. In this procedure we want to take a rational function such as $\dfrac{3x^2 + 1}{x(x^2 + 1)}$ and write it as the sum of simpler rational functions. This is called separating into partial fractions.

Suppose we have a given fraction with a numerator and denominator that are both polynomials, and we wish to separate it into a sum of partial fractions $\left(\text{i.e., separate } \dfrac{f(x)}{g(x)} \text{ into partial fractions}\right)$. Our success in doing so depends on two things.

---

To separate $\dfrac{f(x)}{g(x)}$ into partial fractions:

1. *The degree of* $f(x)$ *must be less than the degree of* $g(x)$. If this is not the case, we can divide $f(x)$ by $g(x)$ using long division and then work with the remaining fractions. (See Section 11-1.)
2. *The factors of the denominator* $g(x)$ must be known.

---

Problems with partial fractions can be classified into three basic types, and we will, by example, show how to solve each type.

---

*Type I.* If $g(x)$, the denominator, has prime linear factors $(x - r_1)$, $(x - r_2)$, $(x - r_3)$, . . . , $(x - r_n)$, all different, then the fraction

$$\frac{f(x)}{g(x)} = \frac{A}{x - r_1} + \frac{B}{x - r_2} + \frac{C}{x - r_3} + \cdots + \frac{D}{x - r_n}$$

*Type II.* If $g(x)$, the denominator, has linear factors $(x - r)^n$, then

$$\frac{f(x)}{g(x)} = \frac{A}{x - r} + \frac{B}{(x - r)^2} + \frac{C}{(x - r)^3} + \cdots + \frac{D}{(x - r)^n}$$

*Type III.* If $g(x)$, the denominator, has quadratic factors $(x^2 + bx + c)^n$ (and $x^2 + bx + c$ is not factorable), then

$$\frac{f(x)}{g(x)} = \frac{Ax + B}{x^2 + bx + c} + \frac{Cx + D}{(x^2 + bx + c)^2} + \cdots + \frac{Ex + F}{(x^2 + bx + c)^n}$$

---

Of course, a given fraction might be a combination of any of the three types.

EXAMPLE 1 Find partial fractions that sum to $\frac{3x - 1}{x^2 + 5x + 6}$.

***Solution*** We see that the denominator has prime linear factors $(x + 2)(x + 3)$. Thus we see that $\frac{f(x)}{g(x)}$ is Type I. We desire to find $A$ and $B$ in the equation

$$\frac{3x - 1}{x^2 + 5x + 6} = \frac{3x - 1}{(x + 2)(x + 3)} = \frac{A}{x + 2} + \frac{B}{x + 3}$$

Adding the two fractions gives

$$\frac{3x - 1}{(x + 2)(x + 3)} = \frac{A(x + 3) + B(x + 2)}{(x + 2)(x + 3)}$$

By removing parentheses and collecting terms, we get

$$\frac{3x - 1}{(x + 2)(x + 3)} = \frac{(A + B)x + 3A + 2B}{(x + 2)(x + 3)}$$

We now have two fractions with the same denominator, and the numerators of these fractions are polynomials. The fractions will be equal if and only if, in the numerator, coefficients of corresponding powers of $x$ are equal. Therefore

$$A + B = 3 \qquad \text{and} \qquad 3A + 2B = -1$$

Solving these two equations simultaneously gives $A = -7$ and $B = 10$. Hence

$$\frac{3x - 1}{(x + 2)(x + 3)} = \frac{-7}{(x + 2)} + \frac{10}{(x + 3)}$$

We can check the solution by adding the fractions $\frac{-7}{x + 2}$ and $\frac{10}{x + 3}$ to get $\frac{3x - 1}{(x + 2)(x + 3)}$. Can you verify this? ■

EXAMPLE 2 Find partial fractions that sum to $\frac{2x + 7}{x^2 + 10x + 25}$.

***Solution*** Since the denominator can be written as $(x + 5)^2$, we see that $\frac{f(x)}{g(x)}$ is Type II.

We desire to find $A$ and $B$ in the equation

$$\frac{2x+7}{x^2+10x+25} = \frac{2x+7}{(x+5)^2} = \frac{A}{x+5} + \frac{B}{(x+5)^2}$$

$$= \frac{A(x+5)+B}{(x+5)^2}$$

$$= \frac{Ax+5A+B}{(x+5)^2}$$

Since $$\frac{2x+7}{(x+5)^2} = \frac{Ax+5A+B}{(x+5)^2}$$

then $$A = 2 \quad \text{and} \quad 5A+B=7$$

So $$A = 2 \quad \text{and} \quad B = -3$$

Hence $$\frac{2x+7}{(x+5)^2} = \frac{2}{x+5} - \frac{3}{(x+5)^2}$$

Can you check by adding $\frac{2}{x+5}$ and $\frac{-3}{(x+5)^2}$ to get $\frac{2x+7}{(x+5)^2}$? ■

**EXAMPLE 3** Find partial fractions that sum to $\frac{3x^2+5x+1}{(x^2+2x+3)^2}$.

***Solution*** Since $x^2+2x+3$ cannot be factored, we see that $\frac{f(x)}{g(x)}$ is Type III.

$$\frac{3x^2+5x+1}{(x^2+2x+3)^2} = \frac{Ax+B}{x^2+2x+3} + \frac{Cx+D}{(x^2+2x+3)^2}$$

$$= \frac{(Ax+B)(x^2+2x+3)+Cx+D}{(x^2+2x+3)^2}$$

$$= \frac{Ax^3+2Ax^2+3Ax+Bx^2+2Bx+3B+Cx+D}{(x^2+2x+3)^2}$$

$$= \frac{Ax^3+(2A+B)x^2+(3A+2B+C)x+3B+D}{(x^2+2x+3)^2}$$

Equating coefficients of like terms in the numerator gives the four equations

$$A = 0 \qquad \textbf{(1)}$$

$$2A+B = 3 \qquad \textbf{(2)}$$

$$3A+2B+C = 5 \qquad \textbf{(3)}$$

$$3B+D = 1 \qquad \textbf{(4)}$$

To solve this system of equations, we substitute $A = 0$ from equation (1) into equation (2) and obtain $B = 3$. If we substitute $A = 0, B = 3$ into equation (3), we obtain $0 + 6 + C = 5$, and thus $C = -1$. If we substitute $B = 3$ into equation (4), we obtain $D = -8$. Thus we have

$$A = 0 \qquad B = 3 \qquad C = -1 \qquad D = -8$$

Therefore

$$\frac{3x^2 + 5x + 1}{(x^2 + 2x + 3)^2} = \frac{3}{x^2 + 2x + 3} + \frac{-x - 8}{(x^2 + 2x + 3)^2}$$

The result may also be written as

$$\frac{3x^2 + 5x + 1}{(x^2 + 2x + 3)^2} = \frac{3}{x^2 + 2x + 3} - \frac{x + 8}{(x^2 + 2x + 3)^2}$$

These results can be checked by combining the partial fractions. ■

**EXAMPLE 4** Find partial fractions that sum to $\dfrac{x^2 + x - 3}{(x + 1)(x + 2)^2}$.

***Solution*** This example is a combination of Types I and II. Hence

$$\begin{aligned}\frac{x^2 + x - 3}{(x + 1)(x + 2)^2} &= \frac{A}{x + 1} + \frac{B}{x + 2} + \frac{C}{(x + 2)^2}\\ &= \frac{A(x + 2)^2 + B(x + 1)(x + 2) + C(x + 1)}{(x + 1)(x + 2)^2}\\ &= \frac{Ax^2 + 4Ax + 4A + Bx^2 + 3Bx + 2B + Cx + C}{(x + 1)(x + 2)^2}\\ &= \frac{(A + B)x^2 + (4A + 3B + C)x + 4A + 2B + C}{(x + 1)(x + 2)^2}\end{aligned}$$

Therefore

$$A + B = 1 \tag{1}$$

$$4A + 3B + C = 1 \tag{2}$$

$$4A + 2B + C = -3 \tag{3}$$

Solving this system yields

$$A = -3 \qquad B = 4 \qquad C = 1$$

Therefore

$$\frac{x^2 + x - 3}{(x+1)(x+2)^2} = \frac{-3}{x+1} + \frac{4}{x+2} + \frac{1}{(x+2)^2} \quad \blacksquare$$

## EXERCISE 11-7

Find partial fractions that sum to each of the following:

1. $\dfrac{1}{(x+1)(x-2)}$
2. $\dfrac{x}{(x+3)(x-1)}$
3. $\dfrac{x^2+1}{(x-2)(x-1)(2x+1)}$
4. $\dfrac{x}{(x-1)(x-2)(x-3)}$
5. $\dfrac{1-x}{x^2+3x+2}$
6. $\dfrac{5x+4}{x^2+2x}$
7. $\dfrac{3x-2}{x^2-x}$
8. $\dfrac{3x+7}{x^2-2x-3}$
9. $\dfrac{11x-11}{6x^2+7x-3}$
10. $\dfrac{3x-13}{6x^2-x-12}$
11. $\dfrac{3x^2+x}{(x-2)(x^2+3)}$
12. $\dfrac{x+10}{(x+1)(x^2+1)}$
13. $\dfrac{x^3-1}{x(x+1)^3}$
14. $\dfrac{1}{(x+1)^3(x-1)}$
15. $\dfrac{x^2+2x+3}{x^3-x}$
16. $\dfrac{x+5}{(x-1)^2(x+2)}$
17. $\dfrac{3x^2+x-2}{(x-1)(x^2+1)}$
18. $\dfrac{x+2}{(x^2+x+1)(x-1)}$
19. $\dfrac{3x^2-x+1}{(x+1)(x^2-x+3)}$
20. $\dfrac{6x^2-3x+1}{(4x+1)(x^2+1)}$
21. $\dfrac{3x^2+x+6}{x^4+3x^2+2}$

★22. $\dfrac{x^5}{(x-1)(x^2+2)^2}$

★23. $\dfrac{3}{x(x+1)(x^2+1)}$

★24. $\dfrac{x^4+1}{x(x^2+1)^2}$

★25. $\dfrac{6x^2-15x+22}{(x+3)(x^2+2)^2}$

★26. $\dfrac{x^4+2x^3+5x^2+5x+3}{(x+1)(x^2+2x+2)^2}$

## COMPUTER PROBLEMS FOR CHAPTER 11

See if you can apply your knowledge of computer programming and of Chapter 11 to do the following.

1. Write a computer program to find the function values for any value of a polynomial of degree 5 or less. The polynomial will be entered in the form $Ax^5 + Bx^4 + Cx^3 + Dx^2 + Ex + F$. The program should ask for input of the values of $A, B, C, D, E, F$. The program should ask for input for a value of $x$. The program should then print the value of $p(x)$ and ask whether other values of $x$ will be entered for that polynomial. If no further values are needed, the program should loop back to the place in the program where the values of $A, B, C, D, E, F$ are entered. Use the program to find the following:
   - **a.** $p(x) = 4x^5 + 3x^4 + 2x^2 - 5x - 8$ evaluated at $p(3)$, $p(2)$, $p(-1)$, $p(0)$, $p(1)$, $p(2.5)$, $p(3.7)$.
   - **b.** $f(x) = -6x^5 - 3x^3 + 5x^2 + 2x - 3$ evaluated at $f(-6.1)$, $f(-3.2)$, $f(-1.1)$, $f(0)$, $f(1.274)$.
   - **c.** $g(x) = x^3 - 5x^2 + 1.3x + 1.62$ evaluated at $g(-5.163)$, $g(-3)$, $g(0)$, $g(1.52)$, $g(28.203)$.

**2.** Modify the program created in Problem 1 so that the function values are printed out for every value between any two end intervals with a specified step between each value. For example, the portion of the program that specified finding $f(x)$ for $f(2.0)$, $f(2.2)$, $f(2.4)$, $f(2.6)$, $f(2.8)$, $f(3.0)$ would look as follows:

```
BEGINNING X VALUE OF INTERVAL?  2
ENDING X VALUE OF INTERVAL?  3
STEP INCREMENT?  0.2
```

Use the program to find the following:

**a.** $p(x) = 9x^4 - 3x^3 + 16x^2 - 6x - 4$ evaluated at $p(-1.0)$, $p(-0.9)$, $p(-0.8)$, . . . , $p(0.8)$, $p(1.0)$.
From your printout, what values of $x$ do you feel are approximate zeros of $p(x)$? Can you verify this?

**b.** $h(x) = x^4 - 6x^3 + 10x^2 - 18x + 21$ evaluated at $h(4.00)$, $h(4.05)$, $h(4.10)$, $h(4.15)$, . . . , $h(4.95)$, $h(5.00)$.
From your printout, what values of $x$ do you feel are approximate zeros of $h(x)$? Can you verify this?

## KEY TERMS AND CONCEPTS

Be sure you understand each of the following and can explain it in your own words.

Real polynomial
Complex polynomial
Degree of a polynomial
Zero of a polynomial
Integral domain
Division algorithm for polynomials
Remainder theorem
Factor theorem
Synthetic division
Descartes' Rule of Signs
Upper bound of a zero of a polynomial
Lower bound of a zero of a polynomial
Intermediate value theorem
Partial fractions

## SUMMARY OF PROCEDURES AND CONCEPTS

The following theorems are often helpful in determining the number of zeros of a polynomial and in determining those zeros.

*Theorem 11-1*

If $f(x)$ and $g(x)$ are polynomials and if $g(x) \neq 0$, then there exist unique polynomials $q(x)$ and $r(x)$ such that $f(x) = g(x) \cdot q(x) + r(x)$ where either $r(x) = 0$ or $r(x)$ has degree less than the degree of $g(x)$.

*Theorem 11-2*

***(Remainder Theorem)*** If a polynomial $f(x)$ is divided by $(x - c)$, the remainder is $f(c)$.

*Theorem 11-3*

***(Factor Theorem)*** A polynomial $f(x)$ has a factor $(x - c)$ if and only if $f(c) = 0$.

*Theorem 11-4*

***(Fundamental Theorem of Algebra)*** If $f(x)$ is a polynomial of degree greater than zero, with complex coefficients, then $f(x)$ has at least one complex zero.

*Theorem 11-5*

If $f(x)$ is a polynomial of degree $n > 0$ with complex coefficients, then $f(x)$ has exactly $n$ zeros. (Note that this does not imply $n$ different zeros. If a zero occurs more than once, it is counted each time.)

*Theorem 11-6*

If a polynomial $f(x)$ of degree $n > 0$ with real coefficients has as a zero $(a + bi)$, then $(a - bi)$ is also a zero.

*Theorem 11-7*

Every polynomial $f(x)$ with real coefficients and of odd degree has at least one real zero.

*Theorem 11-8*

***(Descartes' Rule of Signs)*** If $f(x)$ is a polynomial with real coefficients, the number of positive real zeros of $f(x)$ is either equal to the number of sign variations of $f(x)$ or less than the number of sign variations by a positive even integer. Also, the number of negative real zeros of $f(x)$ is either equal to the number of sign variations of $f(-x)$ or less than the number of sign variations of $f(-x)$ by a positive even integer.

*Theorem 11-9*

If a polynomial $p(x)$ of degree $n > 0$ with rational coefficients has a zero of the form $a + b\sqrt{c}$ where $a$, $b$ are rational numbers and $c$ is not a perfect square, then $a - b\sqrt{c}$ is also a zero.

*Theorem 11-10*

Let the polynomial $f(x) = a_0x^n + a_1x^{n-1} + a_2x^{n-2} + \cdots + a_n$ have only integer coefficients with $a_0 \neq 0$. If $f(x)$ has a rational zero $p/q$ (where $p/q$ is in reduced form), then $p$ is an exact divisor of $a_n$ and $q$ is an exact divisor of $a_0$.

*Corollary 11-11*

An integral zero of the polynomial $f(x)$ must be an exact divisor of the constant term $a_n$.

*Corollary 11-12*

If the polynomial $f(x)$ has the leading coefficient $a_0 = 1$, then any rational zero is an integer.

*Theorem 11-13*

If $f(x) = a_0x^n + a_1x^{n-1} + \cdots + a_n$ and if $a_k$ is the coefficient of greatest absolute value, then no positive zero can exceed $|a_k/a_0| + 1$.

*Theorem 11-14*

If, in the synthetic division of a polynomial $f(x)$ with real coefficients by $(x - a)$ where $a > 0$, all the numbers in the third row are positive, then $f(x)$ has no real zero larger than $a$.

*Theorem 11-15*

If, in the synthetic division of a polynomial $f(x)$ with real coefficients by $(x - a)$ where $a < 0$, the numbers in the third row are alternately positive and negative, then $f(x)$ has no real zero less than $a$.

*Theorem 11-16*

***(Intermediate Value Theorem)*** If for the polynomial $f(x)$ with real coefficients, $f(a)$ and $f(b)$ are opposite in sign, then $f(x)$ has a real zero between $a$ and $b$.

## CHAPTER 11 REVIEW EXERCISE

1. Using direct substitution in $f(x) = x^5 + 3x^3 - 4x + 5$, find $f(-1)$.
2. Is $-3$ a zero of $P(x) = x^3 - 2x^2 - 11x + 12$?
3. State the degree of $g(x) = 4x^3 - 2x^2 + 1$.
4. If $f(x) = x^5 - 2x^4 + x - 1$ and $g(x) = 3x^2 + x - 4$, state the degree of $f(x) \cdot g(x)$.
5. If $f(x) = 4x^3 + x + 2$ and $g(x) = (a - 1)x^3 + (b + 1)x^2 + (c - 10)x + 2$, find values for $a$, $b$, and $c$ so that $f(x) = g(x)$.

Use long division to find the remainder in Problems 6 and 7.

6. $(5x^4 - 2x^3 + x^2 + 2x - 1) \div (x - 2)$
7. $(3x^5 - 12x^3 + x + 4) \div (x + 3)$

Use the remainder theorem to find the remainder in Problems 8 and 9.

8. $(x^4 - 5x^3 + 11x^2 + x - 5) \div (x - 1)$
9. $(x^3 - 4x^2 + 3x + 2) \div (x + 3)$

Use the factor theorem to answer Problems 10 and 11.

10. Is $(x - 3)$ a factor of $4x^4 - 11x^3 - 10x - 3$?
11. Is $(x + 2)$ a factor of $x^5 + 2x^4 - x^3 + 4x$?
12. Use synthetic division to divide $x^4 + 3x^3 - 44x^2 + 19x + 6$ by $(x - 5)$.
13. Use synthetic division to divide $x^5 - 30x^3 + 40x^2 - 103$ by $(x + 6)$.
14. If $f(x) = x^6 - 3x^4 + x - 1$, find $f(2)$ using synthetic division.
15. Use synthetic division to find $f(-4)$ if $f(x) = x^4 + 3x^3 + 12x - 1$.
16. Use synthetic division to determine if 6 is a zero of $f(x) = x^4 - 5x^3 - 8x^2 + 11x - 6$.
17. Use synthetic division to determine if $-3$ is a zero of $g(x) = 5x^4 + 11x^3 + 29x + 6$.
18. Use synthetic division to determine if $(2 + i)$ is a zero of $h(x) = x^3 - 2x^2 - 3x + 10$.
19. How many zeros does $f(x) = 5x^6 - 3x^3 + x^2 - 1$ have?
20. Find a third-degree polynomial that has zeros $-2$, 2, and 5.
21. Find a third-degree polynomial that has zeros 3, $\sqrt{5}$, and $-2\sqrt{5}$.
22. Find a third-degree polynomial that has zeros 1, $i$, and $-i$.
23. Use the polynomial $f(x) = x^3 + 1$ to find the three cube roots of $-1$.

Using Descartes' Rule of Signs, answer Problems 24–29.

24. How many positive real zeros could $f(x) = 6x^4 + x^3 - 2x^2 - x + 1$ have?
25. How many negative real zeros could $f(x) = 6x^4 + x^3 - 2x^2 - x + 1$ have?
26. How many positive real zeros could $f(x) = x^5 + 4x^3 + x - 2$ have?
27. How many negative real zeros could $f(x) = x^5 + 4x^3 + x - 2$ have?
28. How many real zeros does $f(x) = 3x^6 + 2x^4 + 5x^2 + 4$ have?
29. How many real zeros does $f(x) = 2x^5 + 4x^3 + 3x$ have?
30. List all possible rational zeros of $f(x) = 3x^5 - 4x^2 + x - 10$.

Find *all* zeros of the polynomials in Problems 31 to 35.

**31.** $f(x) = x^3 - 7x^2 + 7x + 15$

**32.** $f(x) = 4x^3 - 13x + 6$

**33.** $f(x) = x^4 - 8x^3 + 22x^2 - 24x + 9$

**34.** $f(x) = 8x^4 + 6x^3 - 15x^2 - 12x - 2$

**35.** $f(x) = 12x^5 - 44x^4 - 21x^3 + 21x^2 - 6x + 8$

Locate the intervals for the real zeros of the polynomials in Problems 36 and 37.

**36.** $f(x) = x^3 - 3x^2 - 5$

**37.** $f(x) = x^4 - 16x^3 + 69x^2 - 70x - 42$

Find, to one decimal place, the indicated real zero in each of the polynomials in Problems 38 and 39.

**38.** $f(x) = x^3 + 2x^2 - 7x + 1$, the zero between 0 and 1.

**39.** $f(x) = x^3 - 6x^2 - 8x + 40$, the zero between $-3$ and $-2$.

**40.** Find all real zeros of $f(x) = 8x^3 - 12x^2 + 1$ to one decimal place.

In Problems 41–45 find partial fractions that sum to the given fraction.

**41.** $\dfrac{3x + 29}{x^2 + 3x - 10}$

**42.** $\dfrac{5x^3 + 20x^2 + 18x + 26}{(x^2 + 2)(x + 2)^2}$

**43.** $\dfrac{8x^2 + 9x + 22}{x^3 - 8}$

**44.** $\dfrac{3x^2 + 10x + 11}{(x + 1)^3}$

**45.** $\dfrac{2x^3 + x^2 - 3x - 10}{(x^2 + 2x + 3)^2}$

## PRACTICE TEST FOR CHAPTER 11

**1.** If $f(x) = 3x^4 + x^2 - 5x + 6$ and $g(x) = 2x^3 + x^2 - 7x + 1$, state the degree of $f(x) \cdot g(x)$.

**2.** Find the remainder when $5x^6 - 14x^4 + 11x^3 + x + 4$ is divided by $(x + 2)$.

**3.** Use the factor theorem to show that $(x + 5)$ is a factor of $x^5 + 6x^4 + 5x^3 + 4x^2 + 18x - 10$.

**4.** Use synthetic division to determine if $-1$ is a zero of $f(x) = 3x^5 + 5x^3 + 9x^2 - 1$.

**5.** Find a third-degree polynomial that has zeros $-3$, $-1$, and 2.

**6.** Using Descartes' Rule of Signs, show that $f(x) = 3x^5 - 2x^4 + x^3 - 6x^2 - 1$ has no negative real zeros.

**7.** List all possible rational zeros of $f(x) = 3x^5 + 2x^4 - 5x^3 + x - 5$.

**8.** Find *all* zeros of $f(x) = 3x^3 - 5x^2 - 11x - 3$.

**9.** Find *all* zeros of $f(x) = 2x^3 - x^2 + 8x - 4$.

**10.** Find *all* zeros of $f(x) = 2x^4 + x^3 - 7x^2 - 3x + 3$.

**11.** Give the intervals for the real zeros of $f(x) = x^3 + 2x^2 - 15x - 25$.

**12.** Find partial fractions that sum to give $\dfrac{2x^2 - 34x + 100}{(x + 5)(x - 3)^2}$.

# 12 *Sequences, Series, and Probability*

*So teach us to number our days, that we may apply our hearts unto wisdom.*

*Psalm 90:12*

For over a thousand years people have compiled statistics on the length of life of individuals or the likelihood that some event will happen. Others have studied how to determine the next number in a series of numbers, or how to find the sum of an infinite series of numbers. In this chapter we will investigate each of these topics in detail.

## 12-1 MATHEMATICAL INDUCTION

In order for a statement about an infinite set of numbers to be true, it must be true for every element within the set. Whenever a set is infinite, it is, of course, not possible to check the statement for every element of the set. **Mathematical induction** is a method of proof used when we have a statement concerning the set of positive integers. This method is based on the following theorem.

***The Principle of Mathematical Induction***

If a statement $P$ is true for the integer 1 and if the assumption that $P$ is true for the integer $k$ implies that $P$ is true for $(k + 1)$, then $P$ is true for the set of positive integers.

Let us consider a real-life situation that is easy to visualize. Suppose we line up 100,000 dominoes and space them in such a way that if one falls down, it will always hit the next one. Now if we want to know whether all of the dominoes can be knocked over by tipping over the first one in the line, we can prove it by mathematical induction in the following way.

*Step 1* Determining that the first domino is tipped over ($n = 1$).

*Step 2* Consider any given domino in the line ($n = k$) that is being tipped over.

*Step 3* Determine that by the domino $n = k$ being tipped over, the domino $n = k + 1$ will be tipped over as a result.

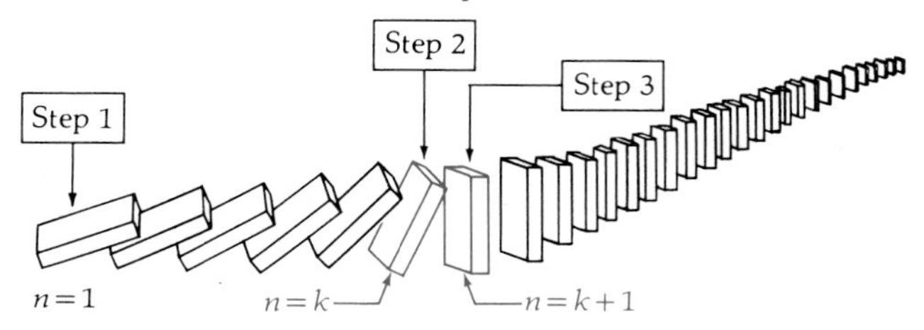


We thus conclude that all 100,000 dominoes can be tipped over by hitting the first one.

**Historical Note:** On June 9, 1979, Michael Cairney, age 23, knocked over a line of 169,713 dominoes by tipping the first one and allowing each domino to tip over the next one in the line. This line of dominoes was not straight, but if it had been, it would have extended for 4.3 miles.

We now look at a mathematical example.

**EXAMPLE 1** Prove that the sum of integers 1 through $n$ is given by the formula $\dfrac{n^2 + n}{2}$.

In other words, show that

$$1 + 2 + 3 + \cdots + n = \frac{n^2 + n}{2}$$

***Solution*** *Step 1* Is the formula true for $n = 1$?

$$\frac{1^2 + 1}{2} = \frac{1 + 1}{2} = 1$$

Thus the formula holds true for $n = 1$.

*Step 2* Assume that the formula is true for $n = k$.

$$1 + 2 + 3 + \cdots + k = \frac{k^2 + k}{2}$$

*Step 3* We must now show that the assumption of Step 2 will imply that the sum of the first $(k + 1)$ integers is $\frac{(k + 1)^2 + (k + 1)}{2}$. We know from Step 2 that

$$1 + 2 + 3 + \cdots + k = \frac{k^2 + k}{2}$$

Now if we add $(k + 1)$ to each side, we obtain

$$1 + 2 + 3 + \cdots k + (k + 1) = \frac{k^2 + k}{2} + (k + 1)$$

$$= \frac{k^2 + k + 2k + 2}{2} = \frac{k^2 + 3k + 2}{2}$$

We now write $k^2 + 3k + 2$ as $k^2 + 2k + 1 + k + 1$ and factor by grouping the first three terms so that we get $(k + 1)^2 + (k + 1)$. We now have the desired result,

$$1 + 2 + 3 + \cdots + (k + 1) = \frac{(k + 1)^2 + (k + 1)}{2}$$

which shows that the formula holds for $n = k + 1$. Thus by the principle of mathematical induction, the formula is true for $n$ equal to any integer. ■

This method of proof can be used to establish De Moivre's Theorem (9-2). See if you can follow each step of the proof.

***De Moivre's Theorem 9-2***

For any complex number $r$ cis $\theta$ and any positive integer $n$,

$$(r \text{ cis } \theta)^n = r^n \text{ cis } n\theta$$

***Proof*** *Step 1* The proof is by mathematical induction. First, the theorem holds for $n = 1$ since

$$(r \text{ cis } \theta)^1 = r^1 \text{ cis } 1\theta$$

*Step 2* Assume that the theorem is true for $n = k$.

$$(r \text{ cis } \theta)^k = r^k \text{ cis } k\theta$$

*Step 3* We must now show that the theorem is true for $n = k + 1$. Multiplying both sides of our previous statement by $(r \text{ cis } \theta)$, we have

$$(r \text{ cis } \theta)(r \text{ cis } \theta)^k = (r \text{ cis } \theta)(r^k \text{ cis } k\theta)$$

$$(r \text{ cis } \theta)^{k+1} = r^{k+1} \text{ cis}(k\theta + \theta)$$

$$= r^{k+1} \text{ cis}[(k + 1)\theta]$$

*Do you see why this step is justified?*

Since the theorem is true for $n = 1$, and the assumption that the theorem is true for $n = k$ implies that it is true for $n = k + 1$, the theorem is therefore true for all $n$. ■

Any and all manipulative skills of algebra can be used to rearrange the right-hand side of Step 3 to obtain the proper form. A great deal of ingenuity is sometimes required.

The method of mathematical induction can be used to prove both equations and inequalities. We will now use the procedure to prove a formula for inequalities. Two of the most useful properties of inequalities are the following:

**1.** If $a < b$ and $b < c$, then $a < c$.
**2.** If $a < b$ and $c < d$, then $a + c < b + d$.

We will use each of these in Example 2.

EXAMPLE 2 Prove $n < 2^n$ for all positive integers $n$.

***Solution*** *Step 1* Show the statement is true for $n = 1$.

$$1 < 2^1 \qquad \text{or} \qquad 1 < 2$$

*Step 2* Assume the statement is true for $n = k$.

$$k < 2^k \qquad \textbf{Inequality (1)}$$

*Step 3* Show that the assumption of Step 3 implies that $k + 1 < 2^{k+1}$ From Step 2 we have

$$k < 2^k$$

Now since $1 < 2$, we know that $1^k < 2^k$ for all positive integers. [This is Theorem (2-4).] Since $1^k = 1$ for all positive integers $k$, we have

$$1 < 2^k \qquad \textbf{Inequality (2)}$$

We now add inequality (1) to inequality (2). We have

| | |
|---|---|
| $k + 1 < 2^k + 2^k$ | *If $a < b$ and $c < d$, then $a + c < b + d$.* |
| $k + 1 < 2^k(1 + 1)$ | *Factor out $2^k$.* |
| $k + 1 < 2^k(2)$ | *Simplify.* |
| $k + 1 < 2^k \cdot 2^1$ | |
| $k + 1 < 2^{k+1}$ | *By the law of exponents.* |

We have thus shown the formula is true for $n = k + 1$. Therefore by mathematical induction the statement is true for all $n$. ■

## EXERCISE 12-1

Prove the statements in Problems 1–20 by using mathematical induction.

**1.** $1 + 3 + 5 + \cdots + (2n - 1) = n^2$

**2.** $2 + 6 + 10 + \cdots + (4n - 2) = 2n^2$

**3.** $1 + 4 + 7 + \cdots + (3n - 2) = \dfrac{3n^2 - n}{2}$

**4.** $2 + 7 + 12 + \cdots + (5n - 3) = \dfrac{n(5n - 1)}{2}$

**5.** $1^2 + 2^2 + 3^2 + \cdots + n^2 = \dfrac{n(n + 1)(2n + 1)}{6}$

**6.** $1^3 + 2^3 + 3^3 + \cdots + n^3 = \dfrac{n^2(n + 1)^2}{4}$

**7.** $1^3 + 3^3 + 5^3 + \cdots + (2n - 1)^3 = n^2(2n^2 - 1)$

**8.** $1^2 + 3^2 + 5^2 + \cdots + (2n - 1)^2 = \dfrac{n(2n - 1)(2n + 1)}{3}$

**9.** $\dfrac{1}{1(2)} + \dfrac{1}{2(3)} + \dfrac{1}{3(4)} + \cdots + \dfrac{1}{n(n + 1)} = \dfrac{n}{n + 1}$

**10.** $\dfrac{1}{2(3)} + \dfrac{1}{3(4)} + \dfrac{1}{4(5)} + \cdots + \dfrac{1}{(n + 1)(n + 2)} = \dfrac{n}{2(n + 2)}$

**11.** $\dfrac{1}{1(2)(3)} + \dfrac{1}{2(3)(4)} + \dfrac{1}{3(4)(5)} + \cdots + \dfrac{1}{n(n + 1)(n + 2)} = \dfrac{n(n + 3)}{4(n + 1)(n + 2)}$

**12.** $\dfrac{1}{1(2)} + \dfrac{5}{2(3)} + \dfrac{11}{3(4)} + \cdots + \dfrac{n^2 + n - 1}{n(n + 1)} = \dfrac{n^2}{n + 1}$

**13.** $1 + 2^1 + 2^2 + 2^3 + \cdots + 2^{n-1} = 2^n - 1$

**14.** $3 + 3^2 + 3^3 + \cdots + 3^n = \dfrac{3(3^n - 1)}{2}$

**15.** $1(2) + 2(3) + 3(4) + \cdots + n(n + 1) = \dfrac{n(n + 1)(n + 2)}{3}$

**16.** $1(4) + 2(9) + 3(16) + \cdots + n(n + 1)^2 = \dfrac{n(n + 1)(n + 2)}{12}$

**17.** $1 + 4n \leq 5^n$

**18.** $1 + 2n \leq 3^n$

★ **19.** $1 + \frac{1}{2} + \frac{1}{2^2} + \frac{1}{2^3} + \cdots + \frac{1}{2^{n-1}} = 2\left(1 - \frac{1}{2^n}\right)$

★ **20.** $1 + x + x^2 + \cdots + x^{n-1} = \frac{x^n - 1}{x - 1}, \quad x \neq 1$

---

## 12-2 ARITHMETIC SEQUENCES AND SERIES

We will now introduce a formal definition of a sequence and the notation that mathematicians usually use to describe a sequence.

*Definition*

A *sequence* is a function whose domain is either the set of positive integers or a finite set of successive positive integers.

Note that the definition of a function would require exactly one term for each positive integer. Thus $f(1)$ would be the first term, $f(2)$ the second term, and so on. Subscripts are generally used to designate the number of a term. Instead of $f(1), f(2), f(3), \ldots, f(n)$, it is generally customary to use $a_1, a_2, a_3, \ldots, a_n$ to designate a sequence with $n$ terms.

*Definition*

A sequence with a finite number of terms is called a *finite sequence,* and a sequence with an infinite number of terms is called an *infinite sequence.*

### Formulas for the *n*th Term of a Sequence

It is most helpful to have a general formula to find the $n$th term of a given sequence. To avoid any misunderstanding about ordering, we will often designate the $n$th term ($a_n$) of a sequence by an algebraic formula.

EXAMPLE 1 Find a formula for the $n$th term of the infinite sequence 3, 6, 9, 12, 15, 18, . . . .

***Solution*** We see that each term is a multiple of 3. Therefore the sequence could be written 3(1), 3(2), 3(3), 3(4), 3(5), 3(6), . . . , and, in general, for the $n$th term, $3n$.

Thus the sequence could be described in the form 3, 6, 9, 12, 15, 18, . . . , $3n$, . . . . ■

EXAMPLE 2 Find a formula for the $n$th term of the infinite sequence 2, 5, 10, 17, 26, 37, . . . .

***Solution*** The sequence is somewhat similar to the squares of each positive integer 1, 4, 9, 16, 25, 36, . . . , but each term of the given sequence is larger by one.

The given sequence could be written as $1^2 + 1$, $2^2 + 1$, $3^2 + 1$, $4^2 + 1$, $5^2 + 1$, $6^2 + 1$, . . . .

In general, the $n$th term could be written as $n^2 + 1$.

The sequence could be described as 2, 5, 10, 17, 26, 37, . . . , $n^2 + 1$, . . . .

■

If you are given a formula for the $n$th term of a sequence, but no terms are given, you can use the formula to obtain each desired term.

EXAMPLE 3 Write the first three terms, the sixth term, and the twentieth term for a sequence whose $n$th term is given by $a_n = 2n^2 + n - 1$.

***Solution***

$$a_1 = 2(1)^2 + 1 - 1 = 2$$

$$a_2 = 2(2)^2 + 2 - 1 = 9$$

$$a_3 = 2(3)^2 + 3 - 1 = 20$$

$$a_6 = 2(6)^2 + 6 - 1 = 77$$

$$a_{20} = 2(20)^2 + 20 - 1 = 800 + 20 - 1 = 819$$ ■

### Recursion Formulas

Sometimes a given term of a sequence is described by a formula relating it to a previous term.

*Definition*

A *recursive* definition of a sequence occurs when we give a specific term and then give a method of finding the $(k + 1)$ term from the $k$th term.

EXAMPLE 4 Find the sixteenth term of the sequence if $a_{15} = 12$ and $a_{k+1} = \frac{a_k}{4} + 1$.

***Solution*** We will use $a_{k+1} = \frac{a_k}{4} + 1$, where $k = 15$ and $k + 1 = 16$.

So we have

$$a_{16} = \frac{a_{15}}{4} + 1 = \frac{12}{4} + 1 = 3 + 1 = 4 \quad \blacksquare$$

### The Series and Sigma Notation

In higher branches of mathematics the concept of a series is very important.

*Definition*

A *series* is the indicated sum of a sequence.

EXAMPLE 5 Find the sum of the series $1^2 + 2^2 + 3^2 + 4^2 + 5^2$.

***Solution*** $1^2 + 2^2 + 3^2 + 4^2 + 5^2 = 1 + 4 + 9 + 16 + 25 = 55 \quad \blacksquare$

A special notation is sometimes used to indicate a series. It is especially useful when we have an expression for the general term of a sequence. The uppercase Greek letter sigma ($\sum$) is used to indicate a sum. To illustrate, $\sum_{n=1}^{3} 2n$ represents the sum $2(1) + 2(2) + 2(3)$. Also consider this sum:

$$\sum_{i=1}^{4} a_i = a_1 + a_2 + a_3 + a_4$$

which represents a general series with four terms. The initial integral value may be a number other than one.

EXAMPLE 6 Find the value of the following expressions.

**a.** $\sum_{n=3}^{6} (3n - 4)$ **b.** $\sum_{k=0}^{3} [5 + (-2)^k]$

***Solution***

**a.** We replace $n$ by 3, 4, 5, 6 in the expression $(3n - 4)$ and obtain the results

$$\begin{aligned} &= [3(3) - 4] + [3(4) - 4] + [3(5) - 4] + [3(6) - 4] \\ &= 5 + 8 + 11 + 14 = 38 \end{aligned}$$

**b.** Observe the alternating positive and negative sign created by raising $-2$ to both even and odd powers.

$$\begin{aligned} &= [5 + (-2)^0] + [5 + (-2)^1] + [5 + (-2)^2] + [5 + (-2)^3] \\ &= [5 + 1] + [5 - 2] + [5 + 4] + [5 - 8] \\ &= 6 + 3 + 9 - 3 = 15 \quad \blacksquare \end{aligned}$$

### Arithmetic Sequences

*Definition*

An *arithmetic sequence* is a sequence in which each term is obtained by adding a constant (we will designate the constant by $d$) to the preceding term.

Note that this definition is recursive since $a_{k+1} = a_k + d$. Given the first term and $d$, we could write the first $n$ terms.

**EXAMPLE 7** Write the first five terms of an arithmetic sequence whose first term is 7 and whose constant difference $d$ is 4.

***Solution*** If the first term of an arithmetic sequence is 7 and $d = 4$, the first five terms are

$$7, 11, 15, 19, 23 \quad \blacksquare$$

If we designate the first term of an arithmetic sequence by $a$ and the constant difference by $d$, then the sequence is

$$a, \quad a + d, \quad (a + d) + d, \quad [(a + d) + d] + d, \quad \text{etc.}$$

or

$$a, \quad a + d, \quad a + 2d, \quad a + 3d, \quad \text{etc.}$$

An inspection of this sequence leads us to a formula for the $n$th term of an arithmetic sequence.

*Theorem 12-1*

If $a_1$ is the first term and $d$ is the constant difference in an arithmetic sequence, then the $n$th term is given by $a_n = a_1 + (n - 1)d$.

A formal proof of this theorem could be given by using mathematical induction.

**EXAMPLE 8** Find the tenth term of an arithmetic sequence where $a_1 = 3$ and $d = 7$.

***Solution***

$$a_n = a_1 + (n - 1)d$$

$$a_{10} = 3 + (10 - 1)7 = 66$$

The advantage of using Theorem 12-1 is obvious. Without it, we would have needed to find ten terms of the sequence

$$3, 10, 17, 24, 31, 38, 45, 52, 59, \underbrace{66}_{\text{tenth term}}, \ldots$$

We would have obtained the same answer but used a lot of unnecessary effort. ■

If you are given the arithmetic sequence of numbers and asked to find a given term, you will first need to determine $a_1$ and $d$ and then proceed to use Theorem 12-1.

**EXAMPLE 9** Find the 89th term of the sequence 7, 12, 17, 22, 27, 32, . . . .

***Solution*** This is an arithmetic sequence in which we see that the common difference between terms is 5. Therefore $d = 5$. The first term $a_1 = 7$.

Thus

$$a_{89} = a_1 + (89 - 1)d$$

$$a_{89} = 7 + (88)(5) = 7 + 440 = 447 \quad ■$$

### Sums of the First $n$ Terms of an Arithmetic Sequence

We often need to sum the terms of a given arithmetic sequence. We can develop a convenient formula to find such a sum.

***Theorem 12-2*** The sum $S_n$ of the first $n$ terms of an arithmetic series is given by the formula $S_n = \frac{n}{2}(a_1 + a_n)$ where $a_1$ is the first term and $a_n$ is the $n$th term.

We can use a slightly different formula if we do not know the last term of the sequence.

***Theorem 12-3*** The sum $S_n$ of the first $n$ terms of an arithmetic series is given by the formula $S_n = \frac{n}{2}[2a_1 + (n - 1)d]$ where $a_1$ is the first term and $d$ is the constant difference.

**EXAMPLE 10** If the first term of an arithmetic sequence is 7 and the constant difference is 4, find the sum of the first ten terms.

***Solution*** We use $S_n = \frac{n}{2}[2a_1 + (n - 1)d]$ (from Theorem 12-3) and we will substitute the values $n = 10$, $a_1 = 7$, and $d = 4$:

$$S_{10} = \frac{10}{2}[2(7) + (9)(4)] = 5(14 + 36) = 250 \quad \blacksquare$$

If you are given only the sequence, you must first determine whether it is an arithmetic sequence and then determine $a_1$ and $d$.

**EXAMPLE 11** Find the sum of the first 20 terms of the sequence $0, -\frac{2}{3}, -\frac{4}{3}, -2, -\frac{8}{3}, \ldots$.

***Solution*** We see that this is an arithmetic sequence. However, we must notice carefully that the value of $d$ is negative. Each subsequent term is *decreasing* by $\frac{2}{3}$. Therefore the value of $d = -\frac{2}{3}$. The first term is zero.

Thus we use $S_n = \frac{n}{2}[2a_1 + (n - 1)d]$ where $n = 20$, $a_1 = 0$, and $d = -\frac{2}{3}$:

$$S_{20} = \frac{20}{2}\left[2(0) + (20 - 1)\left(-\frac{2}{3}\right)\right]$$

$$= 10\left[0 + (19)\left(-\frac{2}{3}\right)\right] = 10\left(-\frac{38}{3}\right)$$

$$= -\frac{380}{3} \quad \text{or} \quad -126\frac{2}{3} \quad \blacksquare$$

An arithmetic series may be indicated by sigma notation. For example, the series 3, 5, 7, . . . , 59, 61 has 30 terms and can be written in sigma notation as $\sum_{k=1}^{30}(2k + 1)$. In general, all indicated sums of the form $\sum_{k=1}^{n}(bk + c)$ are arithmetic series and have a common difference of $b$. Do you see why this is true?

## EXERCISE 12-2

Discover a pattern in each of the sequences in Problems 1–4.

**a.** In each case, find the next two numbers of the sequence.

**b.** Write an expression for the $n$th term of the sequence.

**1.** 12, 16, 20, 24, 28, ?, ?, . . .

**2.** 50, 43, 36, 29, 22, ?, ?, . . .

**3.** $\frac{3}{4}, \frac{4}{9}, \frac{5}{16}, \frac{6}{25}, \frac{7}{36}$, ?, ?, . . .

**4.** $\frac{1}{2}, -\frac{1}{3}, \frac{1}{4}, -\frac{1}{5}, \frac{1}{6}, -\frac{1}{7}$, ?, ?, . . .

Write the first four terms and the tenth term of the sequence whose $n$th term is defined in Problems 5–10.

**5.** $a_n = 3n - 2$

**6.** $a_n = \frac{n}{2} + 1$

**7.** $a_n = n^2 + 3$

**8.** $a_n = 2n^2 - 5n + 1$

**9.** $a_n = \frac{\log n}{2n}$

**10.** $a_n = \sqrt{3n}$

Find the first four terms of each of the sequences defined recursively in Problems 11–14.

**11.** $a_1 = 3, a_{k+1} = -6 + 2a_k$

**12.** $a_1 = 2, a_{k+1} = 3a_k - 4$

**13.** $a_1 = -4, a_{k+1} = 5a_k + 1$

**14.** $a_1 = \frac{1}{2}, a_{k+1} = \frac{a_k}{2}$

Find the value of each of the expressions in Problems 15–18.

**15.** $\sum_{n=1}^{4} 5n$

**16.** $\sum_{k=3}^{5} \frac{1}{2}k$

**17.** $\sum_{i=0}^{4} (2^i + 1)$

**18.** $\sum_{k=2}^{5} \frac{k}{k-1}$

Write each of the series in Problems 19–20 in sigma notation.

**19.** $1 - 1 + 1 - 1 + 1 - 1$

**20.** $1 + 6 + 11 + 16 + 21$

**21.** Write the first six terms of the arithmetic sequence where $a_1 = 3$ and $d = 5$.

**22.** Write the first seven terms of the arithmetic sequence where $a_1 = 8$ and $d = -3$.

Find the indicated term in each of the arithmetic sequences in Problems 23–26.

**23.** $a_1 = 2, d = 5$: 11th term

**24.** $a_1 = 6, d = \frac{1}{2}$: 13th term

**25.** $a_1 = 3, d = -2$: 6th term

**26.** $a_1 = \frac{1}{2}, d = \frac{1}{3}$: 20th term

Find the sum of each of the arithmetic sequences in Problems 27–30.

**27.** $a_1 = 2, d = 4, n = 10$

**28.** $a_1 = 3, d = \frac{1}{2}, n = 9$

**29.** $a_1 = 10, d = -4, n = 8$

**30.** $a_1 = -\frac{1}{2}, d = 5, n = 10$

In Problems 31–34, calculate each sum by using the theorems of this chapter. (*Hint:* As a first step, you may want to write out the first two or three terms of the series.)

**31.** $\sum_{n=1}^{36} (3n - 4)$ **32.** $\sum_{n=1}^{100} (2n + 5)$ **33.** $\sum_{i=3}^{52} (5i + 4)$ **34.** $\sum_{i=4}^{66} (2i - 8)$

★ **35.** For what value of $k$ is the sequence $2k + 4$, $3k - 7$, $k + 12$ an arithmetic sequence?

★ **36.** If the numbers $a$, $b$, $c$, $d$, $e$ form an arithmetic sequence, then $b$, $c$, and $d$ are called *arithmetic means* between $a$ and $e$. Find three arithmetic means between $-12$ and 60.

★ **37.** A sequence of numbers is called a *harmonic progression* if their reciprocals form an arithmetic sequence. The sequence $\frac{1}{3}, \frac{1}{6}, \frac{1}{9}, \frac{1}{12}$ is a harmonic progression, since their reciprocals 3, 6, 9, 12 form an arithmetic sequence. Is the sequence $\frac{6}{7}, \frac{3}{5}, \frac{6}{13}, \frac{3}{8}$ a harmonic progression?

★ **38.** The *harmonic mean* of two numbers is the reciprocal of the arithmetic mean of the reciprocals of the two numbers. Find the harmonic mean of $\frac{1}{5}$ and $\frac{1}{8}$.

---

## 12-3 GEOMETRIC SEQUENCES

Another sequence of special interest is the geometric sequence. Consider the sequence 1, 2, 4, 8, 16, 32, 64, . . . . We observe that multiplying each term by 2 creates the value of the next term. This type of sequence is called a geometric sequence.

*Definition*

> A *geometric sequence* is a sequence in which each term is obtained by multiplying the preceding term by a constant ratio (designated by $r$).

In the sequence 1, 2, 4, 8, 16, 32, 64, . . . we saw that the common ratio was 2. Let us examine the first five terms:

$$a_1 = 1 \qquad a_4 = 8 = a_1(2)^3$$
$$a_2 = 2 = a_1(2) \qquad a_5 = 16 = a_1(2)^4$$
$$a_3 = 4 = a_1(2)^2$$

Note that the definition above is a recursive definition. If we denote the first term as $a_1$ and the constant ratio as $r$, we obtain the expansion of the geometric sequence as

$$a_1, a_1r, a_1r^2, a_1r^3, \ldots, a_1r^{n-1}$$

*Theorem 12-4*

The formula for the $n$th term of a geometric sequence having a first term $a_1$ and a constant ratio $r$ is $a_n = a_1 r^{n-1}$.

EXAMPLE 1 Write the first four terms and the seventh term of a geometric sequence if the first term is 3 and the constant ratio is 2.

***Solution*** We desire to find

$$a_1, \quad a_1r, \quad a_1r^2, \quad a_1r^3$$

when $a_1 = 3$ and $r = 2$.

This gives us 3, 3(2), $3(2)^2$, and $3(2)^3$; thus the first four terms are 3, 6, 12, 24. Using $a_n = a_1r^{n-1}$ when $n = 7$, $a_1 = 3$, and $r = 2$, we find that the seventh term is

$$a_7 = 3(2)^6 = 3(64) = 192 \quad ■$$

The value of $r$ may be negative.

EXAMPLE 2 Find the tenth term of the geometric sequence 3, $-6$, 12, $-24$, 48, . . . .

***Solution*** We see that if we multiply each term by $-2$, we obtain the next term. Thus $r = -2$.

To obtain the tenth term, we use $a_n = a_1r^{n-1}$ where $n = 10$, $a_1 = 3$, and $r = -2$:

$$a_{10} = (3)(-2)^{10-1} = (3)(-2)^9 = (3)(-512) = -1536 \quad ■$$

*Theorem 12-5*

Given a geometric sequence with a first term of $a_1$ and a common ratio $r \neq 1$, the sum of the first $n$ terms of the geometric sequence is

$$S_n = \frac{a_1 - a_1r^n}{1 - r}$$

This formula is useful if we do not know the $a_n$ term. If we already have $a_n$, then another formula is more helpful.

*Theorem 12-6*

> Given a geometric sequence with a first term of $a_1$, an $n$th term of $a_n$, and a common ratio $r \neq 1$, the sum of the first $n$ terms of the geometric sequence is
>
> $$S_n = \frac{a_1 - a_n r}{1 - r}$$

**EXAMPLE 3** Find the sum of the first five terms of the geometric sequence in which $a_1 = 2$ and $r = 3$.

***Solution*** Since we do not know $a_5$, we use Theorem 12-5.

$$S_n = \frac{a_1 - a_1 r^n}{1 - r} \qquad \text{where } a_1 = 2,\ n = 5, \text{ and } r = 3$$

$$S_5 = \frac{2 - 2(3)^5}{1 - 3} = \frac{2 - 2(243)}{-2} = \frac{2 - 486}{-2} = \frac{-484}{-2} = 242$$

The sum of the first five terms is 242. ■

**EXAMPLE 4** Find the sixth term and then find the sum of the first six terms of the geometric sequence in which $a_1 = \frac{2}{3}$ and $r = \frac{1}{2}$.

***Solution*** First we find $a_6$ using Theorem 12-4:

$$a_6 = a_1 r^{(6-1)}$$

$$a_6 = \frac{2}{3}\left(\frac{1}{2}\right)^5 = \frac{2}{3}\left(\frac{1}{32}\right) = \frac{1}{48}$$

Now we find $S_6$ using Theorem 12-6.

$$S_6 = \frac{a_1 - a_6 r}{1 - r}$$

$$= \frac{\frac{2}{3} - \left(\frac{1}{48}\right)\left(\frac{1}{2}\right)}{1 - \frac{1}{2}} = \frac{\frac{2}{3} - \frac{1}{96}}{\frac{1}{2}} = \frac{21}{16}$$

■

*Definition*

> If the numbers $a_1, a_2, a_3, \ldots, a_n$ form a geometric sequence, the numbers $a_2 \cdots a_{n-1}$ are called *geometric means* between $a_1$ and $a_n$.

**EXAMPLE 5** Insert three geometric means between 3 and 48. Find all possible answers.

***Solution*** We are given that $a_1 = 3$ and $a_5 = 48$. We know that

$$a_5 = a_1 r^4$$

so

$$48 = 3r^4$$

$$r^4 = 16$$

$$r = \pm 2$$ *Notice that we must consider both sign possibilities.*

Hence the sequence is

$$3, 6, 12, 24, 48$$

or

$$3, -6, 12, -24, 48$$ ■

### Infinite Geometric Sequences

The formula for $S_n$ gives us the sum of the first $n$ terms in a geometric sequence. The question arises, "Can we find the sum of the terms in an infinite geometric sequence?" The answer is "sometimes," and the determining factor is the size of the constant ratio $r$.

*Theorem 12-7* The sum $S$ of an infinite geometric sequence $a_1, a_1 r, a_1 r^2, \ldots$ with $|r| < 1$ is given by the formula

$$S = \frac{a_1}{1 - r}$$

**EXAMPLE 6** Find the sum of the infinite geometric sequence $125, -25, 5, -1, \frac{1}{5}, \ldots$.

***Solution*** We see that $a_1 = 125$ and $r = -\frac{1}{5}$. Thus by Theorem 12-7,

$$S = \frac{a_1}{1 - r}$$

$$S = \frac{125}{1 - \left(-\frac{1}{5}\right)} = \frac{125}{\frac{5}{5} + \frac{1}{5}} = \frac{125}{\frac{6}{5}}$$

$$S = \frac{625}{6} \quad \text{or} \quad 104\frac{1}{6}$$ ■

## EXERCISE 12-3

In each of the sequences in Problems 1–8, indicate if the sequence is geometric or if it is not. If it is geometric, find the value of the constant ratio $r$.

**1.** $108, 36, 12, 4, \ldots$

**2.** $8, -24, 72, -216, \ldots$

**3.** $\sqrt{7}, -7, 7\sqrt{7}, -49, \ldots$

**4.** $25, 5\sqrt{5}, 5, \sqrt{5}, \ldots$

**5.** $\frac{1}{3}, \frac{1}{9}, \frac{1}{81}, \frac{1}{243}, \ldots$

**6.** $\frac{2}{5}, \frac{2}{25}, \frac{1}{125}, \frac{1}{625}, \ldots$

**7.** $0.12, 0.024, 0.0048, 0.00096, \ldots$

**8.** $1.12, (1.12)^2, (1.12)^3, (1.12)^4, \ldots$

**9.** Write the first five terms of a geometric sequence where the first term is 2 and the constant ratio is 3.

**10.** Write the first four terms of a geometric sequence where the first term is 10 and the constant ratio is $-2$.

Find the indicated term in each of the geometric sequences in Problems 11–14.

**11.** $a_1 = 3, r = 5$: 6th term

**12.** $a_1 = 8, r = \frac{3}{2}$: 5th term

**13.** $a_1 = -3, r = -2$: 8th term

**14.** $a_1 = 4, r = -\frac{1}{2}$: 6th term

Find the sum of the first seven terms of the geometric sequences in Problems 15–18.

**15.** $\frac{4}{5}, \frac{1}{5}, \frac{1}{20}, \frac{1}{80}, \ldots$

**16.** $\frac{1}{3}, \frac{1}{6}, \frac{1}{12}, \frac{1}{24}, \ldots$

**17.** $4, -16, 64, -256, \ldots$

**18.** $27, -9, 3, -1, \ldots$

Find the indicated sums in Problems 19–22 by using one of the theorems of this section.

**19.** $\frac{3}{4} + \frac{3}{8} + \frac{3}{16} + \frac{3}{32} + \frac{3}{64}$

**20.** $4 + \frac{4}{5} + \frac{4}{25} + \frac{4}{125} + \frac{4}{625}$

**21.** $12 - 4 + \frac{4}{3} - \frac{4}{9} + \frac{4}{27} - \frac{4}{81}$

**22.** $-147 + 21 - 3 + \frac{3}{7} - \frac{3}{49}$

Each of the Problems 23–28 refers to geometric sequences.

**23.** If $a_1 = 1$, $r = -2$, and $S_n = -21$, find $n$.

**24.** If $a_7 = 192$ and $r = 2$, find $S_7$.

**25.** If $a_1 = 4$ and $S_4 = -80$, find $r$.

**26.** If $r = -3$ and $S_7 = 2188$, find $a_7$.

**27.** If $a_3 = 3$ and $a_6 = -81$, find $a_8$.

**28.** If $a_2 = -6$ and $a_5 = \frac{3}{4}$, find $a_7$.

**29.** If log 3 is the first term of a geometric sequence and log 9 is the second term, find the third and fourth terms.

**30.** For what value of $k$ is the sequence $k - 2, k - 6, 2k + 3$ a geometric sequence?

**31.** Insert four geometric means between 3 and 96.

**32.** Insert three geometric means between 16 and 625.

Find the sum, if it exists, of each infinite geometric series in Problems 33–40.

**33.** $4 + 2 + 1 + \frac{1}{2} + \cdots$

**34.** $1 - \frac{1}{2} + \frac{1}{4} - \frac{1}{8} + \cdots$

**35.** $\sqrt{3} + 3 + \sqrt{27} + 9 + \cdots$

**36.** $15 - 5 + \frac{5}{3} - \frac{5}{9} + \cdots$

**37.** $4 + 2\sqrt{2} + 2 + \sqrt{2} + \cdots$

**38.** $1 + 0.5 + 0.25 + 0.125 + \cdots$

**39.** $7 - 3 + \frac{9}{7} - \frac{27}{49} + \cdots$

**40.** $4 + \frac{8}{5} + \frac{16}{25} + \frac{32}{125} + \cdots$

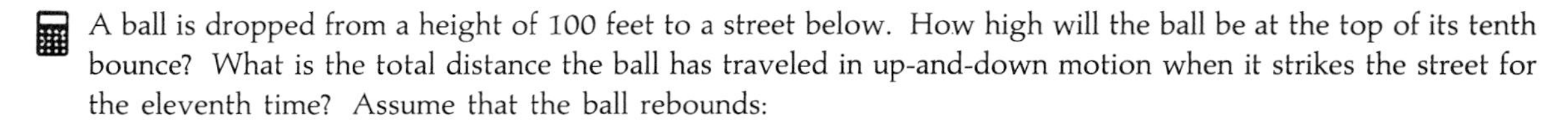

A ball is dropped from a height of 100 feet to a street below. How high will the ball be at the top of its tenth bounce? What is the total distance the ball has traveled in up-and-down motion when it strikes the street for the eleventh time? Assume that the ball rebounds:

★ **41.** to $\frac{3}{5}$ of its original height at each bounce.

★ **42.** to $\frac{1}{3}$ of its original height at each bounce.

---

## 12-4 PERMUTATIONS AND COMBINATIONS

### Permutations

If a succession of acts can occur in more than one way, or if a set of objects can be arranged in more than one way, there is often a need to know the total number of ways these events can occur.

***Theorem 12-8 The Fundamental Counting Principle***

If an act can occur in $r$ ways and, after it has occurred, a second act can occur in $s$ ways, and thereafter a third act can occur in $t$ ways, and so forth, until the last act can occur in $k$ ways, then the successive acts can occur in $(r)(s)(t) \ldots (k)$ ways.

EXAMPLE 1 An apartment building has four entrance doors on the first floor. It has five different stairways to get from the first floor to the second floor. It has four different stairways that lead from the second floor to the third floor. In how many ways can an occupant enter the apartment building and ascend to the third floor?

***Solution*** By Theorem 12-8 we have $4 \cdot 5 \cdot 4 = 80$ different ways. ■

Often we are concerned with the arrangement of a set of objects in *some specified order*. In these cases we will be concerned with a concept in mathematics called *permutations*.

*Definition* The arrangement of a group of objects in a definite order is called a *permutation* of the objects.

EXAMPLE 2 How many permutations are possible with the three letters $a$, $b$, and $c$?

***Solution*** We can solve the problem by listing the arrangements.

$$a, b, c \qquad b, a, c \qquad c, a, b$$
$$a, c, b \qquad b, c, a \qquad c, b, a$$

We see that there are six permutations. ■

In order to provide consistency for future formulas involving factorial notation, we make the following definitions.

*Definition* If $n$ is a positive integer, then the symbol $n!$ (read "$n$ factorial") is defined as $n(n-1)(n-2)\ldots(3)(2)(1)$. If $n = 0$ then $0! = 1$.

EXAMPLE 3 Evaluate the following.

**a.** $4!$ **b.** $8!$

***Solution***

**a.** $4! = (4)(3)(2)(1) = 24$

**b.** $8! = (8)(7)(6)(5)\underbrace{(4)(3)(2)(1)}_{4!} = (8)(7)(6)(5)(4!) = 40{,}320$ ■

On most scientific calculators there is a button labeled $\boxed{x!}$. (Factorial tables are included in books of mathematical tables located in libraries for those without a calculator.)

EXAMPLE 4 Evaluate the following on a calculator.

**a.** $11!$ **b.** $29!$

***Solution***

**a.** On a calculator we enter 11 [x!] 39916800

**b.** On a calculator we enter 29 [x!] 8.84176199 30 ■

In this case the answer is not exact. The value $8.84176199 \times 10^{30}$ is an approximation which is accurate to nine significant digits. ■

***Theorem 12-9*** The number of permutations of $n$ things taken all at a time is equal to $n!$.

Thus if we had a group of eleven different people, the numbers of different ways they could be placed in a line is 11!. From part a of Example 4 we see that they can be arranged in a line 39,916,800 different ways.

The permutation of $n$ things *taken less than n at a time* would of course be less than $n!$.

***Theorem 12-10*** The number of permutations of $n$ things taken $r$ at a time is

$$_nP_r = \frac{n!}{(n-r)!}$$

**EXAMPLE 5** In how many ways can 12 objects be arranged if we use three at a time?

***Solution*** We will use Theorem 12-10:

$$_{12}P_3 = \frac{12!}{(12-3)!} = \frac{12!}{9!} = \frac{(12)(11)(10)(9)(8)(7)(6)(5)(4)(3)(2)(1)}{(9)(8)(7)(6)(5)(4)(3)(2)(1)}$$
$$= (12)(11)(10) = 1320$$

Do you see that we have written out some unnecessary work in Step 3 of our solution? We could have shortened the work somewhat by writing $\frac{12!}{9!}$ as $\frac{(12)(11)(10)(9!)}{9!}$ and then reducing to $(12)(11)(10) = 1320$. ■

The power of a scientific calculator here is most helpful. We will illustrate this in the following example.

**EXAMPLE 6** A college club has 20 members. In how many ways can the club elect a president, vice-president, secretary, and treasurer?

***Solution*** We will use Theorem 12-10 for 20 things taken four at a time.

$$
\begin{aligned}
{}_{20}P_4 &= \frac{20!}{(20-4)!} = \frac{20!}{16!} \\
&= \frac{(20)(19)(18)(17)(16!)}{16!} \\
&= (20)(19)(18)(17) = 116{,}280
\end{aligned}
$$

$\frac{20!}{16!}$ can be evaluated directly on a scientific calculator

20 [x!] [÷] 16 [x!] [=] 116,280 ■

### Combinations

In many instances we are interested in a collection of items, but the order in which they are arranged is not important. In such cases we are dealing with *combinations*.

*Definition*

A set of elements in which order is not considered is called a *combination*.

The context of a problem will distinguish between permutations and combinations. For instance, if we ask, "How many four-digit numbers can be written using the digits $\{1, 2, 3, 4, 5, 6\}$ without repeating any one digit in a given number?", we have a permutation problem, since the different order of the digits would give a different number. If we ask, "How many different four-member committees can be selected from six people?", we have a combination problem, since the order of selection would not make a different committee.

*Theorem 12-11*

The number of combinations of $n$ things taken $r$ at a time is denoted by $\binom{n}{r}$ or by ${}_nC_r$ and is obtained by the formula

$$\binom{n}{r} = \frac{n!}{(n-r)!\,r!}$$

**EXAMPLE 7** How many four-member committees can be formed from a group of nine people?

***Solution***

$$\binom{9}{4} = \frac{9!}{(9-4)!4!} = \frac{9!}{5!4!}$$

$$= \frac{(9)(8)(7)(6)(5)(4)(3)(2)(1)}{(5)(4)(3)(2)(1)(4)(3)(2)(1)} = 126 \quad ■$$

In some applied situations, both our knowledge of the fundamental counting principle and the use of combinations is necessary to solve the problem.

**EXAMPLE 8** A seven-person homecoming committee is to be selected from the student senate. The 14 members on the senate are classified as five seniors, four juniors, three sophomores, and two freshmen. In how many ways can a homecoming committee be selected if it must have three seniors, two juniors, one sophomore, and one freshman?

***Solution*** The choice of three seniors out of five possible seniors can be made in $\binom{5}{3}$ ways. In similar fashion the choice of two juniors can be made in $\binom{4}{2}$ ways. The choice of one sophomore can be made in $\binom{3}{1}$ ways, and the choice of one freshman can be made in $\binom{2}{1}$ ways. Since we must have three seniors *and* two juniors *and* one sophomore *and* one freshman, we use fundamental counting principle. The selection can be made in $\binom{5}{3}\binom{4}{2}\binom{3}{1}\binom{2}{1}$ ways.

Evaluating each combination gives

$$\binom{5}{3} = \frac{5!}{3!2!} = \frac{(5)(4)}{(2)(1)} = 10 \qquad \binom{4}{2} = \frac{4!}{2!2!} = \frac{(4)(3)}{(2)(1)} = 6$$

$$\binom{3}{1} = \frac{3!}{2!1!} = \frac{3}{1} = 3 \qquad \binom{2}{1} = \frac{2!}{1!1!} = 2$$

Thus the selection can be made in $(10)(6)(3)(2) = 360$ ways. ■

It is worthwhile to see how combination problems would be done using a scientific calculator.

**EXAMPLE 9** A quality control inspector must carefully examine in detail a sample of 19 parts of a total of 58 parts that are manufactured in one section of a factory each hour. How many different combinations of parts could be selected for this detailed examination?

***Solution*** The inspector must select 19 out of the 58 parts manufactured each hour. The number of combinations of 58 things taken 19 at a time is

$$\binom{58}{19} = \frac{58!}{19!(58-19)!} = \frac{58!}{19!39!}$$

These are huge numbers, and some slight roundoff error will be incurred when we perform multiplications and divisions of these on the calculator.

The problem can be done directly on the calculator as follows, using parentheses:

58 [x!] [÷] [(] 19 [x!] [×] 39 [x!] [)] [=] 9.473094926 14

Thus we could say that the inspector could inspect any one of 947,309,493,000,000 combinations of parts when the 19 parts are selected. (This answer is an approximate value rounded to the nearest million.) ■

## EXERCISE 12-4

A calculator may be useful in some of the following problems. Evaluate the expressions in Problems 1–8.

**1.** $5!$ **2.** $10!$ **3.** $(4!)(3!)$ **4.** $\dfrac{20!}{18!}$

**5.** ${}_6P_4$ **6.** ${}_8P_3$ **7.** ${}_{10}P_5$ **8.** ${}_{12}P_4$

Write each of Problems 9–12 in factorial notation. Express your answer in a simplified form whenever possible.

**9.** $(k+2)(k+1)(k!)$ **10.** $(k-3)(k-4)(k-5)[(k-6)!]$

**11.** $(9)(9!) + 9!$ **12.** $8! + (8)(8!)$

**13.** In how many ways can the letters in the word "triangle" be arranged?

**14.** How many five-digit numerals can be formed from the digits $\{2, 3, 4, 5, 6\}$ if no digit is repeated?

**15.** How many three-digit numerals can be formed from the digits $\{1, 2, 3, 4, 5, 6\}$ if no digit is repeated?

**16.** In how many ways can six people be arranged in a straight line?

**17.** In how many ways can the letters of the alphabet be arranged in a sequence of five letters if no letter is repeated?

**18.** A disc jockey wishes to play 11 records on a radio program. How many different orders of playing them are there?

A calculator may be useful in some of the following problems. Evaluate the expressions in Problems 19–24.

**19.** $\binom{10}{3}$ **20.** $\binom{10}{7}$ **21.** $\binom{12}{8}$ **22.** $\binom{12}{4}$ **23.** $\binom{22}{19}$ **24.** $\binom{36}{34}$

**25.** Show that $\binom{n}{r} = \binom{n}{n-r}$ **26.** Show that $\binom{n}{n-1} = n.$

**27.** Show that $\binom{n}{0} = 1.$ **28.** Show that $\binom{n}{n} = 1.$

**29.** How many four-member committees can be made from a group of 10 people?

**30.** How many three-member committees can be formed from a group of 12 people?

**31.** A man has a penny, a nickel, a dime, a quarter, and a half-dollar in his pocket, from which he randomly draws three coins. How many ways can he do this?

**32.** A student must select five classes from a schedule of eight. How many ways are there of making the selection?

**33.** A woman has found eight art prints that she likes but can only buy three. How many ways could she select them?

**34.** A district office of a national company is requested to transfer three of its salespeople to the head office. If they have ten salespeople to choose from, in how many ways could they make the selection?

★ **35.** A state house of representatives redistricting committee is to be appointed. It will have three Republicans, three Democrats, and one Independent member. The state currently has 39 members in the House of Representatives. Of these, 31 are Democrats, five are Republicans, and three are Independents. In how many ways can the committee be selected?

★ **36.** There are 12 teams in a town basketball league. How many games will be played by the 12 teams if each team plays each of the other teams just once?

## 12-5 THE BINOMIAL THEOREM

Raising a binomial to a power occurs in many phases of mathematics, and the coefficients of this expansion have a special place in the study of statistics. In this section we will discuss and use the theorem for the binomial expansion.

***Theorem 12-12 (Binomial Theorem)***

For any natural number $n$,

$$(x + y)^n = \sum_{r=0}^{n} \binom{n}{r} x^{n-r} y^r$$

EXAMPLE 1 Expand $(x + y)^5$ by using the binomial theorem.

***Solution*** We will use Theorem 12-12 with $n = 5$. Thus we have

$$(x + y)^5 = \sum_{r=0}^{5} \binom{5}{r} x^{5-r}y^r$$

$$= \binom{5}{0} x^5 + \binom{5}{1} x^4y + \binom{5}{2} x^3y^2 + \binom{5}{3} x^2y^3 + \binom{5}{4} xy^4 + \binom{5}{5} y^5$$

Since $\binom{5}{0}$ and $\binom{5}{5} = 1$ and since $\binom{5}{1}$ and $\binom{5}{4} = 5$, we need to find only

$$\binom{5}{3} = \frac{5!}{2!3!} = \frac{(5)(4)}{(2)(1)} = 10 \text{ and similarly } \binom{5}{2} = \frac{5!}{3!2!} = 10$$

Thus we have

$$(x + y)^5 = x^5 + 5x^4y + 10x^3y^2 + 10x^2y^3 + 5xy^4 + y^5 \quad \blacksquare$$

The great advantage of the binomial theorem is that we can use it to write just one given term or a small number of terms of a long binomial expansion.

**EXAMPLE 2** Find the eighth term of the expansion of $(x + y)^{12}$.

***Solution*** Here we have $n = 12$. Note that in the eighth term, $r = 7$. Thus the eighth term is $\binom{12}{7} x^5y^7$. Since

$$\binom{12}{7} = \frac{12!}{5!7!} = \frac{(\overset{6}{\cancel{12}})(11)(\overset{2}{\cancel{10}})(\overset{3}{\cancel{9}})(\overset{2}{\cancel{8}})}{(\underset{1}{\cancel{5}})(\underset{1}{\cancel{4}})(\underset{1}{\cancel{3}})(\underset{1}{\cancel{2}})(1)}$$

$$= (6)(11)(2)(3)(2) = 792$$

On a scientific calculator we would evaluate this quickly by entering

12 [x!] [÷] 5 [x!] [÷] 7 [x!] [=] 792

We know that the eighth term is $792x^5y^7$. $\blacksquare$

**EXAMPLE 3** Find the third term of $(2x + 3)^8$.

***Solution*** First we write $(2x + 3)^8 = [(2x) + 3]^8$. Note that in the third term, $r = 2$. Thus the third term is

$$\binom{8}{2} (2x)^6(3)^2 = 28(64x^6)(9)$$

$$= 16{,}128x^6 \quad \blacksquare$$

### Pascal's Triangle

A second method of determining the coefficients of the terms of an expansion of a binomial uses an array of numbers called *Pascal's triangle.* This triangular shaped array was named after the famous French mathematician Blaise Pascal (1623–1662), who studied such arrays and wrote a paper on these properties. The pattern of this array is indicated below.

| *Binomial* | *Coefficients of the Binomial* |
|---|---|
| $(a + b)^0$ | 1 |
| $(a + b)^1$ | 1 1 |
| $(a + b)^2$ | 1 2 1 |
| $(a + b)^3$ | 1 3 3 1 |
| $(a + b)^4$ | 1 4 6 4 1 |
| $(a + b)^5$ | 1 5 10 10 5 1 |

This pattern may be continued indefinitely. Notice that each row in the array begins and ends with a 1. The second number in any row is obtained by adding the first and second numbers in the preceding row. The third number is obtained by adding the second and third numbers in the preceding row, and so on.

The numbers in the $n$th row of the triangle are the coefficients of the binomial expansion of $(a + b)^{n-1}$. Notice from the triangle, for example, that in the fifth row the coefficients of the expansion $(a + b)^4$ are 1, 4, 6, 4, 1. Thus

$$(a + b)^4 = a^4 + 4a^3b + 6a^2b^2 + 4ab^3 + b^4$$

**EXAMPLE 4** Using Pascal's triangle, expand $(a + b)^7$.

***Solution*** First we need to write out eight rows of Pascal's triangle.

| *Binomial* | *Coefficients of the Binomial* |
|---|---|
| $(a + b)^0$ | 1 |
| $(a + b)^1$ | 1 1 |
| $(a + b)^2$ | 1 2 1 |
| $(a + b)^3$ | 1 3 3 1 |
| $(a + b)^4$ | 1 4 6 4 1 |
| $(a + b)^5$ | 1 5 10 10 5 1 |
| $(a + b)^6$ | 1 6 15 20 15 6 1 |
| $(a + b)^7$ | 1 7 21 35 35 21 7 1 |

Thus

$$(a+b)^7 = a^7 + 7a^6b + 21a^5b^2 + 35a^4b^3 + 35a^3b^4 + 21a^2b^5 + 7ab^6 + b^7$$

■

In some cases the initial exponents of $a$, $b$ in the expression $(a+b)^n$ are not one. You will need to work with exponents carefully in such cases.

**EXAMPLE 5** Using Pascal's triangle, expand $\left(3x^2 - \frac{2}{y}\right)^5$.

***Solution*** From row 6 of the triangle we know the coefficients are 1, 5, 10, 10, 5, 1. We rewrite $\left(3x^2 - \frac{2}{y}\right)^5$ as $[3x^2 + (-2y^{-1})]^5$. Then we have

$$\begin{aligned}(3x^2 - 2y^{-1})^5 &= (3x^2)^5 + 5(3x^2)^4(-2y^{-1})^1 + 10(3x^2)^3(-2y^{-1})^2 \\ &\quad + 10(3x^2)^2(-2y^{-1})^3 + 5(3x^2)(-2y^{-1})^4 + (-2y^{-1})^5 \\ &= 243x^{10} + 5(81x^8)(-2y^{-1}) + 10(27x^6)(4y^{-2}) \\ &\quad + 10(9x^4)(-8y^{-3}) + 5(3x^2)(16y^{-4}) + (-32y^{-5}) \\ &= 243x^{10} - 810x^8y^{-1} + 1080x^6y^{-2} - 720x^4y^{-3} \\ &\quad + 240x^2y^{-4} - 32y^{-5}\end{aligned}$$

■

## EXERCISE 12-5

Expand the expressions in Problems 1–5 by the binomial theorem.

**1.** $(x+2)^4$ **2.** $(1+2x)^6$ **3.** $(x-y)^6$ **4.** $(2x^2-y)^5$ **5.** $\left(2a + \frac{b}{2}\right)^7$

Use the binomial theorem to find the indicated term or terms in Problems 6–10.

**6.** $(x+y)^{30}$, first four terms **7.** $(a-3b)^{100}$, first four terms **8.** $\left(x - \frac{2}{x}\right)^{10}$, first four terms

**9.** $(2a^2-b^2)^7$, fourth term **10.** $(2x-\sqrt{y})^9$, seventh term

Continue to find the binomial coefficients by extending Pascal's triangle to find the expressions in Problems 11 and 12.

★**11.** $(x-2y)^8$ ★**12.** $(2x+y)^9$

By any method discussed in this section, find the expressions in Problems 13–20.

**13.** $\left(3x - \frac{2}{y}\right)^4$ **14.** $\left(5x + \frac{2}{y^2}\right)^3$ **15.** $(2\sqrt{x}+y^3)^5$ **16.** $(x^2 - 2\sqrt{y})^4$

**17.** The term involving $x^6$ in $(3x-y)^8$. **18.** The term involving $y^5$ in $(2x+3y)^7$.

**19.** The term involving $y^3$ in $(4x^2+y)^9$. **20.** The term involving $x^8$ in $(3x^2-y)^8$.

## 12-6 AN INTRODUCTION TO PROBABILITY (OPTIONAL)

*Definition*

If an event can occur in any one of $n$ equally likely ways and if $m$ of these ways are considered favorable, the *probability* of a favorable event is the ratio $\frac{m}{n}$. The notation is $P(E) = \frac{m}{n}$.

Suppose we designate $A$ as an event; then $P(A)$ is read as "the probability of $A$."

Our definition states

$$P(\text{favorable outcome}) = \frac{\text{number of favorable outcomes}}{\text{total number of outcomes}}$$

Suppose there are no favorable outcomes possible; then the ratio would be $\frac{0}{n}$ or 0. Also, if every outcome is favorable, then the ratio would be $\frac{n}{n}$ or 1. It is obvious that the numerator of the fraction cannot be larger than the denominator. This establishes that the probability of an event is never less than 0 nor greater than 1, that is,

$$0 \leq P(A) \leq 1$$

EXAMPLE 1 Consider a drawing for a prize in which there are ten cards with one name on each card. We want to know the probability that Jay's name will be drawn to receive the prize.

**a.** Find the probability that Jay wins if his name is on one of the cards.
**b.** Find the probability that Jay wins if his name is on three of the cards.

***Solution***

**a.** If Jay's name is on only one of the cards, then the chance of Jay's winning the prize is one in ten or

$$P(\text{Jay wins}) = \frac{1}{10}$$

**b.** If Jay's name is on three of the cards, then

$$P(\text{Jay wins}) = \frac{3}{10}$$ ■

EXAMPLE 2 A card is drawn at random from a bridge deck. What is the probability that the card is the ace of spades?

***Solution*** There are 52 cards in a bridge deck, and only one of them is the ace of spades. Hence

$$P(\text{ace of spades}) = \frac{1}{52}$$ ■

We sometimes want to know the probability that something will not occur.

*Definition*

If the probability that an event will occur is $P(A)$, then the probability that it will *not occur* is $1 - P(A)$. The notation is written as

$$P(\bar{A}) = 1 - P(A).$$

You will need to be clear about the notation $\bar{A}$. We use the notation $\bar{A}$ to designate "not $A$." Note that

$$P(A) + P(\bar{A}) = 1$$

and also

$$P(A) = 1 - P(\bar{A})$$

These relationships are used in later formulas.

EXAMPLE 3 A golf bag contains three red tees, five blue tees, and seven white tees.

**a.** What is the probability that a tee drawn at random will be blue?

**b.** What is the probability that it will not be red?

***Solution***

**a.** Since the bag contains a total of 15 tees and five are blue, the ratio of favorable outcomes (blue tee is drawn) to the total possible outcomes is $\frac{5}{15}$ or $\frac{1}{3}$. Thus $P(\text{blue}) = \frac{1}{3}$.

**b.** The probability of *not obtaining* a red tee is $1 - P(\text{red})$. First we determine the probability of obtaining red:

$$P(\text{red}) = \frac{3}{15} = \frac{1}{5}$$

Thus the probability of not obtaining a red is

$$P(\overline{\text{red}}) = 1 - P(\text{red}) = 1 - \frac{1}{5} = \frac{4}{5}$$ ■

### Independent Events

We must have an understanding of independent events. Some events have an effect on other events, and finding the probability of these is more difficult. In this next section we will consider events that do not affect the outcome of other events and thus are independent.

*Definition*

Two events are *independent* if the outcome of one event does not affect the probability that the other event will take place.

A good example of independent events is tossing a coin several times in succession. The likelihood of getting a head on the second try is not affected by the outcome of the first toss. If the outcome of the first event changed the probability of the second event occurring, then the events would be **dependent**. If tossing the coin the first time bent the coin so that it was more likely to turn up heads, then the second toss and the first toss would be dependent events.

EXAMPLE 4 What is the probability of tossing two heads in succession with a single coin? (Assume that the events are independent.)

***Solution*** If we examine this problem by listing the possibilities, we find that there are four. We can denote the four outcomes as follows:

HH HT TH TT

Out of the four possible outcomes there is one favorable event. Hence by the definition of probability,

$$P(\text{two heads}) = \frac{1}{4}$$ ■

Observe the following interesting property: The probability of obtaining a head on the first toss is $\frac{1}{2}$; the probability of obtaining a head on the second toss is $\frac{1}{2}$; and $\frac{1}{2} \times \frac{1}{2} = \frac{1}{4}$, which is the probability of two heads.

This gives us some intuitive support for an important theorem in probability. We will state the theorem but will not prove it here.

*Theorem 12-13*

If $A$ and $B$ are independent events, then $P(A \text{ and } B) = P(A) \times P(B)$.

EXAMPLE 5 In a small city fire department, the probability that a given fire truck will start when firefighters need it is 0.99. The North Side station has two trucks. A fire alarm is sounded, and both trucks are needed. What is the probability that both will start?

***Solution***

$$P(\text{truck 1 starts and truck 2 starts}) = P(\text{truck 1 starts}) \times P(\text{truck 2 starts})$$
$$= (0.99)(0.99) = 0.9801$$

Thus the probability that both trucks will start is 0.9801. ■

### Dependent Events

When the outcome of one event changes the likelihood that another event happens, the probability is calculated in a somewhat different fashion.

*Definition*

> Two events are *dependent* if the outcome of one event affects the probability that the other event will take place.

For example, suppose a card is drawn from a bridge deck, and then a second card is drawn without replacing the first card. If we wish to determine the probability that both cards are aces, we proceed as follows.

There are two events to consider.

**1.** The probability that the first card drawn is an ace.

$$P(\text{first card is an ace}) = \frac{4}{52} = \frac{1}{13}$$

**2.** Assuming that the first card is an ace (if it is not an ace, there is no need to consider the second card), we need to determine the probability that the second card is an ace. We have 51 cards remaining, and 3 of them are aces. Therefore

$$P(\text{second card is an ace}) = \frac{3}{51} = \frac{1}{17}$$

Thus the probability that both cards are aces is

$$P(\text{first is an ace } \textit{and} \text{ second is an ace}) = \frac{1}{13} \times \frac{1}{17} = \frac{1}{221}$$

This illustrates the following theorem for dependent events.

*Theorem 12-14*

> The probability of two dependent events $A$ and $B$ occurring one after the other is the product of the probability of $A$ and the probability of $B$, assuming $A$ has occurred.
>
> $$P(A \text{ and } B) = P(A) \times P(B, \text{ assuming } A \text{ has occurred})$$

**EXAMPLE 6** A bag contains five red marbles, seven green marbles, and six blue marbles. If two marbles are drawn, what is the probability of both being green?

***Solution*** $P(\text{first green}) = \frac{7}{18}$

We assume that the first one chosen is green. There are six green marbles remaining in a bag that now contains 17 marbles. Thus

$$P(\text{second green, assuming first is green}) = \frac{6}{17}$$

$$P(\text{both green}) = \frac{7}{18} \times \frac{6}{17} = \frac{7}{51} \qquad \textit{By Theorem 12-14.}$$ ■

### Mutually Exclusive Events

*Definition*

If two events cannot occur at the same time, they are said to be *mutually exclusive.*

*Theorem 12-15*

If $A$ and $B$ are mutually exclusive events, then the probability that $A$ or $B$ occurs is $P(A \text{ or } B) = P(A) + P(B)$.

**EXAMPLE 7** The *Daily Journal* did a survey of cars in the city. The likelihood of a given compact car in Jonesville being a German car is 0.21. The likelihood of a compact car being a Japanese car is 0.38. What is the likelihood that any given compact car in Jonesville is manufactured in either Japan or Germany?

***Solution*** We use Theorem 12-15:

$$P(J \text{ or } G) = P(J) + P(G) = 0.38 + 0.21 = 0.59$$ ■

In logic, ($A$ or $B$) is true in three cases: if $A$ is true; if $B$ is true; if both $A$ and $B$ are true. Since this definition of "or" allows both $A$ and $B$ to be true, it is important that we distinguish mutually exclusive events from those that can both happen at the same time.

If we wish to compute the probability of ($A$ or $B$) when $A$ and $B$ are not mutually exclusive, we must be careful not to count an event twice.

*Theorem 12-16*

> If $A$ and $B$ are not mutually exclusive, the probability that $A$ or $B$ occurs is $P(A \text{ or } B) = P(A) + P(B) - P(A \text{ and } B)$.

Notice that this formula will also hold for mutually exclusive events, since we know that $P(A \text{ and } B)$ is zero in that case.

EXAMPLE 8 A die is rolled and three coins are tossed. What is the probability of obtaining a 4 or obtaining three heads?

***Solution*** $P(4) = \frac{1}{6}$ and $P(3 \text{ heads}) = \frac{1}{8}$ (Do you see why?)

Now we know that these are independent events. We can use Theorem 12-13 to find

$$P(4 \text{ and } 3 \text{ heads}) = \frac{1}{6} \times \frac{1}{8} = \frac{1}{48}$$

Now we use Theorem 12-16:

$$P(A \text{ or } B) = P(A) + P(B) - P(A \text{ and } B)$$

$$P(4 \text{ or } 3 \text{ heads}) = P(4) + P(3 \text{ heads}) - P(4 \text{ and } 3 \text{ heads})$$

$$= \frac{1}{6} + \frac{1}{8} - \frac{1}{48} = \frac{13}{48}$$ ■

### The Odds of an Event Occurring

A word closely associated with probability is *odds*. This word designates the relative chances of a favorable event occurring.

*Definition*

> The *odds* in favor of event $A$ is the ratio of the probability of $A$ to the probability of not $A$:
>
> $$\text{Odds in favor of } A = \frac{P(A)}{P(\bar{A})}$$

EXAMPLE 9 If a die is rolled, what are the odds that a 2 will occur on the top face?

***Solution*** $P(2) = \frac{1}{6}$ and $P(\bar{2}) = \frac{5}{6}$

$$\text{Therefore the odds in favor of a } 2 = \frac{\frac{1}{6}}{\frac{5}{6}} = \frac{1}{5}.$$

The odds in favor of rolling a 2 with a single die are 1 to 5, often written as 1:5. ■

**EXAMPLE 10** Professor Hersey will select a graduate assistant in statistics from five candidates. Each candidate is equally likely to be selected. The candidates are Susan, Melissa, Alice, Jane, and Robert. What are the odds in favor of a woman being selected?

***Solution*** The probability of a woman being selected is $P(W) = \frac{4}{5}$. The probability of a man being selected is $P(M) = \frac{1}{5}$.

$$\text{Odds in favor of a woman being selected} = \frac{\frac{4}{5}}{\frac{1}{5}} = \frac{4}{1}$$

Thus the odds in favor of his selecting a woman graduate assistant in statistics are 4:1. $\left(\text{Note that we do not reduce } \frac{4}{1} \text{ to a whole number; 4 to 1 or 4:1 is a proper way to indicate the odds, not 4.}\right)$ ■

## EXERCISE 12-6

1. If a coin is tossed, what is the probability that it will land heads up?
2. A die is rolled. What is the probability that the result will be
   **a.** a 3? **b.** an even number? **c.** a number greater than 4?
3. If a card is drawn at random from a bridge deck, what is the probability that it will be
   **a.** the king of clubs? **b.** an ace? **c.** a heart? **d.** a picture card?
4. Five hundred raffle tickets were sold for a prize, and you purchased ten of them. What is the probability that you will win?

5. A candy jar contains 10 red jelly beans, 14 green jelly beans, 20 white jelly beans, and 18 black jelly beans. If a jelly bean is drawn at random from the jar, what is the probability that it will be
   **a.** red? **b.** black? **c.** red or green? **d.** not white? **e.** not green or black?
6. A box contains 16 defective generators, 10 partially defective generators, and 34 good generators. If a mechanic selects one, what is the probability that it is
   **a.** defective? **b.** defective or partially defective?
   **c.** good? **d.** good or partially defective?
7. What is the probability of rolling two 6's with a pair of dice?
8. What is the probability of tossing five heads in succession with a single coin?
9. A die is rolled and a coin is tossed. What is the probability of obtaining an even number on the die and a tail on the coin?
10. A card is drawn at random from a bridge deck. It is replaced, the deck is shuffled, and a second card is drawn. What is the probability that
    **a.** both cards are aces? **b.** both cards are diamonds?
    **c.** both cards are picture cards? **d.** the first card is an ace and the second card is a club?
    **e.** both cards are greater than 7?
11. In a single roll of two dice, find the probability that the sum of the numbers obtained is even.
12. Two cards are drawn together from a bridge deck. What is the probability that
    **a.** both cards are jacks? **b.** both cards are hearts?
    **c.** both cards are picture cards? **d.** both cards are less than 9?
13. Two people are to be chosen at random from a group of four women and three men. What is the probability that they will both be women?
14. Six cards numbered 1 through 6 are placed in a box, and three cards are drawn at random. What is the probability that they are all even numbers?
15. In a 12-horse race a person picks number 8 to win, number 4 to finish second, and number 10 to finish third. What is the probability that the horses will finish in exactly this order?
16. Five cards are dealt from a bridge deck. What is the probability that they are all hearts?
★ 17. A bridge hand consists of 13 cards. What is the probability of being dealt all the spades?
18. Three positions are open for an election. What is the probability of selecting all three Democrats if all candidates are equally likely to win in a race with eight Democrats, twelve Republicans, five Independents, and two Socialist candidates.
19. A bag contains five white, six green, and twelve red marbles. If a marble is drawn at random, what is the probability that it will be red or green?
20. If a die is tossed, what is the probability that it will show a 3 or 5?
21. A card is drawn from a bridge deck. What is the probability that it is an ace or king?
22. Two dice are rolled. What is the probability that the sum of the numbers obtained is 3?
23. Two dice are rolled. What is the probability that the sum of the numbers is 7?
24. Two coins are tossed. What is the probability of obtaining at least one head?
25. One card is drawn from a bridge deck. What is the probability that it is an 8 or a club?
26. One card is drawn from a bridge deck. What is the probability that it is a heart or a picture card?

**27.** Two dice are rolled and two coins are tossed. What is the probability of obtaining a 12 or two heads?

**28.** A family has two children.

**a.** What is the probability that one of the children is a girl? (Assume that the chances of the child being a girl or a boy are equally likely.)

**b.** If one of the children is a boy, what is the probability that the other is a girl?

**c.** If the older child is a boy, what is the probability that the other is a girl?

**29.** If a die is rolled, what are the odds that a 5 will occur on the top face?

**30.** A die is rolled. What are the odds that an even number will occur on the top face?

**31.** A card is drawn from a bridge deck. What are the odds that the card will be an ace?

**32.** What are the odds of obtaining three heads on a toss of three coins?

**33.** Two dice are rolled. What are the odds that the sum of the faces will be

**a.** 2? **b.** 7?

**34.** A box contains six blue cards, ten red cards, seven green cards, and nine white cards. If a card is drawn at random, what are the odds that it will be

**a.** red? **b.** blue or red? **c.** not blue or green?

**35.** Two cards are drawn from a bridge deck. What are the odds that they both will be aces?

**36.** The odds against a horse winning a race are 9:5. What is the probability that the horse will win the race?

★ **37.** Five cards are dealt from a bridge deck. What are the odds that they will all be clubs?

★ **38.** On a roulette wheel there are 38 positions in which the ball can stop (numbers 1 to 36, plus 0 and 00). There are 18 red numbers and 18 black numbers (the 0 and 00 positions are green). If the wheel is spun, what are the odds that the ball will stop at

**a.** 00? **b.** a red number? **c.** a green number? **d.** a red or black number?

★ **39.** A new drug is found to help 3100 out of 5600 people with a certain disease in a laboratory test. A doctor gives the drug to two people. What is the probability that both will be helped?

★ **40.** Fred is blindfolded. He is asked whether or not each of six people named is smiling. He answers yes or no to each of the six questions. Assume that he merely guesses the correct results in all six cases and is just lucky. What is the probability that Fred could obtain the correct result for all six cases by guessing? (Assume that the likelihood of a given person smiling is 0.5.)

## COMPUTER PROBLEMS FOR CHAPTER 12

See if you can apply your knowledge of computer programming and of Chapter 12 to do the following:

**1.** Write a computer program that determines whether a given sequence of four terms is a geometric sequence. If it is, the program should print out the message: "THIS IS A GEOMETRIC SEQUENCE. THE VALUE OF THE COMMON RATIO IS" and at that point the value of $r$ is printed. If the entered sequence is not a geometric sequence, the program should print out the message: "THIS IS NOT A GEOMETRIC SEQUENCE." Test your program on the following:

**a.** 1.44, 3.60, 9.00, 22.5 **b.** 9, −135, 2025, −303075 **c.** 0.0166, 0.0498, 0.1743, 0.5229

2. Write a computer program that, when you enter the values of $a_1$ and $r$ of a geometric sequence, will type out $a_n$ and $S_n$ for any positive integer $n \geq 1$. Test your program on the following:
   a. $a_1 = 36$, $r = 0.05$; find $a_{20}$, $S_{20}$.
   b. $a_1 = 0.07$, $r = 2.5$; find $a_{16}$, $S_{16}$.
   c. $a_1 = 166{,}000$, $r = -0.4$; find $a_{30}$, $S_{30}$.

## KEY TERMS AND CONCEPTS

Be sure you understand what is meant by these terms and can give an example of each.

Mathematical induction
Sequence of numbers
Series of numbers
Finite sequence
Infinite sequence
Recursive definition of a sequence
Sigma notation
Arithmetic sequence
Constant difference of an arithmetic sequence
Arithmetic series
Geometric sequence
Constant ratio of a geometric sequence
Geometric series
Geometric means
Permutation
Arrangement
Fundamental counting principle
Factorial
Combination
Binomial expansion
Binomial theorem
Pascal's triangle
Probability of an event
Event
Independent events
Equally likely events
Dependent events
Mutually exclusive events
Odds in favor of an event

## SUMMARY OF PROCEDURES AND CONCEPTS

### Sequences

*Theorem 12-1*

If $a_1$ is the first term and $d$ is the constant difference in an arithmetic sequence, then the $n$th term is given by $a_n = a_1 + (n - 1)d$.

*Theorem 12-2*

The sum $S_n$ of the first $n$ terms of an arithmetic series is given by the formula $S_n = \frac{n}{2}(a_1 + a_n)$ where $a_1$ is the first term and $a_n$ is the $n$th term.

*Theorem 12-3*

The sum $S_n$ of the first $n$ terms of an arithmetic series is given by the formula $S_n = \frac{n}{2}[2a_1 + (n - 1)d]$ where $a_1$ is the first term and $d$ is the constant difference.

*Theorem 12-4*

The formula for the $n$th term of a geometric sequence having a first term $a_1$ and a constant ratio $r$ is $a_n = a_1r^{n-1}$.

*Theorem 12-5*

Given a geometric sequence with a first term of $a_1$ and a common ratio $r \neq 1$, the sum of the first $n$ terms of the geometric sequence is $S_n = \frac{a_1 - a_1r^n}{1 - r}$.

*Theorem 12-6*

Given a geometric sequence with a first term of $a_1$, an $n$th term of $a_n$, and a common ratio $r \neq 1$, the sum of the first $n$ terms of the geometric sequence is $S_n = \frac{a_1 - a_nr}{1 - r}$.

*Theorem 12-7*

The sum $S$ of an infinite geometric sequence $a_1, a_1r, a_1r^2, \ldots$ with $|r| < 1$ is given by the formula $S = \frac{a_1}{1 - r}$.

**Permutations**

*Theorem 12-10*

The number of permutations of $n$ things taken $r$ at a time is ${}_nP_r = \frac{n!}{(n - r)!}$.

### Combinations

*Theorem 12-11* The number of combinations of $n$ things taken $r$ at a time is denoted by $\binom{n}{r}$ or by ${}_nC_r$ and is obtained by the formula

$$\binom{n}{r} = \frac{n!}{(n-r)!r!}$$

### Binomial Theorem

*Theorem 12-12* For any natural number $n$,

$$(x+y)^n = \sum_{r=0}^{n} \binom{n}{r} x^{n-r} y^r$$

## CHAPTER 12 REVIEW EXERCISE

Use mathematical induction to prove the statements in Problems 1–6.

1. $\frac{1}{2}(1) + \frac{1}{2}(2) + \frac{1}{2}(3) + \cdots + \frac{1}{2}(n) = \frac{n^2+n}{4}$
2. $4 + 10 + 18 + \cdots + n(n+3) = \frac{n(n+1)(n+5)}{3}$
3. $2 + 2^2 + 2^3 + \cdots + 2^n = 2^{n+1} - 2$
4. $1 + 3 + 6 + \cdots + \frac{n(n+1)}{2} = \frac{n(n+1)(n+2)}{6}$
5. $1^2 + 3^2 + 5^2 + \cdots + (2n-1)^2 = \frac{n(4n^2-1)}{3}$
6. $\frac{1}{1(3)} + \frac{1}{3(5)} + \frac{1}{5(7)} + \cdots + \frac{1}{(2n-1)(2n+1)} = \frac{n}{2n+1}$
7. Write the first four terms and the eighth term of the sequence whose $n$th term is given as $a_n = n^3 - 4$.
8. Write the first and the 112th term of the sequence whose $n$th term is $a_n = (-1)^n + 5$.
9. Find the first four terms of the sequence that is defined recursively as $a_1 = 1$, $a_{k+1} = k(a_k + 1)$.
10. Evaluate $\sum_{i=1}^{5} 2^{i-1}$.

**11.** Express in expanded form: $\sum_{i=1}^{4} \frac{x^{i+2}}{4i}$.

**12.** Express the series $2 + 5 + 10 + 17 + 26$ in sigma notation.

**13.** Express the series $1 + 5 + 9 + 13 + 17 + 21$ in sigma notation.

Indicate whether the sequences in Problems 14–17 are *arithmetic, geometric,* or *neither.*

**14.** $\frac{3}{4}, \frac{13}{12}, \frac{17}{12}, \ldots$

**15.** $-8, 4, -2, \ldots$

**16.** $\frac{1}{5}, \frac{1}{25}, \frac{1}{625}, \ldots$

**17.** $0.0176, 0.0376, 0.0576, \ldots$

**18.** In an arithmetic sequence, if $a_1 = -8$ and $d = 5$, find the 18th term.

**19.** In an arithmetic sequence, if $a_1 = 8$ and $d = -\frac{1}{2}$, find the 29th term.

**20.** In an arithmetic sequence, if $a_1 = x + 2y$ and $d = 2x + y$, find the 12th term.

**21.** Find the sum of the first 100 positive odd integers.

**22.** If the sum of the first nine terms of an arithmetic sequence in which $d = -3$ is 468, find the first term.

**23.** An arithmetic sequence having ten terms has a first term of 18 and a sum of 45. Find the seventh term.

**24.** For what value of $k$ is the sequence $k - 3$, $k + 5$, $2k - 1$ an arithmetic sequence?

**25.** Write the first four terms of a geometric sequence in which the first term is $\sqrt{2}$ and the constant ratio is $\sqrt{2}$.

**26.** Find the eighth term of a geometric sequence in which the first term is 6 and the constant ratio is 2.

**27.** Find the fifth term of a geometric sequence in which the first term is $-8$ and the constant ratio is $-\frac{1}{2}$.

**28.** Find the sum of the first six terms of a geometric sequence in which $a_1 = 16$ and $r = -\frac{1}{4}$.

**29.** For what values of $k$ is the sequence $3k + 4$, $k - 2$, $5k + 1$ a geometric sequence?

**30.** Insert three geometric means between 81 and 16.

A calculator may be helpful in some of the following problems.

**31.** Evaluate $8!$.

**32.** Evaluate $\frac{18!}{15!}$.

**33.** Evaluate ${}_{10}P_3$.

**34.** Evaluate $\binom{12}{4}$.

**35.** In how many ways can the letters of the word "trapezoid" be arranged?

**36.** The three offices of president, vice-president, and secretary are to be filled from a group of 35 people. If each person is eligible for election to any office, how many possible arrangements are there?

37. How many different seven-digit telephone numbers can be formed from the digits 0 through 9 if no digit is used more than once?

38. How many different seven-digit telephone numbers can be formed from the digits 0 through 9 if digits may be repeated?

39. How many triangles are determined by 12 points, no three of which are collinear?

40. In how many ways can a set of 13 cards be selected from a deck of 52 cards?

41. A person is to select one or more books from a set of seven books. How many possible combinations are there?

42. A committee of three is to be selected from a group of ten teachers and 12 parents. If the committee must have at least one teacher as a member, how many possible committees could be formed?

43. Expand $(2x + 3y)^6$ by using the binomial theorem.

44. Expand $(y^3 - 3a)^4$ by using the binomial theorem.

45. Find the fifth term of the expansion of $(y^2 - 2)^{10}$.

46. Find the middle term of the expansion of $\left(x - \frac{2}{x}\right)^8$.

47. A box contains eleven red, four green, six blue, and ten yellow cards. If a card is drawn at random, what is the probability that it is either green or yellow?

48. A card is drawn from a bridge deck. What is the probability that it is a club or greater than 10?

49. Five coins are tossed. What are the odds of obtaining all heads?

50. Five dice are rolled. Find the odds that all faces will be the same value.

51. Five cards are dealt from a poker deck. What are the odds against a "royal flush" (i.e., the 10, jack, queen, king, and ace of the same suit) being dealt?

52. Steakhouse West has eight restaurants. It has 11 people who are eligible to be appointed restaurant manager. In how many ways can the appointments be made? (Assume that each restaurant is distinctively different.)

## PRACTICE TEST FOR CHAPTER 12

1. Use mathematical induction to prove $2 + 4 + 6 + \cdots + 2n = n(n + 1)$.

2. Write the first four terms of the sequence whose $n$th term is defined as $a_n = 3n^2 - n - 2$.

3. Evaluate $\sum_{i=2}^{7} (2i - 1)$.

4. Express the series $-1 + 2 + 7 + 14$ in sigma notation.

5. In an arithmetic sequence, if $a_1 = 3$, $a_n = 27$, and $S_n = 150$, find $n$.

6. The first term of a geometric sequence is 625 and the last term is 1. If the sum of the terms is 521, find the fourth term.

7. Insert three geometric means between $\frac{2}{3}$ and 54.

8. Find the sum of the infinite geometric series $6 - 4 + \frac{8}{3} - \frac{16}{9} + \cdots$.

The use of a calculator may be helpful for some problems.

**9.** Evaluate $\frac{9!\,5!}{10!}$. **10.** Evaluate ${}_{20}P_4$. **11.** Evaluate $\binom{18}{3}$.

**12.** The four positions of president, vice-president, secretary, and treasurer are to be filled from a class of 30. In how many ways can these offices be filled?

**13.** How many different committees of four persons each can be chosen from a group of ten persons?

**14.** Find the tenth term of $\left(x + \frac{1}{2y}\right)^{12}$.

**15.** Use the binomial expansion to write out all the terms of $(x - 2y)^5$.

**16.** Eight cards numbered 1 to 8 are placed in a box, and three cards are drawn at random. What is the probability that they are all even numbers?

# APPENDICES *and* ANSWERS TO ODD-NUMBERED PROBLEMS

APPENDIX

# A Polar Coordinates and Graphs

For some types of curves, it is more convenient to use a coordinate system known as the **polar coordinate system**. In such a system the origin is called the **pole**, and the positive $x$-axis is called the **polar axis**. Any point can be designated by polar coordinates $(r, \theta)$ as shown in the figure below. The value $r$ is the directed distance from $O$ to $P$ and $\theta$ is the directed angle from the polar axis to $OP$. The angle $\theta$ is considered **positive** if the angle is a counterclockwise rotation of the polar axis. The angle $\theta$ is considered **negative** if the angle is a clockwise rotation of the polar axis. The angle $\theta$ may be measured in radians or degrees. Polar coordinate graph paper makes it easier to plot points identified by polar coordinates.

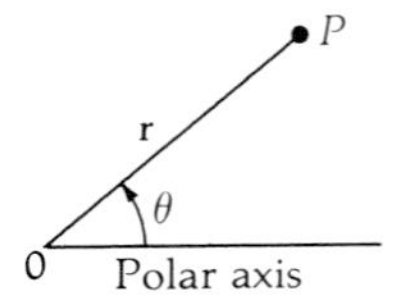


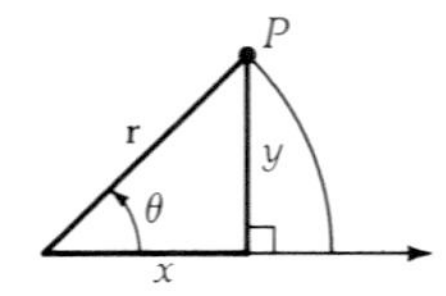


**EXAMPLE 1** Plot the following points on polar coordinate paper.

**a.** $\left(3, \frac{\pi}{6}\right)$ **b.** $\left(2, \frac{5\pi}{3}\right)$

**c.** $\left(4, -\frac{3\pi}{2}\right)$ **d.** $\left(5, -\frac{\pi}{3}\right)$

**e.** $\left(5.5, \frac{3\pi}{4}\right)$

***Solution***

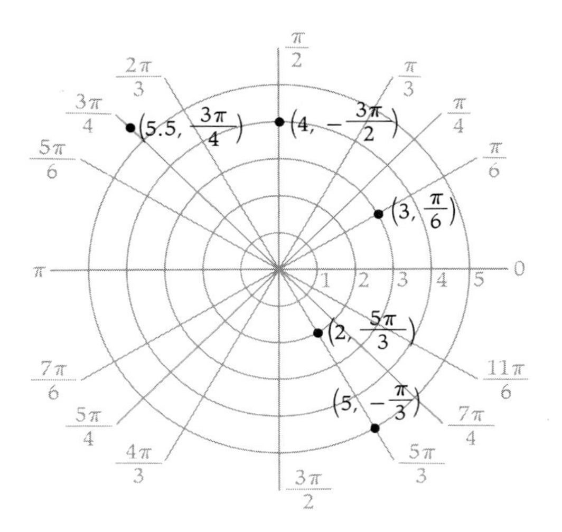


*Figure A-1*

The relationship between the rectangular coordinates of a point and the polar coordinates of a point come directly from the trigonometric functions.

**Polar Coordinate Formulas**

$$x = r\cos\theta \qquad y = r\sin\theta$$

$$\tan\theta = \frac{y}{x} \qquad r = \sqrt{x^2 + y^2}$$

It should be noted that in polar coordinates the value of $r$ can be positive or negative. In the polar coordinate system there are many possible coordinates for the same point.

We now consider polar equations, which are equations in $r$ and $\theta$. The graphs of such equations will be plotted on polar coordinate graph paper.

EXAMPLE 2 Sketch the graph of the equation $r = 6\sin\theta$ for the values $0 \le \theta \le \pi$.

***Solution*** We find a few approximate values and list them in a table.

| $\theta$ | $0$ | $\frac{\pi}{6}$ | $\frac{\pi}{4}$ | $\frac{\pi}{3}$ | $\frac{\pi}{2}$ | $\frac{2\pi}{3}$ | $\frac{3\pi}{4}$ | $\frac{5\pi}{6}$ | $\pi$ |
|---|---|---|---|---|---|---|---|---|---|
| $r$ | 0 | 3.0 | 4.2 | 5.2 | 6.0 | 5.2 | 4.2 | 3.0 | 0 |

See Figure A-2. We see from our graph that we have a circle.

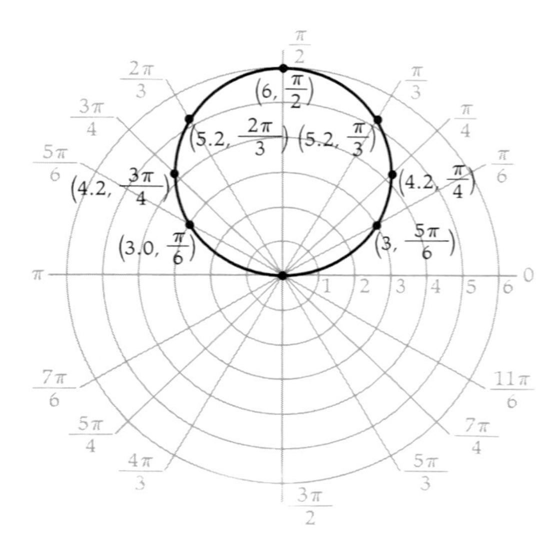


*Figure A-2*

What would happen if we continued graphing the equation $r = 6 \sin \theta$ for $\pi \leq \theta \leq 2\pi$? Do you see we would get the same points again? For example, if $\theta = \frac{7\pi}{6}$, $r = -3.0$. If we plot $\left(-3.0, \frac{7\pi}{6}\right)$, it will coincide with the same point as $\left(3.0, \frac{\pi}{6}\right)$. Because $\sin(\pi + \theta) = -\sin \theta$, if we continue to plot points over an interval greater than $\pi$, we just obtain the same points as we originally obtained. ■

**EXAMPLE 3** Sketch the graph of the equation $r = 6 \cos 2\theta$ for the values $0 \leq \theta \leq 2\pi$.

***Solution***

| $\theta$ | $0$ | $\frac{\pi}{6}$ | $\frac{\pi}{4}$ | $\frac{\pi}{3}$ | $\frac{\pi}{2}$ | $\frac{2\pi}{3}$ | $\frac{3\pi}{4}$ | $\frac{5\pi}{6}$ | $\pi$ |
|---|---|---|---|---|---|---|---|---|---|
| $r$ | 6 | 3 | 0 | −3 | −6 | −3 | 0 | 3 | 6 |

Although these additional values are similar, we will need them in order to obtain a complete graph. (Do you see why?)

| $\theta$ | $\frac{7\pi}{6}$ | $\frac{5\pi}{4}$ | $\frac{4\pi}{3}$ | $\frac{3\pi}{2}$ | $\frac{5\pi}{3}$ | $\frac{7\pi}{4}$ | $\frac{11\pi}{6}$ | $2\pi$ |
|---|---|---|---|---|---|---|---|---|
| $r$ | 3 | 0 | $-3$ | $-6$ | $-3$ | 0 | 3 | 6 |

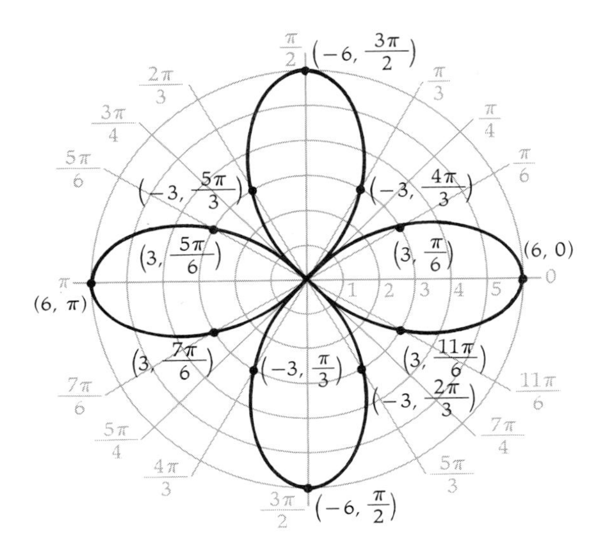


*Figure A-3*

The curve is called a four-leaf *rose*. See Figure A-3. ■

Careful use of one or more of the polar coordinate formulas will allow you to convert from polar equations to rectangular equations and vice-versa.

**EXAMPLE 4** Convert to a polar equation: $x^2 - 4y^2 = 4$.

***Solution*** Since $x = r\cos\theta$ and $y = r\sin\theta$, we have

$$(r\cos\theta)^2 - 4(r\sin\theta)^2 = 4$$

$$r^2\cos^2\theta - 4r^2\sin^2\theta = 4$$

$$r^2(\cos^2\theta - 4\sin^2\theta) = 4 \quad ■$$

**EXAMPLE 5** Convert to a rectangular equation: $r = \dfrac{1}{1 - \cos\theta}$.

***Solution*** Since $\cos\theta = \frac{x}{r}$, we have

$$r = \frac{1}{1 - \frac{x}{r}} \tag{1}$$

The right-hand side of Equation **(1)** is a complex fraction which can be written as $\frac{r}{r - x}$. (Do you see why?) Thus, Equation **(1)** becomes $r = \frac{r}{r - x}$. We therefore conclude that if $r \neq 0$ then

$$r - x = 1 \tag{2}$$

Since $r = \sqrt{x^2 + y^2}$, Equation **(2)** can be written as

$$\sqrt{x^2 + y^2} - x = 1$$

$$\sqrt{x^2 + y^2} = 1 + x$$

$$x^2 + y^2 = x^2 + 2x + 1 \qquad \textit{Squaring each side.}$$

$$y^2 = 2x + 1 \quad ■$$

## APPENDIX A EXERCISE

Sketch the graphs of the equations in Problems 1–16. Use polar coordinate graph paper.

**1.** $\theta = \frac{\pi}{12}$

**2.** $\theta = -\frac{2\pi}{3}$

**3.** $r = -4$

**4.** $r = 2$

**5.** $r = 2\cos\theta$ (circle)

**6.** $r(\cos\theta + \sin\theta) + 2 = 0$ (straight line)

**7.** $r = 2\sin 3\theta$ (three-leaf rose)

**8.** $r = 3\cos 3\theta$ (three-leaf rose)

**9.** $r = 3(1 + \cos\theta)$ (cardiod)

**10.** $r = 4 + 4\sin\theta$ (cardiod)

**11.** $r = 5(\sin 2\theta)$ (four-leaf rose)

**12.** $r = 3(\cos 2\theta)$ (four-leaf rose)

**13.** $r^2 = 4\cos 2\theta$ (lemniscate)

**14.** $r^2 = 9\sin 2\theta$ (lemniscate)

**15.** $r = 2 + 3\cos\theta$ (limaçon)

**16.** $r = 1 + 2\sin\theta$ (limaçon)

Convert to a polar equation in Problems 17–20.

**17.** $y^2 = 9x$ **18.** $x^2 + y^2 = 4$ **19.** $x^2 + 4y^2 = 16$ **20.** $y = 8x$

Convert to a rectangular equation in Problems 21–24.

**21.** $r\cos\theta = 2$ **22.** $r = 8$ **23.** $r = 6\sin\theta$ **24.** $r = 6\cos\theta$

APPENDIX

# B Tables

TABLE A. **Squares, Square Roots, and Reciprocals**

| $x$ | $x^2$ | $\sqrt{x}$ | $\sqrt{10x}$ | $1/x$ |
|---|---|---|---|---|
| 1 | 1 | 1.000 | 3.162 | 1.00000 |
| 2 | 4 | 1.414 | 4.472 | 0.50000 |
| 3 | 9 | 1.732 | 5.477 | 0.33333 |
| 4 | 16 | 2.000 | 6.325 | 0.25000 |
| 5 | 25 | 2.236 | 7.071 | 0.20000 |
| 6 | 36 | 2.449 | 7.746 | 0.16667 |
| 7 | 49 | 2.646 | 8.367 | 0.14286 |
| 8 | 64 | 2.828 | 8.944 | 0.12500 |
| 9 | 81 | 3.000 | 9.487 | 0.11111 |
| **10** | 100 | 3.162 | 10.000 | 0.10000 |
| 11 | 121 | 3.317 | 10.488 | 0.09091 |
| 12 | 144 | 3.464 | 10.954 | 0.08333 |
| 13 | 169 | 3.606 | 11.402 | 0.07692 |
| 14 | 196 | 3.742 | 11.832 | 0.07143 |
| 15 | 225 | 3.873 | 12.247 | 0.06667 |
| 16 | 256 | 4.000 | 12.649 | 0.06250 |
| 17 | 289 | 4.123 | 13.038 | 0.05882 |
| 18 | 324 | 4.243 | 13.416 | 0.05556 |
| 19 | 361 | 4.359 | 13.784 | 0.05263 |
| **20** | 400 | 4.472 | 14.142 | 0.05000 |
| 21 | 441 | 4.583 | 14.491 | 0.04762 |
| 22 | 484 | 4.690 | 14.832 | 0.04545 |
| 23 | 529 | 4.796 | 15.166 | 0.04348 |
| 24 | 576 | 4.899 | 15.492 | 0.04167 |
| 25 | 625 | 5.000 | 15.811 | 0.04000 |
| 26 | 676 | 5.099 | 16.125 | 0.03846 |
| 27 | 729 | 5.196 | 16.432 | 0.03704 |
| 28 | 784 | 5.292 | 16.733 | 0.03571 |
| 29 | 841 | 5.385 | 17.029 | 0.03448 |
| **30** | 900 | 5.477 | 17.321 | 0.03333 |
| 31 | 961 | 5.568 | 17.607 | 0.03226 |
| 32 | 1024 | 5.657 | 17.889 | 0.03125 |
| 33 | 1089 | 5.745 | 18.166 | 0.03030 |
| 34 | 1156 | 5.831 | 18.439 | 0.02941 |
| 35 | 1225 | 5.916 | 18.708 | 0.02857 |
| 36 | 1296 | 6.000 | 18.974 | 0.02778 |
| 37 | 1369 | 6.083 | 19.235 | 0.02703 |
| 38 | 1444 | 6.164 | 19.494 | 0.02632 |
| 39 | 1521 | 6.245 | 19.748 | 0.02564 |
| **40** | 1600 | 6.325 | 20.000 | 0.02500 |
| 41 | 1681 | 6.403 | 20.248 | 0.02439 |
| 42 | 1764 | 6.481 | 20.494 | 0.02381 |
| 43 | 1849 | 6.557 | 20.736 | 0.02326 |
| 44 | 1936 | 6.633 | 20.976 | 0.02273 |
| 45 | 2025 | 6.708 | 21.213 | 0.02222 |
| 46 | 2116 | 6.782 | 21.448 | 0.02174 |
| 47 | 2209 | 6.856 | 21.679 | 0.02128 |
| 48 | 2304 | 6.928 | 21.909 | 0.02083 |
| 49 | 2401 | 7.000 | 22.136 | 0.02041 |
| **50** | 2500 | 7.071 | 22.361 | 0.02000 |
| 51 | 2601 | 7.141 | 22.583 | 0.01961 |
| 52 | 2704 | 7.211 | 22.804 | 0.01923 |
| 53 | 2809 | 7.280 | 23.022 | 0.01887 |
| 54 | 2916 | 7.348 | 23.238 | 0.01852 |
| 55 | 3025 | 7.416 | 23.452 | 0.01818 |
| 56 | 3136 | 7.483 | 23.664 | 0.01786 |
| 57 | 3249 | 7.550 | 23.875 | 0.01754 |
| 58 | 3364 | 7.616 | 24.083 | 0.01724 |
| 59 | 3481 | 7.681 | 24.290 | 0.01695 |
| **60** | 3600 | 7.746 | 24.495 | 0.01667 |
| 61 | 3721 | 7.810 | 24.698 | 0.01639 |
| 62 | 3844 | 7.874 | 24.900 | 0.01613 |
| 63 | 3969 | 7.937 | 25.100 | 0.01587 |
| 64 | 4096 | 8.000 | 25.298 | 0.01562 |
| 65 | 4225 | 8.062 | 25.495 | 0.01538 |
| 66 | 4356 | 8.124 | 25.690 | 0.01515 |
| 67 | 4489 | 8.185 | 25.884 | 0.01493 |
| 68 | 4624 | 8.246 | 26.077 | 0.01471 |
| 69 | 4761 | 8.307 | 26.268 | 0.01449 |
| **70** | 4900 | 8.367 | 26.458 | 0.01429 |
| 71 | 5041 | 8.426 | 26.646 | 0.01408 |
| 72 | 5184 | 8.485 | 26.833 | 0.01389 |
| 73 | 5329 | 8.544 | 27.019 | 0.01370 |
| 74 | 5476 | 8.602 | 27.203 | 0.01351 |
| 75 | 5625 | 8.660 | 27.386 | 0.01333 |
| 76 | 5776 | 8.718 | 27.568 | 0.01316 |
| 77 | 5929 | 8.775 | 27.749 | 0.01299 |
| 78 | 6084 | 8.832 | 27.928 | 0.01282 |
| 79 | 6241 | 8.888 | 28.107 | 0.01266 |
| **80** | 6400 | 8.944 | 28.284 | 0.01250 |
| 81 | 6561 | 9.000 | 28.460 | 0.01235 |
| 82 | 6724 | 9.055 | 28.636 | 0.01220 |
| 83 | 6889 | 9.110 | 28.810 | 0.01205 |
| 84 | 7056 | 9.165 | 28.983 | 0.01190 |
| 85 | 7225 | 9.220 | 29.155 | 0.01176 |
| 86 | 7396 | 9.274 | 29.326 | 0.01163 |
| 87 | 7569 | 9.327 | 29.496 | 0.01149 |
| 88 | 7744 | 9.381 | 29.665 | 0.01136 |
| 89 | 7921 | 9.434 | 29.833 | 0.01124 |
| **90** | 8100 | 9.487 | 30.000 | 0.01111 |
| 91 | 8281 | 9.539 | 30.166 | 0.01099 |
| 92 | 8464 | 9.592 | 30.332 | 0.01087 |
| 93 | 8649 | 9.644 | 30.496 | 0.01075 |
| 94 | 8836 | 9.695 | 30.659 | 0.01064 |
| 95 | 9025 | 9.747 | 30.822 | 0.01053 |
| 96 | 9216 | 9.798 | 30.984 | 0.01042 |
| 97 | 9409 | 9.849 | 31.145 | 0.01031 |
| 98 | 9604 | 9.899 | 31.305 | 0.01020 |
| 99 | 9801 | 9.950 | 31.464 | 0.01010 |
| **100** | 10,000 | 10.000 | 31.623 | 0.01000 |

TABLE B. **Exponential Values**

| $x$ | $e^x$ | $e^{-x}$ |
|---|---|---|
| 0.00 | 1.0000 | 1.0000 |
| 0.01 | 1.0101 | 0.9900 |
| 0.02 | 1.0202 | 0.9802 |
| 0.03 | 1.0305 | 0.9704 |
| 0.04 | 1.0408 | 0.9608 |
| 0.05 | 1.0513 | 0.9512 |
| 0.06 | 1.0618 | 0.9418 |
| 0.07 | 1.0725 | 0.9324 |
| 0.08 | 1.0833 | 0.9231 |
| 0.09 | 1.0942 | 0.9139 |
| 0.10 | 1.1052 | 0.9048 |
| 0.11 | 1.1163 | 0.8958 |
| 0.12 | 1.1275 | 0.8869 |
| 0.13 | 1.1388 | 0.8781 |
| 0.14 | 1.1503 | 0.8694 |
| 0.15 | 1.1618 | 0.8607 |
| 0.16 | 1.1735 | 0.8521 |
| 0.17 | 1.1853 | 0.8437 |
| 0.18 | 1.1972 | 0.8353 |
| 0.19 | 1.2092 | 0.8270 |
| 0.20 | 1.2214 | 0.8187 |
| 0.21 | 1.2337 | 0.8106 |
| 0.22 | 1.2461 | 0.8025 |
| 0.23 | 1.2586 | 0.7945 |
| 0.24 | 1.2712 | 0.7866 |
| 0.25 | 1.2840 | 0.7788 |
| 0.26 | 1.2969 | 0.7711 |
| 0.27 | 1.3100 | 0.7634 |
| 0.28 | 1.3231 | 0.7558 |
| 0.29 | 1.3364 | 0.7483 |
| 0.30 | 1.3499 | 0.7408 |
| 0.35 | 1.4191 | 0.7047 |
| 0.40 | 1.4918 | 0.6703 |
| 0.45 | 1.5683 | 0.6376 |
| 0.50 | 1.6487 | 0.6065 |
| 0.55 | 1.7333 | 0.5769 |
| 0.60 | 1.8221 | 0.5488 |
| 0.65 | 1.9155 | 0.5220 |
| 0.70 | 2.0138 | 0.4966 |
| 0.75 | 2.1170 | 0.4724 |
| 0.80 | 2.2255 | 0.4493 |
| 0.85 | 2.3396 | 0.4274 |
| 0.90 | 2.4596 | 0.4066 |
| 0.95 | 2.5857 | 0.3867 |
| 1.0 | 2.7183 | 0.3679 |
| 1.1 | 3.0042 | 0.3329 |
| 1.2 | 3.3201 | 0.3012 |
| 1.3 | 3.6693 | 0.2725 |
| 1.4 | 4.0552 | 0.2466 |
| 1.5 | 4.4817 | 0.2231 |
| 1.6 | 4.9530 | 0.2019 |
| 1.7 | 5.4739 | 0.1827 |
| 1.8 | 6.0496 | 0.1653 |
| 1.9 | 6.6859 | 0.1496 |
| 2.0 | 7.3891 | 0.1353 |
| 2.1 | 8.1662 | 0.1225 |
| 2.2 | 9.0250 | 0.1108 |
| 2.3 | 9.9742 | 0.1003 |
| 2.4 | 11.023 | 0.0907 |
| 2.5 | 12.182 | 0.0821 |
| 2.6 | 13.464 | 0.0743 |
| 2.7 | 14.880 | 0.0672 |
| 2.8 | 16.445 | 0.0608 |
| 2.9 | 18.174 | 0.0550 |
| 3.0 | 20.086 | 0.0498 |
| 3.1 | 22.198 | 0.0450 |
| 3.2 | 24.533 | 0.0408 |
| 3.3 | 27.113 | 0.0369 |
| 3.4 | 29.964 | 0.0334 |
| 3.5 | 33.115 | 0.0302 |
| 3.6 | 36.598 | 0.0273 |
| 3.7 | 40.447 | 0.0247 |
| 3.8 | 44.701 | 0.0224 |
| 3.9 | 49.402 | 0.0202 |
| 4.0 | 54.598 | 0.0183 |
| 4.1 | 60.340 | 0.0166 |
| 4.2 | 66.686 | 0.0150 |
| 4.3 | 73.700 | 0.0136 |
| 4.4 | 81.451 | 0.0123 |
| 4.5 | 90.017 | 0.0111 |
| 4.6 | 99.484 | 0.0101 |
| 4.7 | 109.95 | 0.0091 |
| 4.8 | 121.51 | 0.0082 |
| 4.9 | 134.29 | 0.0074 |
| 5.0 | 148.41 | 0.0067 |
| 5.5 | 244.69 | 0.0041 |
| 6.0 | 403.43 | 0.0025 |
| 6.5 | 665.14 | 0.0015 |
| 7.0 | 1,096.6 | 0.0009 |
| 7.5 | 1,808.0 | 0.00055 |
| 8.0 | 2,981.0 | 0.00034 |
| 8.5 | 4,914.8 | 0.00020 |
| 9.0 | 8,103.1 | 0.00012 |
| 9.5 | 13,360 | 0.000075 |
| 10 | 22,026 | 0.000045 |
| 11 | 59,874 | 0.000017 |
| 12 | 162,754 | 0.0000061 |
| 13 | 442,413 | 0.0000023 |
| 14 | 1,202,604 | 0.0000008 |
| 15 | 3,269,017 | 0.0000003 |

TABLE C. **Common Logarithms**

| N | 0 | 1 | 2 | 3 | 4 | 5 | 6 | 7 | 8 | 9 |
|---|---|---|---|---|---|---|---|---|---|---|
| 1.0 | 0.0000 | 0.0043 | 0.0086 | 0.0128 | 0.0170 | 0.0212 | 0.0253 | 0.0294 | 0.0334 | 0.0374 |
| 1.1 | 0.0414 | 0.0453 | 0.0492 | 0.0531 | 0.0569 | 0.0607 | 0.0645 | 0.0682 | 0.0719 | 0.0755 |
| 1.2 | 0.0792 | 0.0828 | 0.0864 | 0.0899 | 0.0934 | 0.0969 | 0.1004 | 0.1038 | 0.1072 | 0.1106 |
| 1.3 | 0.1139 | 0.1173 | 0.1206 | 0.1239 | 0.1271 | 0.1303 | 0.1335 | 0.1367 | 0.1399 | 0.1430 |
| 1.4 | 0.1461 | 0.1492 | 0.1523 | 0.1553 | 0.1584 | 0.1614 | 0.1644 | 0.1673 | 0.1703 | 0.1732 |
| 1.5 | 0.1761 | 0.1790 | 0.1818 | 0.1847 | 0.1875 | 0.1903 | 0.1931 | 0.1959 | 0.1987 | 0.2014 |
| 1.6 | 0.2041 | 0.2068 | 0.2095 | 0.2122 | 0.2148 | 0.2175 | 0.2201 | 0.2227 | 0.2253 | 0.2279 |
| 1.7 | 0.2304 | 0.2330 | 0.2355 | 0.2380 | 0.2405 | 0.2430 | 0.2455 | 0.2480 | 0.2504 | 0.2529 |
| 1.8 | 0.2553 | 0.2577 | 0.2601 | 0.2625 | 0.2648 | 0.2672 | 0.2695 | 0.2718 | 0.2742 | 0.2765 |
| 1.9 | 0.2788 | 0.2810 | 0.2833 | 0.2856 | 0.2878 | 0.2900 | 0.2923 | 0.2945 | 0.2967 | 0.2989 |
| 2.0 | 0.3010 | 0.3032 | 0.3054 | 0.3075 | 0.3096 | 0.3118 | 0.3139 | 0.3160 | 0.3181 | 0.3201 |
| 2.1 | 0.3222 | 0.3243 | 0.3263 | 0.3284 | 0.3304 | 0.3324 | 0.3345 | 0.3365 | 0.3385 | 0.3404 |
| 2.2 | 0.3424 | 0.3444 | 0.3464 | 0.3483 | 0.3502 | 0.3522 | 0.3541 | 0.3560 | 0.3579 | 0.3598 |
| 2.3 | 0.3617 | 0.3636 | 0.3655 | 0.3674 | 0.3692 | 0.3711 | 0.3729 | 0.3747 | 0.3766 | 0.3784 |
| 2.4 | 0.3802 | 0.3820 | 0.3838 | 0.3856 | 0.3874 | 0.3892 | 0.3909 | 0.3927 | 0.3945 | 0.3692 |
| 2.5 | 0.3979 | 0.3997 | 0.4014 | 0.4031 | 0.4048 | 0.4065 | 0.4082 | 0.4099 | 0.4116 | 0.4133 |
| 2.6 | 0.4150 | 0.4166 | 0.4183 | 0.4200 | 0.4216 | 0.4232 | 0.4249 | 0.4265 | 0.4281 | 0.4298 |
| 2.7 | 0.4314 | 0.4330 | 0.4346 | 0.4362 | 0.4378 | 0.4393 | 0.4409 | 0.4425 | 0.4440 | 0.4456 |
| 2.8 | 0.4472 | 0.4487 | 0.4502 | 0.4518 | 0.4533 | 0.4548 | 0.4564 | 0.4579 | 0.4594 | 0.4609 |
| 2.9 | 0.4624 | 0.4639 | 0.4654 | 0.4669 | 0.4683 | 0.4698 | 0.4713 | 0.4728 | 0.4742 | 0.4757 |
| 3.0 | 0.4771 | 0.4786 | 0.4800 | 0.4814 | 0.4829 | 0.4843 | 0.4857 | 0.4871 | 0.4886 | 0.4900 |
| 3.1 | 0.4914 | 0.4928 | 0.4942 | 0.4955 | 0.4969 | 0.4983 | 0.4997 | 0.5011 | 0.5024 | 0.5038 |
| 3.2 | 0.5051 | 0.5065 | 0.5079 | 0.5092 | 0.5105 | 0.5119 | 0.5132 | 0.5145 | 0.5159 | 0.5172 |
| 3.3 | 0.5185 | 0.5198 | 0.5211 | 0.5224 | 0.5237 | 0.5250 | 0.5263 | 0.5276 | 0.5289 | 0.5302 |
| 3.4 | 0.5315 | 0.5328 | 0.5340 | 0.5353 | 0.5366 | 0.5378 | 0.5391 | 0.5403 | 0.5416 | 0.5428 |
| 3.5 | 0.5441 | 0.5453 | 0.5465 | 0.5478 | 0.5490 | 0.5502 | 0.5514 | 0.5527 | 0.5539 | 0.5551 |
| 3.6 | 0.5563 | 0.5575 | 0.5587 | 0.5599 | 0.5611 | 0.5623 | 0.5635 | 0.5647 | 0.5658 | 0.5670 |
| 3.7 | 0.5682 | 0.5694 | 0.5705 | 0.5717 | 0.5729 | 0.5740 | 0.5752 | 0.5763 | 0.5775 | 0.5786 |
| 3.8 | 0.5798 | 0.5809 | 0.5821 | 0.5832 | 0.5843 | 0.5855 | 0.5866 | 0.5877 | 0.5888 | 0.5899 |
| 3.9 | 0.5911 | 0.5922 | 0.5933 | 0.5944 | 0.5955 | 0.5966 | 0.5977 | 0.5988 | 0.5999 | 0.6010 |
| 4.0 | 0.6021 | 0.6031 | 0.6042 | 0.6053 | 0.6064 | 0.6075 | 0.6085 | 0.6096 | 0.6107 | 0.6117 |
| 4.1 | 0.6128 | 0.6138 | 0.6149 | 0.6160 | 0.6170 | 0.6180 | 0.6191 | 0.6201 | 0.6212 | 0.6222 |
| 4.2 | 0.6232 | 0.6243 | 0.6253 | 0.6263 | 0.6274 | 0.6284 | 0.6294 | 0.6304 | 0.6314 | 0.6325 |
| 4.3 | 0.6335 | 0.6345 | 0.6355 | 0.6365 | 0.6375 | 0.6385 | 0.6395 | 0.6405 | 0.6415 | 0.6425 |
| 4.4 | 0.6435 | 0.6444 | 0.6454 | 0.6464 | 0.6474 | 0.6484 | 0.6493 | 0.6503 | 0.6513 | 0.6522 |
| 4.5 | 0.6532 | 0.6542 | 0.6551 | 0.6561 | 0.6571 | 0.6580 | 0.6590 | 0.6599 | 0.6609 | 0.6618 |
| 4.6 | 0.6628 | 0.6637 | 0.6646 | 0.6656 | 0.6665 | 0.6675 | 0.6684 | 0.6693 | 0.6702 | 0.6712 |
| 4.7 | 0.6721 | 0.6730 | 0.6739 | 0.6749 | 0.6758 | 0.6767 | 0.6776 | 0.6785 | 0.6794 | 0.6803 |
| 4.8 | 0.6812 | 0.6821 | 0.6830 | 0.6839 | 0.6848 | 0.6857 | 0.6866 | 0.6875 | 0.6884 | 0.6893 |
| 4.9 | 0.6902 | 0.6911 | 0.6920 | 0.6928 | 0.6937 | 0.6946 | 0.6955 | 0.6964 | 0.6972 | 0.6981 |
| 5.0 | 0.6990 | 0.6998 | 0.7007 | 0.7016 | 0.7024 | 0.7033 | 0.7042 | 0.7050 | 0.7059 | 0.7067 |
| 5.1 | 0.7076 | 0.7084 | 0.7093 | 0.7101 | 0.7110 | 0.7118 | 0.7126 | 0.7135 | 0.7143 | 0.7152 |
| 5.2 | 0.7160 | 0.7168 | 0.7177 | 0.7185 | 0.7193 | 0.7202 | 0.7210 | 0.7218 | 0.7226 | 0.7235 |
| 5.3 | 0.7243 | 0.7251 | 0.7259 | 0.7267 | 0.7275 | 0.7284 | 0.7292 | 0.7300 | 0.7308 | 0.7316 |
| 5.4 | 0.7324 | 0.7332 | 0.7340 | 0.7348 | 0.7356 | 0.7364 | 0.7372 | 0.7380 | 0.7388 | 0.7396 |

TABLE C. **Common Logarithms (*continued*)**

| N | 0 | 1 | 2 | 3 | 4 | 5 | 6 | 7 | 8 | 9 |
|---|---|---|---|---|---|---|---|---|---|---|
| 5.5 | 0.7404 | 0.7412 | 0.7419 | 0.7427 | 0.7435 | 0.7443 | 0.7451 | 0.7459 | 0.7466 | 0.7474 |
| 5.6 | 0.7482 | 0.7490 | 0.7497 | 0.7505 | 0.7513 | 0.7520 | 0.7528 | 0.7536 | 0.7543 | 0.7551 |
| 5.7 | 0.7559 | 0.7566 | 0.7574 | 0.7582 | 0.7589 | 0.7597 | 0.7604 | 0.7612 | 0.7619 | 0.7627 |
| 5.8 | 0.7634 | 0.7642 | 0.7649 | 0.7657 | 0.7664 | 0.7672 | 0.7679 | 0.7686 | 0.7694 | 0.7701 |
| 5.9 | 0.7709 | 0.7716 | 0.7723 | 0.7731 | 0.7738 | 0.7745 | 0.7752 | 0.7760 | 0.7767 | 0.7774 |
| 6.0 | 0.7782 | 0.7789 | 0.7796 | 0.7803 | 0.7810 | 0.7818 | 0.7825 | 0.7832 | 0.7839 | 0.7846 |
| 6.1 | 0.7853 | 0.7860 | 0.7868 | 0.7875 | 0.7882 | 0.7889 | 0.7896 | 0.7903 | 0.7910 | 0.7917 |
| 6.2 | 0.7924 | 0.7931 | 0.7938 | 0.7945 | 0.7952 | 0.7959 | 0.7966 | 0.7973 | 0.7980 | 0.7987 |
| 6.3 | 0.7993 | 0.8000 | 0.8007 | 0.8014 | 0.8021 | 0.8028 | 0.8035 | 0.8041 | 0.8048 | 0.8055 |
| 6.4 | 0.8062 | 0.8069 | 0.8075 | 0.8082 | 0.8089 | 0.8096 | 0.8102 | 0.8109 | 0.8116 | 0.8122 |
| 6.5 | 0.8129 | 0.8136 | 0.8142 | 0.8149 | 0.8156 | 0.8162 | 0.8169 | 0.8176 | 0.8182 | 0.8189 |
| 6.6 | 0.8195 | 0.8202 | 0.8209 | 0.8215 | 0.8222 | 0.8228 | 0.8235 | 0.8241 | 0.8248 | 0.8254 |
| 6.7 | 0.8261 | 0.8267 | 0.8274 | 0.8280 | 0.8287 | 0.8293 | 0.8299 | 0.8306 | 0.8312 | 0.8319 |
| 6.8 | 0.8325 | 0.8331 | 0.8338 | 0.8344 | 0.8351 | 0.8357 | 0.8363 | 0.8370 | 0.8376 | 0.8382 |
| 6.9 | 0.8388 | 0.8395 | 0.8401 | 0.8407 | 0.8414 | 0.8420 | 0.8426 | 0.8432 | 0.8439 | 0.8445 |
| 7.0 | 0.8451 | 0.8457 | 0.8463 | 0.8470 | 0.8476 | 0.8482 | 0.8488 | 0.8494 | 0.8500 | 0.8506 |
| 7.1 | 0.8513 | 0.8519 | 0.8525 | 0.8531 | 0.8537 | 0.8543 | 0.8549 | 0.8555 | 0.8561 | 0.8567 |
| 7.2 | 0.8573 | 0.8579 | 0.8585 | 0.8591 | 0.8597 | 0.8603 | 0.8609 | 0.8615 | 0.8621 | 0.8627 |
| 7.3 | 0.8633 | 0.8639 | 0.8645 | 0.8651 | 0.8657 | 0.8663 | 0.8669 | 0.8675 | 0.8681 | 0.8686 |
| 7.4 | 0.8692 | 0.8698 | 0.8704 | 0.8710 | 0.8716 | 0.8722 | 0.8727 | 0.8733 | 0.8739 | 0.8745 |
| 7.5 | 0.8751 | 0.8756 | 0.8762 | 0.8768 | 0.8774 | 0.8779 | 0.8785 | 0.8791 | 0.8797 | 0.8802 |
| 7.6 | 0.8808 | 0.8814 | 0.8820 | 0.8825 | 0.8831 | 0.8837 | 0.8842 | 0.8848 | 0.8854 | 0.8859 |
| 7.7 | 0.8865 | 0.8871 | 0.8876 | 0.8882 | 0.8887 | 0.8893 | 0.8899 | 0.8904 | 0.8910 | 0.8915 |
| 7.8 | 0.8921 | 0.8927 | 0.8932 | 0.8938 | 0.8943 | 0.8949 | 0.8954 | 0.8960 | 0.8965 | 0.8971 |
| 7.9 | 0.8976 | 0.8982 | 0.8987 | 0.8993 | 0.8998 | 0.9004 | 0.9009 | 0.9015 | 0.9020 | 0.9025 |
| 8.0 | 0.9031 | 0.9036 | 0.9042 | 0.9047 | 0.9053 | 0.9058 | 0.9063 | 0.9069 | 0.9074 | 0.9079 |
| 8.1 | 0.9085 | 0.9090 | 0.9096 | 0.9101 | 0.9106 | 0.9112 | 0.9117 | 0.9122 | 0.9128 | 0.9133 |
| 8.2 | 0.9138 | 0.9143 | 0.9149 | 0.9154 | 0.9159 | 0.9165 | 0.9170 | 0.9175 | 0.9180 | 0.9186 |
| 8.3 | 0.9191 | 0.9196 | 0.9201 | 0.9206 | 0.9212 | 0.9217 | 0.9222 | 0.9227 | 0.9232 | 0.9238 |
| 8.4 | 0.9243 | 0.9248 | 0.9253 | 0.9258 | 0.9263 | 0.9269 | 0.9274 | 0.9279 | 0.9284 | 0.9289 |
| 8.5 | 0.9294 | 0.9299 | 0.9304 | 0.9309 | 0.9315 | 0.9320 | 0.9325 | 0.9330 | 0.9335 | 0.9340 |
| 8.6 | 0.9345 | 0.9350 | 0.9355 | 0.9360 | 0.9365 | 0.9370 | 0.9375 | 0.9380 | 0.9385 | 0.9390 |
| 8.7 | 0.9395 | 0.9400 | 0.9405 | 0.9410 | 0.9415 | 0.9420 | 0.9425 | 0.9430 | 0.9435 | 0.9440 |
| 8.8 | 0.9445 | 0.9450 | 0.9455 | 0.9460 | 0.9465 | 0.9469 | 0.9474 | 0.9479 | 0.9484 | 0.9489 |
| 8.9 | 0.9494 | 0.9499 | 0.9504 | 0.9509 | 0.9513 | 0.9518 | 0.9523 | 0.9528 | 0.9533 | 0.9538 |
| 9.0 | 0.9542 | 0.9547 | 0.9552 | 0.9557 | 0.9562 | 0.9566 | 0.9571 | 0.9576 | 0.9581 | 0.9586 |
| 9.1 | 0.9590 | 0.9595 | 0.9600 | 0.9605 | 0.9609 | 0.9614 | 0.9619 | 0.9624 | 0.9628 | 0.9633 |
| 9.2 | 0.9638 | 0.9643 | 0.9647 | 0.9652 | 0.9657 | 0.9661 | 0.9666 | 0.9671 | 0.9675 | 0.9680 |
| 9.3 | 0.9685 | 0.9689 | 0.9694 | 0.9699 | 0.9703 | 0.9708 | 0.9713 | 0.9717 | 0.9722 | 0.9727 |
| 9.4 | 0.9731 | 0.9736 | 0.9741 | 0.9745 | 0.9750 | 0.9754 | 0.9759 | 0.9763 | 0.9768 | 0.9773 |
| 9.5 | 0.9777 | 0.9782 | 0.9786 | 0.9791 | 0.9795 | 0.9800 | 0.9805 | 0.9809 | 0.9814 | 0.9818 |
| 9.6 | 0.9823 | 0.9827 | 0.9832 | 0.9836 | 0.9841 | 0.9845 | 0.9850 | 0.9854 | 0.9859 | 0.9863 |
| 9.7 | 0.9868 | 0.9872 | 0.9877 | 0.9881 | 0.9886 | 0.9890 | 0.9894 | 0.9899 | 0.9903 | 0.9908 |
| 9.8 | 0.9912 | 0.9917 | 0.9921 | 0.9926 | 0.9930 | 0.9934 | 0.9939 | 0.9943 | 0.9948 | 0.9952 |
| 9.9 | 0.9956 | 0.9961 | 0.9965 | 0.9969 | 0.9974 | 0.9978 | 0.9983 | 0.9987 | 0.9991 | 0.9996 |

TABLE D. **Trigonometric Functions of Angles**

| $\theta$ deg | deg min | $\sin\theta$ | $\cos\theta$ | $\tan\theta$ | $\csc\theta$ | $\sec\theta$ | $\cot\theta$ | | |
|---|---|---|---|---|---|---|---|---|---|
| 0.0 | 0 0 | 0.0000 | 1.0000 | 0.0000 | No value | 1.0000 | No value | 90 0 | 90.0 |
| 0.1 | 0 6 | 0.0017 | 1.0000 | 0.0017 | 572.96 | 1.0000 | 572.96 | 89 54 | 89.9 |
| 0.2 | 0 12 | 0.0035 | 1.0000 | 0.0035 | 286.48 | 1.0000 | 286.48 | 89 48 | 89.8 |
| 0.3 | 0 18 | 0.0052 | 1.0000 | 0.0052 | 190.99 | 1.0000 | 190.98 | 89 42 | 89.7 |
| 0.4 | 0 24 | 0.0070 | 1.0000 | 0.0070 | 143.24 | 1.0000 | 143.24 | 89 36 | 89.6 |
| 0.5 | 0 30 | 0.0087 | 1.0000 | 0.0087 | 114.59 | 1.0000 | 114.59 | 89 30 | 89.5 |
| 0.6 | 0 36 | 0.0105 | 0.9999 | 0.0105 | 95.495 | 1.0001 | 95.490 | 89 24 | 89.4 |
| 0.7 | 0 42 | 0.0122 | 0.9999 | 0.0122 | 81.853 | 1.0001 | 81.847 | 89 18 | 89.3 |
| 0.8 | 0 48 | 0.0140 | 0.9999 | 0.0140 | 71.622 | 1.0001 | 71.615 | 89 12 | 89.2 |
| 0.9 | 0 54 | 0.0157 | 0.9999 | 0.0157 | 63.665 | 1.0001 | 63.657 | 89 6 | 89.1 |
| 1.0 | 1 0 | 0.0175 | 0.9998 | 0.0175 | 57.299 | 1.0002 | 57.290 | 89 0 | 89.0 |
| 1.1 | 1 6 | 0.0192 | 0.9998 | 0.0192 | 52.090 | 1.0002 | 52.081 | 88 54 | 88.9 |
| 1.2 | 1 12 | 0.0209 | 0.9998 | 0.0209 | 47.750 | 1.0002 | 47.740 | 88 48 | 88.8 |
| 1.3 | 1 18 | 0.0227 | 0.9997 | 0.0227 | 44.077 | 1.0003 | 44.066 | 88 42 | 88.7 |
| 1.4 | 1 24 | 0.0244 | 0.9997 | 0.0244 | 40.930 | 1.0003 | 40.917 | 88 36 | 88.6 |
| 1.5 | 1 30 | 0.0262 | 0.9997 | 0.0262 | 38.202 | 1.0003 | 38.188 | 88 30 | 88.5 |
| 1.6 | 1 36 | 0.0279 | 0.9996 | 0.0279 | 35.815 | 1.0004 | 35.801 | 88 24 | 88.4 |
| 1.7 | 1 42 | 0.0297 | 0.9996 | 0.0297 | 33.708 | 1.0004 | 33.694 | 88 18 | 88.3 |
| 1.8 | 1 48 | 0.0314 | 0.9995 | 0.0314 | 31.836 | 1.0005 | 31.821 | 88 12 | 88.2 |
| 1.9 | 1 54 | 0.0332 | 0.9995 | 0.0332 | 30.161 | 1.0005 | 30.145 | 88 6 | 88.1 |
| 2.0 | 2 0 | 0.0349 | 0.9994 | 0.0349 | 28.654 | 1.0006 | 28.636 | 88 0 | 88.0 |
| 2.1 | 2 6 | 0.0366 | 0.9993 | 0.0367 | 27.290 | 1.0007 | 27.271 | 87 54 | 87.9 |
| 2.2 | 2 12 | 0.0384 | 0.9993 | 0.0384 | 26.050 | 1.0007 | 26.031 | 87 48 | 87.8 |
| 2.3 | 2 18 | 0.0401 | 0.9992 | 0.0402 | 24.918 | 1.0008 | 24.898 | 87 42 | 87.7 |
| 2.4 | 2 24 | 0.0419 | 0.9991 | 0.0419 | 23.880 | 1.0009 | 23.859 | 87 36 | 87.6 |
| 2.5 | 2 30 | 0.0436 | 0.9990 | 0.0437 | 22.926 | 1.0010 | 22.904 | 87 30 | 87.5 |
| 2.6 | 2 36 | 0.0454 | 0.9990 | 0.0454 | 22.044 | 1.0010 | 22.022 | 87 24 | 87.4 |
| 2.7 | 2 42 | 0.0471 | 0.9989 | 0.0472 | 21.229 | 1.0011 | 21.205 | 87 18 | 87.3 |
| 2.8 | 2 48 | 0.0488 | 0.9988 | 0.0489 | 20.471 | 1.0012 | 20.446 | 87 12 | 87.2 |
| 2.9 | 2 54 | 0.0506 | 0.9987 | 0.0507 | 19.766 | 1.0013 | 19.740 | 87 6 | 87.1 |
| 3.0 | 3 0 | 0.0523 | 0.9986 | 0.0524 | 19.107 | 1.0014 | 19.081 | 87 0 | 87.0 |
| 3.1 | 3 6 | 0.0541 | 0.9985 | 0.0542 | 18.492 | 1.0015 | 18.464 | 86 54 | 86.9 |
| 3.2 | 3 12 | 0.0558 | 0.9984 | 0.0559 | 17.914 | 1.0016 | 17.886 | 86 48 | 86.8 |
| 3.3 | 3 18 | 0.0576 | 0.9983 | 0.0577 | 17.372 | 1.0017 | 17.343 | 86 42 | 86.7 |
| 3.4 | 3 24 | 0.0593 | 0.9982 | 0.0594 | 16.862 | 1.0018 | 16.832 | 86 36 | 86.6 |
| 3.5 | 3 30 | 0.0610 | 0.9981 | 0.0612 | 16.380 | 1.0019 | 16.350 | 86 30 | 86.5 |
| 3.6 | 3 36 | 0.0628 | 0.9980 | 0.0629 | 15.926 | 1.0020 | 15.895 | 86 24 | 86.4 |
| 3.7 | 3 42 | 0.0645 | 0.9979 | 0.0647 | 15.496 | 1.0021 | 15.464 | 86 18 | 86.3 |
| 3.8 | 3 48 | 0.0663 | 0.9978 | 0.0664 | 15.089 | 1.0022 | 15.056 | 86 12 | 86.2 |
| 3.9 | 3 54 | 0.0680 | 0.9977 | 0.0682 | 14.703 | 1.0023 | 14.669 | 86 6 | 86.1 |
| 4.0 | 4 0 | 0.0698 | 0.9976 | 0.0699 | 14.336 | 1.0024 | 14.301 | 86 0 | 86.0 |
| 4.1 | 4 6 | 0.0715 | 0.9974 | 0.0717 | 13.987 | 1.0026 | 13.951 | 85 54 | 85.9 |
| 4.2 | 4 12 | 0.0732 | 0.9973 | 0.0734 | 13.654 | 1.0027 | 13.617 | 85 48 | 85.8 |
| 4.3 | 4 18 | 0.0750 | 0.9972 | 0.0752 | 13.337 | 1.0028 | 13.300 | 85 42 | 85.7 |
| 4.4 | 4 24 | 0.0767 | 0.9971 | 0.0769 | 13.035 | 1.0030 | 12.996 | 85 36 | 85.6 |
| 4.5 | 4 30 | 0.0785 | 0.9969 | 0.0787 | 12.746 | 1.0031 | 12.706 | 85 30 | 85.5 |
| 4.6 | 4 36 | 0.0802 | 0.9968 | 0.0805 | 12.469 | 1.0032 | 12.429 | 85 24 | 85.4 |
| 4.7 | 4 42 | 0.0819 | 0.9966 | 0.0822 | 12.204 | 1.0034 | 12.163 | 85 18 | 85.3 |
| 4.8 | 4 48 | 0.0837 | 0.9965 | 0.0840 | 11.951 | 1.0035 | 11.909 | 85 12 | 85.2 |
| 4.9 | 4 54 | 0.0854 | 0.9963 | 0.0857 | 11.707 | 1.0037 | 11.665 | 85 6 | 85.1 |
| | | $\cos\theta$ | $\sin\theta$ | $\cot\theta$ | $\sec\theta$ | $\csc\theta$ | $\tan\theta$ | deg min | $\theta$ deg |

TABLE D. **Trigonometric Functions of Angles** ***(continued)***

| $\theta$ deg | deg min | $\sin\theta$ | $\cos\theta$ | $\tan\theta$ | $\csc\theta$ | $\sec\theta$ | $\cot\theta$ | | |
|---|---|---|---|---|---|---|---|---|---|
| 5.0 | 5 0 | 0.0872 | 0.9962 | 0.0875 | 11.474 | 1.0038 | 11.430 | 85 0 | 85.0 |
| 5.1 | 5 6 | 0.0889 | 0.9960 | 0.0892 | 11.249 | 1.0040 | 11.205 | 84 54 | 84.9 |
| 5.2 | 5 12 | 0.0906 | 0.9959 | 0.0910 | 11.034 | 1.0041 | 10.988 | 84 48 | 84.8 |
| 5.3 | 5 18 | 0.0924 | 0.9957 | 0.0928 | 10.826 | 1.0043 | 10.780 | 84 42 | 84.7 |
| 5.4 | 5 24 | 0.0941 | 0.9956 | 0.0945 | 10.626 | 1.0045 | 10.579 | 84 36 | 84.6 |
| 5.5 | 5 30 | 0.0958 | 0.9954 | 0.0963 | 10.433 | 1.0046 | 10.385 | 84 30 | 84.5 |
| 5.6 | 5 36 | 0.0976 | 0.9952 | 0.0981 | 10.248 | 1.0048 | 10.199 | 84 24 | 84.4 |
| 5.7 | 5 42 | 0.0993 | 0.9951 | 0.0998 | 10.069 | 1.0050 | 10.019 | 84 18 | 84.3 |
| 5.8 | 5 48 | 0.1011 | 0.9949 | 0.1016 | 9.8955 | 1.0051 | 9.8448 | 84 12 | 84.2 |
| 5.9 | 5 54 | 0.1028 | 0.9947 | 0.1033 | 9.7283 | 1.0053 | 9.6768 | 84 6 | 84.1 |
| 6.0 | 6 0 | 0.1045 | 0.9945 | 0.1051 | 9.5668 | 1.0055 | 9.5144 | 84 0 | 84.0 |
| 6.1 | 6 6 | 0.1063 | 0.9943 | 0.1069 | 9.4105 | 1.0057 | 9.3573 | 83 54 | 83.9 |
| 6.2 | 6 12 | 0.1080 | 0.9942 | 0.1086 | 9.2593 | 1.0059 | 9.2052 | 83 48 | 83.8 |
| 6.3 | 6 18 | 0.1097 | 0.9940 | 0.1104 | 9.1129 | 1.0061 | 9.0579 | 83 42 | 83.7 |
| 6.4 | 6 24 | 0.1115 | 0.9938 | 0.1122 | 8.9711 | 1.0063 | 8.9152 | 83 36 | 83.6 |
| 6.5 | 6 30 | 0.1132 | 0.9936 | 0.1139 | 8.8337 | 1.0065 | 8.7769 | 83 30 | 83.5 |
| 6.6 | 6 36 | 0.1149 | 0.9934 | 0.1157 | 8.7004 | 1.0067 | 8.6428 | 83 24 | 83.4 |
| 6.7 | 6 42 | 0.1167 | 0.9932 | 0.1175 | 8.5711 | 1.0069 | 8.5126 | 83 18 | 83.3 |
| 6.8 | 6 48 | 0.1184 | 0.9930 | 0.1192 | 8.4457 | 1.0071 | 8.3863 | 83 12 | 83.2 |
| 6.9 | 6 54 | 0.1201 | 0.9928 | 0.1210 | 8.3238 | 1.0073 | 8.2636 | 83 6 | 83.1 |
| 7.0 | 7 0 | 0.1219 | 0.9925 | 0.1228 | 8.2055 | 1.0075 | 8.1444 | 83 0 | 83.0 |
| 7.1 | 7 6 | 0.1236 | 0.9923 | 0.1246 | 8.0905 | 1.0077 | 8.0285 | 82 54 | 82.9 |
| 7.2 | 7 12 | 0.1253 | 0.9921 | 0.1263 | 7.9787 | 1.0079 | 7.9158 | 82 48 | 82.8 |
| 7.3 | 7 18 | 0.1271 | 0.9919 | 0.1281 | 7.8700 | 1.0082 | 7.8062 | 82 42 | 82.7 |
| 7.4 | 7 24 | 0.1288 | 0.9917 | 0.1299 | 7.7642 | 1.0084 | 7.6996 | 82 36 | 82.6 |
| 7.5 | 7 30 | 0.1305 | 0.9914 | 0.1317 | 7.6613 | 1.0086 | 7.5958 | 82 30 | 82.5 |
| 7.6 | 7 36 | 0.1323 | 0.9912 | 0.1334 | 7.5611 | 1.0089 | 7.4947 | 82 24 | 82.4 |
| 7.7 | 7 42 | 0.1340 | 0.9910 | 0.1352 | 7.4635 | 1.0091 | 7.3962 | 82 18 | 82.3 |
| 7.8 | 7 48 | 0.1357 | 0.9907 | 0.1370 | 7.3684 | 1.0093 | 7.3002 | 82 12 | 82.2 |
| 7.9 | 7 54 | 0.1374 | 0.9905 | 0.1388 | 7.2757 | 1.0096 | 7.2066 | 82 6 | 82.1 |
| 8.0 | 8 0 | 0.1392 | 0.9903 | 0.1405 | 7.1853 | 1.0098 | 7.1154 | 82 0 | 82.0 |
| 8.1 | 8 6 | 0.1409 | 0.9900 | 0.1423 | 7.0972 | 1.0101 | 7.0264 | 81 54 | 81.9 |
| 8.2 | 8 12 | 0.1426 | 0.9898 | 0.1441 | 7.0112 | 1.0103 | 6.9395 | 81 48 | 81.8 |
| 8.3 | 8 18 | 0.1444 | 0.9895 | 0.1459 | 6.9273 | 1.0106 | 6.8548 | 81 42 | 81.7 |
| 8.4 | 8 24 | 0.1461 | 0.9893 | 0.1477 | 6.8454 | 1.0108 | 6.7720 | 81 36 | 81.6 |
| 8.5 | 8 30 | 0.1478 | 0.9890 | 0.1495 | 6.7655 | 1.0111 | 6.6912 | 81 30 | 81.5 |
| 8.6 | 8 36 | 0.1495 | 0.9888 | 0.1512 | 6.6874 | 1.0114 | 6.6122 | 81 24 | 81.4 |
| 8.7 | 8 42 | 0.1513 | 0.9885 | 0.1530 | 6.6111 | 1.0116 | 6.5350 | 81 18 | 81.3 |
| 8.8 | 8 48 | 0.1530 | 0.9882 | 0.1548 | 6.5366 | 1.0119 | 6.4596 | 81 12 | 81.2 |
| 8.9 | 8 54 | 0.1547 | 0.9880 | 0.1566 | 6.4637 | 1.0122 | 6.3859 | 81 6 | 81.1 |
| 9.0 | 9 0 | 0.1564 | 0.9877 | 0.1584 | 6.3925 | 1.0125 | 6.3138 | 81 0 | 81.0 |
| 9.1 | 9 6 | 0.1582 | 0.9874 | 0.1602 | 6.3228 | 1.0127 | 6.2432 | 80 54 | 80.9 |
| 9.2 | 9 12 | 0.1599 | 0.9871 | 0.1620 | 6.2547 | 1.0130 | 6.1742 | 80 48 | 80.8 |
| 9.3 | 9 18 | 0.1616 | 0.9869 | 0.1638 | 6.1880 | 1.0133 | 6.1066 | 80 42 | 80.7 |
| 9.4 | 9 24 | 0.1633 | 0.9866 | 0.1655 | 6.1227 | 1.0136 | 6.0405 | 80 36 | 80.6 |
| 9.5 | 9 30 | 0.1650 | 0.9863 | 0.1673 | 6.0589 | 1.0139 | 5.9758 | 80 30 | 80.5 |
| 9.6 | 9 36 | 0.1668 | 0.9860 | 0.1691 | 5.9963 | 1.0142 | 5.9124 | 80 24 | 80.4 |
| 9.7 | 9 42 | 0.1685 | 0.9857 | 0.1709 | 5.9351 | 1.0145 | 5.8502 | 80 18 | 80.3 |
| 9.8 | 9 48 | 0.1702 | 0.9854 | 0.1727 | 5.8751 | 1.0148 | 5.7894 | 80 12 | 80.2 |
| 9.9 | 9 54 | 0.1719 | 0.9851 | 0.1745 | 5.8164 | 1.0151 | 5.7297 | 80 6 | 80.1 |
| | | $\cos\theta$ | $\sin\theta$ | $\cot\theta$ | $\sec\theta$ | $\csc\theta$ | $\tan\theta$ | deg min | $\theta$ deg |

TABLE D. **Trigonometric Functions of Angles** (*continued*)

| $\theta$ *deg* | *deg min* | $\sin\theta$ | $\cos\theta$ | $\tan\theta$ | $\csc\theta$ | $\sec\theta$ | $\cot\theta$ | | |
|---|---|---|---|---|---|---|---|---|---|
| 10.0 | 10 0 | 0.1736 | 0.9848 | 0.1763 | 5.7588 | 1.0154 | 5.6713 | 80 0 | 80.0 |
| 10.1 | 10 6 | 0.1754 | 0.9845 | 0.1781 | 5.7023 | 1.0157 | 5.6140 | 79 54 | 79.9 |
| 10.2 | 10 12 | 0.1771 | 0.9842 | 0.1799 | 5.6470 | 1.0161 | 5.5578 | 79 48 | 79.8 |
| 10.3 | 10 18 | 0.1788 | 0.9839 | 0.1817 | 5.5928 | 1.0164 | 5.5027 | 79 42 | 79.7 |
| 10.4 | 10 24 | 0.1805 | 0.9836 | 0.1835 | 5.5396 | 1.0167 | 5.4486 | 79 36 | 79.6 |
| 10.5 | 10 30 | 0.1822 | 0.9833 | 0.1853 | 5.4874 | 1.0170 | 5.3955 | 79 30 | 79.5 |
| 10.6 | 10 36 | 0.1840 | 0.9829 | 0.1871 | 5.4362 | 1.0174 | 5.3435 | 79 24 | 79.4 |
| 10.7 | 10 42 | 0.1857 | 0.9826 | 0.1890 | 5.3860 | 1.0177 | 5.2924 | 79 18 | 79.3 |
| 10.8 | 10 48 | 0.1874 | 0.9823 | 0.1908 | 5.3367 | 1.0180 | 5.2422 | 79 12 | 79.2 |
| 10.9 | 10 54 | 0.1891 | 0.9820 | 0.1926 | 5.2883 | 1.0184 | 5.1929 | 79 6 | 79.1 |
| 11.0 | 11 0 | 0.1908 | 0.9816 | 0.1944 | 5.2408 | 1.0187 | 5.1446 | 79 0 | 79.0 |
| 11.1 | 11 6 | 0.1925 | 0.9813 | 0.1962 | 5.1942 | 1.0191 | 5.0970 | 78 54 | 78.9 |
| 11.2 | 11 12 | 0.1942 | 9.9810 | 0.1980 | 5.1484 | 1.0194 | 5.0504 | 78 48 | 78.8 |
| 11.3 | 11 18 | 0.1959 | 0.9806 | 0.1998 | 5.1034 | 1.0198 | 5.0045 | 78 42 | 78.7 |
| 11.4 | 11 24 | 0.1977 | 0.9803 | 0.2016 | 5.0593 | 1.0201 | 4.9595 | 78 36 | 78.6 |
| 11.5 | 11 30 | 0.1994 | 0.9799 | 0.2035 | 5.0159 | 1.0205 | 4.9152 | 78 30 | 78.5 |
| 11.6 | 11 36 | 0.2011 | 0.9796 | 0.2053 | 4.9732 | 1.0209 | 4.8716 | 78 24 | 78.4 |
| 11.7 | 11 42 | 0.2028 | 0.9792 | 0.2071 | 4.9313 | 1.0212 | 4.8288 | 78 18 | 78.3 |
| 11.8 | 11 48 | 0.2045 | 0.9789 | 0.2089 | 4.8901 | 1.0216 | 4.7867 | 78 12 | 78.2 |
| 11.9 | 11 54 | 0.2062 | 0.9785 | 0.2107 | 4.8496 | 1.0220 | 4.7453 | 78 6 | 78.1 |
| 12.0 | 12 0 | 0.2079 | 0.9781 | 0.2126 | 4.8097 | 1.0223 | 4.7046 | 78 0 | 78.0 |
| 12.1 | 12 6 | 0.2096 | 0.9778 | 0.2144 | 4.7706 | 1.0227 | 4.6646 | 77 54 | 77.9 |
| 12.2 | 12 12 | 0.2113 | 0.9774 | 0.2162 | 4.7321 | 1.0231 | 4.6252 | 77 48 | 77.8 |
| 12.3 | 12 18 | 0.2130 | 0.9770 | 0.2180 | 4.6942 | 1.0235 | 4.5864 | 77 42 | 77.7 |
| 12.4 | 12 24 | 0.2147 | 0.9767 | 0.2199 | 4.6569 | 1.0239 | 4.5483 | 77 36 | 77.6 |
| 12.5 | 12 30 | 0.2164 | 0.9763 | 0.2217 | 4.6202 | 1.0243 | 4.5107 | 77 30 | 77.5 |
| 12.6 | 12 36 | 0.2181 | 0.9759 | 0.2235 | 4.5841 | 1.0247 | 4.4737 | 77 24 | 77.4 |
| 12.7 | 12 42 | 0.2198 | 0.9755 | 0.2254 | 4.5486 | 1.0251 | 4.4374 | 77 18 | 77.3 |
| 12.8 | 12 48 | 0.2215 | 0.9751 | 0.2272 | 4.5137 | 1.0255 | 4.4015 | 77 12 | 77.2 |
| 12.9 | 12 54 | 0.2232 | 0.9748 | 0.2290 | 4.4793 | 1.0259 | 4.3662 | 77 6 | 77.1 |
| 13.0 | 13 0 | 0.2250 | 0.9744 | 0.2309 | 4.4454 | 1.0263 | 4.3315 | 77 0 | 77.0 |
| 13.1 | 13 6 | 0.2267 | 0.9740 | 0.2327 | 4.4121 | 1.0267 | 4.2972 | 76 54 | 76.9 |
| 13.2 | 13 12 | 0.2284 | 0.9736 | 0.2345 | 4.3792 | 1.0271 | 4.2635 | 76 48 | 76.8 |
| 13.3 | 13 18 | 0.2300 | 0.9732 | 0.2364 | 4.3469 | 1.0276 | 4.2303 | 76 42 | 76.7 |
| 13.4 | 13 24 | 0.2317 | 0.9728 | 0.2382 | 4.3150 | 1.0280 | 4.1976 | 76 36 | 76.6 |
| 13.5 | 13 30 | 0.2334 | 0.9724 | 0.2401 | 4.2837 | 1.0284 | 4.1653 | 76 30 | 76.5 |
| 13.6 | 13 36 | 0.2351 | 0.9720 | 0.2419 | 4.2528 | 1.0288 | 4.1335 | 76 24 | 76.4 |
| 13.7 | 13 42 | 0.2368 | 0.9715 | 0.2438 | 4.2223 | 1.0293 | 4.1022 | 76 18 | 76.3 |
| 13.8 | 13 48 | 0.2385 | 0.9711 | 0.2456 | 4.1923 | 1.0297 | 4.0713 | 76 12 | 76.2 |
| 13.9 | 13 54 | 0.2402 | 0.9707 | 0.2475 | 4.1627 | 1.0302 | 4.0408 | 76 6 | 76.1 |
| 14.0 | 14 0 | 0.2419 | 0.9703 | 0.2493 | 4.1336 | 1.0306 | 4.0108 | 76 0 | 76.0 |
| 14.1 | 14 6 | 0.2436 | 0.9699 | 0.2512 | 4.1048 | 1.0311 | 3.9812 | 75 54 | 75.9 |
| 14.2 | 14 12 | 0.2453 | 0.9694 | 0.2530 | 4.0765 | 1.0315 | 3.9520 | 75 48 | 75.8 |
| 14.3 | 14 18 | 0.2470 | 0.9690 | 0.2549 | 4.0486 | 1.0320 | 3.9232 | 75 42 | 75.7 |
| 14.4 | 14 24 | 0.2487 | 0.9686 | 0.2568 | 4.0211 | 1.0324 | 3.8947 | 75 36 | 75.6 |
| 14.5 | 14 30 | 0.2504 | 0.9681 | 0.2586 | 3.9939 | 1.0329 | 3.8667 | 75 30 | 75.5 |
| 14.6 | 14 36 | 0.2521 | 0.9677 | 0.2605 | 3.9672 | 1.0334 | 3.8391 | 75 24 | 75.4 |
| 14.7 | 14 42 | 0.2538 | 0.9673 | 0.2623 | 3.9408 | 1.0338 | 3.8118 | 75 18 | 75.3 |
| 14.8 | 14 48 | 0.2554 | 0.9668 | 0.2642 | 3.9147 | 1.0343 | 3.7849 | 75 12 | 75.2 |
| 14.9 | 14 54 | 0.2571 | 0.9664 | 0.2661 | 3.8890 | 1.0348 | 3.7583 | 75 6 | 75.1 |
| | | $\cos\theta$ | $\sin\theta$ | $\cot\theta$ | $\sec\theta$ | $\csc\theta$ | $\tan\theta$ | *deg min* | $\theta$ *deg* |

TABLE D. **Trigonometric Functions of Angles** (*continued*)

| $\theta$ *deg* | *deg min* | $\sin\theta$ | $\cos\theta$ | $\tan\theta$ | $\csc\theta$ | $\sec\theta$ | $\cot\theta$ | | |
|---|---|---|---|---|---|---|---|---|---|
| 15.0 | 15 0 | 0.2588 | 0.9659 | 0.2679 | 3.8637 | 1.0353 | 3.7321 | 75 0 | 75.0 |
| 15.1 | 15 6 | 0.2605 | 0.9655 | 0.2698 | 3.8387 | 1.0358 | 3.7062 | 74 54 | 74.9 |
| 15.2 | 15 12 | 0.2622 | 0.9650 | 0.2717 | 3.8140 | 1.0363 | 3.6806 | 74 48 | 74.8 |
| 15.3 | 15 18 | 0.2639 | 0.9646 | 0.2736 | 3.7897 | 1.0367 | 3.6554 | 74 42 | 74.7 |
| 15.4 | 15 24 | 0.2656 | 0.9641 | 0.2754 | 3.7657 | 1.0372 | 3.6305 | 74 36 | 74.6 |
| 15.5 | 15 30 | 0.2672 | 0.9636 | 0.2773 | 3.7420 | 1.0377 | 3.6059 | 74 30 | 74.5 |
| 15.6 | 15 36 | 0.2689 | 0.9632 | 0.2792 | 3.7186 | 1.0382 | 3.5816 | 74 24 | 74.4 |
| 15.7 | 15 42 | 0.2706 | 0.9627 | 0.2811 | 3.6955 | 1.0388 | 3.5576 | 74 18 | 74.3 |
| 15.8 | 15 48 | 0.2723 | 0.9622 | 0.2830 | 3.6727 | 1.0393 | 3.5339 | 74 12 | 74.2 |
| 15.9 | 15 54 | 0.2740 | 0.9617 | 0.2849 | 3.6502 | 1.0398 | 3.5105 | 74 6 | 74.1 |
| 16.0 | 16 0 | 0.2756 | 0.9613 | 0.2867 | 3.6280 | 1.0403 | 3.4874 | 74 0 | 74.0 |
| 16.1 | 16 6 | 0.2773 | 0.9608 | 0.2886 | 3.6060 | 1.0408 | 3.4646 | 73 54 | 73.9 |
| 16.2 | 16 12 | 0.2790 | 0.9603 | 0.2905 | 3.5843 | 1.0413 | 3.4420 | 73 48 | 73.8 |
| 16.3 | 16 18 | 0.2807 | 0.9598 | 0.2924 | 3.5629 | 1.0419 | 3.4197 | 73 42 | 73.7 |
| 16.4 | 16 24 | 0.2823 | 0.9593 | 0.2943 | 3.5418 | 1.0424 | 3.3977 | 73 36 | 73.6 |
| 16.5 | 16 30 | 0.2840 | 0.9588 | 0.2962 | 3.5209 | 1.0429 | 3.3759 | 73 30 | 73.5 |
| 16.6 | 16 36 | 0.2857 | 0.9583 | 0.2981 | 3.5003 | 1.0435 | 3.3544 | 73 24 | 74.4 |
| 16.7 | 16 42 | 0.2874 | 0.9578 | 0.3000 | 3.4800 | 1.0440 | 3.3332 | 73 18 | 73.3 |
| 16.8 | 16 48 | 0.2890 | 0.9573 | 0.3019 | 3.4598 | 1.0446 | 3.3122 | 73 12 | 73.2 |
| 16.9 | 16 54 | 0.2907 | 0.9568 | 0.3038 | 3.4399 | 1.0451 | 3.2914 | 73 6 | 73.1 |
| 17.0 | 17 0 | 0.2924 | 0.9563 | 0.3057 | 3.4203 | 1.0457 | 3.2709 | 73 0 | 73.0 |
| 17.1 | 17 6 | 0.2940 | 0.9558 | 0.3076 | 3.4009 | 1.0463 | 3.2506 | 72 54 | 72.9 |
| 17.2 | 17 12 | 0.2957 | 0.9553 | 0.3096 | 3.3817 | 1.0468 | 3.2305 | 72 48 | 72.8 |
| 17.3 | 17 18 | 0.2974 | 0.9548 | 0.3115 | 3.3628 | 1.0474 | 3.2106 | 72 42 | 72.7 |
| 17.4 | 17 24 | 0.2990 | 0.9542 | 0.3134 | 3.3440 | 1.0480 | 3.1910 | 72 36 | 72.6 |
| 17.5 | 17 30 | 0.3007 | 0.9537 | 0.3153 | 3.3255 | 1.0485 | 3.1716 | 72 30 | 72.5 |
| 17.6 | 17 36 | 0.3024 | 0.9532 | 0.3172 | 3.3072 | 1.0491 | 3.1524 | 72 24 | 72.4 |
| 17.7 | 17 42 | 0.3040 | 0.9527 | 0.3191 | 3.2891 | 1.0497 | 3.1334 | 72 18 | 72.3 |
| 17.8 | 17 48 | 0.3057 | 0.9521 | 0.3211 | 3.2712 | 1.0503 | 3.1146 | 72 12 | 72.2 |
| 17.9 | 17 54 | 0.3074 | 0.9516 | 0.3230 | 3.2536 | 1.0509 | 3.0961 | 72 6 | 72.1 |
| 18.0 | 18 0 | 0.3090 | 0.9511 | 0.3249 | 3.2361 | 1.0515 | 3.0777 | 72 0 | 72.0 |
| 18.1 | 18 6 | 0.3107 | 0.9505 | 0.3268 | 3.2188 | 1.0521 | 3.0595 | 71 54 | 71.9 |
| 18.2 | 18 12 | 0.3123 | 0.9500 | 0.3288 | 3.2017 | 1.0527 | 3.0415 | 71 48 | 71.8 |
| 18.3 | 18 18 | 0.3140 | 0.9494 | 0.3307 | 3.1848 | 1.0533 | 3.0237 | 71 42 | 71.7 |
| 18.4 | 18 24 | 0.3156 | 0.9489 | 0.3327 | 3.1681 | 1.0539 | 3.0061 | 71 36 | 71.6 |
| 18.5 | 18 30 | 0.3173 | 0.9483 | 0.3346 | 3.1515 | 1.0545 | 2.9887 | 71 30 | 71.5 |
| 18.6 | 18 36 | 0.3190 | 0.9478 | 0.3365 | 3.1352 | 1.0551 | 2.9714 | 71 24 | 71.4 |
| 18.7 | 18 42 | 0.3206 | 0.9472 | 0.3385 | 3.1190 | 1.0557 | 2.9544 | 71 18 | 71.3 |
| 18.8 | 18 48 | 0.3223 | 0.9466 | 0.3404 | 3.1030 | 1.0564 | 2.9375 | 71 12 | 71.2 |
| 18.9 | 18 54 | 0.3239 | 0.9461 | 0.3424 | 3.0872 | 1.0570 | 2.9208 | 71 6 | 71.1 |
| 19.0 | 19 0 | 0.3256 | 0.9455 | 0.3443 | 3.0716 | 1.0576 | 2.9042 | 71 0 | 71.0 |
| 19.1 | 19 6 | 0.3272 | 0.9449 | 0.3463 | 3.0561 | 1.0583 | 2.8878 | 70 54 | 70.9 |
| 19.2 | 19 12 | 0.3289 | 0.9444 | 0.3482 | 3.0407 | 1.0589 | 2.8716 | 70 48 | 70.8 |
| 19.3 | 19 18 | 0.3305 | 0.9438 | 0.3502 | 3.0256 | 1.0595 | 2.8556 | 70 42 | 70.7 |
| 19.4 | 19 24 | 0.3322 | 0.9432 | 0.3522 | 3.0106 | 1.0602 | 2.8397 | 70 36 | 70.6 |
| 19.5 | 19 30 | 0.3338 | 0.9426 | 0.3541 | 2.9957 | 1.0608 | 2.8239 | 70 30 | 70.5 |
| 19.6 | 19 36 | 0.3355 | 0.9421 | 0.3561 | 2.9811 | 1.0615 | 2.8083 | 70 24 | 70.4 |
| 19.7 | 19 42 | 0.3371 | 0.9415 | 0.3581 | 2.9665 | 1.0622 | 2.7929 | 70 18 | 70.3 |
| 19.8 | 19 48 | 0.3387 | 0.9409 | 0.3600 | 2.9521 | 1.0628 | 2.7776 | 70 12 | 70.2 |
| 19.9 | 19 54 | 0.3404 | 0.9403 | 0.3620 | 2.9379 | 1.0635 | 2.7625 | 70 6 | 70.1 |
| | | $\cos\theta$ | $\sin\theta$ | $\cot\theta$ | $\sec\theta$ | $\csc\theta$ | $\tan\theta$ | *deg min* | $\theta$ *deg* |

TABLE D. **Trigonometric Functions of Angles (*continued*)**

| $\theta$ deg | deg min | sin $\theta$ | cos $\theta$ | tan $\theta$ | csc $\theta$ | sec $\theta$ | cot $\theta$ | | |
|---|---|---|---|---|---|---|---|---|---|
| 20.0 | 20 0 | 0.3420 | 0.9397 | 0.3640 | 2.9238 | 1.0642 | 2.7475 | 70 0 | 70.0 |
| 20.1 | 20 6 | 0.3437 | 0.9391 | 0.3659 | 2.9099 | 1.0649 | 2.7326 | 69 54 | 69.9 |
| 20.2 | 20 12 | 0.3453 | 0.9385 | 0.3679 | 2.8960 | 1.0655 | 2.7179 | 69 48 | 69.8 |
| 20.3 | 20 18 | 0.3469 | 0.9379 | 0.3699 | 2.8824 | 1.0662 | 2.7034 | 69 42 | 69.7 |
| 20.4 | 20 24 | 0.3486 | 0.9373 | 0.3719 | 2.8688 | 1.0669 | 2.6889 | 69 36 | 69.6 |
| 20.5 | 20 30 | 0.3502 | 0.9367 | 0.3739 | 2.8555 | 1.0676 | 2.6746 | 69 30 | 69.5 |
| 20.6 | 20 36 | 0.3518 | 0.9361 | 0.3759 | 2.8422 | 1.0683 | 2.6605 | 69 24 | 69.4 |
| 20.7 | 20 42 | 0.3535 | 0.9354 | 0.3779 | 2.8291 | 1.0690 | 2.6464 | 69 18 | 69.3 |
| 20.8 | 20 48 | 0.3551 | 0.9348 | 0.3799 | 2.8161 | 1.0697 | 2.6325 | 69 12 | 69.2 |
| 20.9 | 20 54 | 0.3567 | 0.9342 | 0.3819 | 2.8032 | 1.0704 | 2.6187 | 69 6 | 69.1 |
| 21.0 | 21 0 | 0.3584 | 0.9336 | 0.3839 | 2.7904 | 1.0711 | 2.6051 | 69 0 | 69.0 |
| 21.1 | 21 6 | 0.3600 | 0.9330 | 0.3859 | 2.7778 | 1.0719 | 2.5916 | 68 54 | 68.9 |
| 21.2 | 21 12 | 0.3616 | 0.9323 | 0.3879 | 2.7653 | 1.0726 | 2.5782 | 68 48 | 68.8 |
| 21.3 | 21 18 | 0.3633 | 0.9317 | 0.3899 | 2.7529 | 1.0733 | 2.5649 | 68 42 | 68.7 |
| 21.4 | 21 24 | 0.3649 | 0.9311 | 0.3919 | 2.7407 | 1.0740 | 2.5517 | 68 36 | 68.6 |
| 21.5 | 21 30 | 0.3665 | 0.9304 | 0.3939 | 2.7285 | 1.0748 | 2.5386 | 68 30 | 68.5 |
| 21.6 | 21 36 | 0.3681 | 0.9298 | 0.3959 | 2.7165 | 1.0755 | 2.5257 | 68 24 | 68.4 |
| 21.7 | 21 42 | 0.3697 | 0.9291 | 0.3979 | 2.7046 | 1.0763 | 2.5129 | 68 18 | 68.3 |
| 21.8 | 21 48 | 0.3714 | 0.9285 | 0.4000 | 2.6927 | 1.0770 | 2.5002 | 68 12 | 68.2 |
| 21.9 | 21 54 | 0.3730 | 0.9278 | 0.4020 | 2.6811 | 1.0778 | 2.4876 | 68 6 | 68.1 |
| 22.0 | 22 0 | 0.3746 | 0.9272 | 0.4040 | 2.6695 | 1.0785 | 2.4751 | 68 0 | 68.0 |
| 22.1 | 22 6 | 0.3762 | 0.9265 | 0.4061 | 2.6580 | 1.0793 | 2.4627 | 67 54 | 67.9 |
| 22.2 | 22 12 | 0.3778 | 0.9259 | 0.4081 | 2.6466 | 1.0801 | 2.4504 | 67 48 | 67.8 |
| 22.3 | 22 18 | 0.3795 | 0.9252 | 0.4101 | 2.6354 | 1.0808 | 2.4383 | 67 42 | 67.7 |
| 22.4 | 22 24 | 0.3811 | 0.9245 | 0.4122 | 2.6242 | 1.0816 | 2.4262 | 67 36 | 67.6 |
| 22.5 | 22 30 | 0.3827 | 0.9239 | 0.4142 | 2.6131 | 1.0824 | 2.4142 | 67 30 | 67.5 |
| 22.6 | 22 36 | 0.3843 | 0.9232 | 0.4163 | 2.6022 | 1.0832 | 2.4023 | 67 24 | 67.4 |
| 22.7 | 22 42 | 0.3859 | 0.9225 | 0.4183 | 2.5913 | 1.0840 | 2.3906 | 67 18 | 67.3 |
| 22.8 | 22 48 | 0.3875 | 0.9219 | 0.4204 | 2.5805 | 1.0848 | 2.3789 | 67 12 | 67.2 |
| 22.9 | 22 54 | 0.3891 | 0.9212 | 0.4224 | 2.5699 | 1.0856 | 2.3673 | 67 6 | 67.1 |
| 23.0 | 23 0 | 0.3907 | 0.9205 | 0.4245 | 2.5593 | 1.0864 | 2.3559 | 67 0 | 67.0 |
| 23.1 | 23 6 | 0.3923 | 0.9198 | 0.4265 | 2.5488 | 1.0872 | 2.3445 | 66 54 | 66.9 |
| 23.2 | 23 12 | 0.3939 | 0.9191 | 0.4286 | 2.5384 | 1.0880 | 2.3332 | 66 48 | 66.8 |
| 23.3 | 23 18 | 0.3955 | 0.9184 | 0.4307 | 2.5282 | 1.0888 | 2.3220 | 66 42 | 66.7 |
| 23.4 | 23 24 | 0.3971 | 0.9178 | 0.4327 | 2.5180 | 1.0896 | 2.3109 | 66 36 | 66.6 |
| 23.5 | 23 30 | 0.3987 | 0.9171 | 0.4348 | 2.5078 | 1.0904 | 2.2998 | 66 30 | 66.5 |
| 23.6 | 23 36 | 0.4003 | 0.9164 | 0.4369 | 2.4978 | 1.0913 | 2.2889 | 66 24 | 66.4 |
| 23.7 | 23 42 | 0.4019 | 0.9157 | 0.4390 | 2.4879 | 1.0921 | 2.2781 | 66 18 | 66.3 |
| 23.8 | 23 48 | 0.4035 | 0.9150 | 0.4411 | 2.4780 | 1.0929 | 2.2673 | 66 12 | 66.2 |
| 23.9 | 23 54 | 0.4051 | 0.9143 | 0.4431 | 2.4683 | 1.0938 | 2.2566 | 66 6 | 66.1 |
| 24.0 | 24 0 | 0.4067 | 0.9135 | 0.4452 | 2.4586 | 1.0946 | 2.2460 | 66 0 | 66.0 |
| 24.1 | 24 6 | 0.4083 | 0.9128 | 0.4473 | 2.4490 | 1.0955 | 2.2355 | 65 54 | 65.9 |
| 24.2 | 24 12 | 0.4099 | 0.9121 | 0.4494 | 2.4395 | 1.0963 | 2.2251 | 65 48 | 65.8 |
| 24.3 | 24 18 | 0.4115 | 0.9114 | 0.4515 | 2.4301 | 1.0972 | 2.2148 | 65 42 | 65.7 |
| 24.4 | 24 24 | 0.4131 | 0.9107 | 0.4536 | 2.4207 | 1.0981 | 2.2045 | 65 36 | 65.6 |
| 24.5 | 24 30 | 0.4147 | 0.9100 | 0.4557 | 2.4114 | 1.0989 | 2.1943 | 65 30 | 65.5 |
| 24.6 | 24 36 | 0.4163 | 0.9092 | 0.4578 | 2.4022 | 1.0998 | 2.1842 | 65 24 | 65.4 |
| 24.7 | 24 42 | 0.4179 | 0.9085 | 0.4599 | 2.3931 | 1.1007 | 2.1742 | 65 18 | 65.3 |
| 24.8 | 24 48 | 0.4195 | 0.9078 | 0.4621 | 2.3841 | 1.1016 | 2.1642 | 65 12 | 65.2 |
| 24.9 | 24 54 | 0.4210 | 0.9070 | 0.4642 | 2.3751 | 1.1025 | 2.1543 | 65 6 | 65.1 |
| | | cos $\theta$ | sin $\theta$ | cot $\theta$ | sec $\theta$ | csc $\theta$ | tan $\theta$ | deg min | $\theta$ deg |

TABLE D. **Trigonometric Functions of Angles (*continued*)**

| $\theta$ deg | deg min | sin $\theta$ | cos $\theta$ | tan $\theta$ | csc $\theta$ | sec $\theta$ | cot $\theta$ | | |
|---|---|---|---|---|---|---|---|---|---|
| 25.0 | 25 0 | 0.4226 | 0.9063 | 0.4663 | 2.3662 | 1.1034 | 2.1445 | 65 0 | 65.0 |
| 25.1 | 25 6 | 0.4242 | 0.9056 | 0.4684 | 2.3574 | 1.1043 | 2.1348 | 64 54 | 64.9 |
| 25.2 | 25 12 | 0.4258 | 0.9048 | 0.4706 | 2.3486 | 1.1052 | 2.1251 | 64 48 | 64.8 |
| 25.3 | 25 18 | 0.4274 | 0.9041 | 0.4727 | 2.3400 | 1.1061 | 2.1155 | 64 42 | 64.7 |
| 25.4 | 25 24 | 0.4289 | 0.9033 | 0.4748 | 2.3314 | 1.1070 | 2.1060 | 64 36 | 64.6 |
| 25.5 | 25 30 | 0.4305 | 0.9026 | 0.4770 | 2.3228 | 1.1079 | 2.0965 | 64 30 | 64.5 |
| 25.6 | 25 36 | 0.4321 | 0.9018 | 0.4791 | 2.3144 | 1.1089 | 2.0872 | 64 24 | 64.4 |
| 25.7 | 25 42 | 0.4337 | 0.9011 | 0.4813 | 2.3060 | 1.1098 | 2.0778 | 64 18 | 64.3 |
| 25.8 | 25 48 | 0.4352 | 0.9003 | 0.4834 | 2.2976 | 1.1107 | 2.0686 | 64 12 | 64.2 |
| 25.9 | 25 54 | 0.4368 | 0.8996 | 0.4856 | 2.2894 | 1.1117 | 2.0594 | 64 6 | 64.1 |
| 26.0 | 26 0 | 0.4384 | 0.8988 | 0.4877 | 2.2812 | 1.1126 | 2.0503 | 64 0 | 64.0 |
| 26.1 | 26 6 | 0.4399 | 0.8980 | 0.4899 | 2.2730 | 1.1136 | 2.0413 | 63 54 | 63.9 |
| 26.2 | 26 12 | 0.4415 | 0.8973 | 0.4921 | 2.2650 | 1.1145 | 2.0323 | 63 48 | 63.8 |
| 26.3 | 26 18 | 0.4431 | 0.8965 | 0.4942 | 2.2570 | 1.1155 | 2.0233 | 63 42 | 63.7 |
| 26.4 | 26 24 | 0.4446 | 0.8957 | 0.4964 | 2.2490 | 1.1164 | 2.0145 | 63 36 | 63.6 |
| 26.5 | 26 30 | 0.4462 | 0.8949 | 0.4986 | 2.2412 | 1.1174 | 2.0057 | 63 30 | 63.5 |
| 26.6 | 26 36 | 0.4478 | 0.8942 | 0.5008 | 2.2333 | 1.1184 | 1.9970 | 63 24 | 63.4 |
| 26.7 | 26 42 | 0.4493 | 0.8934 | 0.5029 | 2.2256 | 1.1194 | 1.9883 | 63 18 | 63.3 |
| 26.8 | 26 48 | 0.4509 | 0.8926 | 0.5051 | 2.2179 | 1.1203 | 1.9797 | 63 12 | 63.2 |
| 26.9 | 26 54 | 0.4524 | 0.8918 | 0.5073 | 2.2103 | 1.1213 | 1.9711 | 63 6 | 63.1 |
| 27.0 | 27 0 | 0.4540 | 0.8910 | 0.5095 | 2.2027 | 1.1223 | 1.9626 | 63 0 | 63.0 |
| 27.1 | 27 6 | 0.4555 | 0.8902 | 0.5117 | 2.1952 | 1.1233 | 1.9542 | 62 54 | 62.9 |
| 27.2 | 27 12 | 0.4571 | 0.8894 | 0.5139 | 2.1877 | 1.1243 | 1.9458 | 62 48 | 62.8 |
| 27.3 | 27 18 | 0.4586 | 0.8886 | 0.5161 | 2.1803 | 1.1253 | 1.9375 | 62 42 | 62.7 |
| 27.4 | 27 24 | 0.4602 | 0.8878 | 0.5184 | 2.1730 | 1.1264 | 1.9292 | 62 36 | 62.6 |
| 27.5 | 27 30 | 0.4617 | 0.8870 | 0.5206 | 2.1657 | 1.1274 | 1.9210 | 62 30 | 62.5 |
| 27.6 | 27 36 | 0.4633 | 0.8862 | 0.5228 | 2.1584 | 1.1284 | 1.9128 | 62 24 | 62.4 |
| 27.7 | 27 42 | 0.4648 | 0.8854 | 0.5250 | 2.1513 | 1.1294 | 1.9047 | 62 18 | 62.3 |
| 27.8 | 27 48 | 0.4664 | 0.8846 | 0.5272 | 2.1441 | 1.1305 | 1.8967 | 62 12 | 62.2 |
| 27.9 | 27 54 | 0.4679 | 0.8838 | 0.5295 | 2.1371 | 1.1315 | 1.8887 | 62 6 | 62.1 |
| 28.0 | 28 0 | 0.4695 | 0.8829 | 0.5317 | 2.1301 | 1.1326 | 1.8807 | 62 0 | 62.0 |
| 28.1 | 28 6 | 0.4710 | 0.8821 | 0.5339 | 2.1231 | 1.1336 | 1.8728 | 61 54 | 61.9 |
| 28.2 | 28 12 | 0.4726 | 0.8813 | 0.5362 | 2.1162 | 1.1347 | 1.8650 | 61 48 | 61.8 |
| 28.3 | 28 18 | 0.4741 | 0.8805 | 0.5384 | 2.1093 | 1.1357 | 1.8572 | 61 42 | 61.7 |
| 28.4 | 28 24 | 0.4756 | 0.8796 | 0.5407 | 2.1025 | 1.1368 | 1.8495 | 61 36 | 61.6 |
| 28.5 | 28 30 | 0.4772 | 0.8788 | 0.5430 | 2.0957 | 1.1379 | 1.8418 | 61 30 | 61.5 |
| 28.6 | 28 36 | 0.4787 | 0.8780 | 0.5452 | 2.0890 | 1.1390 | 1.8341 | 61 24 | 61.4 |
| 28.7 | 28 42 | 0.4802 | 0.8771 | 0.5475 | 2.0824 | 1.1401 | 1.8265 | 61 18 | 61.3 |
| 28.8 | 28 48 | 0.4818 | 0.8763 | 0.5498 | 2.0758 | 1.1412 | 1.8190 | 61 12 | 61.2 |
| 28.9 | 28 54 | 0.4833 | 0.8755 | 0.5520 | 2.0692 | 1.1423 | 1.8115 | 61 6 | 61.1 |
| 29.0 | 29 0 | 0.4848 | 0.8746 | 0.5543 | 2.0627 | 1.1434 | 1.8040 | 61 0 | 61.0 |
| 29.1 | 29 6 | 0.4863 | 0.8738 | 0.5566 | 2.0562 | 1.1445 | 1.7966 | 60 54 | 60.9 |
| 29.2 | 29 12 | 0.4879 | 0.8729 | 0.5589 | 2.0598 | 1.1456 | 1.7893 | 60 48 | 60.8 |
| 29.3 | 29 18 | 0.4894 | 0.8721 | 0.5612 | 2.0434 | 1.1467 | 1.7820 | 60 42 | 60.7 |
| 29.4 | 29 24 | 0.4909 | 0.8712 | 0.5635 | 2.0371 | 1.1478 | 1.7747 | 60 36 | 60.6 |
| 29.5 | 29 30 | 0.4924 | 0.8704 | 0.5658 | 2.0308 | 1.1490 | 1.7675 | 60 30 | 60.5 |
| 29.6 | 29 36 | 0.4939 | 0.8695 | 0.5681 | 2.0245 | 1.1501 | 1.7603 | 60 24 | 60.4 |
| 29.7 | 29 42 | 0.4955 | 0.8686 | 0.5704 | 2.0183 | 1.1512 | 1.7532 | 60 18 | 60.3 |
| 29.8 | 29 48 | 0.4970 | 0.8678 | 0.5727 | 2.0122 | 1.1524 | 1.7461 | 60 12 | 60.2 |
| 29.9 | 29 54 | 0.4985 | 0.8669 | 0.5750 | 2.0061 | 1.1535 | 1.7391 | 60 6 | 60.1 |
| | | cos $\theta$ | sin $\theta$ | cot $\theta$ | sec $\theta$ | csc $\theta$ | tan $\theta$ | deg min | $\theta$ deg |

## TABLE D. Trigonometric Functions of Angles (*continued*)

| $\theta$ deg | deg min | $\sin\theta$ | $\cos\theta$ | $\tan\theta$ | $\csc\theta$ | $\sec\theta$ | $\cot\theta$ | | |
|---|---|---|---|---|---|---|---|---|---|
| 30.0 | 30 0 | 0.5000 | 0.8660 | 0.5774 | 2.0000 | 1.1547 | 1.7321 | 60 0 | 60.0 |
| 30.1 | 30 6 | 0.5015 | 0.8652 | 0.5797 | 1.9940 | 1.1559 | 1.7251 | 59 54 | 59.9 |
| 30.2 | 30 12 | 0.5030 | 0.8643 | 0.5820 | 1.9880 | 1.1570 | 1.7182 | 59 48 | 59.8 |
| 30.3 | 30 18 | 0.5045 | 0.8634 | 0.5844 | 1.9821 | 1.1582 | 1.7113 | 59 42 | 59.7 |
| 30.4 | 30 24 | 0.5060 | 0.8625 | 0.5867 | 1.9762 | 1.1594 | 1.7045 | 59 36 | 59.6 |
| 30.5 | 30 30 | 0.5075 | 0.8616 | 0.5890 | 1.9703 | 1.1606 | 1.6977 | 59 30 | 59.5 |
| 30.6 | 30 36 | 0.5090 | 0.8607 | 0.5914 | 1.9645 | 1.1618 | 1.6909 | 59 24 | 59.4 |
| 30.7 | 30 42 | 0.5105 | 0.8599 | 0.5938 | 1.9587 | 1.1630 | 1.6842 | 59 18 | 59.3 |
| 30.8 | 30 48 | 0.5120 | 0.8590 | 0.5961 | 1.9530 | 1.1642 | 1.6775 | 59 12 | 59.2 |
| 30.9 | 30 54 | 0.5135 | 0.8581 | 0.5985 | 1.9473 | 1.1654 | 1.6709 | 59 6 | 59.1 |
| 31.0 | 31 0 | 0.5150 | 0.8572 | 0.6009 | 1.9416 | 1.1666 | 1.6643 | 59 0 | 59.0 |
| 31.1 | 31 6 | 0.5165 | 0.8563 | 0.6032 | 1.9360 | 1.1679 | 1.6577 | 58 54 | 58.9 |
| 31.2 | 31 12 | 0.5180 | 0.8554 | 0.6056 | 1.9304 | 1.1691 | 1.6512 | 58 48 | 58.8 |
| 31.3 | 31 18 | 0.5195 | 0.8545 | 0.6080 | 1.9249 | 1.1703 | 1.6447 | 58 42 | 58.7 |
| 31.4 | 31 24 | 0.5210 | 0.8536 | 0.6104 | 1.9194 | 1.1716 | 1.6383 | 58 36 | 58.6 |
| 31.5 | 31 30 | 0.5225 | 0.8526 | 0.6128 | 1.9139 | 1.1728 | 1.6319 | 58 30 | 58.5 |
| 31.6 | 31 36 | 0.5240 | 0.8517 | 0.6152 | 1.9084 | 1.1741 | 1.6255 | 58 24 | 58.4 |
| 31.7 | 31 42 | 0.5255 | 0.8508 | 0.6176 | 1.9031 | 1.1753 | 1.6191 | 58 18 | 58.3 |
| 31.8 | 31 48 | 0.5270 | 0.8499 | 0.6200 | 1.8977 | 1.1766 | 1.6128 | 58 12 | 58.2 |
| 31.9 | 31 54 | 0.5284 | 0.8490 | 0.6224 | 1.8924 | 1.1779 | 1.6066 | 58 6 | 58.1 |
| 32.0 | 32 0 | 0.5299 | 0.8480 | 0.6249 | 1.8871 | 1.1792 | 1.6003 | 58 0 | 58.0 |
| 32.1 | 32 6 | 0.5314 | 0.8471 | 0.6273 | 1.8818 | 1.1805 | 1.5941 | 57 54 | 57.9 |
| 32.2 | 32 12 | 0.5329 | 0.8462 | 0.6297 | 1.8766 | 1.1818 | 1.5880 | 57 48 | 57.8 |
| 32.3 | 32 18 | 0.5344 | 0.8453 | 0.6322 | 1.8714 | 1.1831 | 1.5818 | 57 42 | 57.7 |
| 32.4 | 32 24 | 0.5358 | 0.8443 | 0.6346 | 1.8663 | 1.1844 | 1.5757 | 57 36 | 57.6 |
| 32.5 | 32 30 | 0.5373 | 0.8434 | 0.6371 | 1.8612 | 1.1857 | 1.5697 | 57 30 | 57.5 |
| 32.6 | 32 36 | 0.5388 | 0.8425 | 0.6395 | 1.8561 | 1.1870 | 1.5637 | 57 24 | 57.4 |
| 32.7 | 32 42 | 0.5402 | 0.8415 | 0.6420 | 1.8510 | 1.1883 | 1.5577 | 57 18 | 57.3 |
| 32.8 | 32 48 | 0.5417 | 0.8406 | 0.6445 | 1.8460 | 1.1897 | 1.5517 | 57 12 | 57.2 |
| 32.9 | 32 54 | 0.5432 | 0.8396 | 0.6469 | 1.8410 | 1.1910 | 1.5458 | 57 6 | 57.1 |
| 33.0 | 33 0 | 0.5446 | 0.8387 | 0.6494 | 1.8361 | 1.1924 | 1.5399 | 57 0 | 57.0 |
| 33.1 | 33 6 | 0.5461 | 0.8377 | 0.6519 | 1.8312 | 1.1937 | 1.5340 | 56 54 | 56.9 |
| 33.2 | 33 12 | 0.5476 | 0.8368 | 0.6544 | 1.8263 | 1.1951 | 1.5282 | 56 48 | 56.8 |
| 33.3 | 33 18 | 0.5490 | 0.8358 | 0.6569 | 1.8214 | 1.1964 | 1.5224 | 56 42 | 56.7 |
| 33.4 | 33 24 | 0.5505 | 0.8348 | 0.6594 | 1.8166 | 1.1978 | 1.5166 | 56 36 | 56.6 |
| 33.5 | 33 30 | 0.5519 | 0.8339 | 0.6619 | 1.8118 | 1.1992 | 1.5108 | 56 30 | 56.5 |
| 33.6 | 33 36 | 0.5534 | 0.8329 | 0.6644 | 1.8070 | 1.2006 | 1.5051 | 56 24 | 56.4 |
| 33.7 | 33 42 | 0.5548 | 0.8320 | 0.6669 | 1.8023 | 1.2020 | 1.4994 | 56 18 | 56.3 |
| 33.8 | 33 48 | 0.5563 | 0.8310 | 0.6694 | 1.7976 | 1.2034 | 1.4938 | 56 12 | 56.2 |
| 33.9 | 33 54 | 0.5577 | 0.8300 | 0.6720 | 1.7929 | 1.2048 | 1.4882 | 56 6 | 56.1 |
| 34.0 | 34 0 | 0.5592 | 0.8290 | 0.6745 | 1.7883 | 1.2062 | 1.4826 | 56 0 | 56.0 |
| 34.1 | 34 6 | 0.5606 | 0.8281 | 0.6771 | 1.7837 | 1.2076 | 1.4770 | 55 54 | 55.9 |
| 34.2 | 34 12 | 0.5621 | 0.8271 | 0.6796 | 1.7791 | 1.2091 | 1.4715 | 55 48 | 55.8 |
| 34.3 | 34 18 | 0.5635 | 0.8261 | 0.6822 | 1.7745 | 1.2105 | 1.4659 | 55 42 | 55.7 |
| 34.4 | 34 24 | 0.5650 | 0.8251 | 0.6847 | 1.7700 | 1.2120 | 1.4605 | 55 36 | 55.6 |
| 34.5 | 34 30 | 0.5664 | 0.8241 | 0.6873 | 1.7655 | 1.2134 | 1.4550 | 55 30 | 55.5 |
| 34.6 | 34 36 | 0.5678 | 0.8231 | 0.6899 | 1.7610 | 1.2149 | 1.4496 | 55 24 | 55.4 |
| 34.7 | 34 42 | 0.5693 | 0.8221 | 0.6924 | 1.7566 | 1.2163 | 1.4442 | 55 18 | 55.3 |
| 34.8 | 34 48 | 0.5707 | 0.8211 | 0.6950 | 1.7522 | 1.2178 | 1.4388 | 55 12 | 55.2 |
| 34.9 | 34 54 | 0.5721 | 0.8202 | 0.6976 | 1.7478 | 1.2193 | 1.4335 | 55 6 | 55.1 |
| | | $\cos\theta$ | $\sin\theta$ | $\cot\theta$ | $\sec\theta$ | $\csc\theta$ | $\tan\theta$ | deg min | $\theta$ deg |

## TABLE D. Trigonometric Functions of Angles (*continued*)

| $\theta$ deg | deg min | $\sin\theta$ | $\cos\theta$ | $\tan\theta$ | $\csc\theta$ | $\sec\theta$ | $\cot\theta$ | | |
|---|---|---|---|---|---|---|---|---|---|
| 35.0 | 35 0 | 0.5736 | 0.8192 | 0.7002 | 1.7434 | 1.2208 | 1.4281 | 55 0 | 55.0 |
| 35.1 | 35 6 | 0.5750 | 0.8181 | 0.7028 | 1.7391 | 1.2223 | 1.4229 | 54 54 | 54.9 |
| 35.2 | 35 12 | 0.5764 | 0.8171 | 0.7054 | 1.7348 | 1.2238 | 1.4176 | 54 48 | 54.8 |
| 35.3 | 35 18 | 0.5779 | 0.8161 | 0.7080 | 1.7305 | 1.2253 | 1.4124 | 54 42 | 54.7 |
| 35.4 | 35 24 | 0.5793 | 0.8151 | 0.7107 | 1.7263 | 1.2268 | 1.4071 | 54 36 | 54.6 |
| 35.5 | 35 30 | 0.5807 | 0.8141 | 0.7133 | 1.7221 | 1.2283 | 1.4019 | 54 30 | 54.5 |
| 35.6 | 35 36 | 0.5821 | 0.8131 | 0.7159 | 1.7179 | 1.2299 | 1.3968 | 54 24 | 54.4 |
| 35.7 | 35 42 | 0.5835 | 0.8121 | 0.7186 | 1.7137 | 1.2314 | 1.3916 | 54 18 | 54.3 |
| 35.8 | 35 48 | 0.5850 | 0.8111 | 0.7212 | 1.7095 | 1.2329 | 1.3865 | 54 12 | 54.2 |
| 35.9 | 35 54 | 0.5864 | 0.8100 | 0.7239 | 1.7054 | 1.2345 | 1.3814 | 54 6 | 54.1 |
| 36.0 | 36 0 | 0.5878 | 0.8090 | 0.7265 | 1.7013 | 1.2361 | 1.3764 | 54 0 | 54.0 |
| 36.1 | 36 6 | 0.5892 | 0.8080 | 0.7292 | 1.6972 | 1.2376 | 1.3713 | 53 54 | 53.9 |
| 36.2 | 36 12 | 0.5906 | 0.8070 | 0.7319 | 1.6932 | 1.2392 | 1.3663 | 53 48 | 53.8 |
| 36.3 | 36 18 | 0.5920 | 0.8059 | 0.7346 | 1.6892 | 1.2408 | 1.3613 | 53 42 | 53.7 |
| 36.4 | 36 24 | 0.5934 | 0.8049 | 0.7373 | 1.6852 | 1.2424 | 1.3564 | 53 36 | 53.6 |
| 36.5 | 36 30 | 0.5948 | 0.8039 | 0.7400 | 1.6812 | 1.2440 | 1.3514 | 53 30 | 53.5 |
| 36.6 | 36 36 | 0.5962 | 0.8028 | 0.7427 | 1.6772 | 1.2456 | 1.3465 | 53 24 | 53.4 |
| 36.7 | 36 42 | 0.5976 | 0.8018 | 0.7454 | 1.6733 | 1.2472 | 1.3416 | 53 18 | 53.3 |
| 36.8 | 36 48 | 0.5990 | 0.8007 | 0.7481 | 1.6694 | 1.2489 | 1.3367 | 53 12 | 53.2 |
| 36.9 | 36 54 | 0.6004 | 0.7997 | 0.7508 | 1.6655 | 1.2505 | 1.3319 | 53 6 | 53.1 |
| 37.0 | 37 0 | 0.6018 | 0.7986 | 0.7536 | 1.6616 | 1.2521 | 1.3270 | 53 0 | 53.0 |
| 37.1 | 37 6 | 0.6032 | 0.7976 | 0.7563 | 1.6578 | 1.2538 | 1.3222 | 52 54 | 52.9 |
| 37.2 | 37 12 | 0.6046 | 0.7965 | 0.7590 | 1.6540 | 1.2554 | 1.3175 | 52 48 | 52.8 |
| 37.3 | 37 18 | 0.6060 | 0.7955 | 0.7618 | 1.6502 | 1.2571 | 1.3127 | 52 42 | 52.7 |
| 37.4 | 37 24 | 0.6074 | 0.7944 | 0.7646 | 1.6464 | 1.2588 | 1.3079 | 52 36 | 52.6 |
| 37.5 | 37 30 | 0.6088 | 0.7934 | 0.7673 | 1.6427 | 1.2605 | 1.3032 | 52 30 | 52.5 |
| 37.6 | 37 36 | 0.6101 | 0.7923 | 0.7701 | 1.6390 | 1.2622 | 1.2985 | 52 24 | 52.4 |
| 37.7 | 37 42 | 0.6115 | 0.7912 | 0.7729 | 1.6353 | 1.2639 | 1.2938 | 52 18 | 52.3 |
| 37.8 | 37 48 | 0.6129 | 0.7902 | 0.7757 | 1.6316 | 1.2656 | 1.2892 | 52 12 | 52.2 |
| 37.9 | 37 54 | 0.6143 | 0.7891 | 0.7785 | 1.6279 | 1.2673 | 1.2846 | 52 6 | 52.1 |
| 38.0 | 38 0 | 0.6157 | 0.7880 | 0.7813 | 1.6243 | 1.2690 | 1.2799 | 52 0 | 52.0 |
| 38.1 | 38 6 | 0.6170 | 0.7869 | 0.7841 | 1.6207 | 1.2708 | 1.2753 | 51 54 | 51.9 |
| 38.2 | 38 12 | 0.6184 | 0.7859 | 0.7869 | 1.6171 | 1.2725 | 1.2708 | 51 48 | 51.8 |
| 38.3 | 38 18 | 0.6198 | 0.7848 | 0.7898 | 1.6135 | 1.2742 | 1.2662 | 51 42 | 51.7 |
| 38.4 | 38 24 | 0.6211 | 0.7837 | 0.7926 | 1.6099 | 1.2760 | 1.2617 | 51 36 | 51.6 |
| 38.5 | 38 30 | 0.6225 | 0.7826 | 0.7954 | 1.6064 | 1.2778 | 1.2572 | 51 30 | 51.5 |
| 38.6 | 38 36 | 0.6239 | 0.7815 | 0.7983 | 1.6029 | 1.2796 | 1.2527 | 51 24 | 51.4 |
| 38.7 | 38 42 | 0.6252 | 0.7804 | 0.8012 | 1.5994 | 1.2813 | 1.2482 | 51 18 | 51.3 |
| 38.8 | 38 48 | 0.6266 | 0.7793 | 0.8040 | 1.5959 | 1.2831 | 1.2437 | 51 12 | 51.2 |
| 38.9 | 38 54 | 0.6280 | 0.7782 | 0.8069 | 1.5925 | 1.2849 | 1.2393 | 51 6 | 51.1 |
| 39.0 | 39 0 | 0.6293 | 0.7771 | 0.8098 | 1.5890 | 1.2868 | 1.2349 | 51 0 | 51.0 |
| 39.1 | 39 6 | 0.6307 | 0.7760 | 0.8127 | 1.5856 | 1.2886 | 1.2305 | 50 54 | 50.9 |
| 39.2 | 39 12 | 0.6320 | 0.7749 | 0.8156 | 1.5822 | 1.2904 | 1.2261 | 50 48 | 50.8 |
| 39.3 | 39 18 | 0.6334 | 0.7738 | 0.8185 | 1.5788 | 1.2923 | 1.2218 | 50 42 | 50.7 |
| 39.4 | 39 24 | 0.6347 | 0.7727 | 0.8214 | 1.5755 | 1.2941 | 1.2174 | 50 36 | 50.6 |
| 39.5 | 39 30 | 0.6361 | 0.7716 | 0.8243 | 1.5721 | 1.2960 | 1.2131 | 50 30 | 50.5 |
| 39.6 | 39 36 | 0.6374 | 0.7705 | 0.8273 | 1.5688 | 1.2978 | 1.2088 | 50 24 | 50.4 |
| 39.7 | 39 42 | 0.6388 | 0.7694 | 0.8302 | 1.5655 | 1.2997 | 1.2045 | 50 18 | 50.3 |
| 39.8 | 39 48 | 0.6401 | 0.7683 | 0.8332 | 1.5622 | 1.3016 | 1.2002 | 50 12 | 50.2 |
| 39.9 | 39 54 | 0.6414 | 0.7672 | 0.8361 | 1.5590 | 1.3035 | 1.1960 | 50 6 | 50.1 |
| | | $\cos\theta$ | $\sin\theta$ | $\cot\theta$ | $\sec\theta$ | $\csc\theta$ | $\tan\theta$ | deg min | $\theta$ deg |

TABLE D. **Trigonometric Functions of Angles (*continued*)**

| $\theta$ deg | deg min | $\sin\theta$ | $\cos\theta$ | $\tan\theta$ | $\csc\theta$ | $\sec\theta$ | $\cot\theta$ | | |
|---|---|---|---|---|---|---|---|---|---|
| 40.0 | 40 0 | 0.6428 | 0.7660 | 0.8391 | 1.5557 | 1.3054 | 1.1918 | 50 0 | 50.0 |
| 40.1 | 40 6 | 0.6441 | 0.7649 | 0.8421 | 1.5525 | 1.3073 | 1.1875 | 49 54 | 49.9 |
| 40.2 | 40 12 | 0.6455 | 0.7638 | 0.8451 | 1.5493 | 1.3092 | 1.1833 | 49 48 | 49.8 |
| 40.3 | 40 18 | 0.6468 | 0.7627 | 0.8481 | 1.5461 | 1.3112 | 1.1792 | 49 42 | 49.7 |
| 40.4 | 40 24 | 0.6481 | 0.7615 | 0.8511 | 1.5429 | 1.3131 | 1.1750 | 49 36 | 49.6 |
| 40.5 | 40 30 | 0.6494 | 0.7604 | 0.8541 | 1.5398 | 1.3151 | 1.1708 | 49 30 | 49.5 |
| 40.6 | 40 36 | 0.6508 | 0.7593 | 0.8571 | 1.5366 | 1.3171 | 1.1667 | 49 24 | 49.4 |
| 40.7 | 40 42 | 0.6521 | 0.7581 | 0.8601 | 1.5335 | 1.3190 | 1.1626 | 49 18 | 49.3 |
| 40.8 | 40 48 | 0.6534 | 0.7570 | 0.8632 | 1.5304 | 1.3210 | 1.1585 | 49 12 | 49.2 |
| 40.9 | 40 54 | 0.6547 | 0.7559 | 0.8662 | 1.5273 | 1.3230 | 1.1544 | 49 6 | 49.1 |
| 41.0 | 41 0 | 0.6561 | 0.7547 | 0.8693 | 1.5243 | 1.3250 | 1.1504 | 49 0 | 49.0 |
| 41.1 | 41 6 | 0.6574 | 0.7536 | 0.8724 | 1.5212 | 1.3270 | 1.1463 | 48 54 | 48.9 |
| 41.2 | 41 12 | 0.6587 | 0.7524 | 0.8754 | 1.5182 | 1.3291 | 1.1423 | 48 48 | 48.8 |
| 41.3 | 41 18 | 0.6600 | 0.7513 | 0.8785 | 1.5151 | 1.3311 | 1.1383 | 48 42 | 48.7 |
| 41.4 | 41 24 | 0.6613 | 0.7501 | 0.8816 | 1.5121 | 1.3331 | 1.1343 | 48 36 | 48.6 |
| 41.5 | 41 30 | 0.6626 | 0.7490 | 0.8847 | 1.5092 | 1.3352 | 1.1303 | 48 30 | 48.5 |
| 41.6 | 41 36 | 0.6639 | 0.7478 | 0.8878 | 1.5062 | 1.3373 | 1.1263 | 48 24 | 48.4 |
| 41.7 | 41 42 | 0.6652 | 0.7466 | 0.8910 | 1.5032 | 1.3393 | 1.1224 | 48 18 | 48.3 |
| 41.8 | 41 48 | 0.6665 | 0.7455 | 0.8941 | 1.5003 | 1.3414 | 1.1184 | 48 12 | 48.2 |
| 41.9 | 41 54 | 0.6678 | 0.7443 | 0.8972 | 1.4974 | 1.3435 | 1.1145 | 48 6 | 48.1 |
| 42.0 | 42 0 | 0.6691 | 0.7431 | 0.9004 | 1.4945 | 1.3456 | 1.1106 | 48 0 | 48.0 |
| 42.1 | 42 6 | 0.6704 | 0.7420 | 0.9036 | 1.4916 | 1.3478 | 1.1067 | 47 54 | 47.9 |
| 42.2 | 42 12 | 0.6717 | 0.7408 | 0.9067 | 1.4887 | 1.3499 | 1.1028 | 47 48 | 47.8 |
| 42.3 | 42 18 | 0.6730 | 0.7396 | 0.9099 | 1.4859 | 1.3520 | 1.0990 | 47 42 | 47.7 |
| 42.4 | 42 24 | 0.6743 | 0.7385 | 0.9131 | 1.4830 | 1.3542 | 1.0951 | 47 36 | 47.6 |
| 42.5 | 42 30 | 0.6756 | 0.7373 | 0.9163 | 1.4802 | 1.3563 | 1.0913 | 47 30 | 47.5 |
| 42.6 | 42 36 | 0.6769 | 0.7361 | 0.9195 | 1.4774 | 1.3585 | 1.0875 | 47 24 | 47.4 |
| 42.7 | 42 42 | 0.6782 | 0.7349 | 0.9228 | 1.4746 | 1.3607 | 1.0837 | 47 18 | 47.3 |
| 42.8 | 42 48 | 0.6794 | 0.7337 | 0.9260 | 1.4718 | 1.3629 | 1.0799 | 47 12 | 47.2 |
| 42.9 | 42 54 | 0.6807 | 0.7325 | 0.9293 | 1.4690 | 1.3651 | 1.0761 | 47 6 | 47.1 |
| 43.0 | 43 0 | 0.6820 | 0.7314 | 0.9325 | 1.4663 | 1.3673 | 1.0724 | 47 0 | 47.0 |
| 43.1 | 43 6 | 0.6833 | 0.7302 | 0.9358 | 1.4635 | 1.3696 | 1.0686 | 46 54 | 46.9 |
| 43.2 | 43 12 | 0.6845 | 0.7290 | 0.9391 | 1.4608 | 1.3718 | 1.0649 | 46 48 | 46.8 |
| 43.3 | 43 18 | 0.6858 | 0.7278 | 0.9424 | 1.4581 | 1.3741 | 1.0612 | 46 42 | 46.7 |
| 43.4 | 43 24 | 0.6871 | 0.7266 | 0.9457 | 1.4554 | 1.3763 | 1.0575 | 46 36 | 46.6 |
| 43.5 | 43 30 | 0.6884 | 0.7254 | 0.9490 | 1.4527 | 1.3786 | 1.0538 | 46 30 | 46.5 |
| 43.6 | 43 36 | 0.6896 | 0.7242 | 0.9523 | 1.4501 | 1.3809 | 1.0501 | 46 24 | 46.4 |
| 43.7 | 43 42 | 0.6909 | 0.7230 | 0.9556 | 1.4474 | 1.3832 | 1.0464 | 46 18 | 46.3 |
| 43.8 | 43 48 | 0.6921 | 0.7218 | 0.9590 | 1.4448 | 1.3855 | 1.0428 | 46 12 | 46.2 |
| 43.9 | 43 54 | 0.6934 | 0.7206 | 0.9623 | 1.4422 | 1.3878 | 1.0392 | 46 6 | 46.1 |
| 44.0 | 44 0 | 0.6947 | 0.7193 | 0.9657 | 1.4396 | 1.3902 | 1.0355 | 46 0 | 46.0 |
| 44.1 | 44 6 | 0.6959 | 0.7181 | 0.9691 | 1.4370 | 1.3925 | 1.0319 | 45 54 | 45.9 |
| 44.2 | 44 12 | 0.6972 | 0.7169 | 0.9725 | 1.4344 | 1.3949 | 1.0283 | 45 48 | 45.8 |
| 44.3 | 44 18 | 0.6984 | 0.7157 | 0.9759 | 1.4318 | 1.3972 | 1.0247 | 45 42 | 45.7 |
| 44.4 | 44 24 | 0.6997 | 0.7145 | 0.9793 | 1.4293 | 1.3996 | 1.0212 | 45 36 | 45.6 |
| 44.5 | 44 30 | 0.7009 | 0.7133 | 0.9827 | 1.4267 | 1.4020 | 1.0176 | 45 30 | 45.5 |
| 44.6 | 44 36 | 0.7022 | 0.7120 | 0.9861 | 1.4242 | 1.4044 | 1.0141 | 45 24 | 45.4 |
| 44.7 | 44 42 | 0.7034 | 0.7108 | 0.9896 | 1.4217 | 1.4069 | 1.0105 | 45 18 | 45.3 |
| 44.8 | 44 48 | 0.7046 | 0.7096 | 0.9930 | 1.4192 | 1.4093 | 1.0070 | 45 12 | 45.2 |
| 44.9 | 44 54 | 0.7059 | 0.7083 | 0.9965 | 1.4167 | 1.4118 | 1.0035 | 45 6 | 45.1 |
| 45.0 | 45 0 | 0.7071 | 0.7071 | 1.0000 | 1.4142 | 1.4142 | 1.0000 | 45 0 | 45.0 |
| | | $\cos\theta$ | $\sin\theta$ | $\cot\theta$ | $\sec\theta$ | $\csc\theta$ | $\tan\theta$ | deg min | $\theta$ deg |

TABLE E. **Trigonometric Functions of Real Numbers or Radians**

| Real Number $u$ or $\theta$ Radians | $\sin u$ or $\sin\theta$ | $\cos u$ or $\cos\theta$ | $\tan u$ or $\tan\theta$ | $\cot u$ or $\cot\theta$ | $\sec u$ or $\sec\theta$ | $\csc u$ or $\csc\theta$ |
|---|---|---|---|---|---|---|
| 0.00 | 0.0000 | 1.0000 | 0.0000 | No value | 1.000 | No value |
| 0.01 | 0.0100 | 1.0000 | 0.0100 | 99.997 | 1.000 | 100.00 |
| 0.02 | 0.0200 | 0.9998 | 0.0200 | 49.993 | 1.000 | 50.00 |
| 0.03 | 0.0300 | 0.9996 | 0.0300 | 33.323 | 1.000 | 33.34 |
| 0.04 | 0.0400 | 0.9992 | 0.0400 | 24.987 | 1.001 | 25.01 |
| 0.05 | 0.0500 | 0.9988 | 0.0500 | 19.983 | 1.001 | 20.01 |
| 0.06 | 0.0600 | 0.9982 | 0.0601 | 16.647 | 1.002 | 16.68 |
| 0.07 | 0.0699 | 0.9976 | 0.0701 | 14.262 | 1.002 | 14.30 |
| 0.08 | 0.0799 | 0.9968 | 0.0802 | 12.473 | 1.003 | 12.51 |
| 0.09 | 0.0899 | 0.9960 | 0.0902 | 11.081 | 1.004 | 11.13 |
| 0.10 | 0.0998 | 0.9950 | 0.1003 | 9.967 | 1.005 | 10.02 |
| 0.11 | 0.1098 | 0.9940 | 0.1104 | 9.054 | 1.006 | 9.109 |
| 0.12 | 0.1197 | 0.9928 | 0.1206 | 8.293 | 1.007 | 8.353 |
| 0.13 | 0.1296 | 0.9916 | 0.1307 | 7.649 | 1.009 | 7.714 |
| 0.14 | 0.1395 | 0.9902 | 0.1409 | 7.096 | 1.010 | 7.166 |
| 0.15 | 0.1494 | 0.9888 | 0.1511 | 6.617 | 1.011 | 6.692 |
| 0.16 | 0.1593 | 0.9872 | 0.1614 | 6.197 | 1.013 | 6.277 |
| 0.17 | 0.1692 | 0.9856 | 0.1717 | 5.826 | 1.015 | 5.911 |
| 0.18 | 0.1790 | 0.9838 | 0.1820 | 5.495 | 1.016 | 5.586 |
| 0.19 | 0.1889 | 0.9820 | 0.1923 | 5.200 | 1.018 | 5.295 |
| 0.20 | 0.1987 | 0.9801 | 0.2027 | 4.933 | 1.020 | 5.033 |
| 0.21 | 0.2085 | 0.9780 | 0.2131 | 4.692 | 1.022 | 4.797 |
| 0.22 | 0.2182 | 0.9759 | 0.2236 | 4.472 | 1.025 | 4.582 |
| 0.23 | 0.2280 | 0.9737 | 0.2341 | 4.271 | 1.027 | 4.386 |
| 0.24 | 0.2377 | 0.9713 | 0.2447 | 4.086 | 1.030 | 4.207 |
| 0.25 | 0.2474 | 0.9689 | 0.2553 | 3.916 | 1.032 | 4.042 |
| 0.26 | 0.2571 | 0.9664 | 0.2660 | 3.759 | 1.035 | 3.890 |
| 0.27 | 0.2667 | 0.9638 | 0.2768 | 3.613 | 1.038 | 3.749 |
| 0.28 | 0.2764 | 0.9611 | 0.2876 | 3.478 | 1.041 | 3.619 |
| 0.29 | 0.2860 | 0.9582 | 0.2984 | 3.351 | 1.044 | 3.497 |
| 0.30 | 0.2955 | 0.9553 | 0.3093 | 3.233 | 1.047 | 3.384 |
| 0.31 | 0.3051 | 0.9523 | 0.3203 | 3.122 | 1.050 | 3.278 |
| 0.32 | 0.3146 | 0.9492 | 0.3314 | 3.018 | 1.053 | 3.179 |
| 0.33 | 0.3240 | 0.9460 | 0.3425 | 2.919 | 1.057 | 3.086 |
| 0.34 | 0.3335 | 0.9428 | 0.3537 | 2.827 | 1.061 | 2.999 |
| 0.35 | 0.3429 | 0.9394 | 0.3650 | 2.740 | 1.065 | 2.916 |
| 0.36 | 0.3523 | 0.9359 | 0.3764 | 2.657 | 1.068 | 2.839 |
| 0.37 | 0.3616 | 0.9323 | 0.3879 | 2.578 | 1.073 | 2.765 |
| 0.38 | 0.3709 | 0.9287 | 0.3994 | 2.504 | 1.077 | 2.696 |
| 0.39 | 0.3802 | 0.9249 | 0.4111 | 2.433 | 1.081 | 2.630 |

TABLE E. **Trigonometric Functions of Real Numbers or Radians (*continued*)**

| Real Number $u$ or $\theta$ Radians | $\sin u$ or $\sin \theta$ | $\cos u$ or $\cos \theta$ | $\tan u$ or $\tan \theta$ | $\cot u$ or $\cot \theta$ | $\sec u$ or $\sec \theta$ | $\csc u$ or $\csc \theta$ |
|---|---|---|---|---|---|---|
| 0.40 | 0.3894 | 0.9211 | 0.4228 | 2.365 | 1.086 | 2.568 |
| 0.41 | 0.3986 | 0.9171 | 0.4346 | 2.301 | 1.090 | 2.509 |
| 0.42 | 0.4078 | 0.9131 | 0.4466 | 2.239 | 1.095 | 2.452 |
| 0.43 | 0.4169 | 0.9090 | 0.4586 | 2.180 | 1.100 | 2.399 |
| 0.44 | 0.4259 | 0.9048 | 0.4708 | 2.124 | 1.105 | 2.348 |
| 0.45 | 0.4350 | 0.9004 | 0.4831 | 2.070 | 1.111 | 2.299 |
| 0.46 | 0.4439 | 0.8961 | 0.4954 | 2.018 | 1.116 | 2.253 |
| 0.47 | 0.4529 | 0.8916 | 0.5080 | 1.969 | 1.122 | 2.208 |
| 0.48 | 0.4618 | 0.8870 | 0.5206 | 1.921 | 1.127 | 2.166 |
| 0.49 | 0.4706 | 0.8823 | 0.5334 | 1.875 | 1.133 | 2.125 |
| 0.50 | 0.4794 | 0.8776 | 0.5463 | 1.830 | 1.139 | 2.086 |
| 0.51 | 0.4882 | 0.8727 | 0.5594 | 1.788 | 1.146 | 2.048 |
| 0.52 | 0.4969 | 0.8678 | 0.5726 | 1.747 | 1.152 | 2.013 |
| 0.53 | 0.5055 | 0.8628 | 0.5859 | 1.707 | 1.159 | 1.987 |
| 0.54 | 0.5141 | 0.8577 | 0.5994 | 1.668 | 1.166 | 1.945 |
| 0.55 | 0.5227 | 0.8525 | 0.6131 | 1.631 | 1.173 | 1.913 |
| 0.56 | 0.5312 | 0.8473 | 0.6269 | 1.595 | 1.180 | 1.883 |
| 0.57 | 0.5396 | 0.8419 | 0.6410 | 1.560 | 1.188 | 1.853 |
| 0.58 | 0.5480 | 0.8365 | 0.6552 | 1.526 | 1.196 | 1.825 |
| 0.59 | 0.5564 | 0.8309 | 0.6696 | 1.494 | 1.203 | 1.797 |
| 0.60 | 0.5646 | 0.8253 | 0.6841 | 1.462 | 1.212 | 1.771 |
| 0.61 | 0.5729 | 0.8196 | 0.6989 | 1.431 | 1.220 | 1.746 |
| 0.62 | 0.5810 | 0.8139 | 0.7139 | 1.401 | 1.229 | 1.721 |
| 0.63 | 0.5891 | 0.8080 | 0.7291 | 1.372 | 1.238 | 1.697 |
| 0.64 | 0.5972 | 0.8021 | 0.7445 | 1.343 | 1.247 | 1.674 |
| 0.65 | 0.6052 | 0.7961 | 0.7602 | 1.315 | 1.256 | 1.652 |
| 0.66 | 0.6131 | 0.7900 | 0.7761 | 1.288 | 1.266 | 1.631 |
| 0.67 | 0.6210 | 0.7838 | 0.7923 | 1.262 | 1.276 | 1.610 |
| 0.68 | 0.6288 | 0.7776 | 0.8087 | 1.237 | 1.286 | 1.590 |
| 0.69 | 0.6365 | 0.7712 | 0.8253 | 1.212 | 1.297 | 1.571 |
| 0.70 | 0.6442 | 0.7648 | 0.8423 | 1.187 | 1.307 | 1.552 |
| 0.71 | 0.6518 | 0.7584 | 0.8595 | 1.163 | 1.319 | 1.534 |
| 0.72 | 0.6594 | 0.7518 | 0.8771 | 1.140 | 1.330 | 1.517 |
| 0.73 | 0.6669 | 0.7452 | 0.8949 | 1.117 | 1.342 | 1.500 |
| 0.74 | 0.6743 | 0.7385 | 0.9131 | 1.095 | 1.354 | 1.483 |
| 0.75 | 0.6816 | 0.7317 | 0.9316 | 1.073 | 1.367 | 1.467 |
| 0.76 | 0.6889 | 0.7248 | 0.9505 | 1.052 | 1.380 | 1.452 |
| 0.77 | 0.6961 | 0.7179 | 0.9697 | 1.031 | 1.393 | 1.437 |
| 0.78 | 0.7033 | 0.7109 | 0.9893 | 1.011 | 1.407 | 1.422 |
| 0.79 | 0.7104 | 0.7038 | 1.009 | 0.9908 | 1.421 | 1.408 |

TABLE E. **Trigonometric Functions of Real Numbers or Radians (*continued*)**

| Real Number $u$ or $\theta$ Radians | $\sin u$ or $\sin \theta$ | $\cos u$ or $\cos \theta$ | $\tan u$ or $\tan \theta$ | $\cot u$ or $\cot \theta$ | $\sec u$ or $\sec \theta$ | $\csc u$ or $\csc \theta$ |
|---|---|---|---|---|---|---|
| 0.80 | 0.7174 | 0.6967 | 1.030 | 0.9712 | 1.435 | 1.394 |
| 0.81 | 0.7243 | 0.6895 | 1.050 | 0.9520 | 1.450 | 1.381 |
| 0.82 | 0.7311 | 0.6822 | 1.072 | 0.9331 | 1.466 | 1.368 |
| 0.83 | 0.7379 | 0.6749 | 1.093 | 0.9146 | 1.482 | 1.355 |
| 0.84 | 0.7446 | 0.6675 | 1.116 | 0.8964 | 1.498 | 1.343 |
| 0.85 | 0.7513 | 0.6600 | 1.138 | 0.8785 | 1.515 | 1.331 |
| 0.86 | 0.7578 | 0.6524 | 1.162 | 0.8609 | 1.533 | 1.320 |
| 0.87 | 0.7643 | 0.6448 | 1.185 | 0.8437 | 1.551 | 1.308 |
| 0.88 | 0.7707 | 0.6372 | 1.210 | 0.8267 | 1.569 | 1.297 |
| 0.89 | 0.7771 | 0.6294 | 1.235 | 0.8100 | 1.589 | 1.287 |
| 0.90 | 0.7833 | 0.6216 | 1.260 | 0.7936 | 1.609 | 1.277 |
| 0.91 | 0.7895 | 0.6137 | 1.286 | 0.7774 | 1.629 | 1.267 |
| 0.92 | 0.7956 | 0.6058 | 1.313 | 0.7615 | 1.651 | 1.257 |
| 0.93 | 0.8016 | 0.5978 | 1.341 | 0.7458 | 1.673 | 1.247 |
| 0.94 | 0.8076 | 0.5898 | 1.369 | 0.7303 | 1.696 | 1.238 |
| 0.95 | 0.8134 | 0.5817 | 1.398 | 0.7151 | 1.719 | 1.229 |
| 0.96 | 0.8192 | 0.5735 | 1.428 | 0.7001 | 1.744 | 1.221 |
| 0.97 | 0.8249 | 0.5653 | 1.459 | 0.6853 | 1.769 | 1.212 |
| 0.98 | 0.8305 | 0.5570 | 1.491 | 0.6707 | 1.795 | 1.204 |
| 0.99 | 0.8360 | 0.5487 | 1.524 | 0.6563 | 1.823 | 1.196 |
| 1.00 | 0.8415 | 0.5403 | 1.557 | 0.6421 | 1.851 | 1.188 |
| 1.01 | 0.8468 | 0.5319 | 1.592 | 0.6281 | 1.880 | 1.181 |
| 1.02 | 0.8521 | 0.5234 | 1.628 | 0.6142 | 1.911 | 1.174 |
| 1.03 | 0.8573 | 0.5148 | 1.665 | 0.6005 | 1.942 | 1.166 |
| 1.04 | 0.8624 | 0.5062 | 1.704 | 0.5870 | 1.975 | 1.160 |
| 1.05 | 0.8674 | 0.4976 | 1.743 | 0.5736 | 2.010 | 1.153 |
| 1.06 | 0.8724 | 0.4889 | 1.784 | 0.5604 | 2.046 | 1.146 |
| 1.07 | 0.8772 | 0.4801 | 1.827 | 0.5473 | 2.083 | 1.140 |
| 1.08 | 0.8820 | 0.4713 | 1.871 | 0.5344 | 2.122 | 1.134 |
| 1.09 | 0.8866 | 0.4625 | 1.917 | 0.5216 | 2.162 | 1.128 |
| 1.10 | 0.8912 | 0.4536 | 1.965 | 0.5090 | 2.205 | 1.122 |
| 1.11 | 0.8957 | 0.4447 | 2.014 | 0.4964 | 2.249 | 1.116 |
| 1.12 | 0.9001 | 0.4357 | 2.066 | 0.4840 | 2.295 | 1.111 |
| 1.13 | 0.9044 | 0.4267 | 2.120 | 0.4718 | 2.344 | 1.106 |
| 1.14 | 0.9086 | 0.4176 | 2.176 | 0.4596 | 2.395 | 1.101 |
| 1.15 | 0.9128 | 0.4085 | 2.234 | 0.4475 | 2.448 | 1.096 |
| 1.16 | 0.9168 | 0.3993 | 2.296 | 0.4356 | 2.504 | 1.091 |
| 1.17 | 0.9208 | 0.3902 | 2.360 | 0.4237 | 2.563 | 1.086 |
| 1.18 | 0.9246 | 0.3809 | 2.427 | 0.4120 | 2.625 | 1.082 |
| 1.19 | 0.9284 | 0.3717 | 2.498 | 0.4003 | 2.691 | 1.077 |

TABLE E. **Trigonometric Functions of Real Numbers or Radians (*continued*)**

| Real Number $u$ or $\theta$ Radians | $\sin u$ or $\sin \theta$ | $\cos u$ or $\cos \theta$ | $\tan u$ or $\tan \theta$ | $\cot u$ or $\cot \theta$ | $\sec u$ or $\sec \theta$ | $\csc u$ or $\csc \theta$ |
|---|---|---|---|---|---|---|
| 1.20 | 0.9320 | 0.3624 | 2.572 | 0.3888 | 2.760 | 1.073 |
| 1.21 | 0.9356 | 0.3530 | 2.650 | 0.3773 | 2.833 | 1.069 |
| 1.22 | 0.9391 | 0.3436 | 2.733 | 0.3659 | 2.910 | 1.065 |
| 1.23 | 0.9425 | 0.3342 | 2.820 | 0.3546 | 2.992 | 1.061 |
| 1.24 | 0.9458 | 0.3248 | 2.912 | 0.3434 | 3.079 | 1.057 |
| 1.25 | 0.9490 | 0.3153 | 3.010 | 0.3323 | 3.171 | 1.054 |
| 1.26 | 0.9521 | 0.3058 | 3.113 | 0.3212 | 3.270 | 1.050 |
| 1.27 | 0.9551 | 0.2963 | 3.224 | 0.3102 | 3.375 | 1.047 |
| 1.28 | 0.9580 | 0.2867 | 3.341 | 0.2993 | 3.488 | 1.044 |
| 1.29 | 0.9608 | 0.2771 | 3.467 | 0.2884 | 3.609 | 1.041 |
| 1.30 | 0.9636 | 0.2675 | 3.602 | 0.2776 | 3.738 | 1.038 |
| 1.31 | 0.9662 | 0.2579 | 3.747 | 0.2669 | 3.878 | 1.035 |
| 1.32 | 0.9687 | 0.2482 | 3.903 | 0.2562 | 4.029 | 1.032 |
| 1.33 | 0.9711 | 0.2385 | 4.072 | 0.2456 | 4.193 | 1.030 |
| 1.34 | 0.9735 | 0.2288 | 4.256 | 0.2350 | 4.372 | 1.027 |
| 1.35 | 0.9757 | 0.2190 | 4.455 | 0.2245 | 4.566 | 1.025 |
| 1.36 | 0.9779 | 0.2092 | 4.673 | 0.2140 | 4.779 | 1.023 |
| 1.37 | 0.9799 | 0.1994 | 4.913 | 0.2035 | 5.014 | 1.021 |
| 1.38 | 0.9819 | 0.1896 | 5.177 | 0.1913 | 5.273 | 1.018 |
| 1.39 | 0.9837 | 0.1798 | 5.471 | 0.1828 | 5.561 | 1.017 |
| 1.40 | 0.9854 | 0.1700 | 5.798 | 0.1725 | 5.883 | 1.015 |
| 1.41 | 0.9871 | 0.1601 | 6.165 | 0.1622 | 6.246 | 1.013 |
| 1.42 | 0.9887 | 0.1502 | 6.581 | 0.1519 | 6.657 | 1.011 |
| 1.43 | 0.9901 | 0.1403 | 7.055 | 0.1417 | 7.126 | 1.010 |
| 1.44 | 0.9915 | 0.1304 | 7.602 | 0.1315 | 7.667 | 1.009 |
| 1.45 | 0.9927 | 0.1205 | 8.238 | 0.1214 | 8.299 | 1.007 |
| 1.46 | 0.9939 | 0.1106 | 8.989 | 0.1113 | 9.044 | 1.006 |
| 1.47 | 0.9949 | 0.1006 | 9.887 | 0.1011 | 9.938 | 1.005 |
| 1.48 | 0.9959 | 0.0907 | 10.983 | 0.0910 | 11.029 | 1.004 |
| 1.49 | 0.9967 | 0.0807 | 12.350 | 0.0810 | 12.390 | 1.003 |
| 1.50 | 0.9975 | 0.0707 | 14.101 | 0.0709 | 14.137 | 1.003 |
| 1.51 | 0.9982 | 0.0608 | 16.428 | 0.0609 | 16.458 | 1.002 |
| 1.52 | 0.9987 | 0.0508 | 19.670 | 0.0508 | 19.695 | 1.001 |
| 1.53 | 0.9992 | 0.0408 | 24.498 | 0.0408 | 24.519 | 1.001 |
| 1.54 | 0.9995 | 0.0308 | 32.461 | 0.0308 | 32.476 | 1.000 |
| 1.55 | 0.9998 | 0.0208 | 48.078 | 0.0208 | 48.089 | 1.000 |
| 1.56 | 0.9999 | 0.0108 | 92.620 | 0.0108 | 92.626 | 1.000 |
| 1.57 | 1.0000 | 0.0008 | 1255.8 | 0.0008 | 1255.8 | 1.000 |

# Answers to Odd-Numbered Problems

**The completely worked out solutions to all odd-numbered exercises are contained in the *Student Tutorial.* Students may order this helpful resource through the bookstore.**

## CHAPTER 1

### Exercise 1-1 (pp. 7–8)

**1.** Associative property of addition **3.** Multiplicative identity **5.** Distributive property
**7.** Commutative property of addition **9.** Additive inverse **11.** Associative property of multiplication
**13.** Closure under multiplication **15.** Commutative property of addition **17.** $\{3, 5\}$ **19.** set $A$ itself
**21.** $\{\sqrt{2}, \sqrt{7}\}$ **23.** $0.\overline{54}$ (repeating decimal) **25.** 0.375 (terminating decimal)
**27.** Integers are closed under addition and multiplication.
**29.** The set of all positive real numbers is closed under addition and multiplication.
**31.** The even counting numbers are closed under addition and multiplication.
**33.** The set $\{0\}$ is closed under addition and multiplication. **35.** $\frac{22}{7} > \pi$ **37.** 1.83 **39.** 6
**41.** 0 **43.** 3 **45.** $\frac{1}{40}$ **47.** 27 **49.** 4.4 **51.** $4 < 6$ **53.** $-3 < 0$ **55.** $-1 > -5$
**57.** $\frac{3}{2} < 2$ **59.** $\frac{1}{3} > 0.33$ **61.** $-7, -2\sqrt{3}, 0, \frac{1}{5}, |4-2|, |-3|, 8$ **63.** $-9$ **65.** $-24$
**67.** $-4$ **69.** 2 **71.** $-\frac{7}{2}$ **73.** 3 **75.** $\pi$

**77.**
$$\begin{aligned} 10-(8-6) &\neq (10-8)-6 \\ 10-2 &\neq 2-6 \\ 8 &\neq -4 \end{aligned}$$
(There are many possible examples.)

**79.**
$$\begin{aligned} a-(b-c) &= a-[b+(-c)] && \text{Definition of subtraction} \\ &= a+-[b+(-c)] && \text{Definition of subtraction} \\ &= a+(-b)+[-(-c)] && \text{Property 16 of properties of real numbers} \\ &= a+(-b)+c && \text{Property 12 of properties of real numbers} \\ &= a-b+c && \text{Definition of subtraction} \end{aligned}$$

### Exercise 1-2 (pp. 14–15)

**1.** $16x^8$ **3.** $\frac{8}{x^6y^3}$ **5.** $\frac{y^2}{x^2}$ **7.** $8x^6y^6$ **9.** 1 **11.** $-\frac{1}{2x}$ **13.** $2x^8y^4$ **15.** $288x^{22}y^{16}$
**17.** $-\frac{2}{x}$ **19.** $-\frac{1}{10xy^4}$ **21.** $\frac{1}{x^3y^5}$ **23.** 9 **25.** $\frac{1}{a+b}$ **27.** $\frac{y^4}{x^2z}$ **29.** $\frac{x^8}{y^{16}}$ **31.** $\frac{1}{x^{15}}$

**33.** $\frac{1}{y^7}$ **35.** $-\frac{9}{8x^{11}}$ **37.** $\frac{x^5}{y^{25}}$ **39.** $\frac{y^3}{x^3}$ **41.** $-\frac{8x^8z^8}{9y^{13}}$ **43.** $\frac{9x}{8y^6}$ **45.** $3.456 \times 10^7$
**47.** $3.68 \times 10^{-3}$ **49.** $2.7 \times 10^9$ acres **51.** $5.878 \times 10^{12}$ miles **53.** $4.8 \times 10^{-10}$ electrostatic unit
**55.** 311,000 **57.** 8,000,000,000,000 **59.** 0.00051836 **61.** 0.000000000013 **63.** 241,000,000
**65.** $3 \times 10^7$ **67.** $12 \times 10^{-4} = 1.2 \times 10^{-3}$ **69.** $4 \times 10^{-11}$ **71.** $6x^{m+n+3}$ **73.** $\frac{12x^6}{5}$

## Exercise 1-3 (pp. 22–23)

**1.** 12 **3.** $-2$ **5.** $-1$ **7.** $5\sqrt{5}$ **9.** $x^2y^3$ **11.** $x^3y^2$ **13.** $4xy^2\sqrt[3]{x}$ **15.** $2xy\sqrt[5]{2x}$
**17.** $7\sqrt{5} - 3\sqrt{3}$ **19.** $3\sqrt{2} - 3$ **21.** $\sqrt{2} - \sqrt[3]{2}$ **23.** $6\sqrt{14}$ **25.** $30\sqrt{2} + 15\sqrt{5}$
**27.** $\sqrt{21} + \sqrt{10}$ **29.** $-167$ **31.** $\frac{3\sqrt{5}}{10}$ **33.** $\frac{\sqrt[3]{2xy^2}}{x}$ **35.** $\frac{8\sqrt{x-2}}{x-2}$ **37.** $3 + \sqrt{5}$
**39.** $\sqrt{a} + 1$ **41.** $\sqrt[40]{x^7}$ **43.** $\frac{\sqrt[3]{25x}}{5x}$ **45.** $\frac{4x^2\sqrt{2xy}}{y^2}$ **47.** $x^2y\sqrt{15}$ **49.** $xy^2\sqrt[5]{27x^2}$
**51.** $\frac{6x + 4\sqrt{xy} + 9x\sqrt{y} + 6y\sqrt{x}}{9x - 4y}$ **53.** 2 **55.** $2 \times 10^6$ **57.** $\frac{5}{6} \times 10^{-2}$ **59.** $a - 4 + \frac{4}{a}$
**61.** $\sqrt{4} + \sqrt{9} \neq \sqrt{4 + 9}$ since $5 \neq \sqrt{13}$

## Exercise 1-4 (pp. 26–27)

**1.** $x^{\frac{4}{5}}$ **3.** $(ab)^{\frac{3}{4}}$ or $a^{\frac{3}{4}}b^{\frac{3}{4}}$ **5.** $(a + 2b)^{\frac{3}{4}}$ **7.** $\frac{5}{x^{\frac{1}{3}}}$ **9.** $\sqrt[12]{x^7}$ **11.** $\sqrt[5]{2xy^2}$ **13.** $\sqrt{a^2 + 2b}$
**15.** 2 **17.** $\frac{1}{3}$ **19.** 2 **21.** 25 **23.** $\frac{343}{8}$ **25.** $x^{10}y^5z^{15}$ **27.** $\frac{x^2}{y^4}$ **29.** $256x^8y^{20}z^{12}$
**31.** $\frac{4x^2y^4}{9z^6}$ **33.** $(x + 2y)^{\frac{1}{10}}$ **35.** $\frac{10x^3}{y^3}$ **37.** $\sqrt[3]{2x}$ **39.** $\sqrt[4]{x}$ **41.** $bc\sqrt[4]{ac^3}$ **43.** $\sqrt[6]{x^5y^3}$
**45.** $y^2\sqrt[3]{5xy}$ **47.** $x\sqrt[6]{x^5}$ **49.** $x\sqrt[4]{x^2y^3}$ **51.** $\sqrt[4]{x}$ **53.** $\sqrt[4]{x^3y^2}$ **55.** $\sqrt[6]{8x^4y}$ **57.** $\frac{1}{x^{\frac{5}{2}}y^4}$

## Exercise 1-5 (pp. 32–33)

**1.** 20 **3.** $-25$ **5.** 4 **7.** $\frac{16}{3}$ **9.** $\frac{3}{2}$ **11.** 65 **13.** second degree
**15.** fourth degree **17.** first degree **19.** $3x^2 - x^2y + 3xy^2$ **21.** $2x^2 - xy + 6y^2$
**23.** $ax + bx - a - b$ **25.** $x^3 + x^2 - 3x + 9$ **27.** $x^3 + 27$ **29.** $a^2 - b^2$ **31.** $a^3 + b^3$
**33.** $13x^2 + 4x - 11$ **35.** $-x^3 - 3x^2 + 12x + 1$ **37.** $2x^3 + 10x^2 - 3x - 15$
**39.** $6x^3 + 37x^2 + 4x - 12$ **41.** $4x^2 + 20xy + 25y^2$ **43.** $3x^3 + 13x^2 - 22x + 8$
**45.** $2x^2 - 2x - 16$ **47.** $x^5 + 3x^4 + x^3 - x^2 + x - 1$
**49.** $(x + y)^3 = (x + y)(x + y)^2 = (x + y)(x^2 + 2xy + y^2) = x^3 + 3x^2y + 3xy^2 + y^3$
**51.** $6x^{m+n} + 2x^m + 3x^n + 1$ **53.** $5x^{2m} - 17x^my^n + 6y^{2n}$

## Exercise 1-6 (p. 38)

**1.** $5a(a - 2)$ **3.** $(3x - 5)^2$ **5.** $5(x + 3)(x - 3)$ **7.** $(x + 5)(x - 3)$ **9.** $(2x - 3)(x + 1)$
**11.** $(2x + 1)(3x - 4)$ **13.** $2x(x + 2)(3x - 2)$ **15.** prime **17.** $(a - b)(x + 3y)$ **19.** $(x + 4)^2$
**21.** $(2x - 7y)^2$ **23.** $(x - 10)(x + 2)$ **25.** $(6x - 5)(x + 1)$ **27.** $(2y - 5)(y + 6)$
**29.** $(3y + 10z)(3y - 10z)$ **31.** $(3s - 2t)(9s^2 + 6st + 4t^2)$ **33.** $(x - 5)(a - 4)$
**35.** $(2x + 3)(a + 1)(a - 1)$ **37.** $(x - y^2)(x^2 + xy^2 + y^4)$ **39.** $(2x + 5)(x + 9)$
**41.** $(x + y + 3)(x - y - 1)$ **43.** $(3a + 3x + 4y)(3a - 3x - 4y)$

**45.** $(a+3)(a-3)(3b-2)$ or $(3+a)(3-a)(2-3b)$ **47.** $5x^3(1-y)(1+y+y^2)$ **49.** $3(x-6)^2$
**51.** $2x(2x+1)(2x-3)$ **53.** $3y(2y+1)(y+3)$ **55.** $(x^n-1)(x^n+1)$ **57.** $(y^n-2)(y^n+2)(y^{2n}+1)$
**59.** $(x^2+3xy+5y^2)(x^2-3xy+5y^2)$

## Exercise 1-7 (pp. 44–45)

**1.** $\frac{x+3}{x+8}$ **3.** $\frac{3}{5(x-1)}$ **5.** $x+2$ **7.** $\frac{1}{(x+1)(x-1)}$ **9.** $\frac{x+2}{x+4}$ **11.** $\frac{x+2}{x-1}$

**13.** $-\frac{2}{x(x+6)}$ **15.** $\frac{y-1}{4y-5}$ **17.** $\frac{s}{2s+7}$ **19.** $-\frac{7y+5}{3}$ **21.** $\frac{2y}{y-3}$ **23.** $x(x-4)$

**25.** $\frac{9x^2-4x-25}{(x+1)(x^2-4)}$ **27.** $\frac{3}{x-1}$ **29.** $\frac{-17x-25}{(x+2)(x+1)(x-7)}$ **31.** $\frac{(x+y)^2}{xy}$ **33.** $-x^2+x+6$

**35.** $\frac{2x^2+xy+3y}{x^2y}$ **37.** $\frac{6y-4}{y-1}$ **39.** $\frac{x^2-5x+5}{x-4}$ **41.** $\frac{x+1}{x+3}$ **43.** $\frac{3y+10}{y(y+2)}$

**45.** $\frac{3s^2-4s-12}{(s-2)(s-3)(s+2)}$ **47.** $\frac{11x^2+31x+8}{(x-4)(5x+2)}$ **49.** $\frac{13-3x}{(x-1)(x+1)^2}$ **51.** $\frac{x+3}{(x+7)(x-2)}$

**53.** $\frac{x-1}{x+1}$ **55.** $\frac{y(x-2y)(x+y)}{x^2}$ or $\frac{x^2y-xy^2-2y^3}{x^2}$ **57.** $\frac{-y^2}{1-2y}$ **59.** $\frac{3(y+2)(y-1)}{y(y+3)(y-3)}$

## Chapter 1 Review Exercise (pp. 47–48)

**1.** 4 **3.** commutative property of addition **5.** multiplicative inverse **7.** $4x^8y^9z^8$

**9.** $\frac{x^8}{y^{12}}$ **11.** not a real number **13.** $y\sqrt[4]{4x^3y}$ **15.** $6\sqrt{3}-8\sqrt{5}-24$ **17.** $\frac{\sqrt{3}}{6}$

**19.** $-\frac{5}{3}$ **21.** $-x-2$ **23.** $5x^2y(3x^2-4y+1)$ **25.** $(x+15)(x+1)$ **27.** $(x+28)(x-5)$

**29.** $(3x-2)^2$ **31.** $4x(x+5)(x-3)$ **33.** $(x-3)(a^2+1)(a-1)(a+1)$ **35.** $\frac{3a+2}{2a-3}$

**37.** $-\frac{1}{2x+3}$ **39.** $\frac{2a+b}{a^2+2a+ab}$ **41.** $6.148\times10^{-7}$ **43.** 9 **45.** $\frac{25y^2}{x^4}$

## Practice Test for Chapter 1 (p. 49)

**1.** $\frac{11}{56}$ **3.** additive identity **5.** $\frac{1}{729}$ **7.** $12\sqrt{2}-12+12\sqrt{6}$ **9.** $\sqrt{2}$ **11.** $5a^2b+5a^2b^2$

**13.** $x^3-125$ **15.** $(x+3)(y-1)$ **17.** $\frac{x+7}{(x+9)(x-4)}$ **19.** $\frac{xy-x^2}{y}$ or $\frac{x(y-x)}{y}$

# CHAPTER 2

## Exercise 2-1 (pp. 55–56) These answers may vary slightly depending on the model of the calculator used.

**1.** 256.17 **3.** $1.1183204\times10^{10}$ **5.** 5396.986 **7.** 30.7572 **9.** $4.6309859\times10^{11}$
**11.** $-37.224$ **13.** $-1.8350717\times10^{-12}$ **15.** $-0.55172414$ **17.** 0.2670206 **19.** 8.3257245
**21.** 0.99002063 **23.** 11.981752 **25.** 1664.8245 **27.** $-5606.3677$ **29.** 3.5353529
**31.** 2.0698963 **33.** 1.8300896 **35.** 0.21228087 **37.** 15.800924 **39.** $6.9168738\times10^{-15}$
**41.** 2198.4671 **43.** 3.2541609 **45.** 4.3391716

## Exercise 2-2 (pp. 60–62)

**1.** $\frac{8}{5}$ **3.** 0 **5.** $-10$ **7.** 3 **9.** $-1$ **11.** $-4$ **13.** 19 **15.** $\frac{14}{5}$ **17.** $\frac{45}{4}$

**19.** $\frac{25}{6}$ **21.** $-\frac{1}{5}$ **23.** 10 **25.** $-\frac{9}{8}$ **27.** no solution **29.** 4 **31.** $-2$

**33.** no solution **35.** 3 **37.** 0 **39.** $\frac{1}{34}$ **41.** $-2$ **43.** $\frac{V-k}{g}$ **45.** $\frac{9C+160}{5}$

**47.** $\frac{-2x-5ax}{b+6}$ **49.** 0.41 **51.** $-9.13$ **53.** Conditional equation; it is true only for $x \geq 0$.

**55.** If $y \neq -2$, then it is an identity.

**57.** **a.** In going from step 2 to step 3, you cannot divide both sides of the equation by $x-2$ because if $x = 2$, you are dividing both sides by zero (which is not allowed). **b.** $x = 2$

**59.** $b = -\frac{9}{5}$

## Exercise 2-3 (pp. 68–72)

**1.** 45 women **3.** 30 mg of vitamin A, 90 mg of vitamin B, 40 mg of vitamin C

**5.** Carlos lost 87 lb, Tony lost 58 lb, and Joan lost 70 lb.

**7.** The daughter is now $6\frac{1}{2}$ years old. The father is now $32\frac{1}{2}$ years old. **9.** width 12 meters, length 20 meters

**11.** The first side is 24 cm, the second side is 12 cm, and the third side is 15 cm. **13.** 2 meters wide

**15.** width 9 meters, length 12 meters **17.** \$6400 at 6%; \$8600 at 8% **19.** \$1700 at 14%; \$1300 at 12%

**21.** \$12,000 **23.** average speed of bus 60 mph; average speed of car 40 mph

**25.** It will take 20 hours for Robert to overtake Jim.

**27.** The slower jet is traveling at 768 kilometers per hour; the faster jet is traveling at 832 kilometers per hour.

**29.** 7.5 liters **31.** 21 liters

**33.** 50 gallons of the mixture of 48% alcohol and 30 gallons of the mixture of 80% alcohol

**35.** $5\frac{5}{11}$ hours or *approximately* 5 hours, 27 minutes **37.** 90 minutes or $1\frac{1}{2}$ hours **39.** 6 days **41.** 93

**43.** They must win 17 more games. **45.** 625 bears

## Exercise 2-4 (pp. 81–83)

**1.** $\{-3, 0\}$ **3.** $\{-3, 1\}$ **5.** $\{3\}$ **7.** $\left\{\frac{1}{5}, \frac{4}{3}\right\}$ **9.** $\left\{-2y, \frac{y}{3}\right\}$ **11.** $\left\{\frac{3}{2}\right\}$ **13.** $\{-5, 3\}$

**15.** $\left\{0, -\frac{3}{2}\right\}$ **17.** $\left\{\frac{-7 \pm \sqrt{29}}{10}\right\}$ **19.** $\{-1 \pm 2\sqrt{2}\}$ **21.** $\left\{\frac{-4 \pm \sqrt{31}}{3}\right\}$

**23.** 1, unequal, rational **25.** 5, unequal, irrational **27.** $-11$, no real roots **29.** 625, unequal, rational

**31.** 49, unequal, rational **33.** $\{2 \pm 2i\}$ **35.** $\left\{\frac{1 \pm i}{4}\right\}$ **37.** $\left\{\frac{1 \pm i\sqrt{3}}{2}\right\}$ **39.** $\left\{\frac{7 \pm \sqrt{5}}{22}\right\}$

**41.** $\left\{\frac{-2 \pm \sqrt{14}}{2}\right\}$ **43.** $\left\{-\frac{1}{3}, \frac{3}{2}\right\}$ **45.** $\left\{\frac{4 \pm \sqrt{70}}{6}\right\}$ **47.** $\{2 \pm 2i\}$ **49.** $\left\{\frac{3 \pm \sqrt{3}}{6}\right\}$

**51.** $\{-4, 6\}$ **53.** The other root is $\frac{2}{3}$. **55.** $V = \pm\sqrt{\frac{Fr}{m}}$ **57.** $r = \frac{-b \pm \sqrt{b^2 + ahV}}{ah}$

**59.** width 30 ft; length 60 ft **61.** 6 meters, 8 meters **63.** width 5 meters, length 10 meters

**65.** base 8 meters, altitude 10 meters **67.** $\{-2.12, 1.60\}$ **69.** 3.67 meters

## Exercise 2-5 (pp. 87–88)

**1.** $\{\pm 1, \pm 2\}$ **3.** $\{\pm\sqrt{2}, \pm\sqrt{5}\}$ **5.** $\{\pm i\sqrt{3}, \pm i\sqrt{5}\}$ **7.** $\{16, 81\}$ **9.** $\{-27, 1\}$

**11.** $\left\{-27, \frac{1}{27}\right\}$ **13.** $\{1, 729\}$ **15.** $\{-1, 1\}$ **17.** $\{-11, -4\}$ **19.** $\{-2\}$ **21.** $\{25\}$

**23.** $\{4\}$ **25.** $\{9\}$ **27.** $\{0, 4\}$ **29.** $\{7\}$ **31.** $\{7\}$ **33.** $\{-7, 3\}$ **35.** $\left\{\frac{9}{5}\right\}$

**37.** $\{2\}$ **39.** $h = \frac{\pm\sqrt{A^2 - \pi^2 r^4}}{\pi r}$ **41.** $x = \pm\sqrt{a^{\frac{2}{3}} - y^{\frac{2}{3}})^3}$ **43.** $\{9.199\}$

**45.** $\{-0.392\}$ *Note:* The value $-1.980$ does not check.

## Exercise 2-6 (pp. 92–93)

**1.** $[3, 8]$ **3.** $(-1, 6)$ **5.** $(-4, -2]$ **7.**

**9.**

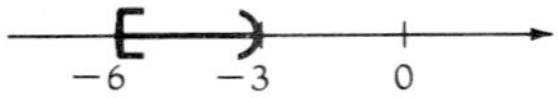


**11.** $(-\infty, 3]$ **13.** $(-\infty, -1]$ **15.** $\left(-\infty, -\frac{1}{2}\right)$

**17.** $(24, +\infty)$ **19.** $(-\infty, 0)$

**21.** $x > 2$

**23.** $x \geq -2$

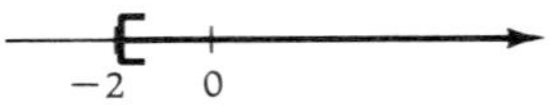


**25.** $x \leq -6$

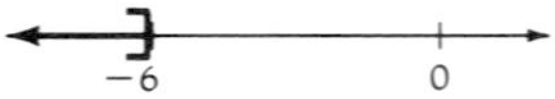


**27.** $-1 < x < 2$

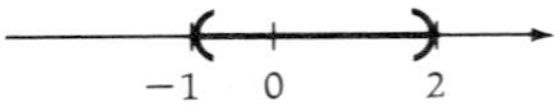


**29.** $2 > x > 1$

**31.** $11 \leq x \leq 14$

**33.** $5 \leq x < \frac{15}{2}$

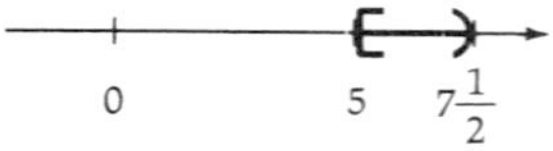


**35.** width $< 2\frac{1}{2}$ feet **37.** base $\leq 11\frac{1}{2}$ meters **39.** $10^\circ \leq$ temperature (in Celsius) $\leq 20^\circ$

**41.** $98 \leq$ test score $\leq 100$ **43.** $x < 1.957$ **45.** 179.641 mm $<$ diameter $<$ 180.899 mm

## Exercise 2-7 (pp. 98–99)

**1.** $(2, 3)$

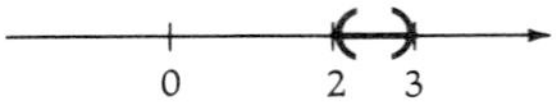


**3.** $(-\infty, -3) \cup (4, \infty)$

**5.** $\left[-\frac{3}{2}, 6\right]$

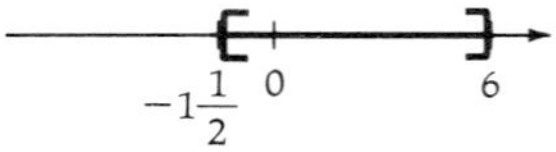


**7.** $(-\infty, -1) \cup \left(\frac{3}{2}, \infty\right)$

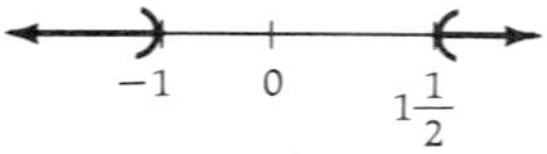


**9.** $\left[-\frac{1}{3}, 2\right]$

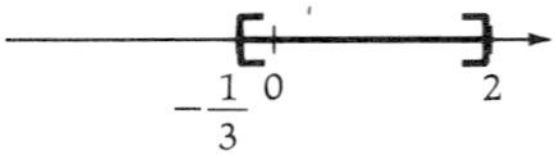


**11.** $(-\infty, -3] \cup [5, \infty)$

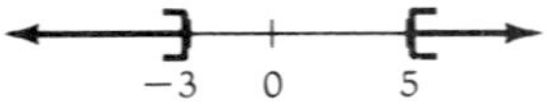


**13.** $\left(-\infty, -\frac{2}{3}\right) \cup (3, \infty)$ **15.** $(-2, 5]$ **17.** $\left(\frac{3}{2}, 5\right)$ **19.** $(-3, -1) \cup (2, \infty)$

**21.** $\left(-\infty, \frac{1}{2}\right] \cup [3, 4]$ **23.** $(-\infty, -1) \cup (2, 4)$ **25.** $(-2, 0) \cup (5, \infty)$ **27.** $(-3, 2) \cup (5, \infty)$

**29.** $(-\infty, 2] \cup (3, 5]$ **31.** no solution **33.** $(-\infty, -1.571) \cup (-1.414, \infty)$ **35.** $(0.354, 5.646)$

## Exercise 2-8 (pp. 103–104)

**1.** $\{-11, 5\}$ **3.** $\{-2, 10\}$ **5.** $\{0, 6\}$ **7.** $\{-3, 2\}$ **9.** $\left\{-\frac{2}{5}, \frac{6}{5}\right\}$ **11.** no solution

**13.** $\left\{-\frac{1}{2}, 5\right\}$ **15.** $\left\{-\frac{7}{5}, \frac{11}{5}\right\}$ **17.** $\{\pm\sqrt{2}, \pm 4\}$ **19.** $\left\{-5, 2, \frac{-3 \pm \sqrt{33}}{2}\right\}$

**21.** $|x| < 3$ **23.** $|x - 1| \leq 5$ **25.** $|x + 3| < 8$ **27.** $(-3, 3)$

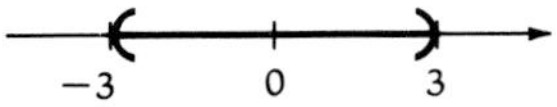


**29.** $\left(-\frac{2}{3}, 2\right)$

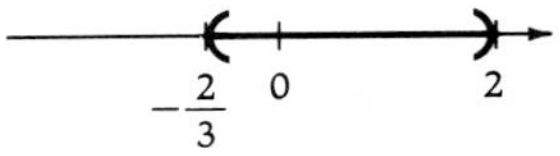


**31.** $[-12, 20]$

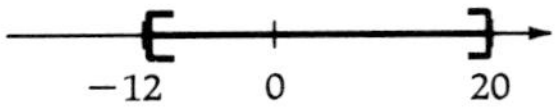


**33.** $\left(-2, \frac{6}{5}\right)$

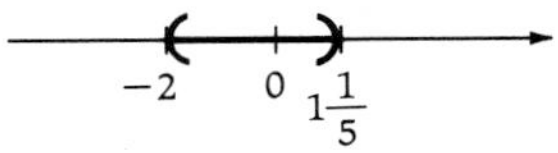


**35.** $[-6, 9]$

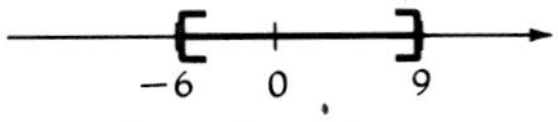


**37.** $(-\infty, -4) \cup (4, \infty)$ **39.** $\left(-\infty, -\frac{3}{2}\right] \cup [0, \infty)$ **41.** $\left(-\infty, -\frac{1}{2}\right) \cup \left(\frac{7}{2}, \infty\right)$

**43.** $(-\infty, -7) \cup (17, \infty)$ **45.** $\left(-\infty, -\frac{3}{4}\right] \cup \left[\frac{7}{4}, \infty\right)$ **47.** $\left(\frac{3}{2}, \infty\right)$

**49.**

$$\left|x - \frac{a+b}{2}\right| < \frac{b-a}{2}$$

$$-\frac{b-a}{2} < x - \frac{a+b}{2} < \frac{b-a}{2}$$

$$-\frac{b-a}{2} + \frac{a+b}{2} < x < \frac{b-a}{2} + \frac{a+b}{2}$$

Finally, adding numerators we have $\frac{2a}{2} < x < \frac{2b}{2}$

$$a < x < b$$

## Chapter 2 Review Exercise (pp. 106–107)

**1.** $\{-3\}$ **3.** $\{1\}$ **5.** no solution **7.** $\{-7, -3\}$ **9.** $\{-2\}$ **11.** 81, real, rational, unequal

**13.** $-7$, no real roots (two complex roots) **15.** $2i\sqrt{3}$ or $2\sqrt{3}\,i$ **17.** $\frac{\sqrt{3}}{2}\,i$ or $\frac{i\sqrt{3}}{2}$ **19.** $\{1 \pm 2i\}$

**21.** $\left\{\frac{3 \pm \sqrt{3}}{2}\right\}$ **23.** $\left\{5, \frac{-5 \pm 5i\sqrt{3}}{2}\right\}$ **25.** $\{16, 625\}$ **27.** $\{-64, 1\}$ **29.** $\{-6\}$

**31.** $\left\{-5, 2, \frac{-3 \pm \sqrt{17}}{2}\right\}$ **33.** $(-\infty, 3]$ **35.** $\left(-\frac{5}{3}, 3\right)$ **37.** $(-7, 3)$ **39.** $(-\infty, 5) \cup (5, +\infty)$

**41.** 57°, 37°, 86° **43.** \$1600 at 12%, \$800 at 10%

## Practice Test for Chapter 2 (pp. 107–108)

**1.** 17, unequal, irrational **3.** $-44$, no real roots (two complex roots) **5.** $\frac{\sqrt{2}}{2}\,i$ or $\frac{i\sqrt{2}}{2}$ **7.** $\{-7\}$

**9.** $\left\{\frac{1}{3}\right\}$ **11.** $\left\{-\frac{3}{2}\right\}$ **13.** $\{1 \pm i\}$ **15.** $\{\pm 1, \pm 3\}$ **17.** $\left\{\frac{-5 \pm \sqrt{5}}{2}, \frac{-5 \pm i\sqrt{11}}{2}\right\}$

**19.** $\left[-\frac{3}{5}, 1\right]$ **21.** 7 articles at \$0.20, 14 articles at \$0.25

## CHAPTER 3

### Exercise 3-1 (pp. 113–114)

**1.**

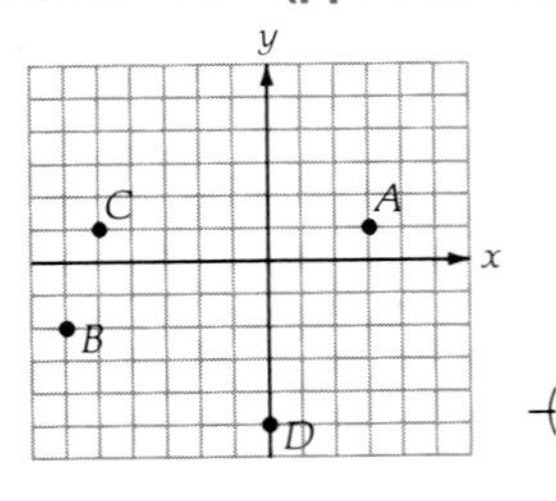


**3.**

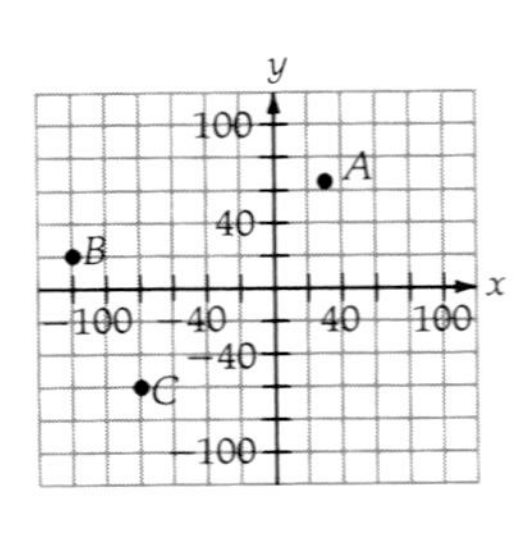


**5.**

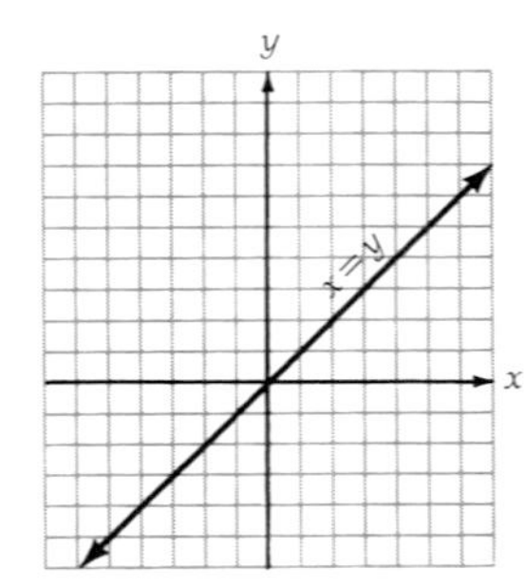


**7.**

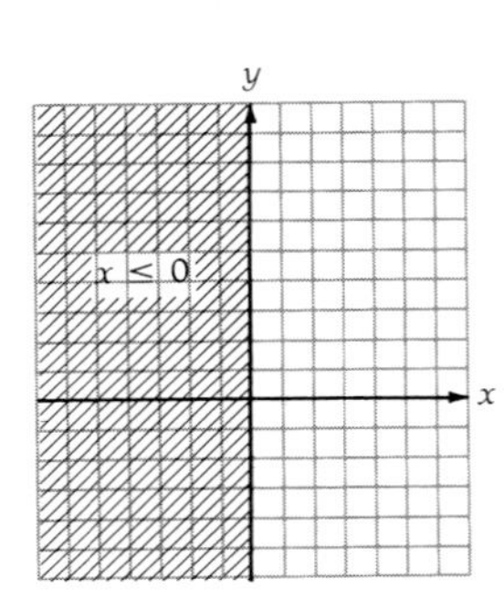


**9.**

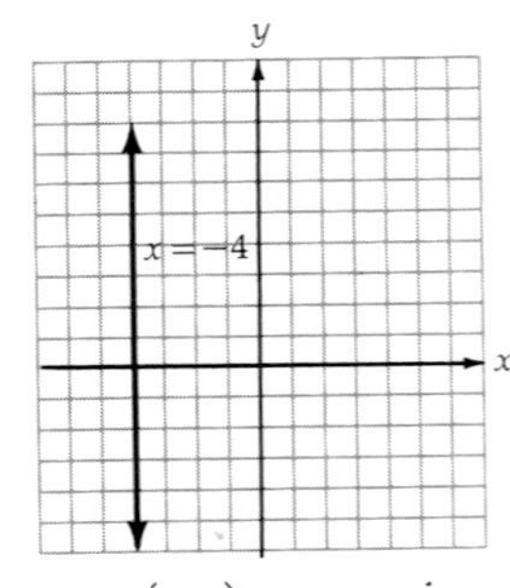


**11.**

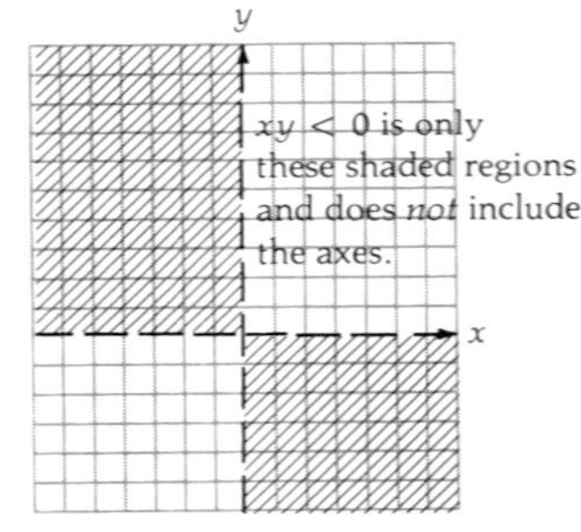


**13.** $d = 8, M = (0, 1)$ **15.** $d = \sqrt{13}, M = \left(4, \frac{1}{2}\right)$ **17.** $d = \sqrt{269}, M = \left(1, \frac{1}{2}\right)$

**19.** $d = 2\sqrt{17}, M = (2, -4)$ **21.** $d = \frac{\sqrt{37}}{6}$ or $\frac{1}{6}\sqrt{37}, M = \left(\frac{5}{12}, -\frac{5}{2}\right)$

**23.** $d \doteq 6.487, M \doteq (1.573, 1.003)$ **25.** $d \doteq 2.806, M \doteq (0.311, 7.621)$ **27.** $B = (-5, 16)$

**29.** Let $A = (3, -4)$, $B = (5, -2)$, and $C = (-2, 1)$. If $\overline{BC}^2 = \overline{AB}^2 + \overline{AC}^2$, then $ABC$ is a right triangle. $\overline{BC} = \sqrt{58}$, $\overline{AB} = \sqrt{8} = 2\sqrt{2}$, $\overline{AC} = \sqrt{50} = 5\sqrt{2}$. $(\sqrt{58})^2 = (2\sqrt{2})^2 + (5\sqrt{2})^2$ Thus it is a right triangle.
$58 = 8 + 50$

**31.** Let $A = (-4, 3)$, $B = (0, 0)$, $C = (3, 4)$, $D = (-1, 7)$. If $\overline{AB} = \overline{BC} = \overline{CD} = \overline{AD}$ and we can demonstrate one right angle, then we establish that $ABCD$ is a square. $\overline{AB} = 5$, $\overline{BC} = 5$, $\overline{CD} = 5$, $\overline{AD} = 5$. Consider triangle $ABC$. $\overline{AC}$ (the diagonal) $= \sqrt{50} = 5\sqrt{2}$. Since $\overline{AB}^2 + \overline{BC}^2 = \overline{AC}^2$, the triangle $ABC$ is a right triangle. Therefore $ABCD$ is a square.

**33.** $\overline{AB} = \sqrt{32} = 4\sqrt{2}$, $\overline{BC} = \sqrt{18} = 3\sqrt{2}$, $\overline{AC} = \sqrt{98} = 7\sqrt{2}$. We see that $\overline{AB} + \overline{BC} = \overline{AC}$. Since the shortest distance between $A$ and $C$ is $7\sqrt{2}$ and is a straight line, it must be that $A$, $B$, and $C$ are collinear.

**35.** $d = a\sqrt{17}, M = \left(3a, \frac{5a}{2}\right)$ **37.** There are *two* values: $x = 1 + 4\sqrt{3}$ or $x = 1 - 4\sqrt{3}$

**39.** Let us draw a rectangle with one corner at (0, 0) as follows:

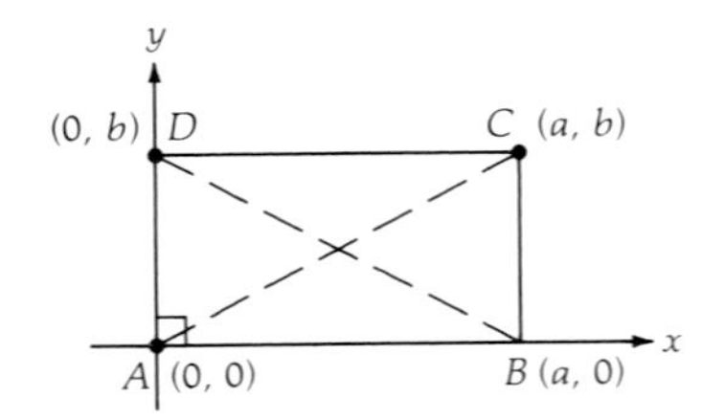


Now we see that $\overline{AC} = \sqrt{a^2 + b^2}$, $\overline{BD} = \sqrt{a^2 + b^2}$. Thus $\overline{AC} = \overline{BD}$.

## Exercise 3-2 (pp. 120–121)

**1.**

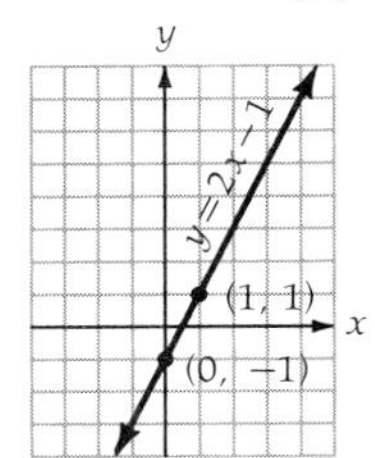


**3.**

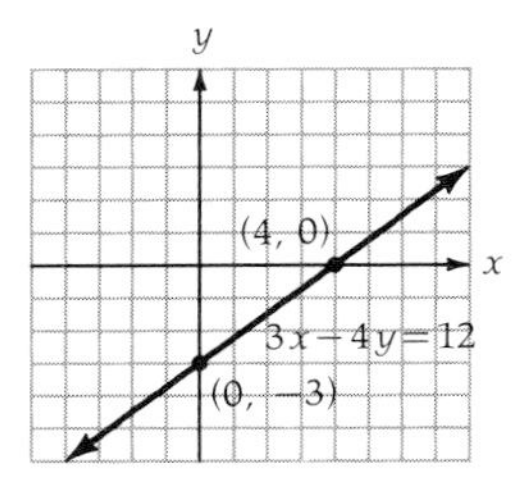


**5.**

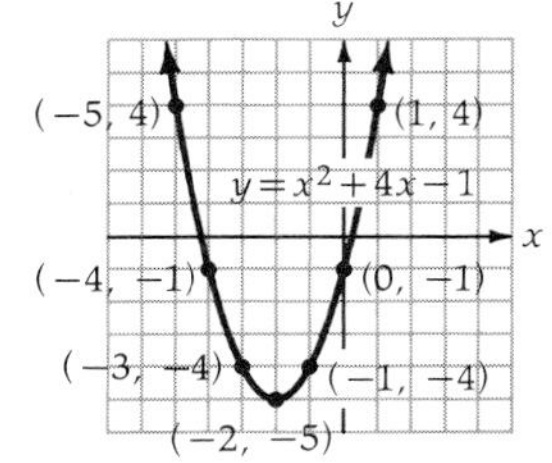


**7.**

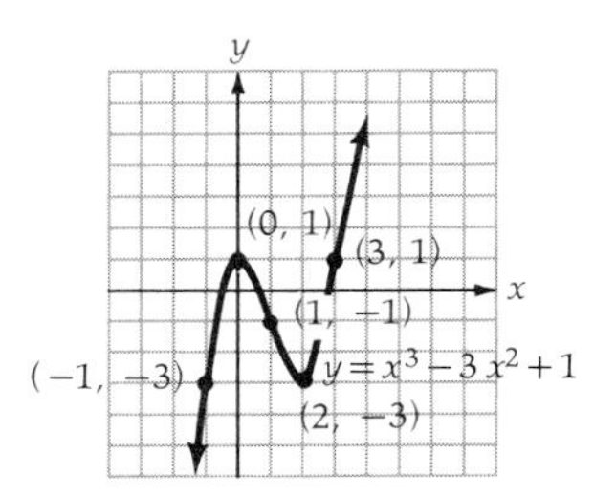


**9.**

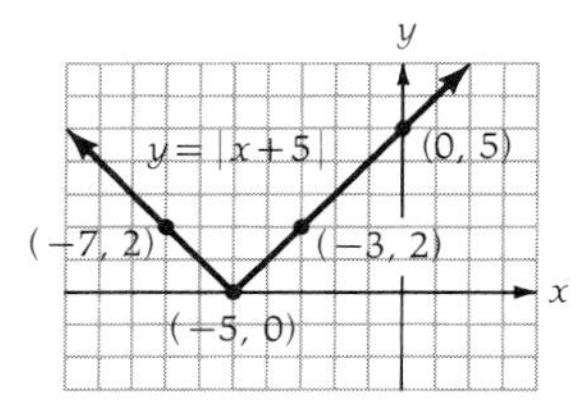


**11.**

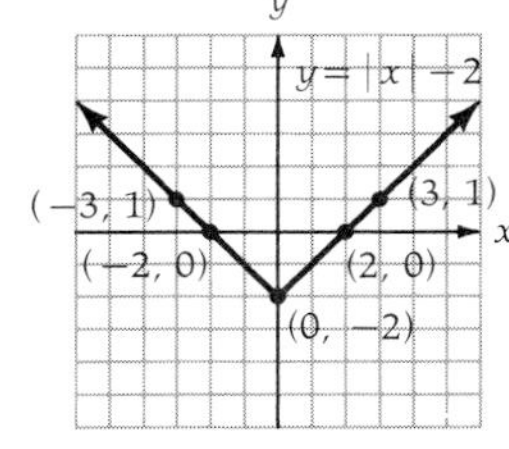


The graph is symmetric with respect to the $y$-axis.

**13.**

$y=|2x+1|$ y

(−3, 5) (2, 5)

(−1, 1) (0, 1) x

$\left(-\frac{1}{2}, 0\right)$

**15.**

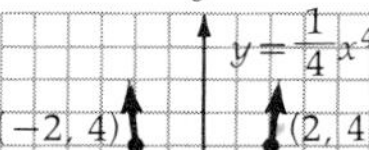


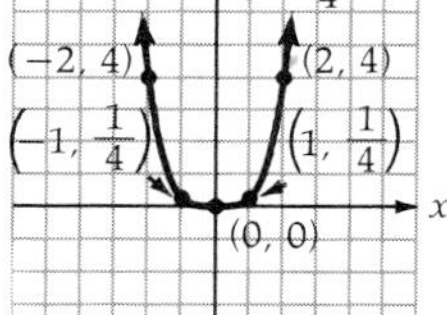


The graph is symmetric with respect to the $y$-axis.

**17.**

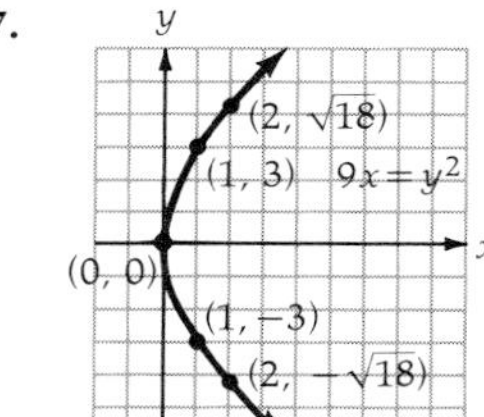


The graph is symmetric with respect to the $x$-axis.

**19.**

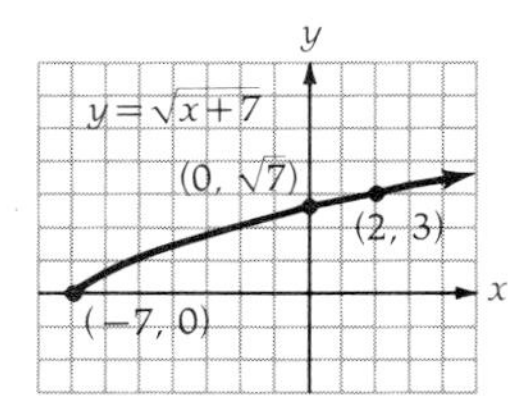


**21.**

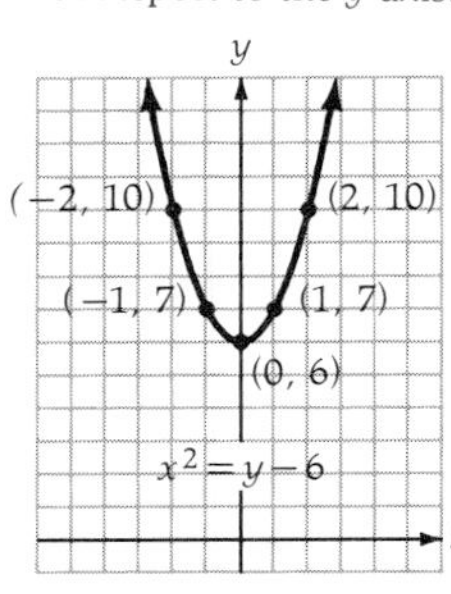


The graph is symmetric with respect to the $y$-axis.

**23.**

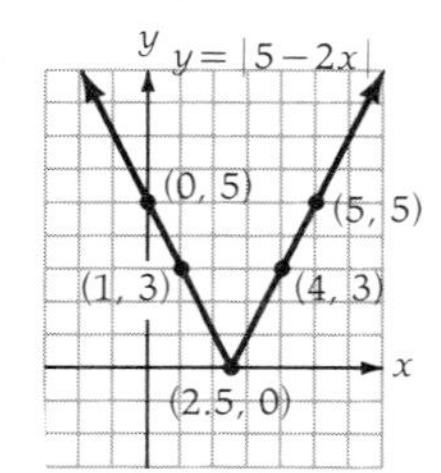

**25.**

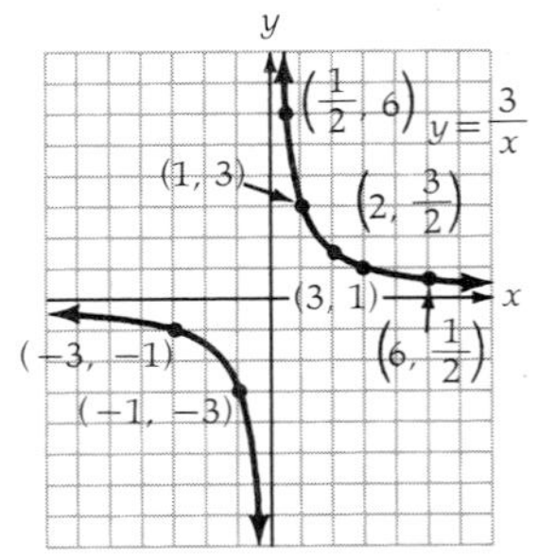


The graph is symmetric with respect to the origin.

**27.**

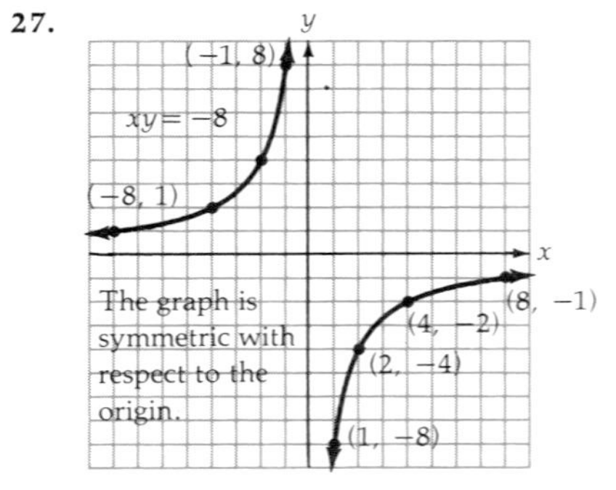


**29.**

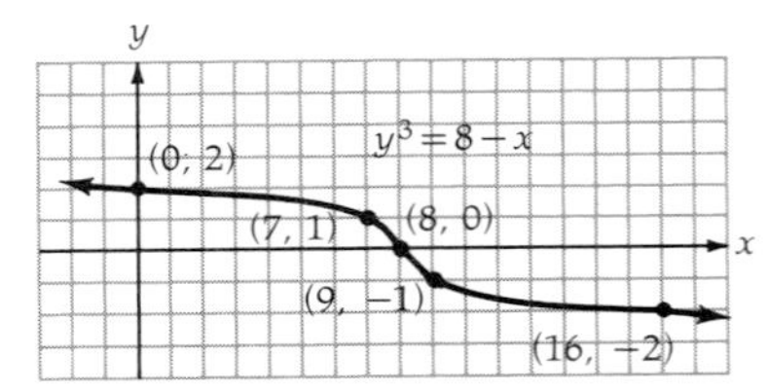


**31.**

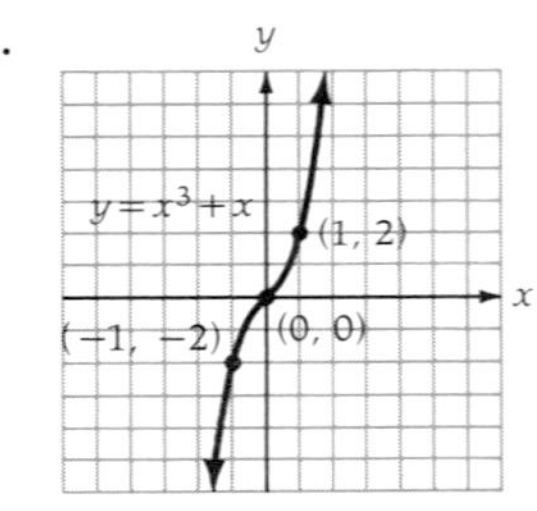


The graph is symmetric with respect to the origin.

**33.**

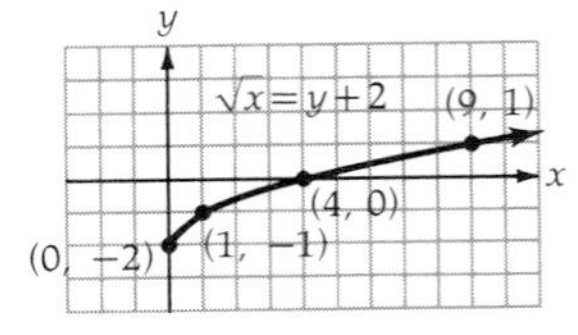


**35.**

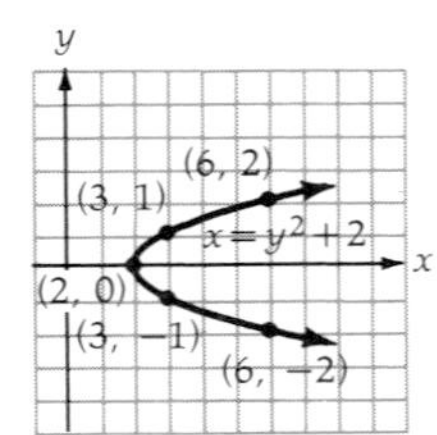


The graph is symmetric with respect to the $x$-axis.

**37.**

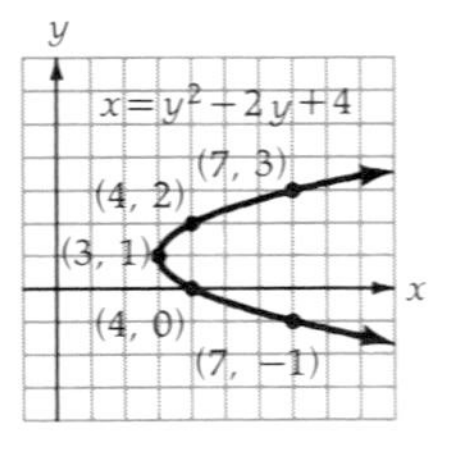


**39.**

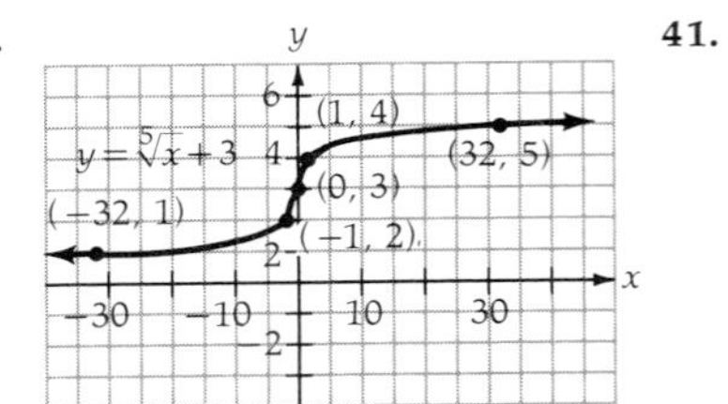


**41.**

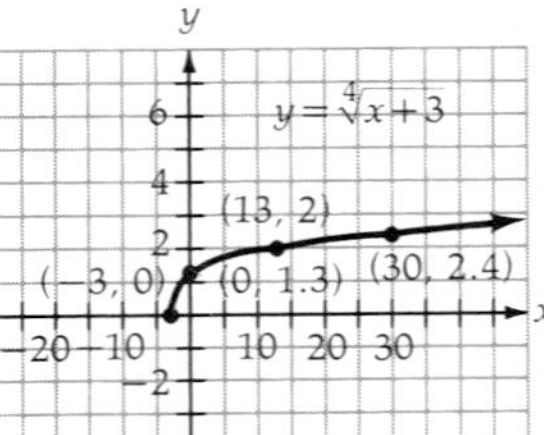


**43.**

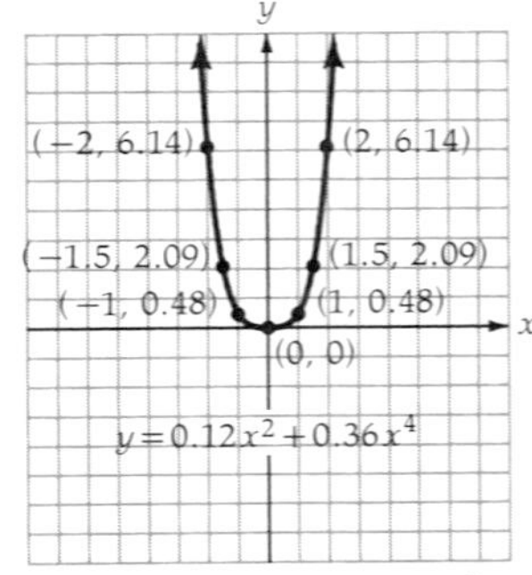


The graph is symmetric with respect to the $y$-axis.

## Exercise 3-3 (pp. 126–128)

**1.** $f(2) = -2$, $f(-3) = 13$, $f\left(\frac{1}{2}\right) = \frac{5}{2}$ **3.** $f(2) = -\frac{1}{3}$, $f(-3) = -\frac{1}{8}$, $f(0) = -\frac{1}{5}$, $f(6) = 1$

**5.** $f(-1) = 7$, $f(-2) = 17$, $f(0) = 1$, $f(3) = 7$ **7.** $f(a) = a^2 - 2a$, $f(2a) = 4a^2 - 4a$, $f(a+2) = a^2 + 2a$

**9.** $g(b) = b^2 + 2$, $g(b+c) = b^2 + 2bc + c^2 + 2$, $g(b) + g(c) = b^2 + c^2 + 4$, $g(b+c) - g(b) = 2bc + c^2$

**11.** $D$: all real numbers; $R$: $f(x) \geq 0$ **13.** $D$: all real numbers; $R$: $f(x) \geq 0$ **15.** $D$: $\{x{:}x \leq 4\}$; $R$: $f(x) \geq 0$ **17.** $D$: all real numbers; $R$: all real numbers

**19.** $\{x{:}x > -4\}$ **21.** $\{x{:}x > 2\}$ **23.** $\{x{:}x \neq -1, x \neq 2\}$ **25.** $\{x{:}x \geq 0, x \neq 16\}$ **27.** $\{x{:}x \geq 0\}$

**29.** neither **31.** odd **33.** even **35.** even **37.** **a.** $f(x) = \frac{3}{x}$ **b.** $g(y) = \frac{3}{y}$

**39.** **a.** $f(x) = 2x^2 + x$ **b.** cannot be done **41.** **a.** cannot be done **b.** $g(y) = |y + 1|$

**43.** **a.** $f(x) = \sqrt[3]{x}$ **b.** $g(y) = y^3$ **45.** **a.** cannot be done **b.** $g(y) = \sqrt{9 - y^2}$

**47.** $f(3.123) = 19.629258$, $f(-0.00026) = -3.000259865$

**49.** $g(1.0008) = 4.198851346$, $g(1576) = 1.639673013 \times 10^{10}$

**51.** First we find $A$ in terms of $s$ and obtain $A = \frac{s^2\sqrt{3}}{4}$. Then solve for $s$: $s(A) = \frac{2\sqrt{A}}{\sqrt[4]{3}}$

**53.** $f\left(\frac{1}{x}\right) = \frac{1}{\frac{1}{x} + 1} = \frac{x}{1+x}$; $\quad x \cdot f(x) = (x)\left(\frac{1}{1+x}\right) = \frac{x}{1+x}$

## Exercise 3-4 (pp. 136–137)

**1.** **a.** no symmetry

**b.**

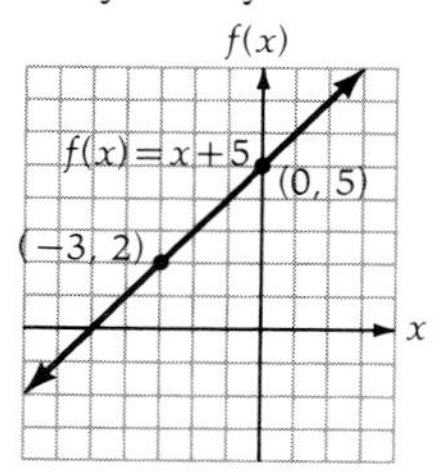


**c.** increasing on $(-\infty, \infty)$

**3.** **a.** no symmetry

**b.**

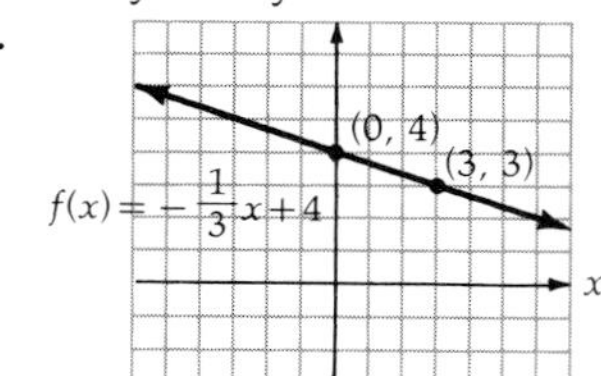


**c.** decreasing on $(-\infty, \infty)$

**5.** **a.** graph is symmetric with respect to $y$-axis

**b.**

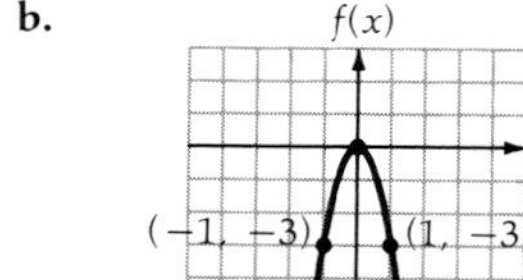


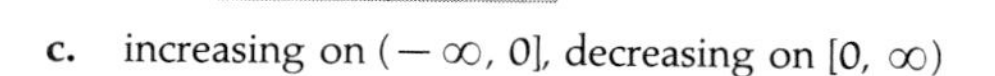

**c.** increasing on $(-\infty, 0]$, decreasing on $[0, \infty)$

**7.** **a.** graph is symmetric with respect to $y$-axis

**b.**

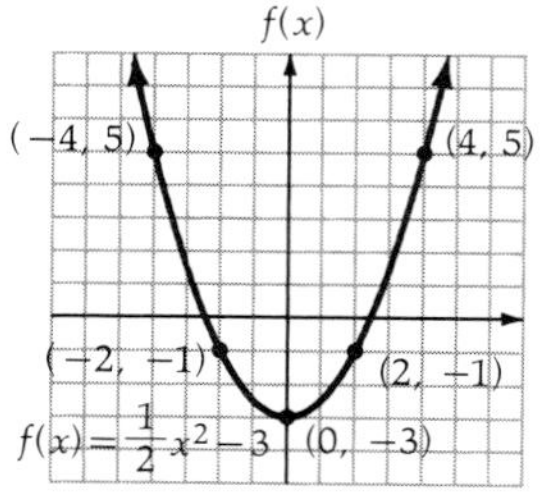


**c.** decreasing on $(-\infty, 0]$, increasing on $[0, \infty)$

**9.** **a.** no symmetry

**b.**

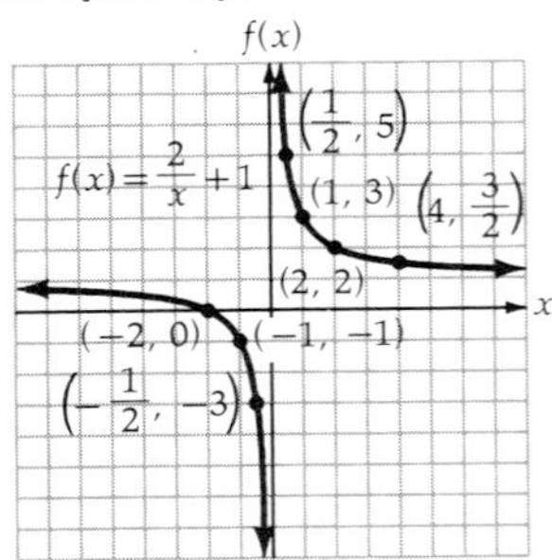


**c.** decreasing on $(-\infty, 0)$, decreasing on $(0, \infty)$

**11.** **a.** graph is symmetric with respect to $y$-axis

**b.**

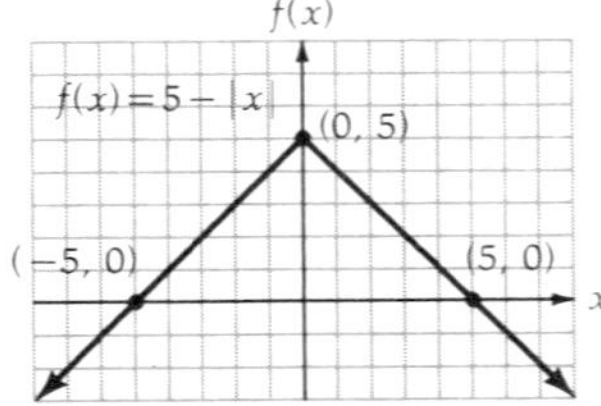


**c.** increasing on $(-\infty, 0]$, decreasing on $[0, \infty)$

**13.** **a.** no symmetry

**b.**

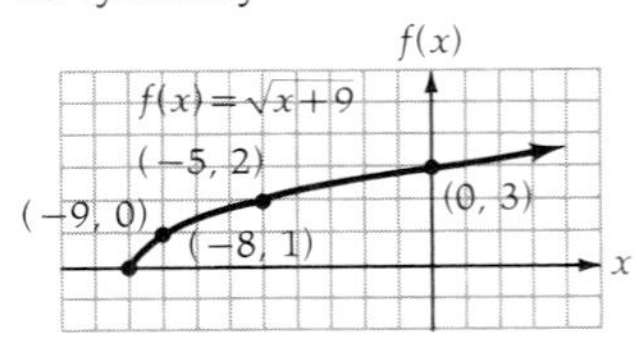


**c.** increasing on $[-9, \infty)$

**15.**

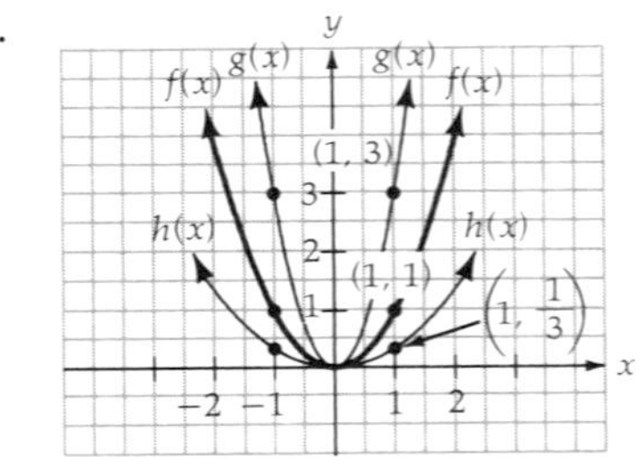


**17.**

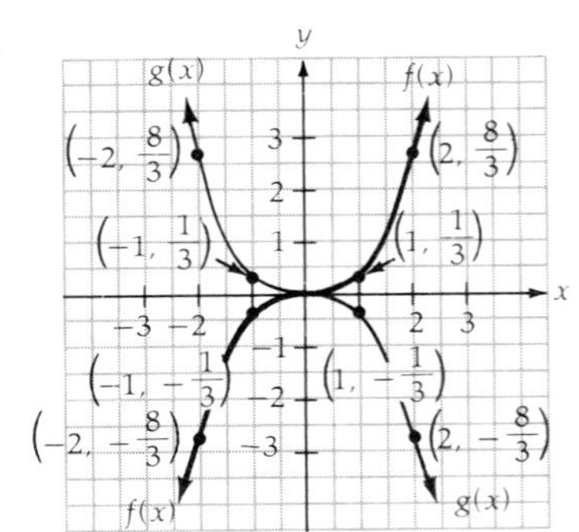


**19.**

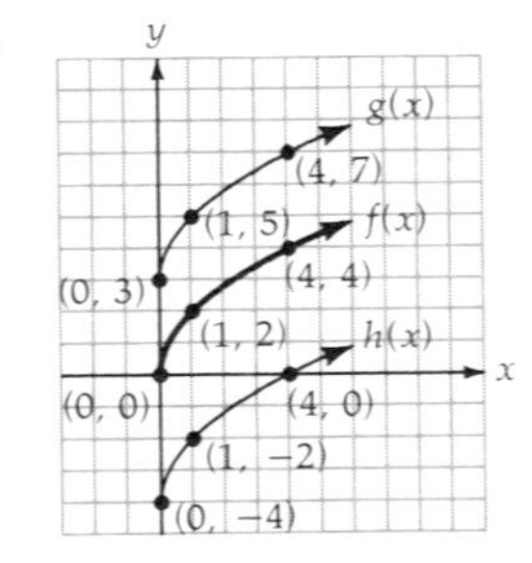


**21.**

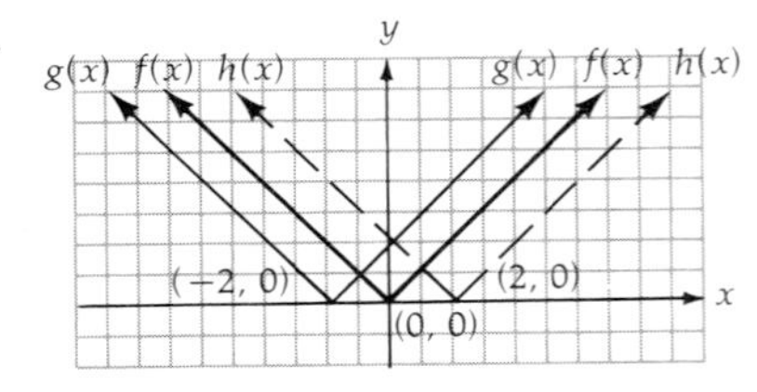


**23.** It is not a function, since a vertical line can cross the curve twice.

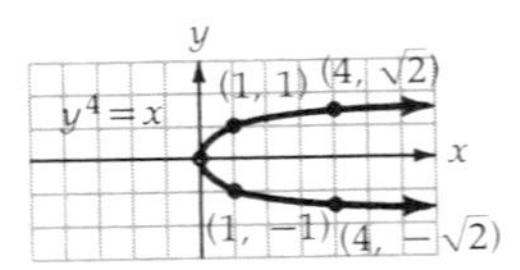

**25.** It is not a function, since a vertical line can cross the curve twice.

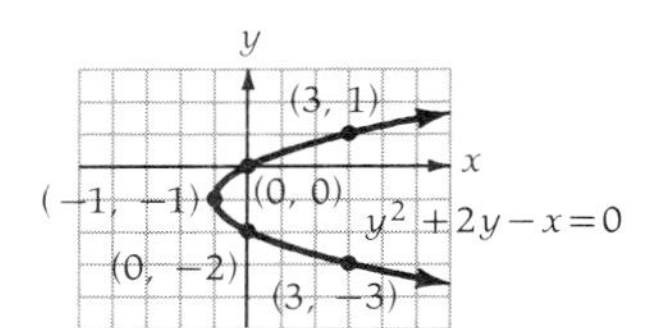


**27.** Note that the graph of $g(x)$ is always a straight line for $x < 0$ and always a curved line for $x \geq 0$.

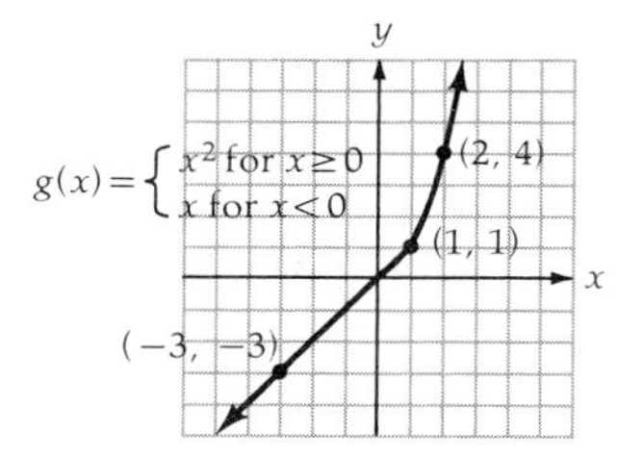


**29.**

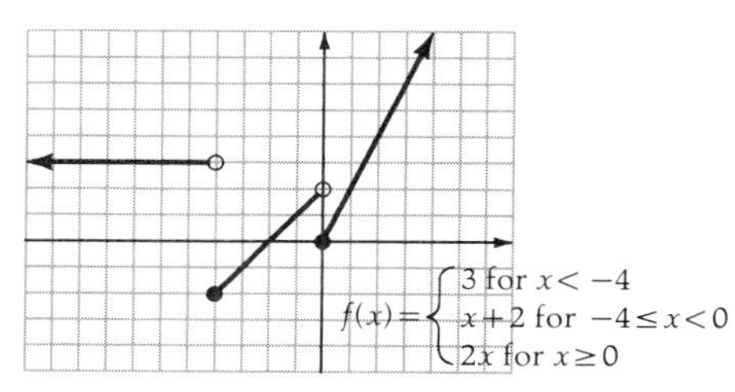

$f(x)=\begin{cases} 3 \text{ for } x<-4 \\ x+2 \text{ for } -4\leq x<0 \\ 2x \text{ for } x\geq 0 \end{cases}$


**31.** $[0] = 0$, $[5.8] = 5$, $[\sqrt[3]{7}] = 1$, $[-17.6] = -18$, $\left[-\frac{\pi}{2}\right] = -2$, $[\pi^2] = 9$

**33.**

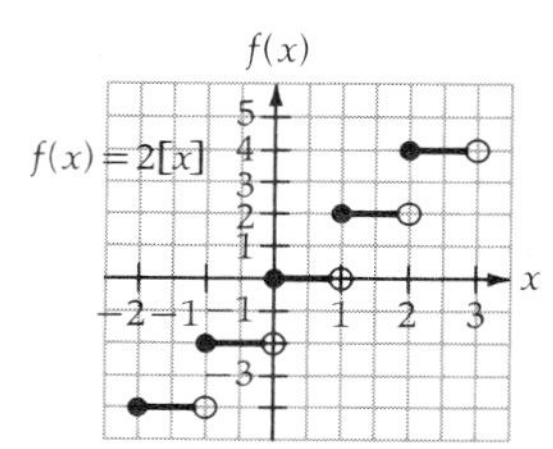


**35.**

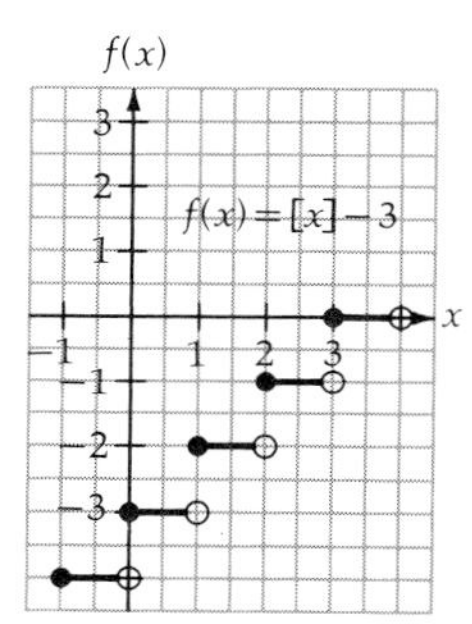


**37.**

$C(0) = 5000$
$C(50) = 5000$
$C(100) = 5000$
$C(150) = 6250$
$C(200) = 7500$
$C(250) = 8000$
$C(300) = 8500$

C(n)
12,000
10,000
8000
6000
4000
2000
50
150
250
n

## Exercise 3-5 (pp. 143–145)

**1.** $8x + 5y = 21$ **3.** $x + 2y = -2$ **5.** $5x + 12y = 20$ **7.** $3x - y = -4$ **9.** $x - 2y = -2$
**11.** $5x - y = -4$ **13.** $2x + 3y = 3$ **15.** $x = -2$

**17.** $y = -2x + 5,\ m = -2,\ b = 5$

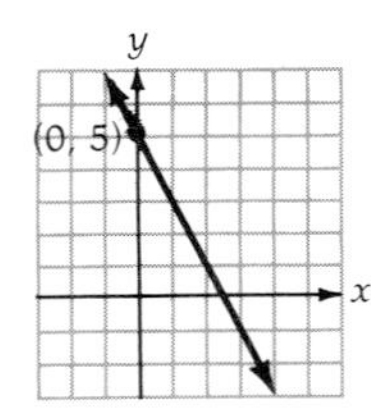


**19.** $y = \frac{1}{2}x - \frac{2}{3},\ m = \frac{1}{2},\ b = -\frac{2}{3}$

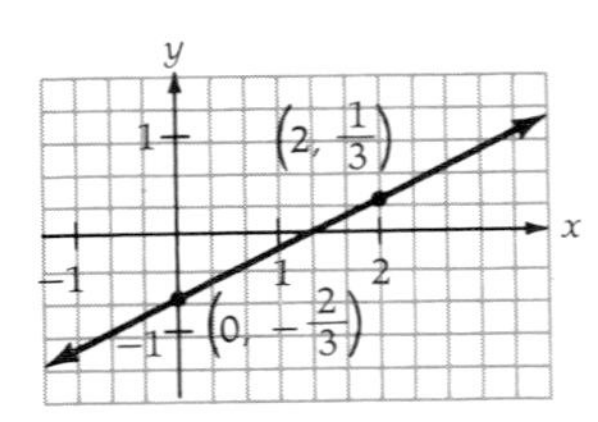


**21.** $y = \frac{5}{2}x,\ m = \frac{5}{2},\ b = 0$

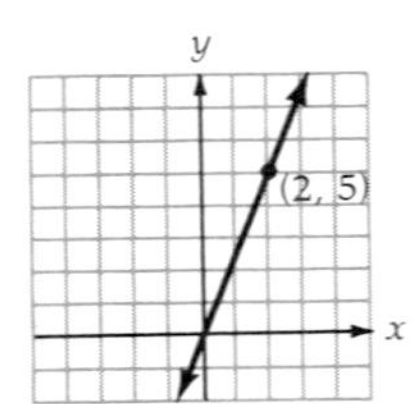


**23.** $y = \frac{2}{5}x - \frac{3}{2},\ m = \frac{2}{5},\ b = -\frac{3}{2}$

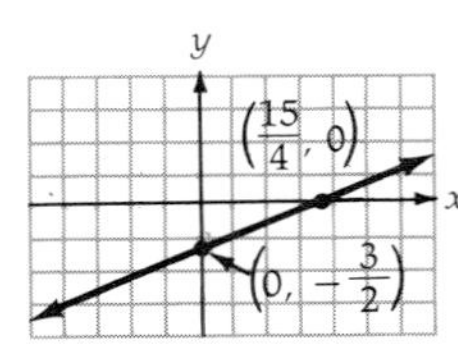


**25.** $7x - 2y = -14$ **27.** $18x - 6y = 5$ **29.** $2x - y = 5$
**31.** $5x - 2y = 27$ **33.** $y = 1$ **35.** $x + 3y = 5$
**37.** $x - 2y = 10$ **39.** $x - 4y = -37$ **41.** $3x - 4y = 12$
**43.** $d = -1$ **45.** $b = \frac{10}{7}$

**47.** Let $A = (2, -1)$, $B = (3, 3)$, $C = (2, 7)$, $D = (1, 3)$. Slope of $AD = -4$, slope of $BC = -4$, slope of $CD = 4$, slope of $AB = 4$. Therefore $AD$ is parallel to $BC$ and $CD$ is parallel to $AB$. Thus $ABCD$ is a parallelogram.

**49.** $f(x) = \begin{cases} \frac{1}{3}x - 2 & \text{if } x < -1 \\ 2x - \frac{1}{3} & \text{if } x \geq -1 \end{cases}$

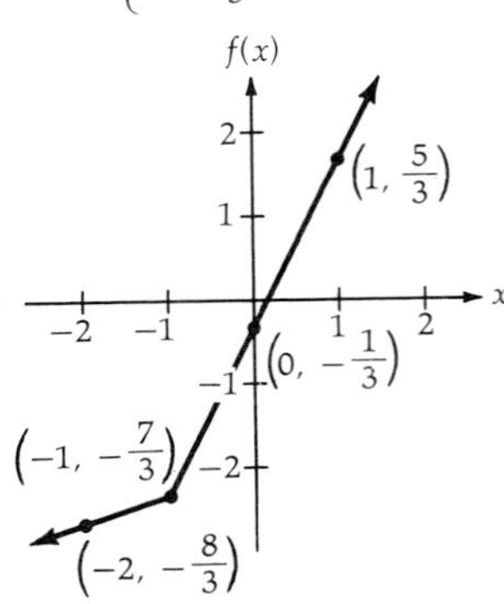


**51.** $y - y_1 = m(x - x_1)$.
Let $(x_1, y_1) = (0, b)$ and $(x_2, y_2) = (a, 0)$.

$$y - b = \frac{b - 0}{0 - a}(x - 0)$$

$$y - b = -\frac{b}{a}(x)$$

$$y + \frac{b}{a}x = b$$

$$\frac{y}{b} + \frac{x}{a} = 1$$

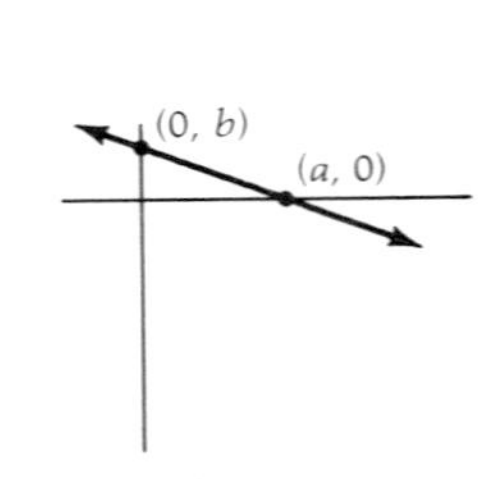


**53.** $f(x) = x + 3$ is a possible answer, and, in general, any equation $f(x) = x + b$ where $b$ is a real number.

**55.** $y = 5.723x - 1010.62$ **57.** $y = 3.2308x - 8.1923$

## Exercise 3-6 (p. 152)

**1.** **a.** $(f + g)(x) = x^2 + 3x + 2,\ D = (-\infty, +\infty)$ **b.** $(f - g)(x) = x^2 - 3x - 2,\ D = (-\infty, +\infty)$

**c.** $(f \cdot g)(x) = 3x^3 + 2x^2,\ D = (-\infty, +\infty)$ **d.** $\left(\frac{f}{g}\right)(x) = \frac{x^2}{3x + 2},\ D = \left(-\infty, -\frac{2}{3}\right) \cup \left(-\frac{2}{3}, +\infty\right)$

**e.** $\left(\frac{g}{f}\right)(x) = \frac{3x + 2}{x^2},\ D = (-\infty, 0) \cup (0, +\infty)$

3. a. $(f+g)(x) = \sqrt{x} + x^2, D = [0, +\infty)$ b. $(f-g)(x) = \sqrt{x} - x^2, D = [0, +\infty)$

c. $(f \cdot g)(x) = x^2\sqrt{x}, D = [0, +\infty)$ d. $\left(\frac{f}{g}\right)(x) = \frac{\sqrt{x}}{x^2}, D = (0, +\infty)$ e. $\left(\frac{g}{f}\right)(x) = \frac{x^2}{\sqrt{x}}, D = (0, +\infty)$

5. $(f \circ g)(x) = 10x - 14, D = (-\infty, +\infty)$
$(g \circ f)(x) = 10x - 1, D = (-\infty, +\infty)$

7. $(f \circ g)(x) = 8x^2 + 6x + 2, D = (-\infty, +\infty)$
$(g \circ f)(x) = 4x^2 - 2x + 3, D = (-\infty, +\infty)$

9. $(f \circ g)(x) = \sqrt{3x^2 + 8}, D = (-\infty, +\infty)$
$(g \circ f)(x) = 3x + 2, D = \left[\frac{1}{3}, +\infty\right)$

11. $(f \circ g)(x) = x, D = [0, +\infty)$
$(g \circ f)(x) = |x|, D = (-\infty, +\infty)$

13. a. yes b. yes c. $f^{-1}(x) = \frac{x+2}{5}$

15. a. yes b. no c. no inverse function can be found unless the domain is restricted

17. a. yes b. no c. no inverse function can be found unless the domain is restricted

19. a. no restrictions b. $f^{-1}(x) = \frac{x+7}{2}$

21. a. no further restrictions are required, since we are given that $x \neq 0$ b. $f^{-1}(x) = \frac{1}{x}, x \neq 0$

23. a. domain restricted to $[0, +\infty)$ b. $f^{-1}(x) = \sqrt{x+3}, x \geq -3$

25. If $y = x^2$ is restricted with $D = (-\infty, 0]$, its inverse is $f^{-1}(x) = -\sqrt{x}$. Both are functions. With that restriction, $f^{-1}[f(x)] = f[f^{-1}(x)]$.

27. $f[g(x)] = (\sqrt{x+5})^2 - 5 = x$
$g[f(x)] = \sqrt{x^2 - 5 + 5} = \sqrt{x^2} = x$ since $x \geq 0$

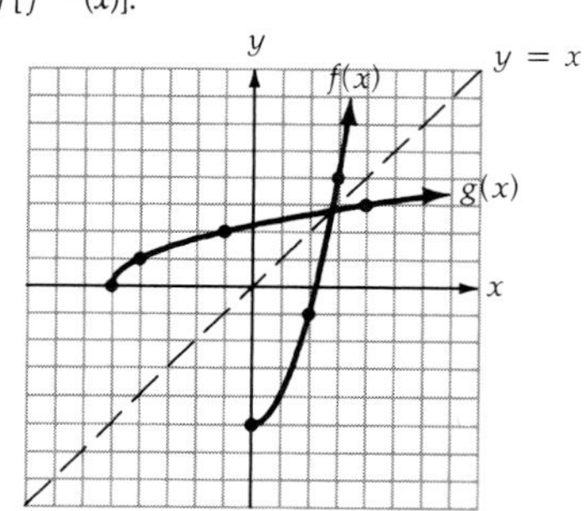


## Chapter 3 Review Exercise (pp. 154–156)

1. $d = 2\sqrt{74}$ 3. $\left(-\frac{7}{4}, -\frac{1}{2}\right)$ 5. $d = \sqrt{64 + 4b^2} = 2\sqrt{16 + b^2}$

7. It is a function
$D = (-\infty, \infty)$
$R = [2, \infty)$

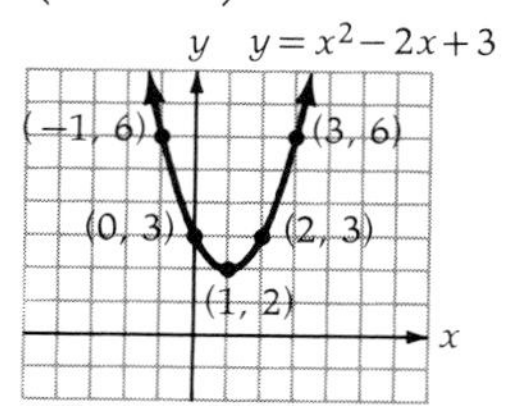


9. Not a function
Domain is only the number 4
$R = (-\infty, +\infty)$

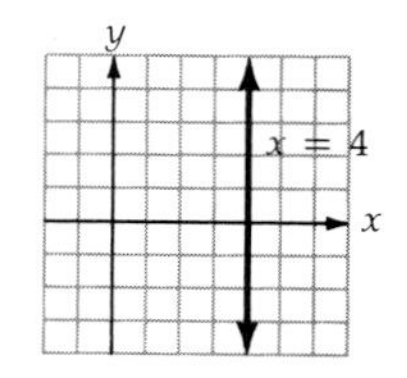

**11.** It is a function
$D = (-\infty, -1] \cup (1, +\infty)$
$R = [0, +\infty)$

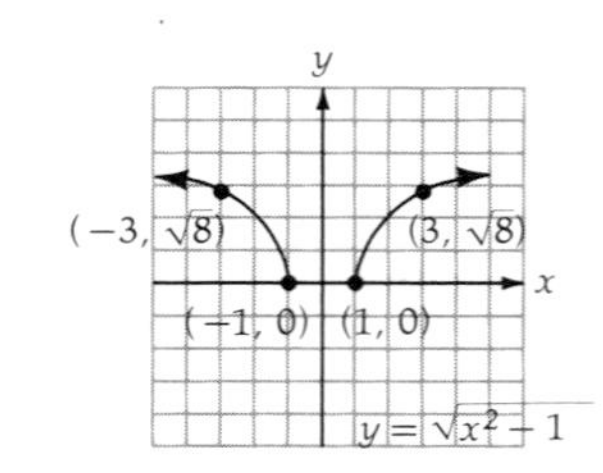


**13.** $f(0) = 0$
$f(-1) = \sqrt{3}$
$f(2) = 0$
$f(a+1) = \sqrt{a^2 - 1}$

**15.** $D = (-\infty, +\infty)$
$y = f(x) = \frac{1}{3}x$, yes

**17.** $D = (-\infty, 0) \cup (0, +\infty)$
$y = f(x) = \frac{3}{x}$, yes

**19.** $D = (-\infty, +\infty)$
$y = f(x) = |x^2 - 1|$, yes

**21.** $D = (-1, \infty)$
$y = f(x) = \frac{3}{\sqrt{x+1}}$, yes

**23.** $D = (-\infty, -1) \cup (-1, 1) \cup (1, +\infty)$
$y = f(x) = \frac{2}{|x| - 1}$, yes

**25.** neither **27.** even

**29.** **a.** graph is symmetric with respect to $y$-axis
**b.**

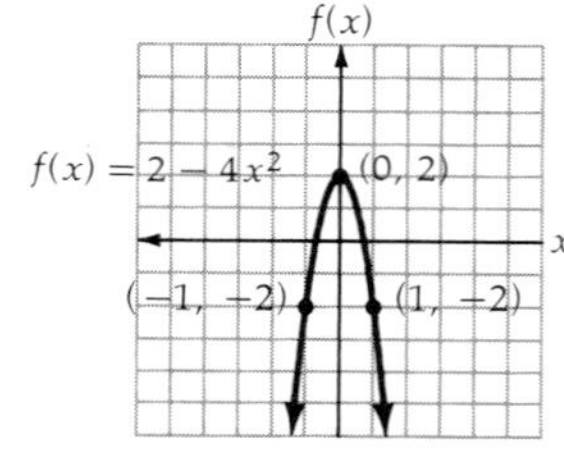


**c.** increasing on $(-\infty, 0]$, decreasing on $[0, +\infty)$

**31.** **a.** no symmetry
**b.**

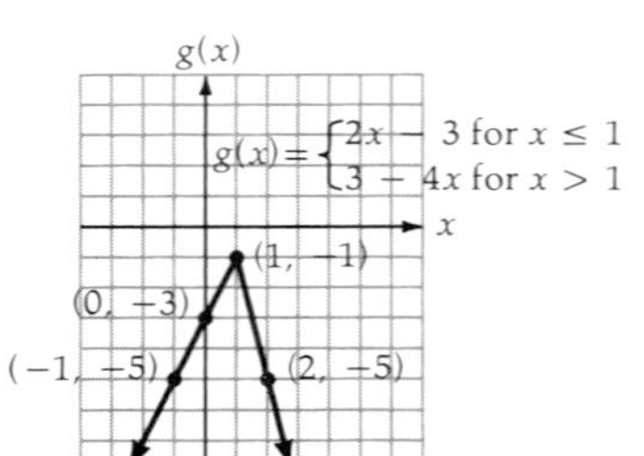


**c.** increasing on $[4, +\infty)$

**33.**

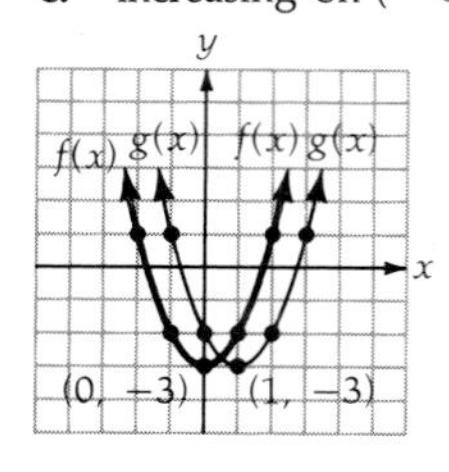


**35.**

g(x)

$g(x) = \begin{cases} 2x - 3 \text{ for } x \leq 1 \\ 3 - 4x \text{ for } x > 1 \end{cases}$

x
(1, −1)
(0, −3)
(−1, −5)
(2, −5)

**37.** $y = \frac{2}{5}x - 2$ **39.** $y = -\frac{3}{8}x + \frac{25}{4}$ **41.** $x$-intercept $= -\frac{8}{7}$

$y$-intercept $= \frac{8}{3}$

**43.** $(f - g)(x) = 3x^2 + x + 1$, $D = (-\infty, +\infty)$ **45.** $\left(\frac{f}{g}\right)(x) = \frac{3x^2 + 2x}{x - 1}$, $D = (-\infty, 1) \cup (1, +\infty)$

**47.** $(f + g)(x) = \frac{3x + 2}{x(x+1)}$, $D = (-\infty, -1) \cup (-1, 0) \cup (0, +\infty)$

**49.** $(f \cdot g)(x) = \frac{2}{x(x+1)}$, $D = (-\infty, -1) \cup (-1, 0) \cup (0, +\infty)$

**51.** $(f \circ g)(x) = \frac{x}{x + 2}$, $D = (-\infty, -2) \cup (-2, 0) \cup (0, +\infty)$

**53.** $f[g(x)] = \sqrt{10 - 3x}$

$g[f(x)] = 3 - \sqrt{3x + 1}$

**55.** $f[g(x)] = \sqrt{5\left(\dfrac{x^2 - 1}{5}\right) + 1} = \sqrt{x^2} = x$, since $x \geq 0$

$g[f(x)] = \dfrac{(\sqrt{5x + 1})^2 - 1}{5} = \dfrac{5x + 1 - 1}{5} = x$

## Practice Test for Chapter 3 (pp. 156–157)

**1.** **a.** $D = \left[\frac{5}{2}, +\infty\right)$

$R = [0, +\infty)$

It is a function

**b.** $D = [-3, 3]$

$R = [-3, 3]$

It is not a function

**c.** $D = (-\infty, +\infty)$

$R = [0, +\infty)$

It is a function

**3.** $f(-3) = 7$

$f(0) = -1$

$f(1) = -1$

$f(3) = 8$

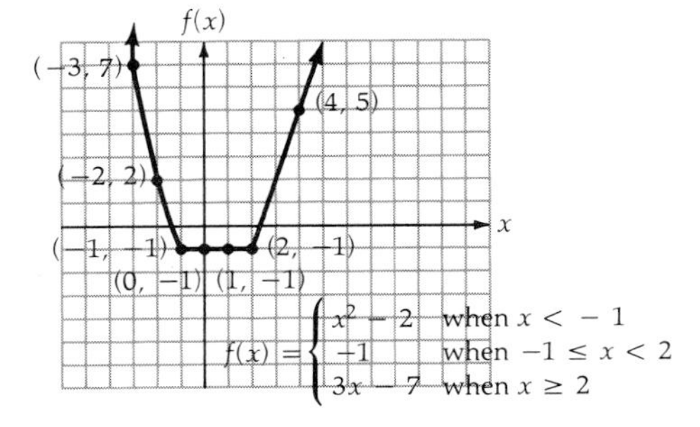


$$f(x) = \begin{cases} x^2 - 2 & \text{when } x < -1 \\ -1 & \text{when } -1 \leq x < 2 \\ 3x - 7 & \text{when } x \geq 2 \end{cases}$$

**5.** $D = [-2, +\infty)$

$R = [0, +\infty)$

It is a function

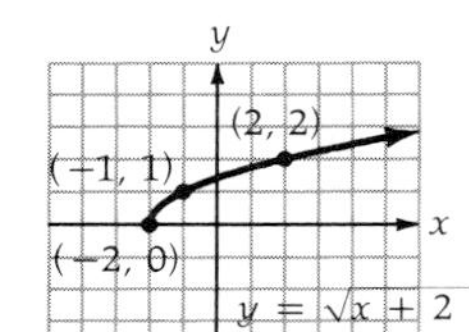


**7.** $y = \frac{8}{3}x - 4$

**9.** **a.** $f^{-1}(x) = \dfrac{x + 5}{6}$

**b.** $f^{-1}(x) = x^2 - 4,\ x \geq 0$

**11.**

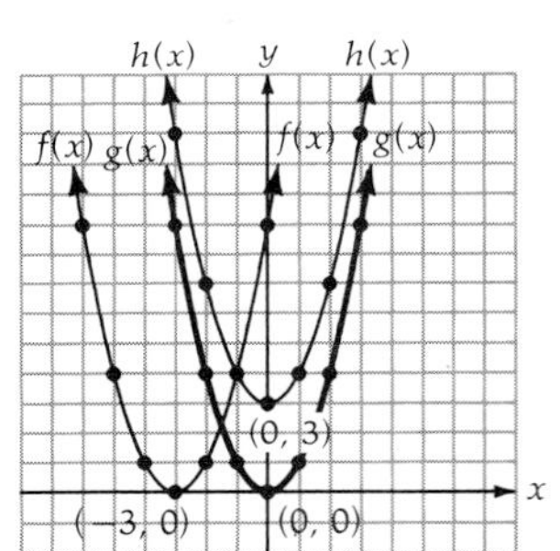


## CHAPTER 4

### Exercise 4-1 (pp. 166–167)

**1.**

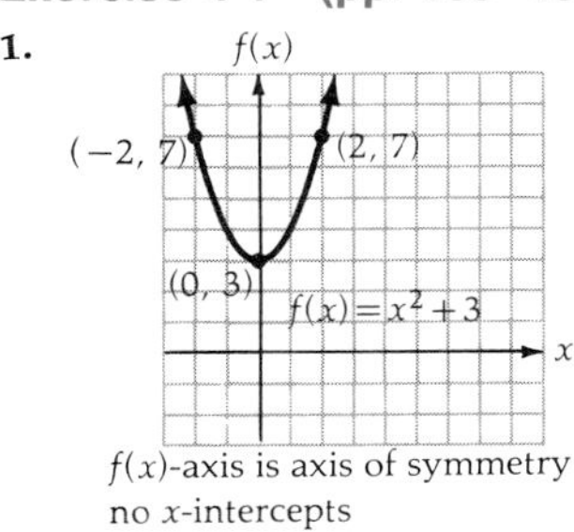


$f(x)$-axis is axis of symmetry
no $x$-intercepts

**3.**

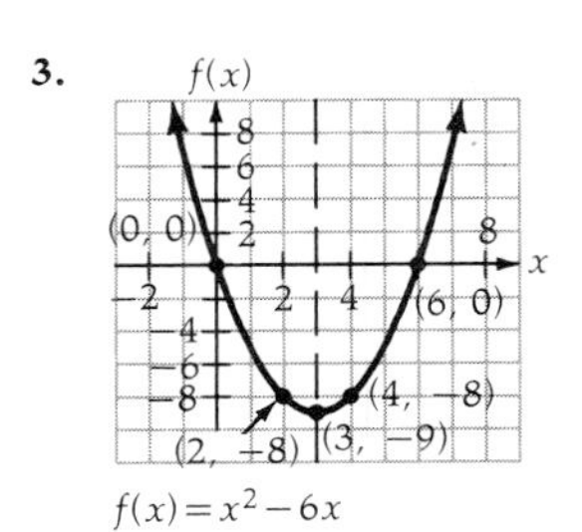


$f(x) = x^2 - 6x$

**5.** $f(x) = x^2 + 7x + 12$

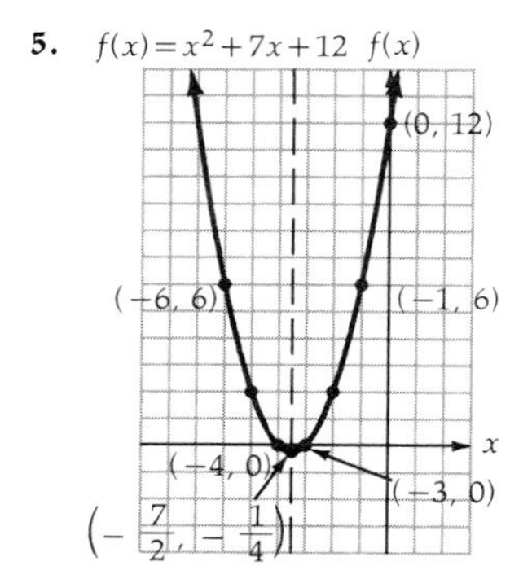

**7.**

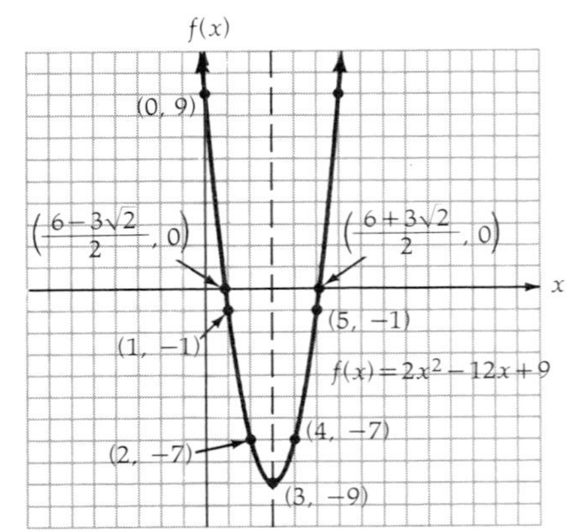


**9.**

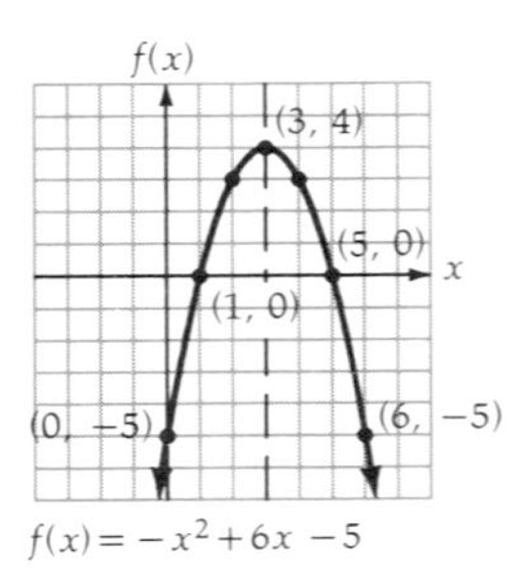


**11.**

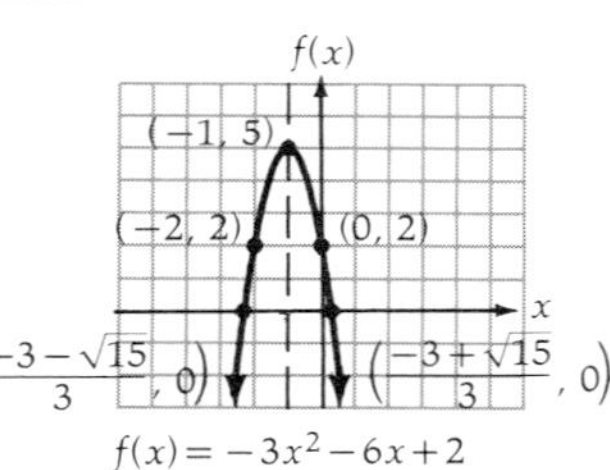


**13.**

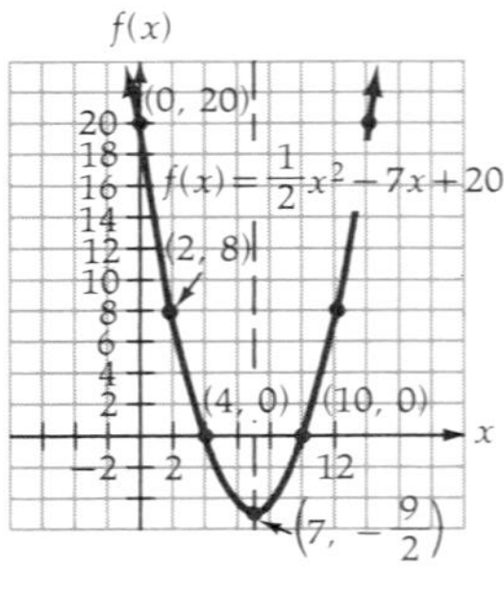


**15.**

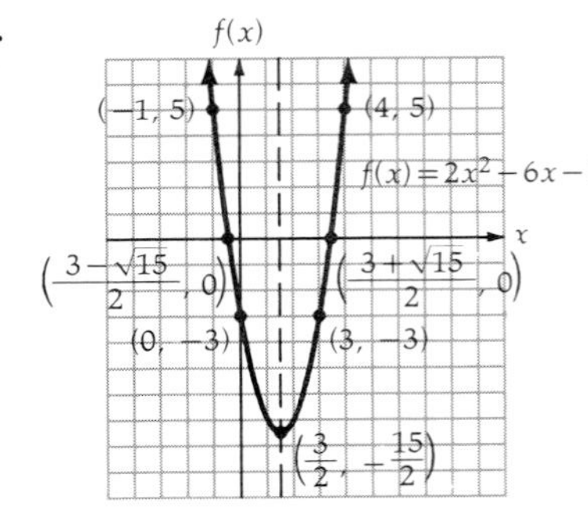


**17.**

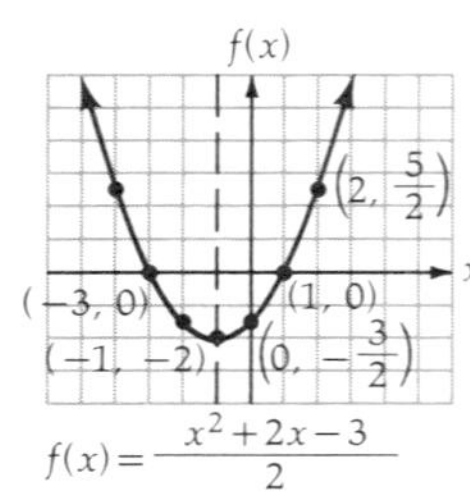


**19.**

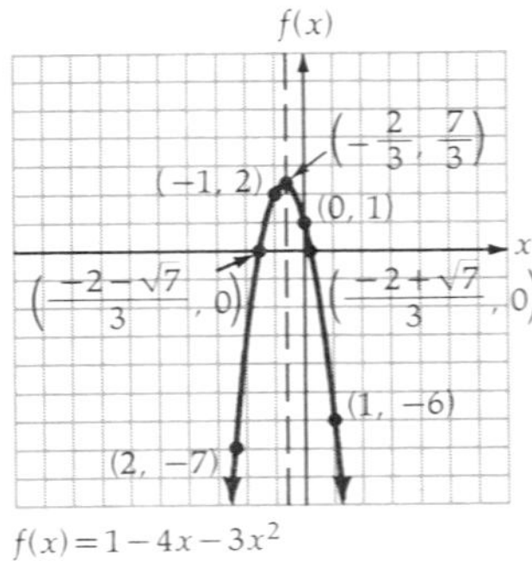


**21.** Side parallel to barn should be 80 meters; two sides perpendicular to barn should each be 40 meters.

**23.** Each number is 18.  **25.** It is a square with each side measuring 150 yards.  **27.** 24 items

**29.** \$3.00

**31.**

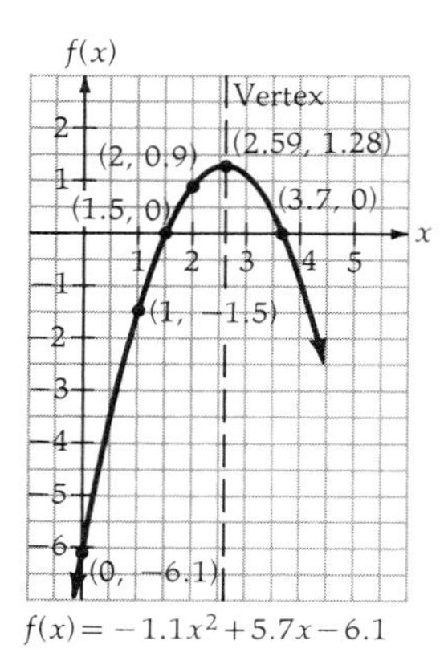


$f(x) = -1.1x^2 + 5.7x - 6.1$

**33.** Maximum height is 1994.7 m; time to achieve max. height is 12.2 s

## Exercise 4-2 (pp. 172–173)

**1.**

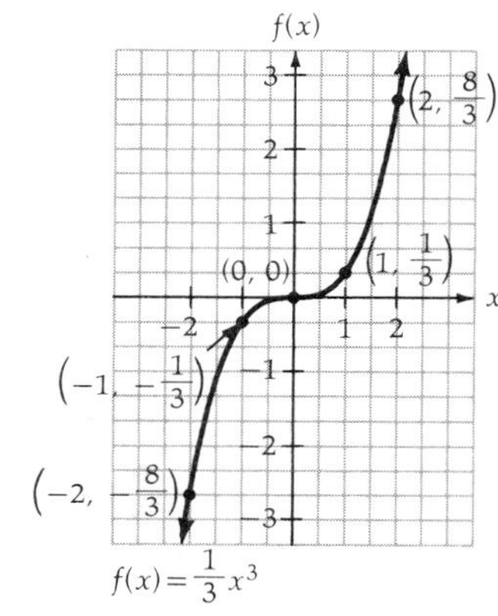


$f(x) = \frac{1}{3}x^3$

**3.**

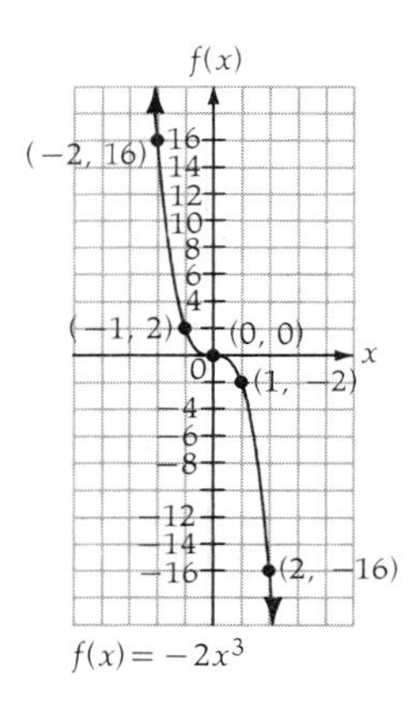


$f(x) = -2x^3$

**5.**

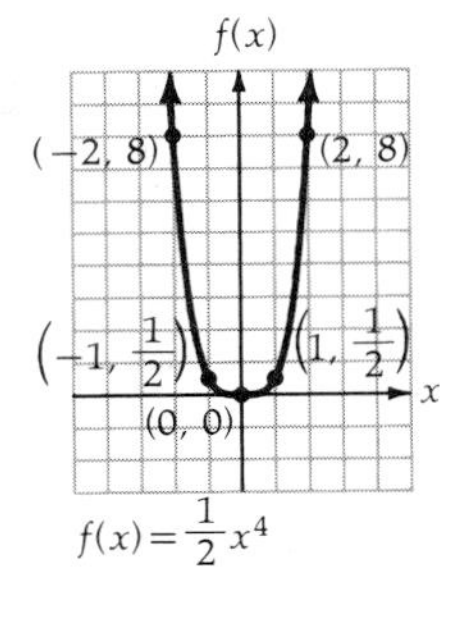


$f(x) = \frac{1}{2}x^4$

**7.**

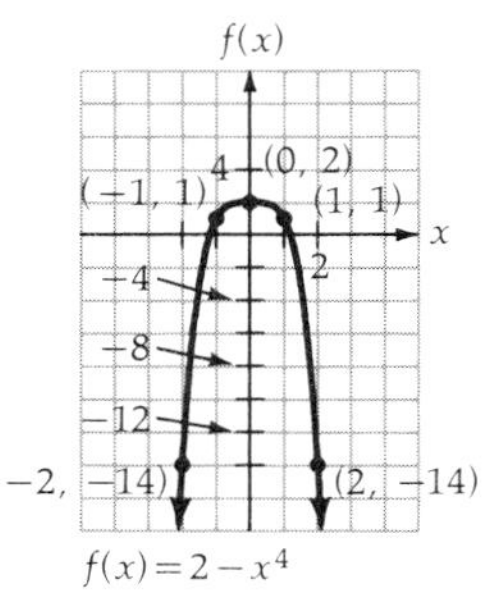


$f(x) = 2 - x^4$

**9.**

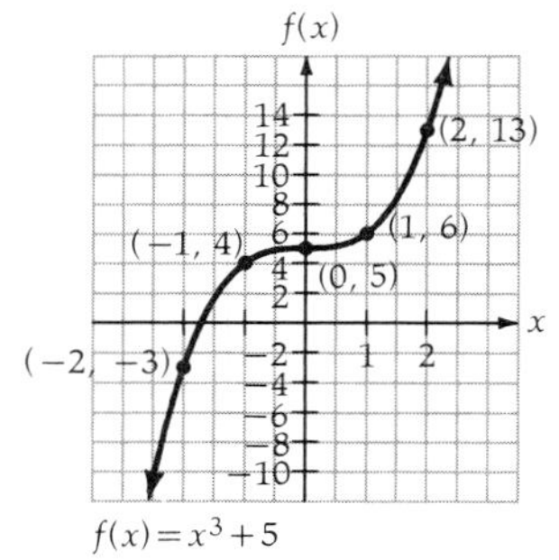


$f(x) = x^3 + 5$

**11.**

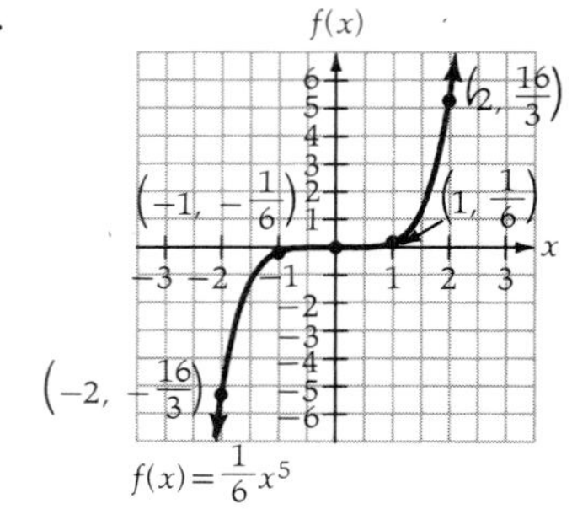


$f(x) = \frac{1}{6}x^5$

**13.**

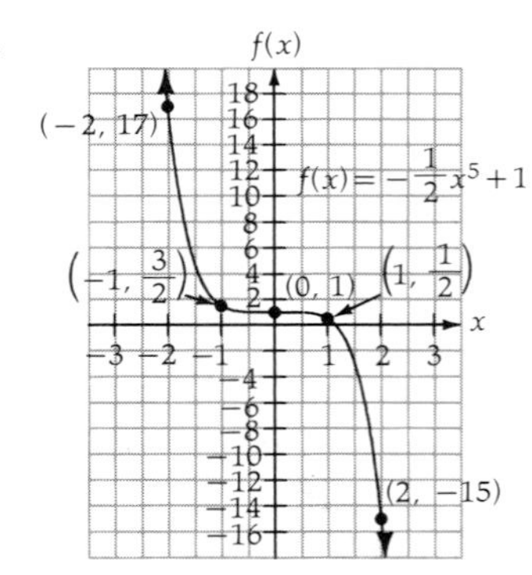


**15.**

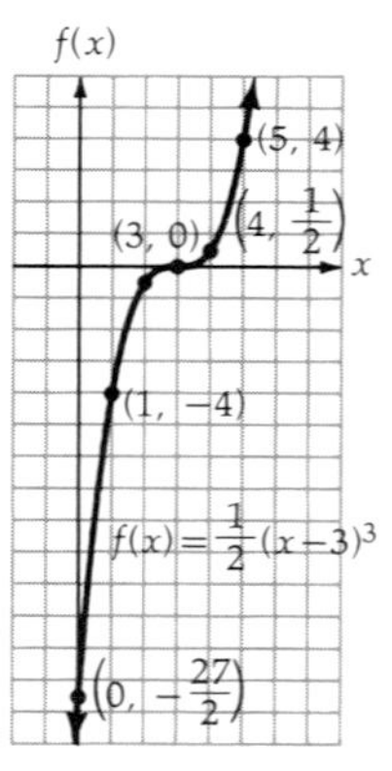


**17.**

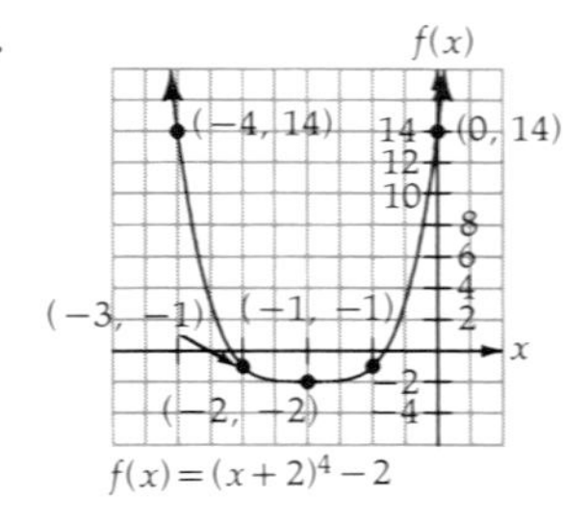


$f(x) = (x+2)^4 - 2$

**19.** Roots $\in \{-2, 0, 2\}$

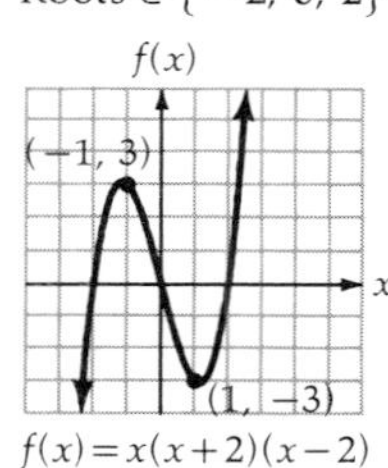


$f(x) = x(x+2)(x-2)$

**21.** Roots $\in \{0, 2, 4\}$

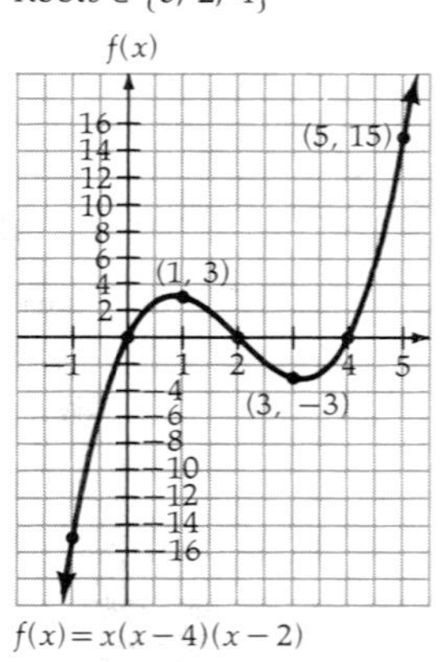


$f(x) = x(x-4)(x-2)$

**23.** Roots $\in \{-3, -1, 2\}$

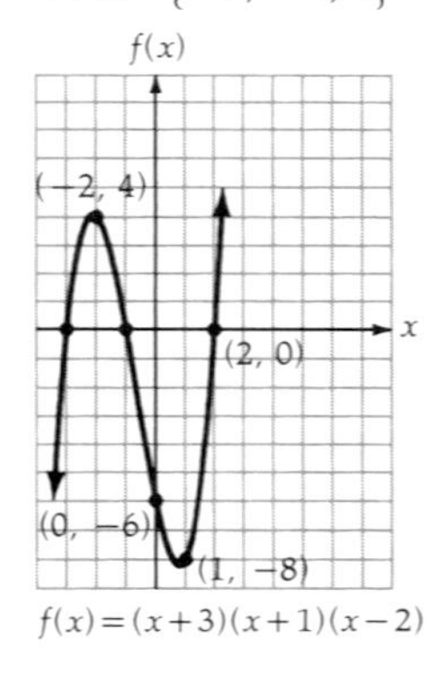


$f(x) = (x+3)(x+1)(x-2)$

**25.** Roots $\in \{-1, 0, 2\}$

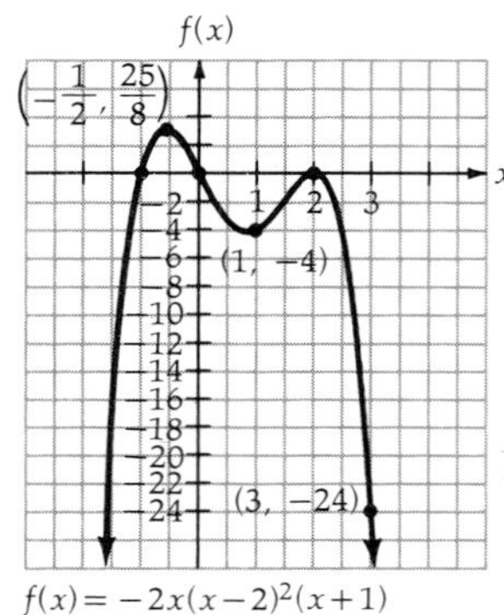


$f(x) = -2x(x-2)^2(x+1)$

**27.** Roots $\in \{-3, 0, 3\}$

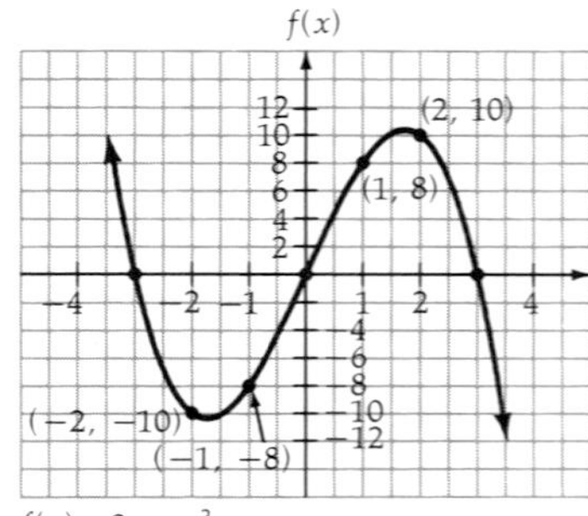


$f(x) = 9x - x^3$

## Exercise 4-2 (continued)

**29.** Roots $\in \left\{-2, 0, \frac{3}{2}\right\}$

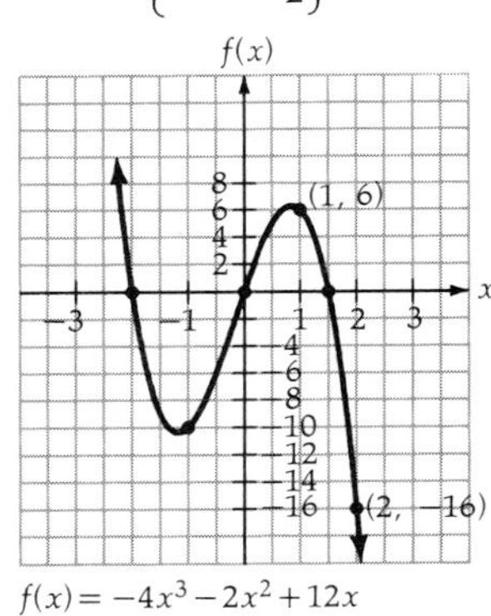


$f(x) = -4x^3 - 2x^2 + 12x$

**31.** Roots $\in \{-2, -1, 2\}$

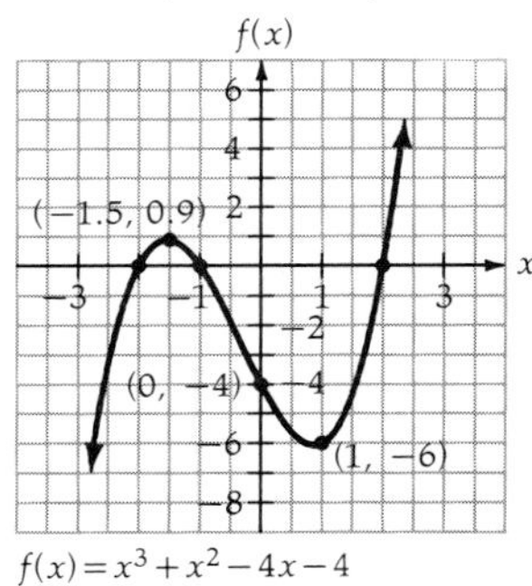


$f(x) = x^3 + x^2 - 4x - 4$

**33.** Roots $\in \{-3, -2, -1, 3\}$

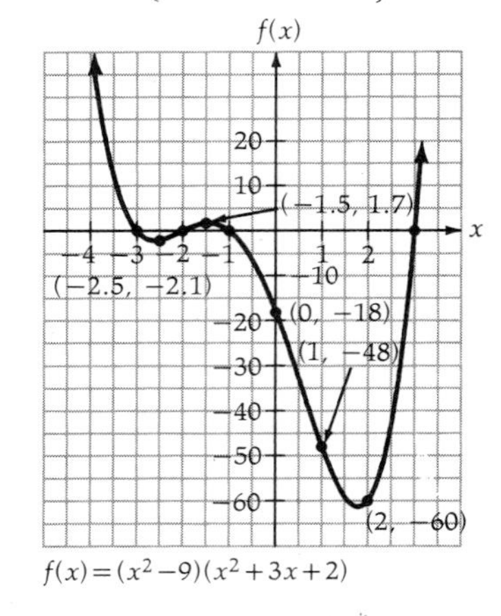


$f(x) = (x^2 - 9)(x^2 + 3x + 2)$

**35.** Roots $\in \left\{-\frac{1}{2}, 2\right\}$

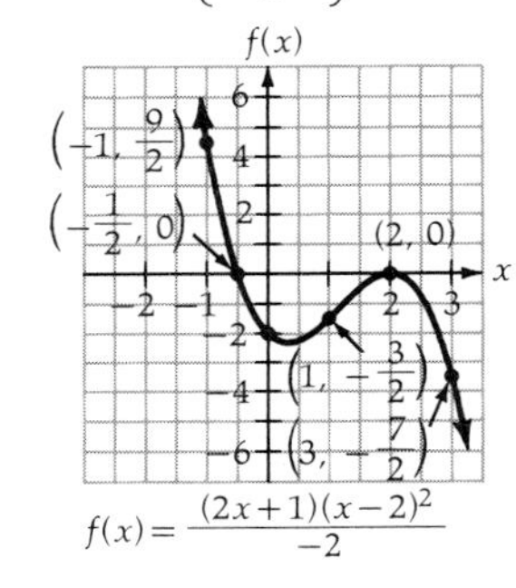


$f(x) = \dfrac{(2x+1)(x-2)^2}{-2}$

**37.**

| $x$ | $-2$ | $-1.5$ | $-1.0$ | $-0.5$ | 0 | 0.5 | 1.0 | 1.5 | 2.0 | 2.5 | 3.0 |
|---|---|---|---|---|---|---|---|---|---|---|---|
| $f(x)$ | $-23$ | $-9.5$ | $-1.4$ | 2.6 | 3.6 | 3.0 | 1.8 | 1.4 | 2.8 | 7.4 | 16.2 |

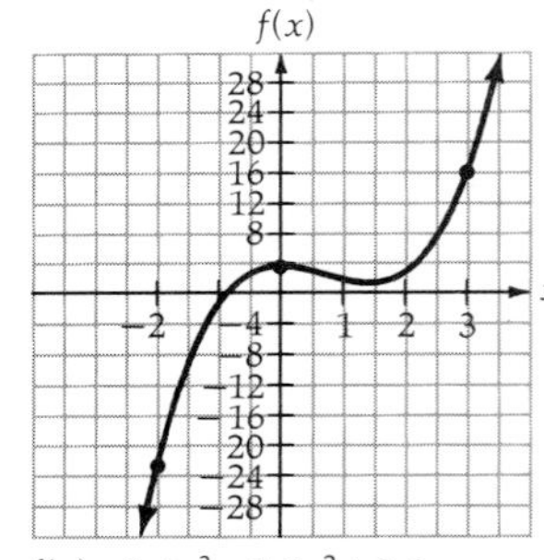


$f(x) = 1.6x^3 - 3.4x^2 + 3.6$

**39.**

| $x$ | 1 | 1.1 | 1.2 | 1.3 | 1.4 | 1.5 | 1.6 | 1.7 | 1.8 | 1.9 | 2.0 |
|---|---|---|---|---|---|---|---|---|---|---|---|
| $g(x)$ | 1.00 | 0.55 | $-0.02$ | $-0.7$ | $-1.5$ | $-2.5$ | $-3.6$ | $-4.9$ | $-6.4$ | $-8.1$ | $-10.0$ |

zero lies between 1.1 and 1.2

## Exercise 4-3 (pp. 178–179)

**1.**

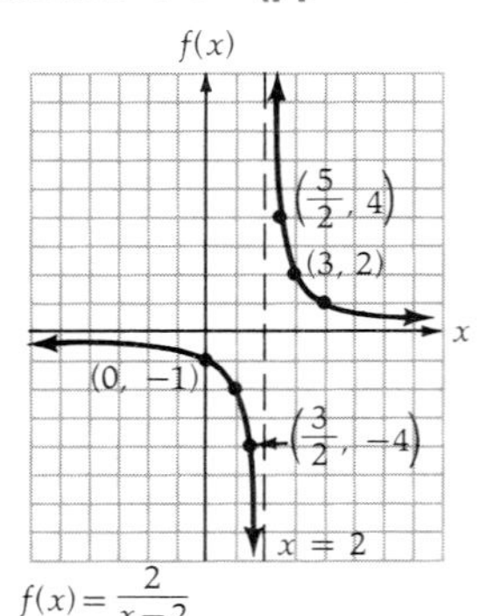


**3.**

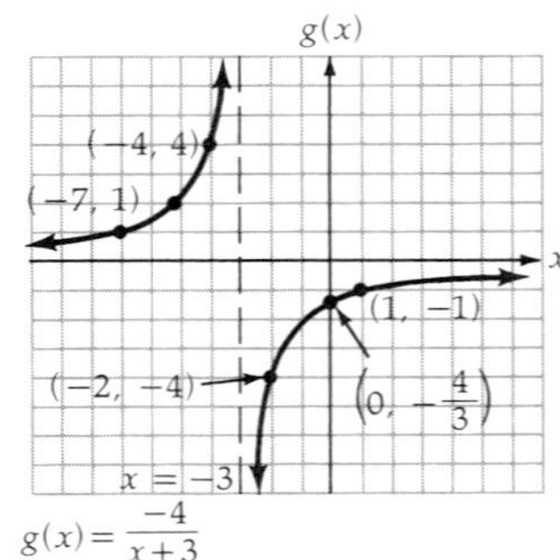


**5.**

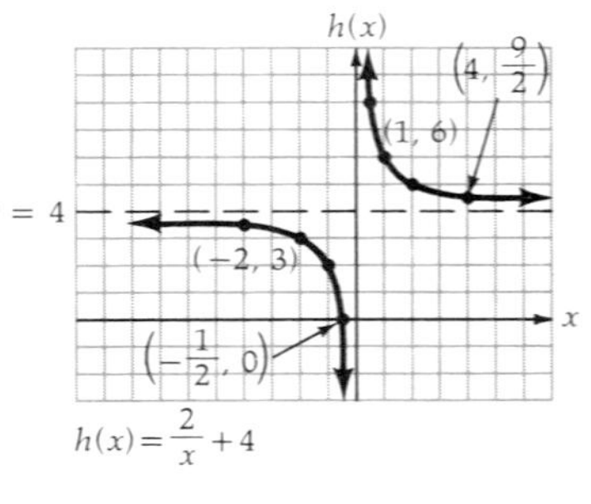


**7.**

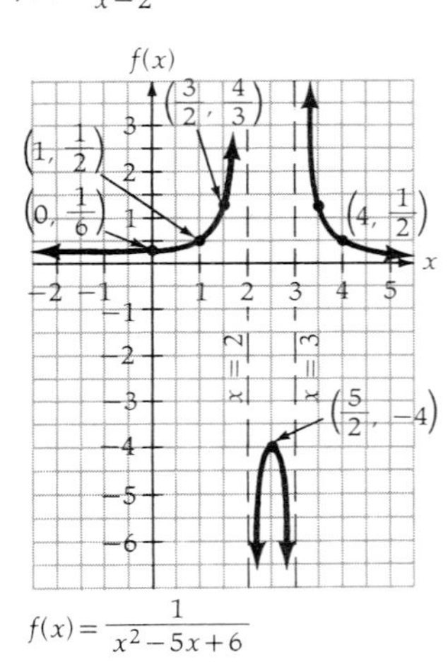


**9.**

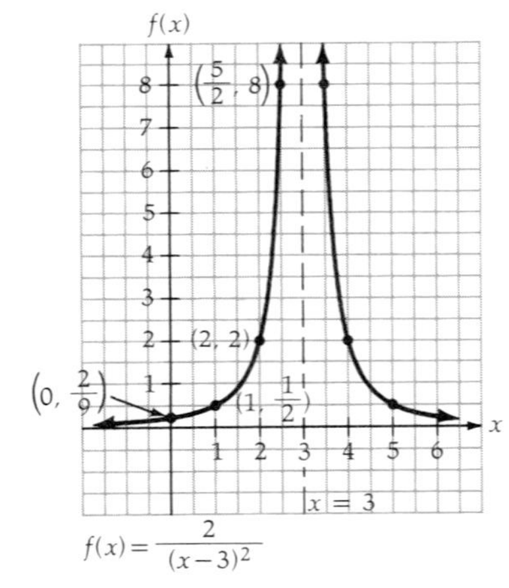


**11.**

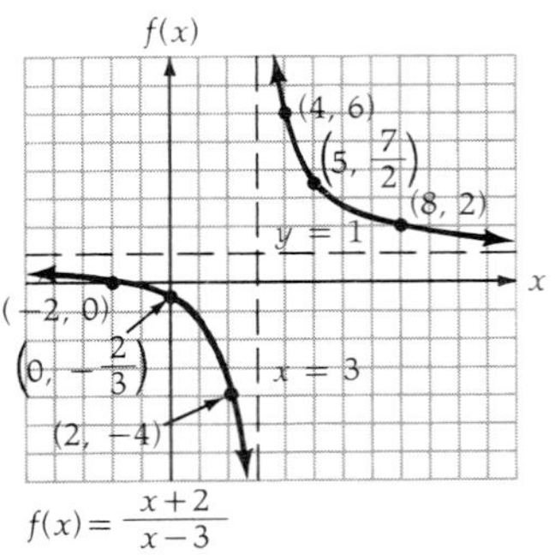


**13.**

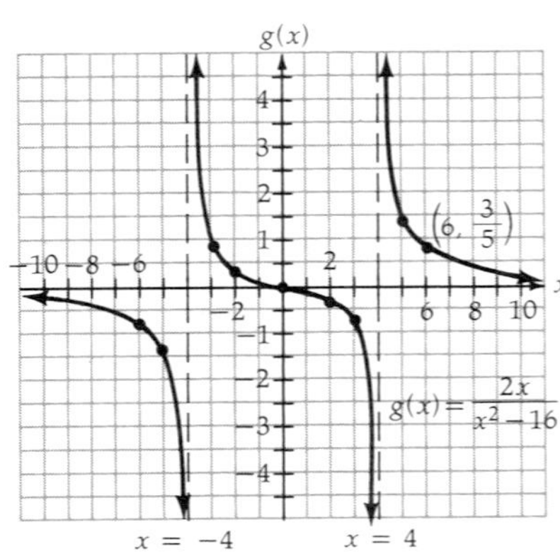

**15.**

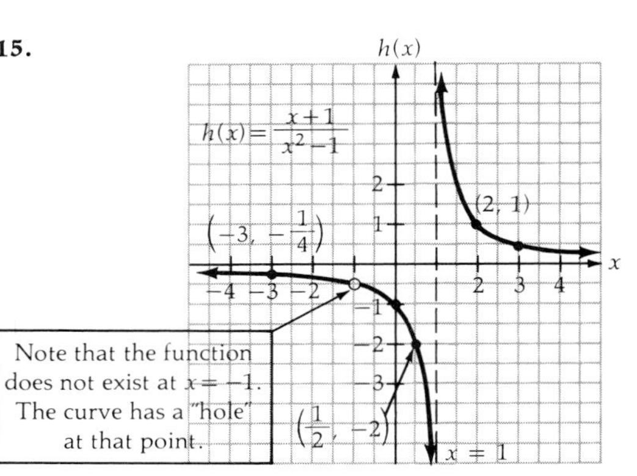
h(x)
h(x) = (x+1)/(x^2−1)
(2, 1)
(−3, −1/4)
(1/2, −2)
x = 1
Note that the function does not exist at x = −1. The curve has a "hole" at that point.

**17.**

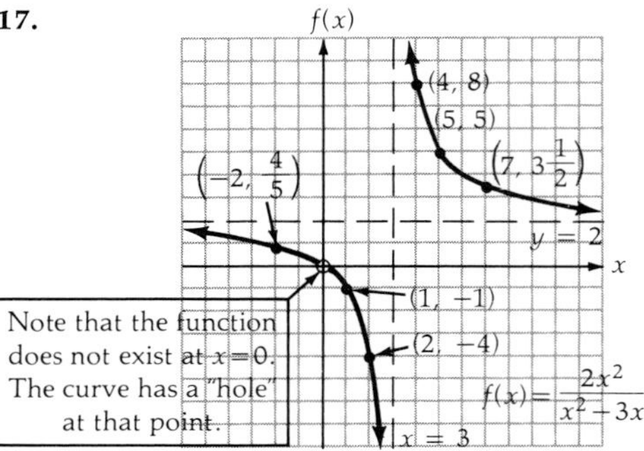
f(x)
(4, 8)
(5, 5)
(7, 3 1/2)
(−2, 4/5)
y = 2
(1, −1)
(2, −4)
f(x) = 2x^2/(x^2 − 3x)
x = 3
Note that the function does not exist at x = 0. The curve has a "hole" at that point.

**19.**

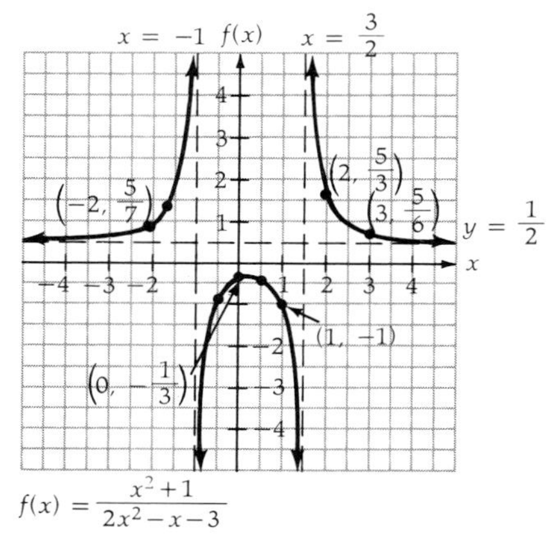
x = −1
f(x)
x = 3/2
(−2, 5/7)
(2, 5/3)
(3, 5/6)
y = 1/2
(1, −1)
(0, −1/3)
f(x) = (x^2+1)/(2x^2 − x − 3)

**21.**

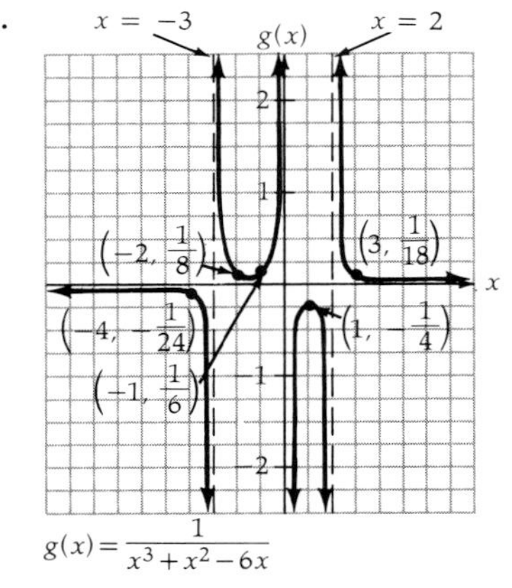
x = −3
g(x)
x = 2
(−2, 1/8)
(3, 1/18)
(−4, −1/24)
(1, −1/4)
(−1, 1/6)
g(x) = 1/(x^3 + x^2 − 6x)

**23.**

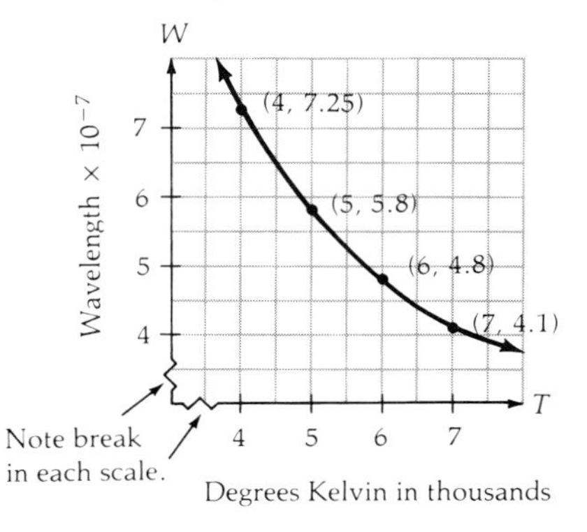
W
Wavelength × 10^−7
(4, 7.25)
(5, 5.8)
(6, 4.8)
(7, 4.1)
T
Note break in each scale.
Degrees Kelvin in thousands

## Exercise 4-4 (pp. 187–188)

**1.** $(x + 2)^2 + (y - 3)^2 = 36$

**3.** $(x + 1)^2 + (y - 6)^2 = 9$; center $= (-1, 6)$, radius $= 3$

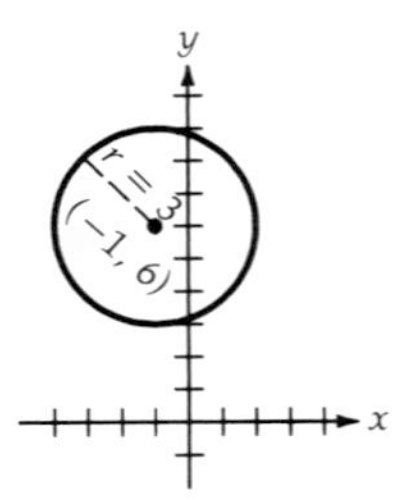


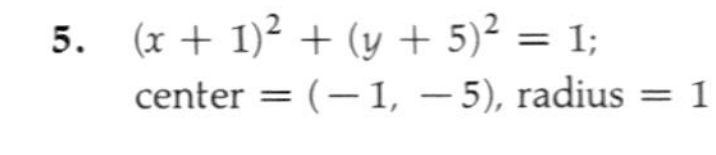

**5.** $(x + 1)^2 + (y + 5)^2 = 1$; center $= (-1, -5)$, radius $= 1$

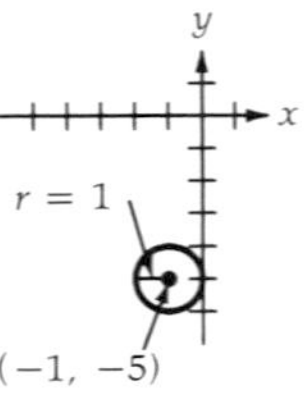


**7.** $(x - 3)^2 + (y + 4)^2 = 25$

**9.** center $= (0, 0)$; vertices $= (7, 0), (-7, 0), (0, 5), (0, -5)$; foci $= (2\sqrt{6}, 0), (-2\sqrt{6}, 0)$

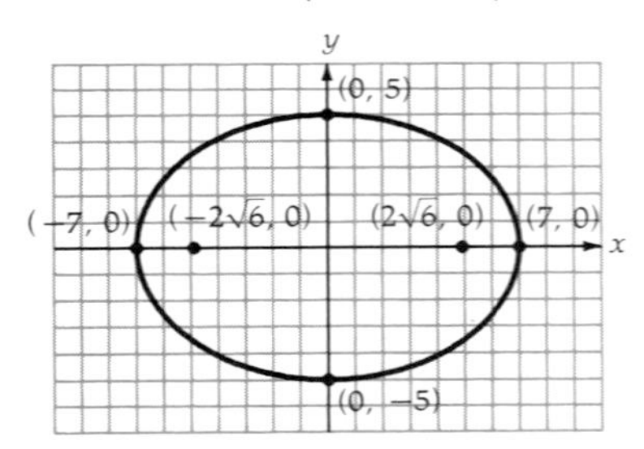


**11.** center $= (0, 0)$; vertices $= (0, 5), (0, -5), (3, 0), (-3, 0)$; foci $= (0, 4), (0, -4)$

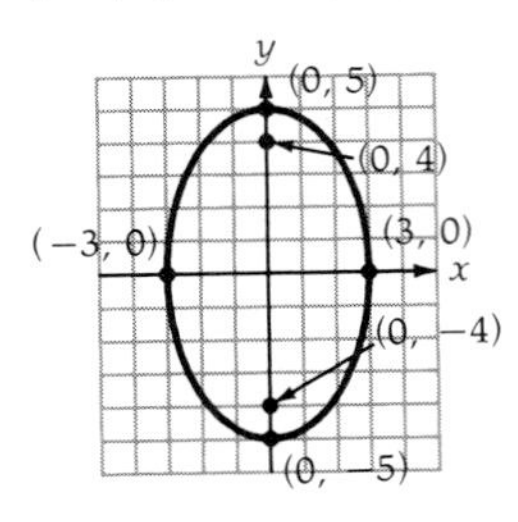


**13.** standard form: $\dfrac{x^2}{(\sqrt{5})^2} + \dfrac{y^2}{5^2} = 1$; center $= (0, 0)$; vertices $= (0, 5), (0, -5), (-\sqrt{5}, 0), (\sqrt{5}, 0)$; foci $= (0, 2\sqrt{5}), (0, -2\sqrt{5})$

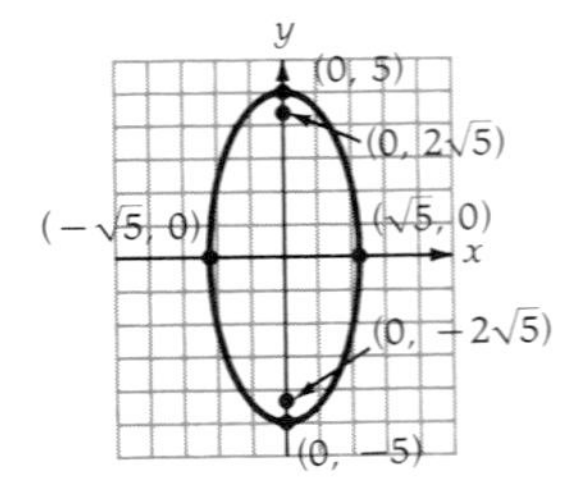


**15.** center $= (3, 1)$; vertices $= (8, 1), (-2, 1), (3, 4), (3, -2)$; foci $= (7, 1), (-1, 1)$

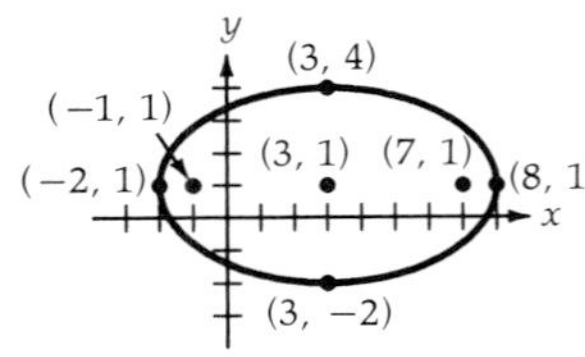

**17.** center $= (-5, -2)$; vertices $= (-5, 0), (-5, -4), (-6, -2), (-4, -2)$; foci $= (-5, -2 + \sqrt{3}), (-5, -2 - \sqrt{3})$

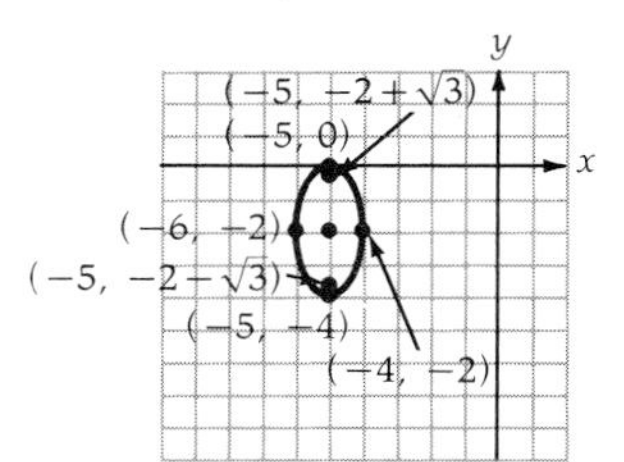


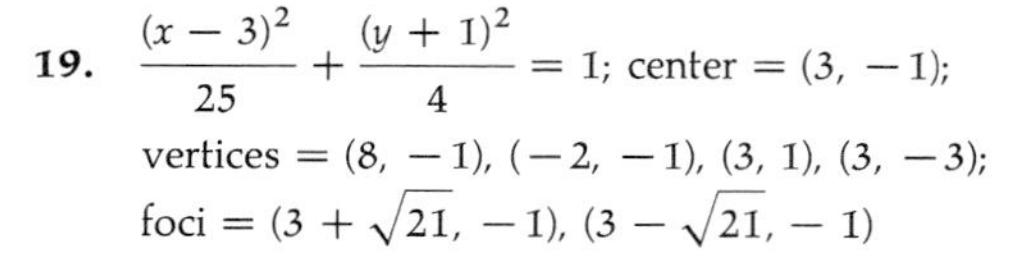

**19.** $\dfrac{(x-3)^2}{25} + \dfrac{(y+1)^2}{4} = 1$; center $= (3, -1)$; vertices $= (8, -1), (-2, -1), (3, 1), (3, -3)$; foci $= (3 + \sqrt{21}, -1), (3 - \sqrt{21}, -1)$

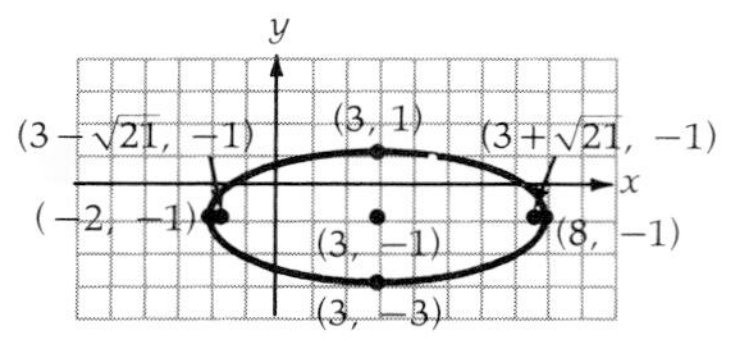


**21.** $\dfrac{(x+3)^2}{16} + \dfrac{(y-3)^2}{25} = 1$; center $= (-3, 3)$; vertices $= (-3, 8), (-3, -2), (1, 3), (-7, 3)$; foci $= (-3, 6), (-3, 0)$

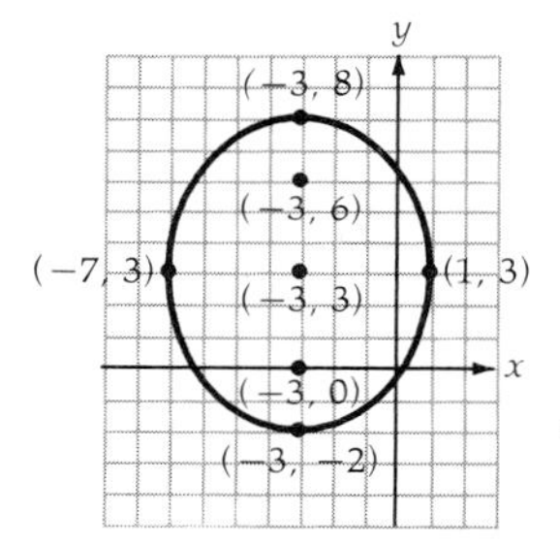


**23.** $\dfrac{(x+1)^2}{25} + \dfrac{(y-3)^2}{16} = 1$

**25.** $\sqrt{(x-h)^2 + [y-(k+c)]^2} + \sqrt{(x-h)^2 + [y-(k-c)]^2} = 2b$. Isolate one radical and square both sides of the equation. Isolate the remaining radical and square both sides of this second equation. Simplify and replace each expression containing $(b^2 - c^2)$ by $a^2$. This will yield $(x-h)^2 b^2 + (y-k)^2 a^2 = a^2 b^2$. Now divide each side by $a^2 b^2$ to obtain $\dfrac{(x-h)^2}{a^2} + \dfrac{(y-k)^2}{b^2} = 1$. For a more detailed solution, please see the *Student Tutorial.*

**27.** $(x-1)^2 + (y+4)^2 = 2$; circle with center at $(1, -4)$ and radius of $\sqrt{2}$

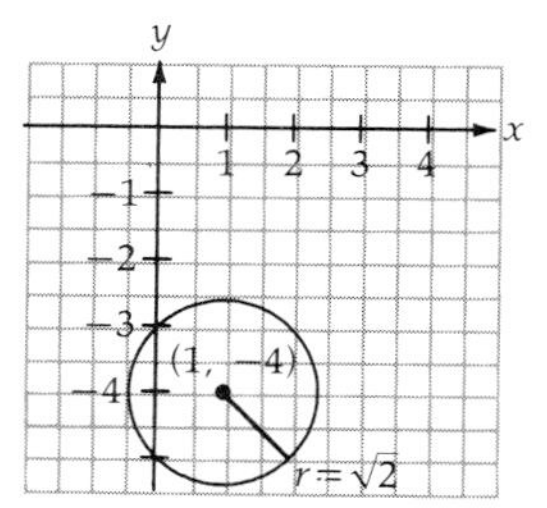


**29.** $\dfrac{(x+3)^2}{4} + \dfrac{(y-5)^2}{16} = 1$; ellipse with center at $(-3, 5)$

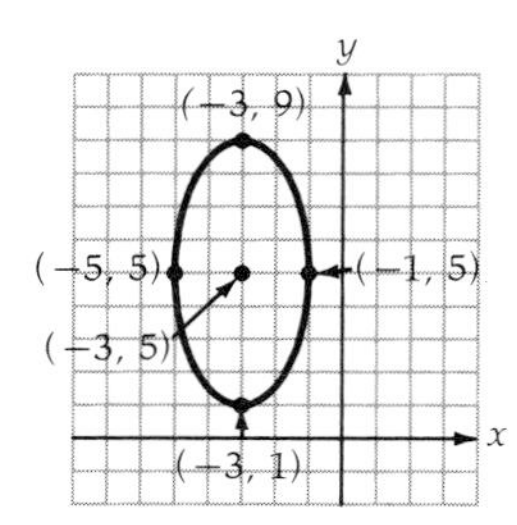

**31.** $\dfrac{(x+3)^2}{2} + \dfrac{y^2}{1} = 1$; ellipse with center at $(-3, 0)$; $a = \sqrt{2}$, $b = 1$

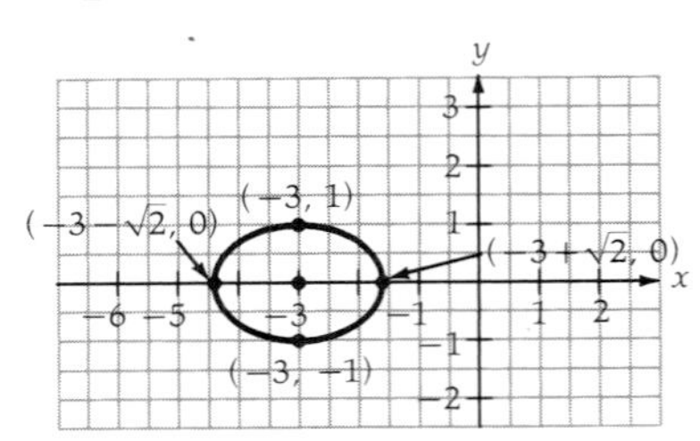


## Exercise 4-5 (p. 195)

**1.**

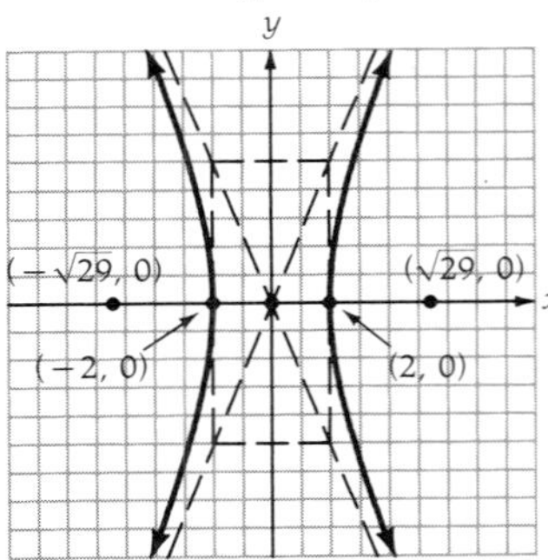


**3.**

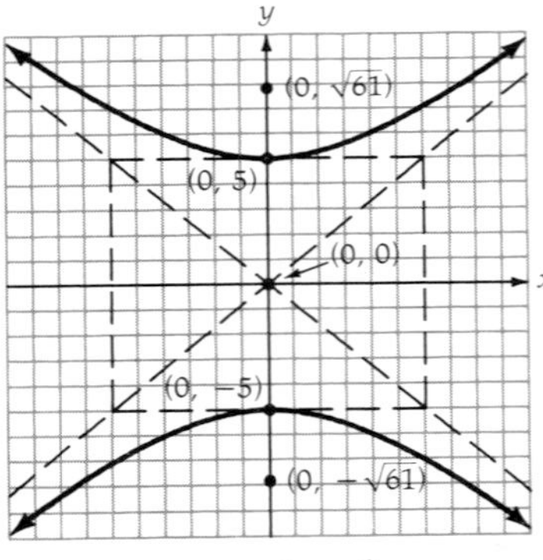


**5.** standard form is $\dfrac{x^2}{36} - \dfrac{y^2}{4} = -1$

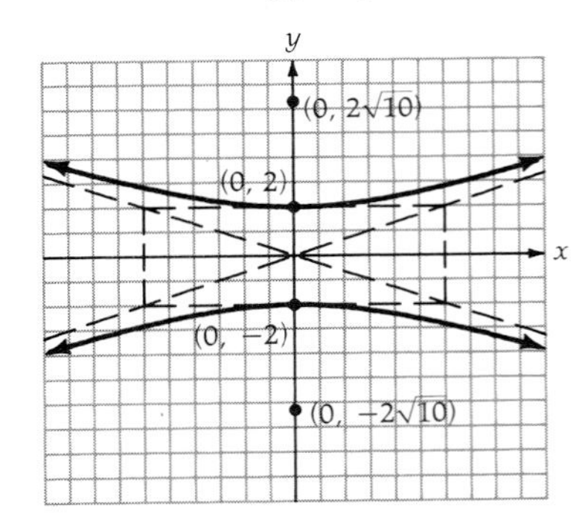


**7.** standard form is $\dfrac{x^2}{1} - \dfrac{y^2}{9} = 1$

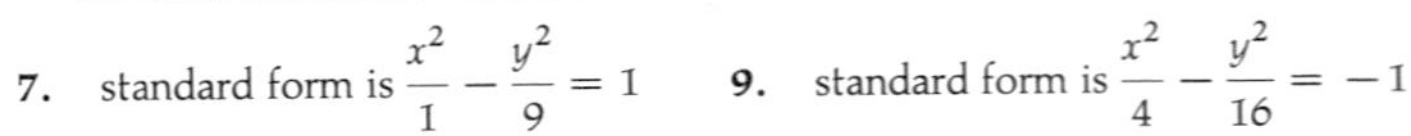

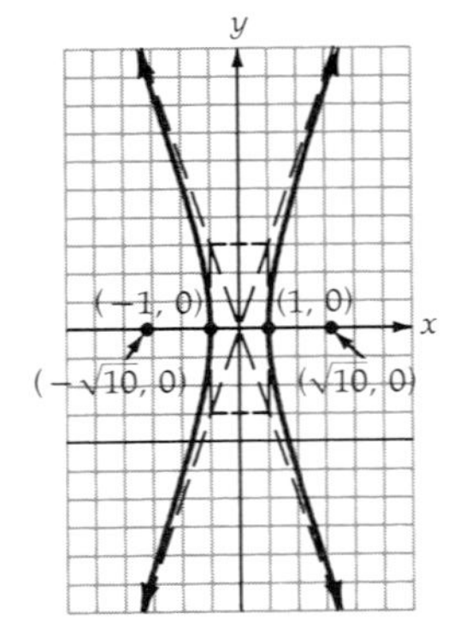


**9.** standard form is $\dfrac{x^2}{4} - \dfrac{y^2}{16} = -1$

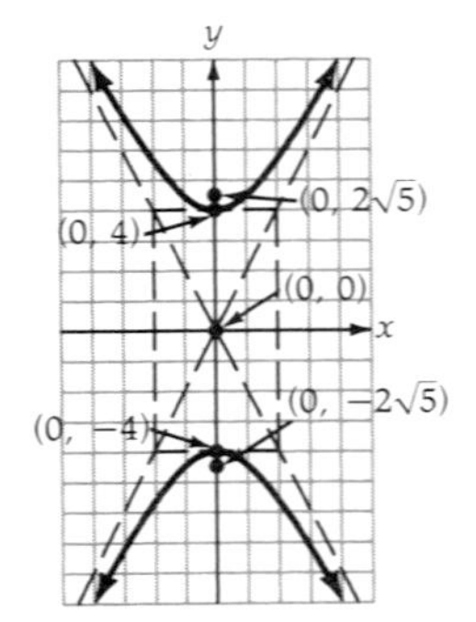

**11.**

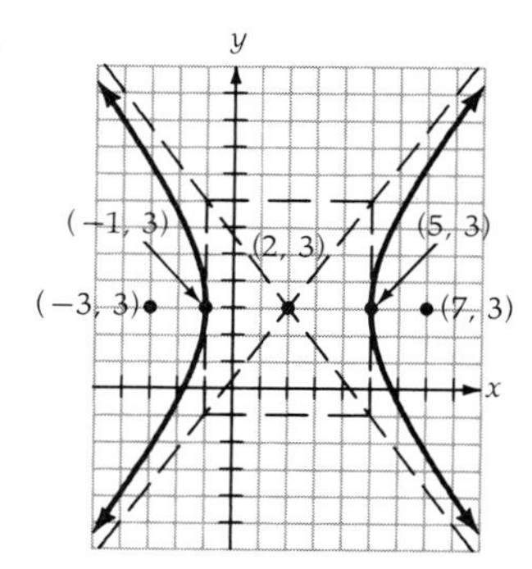


**13.** $(-1, 2+2\sqrt{5})$

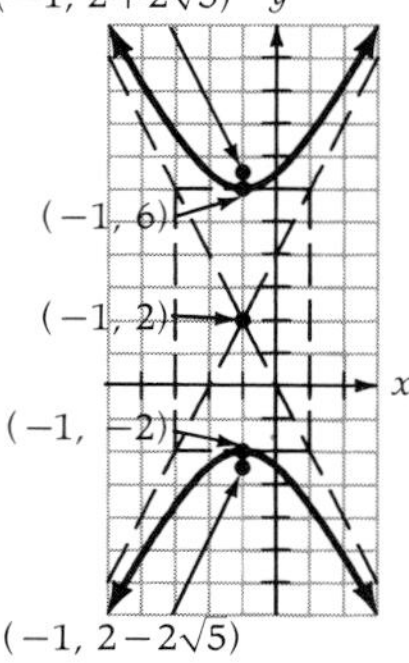


**15.** standard form is $\dfrac{(x-3)^2}{25} - \dfrac{(y+2)^2}{16} = 1$

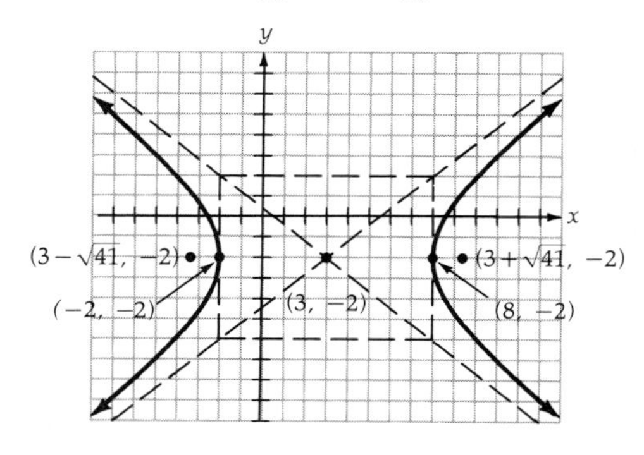


**17.** standard form is $\dfrac{(x+5)^2}{9} - \dfrac{(y-2)^2}{1} = -1$

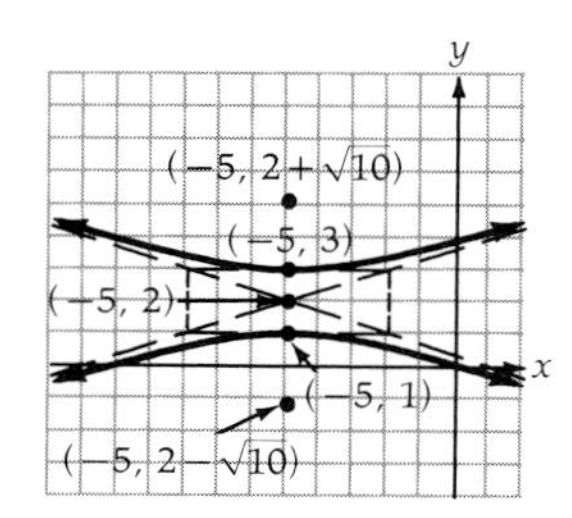


**19.** standard form is $\dfrac{(x-3)^2}{4} - \dfrac{(y+4)^2}{9} = 1$

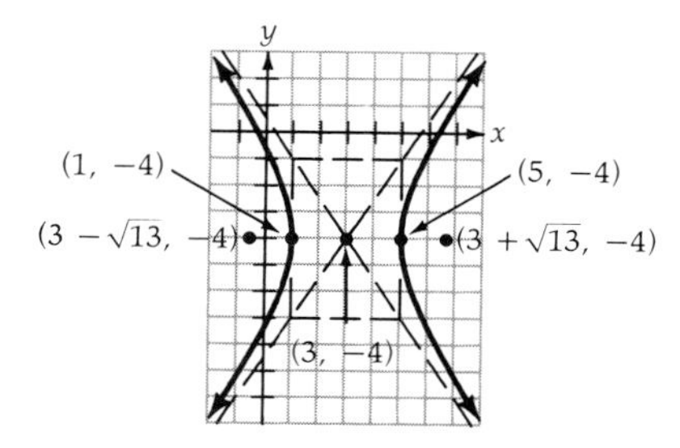


**21.** standard form is $\dfrac{(x+3)^2}{1} - \dfrac{(y-2)^2}{\frac{9}{25}} = -1$; $a = 1$, $b = \dfrac{3}{5}$

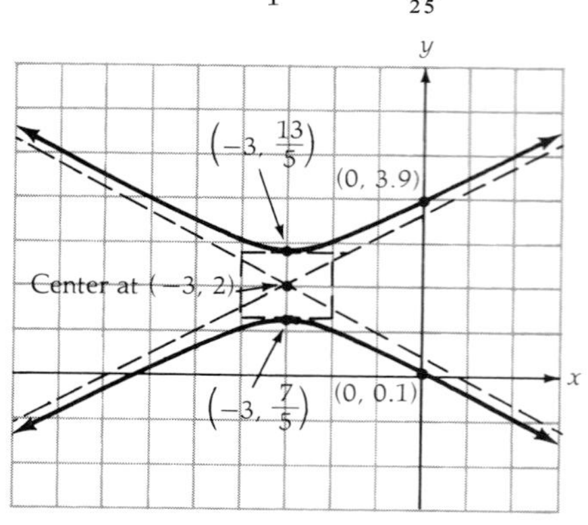


**23.** $\dfrac{(x-1)^2}{9} - \dfrac{(y+5)^2}{16} = -1$

**25.** $\dfrac{(x-1)^2}{9} - \dfrac{(y-2)^2}{7} = 1$

**27.** Use of the distance formula gives

$$\sqrt{(x-h)^2+[y-(k+c)]^2}-\sqrt{(x-h)^2+[y-(k-c)]^2}=2b$$

Isolate one radical and square each side of the equation. Isolate the remaining radical and square each side of the second equation. Simplify the results. Replace $b^2-c^2$ by $a^2$. This yields

$$(x-h)^2b^2-(y-k)^2a^2=-a^2b^2$$

Divide each side by $a^2b^2$ to obtain

$$\frac{(x-h)^2}{a^2}-\frac{(y-k)^2}{b^2}=-1$$

For a more detailed solution, see the *Student Tutorial.*

**29.**

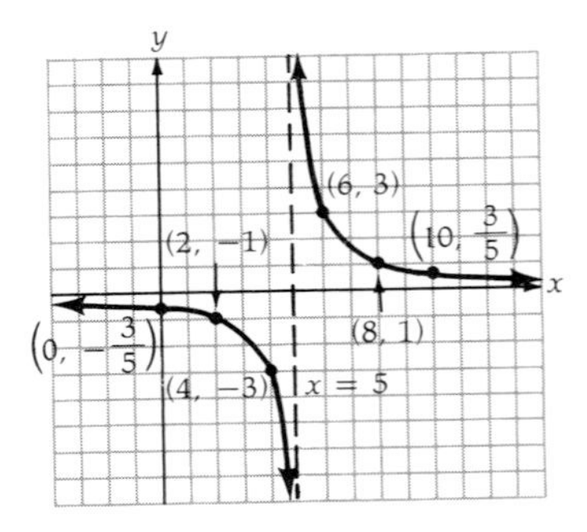


## Exercise 4-6 (pp. 202–203)

**1.**

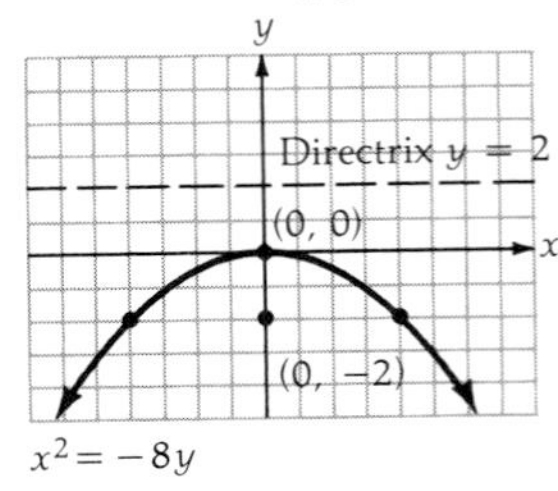


**3.**

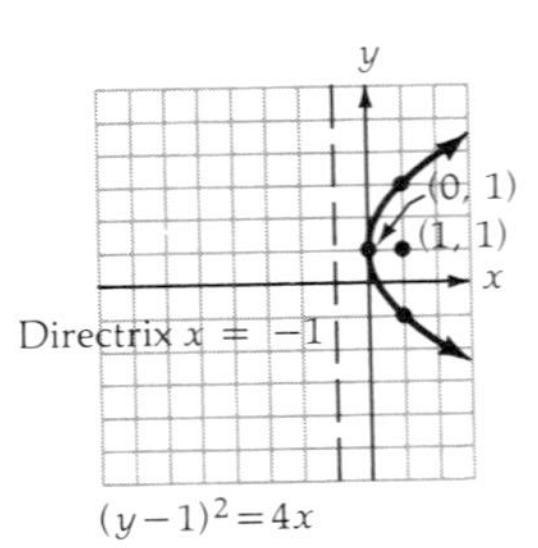


**5.**

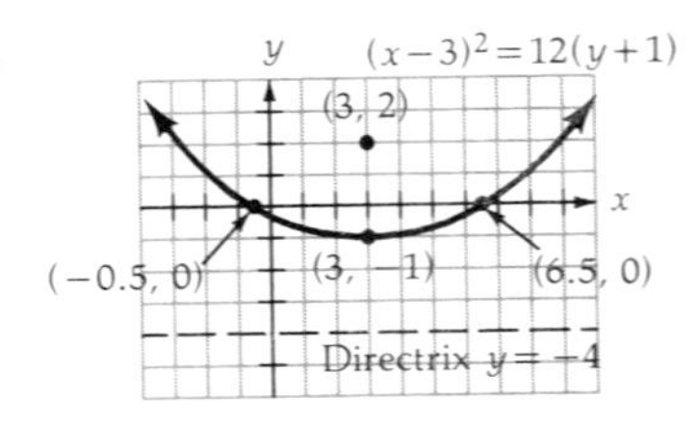


**7.**

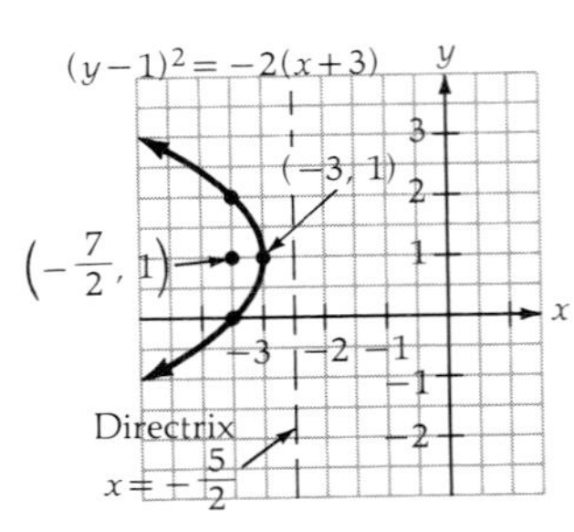


**9.**

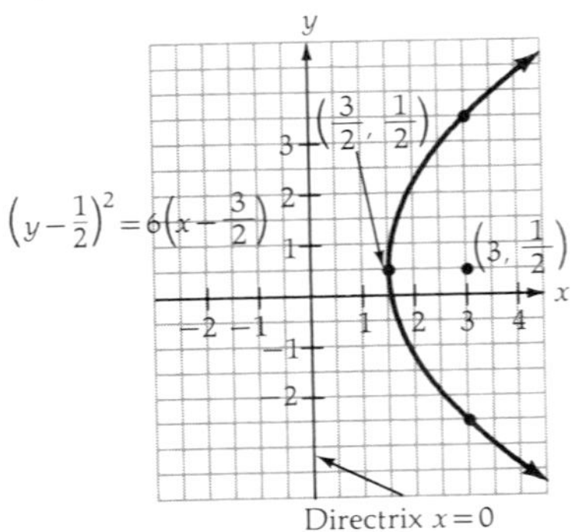

## Exercise 4-6 (continued)

**11.** standard form is $\left(x + \frac{1}{2}\right)^2 = y + \frac{1}{4}$

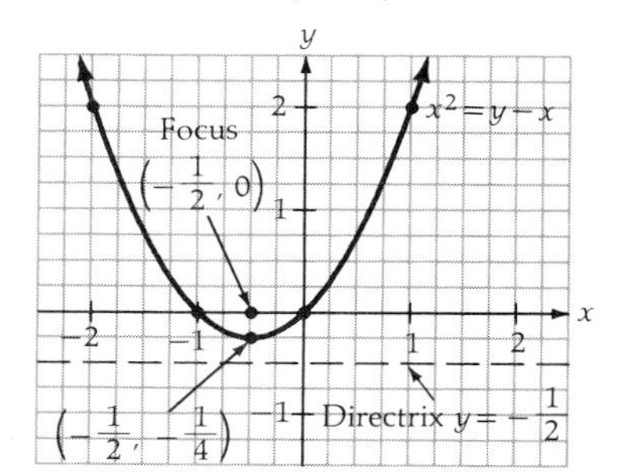


**13.** standard form is $(y - 1)^2 = -8(x - 2)$

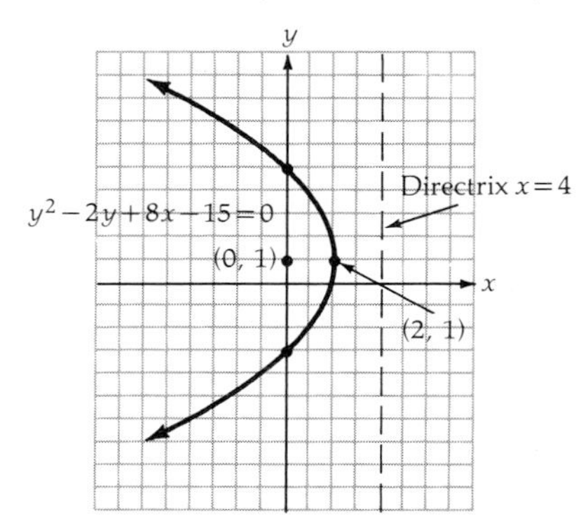


**15.** standard form is $(y + 3)^2 = (x + 4)$

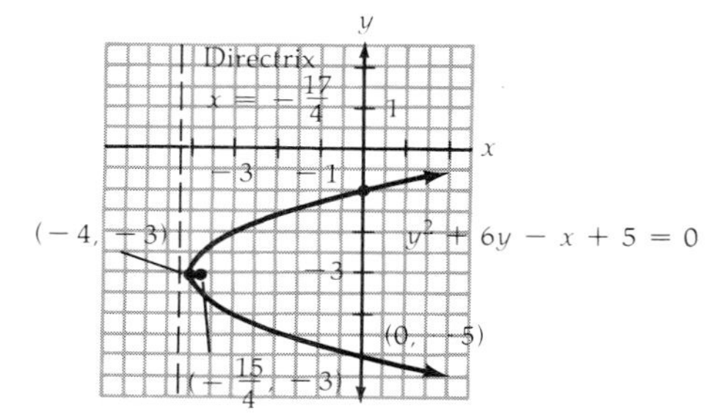


**17.** standard form is $(x - 3)^2 = -4\left(y + \frac{3}{2}\right)$

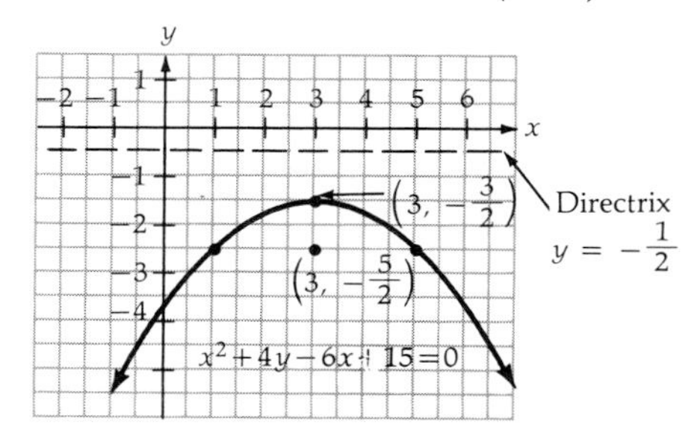


**19.** standard form is $(x - 5)^2 = -4(y + 2)$

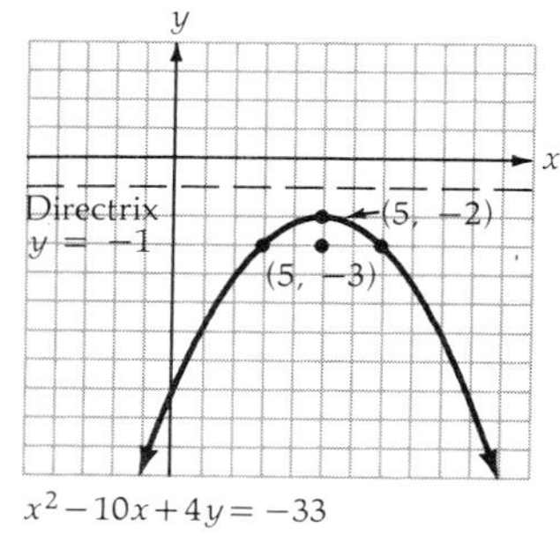


$x^2 - 10x + 4y = -33$

**21.** $(x - 3)^2 = 12(y - 1)$

**23.** $(x + 2)^2 = 12y$

**25.** $x^2 = \frac{2}{3}y$

**27.** focus is at $(h + p, k)$, vertex is at $(h, k)$, directrix is $x = h - p$; using distance formula, we have

$$\sqrt{[x - (h + p)]^2 + (y - k)^2} = [x - (h - p)]$$

Square both sides to obtain

$$(x - h)^2 - 2p(x - h) + p^2 + (y - k)^2 = (x - h)^2 + 2p(x - h) + p^2$$

Simplify. Isolate $(y - k)^2$ on the left-hand side and get all other terms on the right-hand side. Thus we have $y - k)^2 = 4p(x - h)$. For a more detailed solution, see the *Student Tutorial*.

**29.** First find the equation describing the parabolic bowl. If the vertex is located at the origin, the equation is $x^2 = 34y$. Then the distance between the two endpoints of the diameter can be found. The diameter is $2\sqrt{102}$ or approximately 20.2 feet.

**31.** standard form is $(x - 4)^2 = 8(y + 1)$; parabola with vertex at $(4, -1)$

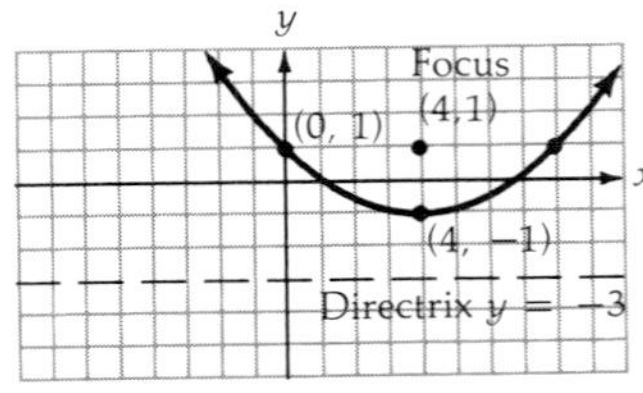


$x^2 - 8x - 8y + 8 = 0$

**33.** standard form is $\frac{(x + 4)^2}{16} - \frac{(y - 3)^2}{9} = 1$; hyperbola with center at $(-4, 3)$ and vertices at $(-8, 3)$ and $(0, 3)$

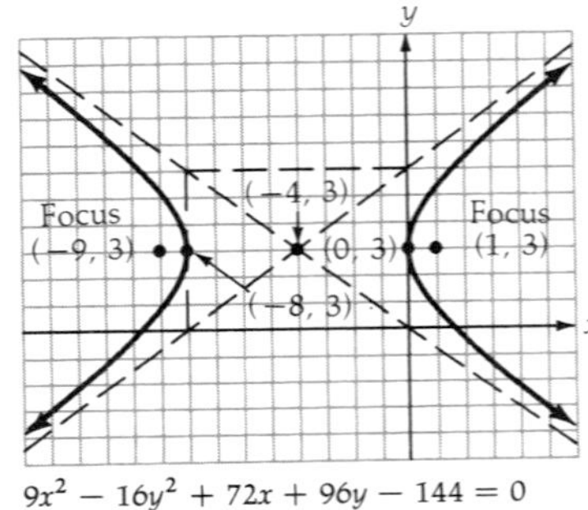


$9x^2 - 16y^2 + 72x + 96y - 144 = 0$

**35.** standard form is $(y + 4)^2 = 12(x - 2)$; parabola with vertex of $(2, -4)$

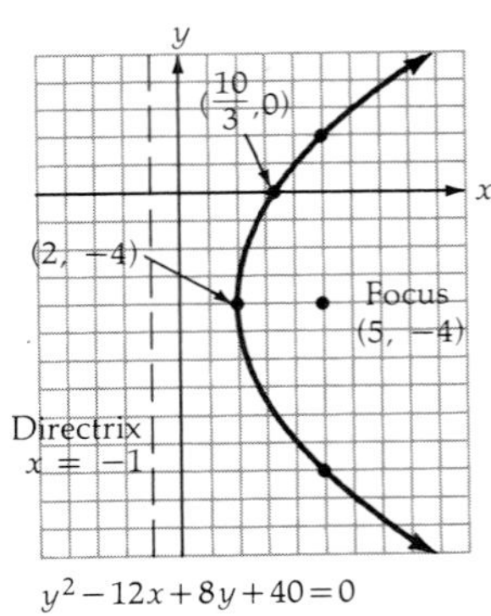


$y^2 - 12x + 8y + 40 = 0$

**37.** standard form is $(x - 6)^2 = 4(y - 3)$; parabola with vertex at $(6, 3)$

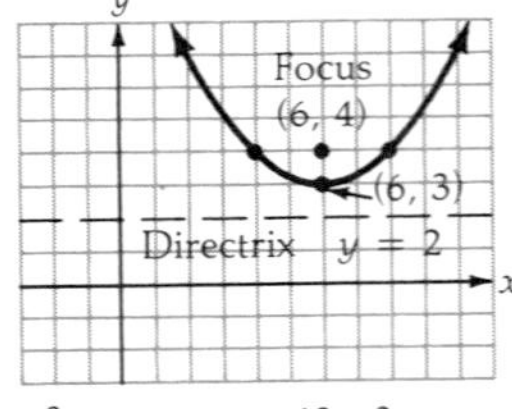


$x^2 - 12x - 4y + 48 = 0$

**39.** standard form is $\frac{(x + 2)^2}{4} - \frac{(y - 5)^2}{9} = 1$; hyperbola with center at $(-2, 5)$ and vertices at $(0, 5)$ and $(-4, 5)$

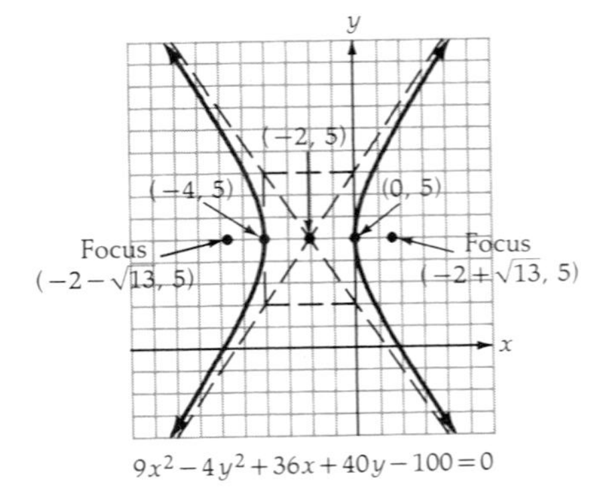


$9x^2 - 4y^2 + 36x + 40y - 100 = 0$

**41.** standard form is $\frac{(x-1)^2}{4} + \frac{(y-4)^2}{16} = 1$; ellipse with center at (1, 4) and vertices at (−1, 4), (1, 8), (3, 4), and (1, 0)

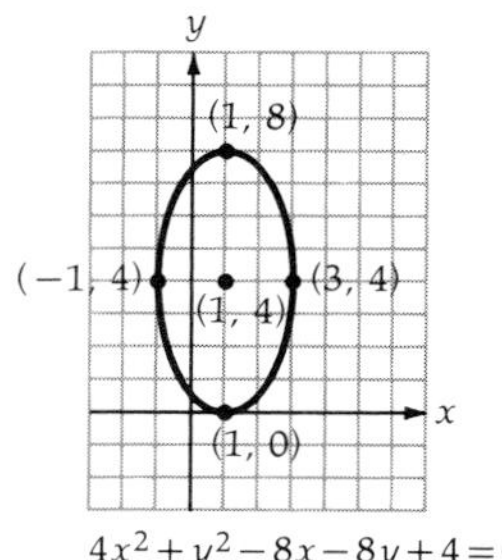

$4x^2 + y^2 - 8x - 8y + 4 = 0$

## Chapter 4 Review Exercise (pp. 207–208)

**1.**

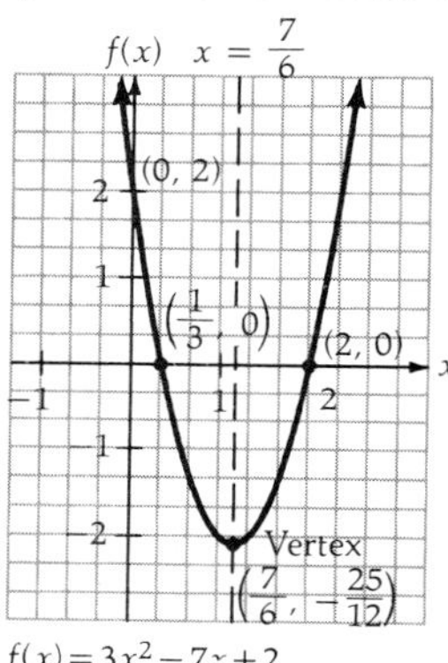

$f(x) = 3x^2 - 7x + 2$

**3.**

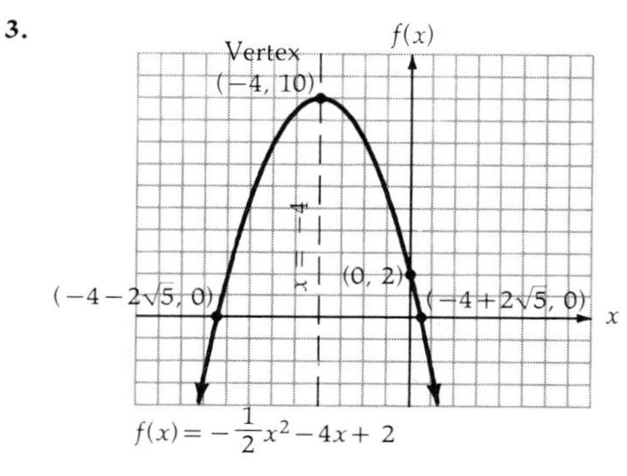

$f(x) = -\frac{1}{2}x^2 - 4x + 2$

**5.** 43 meters of fencing on side parallel to barn, 21.5 meters of fencing on each of 2 sides perpendicular to barn

**7.**

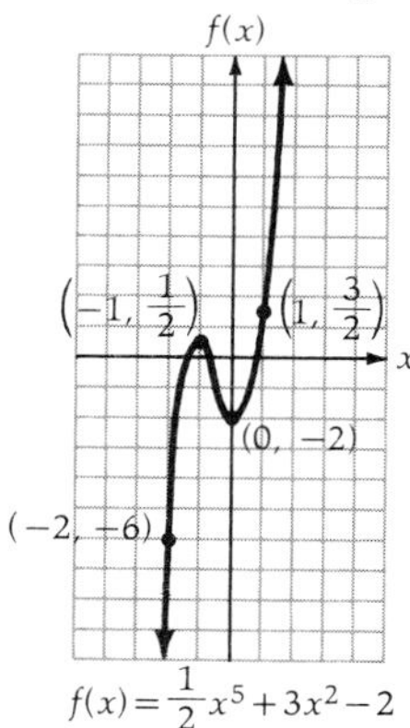

$f(x) = \frac{1}{2}x^5 + 3x^2 - 2$

**9.** roots $\in \{-4, -2, 3\}$

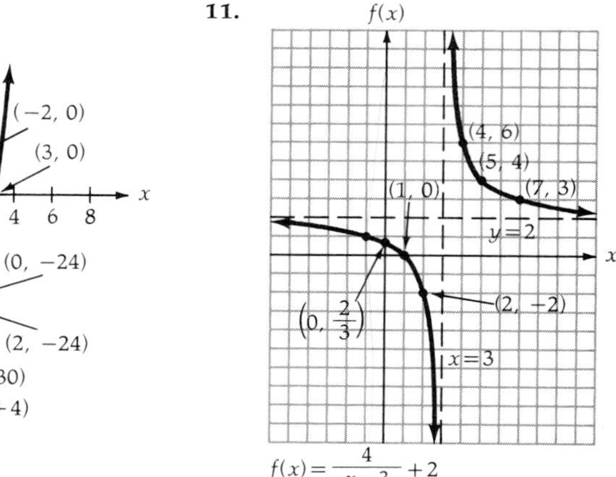

$f(x) = (x-3)(x+2)(x+4)$

**11.**


$f(x) = \frac{4}{x-3} + 2$

**13.**

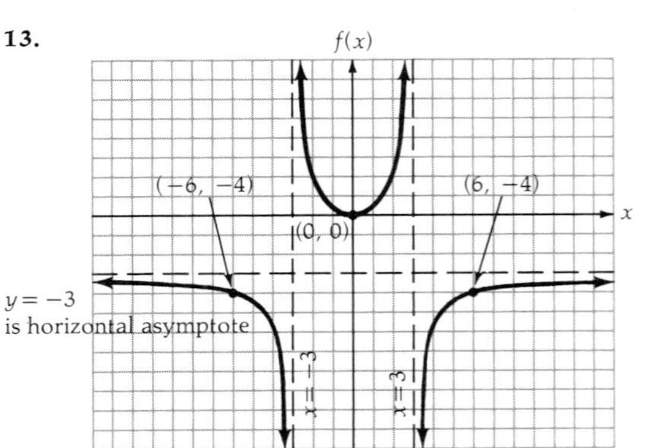


$f(x) = \frac{3x^2}{9 - x^2}$

**15.** This is a circle; in standard form: $(x-3)^2 + (y+1)^2 = 25$; center: $(3, -1)$, radius: 5

**17.**

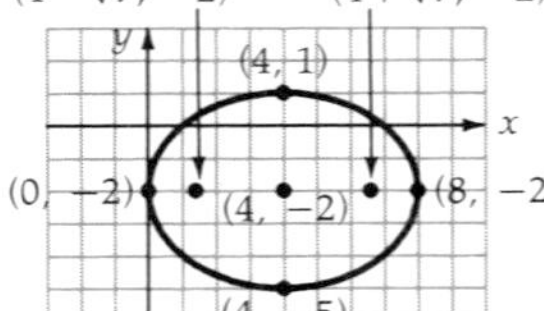


$$\frac{(x-4)^2}{16} + \frac{(y+2)^2}{9} = 1$$

**19.** $$\frac{(x+2)^2}{9} + \frac{(y-1)^2}{16} = 1$$

**21.**

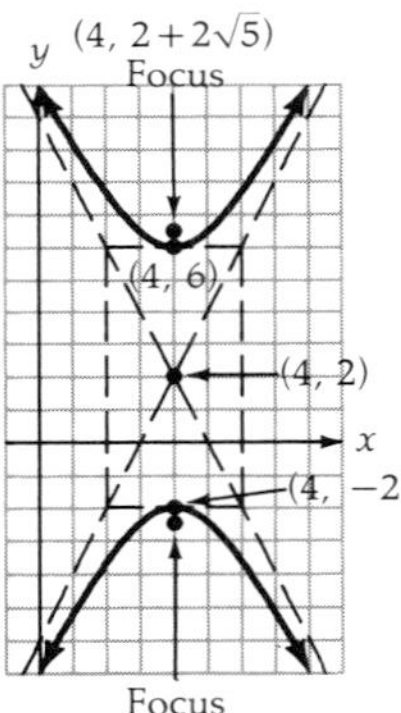


$$\frac{(x-4)^2}{4} - \frac{(y-2)^2}{16} = -1$$

**23.**

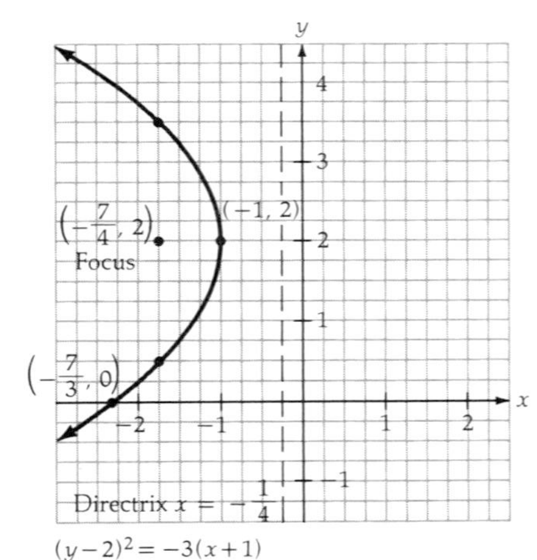


$(y-2)^2 = -3(x+1)$

**25.** $$\frac{(x+3)^2}{9} + \frac{(y-5)^2}{25} = 1,$$
vertical ellipse

**27.** $$\frac{(x-2)^2}{25} - \frac{(y+1)^2}{16} = -1,$$
vertical hyperbola

**29.** $$\frac{(x+3)^2}{4} + \frac{(y-2)^2}{1} = 1,$$
horizontal ellipse

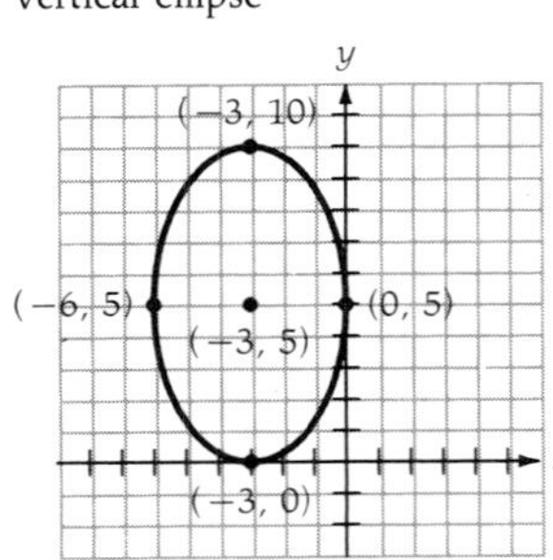


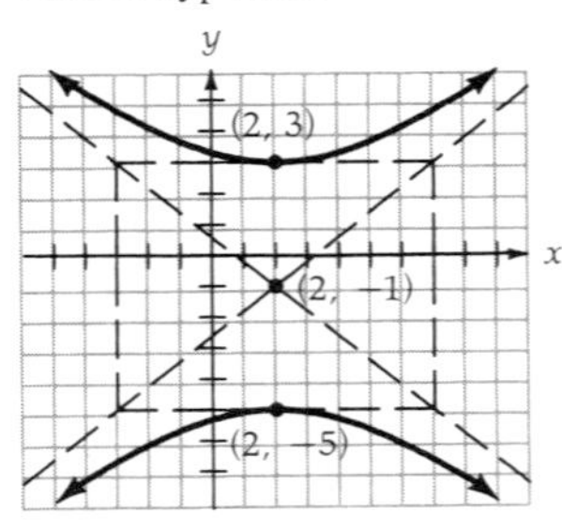


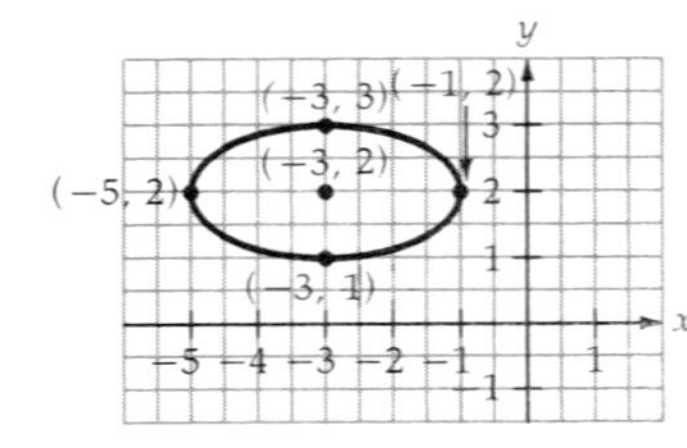

**31.** $(y - 4)^2 = -\frac{1}{2}(x + 3)$, horizontal parabola

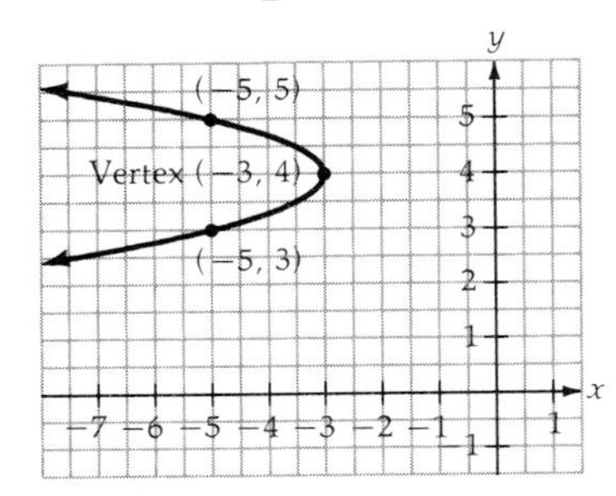


**33.** $\frac{(x - 2)^2}{1} - \frac{(y - 6)^2}{9} = 1$, horizontal hyperbola

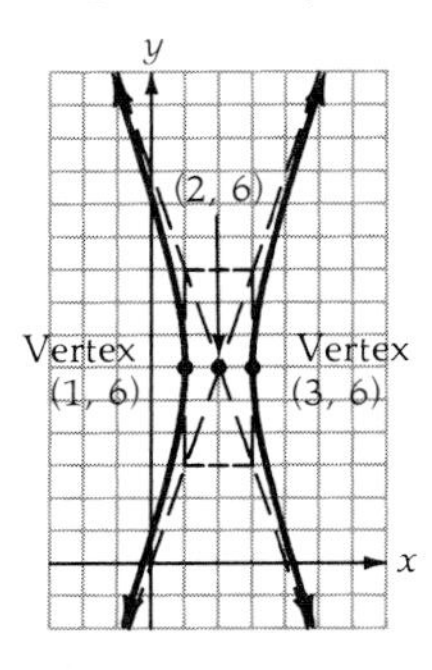


## Practice Test for Chapter 4 (p. 208)

**1.** $(x - 5)^2 + (y + 2)^2 = 36$, center: $(5, -2)$, radius: 6

**3.**

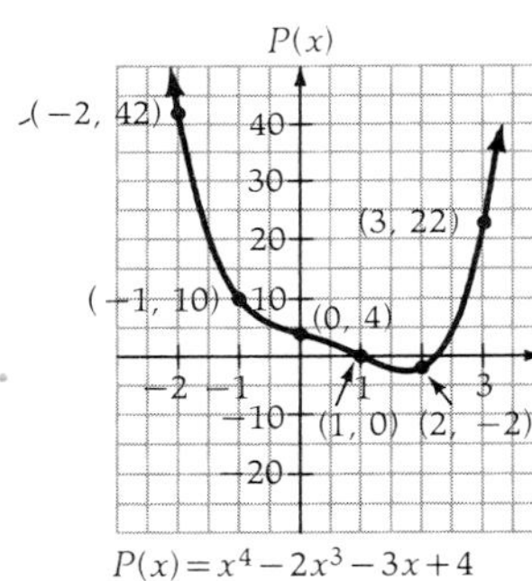


**5.** $\frac{(x - 2)^2}{4} - \frac{(y + 4)^2}{16} = 1$, horizontal hyperbola

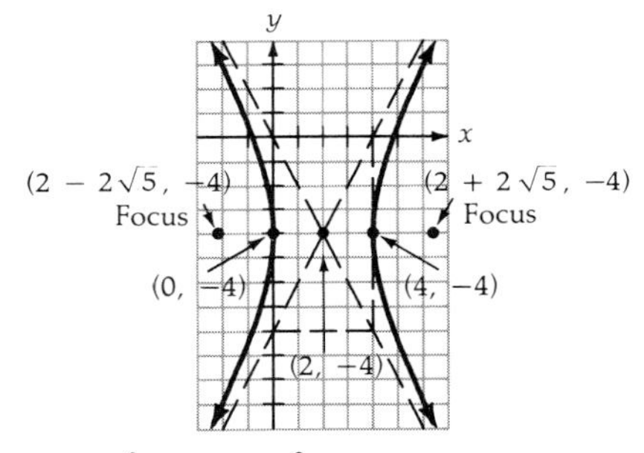


**7.** $(x + 1)^2 + (y - 2)^2 = 4$, circle

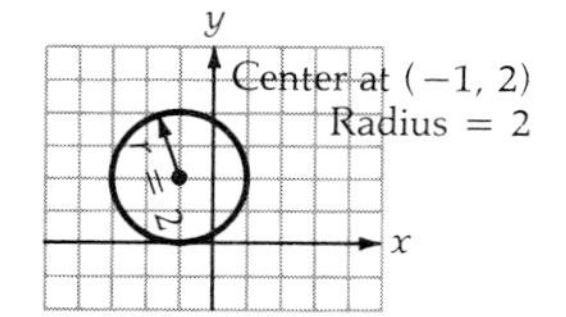


**9.** $\frac{(x - 2)^2}{\frac{9}{25}} - \frac{(y + 3)^2}{1} = 1$, horizontal hyperbola; $a = \frac{3}{5}$, $b = 1$

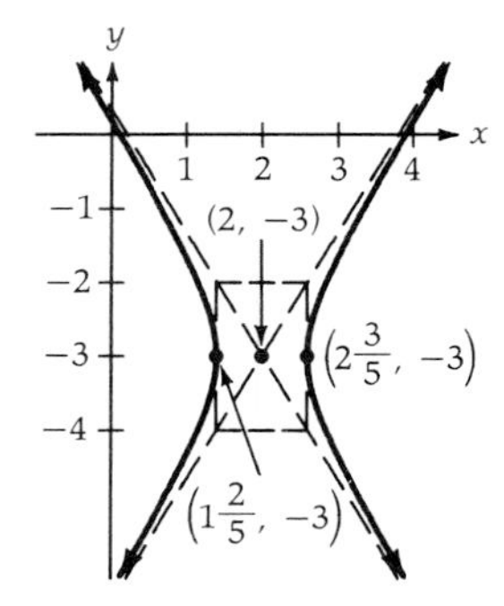

## CHAPTER 5

### Exercise 5-1 (pp. 219–221)

**1.**

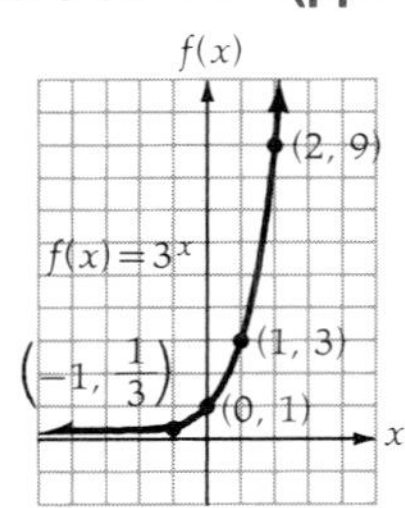


**3.**

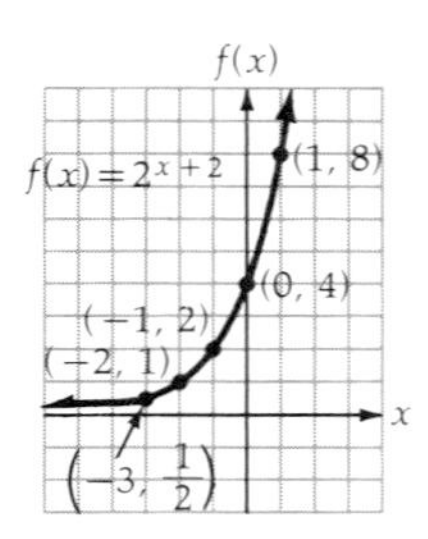


**5.**

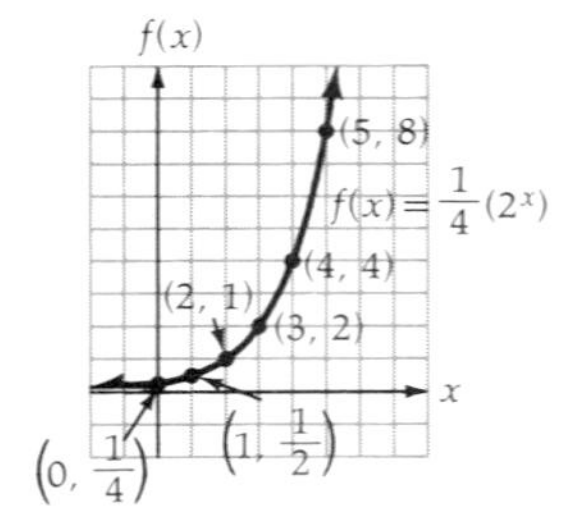


**7.**

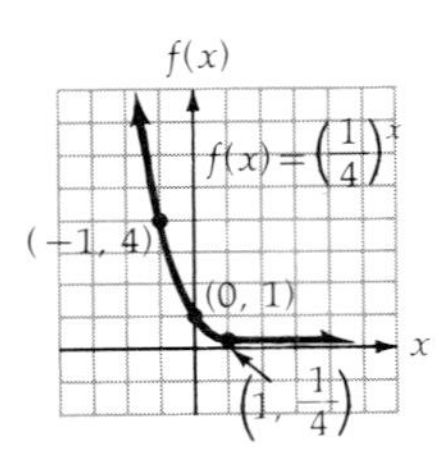


**9.**

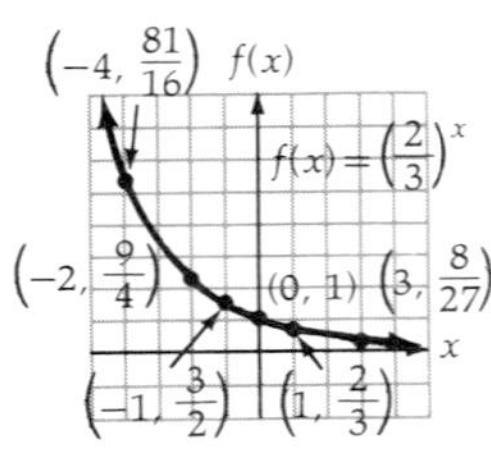


**11.**

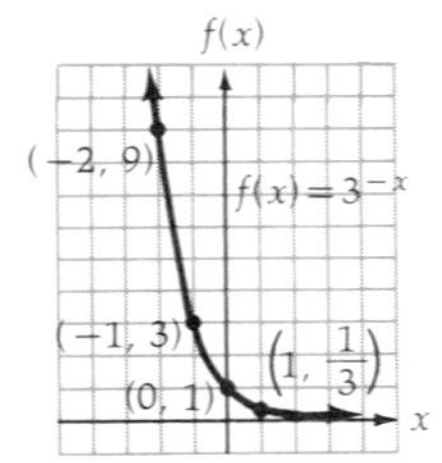


**13.**

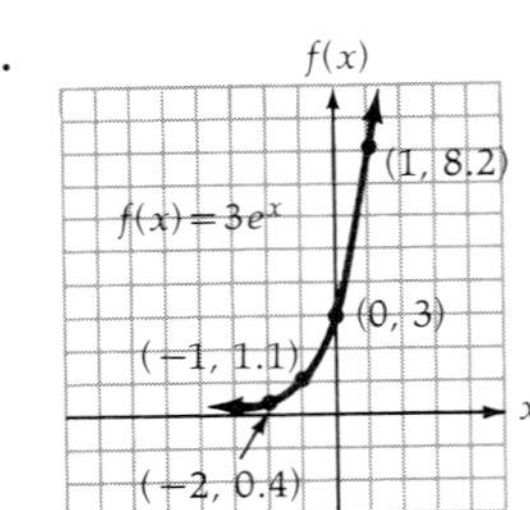


**15.**

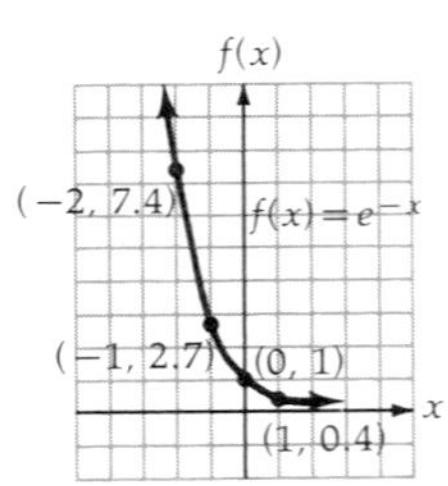


**17.** Curve is symmetric with respect to the $f(x)$-axis.

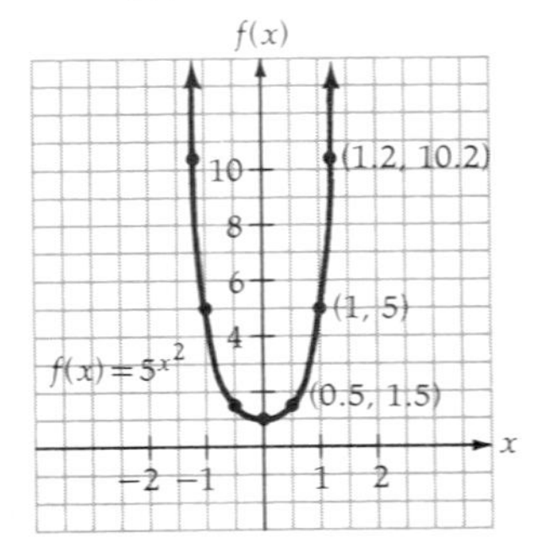


**19.**

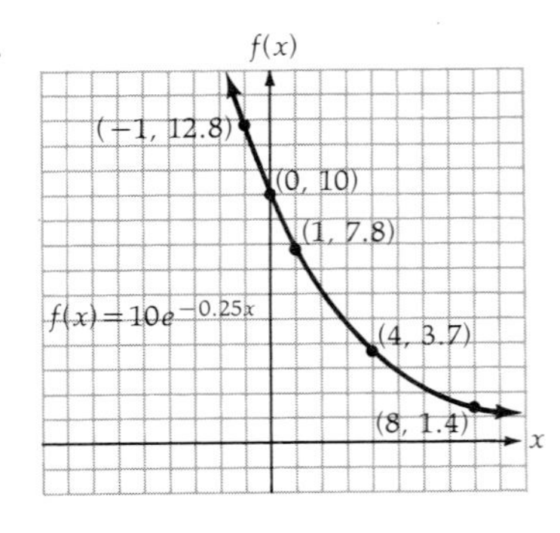


**21.** Curve is symmetric with respect to the $g(x)$-axis. (Note: All plotted values are rounded.)

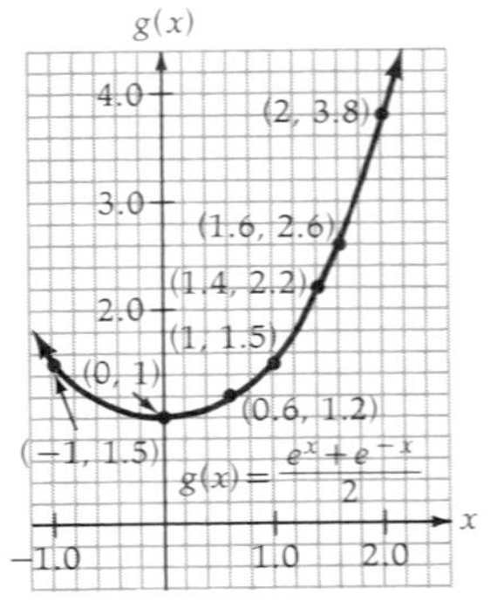

**23.** \$7701.66 **25.** \$4666.36; \$4691.83 **27.** 11,314 (rounded to nearest whole number)
**29.** \$6432.06 − \$6431.66 = \$0.40 additional will be earned. **31.** 0.847 mg **33.** 1.009 mg
**35.** 1.67 microcoulombs

## Exercise 5-2 (pp. 228–229)

**1.** $\log_3 9 = 2$ **3.** $\log_5 125 = 3$ **5.** $\log_3 27 = 3$ **7.** $\log_2 \frac{1}{8} = -3$ **9.** $\log_2 \frac{1}{16} = -4$

**11.** $2^2 = 4$ **13.** $4^2 = 16$ **15.** $3^4 = 81$ **17.** $5^3 = 125$ **19.** 81 **21.** 2 **23.** 5

**25.** $\frac{1}{9}$ **27.** 49 **29.** 1 **31.** 3 **33.** −2

**35.**

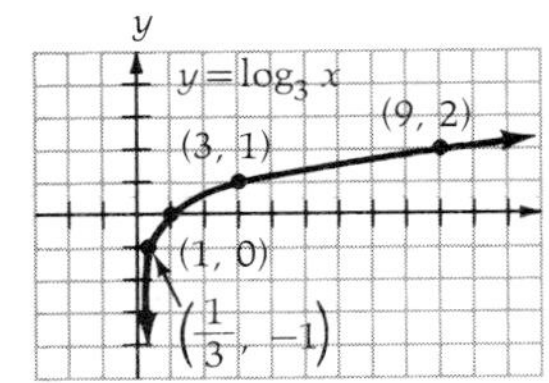


**37.**
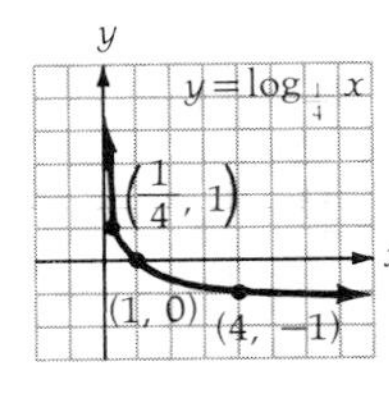


**39.** It is an *exponent.* **41.** 8

**43.** $6^6$ **45.** $\frac{1}{3}$ **47.** $2y$

**49.** $2x + y$ **51.** $x - 2y$

**53.** $2y - 2x$ **55.** 2 **57.** 4

**59.** −2 **61.** 5 **63.** 0.5514 + 2 **65.** 0.5514 − 2 **67.** 0.5514 − 3

**69.** *Prove:* $\log_b \frac{x}{y} = \log_b x - \log_b y$. Let $\log_b x = k$ and $\log_b y = n$; then by definition $b^k = x$ and $b^n = y$.

Therefore $\frac{x}{y} = \frac{b^k}{b^n} = b^{k-n}$ by laws of exponents. Finally, $\log_b\left(\frac{x}{y}\right) = \log_b (b^{k-n}) = k - n = \log_b x - \log_b y$

## Exercise 5-3 (pp. 234–235)

**1.** 0.7505083949 **3.** 2.104248047 **5.** −0.773915884 **7.** −3.773915884 **9.** 4.134909879
**11.** 1.954728298 **13.** −0.2993498773 **15.** 8.136810864 **17.** 12.74198105
**19.** −5.228791304 **21.** 2.350715116 **23.** 4378.244391 **25.** 0.0181217444 **27.** 0.0040963778
**29.** 2.438297378 **31.** 37.40644441 **33.** 1507.488952 **35.** 0.0114335873 **37.** 3.699107862
**39.** 0.1611801934 **41.** 5.870383517 **43.** −4.016249352 **45.** 1.789609235

Problems 47–53 have many possible answers. Owing to roundoff errors, the values may differ slightly in the last decimal place.

**47.** Show that $\log (2 \cdot 5) = \log 2 + \log 5$

$$1 = 0.301029995 + 0.698970004$$

(approximately equal) $1 \doteq 0.999999999$

**49.** Show that $\log 3^6 = 6 \log 3$

$$2.862727528 = (6)(0.477121254)$$

$$2.862727528 = 2.862727528$$

**51.** Show that $\log_3 3^2 = 2$

$$\log_3 9 = 2$$

$$\frac{0.954242509}{0.477121254} = 2$$

(approximately equal) $2.000000002 \doteq 2$

**53.** Show that $2^5 = 10^{5 \log 2}$

$$32 = 10^{1.505149978}$$

$$32 = 32$$

**55.** Show that $10^{\log \sqrt{7}} = \sqrt{7}$

$$10^{0.42254902} = \sqrt{7}$$

$$2.645751311 = 2.645751311$$

## Exercise 5-4 (pp. 244–245)

The answers in this section are obtained by using Table C. Answers obtained on a calculator will vary slightly from these values.

**1.** 0.3711 **3.** 0.2856 **5.** 0.8976 **7.** 0.5694 + 1 **9.** 0.5172 + 2 **11.** 3.5229
**13.** 7580 **15.** 0.045 **17.** 0.00601 **19.** 34,500,000 **21.** 3.9676 **23.** 0.4751 − 1
**25.** Interpolation treats the function between two points as linear. The logarithmic function and the squaring function are not linear.
**27.** 0.01716 **29.** 10,870 **31.** 0.002086 **33.** 1334 **35.** 6.696 **37.** 5.244
**39.** 0.001945 **41.** 0.3084 **43.** $1.583 \times 10^{12}$ **45.** 7.423 **47.** 4.858 **49.** 2.921

## Exercise 5-5 (p. 249)

**1.** 2 **3.** no solution **5.** $\frac{1}{9}$ **7.** no solution **9.** 3, 4 **11.** 4 **13.** $\frac{\log 7}{\log 2}$

**15.** $\frac{\log 9 - \log 16}{\log 8}$ **17.** $\frac{5 \log 3 - \log 2}{\log 2 - 2 \log 3}$ **19.** $\frac{\log 5 - \log 3}{\log 3 + 2 \log 5}$ or $\frac{\log 15 - 2 \log 3}{2 \log 15 - \log 3}$ **21.** $\frac{1006}{3}$

**23.** −0.909 **25.** −1.856 **27.** −270 **29.** 2.261 **31.** 5 **33.** $\frac{5}{2}$ **35.** 4

**37.** 14.217 **39.** 70 **41.** 5.823 **43.** $60.44 \log (a/b)$

## Exercise 5-6 (pp. 254–256)

**1.** 11.9 hours **3.** 51 days **5.** 8.4 years **7.** 9.3 years **9.** 28.25 years **11.** 65.8 years
**13.** $K = -0.00425244$ **15.** 7600 years old **17.** 0.2 **19.** $5 \times 10^{-3}$ moles per liter **21.** 16 times
**23.** $2.5 \times 10^8 I_0$ **25.** $5.75 \times 10^7$

## Chapter 5 Review Exercise (pp. 258–259)

**1.** 64 **3.** 9 **5.** −3 **7.** $3x$ **9.** $3x + 2y$

The answers to Problems 11–19 were obtained on a calculator. If Table C is used, these answers will have fewer significant digits,

**11.** 0.9052560487 **13.** 4.69019608 **15.** −3.806875402 or 0.193124598 − 4 **17.** 908.0296
**19.** −0.346112 or 0.653888 − 1 **21.** 21.87 **23.** 0.08112 **25.** 0.1293 **27.** 0.0001406
**29.** 4.149 **31.** 3.350219859 **33.** 0.509357848 **35.** −0.191160505 **37.** 11

**39.** $\frac{\log 7 - \log 3}{\log 3}$ or 0.771243749 **41.** 14 years **43.** 1.09 mg **45.** 8.25 years

## Practice Test for Chapter 5 (pp. 259–260)

**1.** $\log_3 9 = 2$ **3.** 2 **5.** 3 **7.** 5 **9.** $2x + y$

The answers to Problems 11–15 were obtained on a calculator. If Table C is used, these answers will have fewer significant digits.

**11.** −3.806875402 or 0.193124598 − 4 **13.** 4.151890374 **15.** 2979.888403 **17.** 13,140

**19.** $\frac{1}{2}$ **21.** 100

# CHAPTER 6

## Exercise 6-1 (pp. 266–268)

**1.** $\frac{1}{6}\pi$ radians **3.** $\frac{1}{9}\pi$ radians **5.** $\frac{5}{6}\pi$ radians **7.** $\frac{3}{2}\pi$ radians **9.** $\frac{1}{4}\pi$ radians

The answers to Problems 11–39 assume that the calculation was done without a calculator. Answers are often approximated and rounded to three significant digits. Answers obtained on a calculator will often display more significant digits.

**11.** 0.175 radians **13.** 0.893 radians **15.** 1.75 radians **17.** 1.65 radians **19.** 4.92 radians
**21.** 45° **23.** 30° **25.** 108° **27.** 860° **29.** 573° **31.** 344° **33.** 917°
**35.** 210° **37.** 171° **39.** 247.5° **41.** 0.4778994 radians **43.** 0.0091050 radians
**45.** 3.3797 radians **47.** 7237.185° **49.** 1.4979° **51.** 780.64° **53.** 25.963°
**55.** 3.2 meters **57.** 1.11 inches **59.** The wheel has rotated through an angle of approximately 0.44 radian.

## Exercise 6-2 (pp. 277–280)

**1.** $\sqrt{13}$ **3.** 8 **5.** 24 **7.** $5\sqrt{2}$ **9.** 20.6 meters **11.** $14\sqrt{2}$

**13.** **a.** $\sqrt{3}$ **b.** $3\sqrt{3}$ **c.** 6 **d.** equal **e.** $\frac{1}{2}$ **f.** $\frac{1}{2}$ **g.** $\frac{1}{2}$

**15.** $\sin B = \frac{5}{13}$, $\csc B = \frac{13}{5}$, $\cos B = \frac{12}{13}$, $\sec B = \frac{13}{12}$, $\tan B = \frac{5}{12}$, $\cot B = \frac{12}{5}$

**17.** $\cos D = \frac{3\sqrt{11}}{11}$, $\sec F = \frac{\sqrt{22}}{2}$, $\cot D = \frac{3\sqrt{2}}{2}$ **19.** $g = \sqrt{34}$, $\cos F = \frac{3\sqrt{34}}{34}$, $\sec F = \frac{\sqrt{34}}{3}$, $\cot F = \frac{3}{5}$

**21.**

| *Function* | 30° | 45° | 60° |
|---|---|---|---|
| $\sin\theta$ | $\frac{1}{2}$ | $\frac{\sqrt{2}}{2}$ | $\frac{\sqrt{3}}{2}$ |
| $\cos\theta$ | $\frac{\sqrt{3}}{2}$ | $\frac{\sqrt{2}}{2}$ | $\frac{1}{2}$ |
| $\tan\theta$ | $\frac{\sqrt{3}}{3}$ | 1 | $\sqrt{3}$ |
| $\cot\theta$ | $\sqrt{3}$ | 1 | $\frac{\sqrt{3}}{3}$ |
| $\sec\theta$ | $\frac{2\sqrt{3}}{3}$ | $\sqrt{2}$ | 2 |
| $\csc\theta$ | 2 | $\sqrt{2}$ | $\frac{2\sqrt{3}}{3}$ |

**23.** $\sin A = \frac{2\sqrt{13}}{13}$, $\cos A = \frac{3\sqrt{13}}{13}$, $\cot A = \frac{3}{2}$, $\csc A = \frac{\sqrt{13}}{2}$, $\sec A = \frac{\sqrt{13}}{3}$

**25.** $\frac{1}{2}$ **27.** $\frac{\sqrt{3}}{3}$ **29.** $\frac{\sqrt{61}}{6}$ **31.** $\frac{7\sqrt{33}}{33}$ **33.** sin 45°

**35.** cos 72° **37.** sec 46° **39.** cos 88.5°

## Exercise 6-3 (pp. 285–286)

**1.** $a, b$ **3.** IV **5.** II **7.** I **9.** IV **11.** III **13.** III

**15.** $540^\circ, -180^\circ$

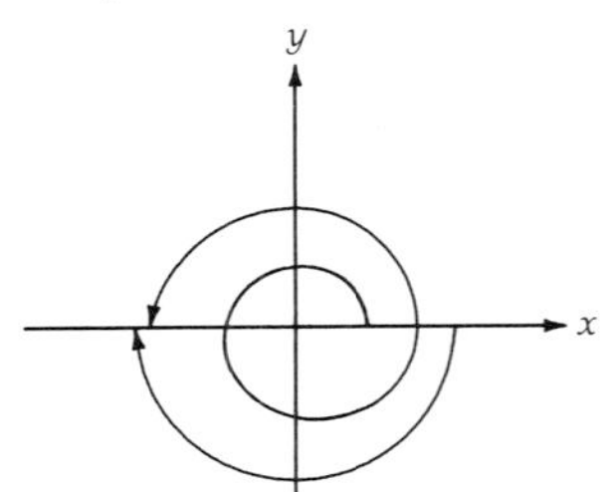


**17.** $517^\circ, -203^\circ$

**19.** 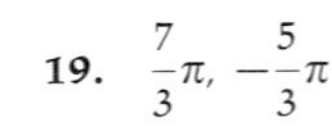 $\frac{7}{3}\pi, -\frac{5}{3}\pi$

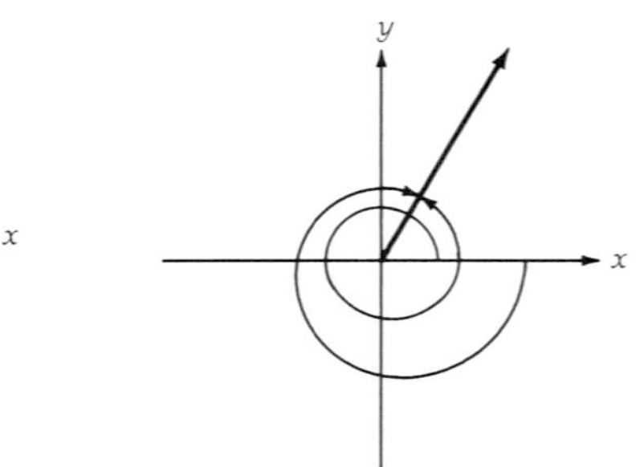


**21.** $\frac{11}{3}\pi, -\frac{\pi}{3}$

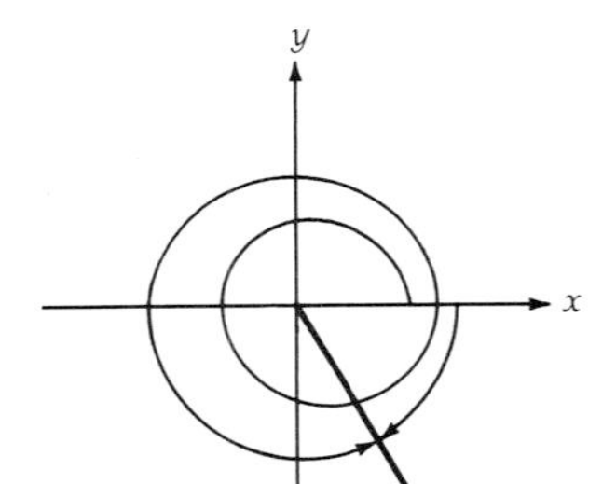


**23.** $\frac{4}{3}\pi, -\frac{8}{3}\pi$

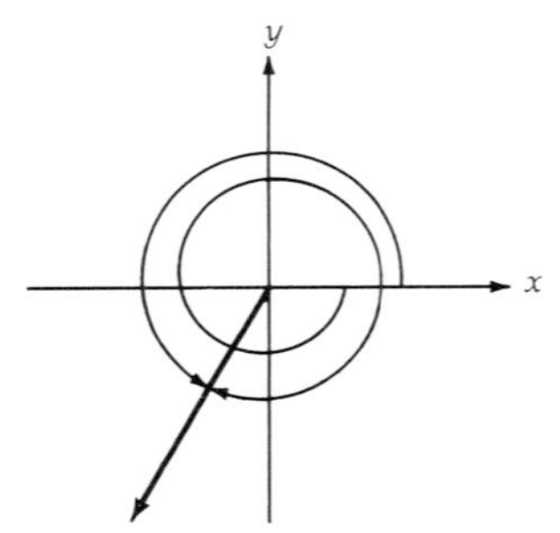


**25.** $\sin\theta = \frac{2\sqrt{13}}{13}$, $\cos\theta = \frac{3\sqrt{13}}{13}$, $\tan\theta = \frac{2}{3}$, $\cot\theta = \frac{3}{2}$, $\sec\theta = \frac{\sqrt{13}}{3}$, $\csc\theta = \frac{\sqrt{13}}{2}$

**27.** $\sin\theta = -\frac{5\sqrt{89}}{89}$, $\cos\theta = \frac{8\sqrt{89}}{89}$, $\tan\theta = -\frac{5}{8}$, $\cot\theta = -\frac{8}{5}$, $\sec\theta = \frac{\sqrt{89}}{8}$, $\csc\theta = -\frac{\sqrt{89}}{5}$

**29.** $\sin\theta = -\frac{2\sqrt{5}}{5}$, $\cos\theta = -\frac{\sqrt{5}}{5}$, $\tan\theta = 2$, $\cot\theta = \frac{1}{2}$, $\sec\theta = -\sqrt{5}$, $\csc\theta = -\frac{\sqrt{5}}{2}$

**31.** $\sin\theta = -\frac{1}{2}$, $\cos\theta = \frac{\sqrt{3}}{2}$, $\tan\theta = -\frac{\sqrt{3}}{3}$, $\cot\theta = -\sqrt{3}$, $\sec\theta = \frac{2\sqrt{3}}{3}$, $\csc\theta = -2$

**33.** $\cos\theta = -\frac{4}{5}$, $\tan\theta = -\frac{3}{4}$, $\cot\theta = -\frac{4}{3}$, $\sec\theta = -\frac{5}{4}$, $\csc\theta = \frac{5}{3}$

**35.** $\sin\theta = -\frac{12}{13}$, $\cos\theta = -\frac{5}{13}$, $\tan\theta = \frac{12}{5}$, $\cot\theta = \frac{5}{12}$, $\csc\theta = -\frac{13}{12}$

**37.** $\sin\theta = \frac{4}{5}$, $\cos\theta = -\frac{3}{5}$, $\tan\theta = -\frac{4}{3}$, $\cot\theta = -\frac{3}{4}$, $\sec\theta = -\frac{5}{3}$

**39.** $\sin\theta = -\frac{5}{13}$, $\tan\theta = -\frac{5}{12}$, $\cot\theta = -\frac{12}{5}$, $\sec\theta = \frac{13}{12}$, $\csc\theta = -\frac{13}{5}$

**41.** $\sin\theta = -\frac{7}{25}$, $\cos\theta = -\frac{24}{25}$, $\tan\theta = \frac{7}{24}$, $\cot\theta = \frac{24}{7}$, $\csc\theta = -\frac{25}{7}$

**43.** $\sin\theta = \frac{8\sqrt{89}}{89}$, $\cos\theta = \frac{5\sqrt{89}}{89}$, $\tan\theta = \frac{8}{5}$, $\sec\theta = \frac{\sqrt{89}}{5}$, $\csc\theta = \frac{\sqrt{89}}{8}$

**45.** $\sin\theta = -\frac{1}{2}$, $\cos\theta = -\frac{\sqrt{3}}{2}$, $\tan\theta = \frac{\sqrt{3}}{3}$, $\cot\theta = \sqrt{3}$, $\sec\theta = \frac{-2\sqrt{3}}{3}$

**47.** There are four possibilities. One of the following is true:
**a.** Angle $A$ = Angle $B$ **b.** Angle $A = 180° +$ Angle $B$ **c.** Angle $A = 180° -$ Angle $B$
**d.** Angle $A = 360° -$ Angle $B$

## Exercise 6-4 (pp. 291–293)

**1.**

| $\theta$ | $\sin\theta$ | $\cos\theta$ | $\tan\theta$ | $\cot\theta$ | $\sec\theta$ | $\csc\theta$ |
|---|---|---|---|---|---|---|
| 0° | 0 | 1 | 0 | No value | 1 | No value |
| 30° | $\frac{1}{2}$ | $\frac{\sqrt{3}}{2}$ | $\frac{\sqrt{3}}{3}$ | $\sqrt{3}$ | $\frac{2\sqrt{3}}{3}$ | 2 |
| 45° | $\frac{\sqrt{2}}{2}$ | $\frac{\sqrt{2}}{2}$ | 1 | 1 | $\sqrt{2}$ | $\sqrt{2}$ |
| 60° | $\frac{\sqrt{3}}{2}$ | $\frac{1}{2}$ | $\sqrt{3}$ | $\frac{\sqrt{3}}{3}$ | 2 | $\frac{2\sqrt{3}}{3}$ |
| 90° | 1 | 0 | No value | 0 | No value | 1 |
| 180° | 0 | $-1$ | 0 | No value | $-1$ | No value |
| 270° | $-1$ | 0 | No value | 0 | No value | $-1$ |
| 360° | 0 | 1 | 0 | No value | 1 | No value |
| 450° | 1 | 0 | No value | 0 | No value | 1 |
| 540° | 0 | $-1$ | 0 | No value | $-1$ | No value |
| 630° | $-1$ | 0 | No value | 0 | No value | $-1$ |
| 720° | 0 | 1 | 0 | No value | 1 | No value |

**3.**

| $\theta$ | $\sin\theta$ | $\cos\theta$ | $\tan\theta$ | $\cot\theta$ | $\sec\theta$ | $\csc\theta$ |
|---|---|---|---|---|---|---|
| 120° | $\frac{\sqrt{3}}{2}$ | $-\frac{1}{2}$ | $-\sqrt{3}$ | $-\frac{\sqrt{3}}{3}$ | $-2$ | $\frac{2\sqrt{3}}{3}$ |
| 135° | $\frac{\sqrt{2}}{2}$ | $-\frac{\sqrt{2}}{2}$ | $-1$ | $-1$ | $-\sqrt{2}$ | $\sqrt{2}$ |
| 150° | $\frac{1}{2}$ | $-\frac{\sqrt{3}}{2}$ | $-\frac{\sqrt{3}}{3}$ | $-\sqrt{3}$ | $-\frac{2\sqrt{3}}{3}$ | 2 |
| 210° | $-\frac{1}{2}$ | $-\frac{\sqrt{3}}{2}$ | $\frac{\sqrt{3}}{3}$ | $\sqrt{3}$ | $-\frac{2\sqrt{3}}{3}$ | $-2$ |
| 225° | $-\frac{\sqrt{2}}{2}$ | $-\frac{\sqrt{2}}{2}$ | 1 | 1 | $-\sqrt{2}$ | $-\sqrt{2}$ |
| 240° | $-\frac{\sqrt{3}}{2}$ | $-\frac{1}{2}$ | $\sqrt{3}$ | $\frac{\sqrt{3}}{3}$ | $-2$ | $-\frac{2\sqrt{3}}{3}$ |
| 300° | $-\frac{\sqrt{3}}{2}$ | $\frac{1}{2}$ | $-\sqrt{3}$ | $-\frac{\sqrt{3}}{3}$ | 2 | $-\frac{2\sqrt{3}}{3}$ |
| 315° | $-\frac{\sqrt{2}}{2}$ | $\frac{\sqrt{2}}{2}$ | $-1$ | $-1$ | $\sqrt{2}$ | $-\sqrt{2}$ |
| 330° | $-\frac{1}{2}$ | $\frac{\sqrt{3}}{2}$ | $-\frac{\sqrt{3}}{3}$ | $-\sqrt{3}$ | $\frac{2\sqrt{3}}{3}$ | $-2$ |

**5.** $-\frac{\sqrt{2}}{2}$ **7.** 1 **9.** $-\frac{\sqrt{3}}{3}$ **11.** $\frac{\sqrt{2}}{2}$ **13.** $\frac{2\sqrt{3}}{3}$ **15.** $-\sqrt{3}$

**17.** We want to show that $\cos(-\theta) = \cos\theta$ for all $\theta$. *Case I:* if $x > 0$. If $-\theta$ is in quadrant IV, then $\theta$ is in quadrant I. Both would have the same positive value for $x$ and $r$. Thus $\cos\theta = \cos(-\theta)$. Similarly, if $-\theta$ is in quadrant I, then $\theta$ is in quadrant IV. The same argument holds.

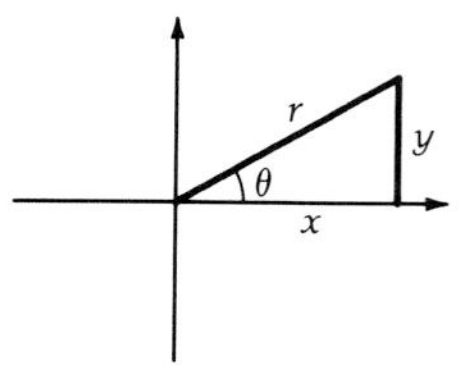

*Case II:* if $x < 0$. If $-\theta$ is in quadrant III, then $\theta$ is in quadrant II. Both would have the same negative value for $x$ and the same positive value for $r$. Thus $\cos\theta = \cos(-\theta)$. Similarly, if $-\theta$ is in quadrant II, then $\theta$ is in quadrant III. The same argument holds. *Case III:* If $x = 0$, then $\theta = 90^\circ$ or $270^\circ$ or is coterminal with those angles. Thus $\cos\theta$ and $\cos(-\theta) = 0$. Thus $\cos\theta = \cos(-\theta)$.

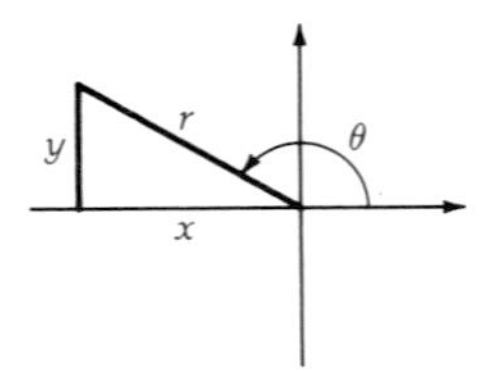


**19.** $\frac{\sqrt{3}}{2}$ **21.** $-1$ **23.** $\frac{1}{2}$ **25.** $-\frac{2\sqrt{3}}{3}$ **27.** $-\frac{\sqrt{3}}{3}$

**29.** $\frac{1}{2} \stackrel{?}{=} 1 - 2\left(\frac{1}{2}\right)^2$
$\frac{1}{2} = \frac{1}{2}$ ✓

**31.** $\left(\frac{\sqrt{2}}{2}\right)\left(\frac{\sqrt{2}}{2}\right) + \left(-\frac{\sqrt{2}}{2}\right)\left(\frac{\sqrt{2}}{2}\right) = 0$ **33.** $1 - 2\left(\frac{\sqrt{2}}{2}\right) = 1 - \sqrt{2}$

## Exercise 6-5 (pp. 297–298)

**1.** 0.2756 **3.** 9.5144 **5.** $-0.9703$ **7.** 1.1223 **9.** 1.2799 **11.** 0.6249 **13.** 0.783477
**15.** 1.101300 **17.** $-0.792077$ **19.** 1.122726 **21.** $-0.021995$ **23.** $-2.391291$

**25.** $\tan 150^\circ = -0.5773503$
$-\frac{\sqrt{3}}{3} = -0.5773503$
(Both answers are rounded to seven decimal places.)

**27.** $(0.5877853)^2 + (0.80901670)^2 \stackrel{?}{=} 1$
$0.345491558 + 0.654508506 \stackrel{?}{=} 1$
Rounded to 7 decimal places $1.0000001 \doteq 1$ ✓

**29.** 4.003102 **31.** 0.694698 **33.** 0.258819 **35.** 1.042217 **37.** $-0.414214$

**39.** $-0.994334$ **41.** $\cot 180^\circ = \frac{-1}{0}$; therefore *no value* **43.** $\tan\frac{3\pi}{2} = \frac{-1}{0}$; therefore *no value*

**45.** $\sec\frac{5\pi}{2} = \sec\frac{\pi}{2} = \frac{1}{0}$; therefore *no value* **47.** $\csc 900^\circ = \csc 180^\circ = \frac{-1}{0}$; therefore *no value*

**49.** $-0.225585$ **51.** 1.178447

## Exercise 6-6 (p. 301)

**1.** 0.2790 **3.** 0.8481 **5.** 0.7869 **7.** 0.4286 **9.** 1.0557 **11.** 1.0251 **13.** 0.3289
**15.** 0.9532 **17.** 1.1918 **19.** $-2.9238$ **21.** $-1.0615$ **23.** 0.4921 **25.** 0.2807
**27.** $-0.9348$ **29.** 0.5542 **31.** 4.6926 **33.** 1.2052 **35.** $-0.8534$ **37.** $-1.0019$
**39.** $-0.7173$

**41.** It should be accurate to only about 3 significant figures. The tangent values are so different that interpolation is not too accurate. The tangent of 88.6° (from Table D) is 40.917; the tangent of 88.7° (from Table D) is 44.066. By interpolation we obtain $\tan 88.66^\circ = 40.917 + 1.889 = 42.806$. From a calculator, a more accurate answer is $\tan 88.66^\circ = 42.75024833$. If we round both answers to the nearest tenth, they agree.

## Exercise 6-7 (pp. 305–306)

**1.** 30° **3.** 135° **5.** 315° **7.** 150° **9.** 270° **11.** 0°, 180° **13.** 120°, 240°
**15.** 35.9°, 215.9° **17.** 34.0°, 326.0° **19.** 248.3°, 291.7° **21.** 138.7°, 318.7°
**23.** 18.4°, 161.6° **25.** 33.2°, 326.8° **27.** 33.8°, 213.8° **29.** 316.82° **31.** 221.57°
**33.** 238.93° **35.** $\tan\theta = 1.459$, $\theta = 55.57°$ **37.** $\sin\theta = 0.9675$, $\theta = 75.35°$ **39.** $\theta = 75.46°$
**41.** $\theta = 44.60°$ **43.** **a.** $x = 3.648$ **b.** $\theta_1 = 26.33°$

## Exercise 6-8 (pp. 313–315)

**1.** $B = 63°$, $b = 6.9$, $c = 7.7$ **3.** $B = 58°$, $a = 11$, $b = 17$ **5.** $B = 42.7°$, $a = 44.8$, $c = 60.9$
**7.** $A = 61.5°$, $b = 10.5$, $c = 22.1$ **9.** $A = 70.5°$, $a = 8.90$, $c = 9.44$ **11.** $B = 57.8°$, $a = 13.7$, $c = 25.7$
**13.** $B = 47.7°$, $a = 16.8$, $b = 18.4$ **15.** $A = 48.8°$, $B = 41.2°$, $c = 21.3$
**17.** $A = 38.6°$, $B = 51.4°$, $a = 22.6$ **19.** $A = 54.44°$, $b = 11.90$, $c = 20.47$ **21.** 12.6 meters
**23.** 21.7 meters **25.** 33.2 meters **27.** 54.2° **29.** S34.6°W **31.** 8.82 meters
**33.** 1740 meters **35.** 44.5 meters **37.** 62.2°, 33.7 kilometers **39.** 13.2 kg
**41.** The angle decreases by 12.2° **43.** 45.1°

## Chapter 6 Review Exercise (pp. 317–319)

**1.** $\frac{1}{18}\pi$ radians **3.** $\frac{5}{3}\pi$ radians **5.** $\frac{5}{4}\pi$ radians **7.** 0.7175 radians **9.** 3.50 radians
**11.** 90° **13.** 114.6° **15.** 540° **17.** 480° **19.** 315° **21.** $\sqrt{106}$ **23.** $\sqrt{51}$ **25.** 6
**27.** $\sin A = \frac{4\sqrt{41}}{41}$, $\cos A = \frac{5\sqrt{41}}{41}$, $\cot A = \frac{5}{4}$, $\sec A = \frac{\sqrt{41}}{5}$, $\csc A = \frac{\sqrt{41}}{4}$ **29.** $\frac{2}{3}$ **31.** 0.3971
**33.** 0.3057 **35.** 18.91° **37.** 60° **39.** quadrant III **41.** quadrant II **43.** quadrant II
**45.** $b, d, f$ **47.** $\sin\theta = \frac{5\sqrt{34}}{34}$, $\cos\theta = \frac{3\sqrt{34}}{34}$, $\tan\theta = \frac{5}{3}$, $\cot\theta = \frac{3}{5}$, $\sec\theta = \frac{\sqrt{34}}{3}$, $\csc\theta = \frac{\sqrt{34}}{5}$
**49.** $\sin\theta = -\frac{7}{25}$, $\cos\theta = \frac{24}{25}$, $\tan\theta = -\frac{7}{24}$, $\cot\theta = -\frac{24}{7}$, $\sec\theta = \frac{25}{24}$, $\csc\theta = -\frac{25}{7}$ **51.** 315°
**53.** 210° **55.** $-0.9004$ **57.** $+1$ **59.** $-\sqrt{3}$ **61.** 0.7859 **63.** 231.17° **65.** 294.49°
**67.** does not exist **69.** $B = 56°50'$, $a = 9.28$, $c = 17.0$ **71.** 10.5 meters **73.** 245 meters
**75.** 48.7 meters

## Practice Test for Chapter 6 (pp. 319–320)

**1.** $\frac{4}{9}\pi$ radians **3.** 40° **5.** $4\sqrt{6}$ **7.** $\frac{5}{6}$ **9.** 0.6191 **11.** quadrant III **13.** $-\frac{5\sqrt{89}}{89}$
**15.** 135° **17.** $-0.5736$ **19.** $-1.4746$ **21.** $A = 55.7°$, $b = 13.4$, $c = 23.7$ **23.** 33.9 meters

## CHAPTER 7

### Exercise 7-1 (pp. 326–328)

**1.** $1 \pm 2\pi$ **3.** $-2 \pm 2\pi$ **5.** $\pm 2\pi$ **7.** $\frac{5}{2}\pi, -\frac{3}{2}\pi$ **9.** $4\pi, -2\pi$ **11.** $\frac{11}{4}\pi, -\frac{5}{4}\pi$

**13.** $2 + \frac{3\pi}{2}, 2 - \frac{5\pi}{2}$ **15.** $10.6974, -1.8690$ **17.** $7.5922, -4.9742$ **19.** $4.3982, -8.1681$

**21.** $(0, 1)$ **23.** $(1, 0)$ **25.** $(0, -1)$

**27.** $\cos u = -\frac{4}{5}, \tan u = -\frac{3}{4}, \cot u = -\frac{4}{3}, \sec u = -\frac{5}{4}, \csc u = \frac{5}{3}$

**29.** $\sin u = \frac{4}{5}, \cos u = -\frac{3}{5}, \tan u = -\frac{4}{3}, \cot u = -\frac{3}{4}, \sec u = -\frac{5}{3}$

**31.** $\sin u = -\frac{7}{25}, \cos u = -\frac{24}{25}, \tan u = \frac{7}{24}, \cot u = \frac{24}{7}, \csc u = -\frac{25}{7}$

**33.** $\sin u = \frac{\sqrt{65}}{9}, \tan u = -\frac{\sqrt{65}}{4}, \cot u = -\frac{4\sqrt{65}}{65}, \sec u = -\frac{9}{4}, \csc u = \frac{9\sqrt{65}}{65}$ **35.** $\left(-\frac{5}{13}, \frac{12}{13}\right)$

**37.** $\sin u = \frac{7\sqrt{65}}{65}, \cos u = -\frac{4\sqrt{65}}{65}, \tan u = -\frac{7}{4}, \sec u = -\frac{\sqrt{65}}{4}, \csc u = \frac{\sqrt{65}}{7}$

**39.** $\sin u = \frac{12}{13}, \cos u = -\frac{5}{13}, \tan u = -\frac{12}{5}, \sec u = -\frac{13}{5}, \csc u = \frac{13}{12}$

**41.** $\sin u = \frac{5\sqrt{34}}{34}, \cos u = \frac{-3\sqrt{34}}{34}, \cot u = -\frac{3}{5}, \sec u = \frac{-\sqrt{34}}{3}, \csc u = \frac{\sqrt{34}}{5}$

**43.** $\sin u = -\frac{1}{5}, \cos u = \frac{-2\sqrt{6}}{5}, \tan u = \frac{\sqrt{6}}{12}, \cot u = 2\sqrt{6}, \sec u = \frac{-5\sqrt{6}}{12}$

**45.** True, since $(\sin u)^2 + (\cos u)^2 = 1$ becomes $y^2 + x^2 = 1$ on the unit circle, which we know to be true.

**47.** On the unit circle, $1 + (\cot u)^2 = 1 + \left(\frac{x}{y}\right)^2$. We may write this as $\frac{y^2 + x^2}{y^2} = \frac{1}{y^2} = \frac{1}{(\sin u)^2} = (\csc u)^2$.

### Exercise 7-2 (pp. 334–335)

**1.**

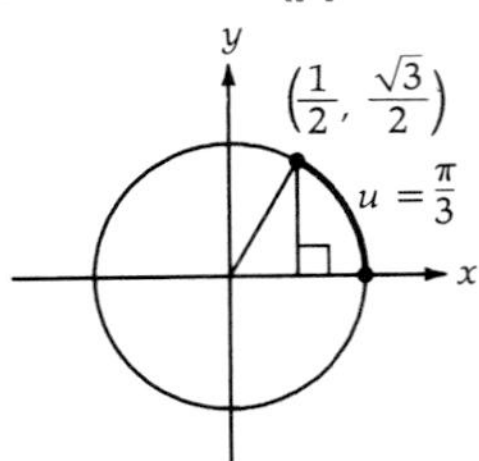


$\sin\frac{\pi}{3} = \frac{\sqrt{3}}{2}, \cos\frac{\pi}{3} = \frac{1}{2}, \tan\frac{\pi}{3} = \sqrt{3}, \cot\frac{\pi}{3} = \frac{\sqrt{3}}{3},$

$\sec\frac{\pi}{3} = 2, \csc\frac{\pi}{3} = \frac{2\sqrt{3}}{3}$

**3.**

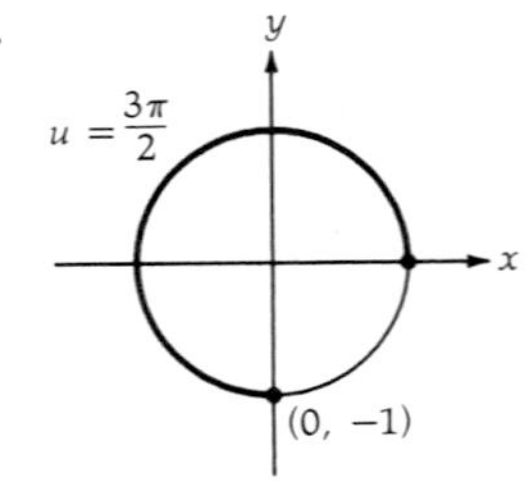


$\sin\frac{3}{2}\pi = -1, \cos\frac{3}{2}\pi = 0, \tan\frac{3}{2}\pi$ does not exist,

$\cot\frac{3}{2}\pi = 0, \sec\frac{3}{2}\pi$ does not exist, $\csc\frac{3}{2}\pi = -1$

## Exercise 7-2 (continued)

**5.**

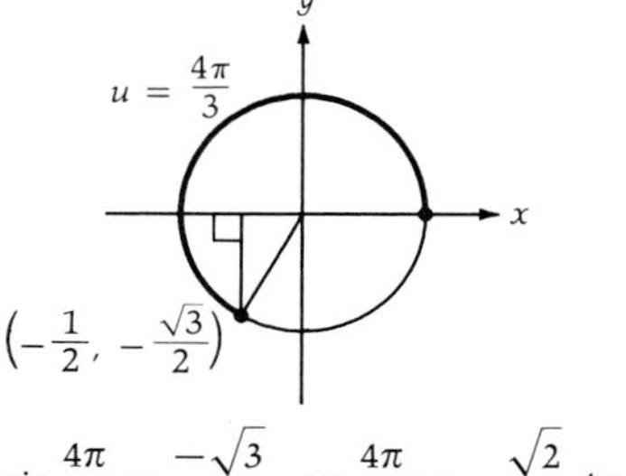


$\sin\frac{4\pi}{3} = \frac{-\sqrt{3}}{2}$, $\cos\frac{4\pi}{3} = -\frac{\sqrt{2}}{2}$, $\tan\frac{4\pi}{3} = \sqrt{3}$,

$\cot\frac{4\pi}{3} = \frac{\sqrt{3}}{3}$, $\sec\frac{4\pi}{3} = -2$, $\csc\frac{4\pi}{3} = -\frac{2\sqrt{3}}{3}$

**7.**

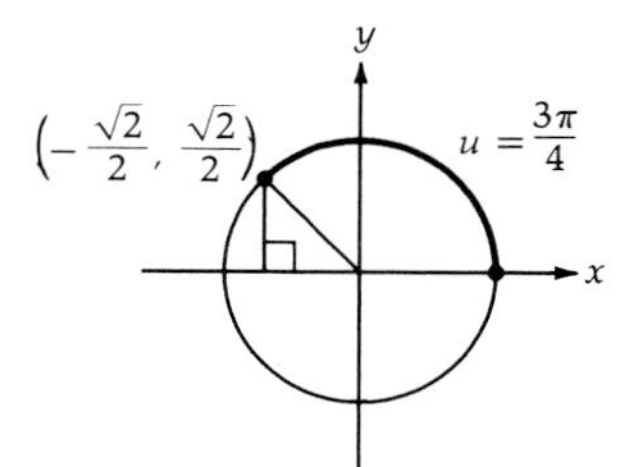


$\sin\frac{3}{4}\pi = \frac{\sqrt{2}}{2}$, $\cos\frac{3}{4}\pi = -\frac{\sqrt{2}}{2}$, $\tan\frac{3}{4}\pi = -1$,

$\cot\frac{3}{4}\pi = -1$, $\sec\frac{3}{4}\pi = -\sqrt{2}$, $\csc\frac{3}{4}\pi = \sqrt{2}$

**9.**

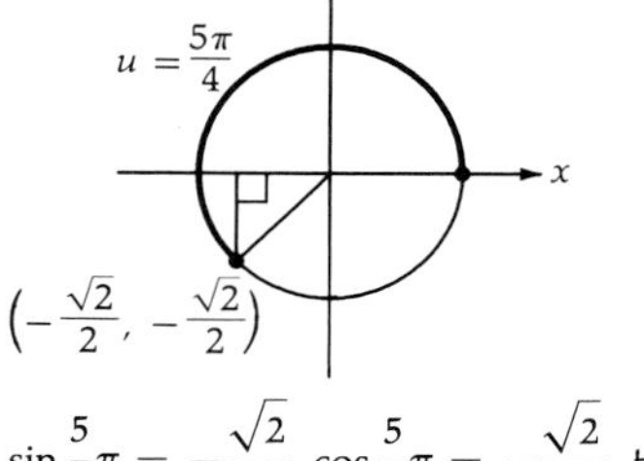


$\sin\frac{5}{4}\pi = -\frac{\sqrt{2}}{2}$, $\cos\frac{5}{4}\pi = -\frac{\sqrt{2}}{2}$, $\tan\frac{5}{4}\pi = 1$,

$\cot\frac{5}{4}\pi = 1$, $\sec\frac{5}{4}\pi = -\sqrt{2}$, $\csc\frac{5}{4}\pi = -\sqrt{2}$

**11.**

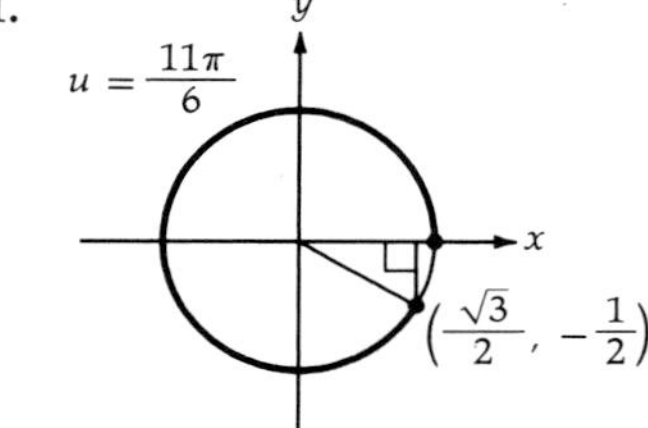


$\sin\frac{11\pi}{6} = -\frac{1}{2}$, $\cos\frac{11\pi}{6} = \frac{\sqrt{3}}{2}$, $\tan\frac{11\pi}{6} = -\frac{\sqrt{3}}{3}$,

$\cot\frac{11\pi}{6} = -\sqrt{3}$, $\sec\frac{11\pi}{6} = \frac{2\sqrt{3}}{3}$, $\csc\frac{11\pi}{6} = -2$

**13.**

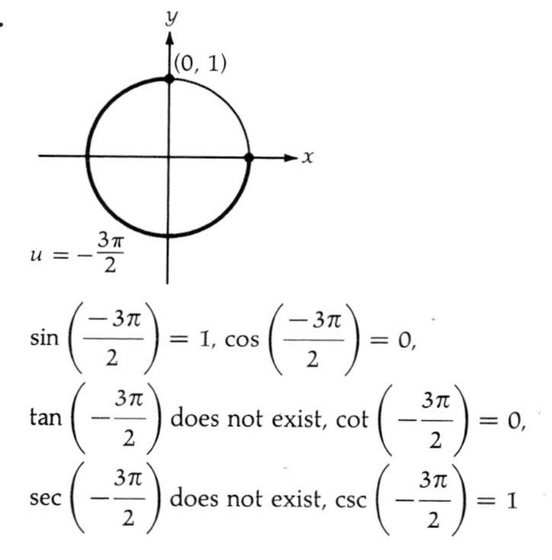


$\sin\left(\frac{-3\pi}{2}\right) = 1$, $\cos\left(\frac{-3\pi}{2}\right) = 0$,

$\tan\left(-\frac{3\pi}{2}\right)$ does not exist, $\cot\left(-\frac{3\pi}{2}\right) = 0$,

$\sec\left(-\frac{3\pi}{2}\right)$ does not exist, $\csc\left(-\frac{3\pi}{2}\right) = 1$

**15.**

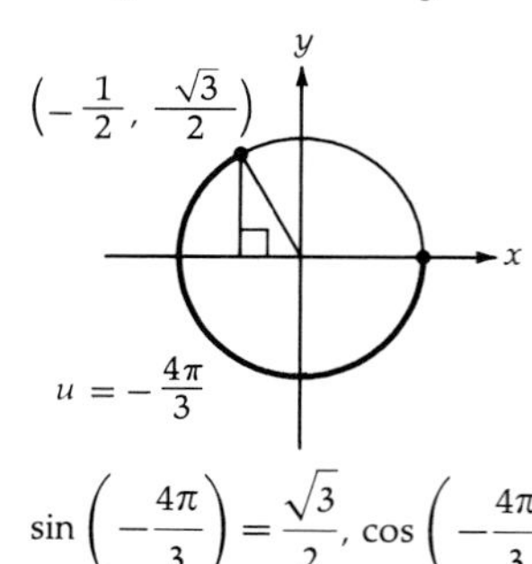


$\sin\left(-\frac{4\pi}{3}\right) = \frac{\sqrt{3}}{2}$, $\cos\left(-\frac{4\pi}{3}\right) = -\frac{1}{2}$,

$\tan\left(-\frac{4\pi}{3}\right) = -\sqrt{3}$, $\cot\left(-\frac{4\pi}{3}\right) = -\frac{\sqrt{3}}{3}$,

$\sec\left(-\frac{4\pi}{3}\right) = -2$, $\csc\left(-\frac{4\pi}{3}\right) = \frac{2\sqrt{3}}{3}$

**17.**

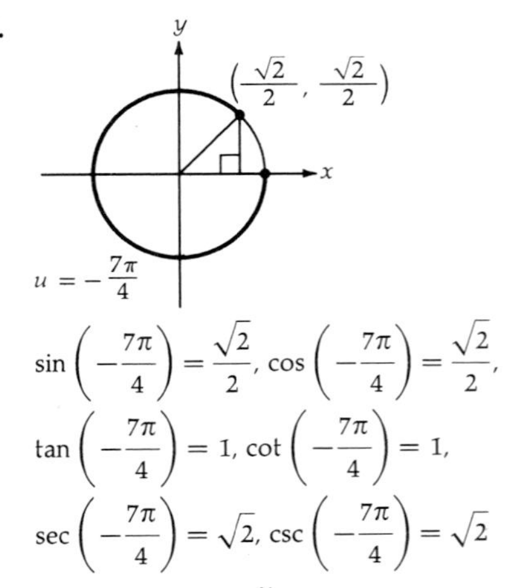


$\sin\left(-\frac{7\pi}{4}\right) = \frac{\sqrt{2}}{2}$, $\cos\left(-\frac{7\pi}{4}\right) = \frac{\sqrt{2}}{2}$,
$\tan\left(-\frac{7\pi}{4}\right) = 1$, $\cot\left(-\frac{7\pi}{4}\right) = 1$,
$\sec\left(-\frac{7\pi}{4}\right) = \sqrt{2}$, $\csc\left(-\frac{7\pi}{4}\right) = \sqrt{2}$

**19.**

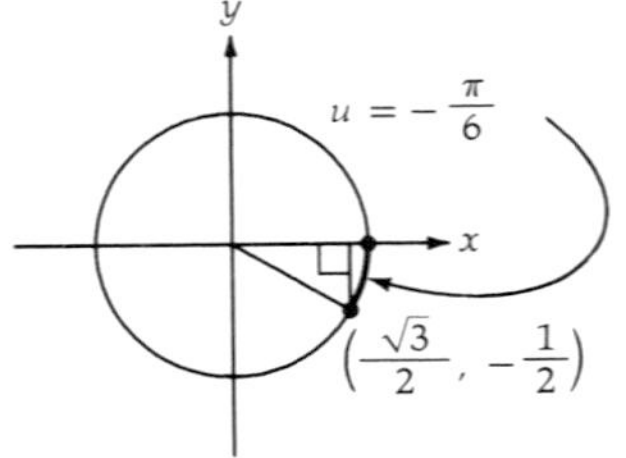


$\sin\left(-\frac{\pi}{6}\right) = -\frac{1}{2}$, $\cos\left(-\frac{\pi}{6}\right) = \frac{\sqrt{3}}{2}$,
$\tan\left(-\frac{\pi}{6}\right) = -\frac{\sqrt{3}}{3}$, $\cot\left(-\frac{\pi}{6}\right) = -\sqrt{3}$,
$\sec\left(-\frac{\pi}{6}\right) = \frac{2\sqrt{3}}{3}$, $\csc\left(-\frac{\pi}{6}\right) = -2$

**21.**

y
x
$\left(-\frac{1}{2}, -\frac{\sqrt{3}}{2}\right)$
$u = -\frac{2\pi}{3}$

$\sin\left(-\frac{2\pi}{3}\right) = -\frac{\sqrt{3}}{2}$, $\cos\left(-\frac{2\pi}{3}\right) = -\frac{1}{2}$,
$\tan\left(-\frac{2\pi}{3}\right) = \sqrt{3}$, $\cot\left(-\frac{2\pi}{3}\right) = \frac{\sqrt{3}}{3}$,
$\sec\left(-\frac{2\pi}{3}\right) = -2$, $\csc\left(-\frac{2\pi}{3}\right) = -\frac{2\sqrt{3}}{3}$

**23.** An arc of $\frac{17\pi}{6}$ is coterminal with an arc of $\frac{5\pi}{6}$;
$\sin\frac{17\pi}{6} = \frac{1}{2}$, $\cos\frac{17\pi}{6} = -\frac{\sqrt{3}}{2}$, $\tan\frac{17\pi}{6} = -\frac{\sqrt{3}}{3}$,
$\cot\frac{17\pi}{6} = -\sqrt{3}$, $\sec\frac{17\pi}{6} = -\frac{2\sqrt{3}}{3}$, $\csc\frac{17\pi}{6} = 2$

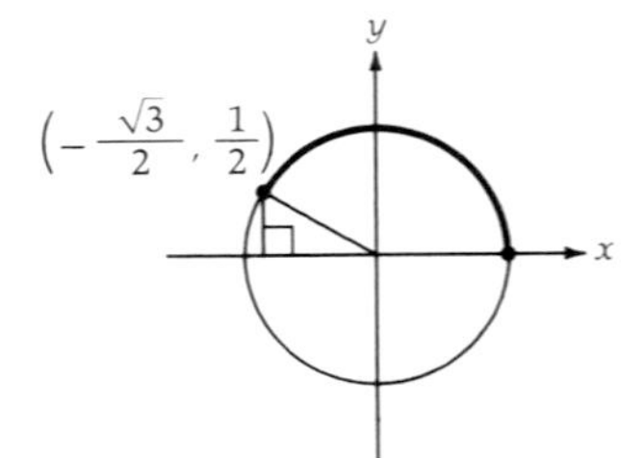


**25.** An arc of $\frac{15\pi}{4}$ is coterminal with an arc of $\frac{7\pi}{4}$;
$\sin\frac{15\pi}{4} = -\frac{\sqrt{2}}{2}$, $\cos\frac{15\pi}{4} = \frac{\sqrt{2}}{2}$, $\tan\frac{15\pi}{4} = -1$,
$\cot\frac{15\pi}{4} = -1$, $\sec\frac{15\pi}{4} = \sqrt{2}$, $\csc\frac{15\pi}{4} = -\sqrt{2}$

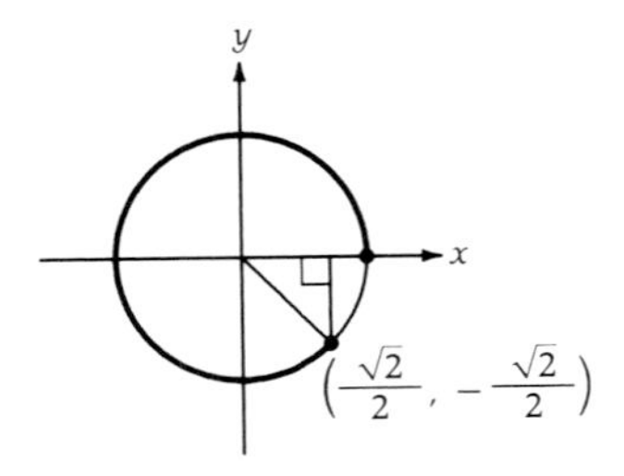


**27.** An arc of $-\frac{11\pi}{3}$ is coterminal with an arc of $-\frac{5\pi}{3}$;
$\sin\left(-\frac{11\pi}{3}\right) = \frac{\sqrt{3}}{2}$, $\cos\left(-\frac{11\pi}{3}\right) = \frac{1}{2}$, $\tan\left(-\frac{11\pi}{3}\right) = \sqrt{3}$.
$\cot\left(-\frac{11\pi}{3}\right) = \frac{\sqrt{3}}{3}$, $\sec\left(-\frac{11\pi}{3}\right) = 2$, $\csc\left(-\frac{11\pi}{3}\right) = \frac{2\sqrt{3}}{3}$

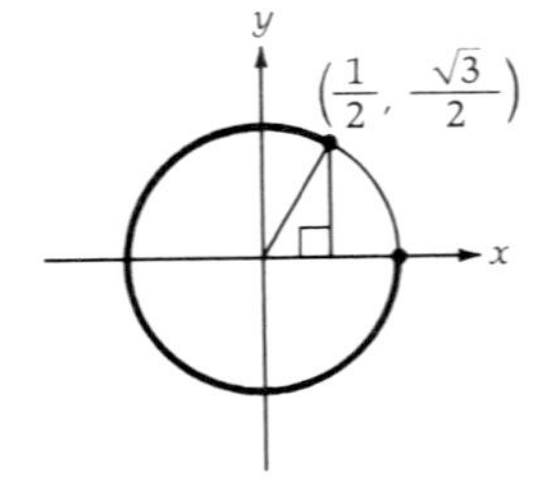

**29.** The details of the proof are to be found in the *Student Tutorial*. The main approach is to locate $u_3 = \pi + u_1$ in quadrant III on a unit circle and to show that the two triangles are congruent. Then $x_1 = |x_3|$ and $y_1 = |y_3|$. Finally, we determine $x_3 = -x_1$ and $y_3 = -y_1$.

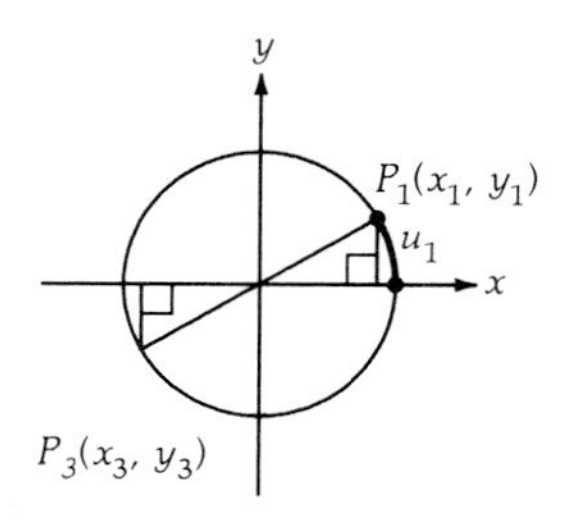


## Exercise 7-3 (pp. 341–342)

**1.** 0.97890 **3.** $-3.20767$ **5.** $-0.99967$ **7.** $-1.00496$ **9.** 2.14409 **11.** $-0.71050$
**13.** 3.73205 **15.** $-0.64279$ **17.** $-34.23253$ for each value **19.** 1.13192 for each value
**21.** $-0.96356$ for each value

In Problems 23 and 25 we use the facts that

If $(-u)$ is in quadrant IV, then $u$ is in quadrant I
If $(-u)$ is in quadrant III, then $u$ is in quadrant II
If $(-u)$ is in quadrant II, then $u$ is in quadrant III
If $(-u)$ is in quadrant I, then $u$ is in quadrant IV

**23.** If $(-u)$ is in quadrant IV, where the sine is negative, then $u$ is in quadrant I, where the sine is positive. It is similar for the other three cases. Thus $\sin(-u) = \sin u$. To illustrate, if $u = -2$, then $\sin(-2) = -\sin 2$ and thus $-0.90930 = -(0.90930)$

**25.** If $(-u)$ is in quadrant IV, where the tangent is negative, then $u$ is in quadrant I, where the tangent is positive. It is similar for the other three cases. Thus $\tan(-u) = -\tan u$. To illustrate, if $u = -3$, then $\tan(-3) = -\tan 3$ and thus $0.14255 = -(-0.14255)$

**27.** 0.7499 **29.** 2.3695 **31.** 2.4846 **33.** 0.6034 **35.** 2.2324

All values listed were obtained by using Table E in Problems 37–55.

**37.** 2.625 **39.** 1.140 **41.** 0.67 **43.** 0.1336 **45.** 0.3844 or 0.3845 **47.** 0.909
**49.** 0.974 **51.** 0.9086 **53.** $-0.1409$ **55.** Using $\pi \doteq 3.14$ we obtain $u = 2.65$

## Exercise 7-4 (pp. 353–354)

**1.** amplitude $= 2$
period $= 2\pi$
phase shift $= 0$

**3.** amplitude $= \frac{1}{2}$
period $= 2\pi$
phase shift $= 0$

**5.** amplitude $= 1$
period $= 2\pi$
phase shift $= 0$

**7.** amplitude $= 4$
period $= \pi$
phase shift $= 0$

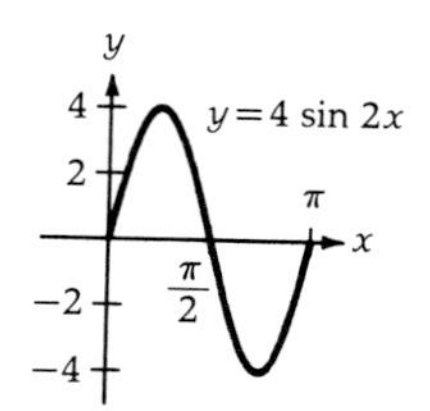

**9.** amplitude $= 0.8$
period $= 6\pi$
phase shift $= 0$

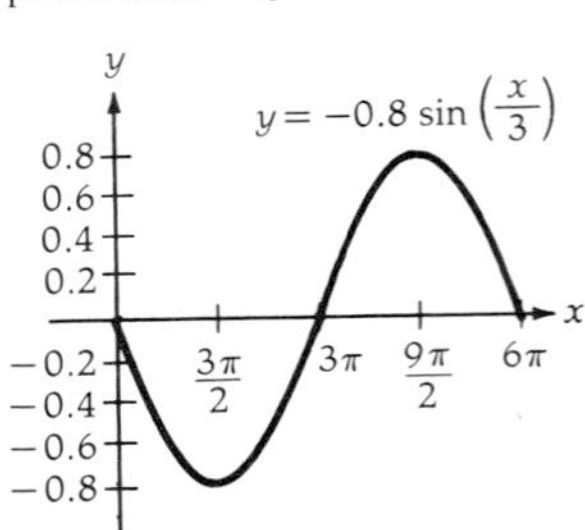


**11.** amplitude $= 1$
period $= 2\pi$
phase shift $= -\frac{\pi}{2}$

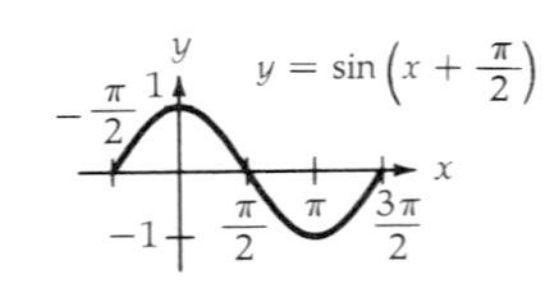


**13.** amplitude $= 1$
period $= \pi$
phase shift $= -\frac{\pi}{2}$

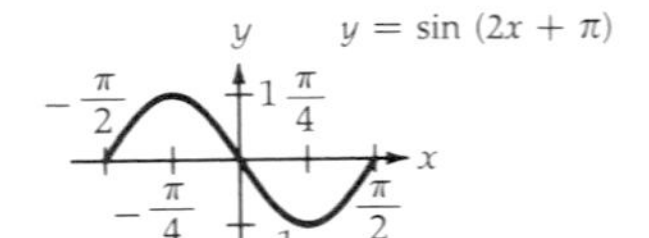


**15.** amplitude $= 2$
period $= 2\pi$
phase shift $= -\frac{3\pi}{2}$

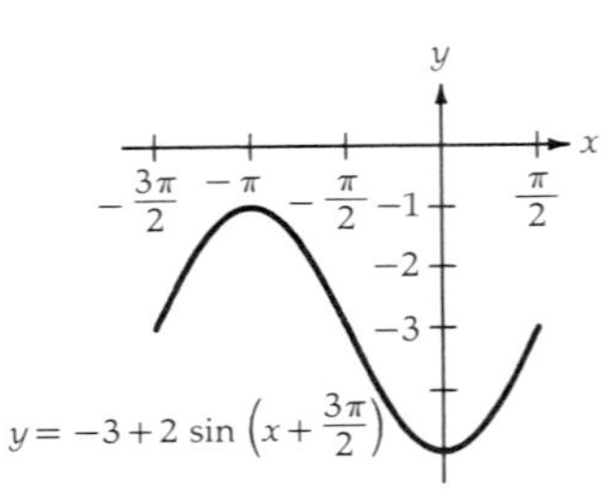


(Note that the curve is shifted downward 3 units lower than the usual sine curve.)

**17.** amplitude $= 2$
period $= 2\pi$
phase shift $= 0$

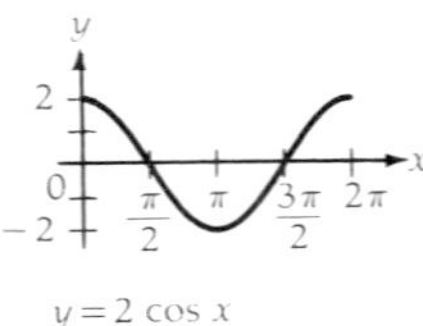


**19.** amplitude $= 1$
period $= 2\pi$
phase shift $= 0$

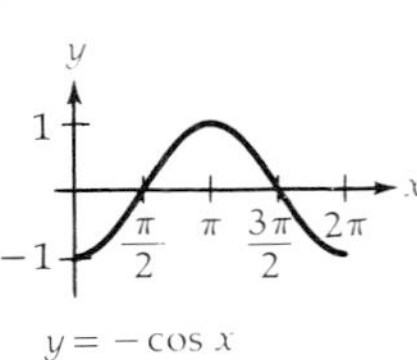


**21.** amplitude $= 1.5$
period $= 3\pi$
phase shift $= 0$

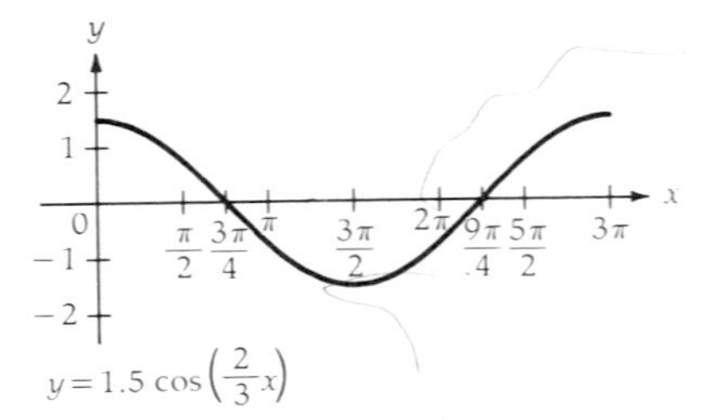


**23.** amplitude $= 2$
period $= 6\pi$
phase shift $= -6\pi$

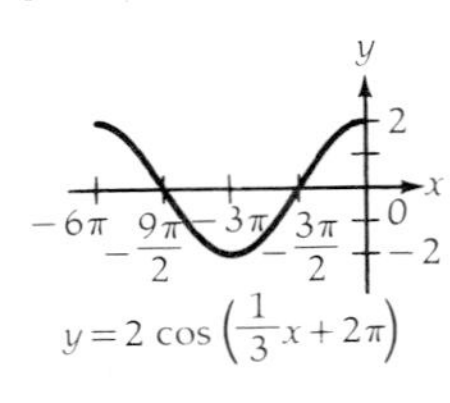


**25.** amplitude $= 3$
period $= \pi$
phase shift $= \frac{\pi}{2}$

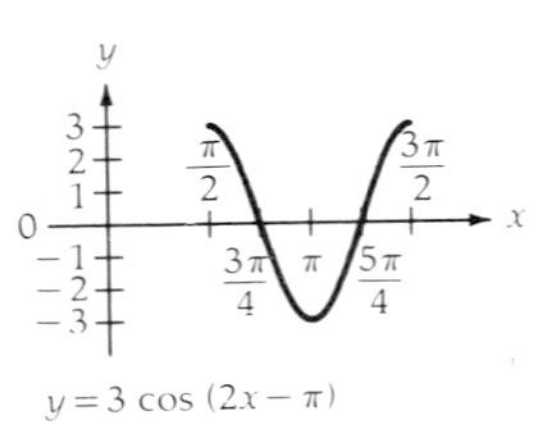


**27.** amplitude $= 0.8$
period $= 6$
phase shift $= -3$

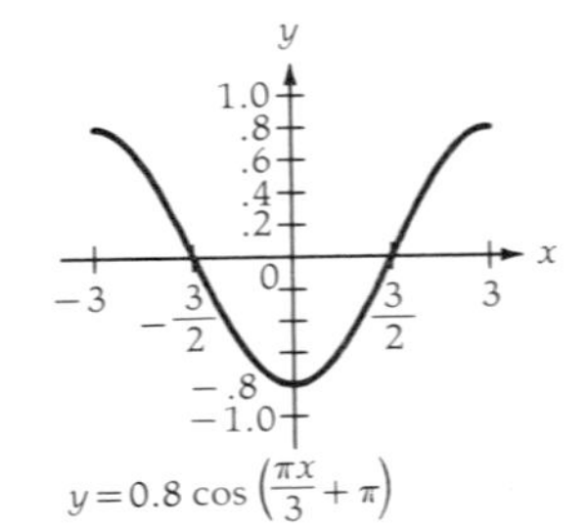

**29.** amplitude $= 100$

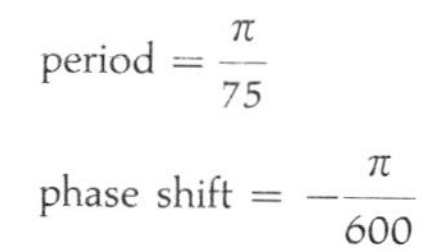

period $= \dfrac{\pi}{75}$

phase shift $= -\dfrac{\pi}{600}$

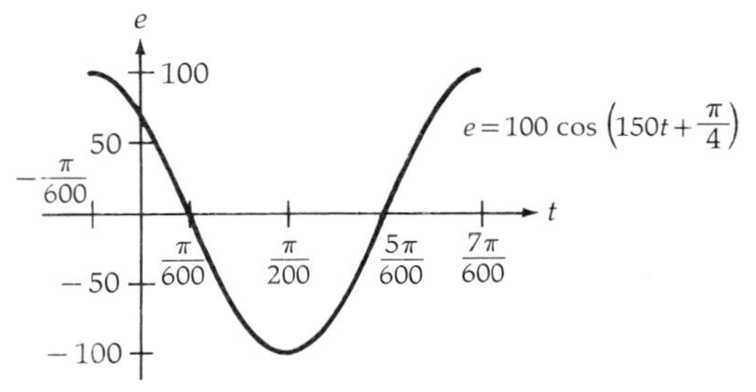


**31.**

| $d$ | $T$ | $d$ | $T$ |
|---|---|---|---|
| 0 | −0.8 | 200 | 64.3 |
| 20 | −4.6 | 220 | 64.6 |
| 40 | −4.3 | 240 | 60.8 |
| 60 | −0.6 | 260 | 53.4 |
| 80 | 7.8 | 280 | 43.2 |
| 100 | 18.2 | 300 | 31.5 |
| 120 | 30.0 | 320 | 19.6 |
| 140 | 41.8 | 340 | 8.9 |
| 160 | 52.2 | 360 | 0.7 |
| 180 | 60.1 | 380 | −4.0 |

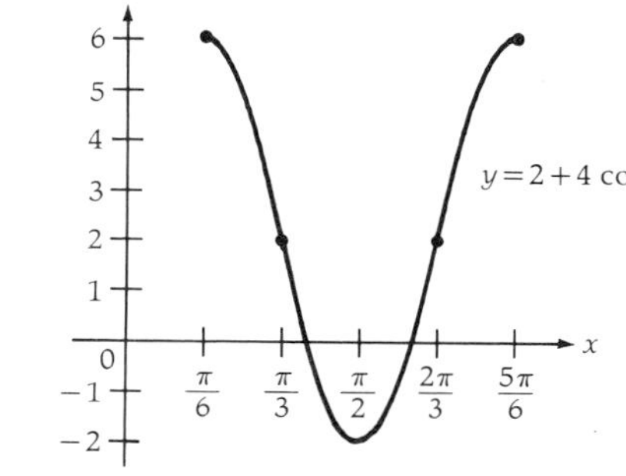


**33.** amplitude $= 4$

period $= \dfrac{2\pi}{3}$

phase shift $= \dfrac{\pi}{6}$



(*Note:* The curve is shifted upward 2 units higher than the usual cosine curve.)

## Exercise 7-5 (pp. 364–365)

**1.**

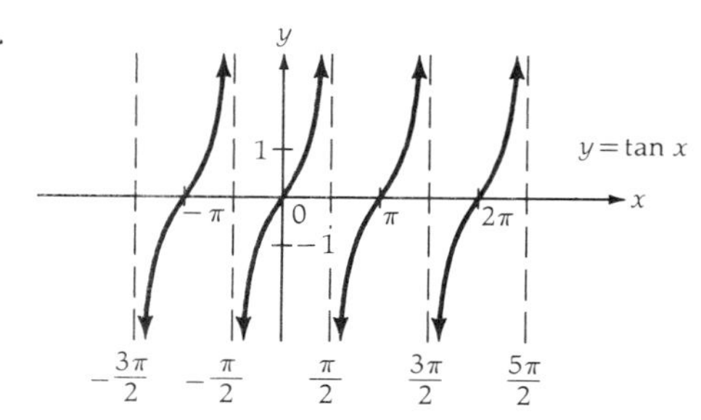


**3.**

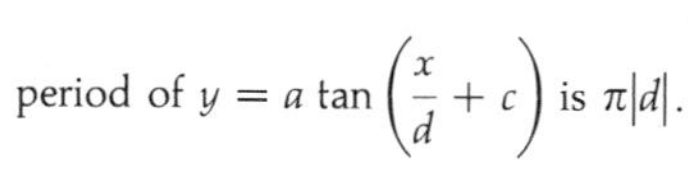

The period of $y = a \tan(bx + c)$ is $\dfrac{\pi}{|b|}$. The period of $y = a \tan\left(\dfrac{x}{d} + c\right)$ is $\pi|d|$.

**5.**

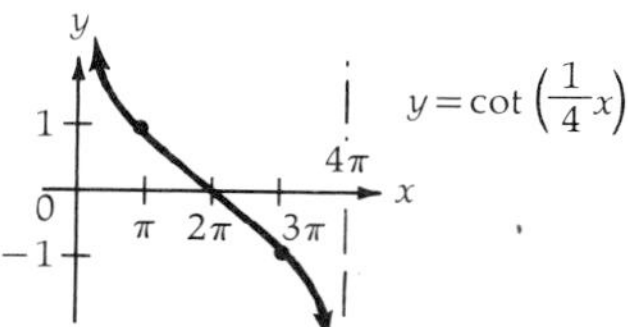

**7.**

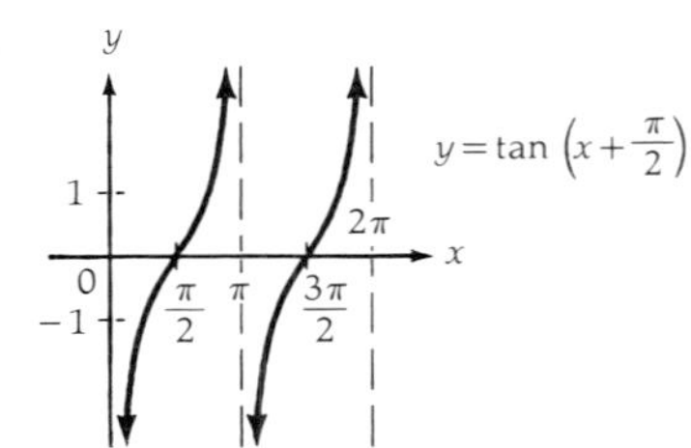


**9.**

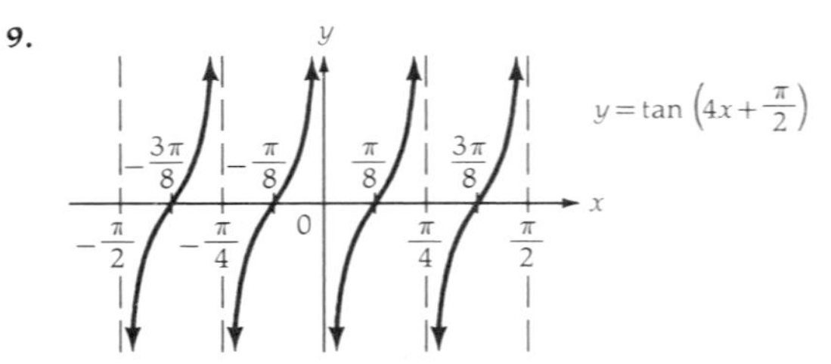


**11.**

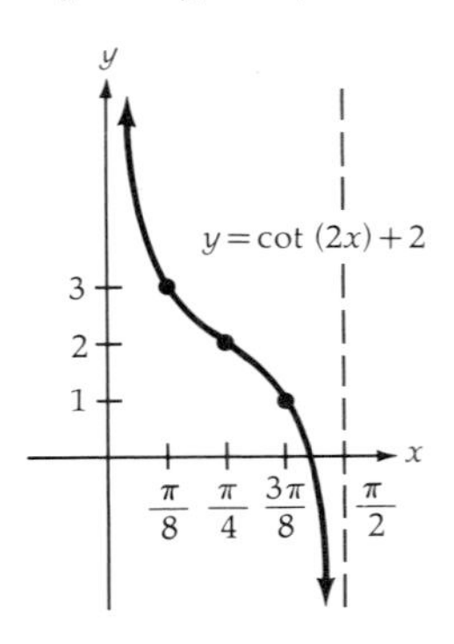


**13.**

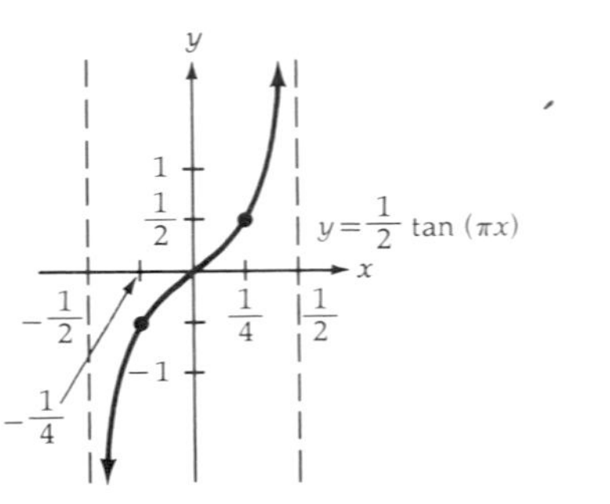


**15.**

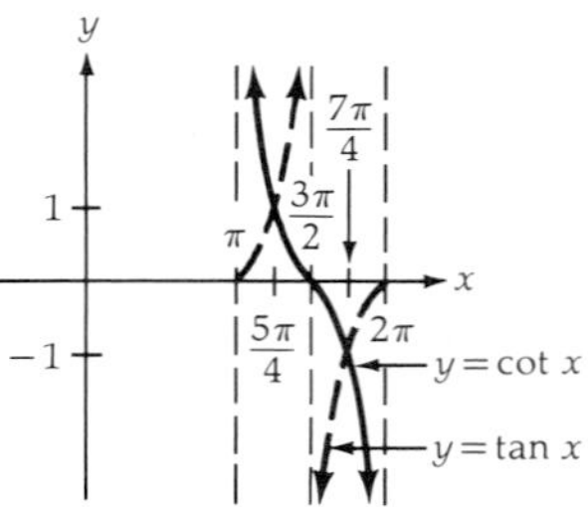


**17.**

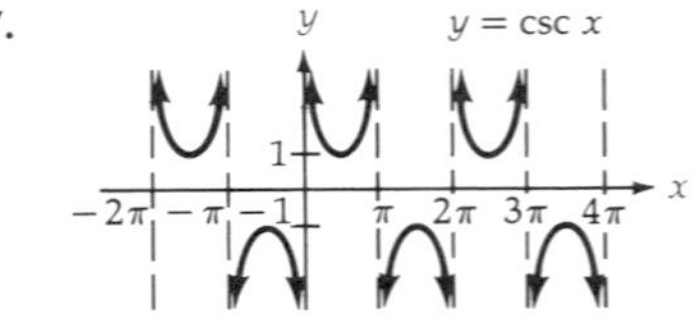


**19.** It does not in the usual sense of the word "amplitude." For functions of the form $y = a \sec(bx + c)$ and $y = a \csc(bx + c)$ the value of $a$ does *not* give a maximum or minimum value in the range of the function. However, the value of $a$ can be used to determine a minimum positive value and a maximum negative value of the function.

**21.**

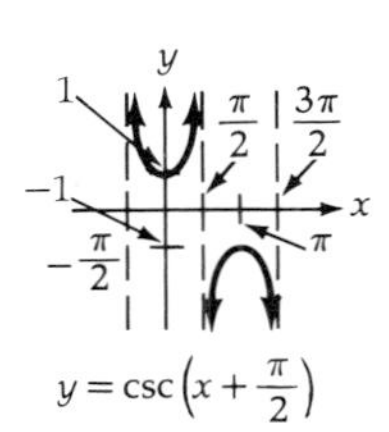


**23.** $\frac{2\pi}{|b|}$

**25.**

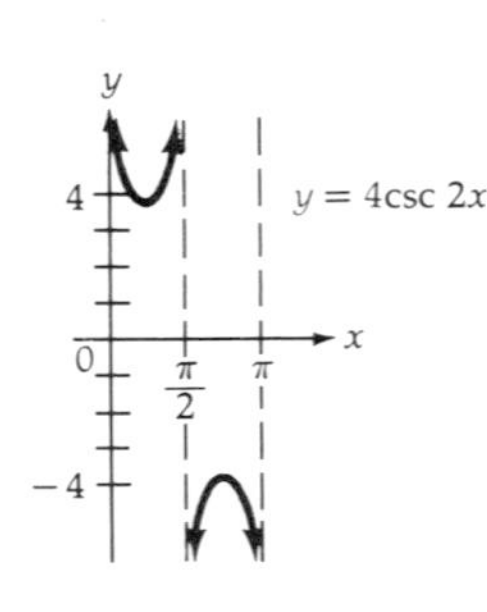


**27.**

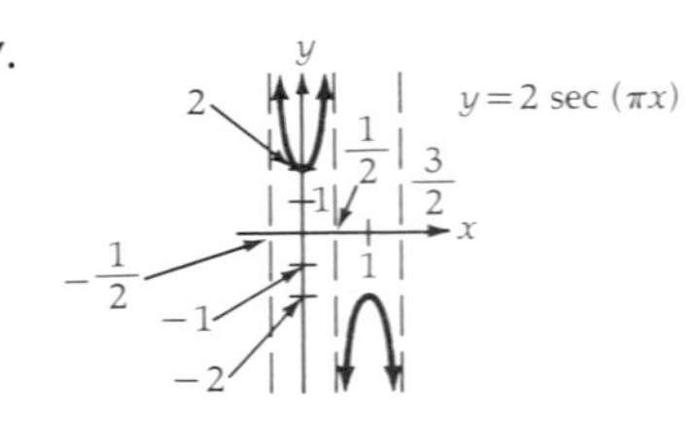

29.

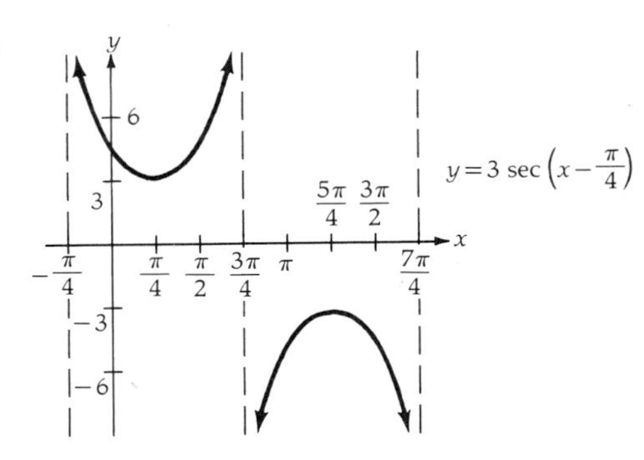


31. $c = -\dfrac{\pi}{2}$

## Exercise 7-6 (p. 368)

1.

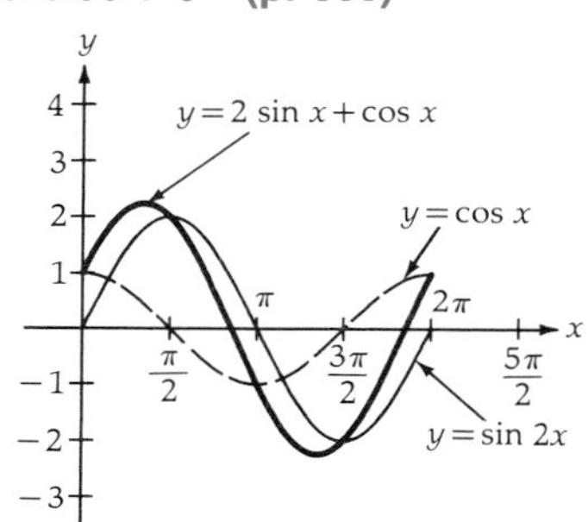


3.

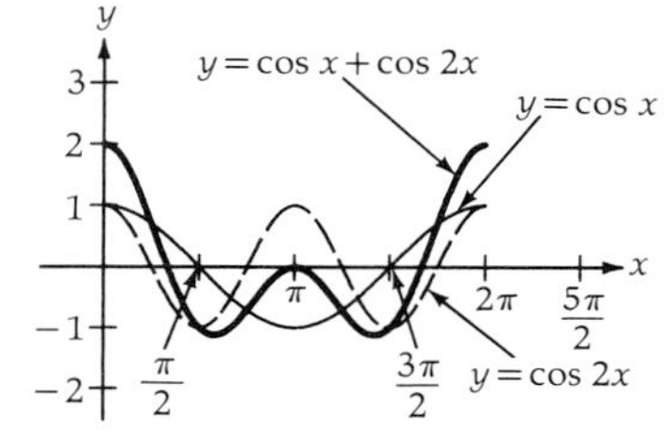


5.

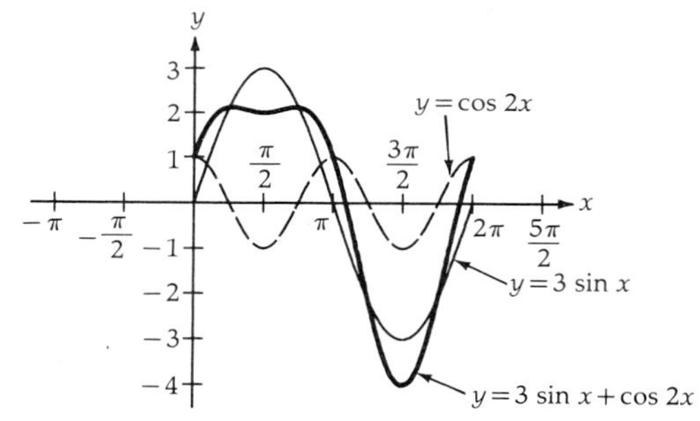


7.

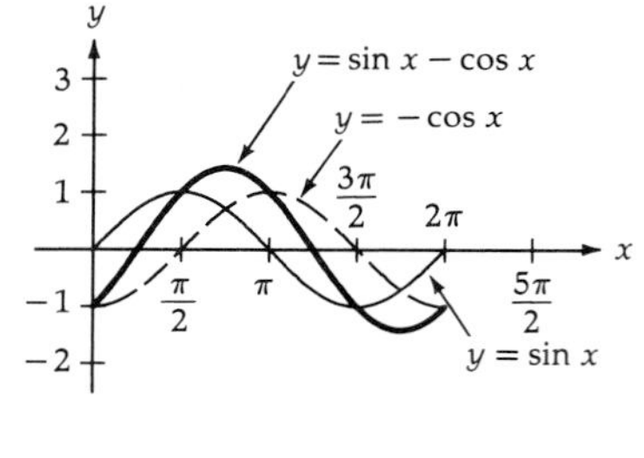


9.

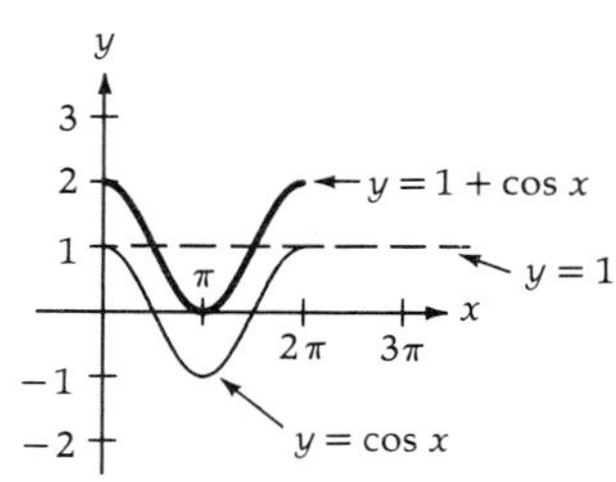


11.

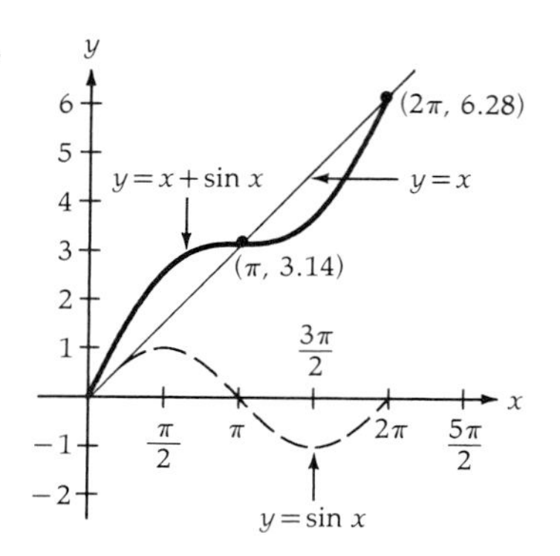

**13.**

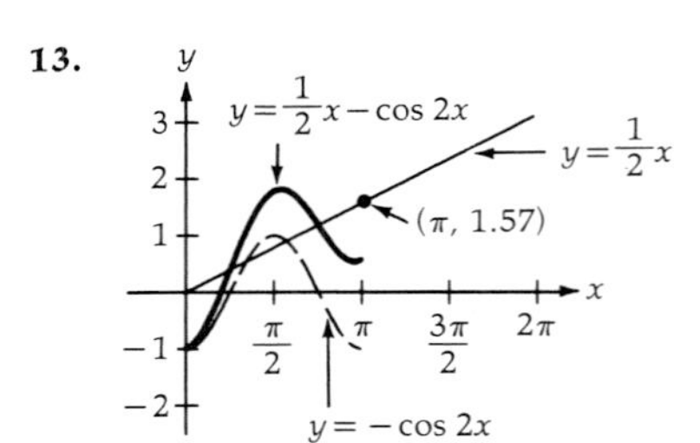


**15.**

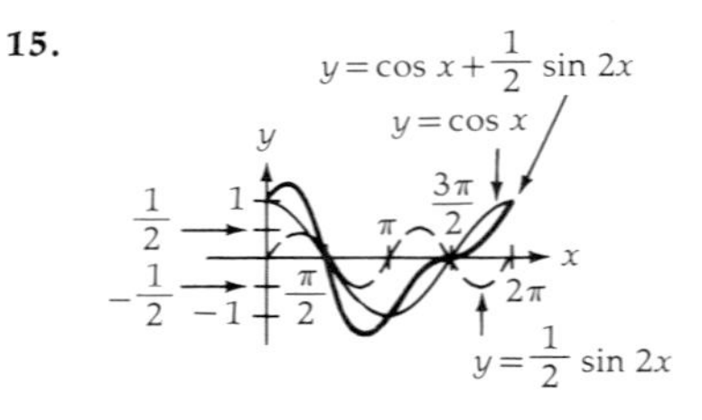


**17.**

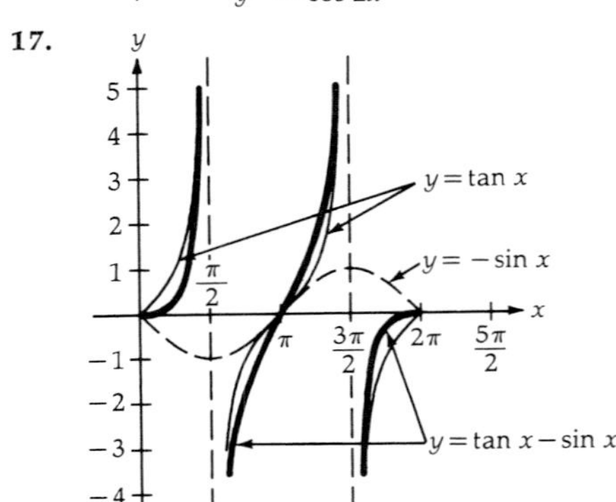


## Exercise 7-7 (pp. 376–377)

**1.** For some values of the domain there would be two images. This would violate the definition of a function.

**3.** No, there would be two images for some elements of the domain. This would violate the definition of a function.

**5.** $\frac{\pi}{6}$ **7.** $\frac{\pi}{3}$ **9.** $n\pi$ **11.** $-\frac{\pi}{3}$ **13.** $\frac{\pi}{2}$ **15.** $\frac{\sqrt{3}}{2}$ **17.** $\frac{\sqrt{2}}{2}$ **19.** $\frac{12}{13}$

**21.** $60^\circ$ or $\frac{\pi}{3}$ **23.** $30^\circ$ or $\frac{\pi}{6}$ **25.** 0 **27.** $\frac{\pi}{3}$ **29.** $\frac{\pi}{3}$ **31.** 0 **33.** $\frac{\sqrt{3}}{3}$

**35.** no solution **37.** $-\frac{1}{2}$ **39.** $-\frac{12}{13}$ **41.** $\sqrt{2}$ **43.** $\frac{4}{5}$ **45.** 0.3713 radians

**47.** $-0.3469$ radians **49.** 0.8242 radians **51.** 0.7752 radians **53.** 0.6396 radians

**55.** no solution **57.** $-0.5245$ **59.** no solution **61.** $\frac{\sqrt{u^2 - 9}}{u}$

## Chapter 7 Review Exercise (pp. 379–381)

**1.** $\left(\frac{\sqrt{2}}{2}, \frac{\sqrt{2}}{2}\right)$ **3.** $(-1, 0)$ **5.** $-\frac{12}{13}$ **7.** $-\frac{\sqrt{5}}{2}$ **9.** $\frac{5}{6}\pi$

The answers for Problems 11–29 were obtained on a scientific calculator. Answers obtained by using Table E will sometimes differ slightly.

**11.** 1.0632 **13.** 0.8473 **15.** 0.224 **17.** 1.290 **19.** $-1.2569$ **21.** $-1.0000$

**23.** $-6.3138$ **25.** 5.733 **27.** 2.392 **29.** 2.272

**31.**

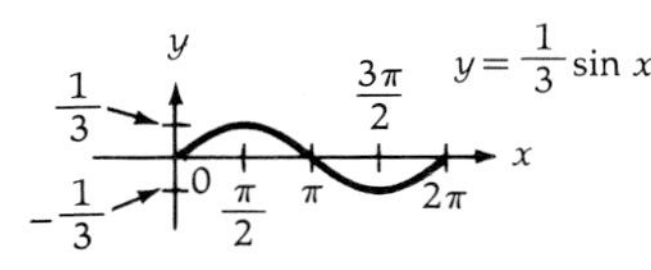


**33.**

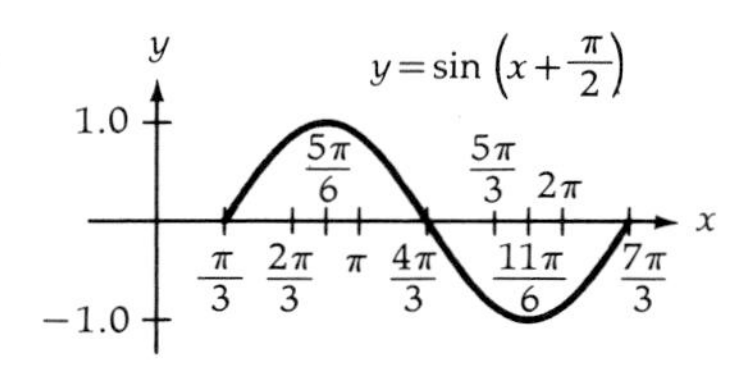


**35.**

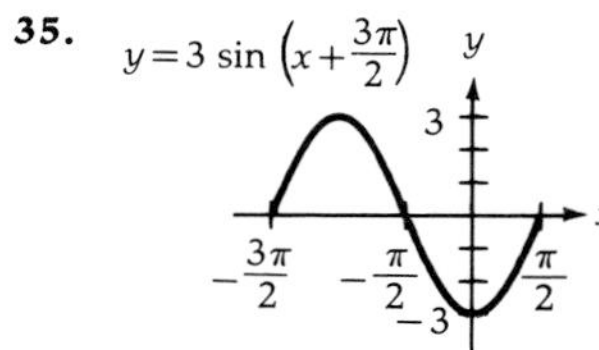
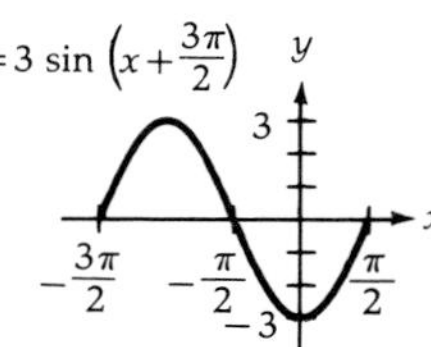


**37.**

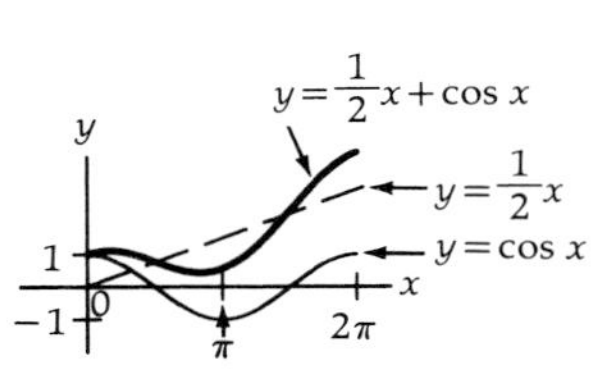


**39.**

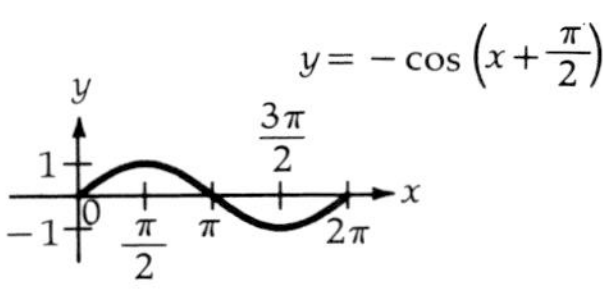


**41.**

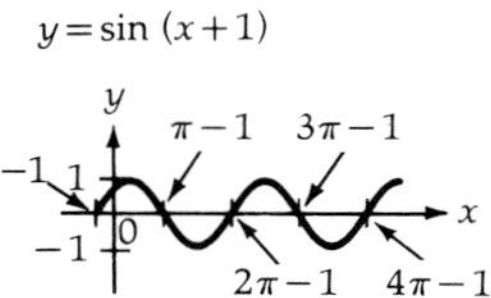


**43.**

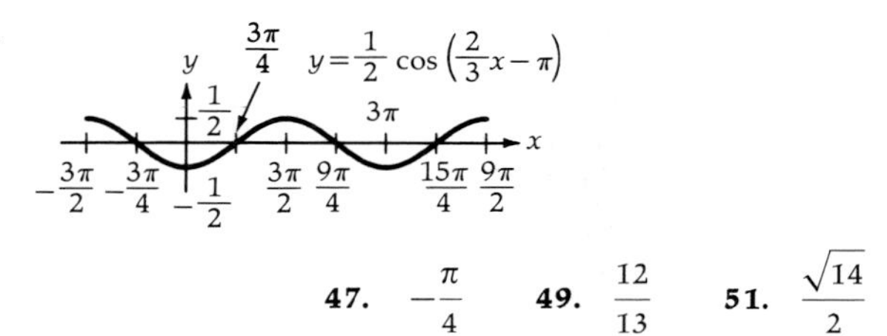


**45.**

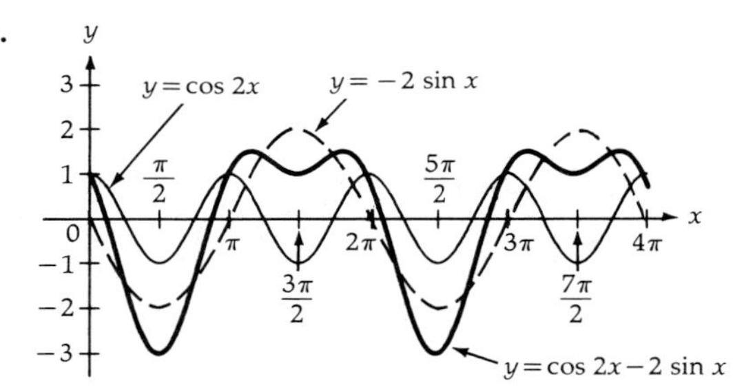


**47.** $-\frac{\pi}{4}$ **49.** $\frac{12}{13}$ **51.** $\frac{\sqrt{14}}{2}$ **55.** $\frac{24}{25}$

## Practice Test for Chapter 7 (pp. 381–382)

**1.** $\left(-\frac{\sqrt{3}}{2}, -\frac{1}{2}\right)$ **3.** $-\frac{3\sqrt{10}}{10}$ **5.** $\frac{5\pi}{3}$

The answers for Problems 7–11 were obtained on a scientific calculator. Answers obtained by using Table E will sometimes differ slightly.

**7.** 1.0669 **9.** 3.4616 **11.** 0.9781 **13.** $\frac{2}{3}\pi$ **15.**

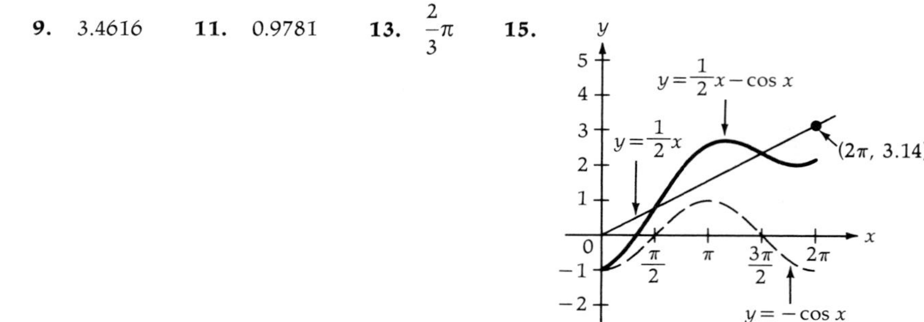


**17.** $-\frac{\pi}{6}$

## CHAPTER 8

### Exercise 8-1 (pp. 388–389)

Some of the steps in verification of each identity are omitted. For more detailed solutions, see the *Student Tutorial*. Each of the problems can be done in more than one way; these solutions are not unique.

**1.** $\dfrac{\cos\theta}{\cot\theta} \equiv \dfrac{\cos\theta}{\dfrac{\cos\theta}{\sin\theta}} \equiv \dfrac{\cos\theta\sin\theta}{\cos\theta} \equiv \sin\theta$

**3.** $\dfrac{\sin x\tan x}{\cos x} \equiv \tan x\tan x \equiv \tan^2 x$

**5.** $\dfrac{\cos\theta}{\tan\theta} \equiv \dfrac{\cos\theta}{\dfrac{\sin\theta}{\cos\theta}} \equiv \dfrac{\cos^2\theta}{\sin\theta} \equiv \dfrac{1-\sin^2\theta}{\sin\theta}$

**7.** $1-2\sin^2 t \equiv 1-\sin^2 t-\sin^2 t \equiv (1-\sin^2 t)-1+(1-\sin^2 t) \equiv \cos^2 t-1+\cos^2 t \equiv 2\cos^2 t-1$

**9.** $\sin\theta\cot\theta+\sin\theta\csc\theta \equiv (\sin\theta)\left(\dfrac{\cos\theta}{\sin\theta}\right)+(\sin\theta)\left(\dfrac{1}{\sin\theta}\right) \equiv \cos\theta+1$

**11.** $(\tan u+\sec u)^2 \equiv \left(\dfrac{\sin u}{\cos u}+\dfrac{1}{\cos u}\right)^2 \equiv \left(\dfrac{\sin u+1}{\cos u}\right)^2 \equiv \dfrac{(\sin u+1)^2}{\cos^2 u} \equiv \dfrac{(\sin u+1)^2}{(1+\sin u)(1-\sin u)} \equiv \dfrac{1+\sin u}{1-\sin u}$

**13.** $\dfrac{1}{\tan^2\theta}-\dfrac{1}{\sec^2\theta} \equiv \dfrac{1}{\dfrac{\sin^2\theta}{\cos^2\theta}}-\cos^2\theta \equiv \dfrac{\cos^2\theta-\sin^2\theta\cos^2\theta}{\sin^2\theta} \equiv \dfrac{\cos^2\theta(1-\sin^2\theta)}{\sin^2\theta} \equiv \dfrac{\cos^4\theta}{\sin^2\theta}$

**15.** $\dfrac{1+\sec\theta}{\tan\theta}-\dfrac{\tan\theta}{\sec\theta} \equiv \dfrac{(\sec\theta)(1+\sec\theta)-\tan^2\theta}{\tan\theta\sec\theta} \equiv \dfrac{\sec\theta+\sec^2\theta-\tan^2\theta}{\sec\theta\tan\theta} \equiv \dfrac{1+\sec\theta}{\sec\theta\tan\theta}$

**17.** $(\tan\theta+\cot\theta)(\cos\theta+\sin\theta) \equiv \left(\dfrac{\sin\theta}{\cos\theta}+\dfrac{\cos\theta}{\sin\theta}\right)(\cos\theta+\sin\theta) \equiv \left(\dfrac{\sin^2\theta+\cos^2\theta}{\cos\theta\sin\theta}\right)(\cos\theta+\sin\theta)$

$\equiv \left(\dfrac{1}{\cos\theta\sin\theta}\right)(\cos\theta+\sin\theta) \equiv \dfrac{1}{\sin\theta}+\dfrac{1}{\cos\theta} \equiv \csc\theta+\sec\theta$

**19.** $\dfrac{\cot^2\theta-1}{1+\cot^2\theta} \equiv \dfrac{\dfrac{\cos^2\theta}{\sin^2\theta}-1}{1+\dfrac{\cos^2\theta}{\sin^2\theta}} \equiv \dfrac{\dfrac{\cos^2\theta-\sin^2\theta}{\sin^2\theta}}{\dfrac{\sin^2\theta+\cos^2\theta}{\sin^2\theta}} \equiv \cos^2\theta-\sin^2\theta \equiv 1-2\sin^2\theta$

**21.** $\dfrac{\sin^3\theta}{1+\cos\theta} \equiv \dfrac{\sin\theta(1-\cos^2\theta)}{1+\cos\theta} \equiv \sin\theta\,(1-\cos\theta) \equiv \left(\dfrac{1}{\csc\theta}\right)(1-\cos\theta) \equiv \dfrac{1-\cos\theta}{\csc\theta}$

**23.** $\cot\theta\sec^2\theta \equiv \cot\theta\,(1+\tan^2\theta) \equiv \cot\theta+\cot\theta\left(\dfrac{1}{\cot\theta}\right)(\tan\theta) \equiv \cot\theta+\tan\theta$

**25.** $\dfrac{1-\tan^2 t}{1+\tan^2 t} \equiv \dfrac{1-(\sec^2 t-1)}{\sec^2 t} \equiv \dfrac{1}{\sec^2 t}-\dfrac{\sec^2 t}{\sec^2 t}+\dfrac{1}{\sec^2 t} \equiv \cos^2 t-1+\cos^2 t \equiv \cos^2 t-\sin^2 t$

**27.** $\dfrac{\sin\alpha}{\csc\alpha-\cot\alpha} \equiv \dfrac{\sin\alpha}{\dfrac{1}{\sin\alpha}-\dfrac{\cos\alpha}{\sin\alpha}} \equiv \dfrac{\sin\alpha}{\dfrac{1-\cos\alpha}{\sin\alpha}} \equiv \dfrac{\sin^2\alpha}{1-\cos\alpha} \equiv \dfrac{\sin^2\alpha}{1-\cos\alpha}\cdot\dfrac{1+\cos\alpha}{1+\cos\alpha} \equiv \dfrac{\sin^2\alpha\,(1+\cos\alpha)}{\sin^2\alpha}$

$\equiv 1+\cos\alpha$

**29.** $\sec u + \cos u + \sin^2 u \sec u \equiv \dfrac{1}{\cos u} + \cos u + (1 - \cos^2 u)\left(\dfrac{1}{\cos u}\right) \equiv \dfrac{1 + \cos^2 u + 1 - \cos^2 u}{\cos u}$

$$\equiv \frac{2}{\cos u} \equiv \frac{2 \sin u}{\cos u \sin u} \equiv \frac{2 \sin u}{\cos u\left(\dfrac{\cos u}{\cos u}\right)(\sin u)} \equiv \frac{2 \sin u}{\cos^2 u \tan u}$$

## Exercise 8-2 (pp. 396–397)

**1.** $\dfrac{\sqrt{6} - \sqrt{2}}{4}$ **3.** $-\dfrac{\sqrt{2}}{2}$ **5.** $-\dfrac{\sqrt{6} + \sqrt{2}}{4}$

**7.** $\cos (A - B) \equiv \cos A \cos B + \sin A \sin B$. Let $A = 0°$ and $B = \theta°$:
$\cos (0° - \theta) \equiv \cos (-\theta) \equiv \cos 0° \cos \theta + \sin 0° \sin \theta \equiv (1)(\cos \theta) + (0)(\sin \theta) \equiv \cos \theta$

**9.** $\cos (360° - \theta) \equiv \cos 360° \cos \theta + \sin 360° \sin \theta \equiv (1)(\cos \theta) + (0)(\sin \theta) \equiv \cos \theta$

**11.** $\cos (270° + \theta) \equiv \cos 270° \cos \theta - \sin 270° \sin \theta \equiv (0)(\cos \theta) - (-1)(\sin \theta) \equiv \sin \theta$

**13.** **a.** $\dfrac{-33}{65}$ **b.** $\dfrac{63}{65}$ **15.** **a.** $\dfrac{2(\sqrt{5} - 1)}{3\sqrt{5}}$ or $\dfrac{2(5 - \sqrt{5})}{15}$ **b.** $\dfrac{2(\sqrt{5} + 1)}{3\sqrt{5}}$ or $\dfrac{2(5 + \sqrt{5})}{15}$

**17.** $\cos 5A$ **19.** $\dfrac{\sqrt{6} - \sqrt{2}}{4}$ **21.** $\dfrac{\sqrt{2} + \sqrt{6}}{4}$ **23.** $-\dfrac{\sqrt{6} + \sqrt{2}}{2}$ **25.** $2 - \sqrt{3}$

**27.** $\dfrac{\sqrt{3} + 1}{1 - \sqrt{3}}$ or $-2 - \sqrt{3}$ **29.** $\dfrac{-1 - \sqrt{3}}{1 - \sqrt{3}}$ or $2 + \sqrt{3}$ **31.** **a.** $\dfrac{56}{65}$ **b.** $-\dfrac{16}{65}$

**33.** **a.** $\dfrac{8}{\sqrt{65}}$ **b.** $-\dfrac{4}{\sqrt{65}}$ **35.** **a.** $-\dfrac{33}{56}$ **b.** $-\dfrac{63}{16}$

**37.** **a.** $\dfrac{3 - 2\sqrt{7}}{6 + \sqrt{7}}$ or $\dfrac{32 - 15\sqrt{7}}{29}$ **b.** $\dfrac{-2\sqrt{7} - 3}{\sqrt{7} - 6}$ or $\dfrac{32 + 15\sqrt{7}}{29}$

**39.** $\sin 2A \equiv \sin (A + A) \equiv \sin A \cos A + \cos A \sin A \equiv 2 \sin A \cos A$

**41.** $\sin (270° + A) \equiv \sin 270° \cos A + \cos 270° \sin A \equiv (-1)(\cos A) + (0)(\sin A) \equiv -\cos A$

**43.** $\tan 2\theta \equiv \tan (\theta + \theta) \equiv \dfrac{\tan \theta + \tan \theta}{1 - \tan \theta \tan \theta} \equiv \dfrac{2 \tan \theta}{1 - \tan^2 \theta}$ **45.** $\dfrac{\sqrt{2}}{2}(\sin \theta - \cos \theta)$ **47.** $-\tan \theta$

**49.** $\dfrac{1 + \tan^2 \theta}{\tan \theta - \tan^2 \theta}$

## Exercise 8-3 (pp. 402–403)

**1.** **a.** $-\dfrac{24}{25}$ **b.** $-\dfrac{7}{25}$ **c.** $\dfrac{24}{7}$ **3.** **a.** $\dfrac{120}{169}$ **b.** $\dfrac{119}{169}$ **c.** $\dfrac{120}{119}$ **5.** $\dfrac{7}{24}$ **7.** $-\dfrac{120}{169}$

**9.** $4 \cos^3 \theta - 3 \cos \theta$

**11.** $\sin 2x \equiv 2 \sin x \cos x \equiv 2 \sin x (\cos x)\left(\dfrac{\cos x}{\cos x}\right) \equiv \left(\dfrac{2 \sin x}{\cos x}\right)(\cos^2 x) \equiv (2 \tan x)\left(\dfrac{1}{\sec^2 x}\right) \equiv \dfrac{2 \tan x}{1 + \tan^2 x}$

**13.** $\tan 2A \equiv \dfrac{2\tan A}{1-\tan^2 A} \equiv \dfrac{\dfrac{2\tan A}{\tan A}}{\dfrac{1-\tan^2 A}{\tan A}} \equiv \dfrac{2}{\dfrac{1}{\tan A}-\tan A} \equiv \dfrac{2}{\cot A - \tan A}$

**15.** $\cos 4A \equiv \cos [2(2A)] \equiv 2\cos^2 (2A) - 1 \equiv 2(2\cos^2 A - 1)^2 - 1$
$\equiv 2[4\cos^4 A - 4\cos^2 A + 1] - 1$
$\equiv 8\cos^4 A - 8\cos^2 A + 1$

**17.** $\cos 2A \equiv \cos^2 A - \sin^2 A \equiv \dfrac{\dfrac{\cos A}{\sin A}(\cos^2 A - \sin^2 A)}{\dfrac{\cos A}{\sin A}} \equiv \dfrac{\dfrac{\cos^3 A}{\sin A} - \cos A \sin A}{\cot A}$

**19.** $2-\sqrt{3}$ **21.** $\dfrac{\sqrt{2-\sqrt{3}}}{2}$ **23.** $\dfrac{\sqrt{2-\sqrt{2}}}{2}$ **25.** $-\dfrac{\sqrt{2+\sqrt{3}}}{2}$ **27.** $\dfrac{2\sqrt{3}}{3}$

**29.** $-\dfrac{3\sqrt{10}}{10}$ **31.** 5 **33.** $\dfrac{\sqrt{17}}{17}$

Some of the steps of verification of each identity in Problems **35–41** are omitted. For more detailed solutions, see the *Student Tutorial.*

**35.** $\tan \dfrac{A}{2} \equiv \dfrac{\sin \dfrac{A}{2}}{\cos \dfrac{A}{2}} \equiv \dfrac{\sin\left(A-\dfrac{A}{2}\right)}{\cos\left(A-\dfrac{A}{2}\right)} \equiv \dfrac{\sin A \cos \dfrac{A}{2} - \cos A \sin \dfrac{A}{2}}{\cos A \cos \dfrac{A}{2} + \sin A \sin \dfrac{A}{2}} \equiv \dfrac{\sin \dfrac{A}{2}\left(\sin A \cos \dfrac{A}{2} - \cos A \sin \dfrac{A}{2}\right)}{\sin \dfrac{A}{2}\left(\cos A \cos \dfrac{A}{2} + \sin A \sin \dfrac{A}{2}\right)}$

$\equiv \dfrac{\sin A\left(\dfrac{\sin A}{2}\right) - \cos A\left(\dfrac{1-\cos A}{2}\right)}{\cos A\left(\dfrac{\sin A}{2}\right) + \sin A\left(\dfrac{1-\cos A}{2}\right)} \equiv \dfrac{\sin^2 A - \cos A + \cos^2 A}{\cos A \sin A + \sin A - \sin A \cos A} \equiv \dfrac{1-\cos A}{\sin A}$

**37.** $\dfrac{1-\tan \dfrac{x}{2}}{1+\tan \dfrac{x}{2}} \equiv \dfrac{1-\dfrac{\sin \frac{x}{2}}{\cos \frac{x}{2}}}{1+\dfrac{\sin \frac{x}{2}}{\cos \frac{x}{2}}} \equiv \dfrac{\cos \dfrac{x}{2} - \sin \dfrac{x}{2}}{\cos \dfrac{x}{2} + \sin \dfrac{x}{2}} \equiv \dfrac{\cos^2 \dfrac{x}{2} - 2\cos \dfrac{x}{2}\sin \dfrac{x}{2} + \sin^2 \dfrac{x}{2}}{\cos^2 \dfrac{x}{2} - \sin^2 \dfrac{x}{2}} \equiv \dfrac{1-\sin x}{\cos x}$

**39.** $\sin \dfrac{A}{2} \cos \dfrac{A}{2} \equiv \dfrac{2\sin \dfrac{A}{2}\cos \dfrac{A}{2}}{2} \equiv \dfrac{\sin\left[2\left(\dfrac{A}{2}\right)\right]}{2} \equiv \dfrac{\sin A}{2}$

**41.** $$\tan\left(\frac{\pi}{4}+\frac{\theta}{2}\right) \equiv \frac{\tan\frac{\pi}{4}+\tan\frac{\theta}{2}}{1-\tan\frac{\pi}{4}\tan\frac{\theta}{2}} \equiv \frac{1+\tan\frac{\theta}{2}}{1-\tan\frac{\theta}{2}} \equiv \frac{1+\dfrac{\sin\frac{\theta}{2}}{\cos\frac{\theta}{2}}}{1-\dfrac{\sin\frac{\theta}{2}}{\cos\frac{\theta}{2}}} \equiv \frac{\left(\cos\frac{\theta}{2}+\sin\frac{\theta}{2}\right)\left(\cos\frac{\theta}{2}+\sin\frac{\theta}{2}\right)}{\left(\cos\frac{\theta}{2}-\sin\frac{\theta}{2}\right)\left(\cos\frac{\theta}{2}+\sin\frac{\theta}{2}\right)}$$

$$\equiv \frac{1+\sin\theta}{\cos\theta} \equiv \sec\theta+\tan\theta$$

## Exercise 8-4 (p. 406)

**1.** $2\sin 40^\circ \cos 20^\circ$ **3.** $2\cos 38.5^\circ \cos 16.5^\circ$ **5.** $2\sin 3x\cos x$ **7.** $2\cos 7.5x\sin 4.5x$

**9.** $-2\sin 5x\sin 2x$ **11.** $\frac{1}{2}[\sin 50^\circ + \sin 10^\circ]$ **13.** $\frac{1}{2}[\cos 78^\circ + \cos 8^\circ]$ **15.** $\sin 6\theta + \sin 4\theta$

**17.** $\frac{1}{2}[\cos 2x - \cos 6x]$ **19.** $\frac{1}{2}[\sin 2A - \sin A]$

Some of the steps of verification of each identity in Problems 21–29 are omitted. For more detailed solutions, see the *Student Tutorial.*

**21.** $$\sin\left(\frac{u+v}{2}\right)\cos\left(\frac{u-v}{2}\right) \equiv \frac{1}{2}\left[\sin\left(\frac{u+v}{2}+\frac{u-v}{2}\right)+\sin\left(\frac{u+v}{2}-\frac{u-v}{2}\right)\right] \equiv \frac{1}{2}[\sin u + \sin v]$$

Therefore $2\sin\left(\frac{u+v}{2}\right)\cos\left(\frac{u+v}{2}\right) \equiv \sin u + \sin v$

**23.** $$\frac{\cos A - \cos B}{\sin A + \sin B} \equiv \frac{-2\sin\frac{1}{2}(A+B)\sin\frac{1}{2}(A-B)}{2\sin\frac{1}{2}(A+B)\cos\frac{1}{2}(A-B)} \equiv \frac{-\sin\frac{1}{2}(A-B)}{\cos\frac{1}{2}(A-B)} \equiv -\tan\frac{1}{2}(A-B)$$

**25.** $\cos(A+B) \equiv \cos A\cos B - \sin A\sin B$; $\cos(A-B) \equiv \cos A\cos B + \sin A\sin B$. By addition of these two equations we have $\cos(A+B)+\cos(A-B) \equiv 2\cos A\cos B$ and therefore

$$\cos A\cos B \equiv \frac{1}{2}[\cos(A+B)+\cos(A-B)]$$

**27.** $$\frac{\sin A + \sin B}{\cos A + \cos B} \equiv \frac{2\sin\frac{1}{2}(A+B)\cos\frac{1}{2}(A-B)}{2\cos\frac{1}{2}(A+B)\cos\frac{1}{2}(A-B)} \equiv \tan\frac{1}{2}(A+B)$$

**29.** $\dfrac{\cot x - \tan x}{\cot x + \tan x} \equiv \dfrac{\dfrac{\cos x}{\sin x} - \dfrac{\sin x}{\cos x}}{\dfrac{\cos x}{\sin x} + \dfrac{\sin x}{\cos x}} \equiv \dfrac{\cos^2 x - \sin^2 x}{\cos^2 x + \sin^2 x} \equiv \cos 2x$

## Exercise 8-5 (pp. 414–416)

The answers to these problems are *rounded off* to the appropriate level of precision for each problem.

**1.** 11 **3.** 18 **5.** 90° **7.** $A = 47°, a = 20, c = 25$ **9.** $C = 61°, a = 8, b = 21$
**11.** no solution **13.** two possible solutions: $B = 54°, C = 92°, c = 39$ or $B = 126°, C = 20°, c = 13$
**15.** two possible solutions: $A = 43°, B = 75°, a = 16$ or $A = 13°, B = 105°, a = 5.4$
**17.** $B = 142.9°, a = 7.06, c = 4.03$ **19.** $A = 91.4°, b = 1.38, a = 3.00$
**21.** two possible solutions: $A = 25.3°, B = 132.2°, b = 9.43$ or $A = 154.7°, B = 2.8°, b = 0.62$
**23.** $B = 9.7°, C = 67.7°, c = 310$ **25.** 1.8 km **27.** 78.1° **29.** 27.8 m wide **31.** 115 cm **33.** 22.0 km

**35.** $\dfrac{\sin A + \sin B}{\sin B} \equiv \dfrac{a \sin A + a \sin B}{a \sin B}$. Now from the law of sines, $a \sin B = b \sin A$. Therefore we have

$$\frac{\sin A + \sin B}{\sin B} \equiv \frac{a \sin A + b \sin A}{b \sin A} \equiv \frac{a + b}{b}$$

## Exercise 8-6 (pp. 420–421)

The answers to these problems are *rounded off* to the appropriate level of precision for each problem.

**1.** $c^2 = a^2 + b^2$, Pythagorean Theorem **3.** 7.3 **5.** 22° **7.** 90° **9.** $c = 12, A = 40°, B = 95°$
**11.** $a = 33, B = 45°, C = 15°$ **13.** $A = 44°, B = 53°, C = 83°$ **15.** $A = 31.2°, B = 52.0°, C = 96.8°$
**17.** $c = 43.4, A = 48.5°, B = 64.1°$ **19.** $A = 45.9°, B = 28.8°, C = 105.3°$ **21.** angle of 55°
**23.** 251 m **25.** angle of 36.1° **27.** 2502 m **29.** 195.113 miles

## Exercise 8-7 (pp. 425–426)

**1.** 0°, 90°, 270° **3.** 60°, 90°, 270°, 300° **5.** 30°, 150°, 270° **7.** 240°, 300° **9.** no solution
**11.** 60°, 300° **13.** 30°, 150°, 210°, 270°, 330° **15.** 15°, 75°, 195°, 255° **17.** 60°, 120°, 240°, 300°
**19.** 124.5°, 235.5° **21.** 30°, 150°, 210°, 330° **23.** 30°, 60°, 210°, 240°
**25.** 109.5°, 120°, 240°, 250.5° **27.** $\frac{\pi}{4}, \frac{3\pi}{4}, \frac{5\pi}{4}, \frac{7\pi}{4}$ **29.** $\frac{3\pi}{4}, \frac{7\pi}{4}$ **31.** $\frac{\pi}{6}, \frac{5\pi}{6}$
**33.** $\frac{4\pi}{9}, \frac{5\pi}{9}, \frac{10\pi}{9}, \frac{11\pi}{9}, \frac{16\pi}{9}, \frac{17\pi}{9}$ **35.** 0.8758, 1.9718, 4.3113, 5.4074 radians **37.** 0°, 180°
**39.** 45°, 90°, 225°, 270° **41.** no solution **43.** 30°, 150°, 210°, 330° **45.** 270°

## Exercise 8-8 (pp. 433–435)

Given vectors **a**, **b**, **c**, **d**, **e**, as shown below.

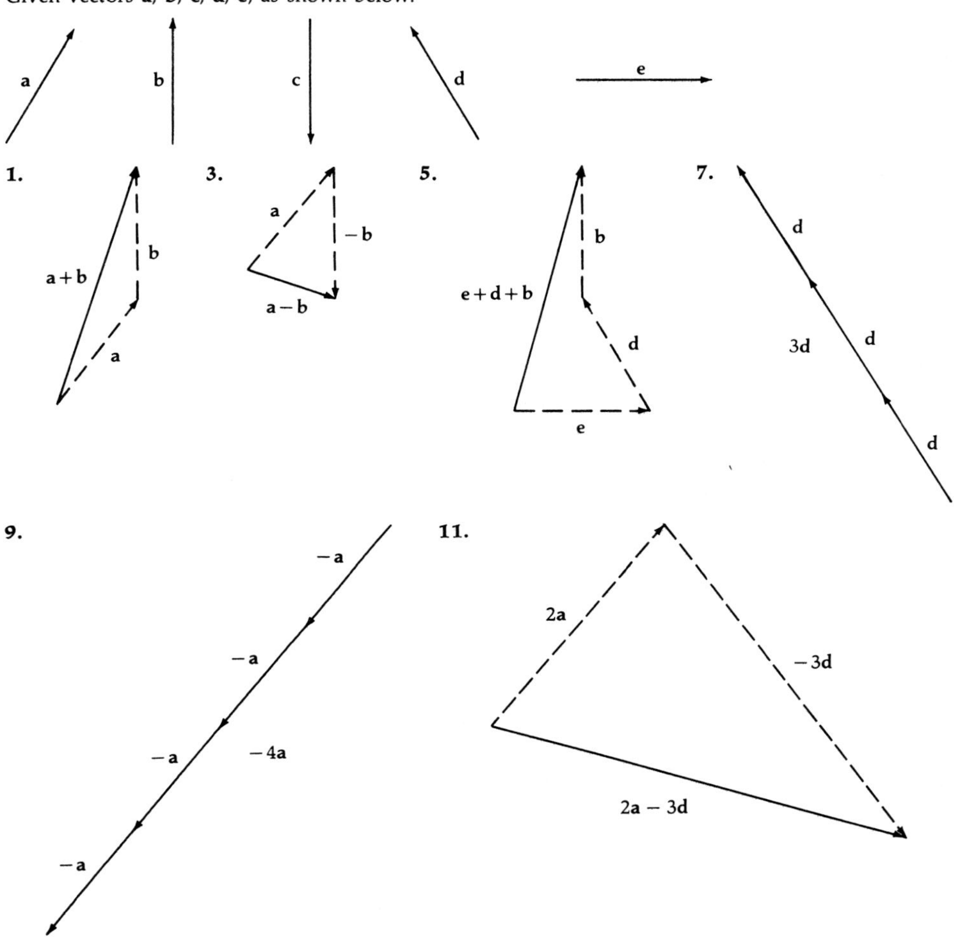


**13.** $\mathbf{a} = \langle 141.27, 75.12 \rangle$ (rounded to nearest hundredth)

The four required resultant vectors for each problem are listed for Problems 15–21 in the order: $2\mathbf{A}$, $\mathbf{A} + \mathbf{B}$, $\mathbf{A} - \mathbf{B}$, $3\mathbf{A} - 2\mathbf{B}$.

**15.** $\langle 6, 14 \rangle, \langle 5, 8 \rangle, \langle 1, 6 \rangle, \langle 5, 19 \rangle$ **17.** $\langle -12, 4 \rangle, \langle -4, 1 \rangle, \langle -8, 3 \rangle, \langle -22, 8 \rangle$

**19.** $\langle 0, 1 \rangle, \langle -3, \frac{9}{2} \rangle, \langle -3, -\frac{7}{2} \rangle, \langle 6, -\frac{13}{2} \rangle$ **21.** $\langle 3.2, -4.6 \rangle, \langle 6.7, 0.3 \rangle, \langle -3.5, -4.9 \rangle, \langle -5.4, -12.1 \rangle$

**23.** $\langle 5, 4 \rangle$ **25.** $23 \angle 156^\circ$ **27.** $88 \angle 196.6^\circ$ **29.** $164 \angle 312^\circ$ **31.** $3\sqrt{5} \angle 70^\circ$

**33.** $4.24 \angle 135^\circ$ **35.** $2 \angle 330^\circ$ **37.** $5.39 \angle 111.8^\circ$ **39.** $1.8 \angle 270^\circ$ **41.** $0.3 \angle 33.7^\circ$

**43.** $1.62 \angle 180^\circ$ **45.** $\langle 1.39, 0.80 \rangle$ **47.** $\langle 5.66, -5.66 \rangle$ **49.** $\langle -8.82, 12.14 \rangle$

**51.** $\langle 0.44, -0.44 \rangle$

## Chapter 8 Review Exercise (pp. 437–440)

**1.** **a.** $\frac{1}{2}$ **b.** $\frac{1}{2}$ **3.** $\frac{\sqrt{6}-\sqrt{2}}{4}$ **5.** $\frac{7}{25}$ **7.** $\frac{\sqrt{10}}{10}$ **9.** 5 **11.** $\frac{56}{65}$ **13.** $-\frac{65}{63}$

**15.** $\frac{2\sqrt{5}}{5}$ **17.** $2 \sin 40° \sin 15°$ **19.** $\cos 99° + \cos 23°$

**21.** $\frac{1+\sin\theta}{\cot^2\theta} \equiv (1+\sin\theta)\left(\frac{\sin^2\theta}{\cos^2\theta}\right) \equiv \frac{(1+\sin\theta)(\sin^2\theta)}{1-\sin^2\theta} \equiv \frac{\sin^2\theta}{1-\sin\theta} \equiv \frac{\sin\theta}{\csc\theta - 1}$

**23.** $\frac{\tan^2 x + 1}{\cot^2 x + 1} \equiv \frac{\sec^2 x}{\csc^2 x} \equiv \frac{\sin^2 x}{\cos^2 x} \equiv \tan^2 x$ **25.** $\frac{1-\cot x}{1+\cot x} \equiv \frac{1-\frac{1}{\tan x}}{1+\frac{1}{\tan x}} \equiv \frac{\tan x - 1}{\tan x + 1}$

**27.** $\frac{\cos^2\beta - \cot\beta}{\sin^2\beta - \tan\beta} \equiv \frac{\cos^2\beta - \frac{\cos\beta}{\sin\beta}}{\sin^2\beta - \frac{\sin\beta}{\cos\beta}} \equiv \frac{\cos^2\beta\,(\sin\beta\cos\beta - 1)}{\sin^2\beta\,(\sin\beta\cos\beta - 1)} \equiv \cot^2\beta$

**29.** $\cos 2\alpha \equiv \cos^2\alpha - \sin^2\alpha \equiv \frac{1}{\sec^2\alpha} - \frac{1}{\csc^2\alpha} \equiv \frac{\csc^2\alpha - \sec^2\alpha}{\sec^2\alpha\csc^2\alpha} \equiv \frac{\frac{\csc^2\alpha}{\sec^2\alpha} - 1}{\csc^2\alpha} \equiv \frac{\cot^2\alpha - 1}{\csc^2\alpha}$

**31.** $\frac{\cos 2\alpha - \cos 2\beta}{\cos\alpha - \cos\beta} \equiv \frac{(2\cos^2\alpha - 1) - (2\cos^2\beta - 1)}{\cos\alpha - \cos\beta} \equiv \frac{2(\cos\alpha - \cos\beta)(\cos\alpha + \cos\beta)}{\cos\alpha - \cos\beta} \equiv 2(\cos\alpha + \cos\beta)$

**33.** $\frac{1+\sin\alpha}{\tan\alpha} \equiv \frac{1}{\tan\alpha} + \frac{\sin\alpha}{\tan\alpha} \equiv \cot\alpha + \cos\alpha$

**35.** $\sin^2 3x - \sin^2 2x \equiv (\sin 3x - \sin 2x)(\sin 3x + \sin 2x) \equiv \left(2\cos\frac{5x}{2}\sin\frac{x}{2}\right)\left(2\sin\frac{5x}{2}\cos\frac{x}{2}\right)$

$\equiv \left(2\sin\frac{x}{2}\cos\frac{x}{2}\right)\left(2\sin\frac{5x}{2}\cos\frac{5x}{2}\right) \equiv \sin x \sin 5x$

**37.** $\frac{1+\sec\beta}{\sec\beta + \tan\beta + 1} \equiv \frac{1+\frac{1}{\cos\beta}}{\frac{1}{\cos\beta} + \frac{\sin\beta}{\cos\beta} + 1} \equiv \frac{1+\cos\beta}{\cos\beta + \sin\beta + 1}$

**39.** $\frac{\sin A \tan\frac{A}{2}}{2} \equiv \frac{2\sin\frac{A}{2}\cos\frac{A}{2}}{2} \cdot \frac{\sin\frac{A}{2}}{\cos\frac{A}{2}} \equiv \sin^2\frac{A}{2}$

In Problems 41–59 the answers are rounded to the appropriate level of precision.

**41.** 38 **43.** 67 **45.** 22 **47.** no solution **49.** $B = 25°$, $C = 125°$, $c = 20$
**51.** $C = 78°$, $a = 30$, $c = 34$ **53.** $B = 27°$, $C = 35°$, $a = 2.5$ **55.** $A = 19°$, $B = 138°$, $C = 23°$
**57.** 71 meters **59.** 1,660 meters **61.** 0°, 180°, 270° **63.** 28.15°, 118.15°, 208.15°, 298.15°
**65.** 210°, 330° **67.** 60°, 180°, 300° **69.** 270° **71.** 0°, 120°, 180°, 240°

**73.** 30°, 90°, 270°, 330° **75.** 0°, 180° **77.** 45°, 90°, 135°, 225°, 270°, 315° **79.** 180°
**81.** $\langle -5, 5.3\rangle$, $\langle 16, -6.1\rangle$ **83.** To nearest tenth, we have $\langle -12.2, -11.8\rangle$

## Practice Test for Chapter 8 (p. 440)

**1.** $\frac{4}{3}$ **3.** $-\frac{3}{2}$ **5.** $-\frac{24}{7}$

**7.** $\cot^2\theta - \cos^2\theta \equiv \dfrac{\cos^2\theta}{\sin^2\theta} - \cos^2\theta \equiv \dfrac{\cos^2\theta - \sin^2\theta\cos^2\theta}{\sin^2\theta} \equiv \dfrac{\cos^2\theta(1-\sin^2\theta)}{\sin^2\theta} \equiv \cot^2\theta\cos^2\theta$

**9.** $C \doteq 48°$ **11.** 0°, 60°, 300° **13.** 0°, 45°, 135°, 180°, 225°, 315°
**15.** To nearest tenth, we have $\langle -14.8, 4.0\rangle$

# CHAPTER 9

## Exercise 9-1 (pp. 445–446)

**1.** $7 + 7i$ **3.** $-3 - i$ **5.** $5 - 4i$ **7.** $10 - 15i$ **9.** $-3 - 18i$ **11.** $5 + 16i$

**13.** $-10 + 11i$ **15.** $-5 + 10i$ **17.** $11 - 60i$ **19.** $x^2 + y^2$ **21.** $\dfrac{6+8i}{25}$ or $\frac{6}{25} + \frac{8}{25}i$

**23.** $\dfrac{4-3i}{5}$ or $\frac{4}{5} - \frac{3}{5}i$ **25.** $\dfrac{1+7i}{10}$ or $\frac{1}{10} + \frac{7}{10}i$ **27.** $\dfrac{14+5i}{17}$ or $\frac{14}{17} + \frac{5}{17}i$ **29.** $\dfrac{11+41i}{34}$ or $\frac{11}{34} + \frac{41}{34}i$

**31.** $\dfrac{-6-9i}{9}$ or $-\frac{2}{3} - i$ **33.** $0 + 4i$ **35.** $2\sqrt{3} + 5i\sqrt{2}$ **37.** $-\sqrt{30} + 0i$ **39.** $i$

**41.** $-1$ **43.** $-i$
**45.** To evaluate $i^n$ where $n$ is odd, use the following procedure: a. If $(n-1)$ is divisible by 4, then $i^n = i$. b. If $(n-1)$ is not divisible by 4, then $i^n = -i$.

**47.** $-44 + 117i$ **49.** $\dfrac{3}{13} + \dfrac{41}{13}i$

## Exercise 9-2 (pp. 449–450)

**1.** $0 - \dfrac{11}{2}i$ **3.** $\dfrac{7}{37} - \dfrac{5}{37}i$ **5.** $-\dfrac{14}{13} - \dfrac{8}{13}i$ **7.** $-\dfrac{5}{2} + \dfrac{1}{2}i$ **9.** $\dfrac{9}{10} - \dfrac{11}{5}i$ **11.** $2 \pm i$

**13.** $-\dfrac{1}{6} \pm \dfrac{\sqrt{23}}{6}i$ **15.** $\dfrac{-1 - i \pm \sqrt{6i}}{2}$ **17.** $0 - 2i$ and $0 + \dfrac{1}{2}i$. **19.** $\dfrac{3}{13} + \dfrac{2}{13}i$

**21.** $2\left(\dfrac{1}{2} + \dfrac{1}{2}i\right)^2 - 2\left(\dfrac{1}{2} + \dfrac{1}{2}i\right) + 1 = 2\left(\dfrac{1}{4} + \dfrac{1}{2}i + \dfrac{1}{4}i^2\right) - 1 - i + 1 = i - 1 - i + 1 = 0$

The answers to Problems 23–33 are not unique.

**23.** $x^2 + 9 = 0$ **25.** $x^2 + 2 = 0$ **27.** $4x^3 - 12x^2 + x - 3 = 0$ **29.** $x^2 - 4x + 13 = 0$
**31.** $x^3 + (-3 - 2i)x^2 + (1 + 5i)x + (2 - 2i) = 0$ **33.** $x^3 + (-6 - i)x^2 + (10 + 6i)x - 10i = 0$
**35.** $-2.02 \pm 1.62i$

## Exercise 9-3 (pp. 453–454)

Answers to Problems 1–5 are graphed below.

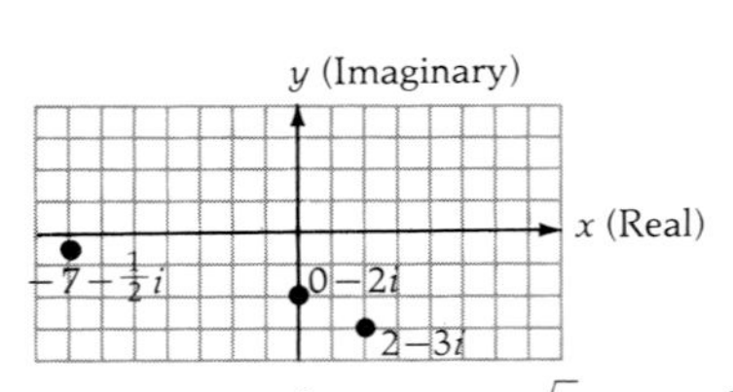


**7.** $\sqrt{41}$ **9.** 13 **11.** $\frac{\sqrt{37}}{2}$ **13.** 315° **15.** 201.8° **17.** 109.5° **19.** $\frac{5\sqrt{2}}{2} + \frac{5\sqrt{2}}{2}i$ **21.** $-3 - 3i$ **23.** $3.544 + 1.855i$ **25.** $-1.854 + 0.749i$ **27.** $6.782 - 7.349i$ **29.** 2 cis 300° **31.** $\sqrt{5}$ cis 26.6° **33.** $\sqrt{13}$ cis 56.3° **35.** 20 cis 150° **37.** 6 cis 90° **39.** 6 cis 330°

## Exercise 9-4 (pp. 456–457)

**1.** 6 cis 53° **3.** 40 cis 98° **5.** $\sqrt{6}$ cis 69.2° **7.** 105 cis 29° **9. a.** $2\sqrt{3} - 2i$ **b.** 4 cis 330° **c.** (2 cis 30°)(2 cis 300°) = 4 cis 330° **11.** 8 cis 135° **13.** cis 40° **15.** $\frac{9}{2} + \frac{9\sqrt{3}}{2}i$ **17.** $73.474 + 101.128i$ **19.** $-16\sqrt{3} - 16i$ **21.** $-\frac{9}{2} + \frac{9\sqrt{3}}{2}i$ **23.** $\frac{2}{5}$ cis 60° **25.** $\frac{3}{4}$ cis 336° **27.** $\frac{27}{4}$ cis 45° **29.** 707281 cis 185.589° = $-703919 - 68883i$

## Exercise 9-5 (p. 462)

**1.** $\pm\left(\frac{\sqrt{2}}{2} + \frac{\sqrt{2}}{2}i\right)$ **3.** $\pm\left(\frac{\sqrt{2}}{2} - \frac{\sqrt{6}}{2}i\right)$ **5.** $-2,\ 1 + i\sqrt{3},\ 1 - i\sqrt{3}$

**7.** Polar form: $\sqrt[6]{2}$ cis 45°, $\sqrt[6]{2}$ cis 165°, $\sqrt[6]{2}$ cis 285°
Rectangular form: $0.794 + 0.794i$, $-1.084 + 0.291i$, $0.291 - 1.084i$

**9.** Polar form: $\sqrt[4]{10}$ cis 30°, $\sqrt[4]{10}$ cis 120°, $\sqrt[4]{10}$ cis 210°, $\sqrt[4]{10}$ cis 300°
Rectangular form: $1.540 + 0.889i$, $-0.889 + 1.540i$, $-1.540 - 0.899i$, $0.889 - 1.540i$

**11.** 2 cis 45°, 2 cis 117°, 2 cis 189°, 2 cis 261°, 2 cis 333°
**13.** $\sqrt{2}$ cis 30°, $\sqrt{2}$ cis 90°, $\sqrt{2}$ cis 150°, $\sqrt{2}$ cis 210°, $\sqrt{2}$ cis 270°, $\sqrt{2}$ cis 330°
**15.** 3 cis 0°, 3 cis 120°, 3 cis 240° **17.** 2 cis 36°, 2 cis 108°, 2 cis 180°, 2 cis 252°, 2 cis 324°
**19.** 2 cis 40°, 2 cis 112°, 2 cis 184°, 2 cis 256°, 2 cis 328°
**21.** 1.710 cis 102.3°, 1.710 cis 222.3°, 1.710 cis 342.3° **23.** 1.800 cis 40.3°, 1.800 cis 160.3°, 1.800 cis 280.3°
**25.** 1.334 cis 62.9°, 1.334 cis 152.9°, 1.334 cis 242.9°, 1.334 cis 332.9°

## Chapter 9 Review Exercise (pp. 464–465)

**1.** $8 - 2i$ **3.** $6 - 6i$ **5.** $2 + 6i$ **7.** $2 - 23i$ **9.** $2 - i$ **11.** $\frac{16 + 15i}{13}$ or $\frac{16}{13} + \frac{15}{13}i$ **13.** $\frac{45 - 40i}{29}$ or $\frac{45}{29} - \frac{40}{29}i$ **15.** $\frac{-1 \pm i\sqrt{3}}{4}$ **17.** $\frac{7 \pm i\sqrt{11}}{6}$ **19.** 3 cis 90° **21.** $2\sqrt{2}$ cis 45° **23.** 5 cis 36.9° **25.** $\sqrt{41}$ cis 218.7° **27.** $-2 - 2i\sqrt{3}$ **29.** $2 + 0i$ **31.** $\frac{3\sqrt{3}}{2} + \frac{3}{2}i$ **33.** $-5 + 5i\sqrt{3}$ **35.** 6 cis 60° **37.** $2\sqrt{5}$ cis 152° **39.** $\frac{1}{2}$ cis 310° **41.** 32 cis 225° **43.** 3 cis 140° **45.** 729 cis 150° **47.** 32 cis 240° **49.** $\frac{1}{3}$ cis 240°

**51.** $1, -\frac{1}{2}+\frac{\sqrt{3}}{2}i, -\frac{1}{2}-\frac{\sqrt{3}}{2}i$ **53.** $\pm(2.187+0.465i), \pm(0.465-2.187i)$

**55.** $\sqrt{3}+i, -1+i\sqrt{3}, -\sqrt{3}-i, 1-i\sqrt{3}$ **57.** $2+0i, 0+2i, -2+0i, 0-2i$

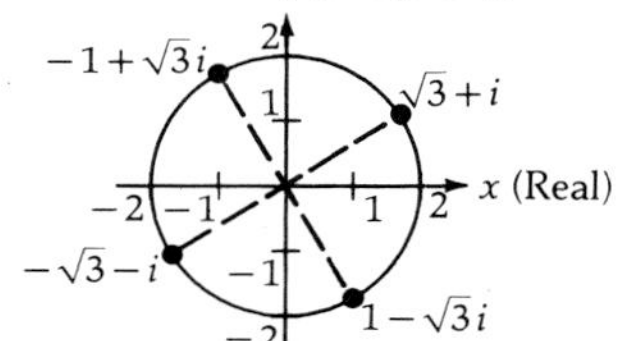


## Practice Test for Chapter 9 (p. 465)

**1.** $-5+8i$ **3.** $31+i$ **5.** $\frac{-3-7i}{2}$ or $-\frac{3}{2}-\frac{7}{2}i$ **7.** $\frac{5\pm i\sqrt{11}}{3}$ **9.** $\sqrt{13}$ cis 146.3°

**11.** 28 cis 273° **13.** $\frac{\sqrt{3}}{2}+\frac{1}{2}i$ **15.** $2+2i\sqrt{3}, -4, 2-2i\sqrt{3}$

# CHAPTER 10

## Exercise 10-1 (pp. 474–475)

**1.** Solution is (1, 1)

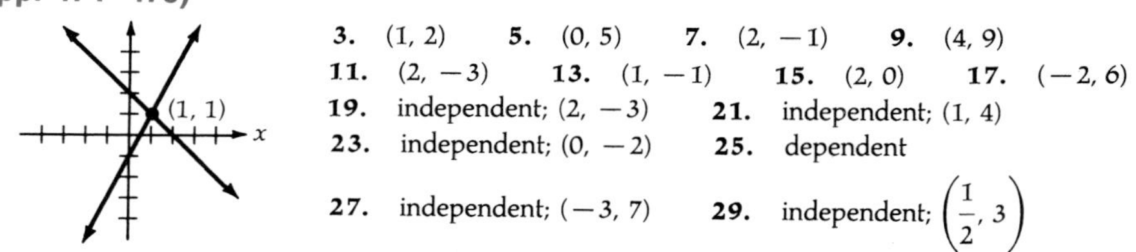


**3.** (1, 2) **5.** (0, 5) **7.** (2, −1) **9.** (4, 9)
**11.** (2, −3) **13.** (1, −1) **15.** (2, 0) **17.** (−2, 6)
**19.** independent; (2, −3) **21.** independent; (1, 4)
**23.** independent; (0, −2) **25.** dependent
**27.** independent; (−3, 7) **29.** independent; $\left(\frac{1}{2}, 3\right)$
**31.** dependent **33.** independent; (3, −8)

**35.** plane (500 mph), wind (100 mph) **37.** boat (12 mph), current (3 mph)
**39.** 5 mg of additive $X$ and 4 mg of additive $Y$

## Exercise 10-2 (pp. 479–481)

**1.** (1, 2, 3) **3.** (−2, 1, 3) **5.** (2, 1, −3) **7.** (−2, 1, 0) **9.** (4, −3, 3) **11.** (1, −2, 3)
**13.** (1, 2, 3, 4) **15.** (3, −1, 2) **17.** inconsistent system of equations; no solution
**19.** infinite number of solutions; they can be described as being of the form $(11-7y, y, 5y-8)$
**21.** 5 nickels, 10 dimes, 13 quarters **23.** 2 liters of 5%, 3 liters of 20%, 4 liters of 50%
**25.** side $x$ is 55 cm, side $y$ is 60 cm, side $z$ is 40 cm

## Exercise 10-3 (pp. 488–490)

**1. a.** $a_{11}=3, a_{22}=-1$ **b.** $a_{21}=2$ **c.** $a_{23}=8$ **3.** $\begin{bmatrix} 7 & -8 & -12 \\ 3 & -2 & -5 \end{bmatrix}$

**5.** $\begin{bmatrix} 3 & 2 & 1 & 5 \\ 9 & -1 & 0 & 8 \\ 6 & -3 & 2 & 5 \end{bmatrix}$ **7.** (4, −1) **9.** (3, −9) **11.** (0, 3) **13.** (2, 2) **15.** (0, 1, −2)

**17.** (2, 5, 0) **19.** $\left(\frac{1}{2}, -1, 5\right)$ **21.** \$10,000 at 5%, \$15,000 at 8% **23.** (1, 3, 4)

**25.** inconsistent system of equations, no solution

**27.** dependent equations, infinite number of solutions of the form $\left(\frac{-3z+24}{7}, \frac{9+5z}{7}, z\right)$ **29.** (2, 0, −3, 6)

## Exercise 10-4 (pp. 494–495)

**1.** 7 **3.** −3 **5.** 11 **7.** −2 **9.** 0 **11.** −101 **13.** 0 **15.** 54 **17.** −1

**19.** 2 **21.** $(-1)^{2+3}\begin{vmatrix} 3 & 6 \\ 7 & 4 \end{vmatrix} = 30$ **23.** $(-1)^{1+2}\begin{vmatrix} 2 & 8 \\ 7 & -3 \end{vmatrix} = 62$ **25.** −9 **27.** 3 **29.** 81

**31.** 46 **33.** 0

**35.** Evaluation of the determinant yields $x(y_1 - y_2) - y(x_1 - x_2) + 1(x_1y_2 - y_1x_2) = 0$
which can be written as $x_1y_1 + yx_2 - x_2y_1 - yx_1 = xy_2 - y_2x_1 - xy_1 + x_1y_1$
If we factor each side, we have $(y - y_1)(x_2 - x_1) = (y_2 - y_1)(x - x_1)$
and therefore $y - y_1 = \frac{y_2 - y_1}{x_2 - x_1}(x - x_1)$

**37.** see *Student Tutorial.* **39.** see *Student Tutorial.*

**41.** $|A| = a\begin{vmatrix} e & f \\ h & i \end{vmatrix} - b\begin{vmatrix} d & f \\ g & i \end{vmatrix} + c\begin{vmatrix} d & e \\ g & h \end{vmatrix} = aei - ahf - bdi + bgf + cdh - cge = aei + bfg + cdh - gec - hfa - idb$

**43.** 1764

## Exercise 10-5 (pp. 501–502)

**1.** (1, 2) **3.** (2, 3) **5.** dependent **7.** (3, 0) **9.** (5, −1) **11.** (−2, 1) **13.** (4, −1)
**15.** (−1, 5) **17.** (1, 2, 3) **19.** (2, 3, −5) **21.** (−3, 1, 0) **23.** (1, −2, 4)

**25.** first number is 3, second number is −10, third number is 7

**27.** $D = \begin{vmatrix} 1 & 2 & -1 \\ 5 & 0 & -1 \\ 3 & -4 & 1 \end{vmatrix} = 0$, so (0, 0, 0) is *not* unique.

## Exercise 10-6 (pp. 509–511)

**1.**

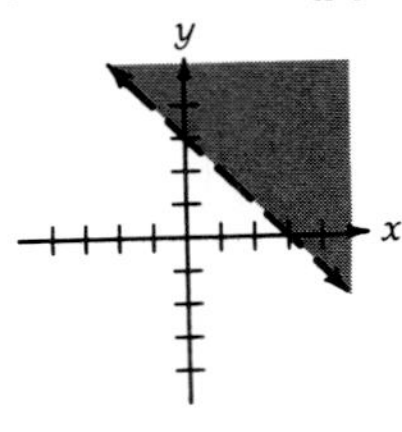


**3.**

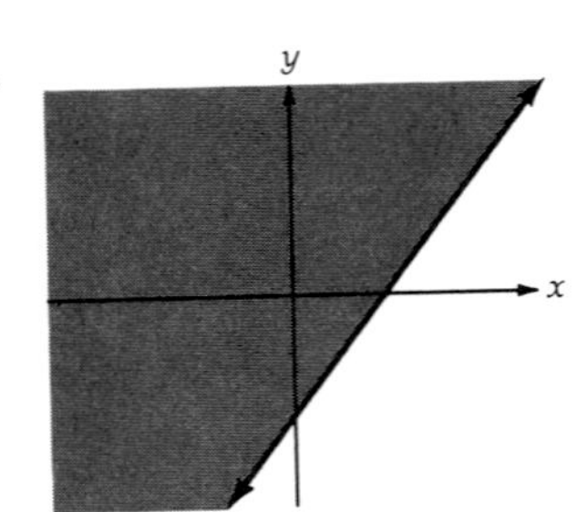


**5.**

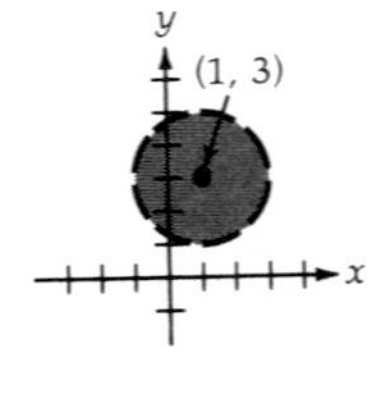

**7.**

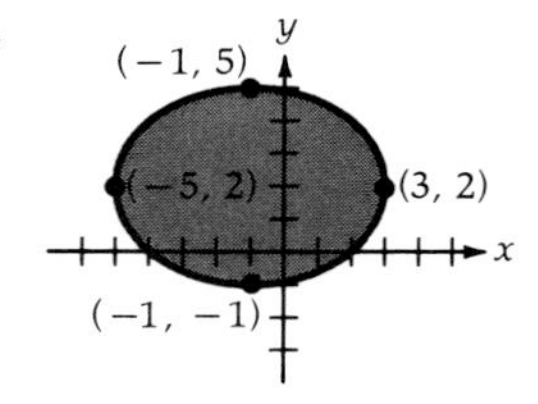


**9.**

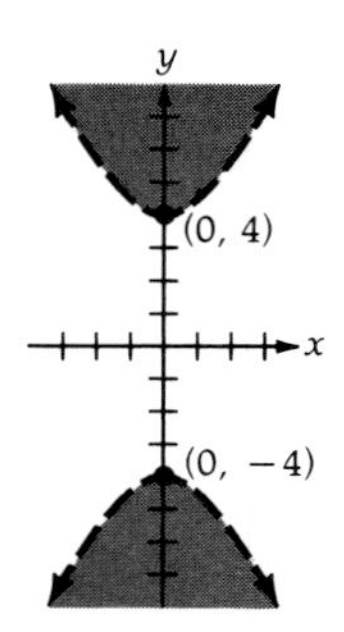


**11.**

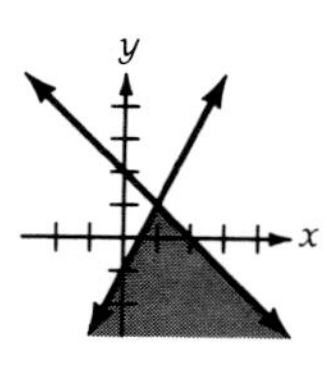


**13.**

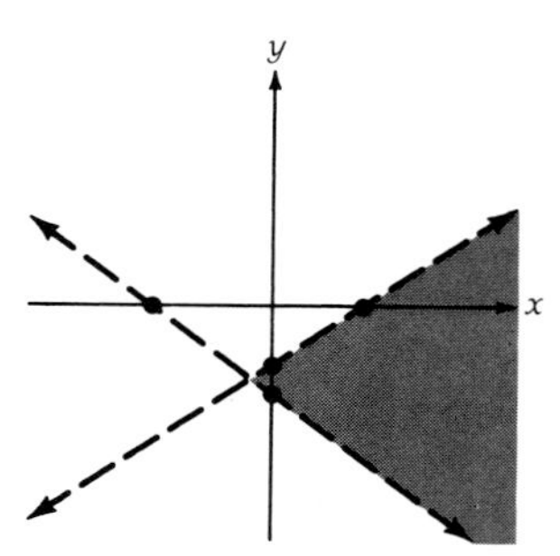


**15.**

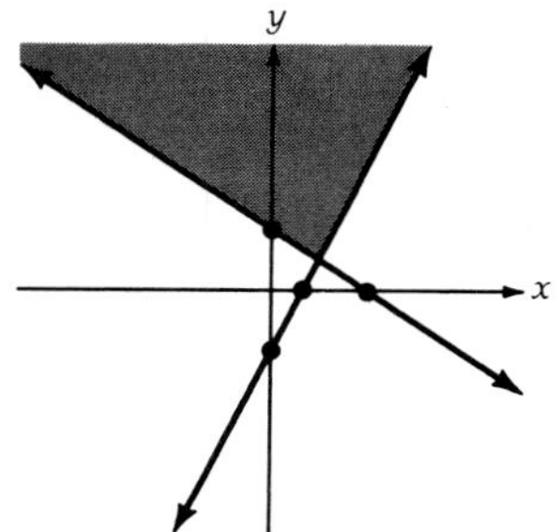


**17.**

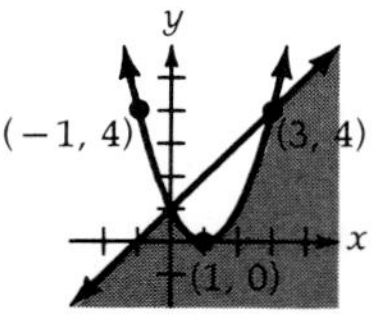


**19.**

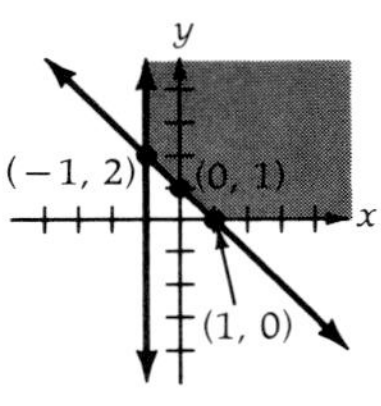


**21.**

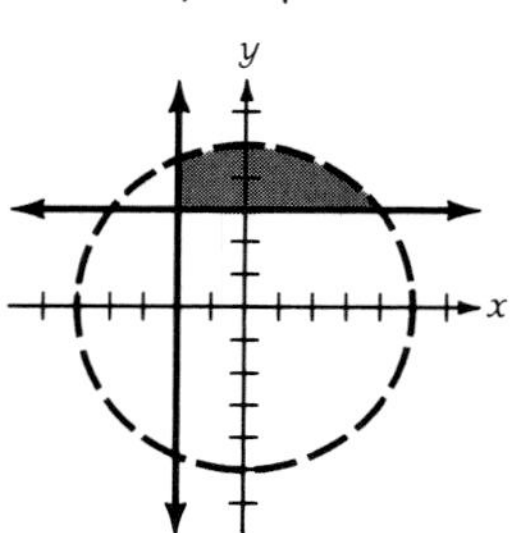


**23.**

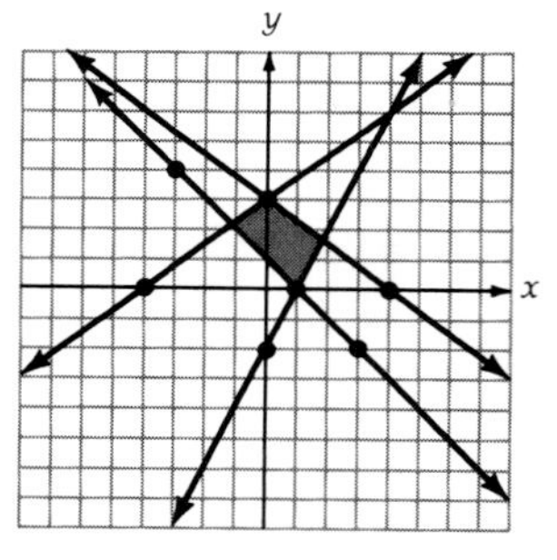


**25.** Minimum value of $c$ is 21. It occurs at $(x, y) = (6, 3)$

**27.** Maximum value of $p$ is 23. It occurs at $(x, y) = \left(6, \frac{5}{2}\right)$ **29.** minimum $(10, -20)$, maximum $\left(\frac{45}{2}, 5\right)$

**31.** He should plant 250 acres of wheat and 50 acres of corn. This will yield a maximum profit of $29,000.

## Exercise 10-7 (pp. 513–514)

**1.** $\{(2, 8), (-1, 2)\}$

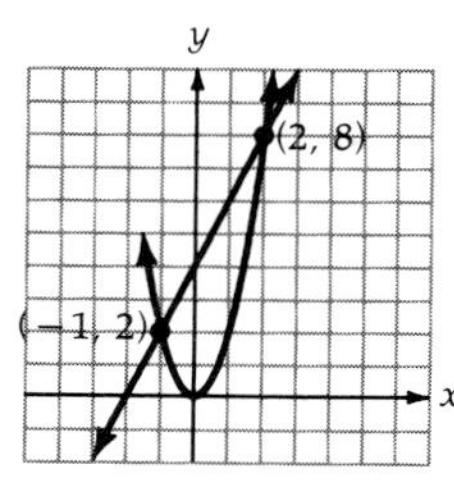


**3.** $\{(0, -1), (-2, 1)\}$

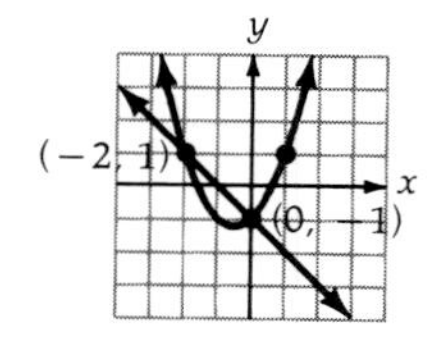


**5.** $\left\{\left(\frac{6}{5}, \frac{12}{5}\right), \left(-\frac{6}{5}, -\frac{12}{5}\right)\right\}$

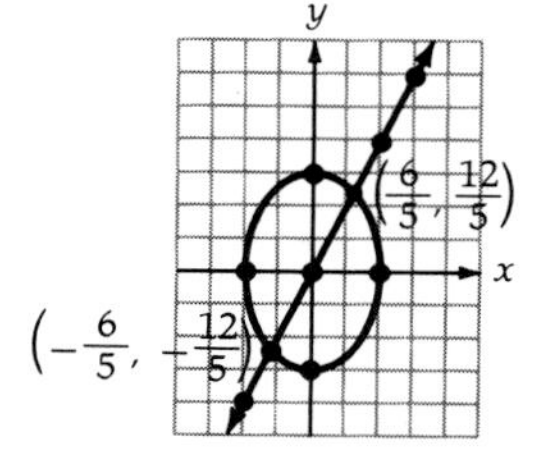

**7.** exact solution: $\left\{\left(\frac{3+\sqrt{23}}{2}, \frac{1-\sqrt{23}}{2}\right), \left(\frac{3-\sqrt{23}}{2}, \frac{1+\sqrt{23}}{2}\right)\right\}$

approximate solution: $\{(3.90, -1.90), (-0.90, 2.90)\}$

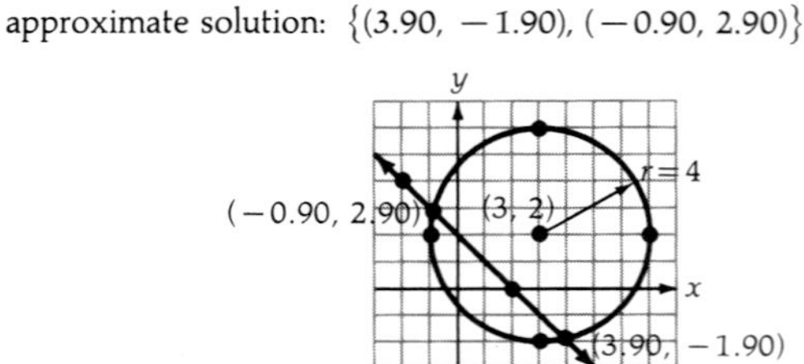


**9.** 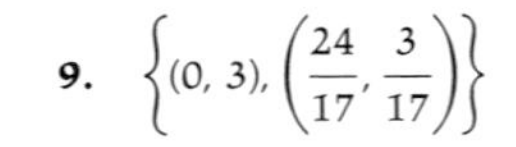 $\left\{(0, 3), \left(\frac{24}{17}, \frac{3}{17}\right)\right\}$

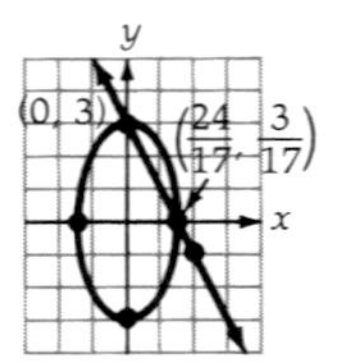


**11.** $\left\{\left(\frac{100}{7}, \frac{72}{7}\right), (4, 0)\right\}$

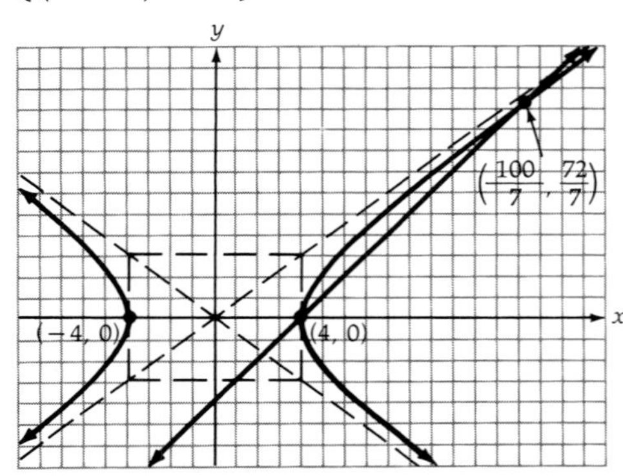


**13.** $\{(4 - 2\sqrt{2}, 2 + \sqrt{2}), (4 + 2\sqrt{2}, 2 - \sqrt{2})\}$

**15.** no real solution

**17.** $\left\{\left(\frac{\sqrt{6}}{6}, \frac{\sqrt{6}}{6}\right), \left(\frac{\sqrt{6}}{6}, -\frac{\sqrt{6}}{6}\right), \left(-\frac{\sqrt{6}}{6}, \frac{\sqrt{6}}{6}\right), \left(-\frac{\sqrt{6}}{6}, -\frac{\sqrt{6}}{6}\right)\right\}$

**19.** no real solution

**21.** $\left\{(2\sqrt{5}, 2\sqrt{5}), (-2\sqrt{5}, -2\sqrt{5}), \left(-3\sqrt{\frac{20}{13}}, \sqrt{\frac{20}{13}}\right), \left(3\sqrt{\frac{20}{13}}, -\sqrt{\frac{20}{13}}\right)\right\}$

**23.** A straight line and a parabola may not intersect at all or they may intersect at one point or at two points.

**25.** An ellipse and a hyperbola may not intersect at all or they may intersect at one, two, three, or four points.

## Chapter 10 Review Exercise (pp. 516–518)

**1.** independent; $(1, -2)$ **3.** independent; $(-2, 1)$ **5.** independent; $(1, -3)$ **7.** inconsistent; no solution **9.** dependent; infinite number of solutions **11.** $(1, 2, -1)$ **13.** $(1, 0, 3)$ **15.** $(-3, 1, -2)$ **17.** $(0, 2)$ **19.** $(0, 0, 4)$ **21.** 12 **23.** 0 **25.** 36 **27.** $(1, -5)$ **29.** $(0, -3)$ **31.** $(1, 2, -1)$ **33.** $(1, 0, 3)$ **35.** $(-3, 1, -2)$ **37.** 5

**39.**

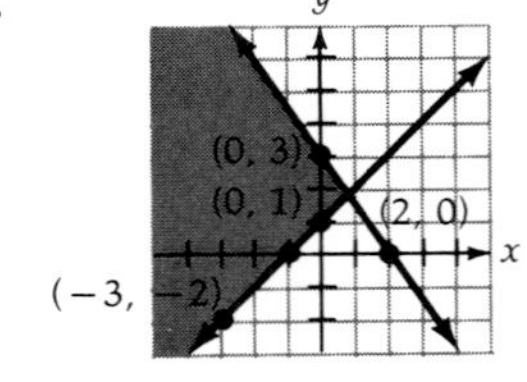


**41.**

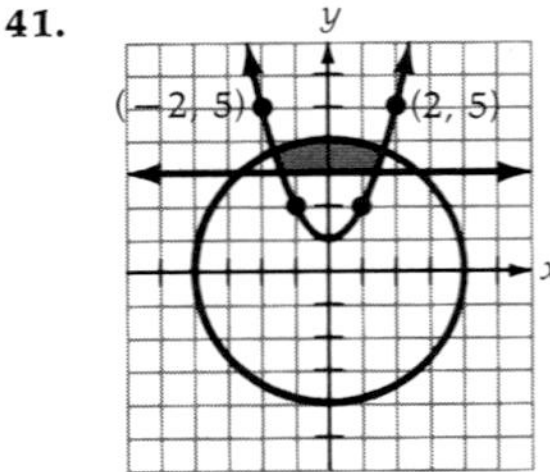


**43.** \$8000 in bond *A*, \$4000 in bond *B*

**45.** 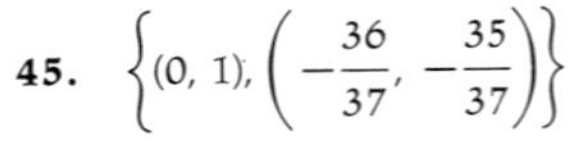 $\left\{(0, 1), \left(-\frac{36}{37}, -\frac{35}{37}\right)\right\}$

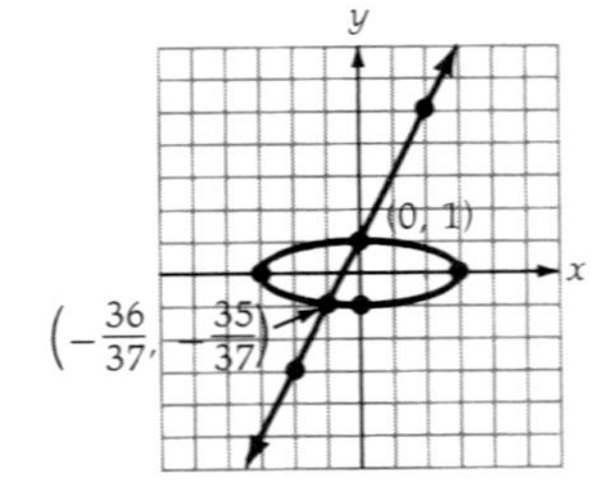

### Practice Test for Chapter 10 (pp. 518–519)

**1.** (3, 1) **3.** (3, −2, 6) **5.** 6 **7.** (5, −4)

**9.**

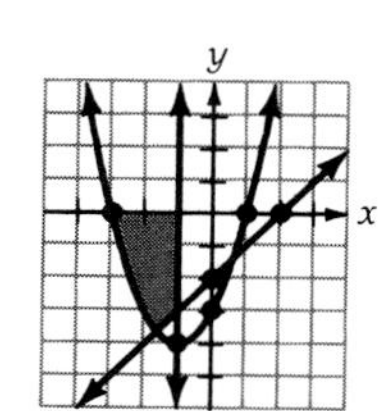


**11.** exact solution: $\left\{\left(\frac{4+2\sqrt{29}}{5}, \frac{3+4\sqrt{29}}{5}\right), \left(\frac{4-2\sqrt{29}}{5}, \frac{3-4\sqrt{29}}{5}\right)\right\}$

approximate solution: {(2.95, 4.91), (−1.35, −3.71)}

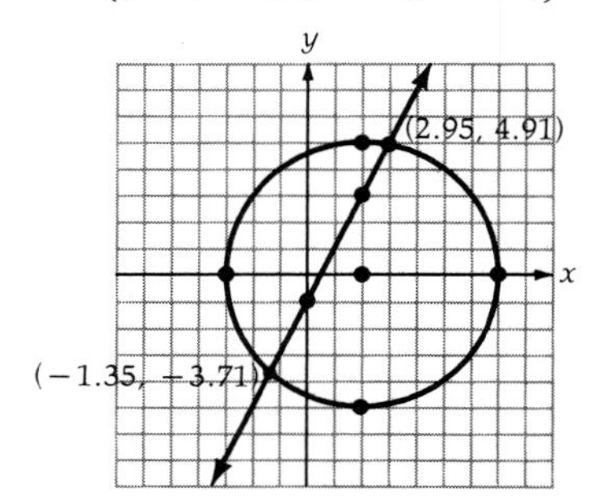


## CHAPTER 11

### Exercise 11-1 (pp. 526–527)

**1.** **a.** −18 **b.** 0 **c.** 12 **d.** 0 **e.** 0 **f.** 0 **g.** yes; the zeros of $P(x)$ are −1, 3, −4, 1
**3.** no **5.** $a = 1, b = 7$ **7.** $3x^2 - 4x + 8, r = -10$ **9.** $x^4 - 4x^3 + 3x^2 - x + 4, r = -1$
**11.** $x^3 - x^2 + x + 1, r = -2x + 9$ **13.** $x^3 + 4x^2 + 19x + 76, r = -36$ **15.** $2x^2 + 3x - 2, r = 0$
**17.** 80 **19.** 110 **21.** 0 **23.** 0 **25.** no **27.** yes **29.** yes
**31.** yes, $(x + 3)$ is a factor of $P(x)$; therefore $x = -3$ is a zero **33.** $b = -24$

### Exercise 11-2 (pp. 530–531)

**1.** $x^2 - x - 1, r = 1$ **3.** $2x^3 - 3x^2 + 6x - 12, r = 23$ **5.** $x^5 + 2x^2 - x, r = 1$
**7.** −9 **9.** 11 **11.** no **13.** yes **15.** no **17.** 0 **19.** $\frac{23}{64}$ **21.** yes
**23.** 203.51571 **25.** no

### Exercise 11-3 (pp. 534–535)

**1.** 2 **3.** 3 **5.** 1
The answers to Problems 7–15 are not unique.
**7.** $6x^3 + 7x^2 - 9x + 2$ **9.** $x^3 - (1 + 4\sqrt{2})x^2 + (6 + 4\sqrt{2})x - 6$
**11.** $x^3 + (3 + \sqrt{7})x^2 + (-14 + 3\sqrt{7})x - 42$ **13.** $x^3 - 2ix^2 + x - 2i$ **15.** $x^4 - 29x^2 + 100$
**17.** −5, 6 **19.** $\pm 1, \pm i$ **21.** $\pm\frac{3}{2}, \pm\frac{3}{2}i$ **23.** ±6 **25.** $\pm\sqrt{3}$ **27.** $\pm i$
**29.** $1, -\frac{1}{2} + \frac{\sqrt{3}}{2}i, -\frac{1}{2} - \frac{\sqrt{3}}{2}i$
The answers to Problems 31 and 33 are not unique.
**31.** $x^4 - 12x^3 + 54x^2 - 108x + 81$ **33.** $4x^3 + 8x^2 - 11x + 3$

## Exercise 11-4 (pp. 537–538)

The answers to Problems 1 and 3 are not unique.

**1.** $x^3 - 4x^2 - 2x + 20$ **3.** $x^4 - 5x^3 + 10x^2 - 16$ **5.** $-i, i, -\frac{3}{2}$ **7.** $-1 + 2i, -1 - 2i, 2$

**9.** **a.** 1 **b.** 3 or 1 **c.** no, $f(0) = -4$ **11.** The constant term must be zero.

**13.** We see that $f(x)$ has one sign variation and $f(-x)$ has one sign variation. Therefore there is one positive real zero and one negative real zero.

**15.** We see that $f(x)$ and $f(-x)$ have no sign variations and $f(0) \neq 0$. Therefore there are no real zeros.

**17.** We see that there is one negative real zero and the real value 0. There are therefore *two* zeros that are not real.

## Exercise 11-5 (p. 542)

The answers to Problems 1 and 3 are not unique.

**1.** $x^3 - 6x^2 + 9x - 2$ **3.** $x^4 - 2x^3 - 13x^2 + 18x - 36$ **5.** $\sqrt{5}, -\sqrt{5}, -\frac{1}{2}$ **7.** $2 - \sqrt{5}, 2 + \sqrt{5}, \frac{1}{3}$

**9.** $\pm\frac{1}{5}, \pm 1, \pm 5, \pm 25$ **11.** $\pm\frac{1}{3}, \pm\frac{2}{3}, \pm 1, \pm\frac{4}{3}, \pm 2, \pm\frac{8}{3}, \pm 3, \pm 4, \pm 6, \pm 8, \pm 12, \pm 24$ **13.** $\pm 1, 2$

**15.** $-2, \frac{1}{2}, 3$ **17.** $-\frac{1}{2}, \frac{1}{3}, \frac{2}{3}$ **19.** $-\frac{3}{2}, -1 \pm \sqrt{3}$ **21.** $\pm 1, -3, \frac{1}{2}$ **23.** $-1, \frac{2}{3}, 2 \pm i$

**25.** $-2, \pm\sqrt{5}$

## Exercise 11-6 (p. 549)

**1.** The intervals are $(-2, -1), (-1, 0), (1, 2)$. **3.** The intervals are $(-7, -6), (0, 1), (5, 6)$.

**5.** Either three real zeros in $(1, 2)$, or one real in $(1, 2)$ and two nonreals.

**7.** There is one real zero in the interval $(1, 2)$. **9.** 2.3 **11.** 3.7 **13.** 4.87 **15.** 2.06

**17.** $-1.54$ **19.** $-1.17, -0.36, 3.53$ **21.** $-0.50, 0.68$

## Exercise 11-7 (p. 554)

**1.** $\dfrac{-\frac{1}{3}}{x + 1} + \dfrac{\frac{1}{3}}{x - 2}$ or $\dfrac{-1}{3(x + 1)} + \dfrac{1}{3(x - 2)}$

**3.** $\dfrac{1}{x - 2} + \dfrac{-\frac{2}{3}}{x - 1} + \dfrac{\frac{1}{3}}{2x + 1}$ or $\dfrac{1}{x - 2} - \dfrac{2}{3(x - 1)} + \dfrac{1}{3(2x + 1)}$ **5.** $\dfrac{2}{x + 1} - \dfrac{3}{x + 2}$ **7.** $\dfrac{2}{x} + \dfrac{1}{x - 1}$

**9.** $\dfrac{5}{2x + 3} - \dfrac{2}{3x - 1}$ **11.** $\dfrac{2}{x - 2} + \dfrac{x + 3}{x^2 + 3}$ **13.** $-\dfrac{1}{x} + \dfrac{2}{x + 1} - \dfrac{1}{(x + 1)^2} + \dfrac{2}{(x + 1)^3}$

**15.** $-\dfrac{3}{x} + \dfrac{3}{x - 1} + \dfrac{1}{x + 1}$ **17.** $\dfrac{1}{x - 1} + \dfrac{2x + 3}{x^2 + 1}$ **19.** $\dfrac{1}{x + 1} + \dfrac{2x - 2}{x^2 - x + 3}$ **21.** $\dfrac{x + 3}{x^2 + 1} - \dfrac{x}{x^2 + 2}$

**23.** $\dfrac{3}{x} - \dfrac{3}{2(x + 1)} - \dfrac{3x + 3}{2(x^2 + 1)}$ or $\dfrac{3}{x} + \dfrac{-\frac{3}{2}}{x + 1} + \dfrac{-\frac{3}{2}(x + 1)}{x^2 + 1}$

**25.** $\dfrac{1}{x + 3} - \dfrac{x - 3}{x^2 + 2} - \dfrac{5x}{(x^2 + 2)^2}$ or $\dfrac{1}{x + 3} + \dfrac{-x + 3}{x^2 + 2} + \dfrac{-5x}{(x^2 + 2)^2}$

### Chapter 11 Review Exercise (pp. 558–559)

**1.** 5 **3.** 3 **5.** $a = 5, b = -1, c = 11$ **7.** $-404$ **9.** $-70$ **11.** yes.
**13.** $x^4 - 6x^3 + 6x^2 + 4x - 24, r = 41$ **15.** 15 **17.** no **19.** 6
**21.** $x^3 - (3 - \sqrt{5})x^2 - (10 + 3\sqrt{5})x + 30$ **23.** $-1, \frac{1 \pm i\sqrt{3}}{2}$ **25.** 2 or 0
**27.** 0 **29.** 1 **31.** $-1, 3, 5$ **33.** 1 (multiplicity 2), 3 (multiplicity 2) **35.** $-1, \frac{2}{3}, 4, \pm\frac{i}{2}$
**37.** The intervals are $(-1, 0), (2, 3), (4, 5), (9, 10)$. **39.** $-2.7$ **41.** $\frac{5}{x-2} - \frac{2}{x+5}$
**43.** $\frac{6}{x-2} + \frac{2x+1}{x^2+2x+4}$ **45.** $\frac{2x-3}{x^2+2x+3} - \frac{3x+1}{(x^2+2x+3)^2}$

### Practice Test for Chapter 11 (p. 559)

**1.** 7 **3.** $f(-5) = 0$ **5.** $x^3 + 2x^2 - 5x - 6$ **7.** $\pm 1, \pm 5, \pm\frac{1}{3}, \pm\frac{5}{3}$ **9.** $\frac{1}{2}, \pm 2i$
**11.** The intervals are $(-5, -4), (-2, -1), (3, 4)$.

## CHAPTER 12

### Exercise 12-1 (pp. 565–566)

The proofs for Problems 1–19 are shown in the *Student Tutorial.*

### Exercise 12-2 (pp. 571–573)

**1.** $32, 36, \ldots, 8 + 4n$ **3.** $\frac{8}{49}, \frac{9}{64}, \ldots, \frac{n+2}{(n+1)^2}$ **5.** $a_1 = 1, a_2 = 4, a_3 = 7, a_4 = 10, a_{10} = 28$
**7.** $a_1 = 4, a_2 = 7, a_3 = 12, a_4 = 19, a_{10} = 103$
**9.** Exact values: $a_1 = \frac{\log 1}{2}, a_2 = \frac{\log 2}{4}, a_3 = \frac{\log 3}{6}, a_4 = \frac{\log 4}{8}, a_{10} = \frac{\log 10}{20}$ Values expressed as decimals ($a_1$ and $a_{10}$ are exact, all others are approximations): $a_1 = 0, a_2 = 0.0753, a_3 = 0.0795, a_4 = 0.0753, a_{10} = 0.05$
**11.** $a_1 = 3, a_2 = 0, a_3 = -6, a_4 = -18$ **13.** $a_1 = -4, a_2 = -19, a_3 = -94, a_4 = -469$ **15.** 50
**17.** 36 **19.** $\sum_{i=1}^{6} (-1)^{i+1}$ (The answer to Problem 19 is not unique.) **21.** 3, 8, 13, 18, 23, 28
**23.** 52 **25.** $-7$ **27.** 200 **29.** 45 **31.** 1854 **33.** 7075 **35.** $k = 10$ **37.** yes

### Exercise 12-3 (pp. 577–578)

**1.** yes, $r = \frac{1}{3}$ **3.** yes, $r = -\sqrt{7}$ **5.** no **7.** no **9.** 2, 6, 18, 54, 162 **11.** 9375 **13.** 384
**15.** $\frac{5461}{5120}$ **17.** 13,108 **19.** $\frac{93}{64}$ **21.** $\frac{728}{81}$ **23.** $n = 6$ **25.** $r = -3$ **27.** $a_8 = -729$
**29.** log 81, log 6561 **31.** 6, 12, 24, 48

**33.** 8 **35.** sum does not exist **37.** $4(2 + \sqrt{2})$ **39.** $\frac{49}{10}$

**41.** Theoretically, it will be 0.6046618 feet high at its tenth bounce. The total distance is 398.18601 feet.

## Exercise 12-4 (pp. 583–584)

**1.** 120 **3.** 144 **5.** 360 **7.** 30,240 **9.** $(k + 2)!$ **11.** $10!$ **13.** 40,320 **15.** 120
**17.** 7,893,600 **19.** 120 **21.** 495 **23.** 1540

**25.** $\binom{n}{r} = \frac{n!}{r!(n-r)!}\ \binom{n}{n-r} = \frac{n!}{(n-r)![n-(n-r)]!} = \frac{n!}{(n-r)!r!}$ **27.** $\binom{n}{0} = \frac{n!}{0!(n-0)!} = \frac{n!}{(1)n!} = 1$

**29.** 210 **31.** 10 **33.** 56 **35.** 134,850

## Exercise 12-5 (p. 587)

**1.** $x^4 + 8x^3 + 24x^2 + 32x + 16$ **3.** $x^6 - 6x^5y + 15x^4y^2 - 20x^3y^3 + 15x^2y^4 - 6xy^5 + y^6$

**5.** $128a^7 + 224a^6b + 168a^5b^2 + 70a^4b^3 + \frac{35}{2}a^3b^4 + \frac{21}{8}a^2b^5 + \frac{7}{32}ab^6 + \frac{b^7}{128}$

**7.** $a^{100} - 300a^{99}b + 44{,}550a^{98}b^2 - 4{,}365{,}900a^{97}b^3$ **9.** $-560a^8b^6$

**11.** $x^8 - 16x^7y + 112x^6y^2 - 448x^5y^3 + 1120x^4y^4 - 1792x^3y^5 + 1792x^2y^6 - 1024xy^7 + 256y^8$

**13.** $81x^4 - 216\frac{x^3}{y} + 216\frac{x^2}{y^2} - 96\frac{x}{y^3} + \frac{16}{y^4}$ **15.** $32x^2\sqrt{x} + 80x^2y^3 + 80x\sqrt{x}y^6 + 40xy^9 + 10\sqrt{x}y^{12} + y^{15}$

**17.** $20{,}412x^6y^2$ **19.** $344{,}064x^{12}y^3$

## Exercise 12-6 (pp. 594–596)

**1.** $\frac{1}{2}$ **3.** **a.** $\frac{1}{52}$ **b.** $\frac{1}{13}$ **c.** $\frac{1}{4}$ **d.** $\frac{3}{13}$

**5.** **a.** $\frac{5}{31}$ **b.** $\frac{9}{31}$ **c.** $\frac{12}{31}$ **d.** $\frac{21}{31}$ **e.** $\frac{15}{31}$ **7.** $\frac{1}{36}$ **9.** $\frac{1}{4}$ **11.** $\frac{1}{2}$ **13.** $\frac{2}{7}$

**15.** $\frac{1}{1{,}320}$ **17.** $P(\text{all spades}) = \frac{1}{\binom{52}{13}} = \frac{13!39!}{52!} = \frac{1}{635{,}013{,}559{,}600} \doteq 1.574769522 \times 10^{-12}$ **19.** $\frac{18}{23}$

**21.** $\frac{2}{13}$ **23.** $\frac{1}{6}$ **25.** $\frac{4}{13}$ **27.** $\frac{13}{48}$ **29.** 1:5 **31.** 1:12 **33.** **a.** 1:35 **b.** 1:5

**35.** 1:220 **37.** 33:66,607 **39.** $\frac{961}{3136}$ or approximately 0.3064

## Chapter 12 Review Exercise (pp. 599–601)

The proofs for Problems 1, 3, and 5 are only briefly outlined. For a more complete proof, see the *Student Tutorial*.

**1.** **a.** Prove that it is true for $n = 1$. **b.** Assume that it is true for $n = k$.

**c.** Add $\frac{1}{2}(k + 1)$ to both sides of the equation in step b and show that it becomes

$$\frac{1}{2}(1) + \frac{1}{2}(2) + \cdots + \frac{1}{2}(k) + \frac{1}{2}(k+1) = \frac{(k+1)^2 + (k+1)}{4}$$

**3.** **a.** Prove that it is true for $n = 1$. **b.** Assume that it is true for $n = k$.
**c.** Add $2^{k+1}$ to both sides of the equation in step b and show that it becomes

$$2 + 2^2 + 2^3 + \cdots + 2^k + 2^{k+1} = 2^{k+2} - 2$$

**5.** **a.** Prove that it is true for $n = 1$. **b.** Assume that it is true for $n = k$.
**c.** Add $(2k + 1)^2$ to both sides of the equation in step b and show that it becomes

$$1^2 + 3^2 + \cdots + (2k - 1)^2 + (2k + 1)^2 = \frac{(k + 1)[4(k + 1)^2 - 1]}{3}$$

**7.** $a_1 = -3, a_2 = 4, a_3 = 23, a_4 = 60, a_8 = 508$ **9.** $a_1 = 1, a_2 = 2, a_3 = 6, a_4 = 21$

**11.** $\frac{x^3}{4} + \frac{x^4}{8} + \frac{x^5}{12} + \frac{x^6}{16}$ **13.** $\sum_{i=1}^{6} (4i - 3)$ (This answer is not unique.) **15.** geometric **17.** arithmetic

**19.** $a_{29} = -6$ **21.** $S_{100} = 10{,}000$ **23.** $a_7 = 0$ **25.** $\sqrt{2}, 2, 2\sqrt{2}, 4$ **27.** $-\frac{1}{2}$

**29.** for $k = 0$ or $k = -\frac{27}{14}$ **31.** 40,320 **33.** 720 **35.** 9! or 362,880 **37.** ${}_{10}P_7$ or 604,800

**39.** 220 **41.** 127 **43.** $64x^6 + 576x^5y + 2160x^4y^2 + 4320x^3y^3 + 4860x^2y^4 + 2916xy^5 + 729y^6$

**45.** $3360y^{12}$ **47.** $\frac{14}{31}$ **49.** 1:31 **51.** 649,739:1

## Practice Test for Chapter 12 (pp. 601–602)

**1.** **a.** It is true for $n = 1$, since $2(1) = 1(1 + 1)$
**b.** Assume that it is true for $n = k$: $2 + 4 + 6 + \cdots + 2k = k(k + 1)$.
**c.** Add $2(k + 1)$ to each side of the equation in step b. The right-hand side is

$$k(k + 1) + 2(k + 1) = (k + 1)(k + 2) = (k + 1)[(k + 1) + 1]$$

This is exactly the formula for $n = k + 1$. Thus we have completed the proof by mathematical induction.

**3.** 48 **5.** $n = 10$ **7.** 2, 6, 18 or $-2, 6, -18$ **9.** 12 **11.** 816 **13.** 210

**15.** $x^5 - 10x^4y + 40x^3y^2 - 80x^2y^3 + 80xy^4 - 32y^5$

## APPENDIX A (p. A-5)

**1.**

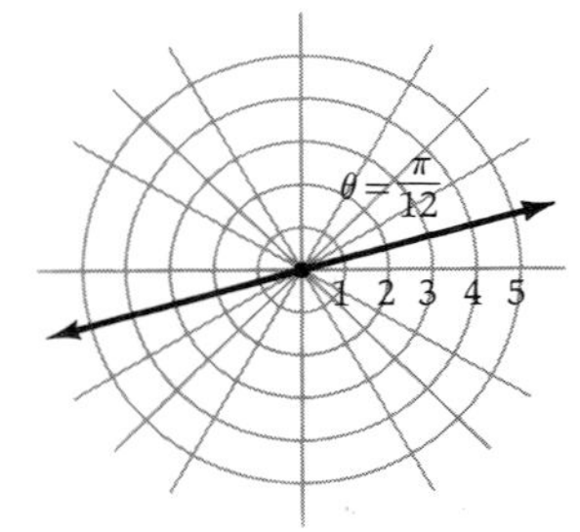


**3.**

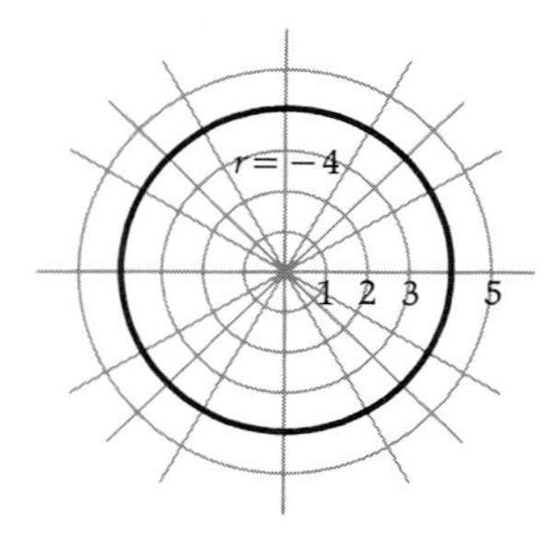

**5.** $r = 2 \cos \theta$

| $\theta$ | $0$ | $\frac{\pi}{6}$ | $\frac{\pi}{4}$ | $\frac{\pi}{3}$ | $\frac{\pi}{2}$ | $\frac{7\pi}{6}$ | $\frac{5\pi}{4}$ | $\frac{4\pi}{3}$ | $\pi$ | etc. |
|---|---|---|---|---|---|---|---|---|---|---|
| $r$ | 2 | 1.7 | 1.4 | 1 | 0 | $-1.7$ | $-1.4$ | $-1$ | $-2$ | |

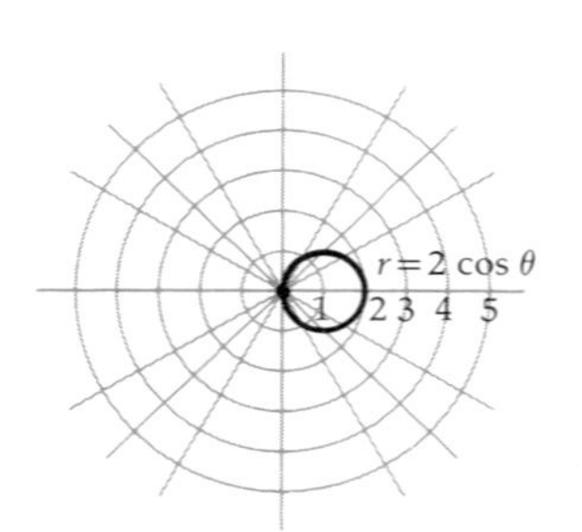


**7.** $r = 2 \sin 3\theta$

| $\theta$ | $0$ | $\frac{\pi}{12}$ | $\frac{\pi}{6}$ | $\frac{\pi}{4}$ | $\frac{\pi}{3}$ | $\frac{\pi}{2}$ | $\frac{2\pi}{3}$ | $\frac{5\pi}{6}$ |
|---|---|---|---|---|---|---|---|---|
| $r$ | 0 | 1.41 | 2 | 1.41 | 0 | $-2$ | 0 | 2 |

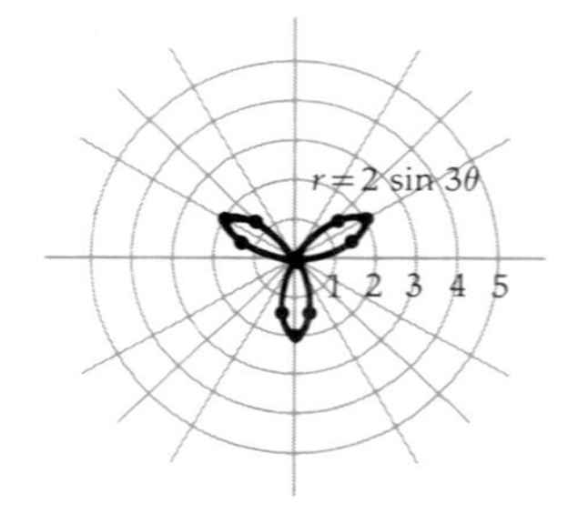


**9.** $r = 3(1 + \cos \theta)$

| $\theta$ | $0$ | $\frac{\pi}{6}$ | $\frac{\pi}{4}$ | $\frac{\pi}{3}$ | $\frac{\pi}{2}$ | $\frac{2\pi}{3}$ | $\frac{3\pi}{4}$ | $\pi$ | $\frac{5\pi}{4}$ | $\frac{3\pi}{2}$ | $\frac{7\pi}{4}$ |
|---|---|---|---|---|---|---|---|---|---|---|---|
| $r$ | 6 | 5.6 | 5.1 | 4.5 | 3 | 1.5 | .9 | 0 | .9 | 3 | 5.1 |

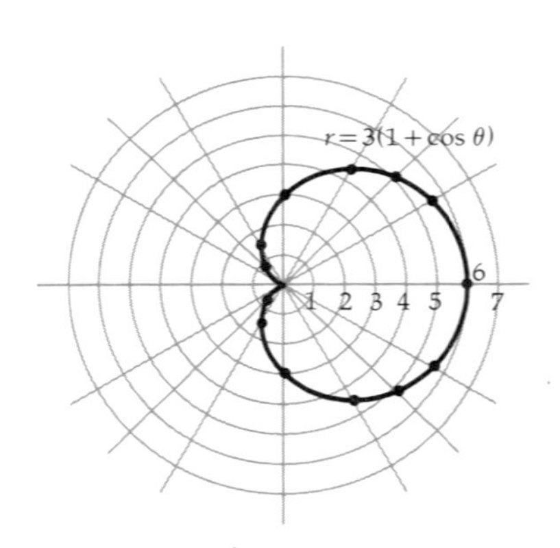


**11.** $r = 5 \sin 2\theta$

| $\theta$ | $0$ | $\frac{\pi}{8}$ | $\frac{\pi}{4}$ | $\frac{3\pi}{8}$ | $\frac{\pi}{2}$ | $\frac{3\pi}{4}$ | $\pi$ | $\frac{5\pi}{4}$ | $\frac{3\pi}{2}$ | $\frac{7\pi}{4}$ | $2\pi$ |
|---|---|---|---|---|---|---|---|---|---|---|---|
| $r$ | 0 | 3.5 | 5 | 3.5 | 0 | $-5$ | 0 | 5 | 0 | $-5$ | 0 |

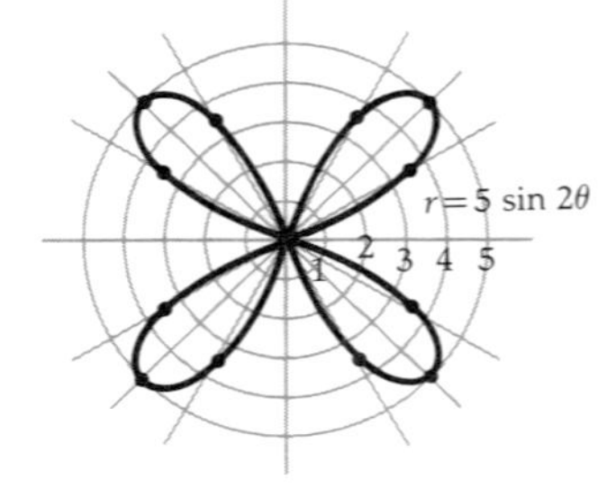

## APPENDIX A (continued)

**13.** $r^2 = 4 \cos 2\theta$

| $\theta$ | 0 | $\frac{\pi}{8}$ | $\frac{\pi}{4}$ | $\frac{3\pi}{8}$ | $\frac{\pi}{2}$ | $\frac{3\pi}{4}$ | $\frac{7\pi}{8}$ | $\pi$ |
|---|---|---|---|---|---|---|---|---|
| $r$ | $\pm 2$ | $\pm 1.7$ | 0 | undef. | undef. | 0 | $\pm 1.7$ | $\pm 2$ |

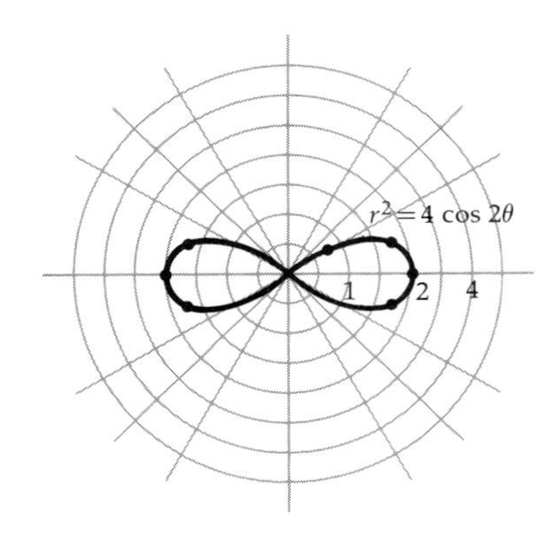


**15.** $r = 2 + 3 \cos \theta$

| $\theta$ | 0 | $\frac{\pi}{12}$ | $\frac{\pi}{6}$ | $\frac{\pi}{4}$ | $\frac{\pi}{3}$ | $\frac{5\pi}{12}$ | $\frac{\pi}{2}$ | $\frac{7\pi}{12}$ | $\frac{2\pi}{3}$ | $\frac{3\pi}{4}$ | $\frac{5\pi}{6}$ | $\frac{11\pi}{12}$ | $\pi$ |
|---|---|---|---|---|---|---|---|---|---|---|---|---|---|
| $r$ | 5 | 4.9 | 4.6 | 4.1 | 3.5 | 2.7 | 2 | 1.2 | .5 | $-.1$ | $-.6$ | $-.9$ | $-1$ |

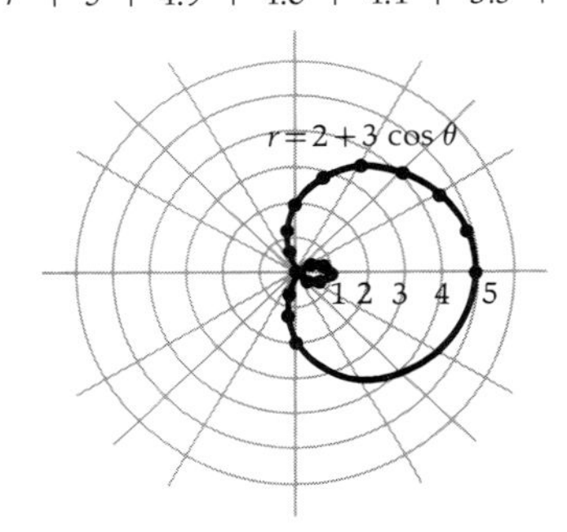


**17.** $r = \dfrac{9 \cos \theta}{\sin^2 \theta}$ **19.** $r^2 = \dfrac{16}{\cos^2 \theta + 4 \sin^2 \theta}$ **21.** $x = 2$ **23.** $x^2 + y^2 - 6y = 0$

# INDEX

**D**

## E



## F

## G



## H



## I

**R**



**S**

**T**



**U**



**V**



**W**

# TRIGONOMETRIC PROPERTIES AND IDENTITIES

## BASIC IDENTITIES

$$\csc A \equiv \frac{1}{\sin A} \qquad \tan A \equiv \frac{\sin A}{\cos A}$$

$$\sec A \equiv \frac{1}{\cos A} \qquad \cot A \equiv \frac{\cos A}{\sin A}$$

$$\cot A \equiv \frac{1}{\tan A}$$

$$\sin^2 A + \cos^2 A \equiv 1$$

$$1 + \tan^2 A \equiv \sec^2 A$$

$$1 + \cot^2 A \equiv \csc^2 A$$

## NEGATIVE IDENTITIES

$$\sin(-A) \equiv -\sin A \qquad \cos(-A) \equiv \cos A \qquad \tan(-A) \equiv -\tan A$$

## COFUNCTION IDENTITIES

$$\sin\left(\frac{\pi}{2} - A\right) \equiv \cos A \qquad \cos\left(\frac{\pi}{2} - A\right) \equiv \sin A \qquad \tan\left(\frac{\pi}{2} - A\right) \equiv \cot A$$

## SUM AND DIFFERENCE OF TWO ARGUMENTS

$$\sin(A + B) \equiv \sin A \cos B + \cos A \sin B$$

$$\sin(A - B) \equiv \sin A \cos B - \cos A \sin B$$

$$\tan(A + B) \equiv \frac{\tan A + \tan B}{1 - \tan A \tan B}$$

$$\tan(A - B) \equiv \frac{\tan A - \tan B}{1 + \tan A \tan B}$$

$$\cos(A + B) \equiv \cos A \cos B - \sin A \sin B$$

$$\cos(A - B) \equiv \cos A \cos B + \sin A \sin B$$